Low Reynolds number hydrodynamics

MECHANICS OF FLUIDS AND TRANSPORT PROCESSES
Editors: R.J. Moreau and G.Æ. Oravas

J. Happel and H. Brenner, Low Reynolds number hydrodynamics. 1983 ISBN 90-247-2877-0.
S. Zahorski, Mechanics of viscoleastic fluids. 1982. ISBN 90-247-2687-5
J.A. Sparenberg, Elements of hydrodynamic propulsion. 1984. ISBN 90-247-2871-1.
B.K. Shivamoggi, Theoretical fluid dynamics. 1984. ISBN 90-247-2999-8
R. Timman, A.J. Hermans and G.C. Hsiao, Water waves and ship hydrodynamics: An introduction. 1985. ISBN 90-247-3218-2.

Low Reynolds number hydrodynamics

with special applications to particulate media

John Happel

Department of Chemical Engineering and Applied Chemistry
Columbia University
New York, New York

Howard Brenner

Department of Chemical Engineering
Massachusetts Institute of Technology
Cambridge, Massachusetts

KLUWER ACADEMIC PUBLISHERS
DORDRECHT / BOSTON / LONDON

Library of Congress Catalogue Card Number: 72-97238

First edition 1963
Second revised edition 1973
Third printing 1983
Fourth printing 1986
Fifth printing 1991

ISBN 90-247-2877-0

Published by Kluwer Academic Publishers,
P.O. Box 17, 3300 AA Dordrecht, The Netherlands.

Kluwer Academic Publishers incorporates
the publishing programmes of
D. Reidel, Martinus Nijhoff, Dr W. Junk and MTP Press.

Sold and distributed in the U.S.A. and Canada
by Kluwer Academic Publishers,
101 Philip Drive, Norwell, MA 02061, U.S.A.

In all other countries, sold and distributed
by Kluwer Academic Publishers Group,
P.O. Box 322, 3300 AH Dordrecht, The Netherlands.

Printed on acid-free paper

Printed in the Netherlands

Contents

Preface

One studying the motion of fluids relative to particulate systems is soon impressed by the dichotomy which exists between books covering theoretical and practical aspects. Classical hydrodynamics is largely concerned with perfect fluids which unfortunately exert no forces on the particles past which they move. Practical approaches to subjects like fluidization, sedimentation, and flow through porous media abound in much useful but uncorrelated empirical information. The present book represents an attempt to bridge this gap by providing at least the beginnings of a rational approach to fluid-particle dynamics, based on first principles.

From the pedagogic viewpoint it seems worthwhile to show that the Navier-Stokes equations, which form the basis of all systematic texts, can be employed for useful practical applications beyond the elementary problems of laminar flow in pipes and Stokes law for the motion of a single particle. Although a suspension may often be viewed as a continuum for practical purposes, it really consists of a discrete collection of particles immersed in an essentially continuous fluid. Consideration of the actual detailed boundary-value problems posed by this viewpoint may serve to call attention to the limitation of idealizations which apply to the overall transport properties of a mixture of fluid and solid particles.

It is hoped that the research worker in this and related fields will be stimulated by noting that not only does the hydrodynamic viewpoint lead to a clearer correlation of much existing work, but that, at every turn, there exists a host of new problems awaiting solution. Among those which seem especially intriguing are the effect of variation in particle size and arrangement on the dynamic behavior of particulate systems, and the possibilities of extending the present treatments to higher Reynolds numbers. Engineers may be interested in the availability of fluid dynamic models which can serve as the framework for more extended investigations involving other transport processes, coupled perhaps with chemical reactions.

The treatment developed here is based almost entirely on the linearized form of the equations of motion which results from omitting the inertial terms from the Navier-Stokes equations, giving the so-called creeping motion or Stokes equations. This is tantamount to assuming that the particle Reynolds numbers are very small. Many systems which involve bulk flow relative to external boundaries at high Reynolds numbers are still characterized by low Reynolds numbers as regards the movement of particles relative to fluid. Also, inertial effects are less important for systems consisting of a number of particles in a bounded fluid medium than they are for the motion of a single particle in an unbounded fluid.

The subject matter is largely confined to a development of the macroscopic properties of fluid-particle systems from first principles. General hydrodynamic and mathematical concepts are not treated in detail beyond what is required for further development. Most of the experimental data reported are confined to critical experiments aimed at demonstrating the applicability of the theoretical results to actual physical systems. Following the first few introductory chapters, subsequent material is organized on the basis of the class of boundary-value problems involved, similar to the approach used by C. W. Oseen in his classical "Hydrodynamik." Starting with the motion of a single particle in an unbounded medium, the problem of the motion of several particles interacting with each other, of particles moving in the presence of bounding walls and finally of combinations of these factors are considered successively. Final chapters deal with the movement of fluids relative to particulate assemblages and with the viscosity of suspensions of particles. The latter treatment is more provisional and includes comparison of theory with empirical equations and data.

Much of the material presented is based on original investigations of the authors and their students, especially Jack Famularo. We were also fortunate in having the advice of O. H. Faxen, who provided us with many early papers by himself and C. W. Oseen, and who also carefully read the entire manuscript. S. Wakiya was also kind enough to review the manuscript and to provide additional useful suggestions. Support in the form of grants and fellowships from the Petroleum Research Fund of the American Chemical Society, The National Science Foundation, The Institute of Paper Chemistry, The Pulp and Paper Research Institute of Canada, and The Texas Company is gratefully acknowledged. The Courant Institute of Mathematical Sciences of New York University graciously provided considerable computer time.

The authors are convinced that the foundation of a scientifically sound development of the fluid dynamics of particulate media at low Reynolds numbers is now available, which should serve as a sound basis for future investigations. They hope that this book will serve to illustrate the versatility of this field of study.

JOHN HAPPEL
HOWARD BRENNER

SYMBOLS

The following is a list of the most frequently occurring symbols used in the book. Symbols not defined here are defined at their first place of use. A few of these symbols are occasionaliy used in other contexts.

a Sphere radius
A Cross-sectional area
b Distance to cylinder axis
B Center of buoyancy
c Particle dimension
C_D Drag coefficient
$\mathbf{C}$; $(\mathbf{C})$; C_{ij} Coupling dyadic; matrix; tensor
$\mathbf{C}_{jk}$ Coupling dyadic in multiparticle system
d Diameter
D/Dt Material (convected) time derivative
$(\mathbf{e}_1, \mathbf{e}_2, \mathbf{e}_3) \equiv \mathbf{e}_j$ Triad of right-handed orthonormal eigenvectors
E Total energy dissipation rate
E^2 Stokes stream function operator
$f_k(\theta, \phi)$ Surface spherical harmonic
F Force
$\mathbf{F}, F_i$ Vector force
$(\mathscr{F})$ Force matrix
$\mathbf{g}$ Local acceleration of gravity vector
$(h_1, h_2, h_3) \equiv h_j$ Metrical coefficients
$\mathscr{H}_n, \mathscr{I}_n, \mathscr{K}_n$ Gegenbauer functions
$i \equiv \sqrt{-1}$ Unit imaginary number
$(\mathbf{i}_1, \mathbf{i}_2, \mathbf{i}_3) \equiv \mathbf{i}_j$ Unit vectors
$\mathbf{i}, \mathbf{j}, \mathbf{k}$, or $\mathbf{i}_x, \mathbf{i}_y, \mathbf{i}_z$ Cartesian unit vectors
I_n, K_n Modified Bessel functions
$\mathbf{I}$ Dyadic idemfactor
k Wall-effect constant; Kozeny constant
$\mathbf{k}$ Wall-effect dyadic
K Translational resistance coefficient for isotropic particle; Darcy permeability constant
$\bar{K}$ Average translational resistance coefficient
K_1, K_2, K_3 Eigenvalues of translation dyadic
$(\mathscr{K})$ Resistance matrix
$\mathbf{K}$; $(\mathbf{K})$; K_{ij} Translation resistance dyadic; matrix; tensor
$\mathbf{K}_{jk}$ Translation dyadic in multiparticle system
l Length; distance to boundary; distance between particles
m Hydraulic radius
m_f; m_p Mass of displaced fluid; mass of particle
M Center of mass
δn Distance measured normal to surface
$\mathbf{n}$ Unit normal vector
N_{Re} Reynolds number
O Arbitrary point fixed in particle
p Local pressure
p_n Solid spherical harmonic
$P_n(\cos\theta)$; $P_n^m(\cos\theta)$ Legendre function; associated Legendre function
ΔP or Δp Pressure drop
$\mathbf{P}$ Intrinsic vector "pressure" field
$\Delta\mathbf{P}$ Vector pressure drop
$\mathscr{P}$ Intrinsic triadic "stress" field
$(q_1, q_2, q_3) \equiv q_j$ Curvilinear coordinates
Q Volumetric flow rate or volume
(r, θ, ϕ) Spherical polar coordinates
$\mathbf{r}$ Position vector
$\mathbf{r}_O$ Position vector relative to origin at O
$\mathbf{r}_{OP}$ Vector from O to P
R Center of reaction
R_o Radius of circular cylinder
$\mathbf{R}$ Position vector
$\mathbf{R}_R$ Vector defined in Eq. (5–7.14)
δs Distance measured along a surface
$\mathbf{s}$ Unit tangent vector in intrinsic coordinates
S Surface area
S_p Particle surface
$d\mathbf{S}$ Directed element of surface area
t Time
$\mathbf{t}$ Unit tangent vector

T Absolute temperature or torque component

$\mathbf{T}, T_i$ Vector torque

u, v, w Component velocities in cartesian coordinates

$(u_1, u_2, u_3) \equiv u_j$ Components of vector $\mathbf{u}$

$\mathbf{u}$ Local fluid velocity

U Particle speed or superficial fluid velocity

U_0 Settling velocity of a single particle, or centerline velocity

U_{MF} Mean fluid velocity

U_{OF} Centerline velocity of fluid

U_{TS} Terminal settling velocity of sphere

$(\mathcal{U})$ Velocity matrix

$\mathbf{U}, U_i$ Particle velocity vector

$\mathbf{U}_m ; U_m$ Superficial velocity vector; speed

$(v_1, v_2, v_3) \equiv v_j$ Components of vector $\mathbf{v}$

$\mathbf{v}$ Local fluid velocity vector

$\mathbf{v}_\infty$ Local fluid velocity vector at infinity

V Volume or velocity

V_m Mean velocity of flow

$\mathbf{V}$ Intrinsic dyadic "velocity" field

W Rate of doing work

x, y, z Cartesian coordinates

$(x_1, x_2, x_3) \equiv x_j$ Cartesian coordinates

$X_n(\theta, \phi), Y_n(\theta, \phi), Z_n(\theta, \phi)$ Surface spherical harmonics

z' Complex variable

β Slip coefficient

δ_{jk} Kronecker delta

$\delta(\mathbf{r} - \mathbf{r}_n)$ Dirac delta function

$\mathbf{\Delta}, \Delta_{ij}$ Rate of strain dyadic, tensor

ϵ Small deformation parameter or fractional void volume

ϵ_{jkl} Permutation symbol

$\boldsymbol{\epsilon}$ Alternating isotropic triadic

$\boldsymbol{\zeta}$ Vorticity vector

κ Bulk viscosity

μ (Shear) viscosity

μ_r Relative viscosity

μ_{sp} Specific viscosity

$[\mu]$ Intrinsic viscosity

ν Kinematic viscosity

$\mathbf{\Pi}, \Pi_{ij}$ Pressure (stress) dyadic, tensor

$\mathbf{\Pi}_n$ Stress vector

ρ, ρ_f Fluid density

ρ_p, ρ' Mean particle density

$\Delta\rho$ Density difference

(ρ, ϕ, z) Circular cylindrical coordinates

σ Internal-external viscosity ratio

τ Volume

ϕ Fractional volume of solids

$\boldsymbol{\phi}_\infty$ Dimensionless translation dyadic in unbounded fluid

Φ Local energy dissipation rate

Φ_n, ϕ_n Solid spherical harmonic

χ_n Solid spherical harmonic

ψ Stream function

$\boldsymbol{\omega}, \omega$ Angular velocity vector, speed

$\mathbf{\Omega}$ Angular velocity vector

$\mathbf{\Omega}; (\mathbf{\Omega}); \Omega_{ij}$ Rotational resistance dyadic; matrix; tensor

$\mathbf{\Omega}_{jk}$ Rotational resistance dyadic in multiparticle system

ϖ Radial cylindrical coordinate

∇ Nabla operator

∇^2 Laplace operator

$\dagger$ Transposition operator

-1 Reciprocal dyadic

$\cdot$ Dot product

$:$ Double-dot product

$\times$ Cross product

$\times\atop\times$ Double-cross product

To our wives and parents

"... e tutto il frutto ricolto del girar di queste spere."

Dante Alighiere

Introduction **1**

1-1 Definition and Purpose

The behavior of systems involving the motion of aggregates of small particles relative to fluids in which they are immersed covers a wide range of phenomena of interest to both scientists and engineers. Broadly speaking we may assign these processes to several classes. Particles may move together in bulk through a fluid, as in sedimentation. In turn, the particles may remain more or less stationary as in a packed bed. The relative particle-fluid motions may be more complex, as in fluidized systems. Finally, the phenomenon of suspension viscosity or resistance to shear is encountered when solid particles move relative to each other owing to shearing motion of the suspending fluid, as contrasted with situations where the fluid moves relative to the entire particle system. Many processes involving these types of motion are found in nature and technology. It is the basic purpose of this book to develop an understanding of such behavior of multiparticle systems, starting with the dynamics of single particles.

One might suppose that all the basic problems and important applications had been solved long ago. For it is fashionable for much of present-day science to probe the behavior of objects of extremely small size by means of cyclotrons or those of much larger size by radio telescopes. Only recently, however, have we begun to establish a satisfactory synthesis of the basic principles of slow viscous flow. Before proceeding to other matters of general interest, it is desirable to form an idea of the size range of objects with which we shall be concerned. The term *particle* is a loose one, but for convenience we shall assume that any solid object less than 10 cm in diameter (2.54 cm = 1 in.) falls into this category. Particles may vary downward from this size until molecular dimensions are reached, the lower limit of size of interest to us here being that in which the fluid surrounding our particles can be regarded as continuous. This will vary depending on the system being studied, but interesting results within the purview of the present treatment are obtained down to sizes of 10^{-7} cm (10 Ångstrom units). This range of size may be visualized by noting that the diameter of Earth is about 12×10^{8} cm (8000 miles). Thus, Earth is just about as many times as big as the largest particle we shall consider, as this particle in turn is larger than the largest organic molecules. Figure 1-1.1 gives an idea of the different types of materials which may be encountered. Note that the scale is plotted logarithmically.

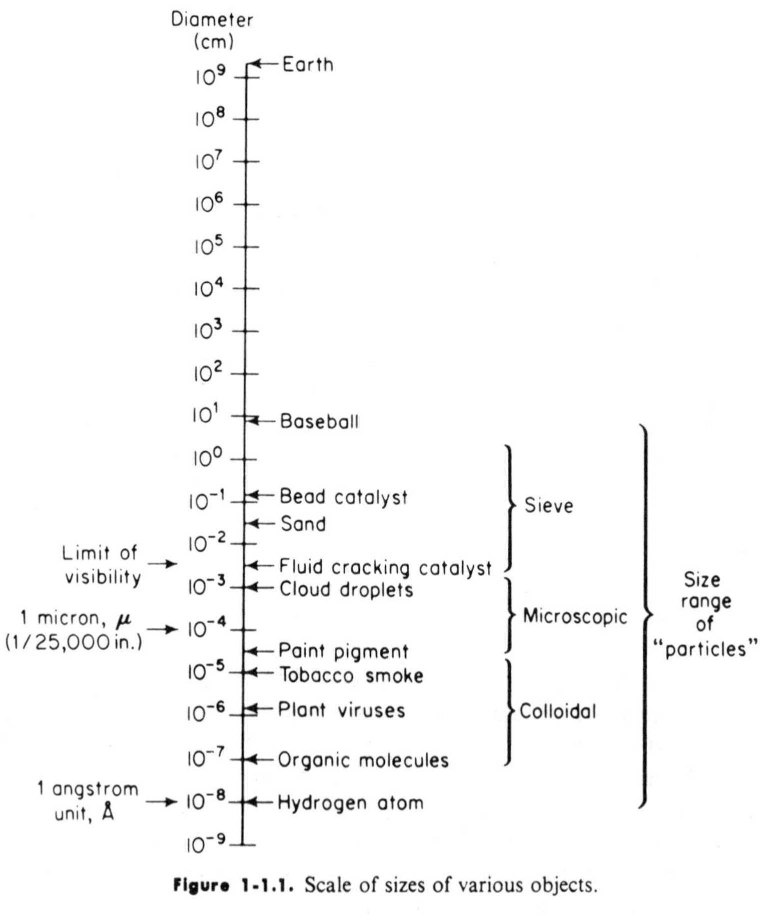

Figure 1-1.1. Scale of sizes of various objects.

In order to construct tractable mathematical models of the flow systems involving particles, it is necessary to resort to a number of simplifications. In this book it is assumed that the flow is laminar and, further, that it is sufficiently "slow" that inertial effects need not be considered in arriving at a solution of the equations of motion which describe the passage of fluid relative to particles in these systems. This simplification is justified, since many multiparticle systems do involve sufficiently slow motions for this assumption to be valid. Often, systems of interest will consist of very small particles, and even when the particles move rapidly with respect to container walls, they will move slowly with respect to fluid passing through them or with respect to each other. The systems which we shall treat are those which exhibit "strongly viscous" behavior, to use an expression coined by Shapiro[40]. Their behavior is governed by the so-called creeping motion or Stokes equations.

A dimensionless criterion which determines the relative importance of inertial and viscous effects is the Reynolds number:

$$\text{Reynolds number} = \frac{\text{fluid density} \times \text{speed} \times \text{size}}{\text{viscosity}}$$

Situations in which the Reynolds number is small are called *slow viscous flows*, because viscous forces arising from shearing motions of the fluid predominate over inertial forces associated with acceleration or deceleration of fluid particles. But the Reynolds number may be small for reasons other than the slowness of the motion or the high viscosity of the fluid. Thus flight of an object through rarefied air high above Earth's surface may represent a very viscous flow even though the air through which the object passes has a very low viscosity, because its density is correspondingly much lower. Of course, in this case, the dimensions of the object must be large compared with the mean-free path of the air molecules. Otherwise the continuum hypothesis is invalid. A speck of dust or a mist settling slowly through normal air, if sufficiently small, may represent a more viscous situation than a steel ball falling in molasses. In many practical situations involving sedimentation and fluidization, the Reynolds number (based on particle diameter) will be less than about 5. These phenomena are then amenable to treatment by the creeping motion equations.

Complications often arise because of the complex geometry encountered in assemblages composed of particles of arbitrary shape. Though the basic differential equations of motion may be well understood, it is still very difficult to arrive at exact or even approximate solutions in all but the simplest cases. Essentially, two techniques have been employed for handling boundary value problems involving a number of particles, namely, the *method of reflections* and the *unit cell* technique. In the method of reflections the boundary conditions are satisfied successively on each of the separate bounding surfaces involved, including the container walls confining the suspension when the fluid-particle system is bounded in extent. If the system is sufficiently dilute, that is, the number of particles per unit volume is small, rapid convergence is possible. This technique is thus especially adapted to determining the interaction effects among a few particles or between single particles and container walls. The unit cell technique, on the other hand, involves the concept that an assemblage can be divided into a number of identical cells, one particle (usually a sphere) occupying each cell. The boundary value problem is thus reduced to consideration of a single particle and its bounding envelope. This technique applies strictly only to periodic arrays. It can, however, also be applied in some stochastic sense to *random* particle arrays. The cell model is of greatest applicability in concentrated assemblages, where the effect of container walls can be neglected.

The cell technique may be employed to show schematically, and in a highly idealized form, the nature of the flow pattern involved in the basic types of motion just discussed. Various investigators have employed different shapes of cells, but the assumption of a spherical shape both for each particle and for a fictitious envelope of fluid surrounding it is of great convenience. The spherical form is of interest mathematically because it enables a surface to be described in terms of a single parameter. It is also of unusual

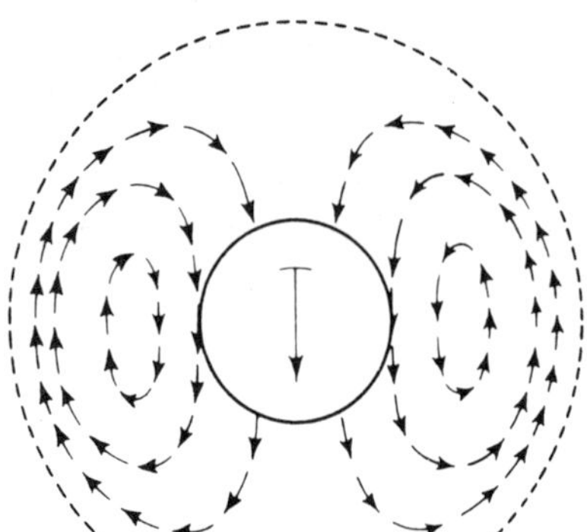

Figure 1-1.2. *Sedimentation.* The solid sphere at the center of the envelope moves downward.

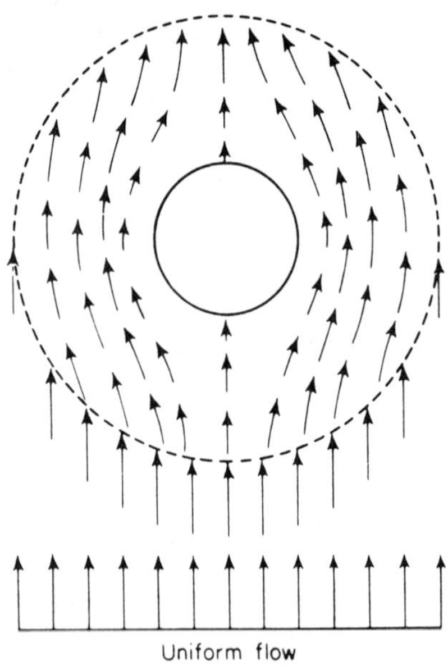

Figure 1-1.3. *Permeability.* The solid sphere is stationary. The undisturbed field entering the fluid envelope is a uniform flow.

practical interest because many particles approximate the spherical form. For illustration, we describe briefly the fluid patterns associated with the assumption of a concentric spherical cell model.

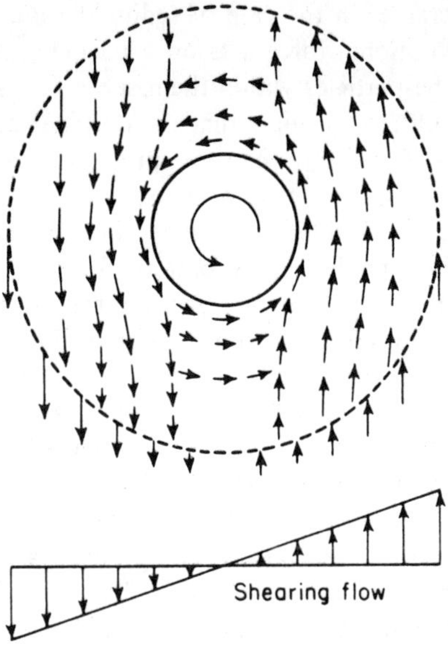

Figure 1-1.4. *Viscosity.* The solid sphere rotates. The undisturbed pattern entering the fluid envelope is a shearing field.

Figures 1-1.2–4 schematically illustrate the spherical "free surface"[23] cell model for the cases of sedimentation, flow through porous media, and suspension viscosity. In the case of sedimentation a number of particles are assumed to be settling with equal velocity under the influence of gravity through a fluid. Attention is focused on one particle, which is surrounded by the dotted line which constitutes the fluid envelope surrounding it. The radius of this fluid envelope is established by assuming that the cell contains the same volumetric proportion of solid to fluid as exists in the entire assemblage. Naturally, the hypothetical envelopes or cells surrounding each particle in an actual assemblage will be distorted, and some "leakage" of fluid from one cell to another will occur, but it is assumed that on the average a spherical cell may be employed, owing to the random arrangement. The entire disturbance due to each particle is thus confined within the cell of fluid with which the particle is associated. The outside surface, represented by the dotted line, is frictionless (that is, the tangential stresses are zero), so that fluid is free to pass over the surface, as if the cell were a raindrop freely suspended in the atmosphere. Instantaneous streamlines representing the path of flow of fluid within this envelope are shown schematically. They depict a circulatory type of motion. The direction and magnitude of motion at the surface of the fluid envelope is such that, if a similar fluid envelope surrounding another sphere approached the one depicted at the point where they touched, it would be found that the motion would be equal in magnitude and direction for the two spheres. Thus no friction is supposed to occur between adjacent fluid envelopes.

From the fluid velocity pattern shown in Fig. 1-1.2 it is possible to compute the drag force exerted by the fluid on the particle contained in the cell. As the particles come closer together the fluid surrounding each particle will be confined to a smaller envelope and resistance to its motion will increase. In the case of sedimentation, this means that, when a constant gravitational force acts on a particle, as concentration increases, the cells and the particles within them settle at a slower rate than in a system which was sufficiently dilute that the particles did not influence each other.

In exactly the same proportion as the settling velocity of an assemblage of given porosity decreases, resistance to flow of a fluid through a fluidized or packed bed of equal porosity increases. Figure 1-1.3 shows an idealized fluid velocity pattern for flow through a stationary cell. Knowledge of this field enables one to estimate, with considerable precision, the pressure drop resulting from passage of fluid through a dense bed of particles.

Figure 1-1.4 depicts the flow pattern which results when an otherwise uniform shearing field is disturbed by the presence of particles. The flow pattern shown permits us to calculate the energy dissipated by fluid friction in the envelope surrounding the particle. A spherical particle carried along by a shearing field will rotate to accommodate itself to the fluid motion as shown. The rate of energy dissipation in the cell occupied by the particle may be compared with that experienced by the undisturbed fluid when sheared at the same rate in the absence of the particle. This comparison provides a reliable estimate of the relative viscosity of a suspension of particles in fairly concentrated systems.

It should be emphasized that although cell models of this type apparently give a satisfactory approximation of the actual average flow pattern close to the particles in a real physical system, they cannot be expected to be valid for points close to the imaginary cell boundaries. Thus, such models will not predict effects such as convective transfer of fluid from one cell to others. For this purpose an approximate solution of the boundary value problem involved, using a technique like the reflection procedure, leads to more satisfactory results.

In addition to the significant simplification arising from the linearization of the equations of motion, and from further simplification of the geometry involved, it is necessary in many cases to make other additional assumptions in order to bring the analysis to fruition. These include such items as the assumption of uniformly sized particles, neglect of Brownian motion of the particles, and except for gravity, the absence of extraneous forces, such as electrostatic or interfacial forces. Good agreement of theoretical results with available experimental data is usually found in situations where these assumptions are warranted.

This book begins with a discussion of the basic equations of motion and their limitations. The behavior of single particles is then considered. Subsequently, the effect of hydrodynamic interactions among individual particles and between particles and walls is examined. Finally, these effects are superposed to treat the behavior of particle assemblages. The basic principles seem clearly established. They serve to illustrate the versatility of the creeping motion equations for handling problems where the viscous effects in particulate motion are dominant.

The treatment in the following chapters is largely mathematical. Experimental data are studied mostly in order to verify that the solutions obtained apply to real physical situations. In all cases treated, however, emphasis is on the hydrodynamic factors involved.

Familiarity with basic vector calculus is assumed. To a lesser extent, some knowledge of tensor and polyadic analysis is also required. Those not familiar with these topics and with the other aspects of engineering mathematics involved, should consult standard treatises. Bird, Stewart, and Lightfoot[5] provide a good introduction to the field of transport phenomena. A brief exposition of the vector and polyadic concepts useful in studying the theory of diffusion, fluid dynamics, and related topics is given in Drew's[13] handbook. Special mention should also be made of Gibbs' text* on vector and polyadic analysis. Aris[1] has published a useful text which devotes special attention to the application of vectors and tensors to problems in fluid mechanics.

At the end of this book, Appendix A provides a useful tabulation of the properties of several important curvilinear coordinate systems. Appendix B briefly summarizes the vector and tensor notation employed in this book.

The following chapter contains references to a number of texts in the general field of hydrodynamics. Closest to the present treatment in objective is the first section of the classic treatise by Oseen[35], which considers, among other things, situations involving the slow motion of particles in the presence of bounding walls.†

As for flow through assemblages of particles, numerous references are cited in Chapter 8. Scheidegger's[41] monograph on the physics of flow through porous media offers a good survey of both English and foreign research papers and texts, covering fundamental subjects and especially systems where the particulate medium is not dispersed.

*J. W. Gibbs, and E. B. Wilson, *Vector Analysis*, Dover reprint (New York: Dover, 1960).

†Other books devoted exclusively or almost exclusively to low Reynolds number flows are H. Villat, *Leçons sur les Fluides Visquex* (Paris: Gauthier-Villars, 1943); W. E., Langlois, *Slow Viscous Flow* (New York: Macmillan, 1964); O. A. Ladyzhenskaya, *The Mathematical Theory of Viscous Incompressible Flow* (New York: Gordon and Breach, 1964). In addition to these books, the following chapters in other books contain extensive summaries of various facets of low Reynolds number flows: I-Dee Chang *in* "Handbook of Engineering Mechanics" (W. Flugge, ed.). New York: McGraw Hill, 1962; C. R. Illingworth *in* "Laminary Boundary Layers" (L. Rosenhead, ed.). London: Oxford, 1963; R. Berker *in* "Encyclopedia of Physics: Fluid Dynamics II" (S. Flugge and C. Truesdell, eds.) Vol. 8 2. Berlin: Springer-Verlag, 1963; P. A. Lagerstrom *in* "Theory of Laminar Flows" (F. K. Moore, ed.). Princeton, N. J.: Princeton University Press, 1964; H. Brenner *in* "Advances in Chemical Engineering" Vol. 6 (T. B. Drew, J. W. Hoopes, Jr., and T. Vermeulen, eds.), New York: Academic Press, 1966.

The remainder of the present chapter is devoted to subjects of general interest in the study of particulate matter; namely, a brief historical review of how the subject has developed and an outline of the various applications in science and technology which supply motivation for research in this area.

1-2 Historical Review

The general history of experimental hydraulics and of theoretical hydrodynamics is interestingly treated by Rouse and Ince[39]. The treatises by Basset[4] and Dryden, Murnaghan, and Bateman[14] also include numerous references to the work of early investigators, especially with regard to laminar flows. Although practical hydraulics had its origins in antiquity, scientific attention to flow relative to particulate media began scarcely one hundred years ago. We do not intend to repeat the names, or summarize the work of all who have contributed to this field, but simply to call attention to those scientists and engineers who have been especially identified with fluid flow relative to particles and porous systems. Thus, we shall not enter into detail regarding the work of such leaders as Navier (1785–1836) who, together with Stokes, is credited with the formulation of the equations of motion used in hydrodynamics; Poiseuille (1799–1869), whose careful experiments on laminar flow established the law which bears his name; Lord Rayleigh (1842–1919), who investigated many phases of hydrodynamics, including energy dissipation; and Boussinesq (1842–1929), who made comprehensive mathematical treatments of laminar flow in pipes and channels. References to pertinent investigations of early contributors not discussed here do appear at appropriate sections of later chapters.

Studies of flow through porous media first engaged the attention of several engineers of the famous Corps des Ponts et Chaussées during the second half of the nineteenth century. Henry P. G. Darcy (1803–1858), à native of Dijon, after being educated in Paris, returned to the city of his birth where he became director of public works. There his major accomplishment was the design and execution of a municipal water supply system. This system not only functioned admirably, but also gave rise to a series of researches which he conducted on the flow of water through sand bed filters. He published the results of these studies along with much other information on the development of water supply systems in 1856[12]. (See Fig. 1-2.1.) The law which Darcy discovered, namely, that the rate of flow is proportional to pressure drop through a bed of fine particles, bears his name. It is widely employed for investigating the behavior of all types of water flow through porous media, such as underground flow to wells, flow in soils being irrigated, and the permeability of dam foundations. In addition, the flow of oil in underground substructures has been found to follow Darcy's

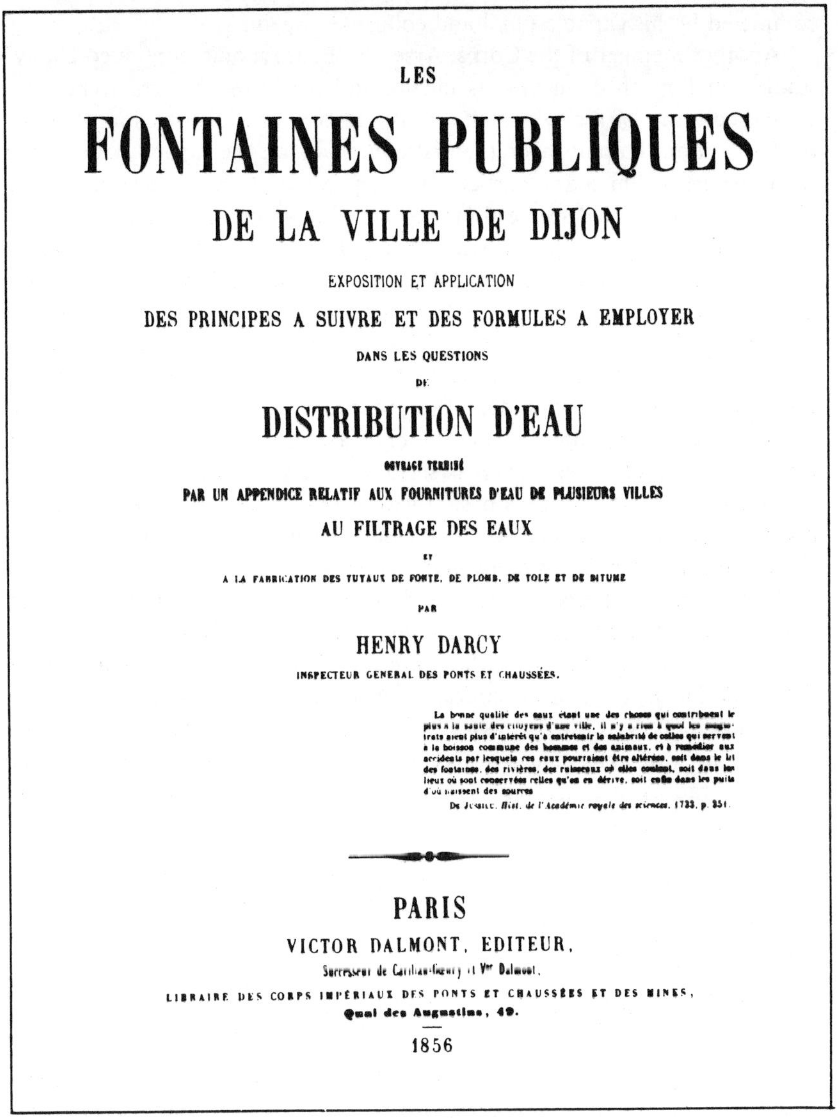

LES

FONTAINES PUBLIQUES

DE LA VILLE DE DIJON

EXPOSITION ET APPLICATION

DES PRINCIPES A SUIVRE ET DES FORMULES A EMPLOYER

DANS LES QUESTIONS

DE

DISTRIBUTION D'EAU

OUVRAGE TERMINÉ

PAR UN APPENDICE RELATIF AUX FOURNITURES D'EAU DE PLUSIEURS VILLES

AU FILTRAGE DES EAUX

ET

A LA FABRICATION DES TUYAUX DE FONTE, DE PLOMB, DE TOLE ET DE BITUME

PAR

HENRY DARCY

INSPECTEUR GENERAL DES PONTS ET CHAUSSÉES.

La bonne qualité des eaux étant une des choses qui contribuent le plus à la santé des citoyens d'une ville, il n'y a rien à quoi les magistrats aient plus d'intérêt qu'à entretenir la salubrité de celles qui servent à la boisson commune des hommes et des animaux, et à remédier aux accidents par lesquels ces eaux pourraient être altérées, soit dans le lit des fontaines, des rivières, des ruisseaux où elles coulent, soit dans les lieux où sont conservées celles qu'on en dérive, soit enfin dans les puits d'où naissent des sources

De Jussieu. *Hist. de l'Académie royale des sciences*, 1733, p. 351.

PARIS

VICTOR DALMONT, EDITEUR,

Successeur de Carilian-Gœury et Vve Dalmont,

LIBRAIRE DES CORPS IMPÉRIAUX DES PONTS ET CHAUSSÉES ET DES MINES,

Quai des Augustins, 49.

1856

Figure 1-2.1. Cover of Darcy's book

law, and a unit of permeability designated as the *darcy* is quite generally used in the oil industry today. Literally hundreds of studies have been devoted to experimental determination of the Darcy permeability for different types of porous media. Darcy's health had been failing for some time when he published his treatise. He died in 1858 while engaged on a project involving the movement of running water in open canals. This work was later completed by his capable pupil and colleague, Bazin.

Another member of the Corps, Arsene J. E. J. Dupuit, continued Darcy's studies on flow through porous media, and published his researches in a second edition[15] of a book on the transportation and distribution of water. In that edition, Dupuit made the first comments on the capacity of a stream to transport sedimentary particles in suspension. Other successors in the Corps carried forward studies in transport of sediment by rivers.

The first study of geometrical arrangements of spheres was made in 1899[42] by Slichter, who was especially interested in the flow of water through soil. He was led to a study of sphere arrangements in order to reduce the hydraulics of a complex soil to an idealized system. He also made the first attempt to derive a porosity function for beds of uniform spheres, by making the simplifying assumption that the average cross-sectional area for flow would be triangular. By applying the equivalent of Poiseuille's law for flow through a tube of triangular cross section, he obtained an appropriate permeability equation. Slichter's basic formula was proved inadequate because his generalized model was oversimplified. Nevertheless, his treatment provided the starting point of many subsequent studies.

One of the most fruitful ideas leading to the type of permeability equations currently employed for predicting resistance to flow through porous media was the utilization of the empirical hydraulic radius concept[6] by Blake (1922), an American chemical engineer. It had been found earlier that resistance to flow through passages of various noncircular conduits could be brought into agreement with that for circular pipes by employing the hydraulic radius to characterize the respective cross sections. This quantity is defined by the expression,

$$\text{Hydraulic radius} = \frac{\text{cross-sectional area normal to flow}}{\text{wetted perimeter}}$$

An equivalent form is

$$\text{Hydraulic radius} = \frac{\text{volume filled with liquid}}{\text{wetted surface}}$$

Blake was apparently the first to realize that a packed bed might be regarded as a single pipe with a very complicated cross section, and that the interstitial volume could be divided by the wetted packing area to obtain a hydraulic radius. Kozeny (1927)[26] extended this treatment. Subsequently, Carman (1937)[8] suggested modifications which resulted in a semiempirical equation, now widely used to correlate data for flow through packed beds, see Eq. (8-5.10).

During the same period that the initial experimental studies were being conducted on flow through porous media, attention was first devoted to the theoretical aspects of flow in particulate systems. The earliest study on resistance of a solid body moving relative to a fluid, in which viscosity was taken into account, was published by Sir George G. Stokes (1819–1903). Born in Skreen, Ireland, Stokes received his education at Cambridge. He became a professor and remained in Cambridge the remainder of his life, making many important contributions to theoretical physics. His early interests centered around the subject of hydrodynamics, which was then rather neglected in the general researches being undertaken. His early papers, mostly in theoretical hydrodynamics, appeared in the Cambridge *Philosophical Transactions.* Of greatest interest to us is the paper in which he linearized the general equations of motion of a viscous incompressible fluid and thus obtained a time-dependent form of the creeping motion equations. He applied these linearized equations to estimate the frictional damping of the motion of a spherical pendulum bob due to air resistance (1851)[47]. As the frequency of oscillation of the pendulum approaches zero, it moves with essentially constant velocity through the air. The resistance to fall of such a body of spherical shape was developed in this paper, and the relationship he discovered is known as Stokes' law, see Eq. (2–6.3.) It has been found to apply to the sedimentation of all manner of small particles which are moving slowly. The derivation he presented is an elegant one mathematically and appears in many text books on hydrodynamics. It applies in the situation where the particles are far enough apart so that the motion of each one of them is not affected by the motion of its neighbors. Stokes made many other important contributions to the Cambridge school of mathematical physics before his death at the age of eighty-four.

Lamb's[30] classic treatise on hydrodynamics which first appeared in 1879, and subsequently passed through six editions, contains much historical and technical information on the development of solutions of the creeping motion equations, though it is devoted mainly to potential flows. Worth special mention also is the solution of the steady translation of an ellipsoid parallel to a principal axis in a viscous fluid by Oberbeck (1876)[34].

As for the influence of bounding surfaces on the motion of a single body, H. A. Lorentz (1896)[31], following the method developed by Stokes (1845)[46], determined the motion of a sphere in the presence of a plane wall. The technique used involves "reflection" of the original motion produced by the body from the surface of the wall and back again to the body. Soon after, R. Ladenburg (1907)[29] exploited the same technique to determine the effect of a cylindrical tube on the axial motion of a centrally positioned sphere.

The same method was employed by the Polish mathematician M. von Smoluchowski to determine the effects of hydrodynamic interaction between two spheres moving in a viscous fluid (1911)[43]. Shortly afterwards, he again employed the "method of reflections" to study the sedimentation of an assemblage of spheres (1912)[44]. E. Cunningham (1910)[10] considered the sedimentation of a cloud of particles in a closed vessel, employing a cell model. His estimate of the decrease in terminal settling velocity due to particle interaction was based on the approximate supposition that each particle moves, on the average, as if it were contained in a rigid spherical envelope, of radius equal to half the distance to its nearest neighbors.

These basic methods are still of considerable utility. For use of the method of reflections see Chapters 6 and 7; various cell models for sedimentation problems are developed in Section 8-4.

Many of the early contributions to low Reynolds number hydrodynamics are summarized in the book by the Swedish physicist, Carl W. Oseen (1927)[35]. Of special interest are the contributions of Hilding Faxen, his coworker, whose early papers are discussed by Oseen. Later researches of Faxen are mentioned in various chapters of the present book. Faxen's work has been of signal importance in advancing the theory of particulate systems.

The problem of finding the disturbance caused by the presence of a particle suspended in an otherwise uniform shearing flow was undertaken somewhat later than that for uniform motion. It is interesting that this problem was first solved as the doctoral dissertation of Albert Einstein (1879–1955). Einstein was born in Germany, but studied physics at the Zurich Polytechnic Institute. When he obtained his doctorate in 1905, he had become a Swiss subject. His thesis was concerned, among other things, with a new method for determining the size of molecules of chemical substances. In order to accomplish this, he developed a theory for the resistance to shear of a suspension of small spherical particles immersed in a continuous fluid, as a model for large molecules in solution. He showed theoretically that the apparent increase in viscosity of the suspending liquid could be related to the volumetric concentration of solid particles (or solute molecules) by a simple proportionality constant (1906, 1911)[16]. The Einstein law for suspension viscosity has been used since as the basis for almost all theories of the behavior of suspensions in shearing fields of flow. (See Section 9-6 for a discussion of his findings on the size of the sugar molecule.) Like Stokes' law, Einstein's applies to the case where the suspended particles are far enough apart on the average that their motion is not influenced by mutual interaction of the disturbances produced by individual particles. As is well known, Einstein's interests soon turned to relativity and quantum theory. He spent the last years of his life in the United States, of which he became a citizen.

Einstein's viscosity theory for suspensions of spherical particles and Jeffery's[25] extension of it to particles of ellipsoidal shape were further extended in several directions in a series of papers by Guth and his coworkers at the University of Vienna in 1936. Guth[21] has given a brief account of these studies. Especially interesting is Guth and Simha's paper[22] which considers wall effects and interaction between particles on the apparent viscosity of suspensions.

Developments during the last thirty years have been more varied and, in common with other branches of science, have become more numerous. Many of these developments were motivated either by other scientific or technological problems or because they found application in applied fields. It is believed that these advances are of interest not only to those interested in engineering applications, but also, hopefully, to the research worker, who may find in these advances sources of ideas from apparently unrelated disciplines.

1-3 Application in Science and Technology

The impact of the science of small particles in many directions has been well documented by DallaValle[11], whose book *Micromeritics* covers many aspects of particle technology other than those of a purely hydrodynamic nature. It includes such topics as the geometry of packing of particles, size measurement, sieving and grading, as well as electrical, optical, sonic, and surface properties of particles. In the survey which follows, we consider mostly those applications which relate in some way to the basic hydrodynamic theme of this book. We consider engineering applications first, in keeping with the idea that hydrodynamics may be approached as an engineering science in its own right, rather than as merely furnishing peripheral information to the physical sciences.

Chemical engineering

Perhaps the most obvious application of particle dynamics occurs in situations where mutual interaction of particles can be neglected, so that the fundamentals of single particle motion apply. Dust and mist collection from dilute suspensions of finely divided solids, and from liquids in gases, furnish simple practical examples of such idealizations[36]. One common application is in the elimination of atmospheric pollution, as in the cleaning of ventilating air or in the reduction of industrial health hazards, such as toxic fumes from chemical operations. The recovery of valuable by-products from the dusts leaving dryers and smelters represents another important application. Finally, many industrial operations involve the production of a powdered product which must be separated from the gas or liquid in which it is suspended, as, for example, the spray drying of milk and soap and the manufacture of zinc oxide and carbon black. Often, very fine particles are involved, permitting the laws of slow viscous flow for bodies falling in infinite media to be applied.

Gravitational and centrifugal sedimentation of dense suspensions of fine particles, in the form of slurries in liquids, constitutes a closely allied application, which arises in many industries. Concentrating and thickening devices are usually still designed on a fairly empirical basis. But it seems reasonable to suppose that their design can be made more rational as we acquire further knowledge of the laws of slow viscous flow.

In the more concentrated range, systems involving towers packed with special shapes or crushed solids have been widely used in chemical processing for many years. Such towers serve as contacting devices: to bring together gases and liquids for the purpose of absorption or desorption; to contact liquids with liquids for extraction; and to transfer heat to or from gases, as in kilns and gas producers. In some of these processes a moving bed of solids is employed, as in a blast furnace.

The past twenty-five years have seen numerous developments of fluid-particle operations in the chemical process industries[51,28,37], stemming from the commercial exploitation of continuous fluids-solids processing in the catalytic cracking of petroleum to produce high octane-number gasoline. An obvious development employed a moving bed in which a pelleted catalyst flowed downward through a reactor against an ascending stream of vapor. The spent catalyst was then subsequently conveyed either by elevator or pneumatically to a regenerator. These endeavors stimulated further developments in moving bed techniques[50]. Among the more recent applications of interest may be mentioned the moving bed oil-shale retort and the use of moving bed reactors for the production of uranium tetrafluoride.

A more striking application followed further developmental work, in which the catalyst was not only conveyed pneumatically, but suspended by vapor in the reactor and regeneration chambers. This led to the fluidized-bed technique. In this contacting method, catalyst is used in powder form. Small particles suspended by the proper velocity of vapors passing through them behave much like true liquids and can be transported or contacted with vapors, as if they were homogeneous fluids.

Figure 1-3.1 illustrates a modern fluid-solid catalytic cracking unit. In these processes, oil-feed vapors carry hot regenerated catalyst particles, varying from 20 to 180 microns, into a fluidized bed in the reactor. The gasified feed is cracked to produce gasoline. The cracked components pass to a separation system and the carbon-coated catalyst passes to the regenerator, flowing by gravity, just as if it were a liquid. The regenerator bed is kept in a fluid condition by means of an air stream which burns off the deposit on the catalyst, and the hot regenerated catalyst is again ready for contact with oil vapors. In a modern unit of the type depicted, 5000 tons/hr of catalyst particles contact the oil fractions being cracked. This is a solid moving process comparable to a major mining operation, such as that conducted by the Kennecott Copper Co. in Utah (300,000 tons/day). Yet it is

Figure 1-3.1. A fluid bed catalytic cracking unit. (Courtesy of M. W. Kellog Co.)

handled by only a handful of men simply watching control instruments, since all the particles are transported as suspensions and maintained in the fluidized state in the reactor and combustion zones.

Many other applications have followed the development of fluidization in the petroleum industry. Among the more interesting are metallurgical applications, such as the use of a fluidized iron ore bed to prepare iron by direct reduction with hydrogen, and the fluid-bed roasting of pyritic ores and limestone. The application of the fluid-bed technique to coal gasification in the production of water-gas appeared in the patent literature as early as 1922 as a development by Winkler. A number of plants were built using this process. It seems likely that this technique will find further wide industrial application.

The dynamics of particles suspended in fluidized beds is very complicated[51]. It is too much, as yet, to expect theoretical relationships to predict their behavior in detail. As noted by Zenz and Othmer[51], however, Reynolds numbers of particles encountered in catalytic cracking systems normally lie in the range $N_{Re} = 0.01–5.0$. The equations for slow viscous flow ought therefore to provide a meaningful basis for interpretation. In this range, pressure drop relationships of the type developed later in this book (see Chapter 8) are in very good agreement with data obtained for the case of particulate fluidization, where smooth uniform expansion of the bed of particles occurs. This type of fluidization normally occurs only when the continuous fluid is a liquid. When gases are employed, aggregative fluidization usually results. Here, nonuniform bed expansion accompanied by bubble formation occurs, making quantitative correlation difficult. This area presents a continuing challenge for both theoretical and experimental work.

The flow properties of many plastic materials and paint products involve the hydrodynamics of shearing flows of suspensions, as do also the flow and aggregating properties of paper-making fibers[45].

Civil engineering

There are many fluid-solids dynamical processes controlling man's environment, which consists, roughly speaking, of earth, water, and air. Earth is to a large extent composed of soil formed from sedimentary deposits of minerals like sand and clay. Soil has many of the properties of packed beds, but is more complicated owing to variations in particle size and shape as well as to nonisotropic distribution in strata.

The infiltration of water and oil through soil finds application in many fields, typical of which are petroleum engineering, soil mechanics, ground water hydrology, and sanitary engineering. Collins[9] presents a unified treatment of flow through porous media, starting with the basic physical characteristics of porous materials and leading ultimately to a treatment of some of the more complex flow problems, among these being simultaneous laminar flow of miscible fluids and flow accompanied by a phase transition.

More dilute systems, where the particles do not touch each other, occur in the case of silt which is carried by rivers. This silt may end by being distributed over wide areas of ocean bottom or confined to the formation of deltas as in the case of the Mississippi and the Nile. Such silt deposits often impart important agricultural advantages, but may, on the other hand, present problems when they accumulate in reservoirs behind dams or in navigation channels. The basic hydrodynamic principles involved are much the same as those encountered in the problems of the process industries.

A different category of problems is that involving dust, smoke, and mist[20], where the particles are smaller and are suspended in air. Though the basic hydrodynamic laws are the same for aerosols and hydrosols, the former have many special properties due to their very fine state of subdivision. Many problems of industrial hygiene derive from contamination of the atmosphere by smoke, bacteria, and other pollutants.

Mining engineering

Knowledge of the behavior and characteristics of suspensions of small particles in liquids is important in the separation of minerals. After ores are crushed to prescribed sizes, various flotation and settling processes are often employed for separation. Indeed, the movement of particles through fluids plays a role in all mineral-dressing processes[19].

Flow and seepage of water and solutions of minerals through permeable soil structures was important in early prehistory in forming deposits of water-soluble materials, and knowledge of such phenomena is important to the mining engineer. Control of the rheological properties of oil well drilling muds is important in petroleum production.

Problems of mine ventilation and prevention of dust explosions from combustible materials, such as coal dust, depend upon a knowledge of the dynamics of dilute suspensions of particles in the atmosphere.

Physical sciences

The flow properties of many disperse systems (suspensions, emulsions, gels, liquid sprays, foams, and powders) are important in their scientific study and characterization[24, 17]. Innumerable rheological investigations on suspensions of colloidal materials and macromolecules have been carried out with the objective of obtaining information as to the physicochemical nature of the particles themselves. Disperse systems, either as emulsions or dispersions, comprise a large group of materials of industrial importance[32], including cement, plaster, paint, ink, and paper coatings.

The falling-ball viscometer, which consists of a sphere falling in a circular cylinder, represents one of the standard fundamental methods[18] for determination of fluid viscosity. The exact theory of this apparatus is a direct application of the appropriate particle-wall interaction dynamics.

Biology

Normal blood is a suspension of particulate matter (red cells, white cells, and platelets) in a continuous medium, the plasma. Of the particulate matter present in whole blood, the red cells dominate in volume, occupying on the average 40 volume per cent. A comprehensive review of the rheological properties of blood, blood plasma, and certain mucous fluids has been given by Merrill and Wells[33]. A more detailed treatment is given in the symposium *Flow Properties of Blood*, edited by Copley and Stainsby.* The biologic functions of blood, plasma, and body fluids are shown to be intimately related to their rheological behavior, which appears far more complicated

*A. L. Copley and G. Stainsby (eds.), *Flow Properties of Blood* (New York: Pergamon Press, 1960).

than man-engineered systems, exhibiting non-newtonian behavior when tested in a special viscometer operating on whole blood samples before clotting occurred. Suspensions of red blood corpuscles in brine, however, appear to exhibit newtonian behavior up to higher concentrations (see Table 9-6.3).

As Merrill and Wells point out, this is but one example of increasingly effective collaboration between engineers, physicians, and scientists on medically important problems.

Earth sciences

Soil permeability has already been discussed under engineering applications. It also has an important bearing on various agricultural problems involving such matters as crop yield, soil accumulation, and vegetation as related to drainage of water supplied by natural and irrigation sources.

Geologists and geographers have always been concerned with the formation of sediments and various types of rock formation produced by them. Twenhofel's[49] treatise provides a source of basic information of many aspects of historical sedimentology. Sedimentologists have been much concerned in recent years with the role of turbidity currents.

Such currents consist of suspensions of sediment, flowing along the inclined floor of a quiescent body of clear water, propelled by gravitational forces resulting from the greater density of the sedimentary suspension. Workers on recent marine sediments are convinced of the important role of turbidity currents in accounting for many topographical features of the continental shelves, such as submarine canyons. One of the largest of these canyons is that formed by the Hudson River. This canyon extends for more than 100 miles across the continental shelf, originating at the entrance to New York Harbor. Its topological structure probably presents an underwater panorama as impressive and spectacular as the familiar Palisades, a few miles up the river. Many geologists believe that these canyons were created by heavy mud flows during periods when the glacial epic lowered the ocean level by several hundred feet. It is generally believed that the fine earth sediments of particles on the continental shelf were churned by storm waves into suspensions which then flowed down the continental slopes at great velocities without mixing with the water above. It is clear that, though inertial and turbulent effects are involved, the slow motion and shear properties of these suspensions will also be important[27].

These turbidity currents display similarities to other natural phenomena. Thus, Kuenen likens the part played by the silt on the shelf to that of a ringing voice bringing down an avalanche which gains in force as it advances. Another interesting example is presented by the clouds which sometimes emerge from volcanic lavas. These are so heavily laden with ashes that they flow down the slopes of the mountain with great velocities[7]. Suspensions of volcanic ash in water can also move rapidly down mountain sides, thus behaving similarly to fluidized beds.

Quicksand provides a dramatic illustration of a fluidization process found in nature. Here, finely divided particles of sand are suspended in the fluidized state by the passage of an underground spring of water through them. The density of the suspension is not very much greater than water, but the suspended material vastly increases the viscosity of the water, making the resistance of the suspension to shear very high. Hence the difficulty in swimming or other movement experienced by those trapped in such a deposit.

Another interesting form of fluid particle interaction in nature is the transportation of particles of solid by *saltation*—a series of leaps and bounds, alternating with impacts on the ground surface. This type of action is responsible for the formation of snow drifts, sand dunes, and various forms of beach erosion. Bagnold[3, 2] has described and analyzed many problems related to the formation of ripples and sand ridges and the growth and movement of sand dunes. In the case of sand storms, particles are not lifted high in the air and carried over great distances, as normally supposed, but move in a relatively dense suspension with saltation.

Meteorologists are concerned with more dilute suspensions, such as are involved in raindrop formation[38] and in the presence of fine particles in the atmosphere. Here, problems are encountered which involve the motion of different sizes of droplets and the possibilities of their agglomeration. Hydrodynamic forces play an important role in these phenomena.

To summarize, there is a whole world of fine particles whose size ranges from those of small molecules to ordinary dust and sand visible with the naked eye. Protein molecules, viruses, synthetic polymers, colloidal particles, cosmic dust, soot, and fly ash particles from atomic tests are part of a domain in which these materials have characteristic size, shape, and properties. Observation of the particles themselves[48] by modern tools, such as the electron microscope, is one phase of their study. Another is their mutual interaction and their movement in the presence of boundaries. In this area the study of hydrodynamics presents a remarkably unified understanding of much of their behavior. True, their hydrodynamic behavior often involves inertial and turbulent flow; other factors, such as particle collisions, electrical charges, and surface properties, may complicate matters further. A study of the effects produced by viscous forces, such as are exerted in slow flows, will, however, usually be advantageous and will very often account to a major extent for the phenomena observed.

BIBLIOGRAPHY

1 Aris, R., *Vectors, Tensors, and the Basic Equations of Fluid Mechanics.* Englewood Cliffs, N. J.: Prentice-Hall, Inc., 1962.

2 Bagnold, R. A., Phil. Trans. Roy. Soc. (London) **249A** (1956), 235.

3 ———, *The Physics of Blown Sand and Desert Dunes.* London: Methuen, 1941.

4 Basset, A. B., *A Treatise On Hydrodynamics*, 2 vols. (1888); New York: Dover, 1961.

5 Bird, R. B., W. E. Stewart, and E. N. Lightfoot, *Transport Phenomena*, New York: Wiley, 1960.

6 Blake, F. C., Trans. Amer. Inst. Chem. Engrs. **14** (1922), 415.

7 Burgers, J. M., Proc. Koningl. Akad. Wetenschap. **45** (1942), 9.

8 Carman, P. C., Trans. Inst. Chem. Engrs. (London) **15** (1937), 150. See also his book *Flow of Gases through Porous Media.* New York: Academic Press, 1956.

9 Collins, E. R., *Flow of Fluids through Porous Materials.* New York: Reinhold, 1961.

10 Cunningham, E., Proc. Roy. Soc. (London) **A83** (1910), 357.

11 DallaValle, J. M., *Micromeritics—The Technology of Fine Particles*, 2nd ed. New York: Pitman, 1948.

12 Darcy, H. P. G., *Les Fontaines Publiques de la Ville de Dijon.* Paris: Victor Dalmont, 1856.

13 Drew, T. B., *Handbook of Vector and Polyadic Analysis.* New York: Reinhold, 1961.

14 Dryden, H. L., F. P. Murnaghan, and H. Bateman, *Hydrodynamics*, reprint, *Bull.* No. 84 Nat. Res. Counc. New York: Dover, 1956.

15 Dupuit, A. J., *Traite theoretique et practique de la conduite et de la distribution des eaux.* Paris: 1865.

16 Einstein, A., Ann. Physik **19** (1906), 289; with correction **34** (1911) 591. For English translation of these and related papers, see Einstein, A., *The Theory of Brownian Movement*, New York: Dover, 1956.

17 Everett, D. H., and F. S. Stone, *The Structure and Properties of Porous Materials.* New York: Academic Press; Butterworth's, London, 1958.

18 Fox, T. G., S. Gratch, and S. Roshaek, in *Rheology—Theory and Applications*, Vol. I, ed. F. R. Eirich. New York: Academic Press, 1956, Chap. 12.

19 Gaudin, A. M., *Principles of Mineral Dressing.* New York: McGraw-Hill, 1939.

20 Green, H. L., and W. R. Lane, *Particulate Clouds, Dusts, Smokes, and Mists.* London: Spon, 1957.

21 Guth, E., Proc. International Congr. Appl. Mech. (1938), pp. 448–55.

22 ———, and R. Simha, Kolloid. Z. **74** (1936), 266.

23 Happel, J., Trans. N. Y. Acad. Sci. **20** (1958), 404.

24 Hermans, J. J., *Flow Properties of Disperse Systems*. Amsterdam: North-Holland Publishing Co.; New York: Interscience, 1953.

25 Jeffery, G. B., Proc. Roy. Soc. (London) **A102** (1922), 161.

26 Kozeny, J., Sitz-Ber. Wiener Akad., Abt. IIa, **136** (1927), 271; see also *Hydraulik*. Wien: Springer, 1953.

27 Kuenen, P. H., Society of Economic Paleontologists and Mineralogists, *Special Publication* No. 2, pp. 14–33, November, 1951.

28 Leva, M., *Fluidization*. New York: McGraw-Hill, 1959.

29 Ladenburg, R., Ann. d. Phys. **23** (1907), 447.

30 Lamb, H., *Hydrodynamics*, 6th ed. Cambridge: Cambridge Univ. Press, 1932; New York: Dover, 1945.

31 Lorentz, H. A., Zittingsverl. Akad. van Wet. **5** (1896), 168; neu bearb.: Abhandl. theoret. phys. **1** (1907), 23.

32 Mill, C. C., *Rheology of Disperse Systems*. New York: Pergamon, 1959.

33 Merrill, E. W., and R. E. Wells, Jr., Appl. Mech. Reviews **14** (1961), 663.

34 Oberbeck, H. A., Crelle, **81** (1876), 62.

35 Oseen, C. W., *Hydrodynamik*. Leipzig: Akademische Verlag, 1927.

36 Perry, J. H., *Chemical Engineers' Handbook*, 3rd ed. New York: McGraw-Hill, 1950.

37 Reboux, P., *Phenomenes de Fluidization*. Paris: Association Française de Fluidization, 28, rue St. Dominique, 1954.

38 Richardson, E. G., *Aerodynamic Capture of Particles*. New York: Pergamon, 1960.

39 Rouse, H., and S. Ince, "History of Hydraulics," State Univ. Iowa Inst. Hydraulic Res. 1957.

40 Shapiro, A. H., *Shape and Flow—The Fluid Dynamics of Drag*. Anchor Books: Science Study Series. New York: Doubleday, 1961.

41 Scheidegger, A. E., *The Physics of Flow through Porous Media*, 2nd ed. New York: Macmillan, 1960.

42 Slichter, C. S., "Theoretical Investigation of the Motion of Ground Waters," U. S. Geological Survey, 19*th Ann. Rep.*, Part 2 (1899), pp. 301–384.

43 Smoluchowski, M., Bull. Intern. acad. polonaise sci. lettres, **1A** (1911), 28.

44 ———, 5th Intern. Congr. Math. **2** (1912), 192.

45 Steenberg, B., and B. Johansson, Svensk Papperstidning **61** (1958), 696.

46 Stokes, G. G., Trans. Cambr. Phil. Soc. **8** (1845), 287.

47 ———, Trans. Cambr. Phil. Soc. **9**, pt. II (1851), 8.

48 Turkevich, J., Amer. Scientist, **47** (1959), 97.

49 Twenhofel, W. H., *Treatise on Sedimentation.* New York: Dover, 1960.

50 Vener, R. E., Chem. Eng. **62** (1955), 175.

51 Zenz, F. A., and D. F. Othmer, *Fluidization and Fluid Particle Systems.* New York: Reinhold, 1960.

The Behavior of Fluids in Slow Motion

2

2-1 The Equations of Change for a Viscous Fluid

In this book we are concerned exclusively with the equations of change for the *isothermal flow* of a homogeneous viscous fluid. These equations express the classical principles of conservation of mass and momentum and are given in detail in standard treatises[48, 39, 6]. For nonisothermal fluids and for inhomogeneous multicomponent fluid mixtures, additional equations are needed to take into account the principles of conservation of energy and the conservation of individual chemical species. Aris[3] presents a careful development of the basic equations from a general viewpoint.

The equation of continuity may be obtained by applying the law of conservation of mass to a small stationary volume element within a flowing fluid. In vector form it may be written as

$$\frac{\partial \rho}{\partial t} = - \nabla \cdot (\rho \mathbf{v}) \tag{2-1.1}$$

where ρ is the local density of the fluid, $\mathbf{v}$ is the local mass average fluid velocity, and $\partial/\partial t$ refers to the time rate of change at a fixed point in the fluid. The quantity $\nabla\cdot(\rho\mathbf{v})$ is called the *divergence* of $\rho\mathbf{v}$ and is sometimes written as div $\rho\mathbf{v}$. In turn, the vector $\rho\mathbf{v}$ is the mass flux density and its divergence is the net rate of mass efflux per unit volume. Thus, Eq. (2–1.1) states that the rate of increase of density in a differential volume element fixed in space is equal to the net rate of mass influx into the element divided by its volume.

The continuity equation may sometimes be used to advantage by writing it in the alternative, but equivalent, form

$$\frac{D\rho}{Dt} = -\rho\nabla\cdot\mathbf{v} \tag{2–1.2}$$

in which the operator D/Dt is the time derivative along a path following the fluid motion. The operator D/Dt, called the *substantial derivative*, or Stokes operator, is defined as follows:

$$\frac{D}{Dt} = \frac{\partial}{\partial t} + \mathbf{v}\cdot\nabla \tag{2–1.3}$$

The equation of continuity in the form Eq. (2–1.2) describes the rate of change of density as measured by an observer moving along with the fluid.

All applications in this book are concerned with incompressible fluids, which have constant density; ρ is therefore constant and the equation of continuity reduces to

$$\nabla\cdot\mathbf{v} = 0 \tag{2–1.4}$$

We consider next the equation of linear momentum for a continuous fluid. This may be obtained by application of Newton's laws of motion to a differential volume of fluid. Newton's laws may be interpreted as stating that the external forces exerted by the surroundings on a stationary fluid element are equal to the time rate at which momentum is being created within the element. The external forces are two in number: (a) the surface or contact forces exerted by the fluid stresses acting over the surface of the element: (b) the volume or body forces, for example, gravity, exerted on the element. Now, the rate of creation of momentum in the volume is simply equal to the rate of accumulation or increase of momentum in the volume plus the net rate of efflux of momentum out of the volume element through its surface. Thus, we obtain

$$\underbrace{\underbrace{\frac{\partial(\rho\mathbf{v})}{\partial t}}_{\substack{\text{Rate of increase}\\ \text{of momentum per}\\ \text{unit volume}}} + \underbrace{\nabla\cdot(\rho\mathbf{v}\mathbf{v})}_{\substack{\text{Rate of}\\ \text{momentum loss by}\\ \text{convection through}\\ \text{the surface, per}\\ \text{unit volume}}}}_{\substack{\text{Rate of creation of}\\ \text{momentum per unit volume}}} = \underbrace{\underbrace{\nabla\cdot\boldsymbol{\Pi}}_{\substack{\text{Stresses on}\\ \text{the surface,}\\ \text{per unit}\\ \text{volume}}} + \underbrace{\rho\mathbf{F}}_{\substack{\text{External body}\\ \text{force on}\\ \text{element, per}\\ \text{unit volume}}}}_{\substack{\text{External forces per}\\ \text{unit volume}}} \tag{2–1.5}$$

The pressure or stress tensor $\mathbf{\Pi}$ is a second-rank tensor (dyadic). It is defined according to the usual convention that if $d\mathbf{S}$ is a directed element of surface area, $d\mathbf{S}\cdot\mathbf{\Pi}$ gives the contact force exerted by the fluid into which the vector $d\mathbf{S}$ is directed on the fluid on the opposite side of the surface element. For the structureless fluids* considered in this book, external angular momentum is conserved. In consequence of this, $\mathbf{\Pi}$ is symmetric.[3] Thus, of the nine components of Π_{jk} $(j, k = 1, 2, 3)$ only six are independent by virtue of the symmetry relation $\Pi_{jk} = \Pi_{kj}$.

In Eq. (2–1.5), $\mathbf{F}$ is the external body force per unit *mass*. Typically, it arises from the action of gravity. If $\mathbf{g}$ is the local acceleration of gravity vector, directed vertically downward, then $\mathbf{F} = \mathbf{g}$.

Note that $\nabla\cdot(\rho\mathbf{v}\mathbf{v})$ and $\nabla\cdot\mathbf{\Pi}$ are not simple divergences because of the tensorial nature of $\rho\mathbf{v}\mathbf{v}$ and $\mathbf{\Pi}$. Eq. (2–1.5) may be rearranged, using the continuity equation, and the definition of the substantial derivative of a vector field, namely,

$$\frac{D\mathbf{v}}{Dt} = \frac{\partial \mathbf{v}}{\partial t} + \mathbf{v}\cdot\nabla\mathbf{v} \tag{2–1.6}$$

where $\nabla\mathbf{v}$ may be taken as a dyadic. The equation of momentum is then obtained in the alternative form

$$\underbrace{\rho\frac{D\mathbf{v}}{Dt}}_{\substack{\text{Mass per unit}\\\text{volume times}\\\text{acceleration}}} = \underbrace{\nabla\cdot\mathbf{\Pi}}_{\substack{\text{Stresses on}\\\text{element per}\\\text{unit volume}}} + \underbrace{\rho\mathbf{F}}_{\substack{\text{Body forces on}\\\text{element per}\\\text{unit volume}}} \tag{2–1.7}$$

In this form, the equation of momentum is referred to a small volume element moving with the fluid and accelerated by forces acting upon it.

The vector velocity $\mathbf{v}$ appearing in these equations is the *mass average* local velocity. Since we are restricting ourselves to homogeneous fluids, where no relative diffusion of individual species occurs, it is unnecessary to distinguish this velocity from other possible velocities. Because of the particular form we have adopted for Newton's laws of motion, however, it should be clearly remembered that $\mathbf{v}$ must be measured relative to an observer in an inertial reference frame.

These equations are applicable to any continuous fluid medium. In order to use them, it is necessary to express the pressure tensor in terms of measurable pressures and velocities. The theory of newtonian fluids leads to the following relation:

$$\mathbf{\Pi} = -p\mathbf{I} + \kappa(\nabla\cdot\mathbf{v})\mathbf{I} + 2\mu\mathbf{\Delta} \tag{2–1.8}$$

*For "structured continua" the pressure tensor is no longer symmetric since external angular momentum may be converted into internal (intrinsic) angular momentum and vice versa. And it is only the *total* angular momentum which need be conserved. For a comprehensive discussion of this point see J. Dahler and L. E. Scriven, *Proc. Roy. Soc.* **A275** (1963), 504.

where p = the hydrostatic pressure the fluid would be supporting if it were at rest at its local density and temperature, T.

$\mathbf{I}$ = the idemfactor (unit tensor δ_{jk}).

κ = bulk or volume viscosity, sometimes written as $\lambda + (2\mu/3)$.

μ = shear viscosity.

$\boldsymbol{\Delta}$ = rate of deformation tensor, defined in Eq. (2–1.12).

The two phenomenological coefficients, κ and μ, are intrinsic properties of the fluid—independent of its state of motion. In general they depend on ρ and T.

The two terms in Eq. (2–1.8) which are multiplied by the idemfactor result in contributions to the normal stresses. For compressible fluids a distinction must be made between the so-called mean normal pressure, $\bar{p}$, at a point and the pressure p which appears in this equation. The mean pressure $\bar{p}$ is defined as follows:

$$\bar{p} = -\tfrac{1}{3}\boldsymbol{\Pi} : \mathbf{I}^* \tag{2–1.9}$$

$$= -\tfrac{1}{3}(\Pi_{xx} + \Pi_{yy} + \Pi_{zz}) \tag{2–1.10}$$

$$= p - \kappa(\nabla \cdot \mathbf{v}) \tag{2–1.11}$$

Thus the pressure, p, at any point in a fluid is larger than the mean normal pressure by an additive term proportional to the local rate of expansion, $\nabla \cdot \mathbf{v}$. The proportionality constant is the bulk viscosity coefficient, which relates stress to volumetric deformation rate in the same way that shear viscosity relates stress to linear deformation rate. The bulk viscosity is important in the case of fluids subjected to rapidly varying forces such as those caused by ultrasonic vibrations. In the case of low density monatomic gases, $\kappa = 0$. Formulas for estimating κ in the case of dilute polyatomic gas and dense gases are available[28]. For a further discussion see Aris[3], and Landau and Lifshitz[35].

The remaining term in Eq. (2–1.8) is based on a linear relationship between the viscous portion of the pressure tensor and the rate of deformation (shear) tensor. This latter is defined as follows:

$$\boldsymbol{\Delta} = \tfrac{1}{2}[\nabla\mathbf{v} + (\nabla\mathbf{v})^{\dagger}] - \tfrac{1}{3}\mathbf{I}(\nabla \cdot \mathbf{v}) \tag{2–1.12}$$

where $(\nabla\mathbf{v})^{\dagger}$ = the transpose or conjugate of $\nabla\mathbf{v}$, sometimes written $(\nabla\mathbf{v})^{t}$ or $\widetilde{\nabla\mathbf{v}}$.

The newtonian hypothesis, Eq. (2–1.8), has been theoretically shown[13] to be a complete first-order correction to the theory of perfect fluids in the limit where all gradients are small. Thus newtonian fluids should approximate the behavior of many real fluids in the limit of slow motions.

*For the double dot product of two dyadics we follow Gibbs'[14] notation, $\mathbf{ab} : \mathbf{cd} = (\mathbf{a} \cdot \mathbf{c})(\mathbf{b} \cdot \mathbf{d})$.

Substitution of the resulting expression for $\mathbf{\Pi}$ into Eq. (2–1.7) ultimately gives the following general equation of motion for a newtonian fluid:

$$\begin{aligned}\rho \frac{D\mathbf{v}}{Dt} = & -\nabla p + \mu\nabla^2\mathbf{v} + \tfrac{1}{3}\mu\nabla(\nabla\cdot\mathbf{v}) \\ & + 2(\nabla\mu)\cdot\nabla\mathbf{v} + (\nabla\mu)\times(\nabla\times\mathbf{v}) \\ & - \tfrac{2}{3}(\nabla\mu)(\nabla\cdot\mathbf{v}) + \kappa\nabla(\nabla\cdot\mathbf{v}) \\ & + (\nabla\kappa)(\nabla\cdot\mathbf{v}) + \rho\mathbf{F}\end{aligned} \tag{2–1.13}$$

This equation, along with the equation of state $p = p(\rho, T)$, the density dependence of shear viscosity $\mu = \mu(\rho, T)$, the density dependence of bulk viscosity $\kappa = \kappa(\rho, T)$, and the boundary and initial conditions, determines completely the pressure, density, and velocity components of a flowing isothermal fluid.

The equation is also applicable to nonisothermal flow of a homogeneous fluid, but it is then also necessary to utilize the equation of energy, in addition to boundary conditions involving temperature or heat flux, in order to determine the dependent variables. Viscosity and density will then be functions of temperature too. Eq. (2–1.13) is not usually used in its complete form, but rather is specialized for individual flow problems.

Thus, κ is often taken equal to zero. If this is done, substitution into Eq. (2–1.13) of the vector identity,

$$\nabla^2\mathbf{v} = \nabla(\nabla\cdot\mathbf{v}) - \nabla\times(\nabla\times\mathbf{v}) \tag{2–1.14}$$

yields a general equation consistent with that given by Milne-Thomson.[39]

When gradients of temperature and pressure are small, terms in $\nabla\mu$ and $\nabla\kappa$ may be omitted. If these terms and those containing κ itself are omitted, we obtain a form frequently used for compressible fluids.[48]

For constant ρ (incompressible fluids, $\nabla\cdot\mathbf{v} = 0$) and constant μ, Eq. (2–1.13) reduces to the one form of the Navier-Stokes equations,

$$\rho\frac{D\mathbf{v}}{Dt} = \rho\left(\frac{\partial\mathbf{v}}{\partial t} + \mathbf{v}\cdot\nabla\mathbf{v}\right) = -\nabla p + \mu\nabla^2\mathbf{v} + \rho\mathbf{F} \tag{2–1.15}$$

first derived by Navier[40] in 1827.

In order to employ the equations of continuity and motion, they must be expressed in coordinate systems appropriate to the shape of the boundaries involved in specific problems. Appendix A outlines general rules for accomplishing this.

In particular, consider a system of Cartesian coordinates (x, y, z). Set

$$\mathbf{v} = \mathbf{i}u + \mathbf{j}v + \mathbf{k}w \quad \text{and} \quad \mathbf{F} = \mathbf{i}X + \mathbf{j}Y + \mathbf{k}Z$$

where $(\mathbf{i}, \mathbf{j}, \mathbf{k})$ are unit vectors parallel to the respective coordinate axes. For incompressible fluids, the continuity equation and Navier-Stokes equations become, respectively,

$$\frac{\partial u}{\partial x} + \frac{\partial v}{\partial y} + \frac{\partial w}{\partial z} = 0 \tag{2–1.16a}$$

and

$$\begin{aligned}
\frac{\partial u}{\partial t} + u\frac{\partial u}{\partial x} + v\frac{\partial u}{\partial y} + w\frac{\partial u}{\partial z} &= -\frac{1}{\rho}\frac{\partial p}{\partial x} + \frac{\mu}{\rho}\left(\frac{\partial^2 u}{\partial x^2} + \frac{\partial^2 u}{\partial y^2} + \frac{\partial^2 u}{\partial z^2}\right) + X \\
\frac{\partial v}{\partial t} + u\frac{\partial v}{\partial x} + v\frac{\partial v}{\partial y} + w\frac{\partial v}{\partial z} &= -\frac{1}{\rho}\frac{\partial p}{\partial y} + \frac{\mu}{\rho}\left(\frac{\partial^2 v}{\partial x^2} + \frac{\partial^2 v}{\partial y^2} + \frac{\partial^2 v}{\partial z^2}\right) + Y \\
\frac{\partial w}{\partial t} + u\frac{\partial w}{\partial x} + v\frac{\partial w}{\partial y} + w\frac{\partial w}{\partial z} &= -\frac{1}{\rho}\frac{\partial p}{\partial z} + \frac{\mu}{\rho}\left(\frac{\partial^2 w}{\partial x^2} + \frac{\partial^2 w}{\partial y^2} + \frac{\partial^2 w}{\partial z^2}\right) + Z
\end{aligned} \tag{2–1.16b}$$

The components of the pressure tensor for a viscous incompressible fluid, corresponding to Eq. (2–1.8) with $\kappa = 0$ and $\nabla \cdot \mathbf{v} = 0$, are, in Cartesian coordinates,

$$\begin{aligned}
\Pi_{xx} &= -p + 2\mu\frac{\partial u}{\partial x} \\
\Pi_{yy} &= -p + 2\mu\frac{\partial v}{\partial y} \\
\Pi_{zz} &= -p + 2\mu\frac{\partial w}{\partial z} \\
\Pi_{xy} &= \Pi_{yx} = \mu\left(\frac{\partial u}{\partial y} + \frac{\partial v}{\partial x}\right) \\
\Pi_{yz} &= \Pi_{zy} = \mu\left(\frac{\partial v}{\partial z} + \frac{\partial w}{\partial y}\right) \\
\Pi_{zx} &= \Pi_{xz} = \mu\left(\frac{\partial w}{\partial x} + \frac{\partial u}{\partial z}\right)
\end{aligned} \tag{2–1.17}$$

When the density ρ is a constant, the terms ∇p and $\rho\mathbf{F}$ can be combined when $\mathbf{F}$ is expressible as the gradient of a potential. Thus, if the body force is due to gravity, $\mathbf{F} = -\nabla gz$, where z is the vertical height of the point above any standard datum plane and $g = |\mathbf{g}| =$ constant. This assumes that variations in the gravitational field with elevation are negligible over the vertical distances of interest. Equation (2–1.15) then becomes

$$\rho\frac{D\mathbf{v}}{Dt} = -\nabla(p + \rho gz) + \mu\nabla^2\mathbf{v} \tag{2–1.18}$$

Normally, this equation will be employed with p written in place of $p + \rho gz$. The pressure p appearing therein is then termed the *dynamic* or *hydrodynamic* pressure. The latter vanishes when the fluid is at rest.

When $\mu = 0$, Eq. (2–1.18) reduces to the well-known Euler equation for a frictionless or ideal fluid. In the case of irrotational motion, $\nabla \times \mathbf{v} = \mathbf{0}$, one obtains the equations of potential flow. These form the basis for much of classical hydrodynamic theory. Since steady potential streaming flows exert no forces on stationary solid bodies, the theory is useful mostly for predicting fluid flow patterns at a distance from boundaries.

In the case of a viscous fluid in contact with a solid, the relative tangential velocity is experimentally observed to be zero (no-slip). In addition, of course, one must satisfy the kinematical condition that the normal velocity of the fluid be the same as that of the boundary. This latter condition holds whether the surface is fluid or solid, or whether the fluid is viscous or not. Thus, for the boundary at a stationary solid surface one has the vector boundary condition

$$\mathbf{v} = \mathbf{0} \tag{2–1.19}$$

for the relative velocity.

The assumption of no slippage is valid unless the mean free path of the molecules becomes large compared with the dimensions of the bounding surfaces. Supersonic aerodynamics and high vacuum flows through small capillaries require, in many cases, a molecular rather than continuum approach[44].

2-2 Mechanical Energy Dissipation in a Viscous Fluid

If the directed element of surface area $d\mathbf{S}$ upon which a stress is acting moves with the fluid at a velocity $\mathbf{v}$, the quantity

$$\mathbf{v}\cdot\mathbf{\Pi}\cdot d\mathbf{S} \tag{2–2.1}$$

gives the instantaneous rate at which fluid on the side of the element into which $d\mathbf{S}$ is directed is doing work on the matter lying on the opposite side of the element, owing to the action of the stresses. The vector $\mathbf{v}\cdot\mathbf{\Pi}$ may therefore be regarded as a flux of mechanical energy. The rate unit time per unit volume at which energy is being supplied by the work done by the stresses on an element of volume (neglecting the effect of body forces) is thus $\nabla\cdot(\mathbf{v}\cdot\mathbf{\Pi})$.

This work may be separated into three parts as follows:

$$\underset{\substack{\text{Work done by}\\ \text{stresses per unit}\\ \text{volume on an}\\ \text{element of volume}}}{\nabla\cdot(\mathbf{v}\cdot\mathbf{\Pi})} = \underset{\substack{\text{Work done on}\\ \text{element by}\\ \text{changing its}\\ \text{volume}}}{-p\nabla\cdot\mathbf{v}} + \underset{\substack{\text{Work done on}\\ \text{element in}\\ \text{overcoming in-}\\ \text{ternal friction}}}{\Phi} + \underset{\substack{\text{Work done in}\\ \text{motion of the}\\ \text{element as a}\\ \text{whole}}}{\mathbf{v}\cdot(\nabla\cdot\mathbf{\Pi})} \tag{2–2.2}$$

The quantity Φ is the rate of mechanical energy dissipation. It expresses the local rate per unit time per unit volume at which mechanical energy is dissipated in deforming a fluid element against the opposing stresses. For a newtonian fluid, it is given by the following expression:

$$\Phi = 2\mu\mathbf{\Delta}:\mathbf{\Delta} + \kappa(\nabla\cdot\mathbf{v})^2 \tag{2–2.3}$$

The coefficients μ and κ are nonnegative, whence it follows that Φ can never be negative. It is zero only when $\mathbf{\Delta} = \mathbf{0}$ and $\nabla \cdot \mathbf{v} = 0$, corresponding to a rigid body motion. The second term in Eq. (2–2.3) vanishes for an incompressible fluid.

The rate of mechanical energy dissipation in any fluid region V may be found by integration of the local dissipation rate over the fluid domain:

$$E = \int_V \Phi \, dV \tag{2–2.4}$$

$\frac{1}{2}E$ is sometimes called the Rayleigh dissipation function[34].

2-3 Force and Couple Acting on a Body Moving in a Viscous Fluid

Consider some body immersed in the fluid, and let $d\mathbf{S}$ be a directed element of surface area normal to the body and pointing into the fluid surrounding the body. The resultant force, due to the stresses, exerted by the surrounding fluid on the body is

$$\mathbf{F} = \int_{\text{body}} \mathbf{\Pi} \cdot d\mathbf{S} \tag{2–3.1}$$

The pressure tensor $\mathbf{\Pi}$ is obtained from Eq. (2–1.8). Thus, if the fluid is incompressible, and μ is constant

$$\mathbf{F} = -\int_{\text{body}} p \, d\mathbf{S} + 2\mu \int_{\text{body}} \mathbf{\Delta} \cdot d\mathbf{S} \tag{2–3.2}$$

Instead of summing the stresses over the surface of the body, an alternative procedure involves applying the theorem of momentum to an infinite mass of fluid surrounding the body[2].

The torque, if any, experienced by the body may be similarly obtained by summing the torques on the surface elements $d\mathbf{S}$. Its value depends upon the choice of origin. Consider an arbitrary origin labeled O, and let $\mathbf{r}_O$ be the position vector of a point relative to an origin at O. Since the force on $d\mathbf{S}$ is $\mathbf{\Pi} \cdot d\mathbf{S}$, the moment of this force about O is $\mathbf{r}_O \times \mathbf{\Pi} \cdot d\mathbf{S}$. Hence, the torque about O due to the stresses exerted by the surrounding fluid is

$$\mathbf{T}_O = \int_{\text{body}} \mathbf{r}_O \times \mathbf{\Pi} \cdot d\mathbf{S} \tag{2–3.3}$$

Since we shall be concerned with incompressible fluids, it is not necessary for us to concern ourselves explicitly with gravitational forces acting on the fluid. Thus, we shall interpret p as the hydrodynamic, rather than the total pressure. The former does not include the hydrostatic pressure. With this convention assigned to p, it is convenient to refer to the force $\mathbf{F}$ in Eq. (2–3.1) as the *hydrodynamic* force exerted by the fluid on the body. It vanishes for a fluid at rest. Since, in reality, gravity always acts on the fluid, in order to obtain the total force exerted by the fluid on the body one must add to Eq. (2–3.1) the *buoyant* force on the body. In accordance with Archimedes' law, this additional force is equal to the weight of fluid displaced by the body. If $\mathbf{g}$ is the acceleration of gravity vector, directed vertically downward, (assumed constant), and m_f is the mass of displaced fluid, the buoyant force is

$$\mathbf{F}_{\text{buoyant}} = -m_f \mathbf{g} \tag{2–3.4}$$

Similar additions must be made to the torque formula, for with this interpretation of the pressure, Eq. (2–3.3) gives only the hydrodynamic torque. One must, in general, add a buoyant torque to this in order to obtain the total torque exerted by the fluid stresses on the body. If $\mathbf{r}_{OB}$ is a vector drawn from O to the center of buoyancy B of the body, then the buoyant torque about O, which the fluid exerts on the body, is

$$(\mathbf{T}_O)_{\text{buoyant}} = -m_f \mathbf{r}_{OB} \times \mathbf{g} \tag{2–3.5}$$

Westberg[62] gives general formulas for the resultant forces and torques exerted by a viscous incompressible fluid in unsteady motion on a body which is not necessarily rigid. These formulas involve various transformations of Eqs. (2–3.2) and (2–3.3).

2-4 Exact Solutions of the Equations of Motion for a Viscous Fluid

Owing to the generally nonlinear nature of the Navier-Stokes equations, exact solutions are difficult to obtain. Only a very few are known, except for the relatively trivial cases where the nonlinear terms vanish identically, as in duct flows. Thus, it has not yet proved possible to carry out a complete investigation for the steady streaming flow of a viscous fluid around a body at large Reynolds numbers. The few available solutions are treated at length in standard texts on fluid mechanics[52, 58] and will only briefly be referred to here. It is worth noting that all the exact solutions known substantiate the assumptions of boundary layer theory, which is widely used for approximate or asymptotic solutions valid at large Reynolds numbers.

Flow between nonparallel plates is a two-dimensional problem in which the streamlines are straight lines converging at the point of intersection of the plates. Convergent laminar streams between nonparallel plates are always free from separation, whereas divergent laminar streams will display separation of flow when the angle between the plates exceeds a limit depending on a properly defined Reynolds number. An exact solution is *not* known for the closely related problem of axisymmetric radial flow in a cone. The comparable low Reynolds number flow in a cone is discussed in Section 4–24.

Laminar flow due to a rotating disk in a viscous fluid filling the semi-infinite space above it illustrates the role of centrifugal forces. Fluid near the plate is hurled away from the axis of rotation whereas fluid from more remote regions is drawn inward toward the center of the plate.

The problem of two-dimensional flow against a plane wall was completely solved by Blasius and Hiemenz. The solution obtained is of interest because of its close relation to solutions of the boundary layer type for flows past a blunt body.

Laminar circular motion of fluid confined between rotating concentric circular cylinders attracted the attention of early investigators. The flow of an incompressible fluid caused by the relative rotation of the two cylinders is known as *Couette flow*. Since the streamlines lie in circular paths, the fluid particles are being accelerated; hence, the inertial terms in the Navier-Stokes equations are non-zero. These nonlinear terms are, however, exactly balanced by a radial pressure gradient, making it a relatively simple matter to solve the resulting equations. In particular, if (r, ϕ, x) are cylindrical coordinates, the only non-zero velocity component is the tangential component, v_ϕ, which will be a function only of the radial distance r. Thus, the continuity equation is automatically satisfied, whereas the Navier-Stokes equations reduce to the two total differential equations

$$\frac{1}{r}\frac{d}{dr}\left(r\frac{dv_\phi}{dr}\right) - \frac{v_\phi}{r^2} = 0 \qquad (2\text{–}4.1\text{a})$$

and

$$\frac{dp}{dr} = \rho\frac{v_\phi^2}{r} \qquad (2\text{–}4.1\text{b})$$

which are to be solved for $v_\phi(r)$ and $p(r)$. If r_i and r_o denote the respective radii of the inner and outer cylinders, and ω_i and ω_o, their angular velocities, the boundary conditions are then

$$v_\phi = \omega_i r_i \qquad \text{at } r = r_i$$

$$v_\phi = \omega_o r_o \qquad \text{at } r = r_o$$

Equation (2–4.1a) has the general solution

$$v_\phi = C_1 r + \frac{C_2}{r} \qquad (2\text{–}4.2)$$

where C_1 and C_2 are constants. The boundary conditions are satisfied by choosing

$$C_1 = \frac{\omega_o r_o^2 - \omega_i r_i^2}{r_o^2 - r_i^2}$$

$$C_2 = \frac{(\omega_o - \omega_i) r_o^2 r_i^2}{r_o^2 - r_i^2} \qquad (2\text{–}4.3)$$

With v_ϕ now known, one may integrate Eq. (2–4.1b) to obtain the pressure. If p_i denotes the constant pressure at the inner cylinder wall, we obtain

$$p = p_i + \frac{\rho}{(r_o^2 - r_i^2)^2}\left[(\omega_o r_o^2 - \omega_i r_i^2)^2 \tfrac{1}{2}(r^2 - r_i^2) - 2r_i^2 r_o^2(\omega_o - \omega_i)(\omega_o r_o^2 - \omega_i r_i^2)\ln\frac{r}{r_i} - r_i^4 r_o^4(\omega_o - \omega_i)^2\left(\frac{1}{r_i^2} - \frac{1}{r^2}\right)\right] \quad (2\text{–}4.4)$$

These results may be checked by observing that, when the two cylinders rotate in the same direction at the same rate, say $\omega_o = \omega_i = \omega$, these reduce to

$$v_\phi = \omega r \quad \text{and} \quad p = p_i + \tfrac{1}{2}\rho\omega^2(r^2 - r_i^2)$$

corresponding to a rigid body rotation.

To obtain the couple on either cylinder, we note that the only non-zero tangential stress component is

$$\Pi_{r\phi} = \mu r \frac{\partial}{\partial r}\left(\frac{v_\phi}{r}\right) \quad (2\text{–}4.5)$$

The torque per unit length of cylinder is obtained by multiplying the foregoing by the area per unit length, $2\pi r$. Division by r gives the moment M of the torque. Thus, for the magnitude of the moment per unit length on either cylinder, we obtain the Couette viscometer formula

$$M = \frac{4\pi\mu|\omega_o - \omega_i|}{(1/r_i^2) - (1/r_o^2)} \quad (2\text{–}4.6)$$

2-5 Laminar Flow in Ducts

Although the Navier-Stokes equations are in general nonlinear, there exist flows for which the nonlinear $\mathbf{v}\cdot\nabla\mathbf{v}$ terms vanish identically. Exact solutions of the resulting linear equations are then relatively easy to obtain, especially if the flow is steady.

The steady motion of an incompressible viscous fluid, when the flow occurs in straight parallel lines, is of special interest in our work. Even when we are concerned predominantly with particulate systems, an external boundary surrounding the fluid always exists in real systems and influences the type of motion involved. Therefore, it is important to have available solutions for flow through pipes and other conduits of constant cross section.

If we assume flow only in the x direction, the Navier-Stokes equations reduce to a single scalar equation,

$$\nabla^2 u = \frac{1}{\mu}\frac{dp}{dx} \quad (2\text{–}5.1)$$

where u, the velocity component in the x direction, is independent of x. The continuity equation is then automatically satisfied: dp/dx will be a constant along the pipe length for steady state conditions and may be written $-\Delta p/l$, where $\Delta p > 0$ is the pressure difference between the ends of the pipe and l is its length. End effects are neglected.

For flow between two infinite parallel plates, spaced a distance h units apart, we take y as the distance from one of the plates. Equation (2–5.1) becomes

$$\frac{d^2u}{dy^2} = -\frac{\Delta p}{\mu l} \tag{2–5.2}$$

The boundary conditions are that $u = 0$ for $y = 0$ and $y = h$. Integration yields

$$u = \frac{\Delta p}{2\mu l} y(h - y) \tag{2–5.3}$$

A flow with this parabolic distribution is known as a *plane Poiseuille flow*. The volumetric flow rate per unit width of gap in a lateral direction is obtained by integration, the result being

$$Q = \frac{h^3 \Delta p}{12\mu l} \tag{2–5.4}$$

In cases of circular symmetry, Eq. (2–5.1) assumes the form,

$$\frac{1}{r}\frac{d}{dr}\left(r\frac{du}{dr}\right) = -\frac{\Delta p}{\mu l} \tag{2–5.5}$$

Upon integration we obtain

$$u = -\frac{\Delta p}{4\mu l} r^2 + a \ln r + b \tag{2–5.6}$$

When the circular pipe is hollow, the constant a must be zero in order for the velocity at the tube center to remain finite. The constant b is obtained from the boundary condition, $u = 0$ at the pipe wall, $r = r_1$. Hence,

$$u = \frac{\Delta p}{4\mu l}(r_1^2 - r^2) \tag{2–5.7}$$

The velocity distribution across the pipe is again parabolic, and known as Hagen-Poiseuille flow. The volumetric discharge is obtained by integrating the differential flow $2\pi ru\,dr$ passing through an annular element $2\pi r\,dr$ of cross-sectional area. Thus,

$$Q = 2\pi \int_0^{r_1} ru\,dr = \frac{\pi r_1^4 \Delta p}{8\mu l} \tag{2–5.8}$$

which is Poiseuille's law for laminar flow in a circular tube.

If the pipe is not hollow, but possesses a centrally located core, we may still employ Eq. (2–5.6). The boundary conditions are now $u = 0$ at r_1, the pipe radius, and r_2, the core radius. Thus,

$$u = \frac{\Delta p}{4\mu l}\left[r_1^2 - r^2 + \frac{r_1^2 - r_2^2}{\ln(r_1/r_2)} \ln\frac{r}{r_1}\right] \tag{2–5.9}$$

The discharge may be computed as previously, giving

$$Q = \frac{\pi \Delta p}{8\mu l}\left[r_1^4 - r_2^4 - \frac{(r_1^2 - r_2^2)^2}{\ln (r_1/r_2)}\right] \qquad (2\text{–}5.10)$$

Equation (2–5.1) may be applied to a cross section of any shape. The cross section may contain one or more cores of arbitrary cross section completely surrounded by fluid, as long as the entire cross section remains the same along the path of flow (Fig. 2-5.1). Since the right-hand side of Eq. (2–5.1) is constant, the entire expression can be transformed into an alternative form by the substitution:

$$u = \psi + \frac{1}{4\mu}\frac{dp}{dx}(y^2 + z^2) \qquad (2\text{–}5.11)$$

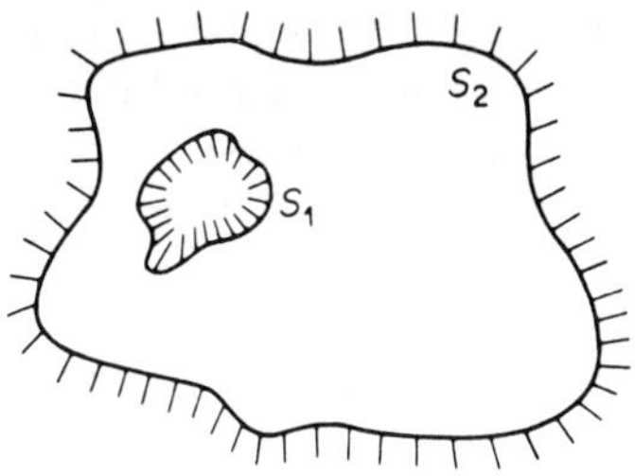

Figure 2-5.1. General cross section for laminar parallel flow.

The function ψ satisfies the two-dimensional form of Laplace's equation,

$$\frac{\partial^2 \psi}{\partial y^2} + \frac{\partial^2 \psi}{\partial z^2} = 0 \qquad (2\text{–}5.12)$$

The no-slip condition requires that

$$\psi = \frac{\Delta p}{4\mu l}(y^2 + z^2) \qquad (2\text{–}5.13)$$

on the stationary bounding surfaces $S_1(y, z)$, $S_2(y, z)$, etc.

The solution of Laplace's equation in two dimensions can be effected by a variety of techniques. If the arbitrary cross section can be mapped onto a circle by an appropriate conformal transformation, the analysis is greatly simplified. Direct numerical solution using relaxation techniques is also applicable. It is also of interest that the solution of Laplace's equation in this form corresponds to several other physical problems, among which may be mentioned the rotation of an ideal fluid, the distortion of a prismatic beam, and the displacement of a thin membrane stretched over the cross section.

The problem of flow through a circular pipe containing an eccentrically located circular core[47,9] has been solved by a conformal transformation of Eqs. (2–5.12) and (2–5.13). The transformation employed is $z' = M \tan(\zeta/2)$, where $z' = y + iz$, $\zeta = \xi + i\eta$, M is a constant, and $i = \sqrt{-1}$. This yields

$$\frac{\partial^2 \psi}{\partial \xi^2} + \frac{\partial^2 \psi}{\partial \eta^2} = 0 \qquad (2\text{–}5.14)$$

which is to be solved for $\psi(\xi, \eta)$ so as to satisfy the boundary conditions

$$\psi = \frac{\Delta p}{4\mu l}\left(\frac{2M^2 \cosh \beta}{\cosh \beta + \cos \xi} - M^2\right) \quad \text{at the inner boundary}$$

$$\psi = \frac{\Delta p}{4\mu l}\left(\frac{2M^2 \cosh \alpha}{\cosh \alpha + \cos \xi} - M^2\right) \quad \text{at the outer boundary} \qquad (2\text{–}5.15)$$

where α and β are the values of η at the outer and inner boundaries, respectively. A complicated expression is obtained for u and the volumetric discharge per second obtained by integration of the expression

$$Q = 2\int_0^\pi \int_\alpha^\beta u \left|\frac{dz'}{d\zeta}\right|^2 d\xi\, d\eta \tag{2–5.16}$$

The final formula for discharge through a tube with pipe radius r_1, core radius r_2, and distance b between the centers of pipe and core is given by the following expression:

$$Q = \frac{\pi\,\Delta p}{8\mu l}\left\{r_1^4 - r_2^4 - \frac{(r_1+r_2+b)(r_1+r_2-b)(r_1-r_2+b)(r_1-r_2-b)}{(\beta-\alpha)} - 4b^2\left[r_2^2 + \frac{r_2^4 r_1^2}{(r_1^2-b^2)^2} + \frac{r_2^6 r_1^4}{\{(r_1^2-b^2)^2 - r_2^2 b^2\}^2} + \cdots\right]\right\} \tag{2–5.17}$$

α and β may be obtained as follows:

$$\alpha = \frac{1}{2}\ln\frac{F+M}{F-M}; \qquad \beta = \frac{1}{2}\ln\frac{F-b+M}{F-b-M} \tag{2–5.18}$$

where $$F = \frac{r_1^2 - r_2^2 + b^2}{2b}; \qquad M = \sqrt{F^2 - r_1^2}$$

In the degenerate case $b = 0$, where the core is centrally located, the last term in Eq. (2–5.17) vanishes. Furthermore,

$$\beta - \alpha = \ln\frac{r_1}{r_2} \tag{2–5.19}$$

whence we obtain

$$Q_c = \frac{\pi\,\Delta p}{8\mu l}\left[r_1^4 - r_2^4 - \frac{(r_1^2 - r_2^2)^2}{\ln(r_1/r_2)}\right] \tag{2–5.20}$$

which is identical to Eq. (2–5.10).

The significance of Eq. (2–5.17) is most conveniently examined by comparing the variation of Q/Q_c, with $b/(r_1 - r_2)$ where Q_c is the discharge with the core centrally located. The results of a number of such calculations by Piercy, Hooper, and Winny[47] are collected in Fig. 2–5.2. A parabolic approximation,

$$\frac{Q}{Q_c} = 1 + \frac{3b^2}{2(r_1 - r_2)^2} \tag{2–5.21}$$

based on a conventional lubrication theory-type approximation, is also plotted for comparison. It is seen that this simple expression approximates the exact solution for values of $r_2/r_1 > 0.5$.

The problem of flow through a circular pipe with eccentric core has also been treated by Redberger and Charles[50], who used a different conformal transformation. These authors reported numerical results which are not in exact agreement with those shown in Fig. 2–5.2. For example, Redberger and

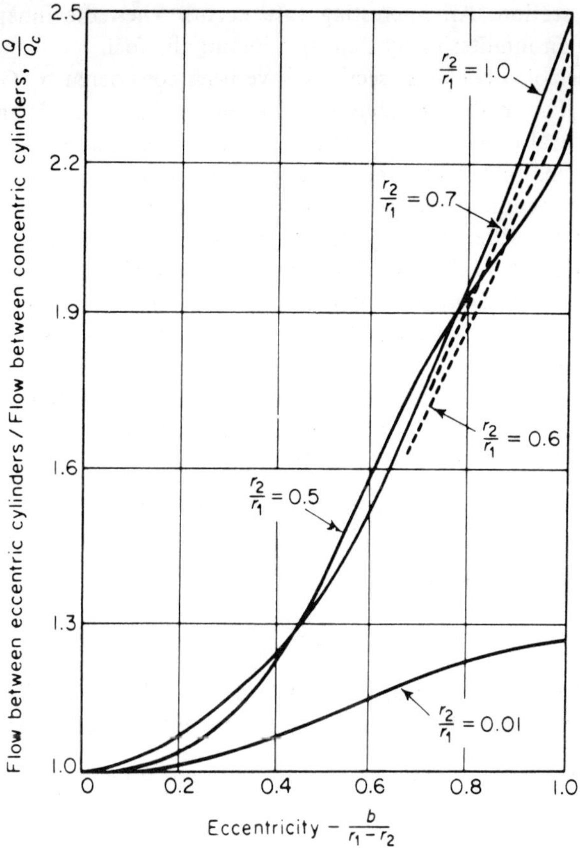

Figure 2-5.2. Variation of discharge with eccentricity of core.

Charles report $Q/Q_c = 2.25$, for an eccentricity of unity and $r_1/r_2 = 0.7$, whereas Fig. 2–5.2 gives $Q/Q_c = 2.4$ for the same conditions.

These studies confirm the general conclusion that a very great increase in flow due to eccentricity occurs in the case of narrow annuli. Engineers have long recognized the large increase in leakage which results when a nearly closed valve binds at one side. Another interesting observation, though hardly unexpected, is that a small core at maximum eccentricity scarcely affects the total resistance to flow.

Steady flow through a long pipe of elliptic cross section containing a confocal elliptic core[47] has also been treated by conformal transformation methods. When the cross-sectional area of the core exceeds one-third that of the pipe, the increase in pressure drop due to the core closely approximates that which would be observed in the case of a concentric circular core of the same cross section. An interesting case occurs when the inner boundary contracts to an infinitesimally thin strip joining the foci.

A number of other cross sections have been considered by Graetz[19] and Greenhill[21]. The method of solution is based on Eqs. (2–5.12) and (2–5.13)., For an elliptical cross section it is possible to express the result in an exact algebraic form. In the case of a rectangular cross section, transcendental functions are required. A second method involves setting up expressions which satisfy the differential equation (2–5.12) and then determining the cross sections which satisfy the boundary conditions, Eq. (2–5.13). As an example of this method it is possible to consider flow through tubes whose cross sections are triangular or quadrilateral with curved sides possessing rounded or pointed corners.

For flow through a pipe of elliptical cross section of major and minor semiaxes a and b, the solution satisfying the condition $u = 0$ on the circumference of the ellipse $y^2/a^2 + z^2/b^2 = 1$ is

$$u = \frac{\Delta p a^2 b^2}{2\mu l(a^2 + b^2)}\left(1 - \frac{y^2}{a^2} - \frac{z^2}{b^2}\right) \tag{2–5.22}$$

This discharge is found to be

$$Q = \frac{\pi \Delta p}{4\mu l}\frac{a^3 b^3}{a^2 + b^2} \tag{2–5.23}$$

For a rectangular cross section with sides a and b in the directions of y and z, respectively, we have as a solution of the differential equation,

$$u = -\frac{\Delta p}{2\mu l} y(y - a) + \sum_{m=1}^{\infty} \sin\left(\frac{m\pi y}{a}\right)\left(A_m \cosh\frac{m\pi z}{a} + B_m \sinh\frac{m\pi z}{a}\right) \tag{2–5.24}$$

This equation satisfies the boundary condition $u = 0$ at $y = 0$ and $y = a$. The coefficients A_m and B_m are determined in accordance with the boundary conditions $u = 0$ at $z = 0$ and $z = b$. The condition at $z = 0$ requires

$$0 = -\frac{\Delta p}{2\mu l} y(y - a) + \sum_{m=1}^{\infty} A_m \sin\frac{m\pi y}{a} \tag{2–5.25}$$

which enables the coefficients A_m to be evaluated by means of Fourier's formula

$$A_m = \frac{\Delta p}{2\mu a l}\int_0^a y(y - a)\sin\left(\frac{m\pi y}{a}\right) dy = \frac{a^2 \Delta p}{\mu m^3 \pi^3 l}(\cos m\pi - 1) \tag{2–5.26}$$

The boundary condition at $z = b = \eta a$, say, requires that

$$0 = -\frac{\Delta p}{2\mu l} y(y - a) + \sum_{m=1}^{\infty} \sin\left(\frac{m\pi y}{a}\right)(A_m \cosh m\eta\pi + B_m \sinh m\eta\pi) \tag{2–5.27}$$

which gives

$$B_m = -\frac{A_m(\cosh m\eta\pi - 1)}{\sinh m\eta\pi} \tag{2–5.28}$$

By integration, we obtain for the discharge

$$Q = \frac{\Delta p}{24\mu l}ab(a^2 + b^2) - \frac{8\Delta p}{\pi^5 \mu l}\sum_{n=1}^{\infty}\frac{1}{(2n-1)^5}\left[a^4 \tanh\left(\frac{2n-1}{2a}\pi b\right) + b^4 \tanh\left(\frac{2n-1}{2b}\pi a\right)\right] \tag{2–5.29}$$

In the case of a square cross section of side a this reduces to

$$Q = \frac{\Delta p a^4}{12\mu l}\left[1 - \frac{192}{\pi^5}\sum_{n=1}^{\infty}\frac{1}{(2n-1)^5}\tanh\left(\frac{2n-1}{2}\pi\right)\right] \tag{2–5.30}$$

or, upon summing the series,

$$Q = \frac{0.035144\, \Delta p a^4}{\mu l} \tag{2–5.31}$$

If we take Q_0 to be the discharge from a circular tube with the same cross-sectional area as a square duct, we find with the help of Eq. (2–5.8) that

$$\frac{Q}{Q_0} = 0.88327 \tag{2–5.32}$$

The reduced discharge is due to the sharp corners on the square duct.

For an equilateral triangle with sides of length b, whose origin is at the center of the cross section, and with the y axis parallel to one of the sides, we have

$$u = \frac{\sqrt{3}\,\Delta p}{6\mu b l}\left(z - \frac{b}{2\sqrt{3}}\right)\left(z + \sqrt{3}\,y - \frac{b}{\sqrt{3}}\right)\left(z - \sqrt{3}\,y - \frac{b}{\sqrt{3}}\right) \tag{2–5.33}$$

The volumetric flow rate is given by

$$Q = \frac{\sqrt{3}\,\Delta p b^4}{320\mu l} \tag{2–5.34}$$

This may be compared with that from a circular tube of equal cross-sectional area:

$$\frac{Q}{Q_0} = 0.72552 \tag{2–5.35}$$

The efflux rate from the equilateral triangular duct is thus substantially lower than from either a square or circular cross section of the same area.

In a later chapter we shall consider the problem of the additional pressure drop created by placing a particle or a dilute system of particles in a duct of any shape. (See Sections 3–6 and 7–1.)

2-6 Simplifications of the Navier-Stokes Equations, Especially for Slow Motion

As we have remarked, the number of solutions of the complete Navier-Stokes equations are few. A still more serious difficulty arises because at sufficiently high velocities turbulent flow occurs. Dependent variables, such as velocity and pressure, are no longer unique functions of space and time coordinates, but must be described by stochastic laws. The occurrence of turbulence[26] is associated with inherent instability of the steady, laminar flow pattern. For flow to be completely stable, it is necessary that small disturbances to the prevailing flow, should they arise, decay with time. If these perturbations, which are ever-present in all real flows, tend to be amplified with time, the flow tends to instability and turbulence normally results. Although the time-dependent forms of the equations of continuity and motion presumably apply, their solution for the instantaneous velocities and pressures is not possible; rather, various approximation schemes must be invoked.

Even with simplifications, the theoretical problem of stability of steady flow past bodies of finite dimensions has not been solved. But it seems certain that steady flow is stable for sufficiently small Reynolds numbers. Experimental data indicate that laminar flow is stable for sufficiently small Reynolds numbers. Experimental data also indicate that laminar flow is always stable in circular ducts up to $N_{\mathrm{Re}} = dU\rho/\mu = 2100$, d being the diameter of the pipe and U the mean velocity. By carefully preventing perturbations at the inlet, however, laminar flows have been attained at much higher Reynolds numbers. In the case of streaming flow past objects immersed in a fluid, the critical Reynolds number is much lower, especially for bodies which are not streamlined to prevent separation of flow. Critical values of the order of 10 to 100 have been reported; for example[35], during flow across a cylinder, undamped nonsteady flow has been observed at $N_{\mathrm{Re}} = dU\rho/\mu = 34$, d being the diameter of the cylinder. The critical Reynolds number at which separation first occurs in flow around a sphere has been reported by Jenson[29] to be $N_{\mathrm{Re}} = 17$, based on a solution of the complete Navier-Stokes equations using relaxation methods.

One common way of simplifying the Navier-Stokes equations at high Reynolds numbers is by complete or partial omission of the viscous terms, $\mu\nabla^2\mathbf{v}$, in comparison to the inertial terms, $\rho\mathbf{v}\cdot\nabla\mathbf{v}$. If the viscous terms are neglected completely, and irrotational motion assumed, one obtains the potential flow equations, which form the basis for much of classical hydrodynamic theory[39]. This theory unfortunately furnishes no information on the drag experienced by bodies immersed in steady flows in unbounded media or on the resistance to flow of fluids through ducts. Because the omission of viscous terms reduces the order of the equations of motion, from four to three, the solution cannot be made to satisfy the no-slip boundary condition at solid boundaries. For this reason, steady flows of this type are devoid of physical meaning, at least in the immediate proximity of solid boundaries, for even very large Reynolds numbers.

In the limiting case of large Reynolds numbers, however, it is possible to subdivide the field of flow around a body into an external region, where the flow is usually irrotational, and a thin layer near the body, together with a wake behind it, where viscous effects are not negligible. In the outer region the flow may be expected approximately to satisfy the equations of ideal flow, whereas in the inner region viscous and inertial effects are of essentially equal importance, and some form of the full Navier-Stokes equations must be employed. This latter region consists of a *boundary layer* immediately adjacent to the body, and a *wake* downstream from it. The concept of a boundary layer was introduced into fluid mechanics by Prandtl at the beginning of the present century and has proved to be very useful in dealing with problems in hydraulics and aerodynamics[54]. Briefly, boundary layer theory involves simplification of the Navier-Stokes equations based on the small order of magnitude of certain terms near a solid surface, and the proper matching of the flow near the surface with the external flow. This theory has been employed in turbulent as well as laminar flow problems, and a wide variety of solutions are available[45, 38]. Unfortunately, the flow on the posterior of submerged objects, after the point of separation, is not amenable to boundary layer treatment. Thus, no exact theoretical evaluation of the drag on submerged objects has yet been possible at Reynolds numbers beyond those at which separation of flow occurs.

Another method of simplification, which will be exploited in this book, involves modification or omission of the inertial terms, $\rho \mathbf{v} \cdot \nabla \mathbf{v}$, compared with the viscous terms, $\mu \nabla^2 \mathbf{v}$, in Eq. (2–1.15). Complete omission of the inertial terms results in the so-called creeping motion or Stokes equations. Thus, for the steady motion of an incompressible fluid, omission of inertia terms and absorption of any conservative extraneous volume forces into the pressure p reduces Eq. (2–1.15) to the following:

$$\nabla^2 \mathbf{v} = \frac{1}{\mu} \nabla p \tag{2–6.1}$$

which, together with the continuity equation,

$$\nabla \cdot \mathbf{v} = 0 \tag{2–6.2}$$

constitutes the creeping motion equations for this situation. Where applicable, neglect of the quadratic inertial terms constitutes a very substantial simplification, for the equations of motion are now linear. That is, if ($\mathbf{v}_1$, p_1) and ($\mathbf{v}_2$, p_2) each separately satisfies Eqs. (2–6.1)–(2–6.2), then so does ($\mathbf{v}_1 + \mathbf{v}_2$, $p_1 + p_2$). Exact solutions are then feasible, even for relatively complex geometries, via classical superposition techniques applicable to linear partial differential equations.

The creeping motion equations may generally be expected to apply for situations where $\rho\mathbf{v}\cdot\nabla\mathbf{v}$ is small compared with $\mu\nabla^2\mathbf{v}$ at each point of the fluid. If, in a given flow problem, l and V are taken to represent a characteristic linear dimension and velocity, respectively, these expressions will be proportional to $\rho V^2/l$ and $\mu V/l^2$, respectively. The ratio of inertial to viscous forces is described in a global sense by the dimensionless parameter $lV\rho/\mu$, a characteristic Reynolds number. Thus, the smaller the Reynolds number, the better will be the approximate solution of the Navier-Stokes equations obtained by retaining only viscous terms. The exact Reynolds number above which the neglect of fluid inertia constitutes a poor approximation depends, in the final analysis, on the accuracy required. The drag on a sphere of radius a moving steadily with velocity U through an unbounded fluid is, by Stokes' law, which derives from the creeping motion equations,

$$F = 6\pi\mu a U \tag{2–6.3}$$

At a Reynolds number $N_{Re} = aU\rho/\mu$ of 0.05, the drag predicted by Stokes' law is only 2 per cent less than the presumably more correct value obtained by Proudman and Pearson[49], based on a rigorous small Reynolds number expansion of the Navier-Stokes equations in which inertial effects are taken into account. Their result is given in Eq. (2–6.6).

When more than a single characteristic dimension enters into a given problem, the Reynolds number may be significantly greater than unity without inertial effects being appreciable. For example, according to the experimental data of McNown, *et al.*[37] on the axial settling of a close-fitting sphere in a circular cylinder filled with viscous fluid, a sphere Reynolds number of 70 must be exceeded before inertial effects become sensible. Similarly, for flow through packed beds of spheres, sphere Reynolds numbers of about 5 (based on superficial fluid velocity) must exist before appreciable deviations from linearity are observed in a plot of pressure drop versus superficial velocity[22]. Here, in addition to sphere radius, the average distance between particle centers serves as a second characteristic length dimension. Similar effects are noted for flow through dense assemblages of cylinders[23].

Equation (2–6.1) also provides a valid approximation of the Navier-Stokes equations for certain classes of *unsteady* motions, as we shall discuss in detail in Section 2–10.

Phenomena arising from the inertia of the fluid are, of course, irrevocably lost in the creeping motion approximation. For example, according to the latter, two identical spheres settling parallel to their line-of-centers experience the same resistance when moving at the same velocity. Hence, they should maintain a fixed distance between them as they fall[60]. At any non-zero Reynolds number, it can, however, be shown that the trailing sphere experiences a smaller resistance than the leading one, so that the former ultimately overtakes the latter[24]. In a similar vein, a neutrally buoyant sphere being transported in a nonaxial location in a vertical circular cylinder through which a viscous fluid is flowing in Poiseuille flow will maintain a fixed position relative to the axis, according to the creeping motion equations[7]. If, however, inertial terms are taken in account, a lateral force arises tending to move the sphere across the streamlines[53]. The smaller the Reynolds number, the smaller will be these inertial effects, all other things being equal. But since no actual flow can occur at a Reynolds number which is identically zero, inertial effects must exist to some extent in all real systems.

Effects which arise when the Reynolds number is small but not wholly negligible have been treated by techniques which attempt to approximate the inertial terms in the Navier-Stokes equations. The first effort in this direction is due to Whitehead[63] who, in 1889, attempted to extend Stokes' original solution for a translating sphere to higher Reynolds numbers, using an apparently straightforward perturbation scheme. Whitehead proposed that Stokes' original solution of the creeping motion equations

$$\mu \nabla^2 \mathbf{v}_0 - \nabla p_0 = \mathbf{0}$$
$$\nabla \cdot \mathbf{v}_0 = 0$$

for streaming flow past a sphere, that is, satisfying the boundary conditions

$$\mathbf{v}_0 = \mathbf{0} \quad \text{at } r = a \qquad \mathbf{v}_0 \to \mathbf{U} \quad \text{as } r \to \infty$$

be employed iteratively to approximate the inertial terms in the Navier-Stokes equations. Thus, in place of the Navier-Stokes equations, Whitehead utilized the relations

$$\mu \nabla^2 \mathbf{v}_1 - \nabla p_1 = \rho \mathbf{v}_0 \cdot \nabla \mathbf{v}_0$$
$$\nabla \cdot \mathbf{v}_1 = 0$$

which he then attempted to solve for $(\mathbf{v}_1, p_1)$, subject to the boundary conditions:

$$\mathbf{v}_1 = \mathbf{0} \quad \text{at } r = a \qquad \mathbf{v}_1 \to \mathbf{U} \quad \text{as } r \to \infty$$

The field $\mathbf{v}_1$ represents an improvement over Stokes' original solution, for it clearly takes account of the first-order effects of Reynolds number. Whitehead's scheme has the obvious advantage that one has now to solve only a *linear*, inhomogeneous equation, rather than a nonlinear equation. What is more, the perturbation scheme can, in principle, be extended indefinitely, by using $\rho \mathbf{v}_1 \cdot \nabla \mathbf{v}_1$ as the next approximation to the inertial terms, proceeding by an obvious iteration scheme to higher-order approximations. Unfortunately, as Whitehead himself discovered, there exists no solution of the preceding equations for $\mathbf{v}_1$ capable of satisfying the condition of uniform flow at infinity. What is more, the next approximation, say, $\mathbf{v}_2$, can be shown to become infinite at infinity. The inability to extend Stokes' solution by the iteration scheme just outlined is known as *Whitehead's paradox*.

It remained for Oseen in 1910 to point out the origin of Whitehead's paradox and to suggest a scheme for its resolution. The details are set forth at great length in Oseen's book[43], along with a variety of applications. As pointed out by Oseen, Stokes' original solution of the creeping motion equations is of the form $\mathbf{v}_0 = \mathbf{U} + UaO(r^{-1})$ at large distances from the sphere. Thus, at great distances, $\nabla^2\mathbf{v}_0 = UaO(r^{-3})$ and $\mathbf{v}_0\cdot\nabla\mathbf{v}_0 = U^2aO(r^{-2})$. The ratio of inertial to viscous terms far from the sphere is therefore

$$\frac{\rho|\mathbf{v}_0\cdot\nabla\mathbf{v}_0|}{\mu|\nabla^2\mathbf{v}_0|} = O\left(\frac{rU\rho}{\mu}\right)$$

which has the form of a *local* Reynolds number. Because of the appearance of the distance r in the foregoing, it is clear that Stokes' implicit assumption that the inertial terms are *everywhere* small compared with the viscous terms is inconsistent with the form of his velocity field. Accordingly, Stokes' velocity field is not a uniformly valid approximation, but breaks down at distances r from the sphere such that $rU\rho/\mu = O(1)$. For at such distances the local inertial and viscous terms are of the same order of magnitude, which is inconsistent with Stokes' complete neglect of inertia. Since $\mathbf{v}_0$ is not uniformly valid, it does not provide a correct estimate of the inertial terms at great distances, and thus cannot serve its proposed role as a first approximation to the inertial terms in Whitehead's iterative scheme. Thus did Oseen point out the source of Whitehead's paradox.

In order to rectify the difficulty, Oseen went on to make the following additional observations. In the limit where the particle Reynolds number $aU\rho/\mu \to 0$, Stokes' approximation becomes invalid only when $r/a \to \infty$. But at such enormous distances, the local velocity $\mathbf{v}$ differs only imperceptibly from a uniform stream of velocity $\mathbf{U}$. Thus, Oseen was inspired to suggest that the inertial term $\rho\mathbf{v}\cdot\nabla\mathbf{v}$ could be uniformly approximated by the term $\rho\mathbf{U}\cdot\nabla\mathbf{v}$. By such arguments, he proposed that uniformly valid solutions of the problem of steady streaming flow past a body at small particle Reynolds numbers could be obtained by solving the *linear* equations:

$$\begin{gathered}\mu\nabla^2\mathbf{v} - \nabla p = \rho\mathbf{U}\cdot\nabla\mathbf{v} \\ \nabla\cdot\mathbf{v} = 0\end{gathered} \tag{2-6.4}$$

These are known as Oseen's equations.

Oseen himself obtained an approximate solution of his equations for flow past a sphere, from which he obtained the drag formula

$$F = 6\pi\mu aU[1 + \tfrac{3}{8}N_{\mathrm{Re}} + O(N_{\mathrm{Re}}^2)] \tag{2-6.5}$$

where $N_{Re} = aU\rho/\mu$. In the limit as $N_{Re} \to 0$ this reduces to Stokes' formula. Resistance formulas comparable in structure to the foregoing have been obtained for a variety of other bodies, such as ellipsoids[2], via approximate solution of the Oseen equations. Many of these results are summarized in Oseen's treatise[43].

Much controversy has centered around the proper interpretation of the relationship between Oseen's differential equations and the Navier-Stokes equations. For though Oseen's term $\mathbf{U}\cdot\nabla\mathbf{v}$ seems a satisfactory approximation to the true inertial term $\mathbf{v}\cdot\nabla\mathbf{v}$ at great distances from the body, it would appear to be a poor approximation in the neighborhood of the body where the boundary condition $\mathbf{v} = \mathbf{0}$ requires that the true inertial term be small. In particular, it is not entirely clear from Oseen's analysis that the inertial correction factor, $3N_{Re}/8$, to the drag on the sphere is really correct; nor does Oseen's analysis furnish us with a systematic perturbation scheme for obtaining higher-order approximations to the solution of the Navier-Stokes equations. On the other hand, Oseen's analysis does place Stokes' law on a firm theoretical basis, as well as pointing out that the relationship of the creeping motion equations to the Navier-Stokes equations at small particle Reynolds numbers is not quite so obvious as elementary dimensional arguments would otherwise make it appear. Oseen's arguments also provide the resolution of Stokes' paradox (Section 2–7), according to which no creeping flow solution exists to the problem of two-dimensional streaming flow perpendicular to the axis of a cylinder in an infinite medium.

The unsatisfactory status of the Oseen equations, relative to inertial effects, persisted until the work of Lagerstrom, Cole, and especially Kaplun at the California Institute of Technology in the mid-1950's. Their ideas were brought to fruition and further extended by the work of Proudman and Pearson[49]. Interestingly enough, the impetus for this work had its origins in problems of laminar boundary layer theory at *large* Reynolds numbers, where attempts to obtain higher-order corrections to Prandtl's theory, and thus extend it to lower Reynolds numbers, were stymied by a lack of clear comprehension of the status of Prandtl's equations relative to the complete Navier-Stokes equations.

Proudman and Pearson[49] point out that Oseen's solution should be interpreted as providing a uniformly valid *zeroth* approximation to the Navier-Stokes equations at small Reynolds numbers. As such it can be employed to justify Stokes' law, but cannot be directly employed to obtain a first-order correction to Stokes' law, of the type appearing in Eq. (2–6.5). If we denote the solution of Oseen's equation by $(\mathbf{v}_0, p_0)$, Proudman and Pearson point out that this field, rather than Stokes' field, when employed in a Whitehead-type perturbation scheme, ought to provide a satisfactory starting point for determining inertial effects at small Reynolds numbers. Owing to the complex structure of the Oseen equations, this promising point of view seems not to have been followed up. Rather, Proudman and Pearson advocate an alternative perturbation scheme for solving the Navier-Stokes equations at low Reynolds numbers. This *singular perturbation* scheme, though conceptually more difficult than the combined Whitehead-Oseen scheme, is easier to carry out in practice. In it, one abandons the attempt to obtain perturbation fields which are uniformly valid throughout the fluid, and seeks instead to find separate solutions which are locally valid in the separate regions near to, and far from, the body. These "inner" and "outer" solutions are each uniquely determined by asymptotically matching them in their common domain of validity.

The major result of their calculation is that the drag on a sphere at low Reynolds numbers is given by the expression

$$F = 6\pi\mu a U[1 + \tfrac{3}{8} N_{\text{Re}} + \tfrac{9}{40} N_{\text{Re}}^2 \ln N_{\text{Re}} + O(N_{\text{Re}}^2)] \qquad (2\text{–}6.6)$$

Owing to the complex nature of the underlying logic, it has not yet proved feasible to establish the range of validity of this asymptotic formula. The agreement with Oseen's formula, Eq. (2–6.5), in the $O(N_{\text{Re}})$ approximation, is fortuitous. For, as has been indicated, Oseen's theory by itself does not suffice to derive a drag formula correct to terms beyond the zeroth order in the Reynolds number, that is, beyond Stokes' law.

Jenson[29] reviews numerical methods for treating streaming flows past spheres and circular cylinders, in the Reynolds number range intermediate between creeping flows and boundary layer flows. He provides a detailed solution to the problem of streaming flow around spheres at intermediate Reynolds numbers utilizing relaxation methods. In this procedure, the Navier-Stokes and continuity equations are condensed into a single non-linear partial differential equation involving the stream function (Chapter 4) as the independent variable, and containing the particle Reynolds number as a free parameter. This, in turn, is replaced by a finite difference equation which is then solved by a graphical numerical procedure. The space external to the sphere is represented by a lattice at whose nodes the stream function is computed by a trial-and-error scheme. Successive trials give better approximations, depending on the fineness of the lattice mesh chosen. Such finite difference equations are also amenable to solution by digital computers. Although neither creeping flow nor boundary layer theory applies in the region considered, predictions of the drag from both theories fit smoothly into Jenson's calculated results, indicating a steady transition from Stokes flow to the comparable boundary layer type of flow as the Reynolds number increases. Figure 2–6.1 is a plot of the drag coefficient, $C_D = F/\frac{1}{2}\rho U^2 \pi a^2$, versus the sphere Reynolds number, $N_{\text{Re}_d} = 2aU\rho/\mu$, based on Stokes' law ($C_D = 24/N_{\text{Re}_d}$), Jenson's calculated results, and actual experimental data[46].

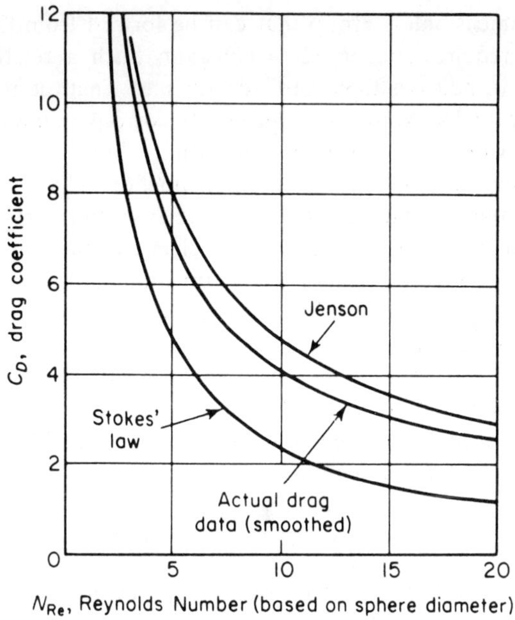

Figure 2-6.1. Comparison of several methods of determining the drag coefficient for a sphere.

In summary, it may be said that the creeping motion equations provide, in many cases, physically meaningful solutions of boundary-value problems involving flow at sufficiently low Reynolds numbers.

2-7 Paradoxes in the Solution of the Creeping Motion Equations

For two-dimensional streaming motion perpendicular to the axis of a circular cylinder there exists no solution of the creeping motion equations vanishing on the cylinder that remains finite at infinity. This is, of course, in marked contrast to the comparable three-dimensional problem of flow past a sphere. This situation is sometimes referred to as Stokes' paradox. That such a paradox should arise for two-dimensional streaming flows is easily demonstrated by invoking simple dimensional arguments. Thus, for flow past a circular cylinder of radius a, one must deal not with a force on the body as in three-dimensional flows, but rather with a force *per unit length* on the body, say, F. Now in creeping flows, where the fluid density ρ is not a parameter, the only variables on which F can depend are μ, a, and U.

But the only dimensionless group that can be formed from these variables is $F/\mu U$. This requires that $F/\mu U =$ constant. Such a relation is clearly impossible, for it indicates that the force per unit length is independent of the size of the cylinder. Were we to let $a \to 0$, corresponding to no cylinder at all, the force would not vanish, as it must in reality.

These arguments show the essential distinction which exists between two- and three-dimensional creeping flows. If a second surface is present, thereby introducing at least one additional characteristic length dimension, say, b, dimensional analysis yields $F/\mu U =$ function (a/b), which is no longer necessarily in conflict with intuition. Alternatively, if the inertial terms are not negligible, so that ρ now enters as a parameter, we obtain $F/\mu U =$ function $(aU\rho/\mu)$. This relation no longer conflicts with physical plausibility, providing that the function vanishes when $aU\rho/\mu \to 0$.

Lamb[34] employed Oseen's equation to obtain a suitable first approximation for the drag per unit length of cylinder,

$$F = \frac{4\pi\mu U}{1/2 - \gamma - \ln\,(Ua\rho/4\mu)} \tag{2-7.1}$$

where $\gamma \doteq 0.577$ is Euler's constant and a is the cylinder radius. Rigorous justification of this result is provided by the singular perturbation techniques of Proudman and Pearson[49].

Krakowski and Charnes[33] have generalized Stokes' paradox to include any two-dimensional flow which is unbounded externally in *all* directions. They show that there exists no solution of the creeping motion equations, aside from the trivial one $\mathbf{v} = \mathbf{0}$ everywhere, which is bounded in the region of flow. The boundary of the flow region is assumed to consist of any number of obstacles, which may be closed or open surfaces or even single points, on which the velocity vanishes.

For two-dimensional regions which are only partly bounded at infinity the creeping motion equations may yield bounded solutions which at low Reynolds numbers are good approximations to the actual flows. Thus, Bairstow, *et al.*[5] and Faxen[18] successfully treated the flow perpendicular to the axis of a circular cylinder between two parallel walls on the basis of the creeping motion equations. Existence theorems are available for such regions[33].

When more than one surface is present in the fluid, it is necessary carefully to consider the nature of the boundary in order to determine whether a physically meaningful result will be obtained by employing the creeping motion equations. Thus, Emersleben[15] obtained a theoretical solution for the one-dimensional problem of viscous flow parallel to the longitudinal axes of equal diameter cylinders in a square array extending indefinitely in the plane perpendicular to the axes. Similarly, Hasimoto[25] presented a theoretical solution of the two-dimensional problem of flow perpendicular to the longitudinal axes of a square array of cylinders extending indefinitely in the plane perpendicular to the cylinder axes. In a periodic array of this type, it is evidently not possible to consider the flow region as unbounded externally. Tamada and Fujikawa[61] investigated the two-dimensional flow past a single infinite row of parallel circular cylinders on the basis of Oseen's equations for the general case where the direction of streaming makes an arbitrary angle relative to the direction of the row. They showed that, for flow perpendicular to the row, the drag on each cylinder tended in the limit of zero Reynolds number to the result obtained from the creeping motion equations. For flow parallel to the row of cylinders (but still perpendicular to the longitudinal axis of each cylinder in the row) a bounded creeping flow solution was not obtainable, as would be surmised from the result of Krakowski and Charnes. Thus any flow oblique to a plane grid of equal-sized parallel cylinders will not possess a creeping motion solution. It is possible, however, to obtain a satisfactory approximation based on the solution of Oseen's equations or, more precisely, by the use of singular perturbation techniques.

Three-dimensional problems involving a number of objects may also lead to paradoxes similar to that of Stokes for the two-dimensional case. Thus, in the case of an unbounded infinite row or column of identical, equally spaced spheres, the creeping motion equations appear to lead to an infinite sedimentation rate. In fact, for the general case of a swarm of an infinite number of particles unbounded at an infinite distance externally, Smoluchowski[59] showed that no bounded solution exists for the field of flow. On the other hand, for flow relative to a periodic array of spheres extending indefinitely in all directions, or indefinitely in two directions perpendicular to the direction of flow, a solution of the creeping motion equations is possible. Hasimoto[25] has considered cubic arrays and cylindrical grids of these types.

2-8 Molecular Effects in Fluid Dynamics

The Navier-Stokes equations are based on the assumption that the fluid may be treated as a continuum. Moreover, it is usually assumed that no slippage occurs at solid surfaces. Actually, these are idealizations of the transport processes occurring. It is convenient, in considering the limitations of the Navier-Stokes equations due to the failure of either of these assumptions, to distinguish two different effects—namely, collisions between fluid molecules themselves and collisions with bounding surfaces.

Chapman[12] has reviewed a number of aspects of gas transport theory in which suspended particles are involved. In the case of gases at sufficiently low pressure, or with sufficiently small particle or container dimensions, the mean free molecular path, l, may become large with respect to either of these macroscopic lengths, d. Under these conditions the dimensionless Knudsen number $K = l/d$ becomes large, intermolecular collisions will be rare, and gas transport will depend to an increasing extent on collisions of molecules with the bounding surfaces. Refinements of theoretical continuum treatments distinguish between *specular* elastic rebound, as from a wall with a perfectly smooth, rigid, elastic surface, and *diffuse* elastic reflection, as from a surface made up of a collection of irregularly disposed, fixed, elastic molecules. In addition, the collisions with walls may be inelastic; a molecule may enter some cavity in the surface and later emerge with an energy different from that on entering. The difference may have a random character, or it may be systematic, as when the wall, or the stratum of it to which the molecules penetrate, is hotter or cooler than the gas[12]. These considerations lead to the conception of an *accommodation* coefficient.

Knudsen studied both flow in ducts[31] and relative to small suspended particles[32] under conditions where these molecular effects are of primary importance. We shall be especially concerned with situations where molecular effects result in only small deviations from continuum behavior.

For Knudsen numbers of $K = 0.01$ or less, Poiseuille's law, Eq. (2–5.8), applies. In the pressure region where the mean free path, though small, is not entirely negligible ($0.01 < K < 0.1$) it is still possible to employ the Navier-Stokes equation solution as given by Poiseuille's law, but a correction must be made to allow for "slippage" between the gas and the solid boundary[30]. No satisfactory theoretical treatment is available for the intermediate region, roughly placed at $0.1 < K < 10$, though empirical relationships have been presented[8] for flow in ducts.

Epstein[16] developed a detailed theory of motion of small spheres relative to gases from the kinetic theory viewpoint. He considered both cases where a particle is small with respect to the mean free path of a gas molecule and those in which a spherical particle may be relatively large. The latter case, as in the problem of flow through ducts, corresponds to the situation where deviations from the creeping motion equations are sufficiently small to be treated by a first-order perturbation scheme. For this circumstance the drag on a sphere, based on a treatment of elastic collisions, both diffuse and specular, is given by [cf. Eq. (4–20.10)]

$$F = 6\pi\mu Ua\left(1 - \frac{\mu}{\beta a}\right) \tag{2–8.1}$$

where β is a coefficient of proportionality, designated as the "coefficient of sliding friction." The "slip coefficient" μ/β has been determined experimentally and found to agree with the general kinetic theory formulation based on the assumption that there is only a small proportion of specular reflection. Schmitt[55], in a paper which includes experimental data, recommends that the slip coefficient be calculated from the formula

$$\frac{\mu}{\beta} = Al \tag{2–8.2}$$

where $A = 1.3 - 1.4$ and l the mean free path of a gas molecule.

Some of the considerations applicable to gas flow through cylindrical tubes have been extended to the case of flow through porous media, on the assumption that the latter may be regarded as bundles of capillary tubes. On this basis, the apparent permeability (flow rate per unit area per unit pressure difference) of a medium to gas flow is shown to be a linear function of the mean gas pressure. This kind of relationship has been observed experimentally by a number of observers[10,4]. In the low pressure domain, however (mean pressure usually of the order of a few mm Hg) discrepancies have been noted. Hirsch[27] has reviewed some of these papers and presented new experimental data indicating a nonlinear dependence of permeability on pressure. In order to account for these effects it would seem necessary to develop a more elaborate model for wall interaction than that employed in the usual kinetic theory, which assumes that the walls are rigid. In actual fact, a solid surface is rapidly covered by a layer of adsorbed gas molecules which are capable of migrating along the surface.

For liquid flow most of the available experimental data indicate that no slippage occurs, although a review which includes new data[56] indicates that, if a solid surface is chemically treated to make it nonwettable by the flowing liquid, appreciable slipping may occur. In the case of non-newtonian liquids displaying a finite yield value, for example, Bingham bodies, slippage may also exist under flow conditions[51].

2-9 Non-newtonian Flow

Fortunately, most liquids which do not possess or form a threadlike or crystalline structure during flow follow the newtonian hypothesis[1]. In flow problems involving polymers and polymer solutions (amorphous, not cross linked), where non-newtonian effects may become important, the flow itself often occurs under conditions which permit the inertial terms to be neglected in the momentum equation, as is done in the derivation of the usual newtonian creeping motion equations.

Several authors, notably Ericksen and Rivlin[17], Oldroyd[42], Noll[41], and Green and Rivlin[20], have introduced simple, theoretically sound, nonlinear models relating the stress tensor to the time rate of deformation in non-newtonian fluids. That of Ericksen and Rivlin has been applied to the creeping flow of non-newtonian fluids by Caswell and Schwarz[11]. The latter solved the equations of motion for slow motion using the matching procedure suggested by Proudman and Pearson[49], but included appropriate non-newtonian terms in both the "inner" and "outer" expansions. Their final expression for drag experienced by a sphere in a streaming flow is

$$F = 6\pi\mu Ua\left(1 + \frac{3}{8}N_{\mathrm{Re}}\right) + \frac{27\pi U^3}{a}\Phi \tag{2–9.1}$$

The coefficient Φ is a complicated expression involving non-newtonian parameters, which should be constants for any given fluid under isothermal conditions. Unfortunately, these cannot as yet be estimated from available experimental data. Leslie[36] also investigated flow around a sphere in creeping motion using the Oldroyd[42] model for non-newtonian behavior. He also obtained a non-newtonian term proportional to U^3. Both models in the limit of very slow shear rates display complete newtonian behavior, and Stokes' law is recovered. An experimental treatment of non-newtonian flow past a sphere by Slattery and Bird[57] relies on empirical models for correlation of data on aqueous solutions of carboxymethylcellulose. Much additional theoretical and experimental work remains before an exact approach to the flow of non-newtonian fluids in particulate systems will be possible.

2-10 Unsteady Creeping Flows

The Cauchy linear momentum equation (2–1.7) may be written as

$$\rho\mathbf{a} = \nabla\cdot\Pi + \rho\mathbf{F}$$

where

$$\mathbf{a} = \frac{D\mathbf{v}}{Dt} = \frac{\partial\mathbf{v}}{\partial t} + \mathbf{v}\cdot\nabla\mathbf{v}$$

is the acceleration of a fluid particle. As this relation is derived by application of Newton's laws of motion to a small fluid particle of mass δm in the elementary form $\delta\mathbf{F}_{\mathrm{external}} = \delta m\mathbf{a}$, it follows that $\mathbf{a}$ and $\mathbf{v}$ must be measured relative to an observer fixed in an inertial reference frame. Though it is not strictly necessary, it is usually convenient also to measure the velocity $\mathbf{v}$ appearing in the newtonian stress relation, Eqs. (2–1.8) and (2–1.12), and in the general continuity equation, Eq. (2–1.1), relative to this same reference frame.

For incompressible fluids in the presence of conservative body force fields, the Navier-Stokes and continuity equations thus become

$$\rho\left(\frac{\partial\mathbf{v}}{\partial t} + \mathbf{v}\cdot\nabla\mathbf{v}\right) = -\nabla p + \mu\nabla^2\mathbf{v} \tag{2–10.1a}$$

$$\nabla\cdot\mathbf{v} = 0 \tag{2–10.1b}$$

where p is the dynamic pressure and $\mathbf{v}$ is measured relative to an inertial frame. As has been argued for *steady* motion at small Reynolds numbers, these reduce to the creeping motion equations,

$$\mathbf{0} = -\nabla p + \mu \nabla^2 \mathbf{v} \tag{2–10.2a}$$

$$\nabla \cdot \mathbf{v} = 0 \tag{2–10.2b}$$

We wish to show that these same relations may also be applied to a large number of *unsteady* low Reynolds number flows.

Typically, consider the problem of the "slow" rotation of an ellipsoid about a principal axis at an angular velocity $\boldsymbol{\omega}$. Such a motion is clearly unsteady since $\partial \mathbf{v}/\partial t \neq \mathbf{0}$. In fact, it is obvious that the motion is periodic in the sense that the local fluid velocity and pressure at any fixed point in space vary periodically in time. Let l denote a characteristic particle dimension and define the following dimensionless variables and parameters:

$\tilde{t} = \omega t =$ dimensionless time
$\tilde{\mathbf{r}} = \mathbf{r}/l =$ dimensionless position vector
$\tilde{\mathbf{v}} = \mathbf{v}/l\omega =$ dimensionless velocity
$\tilde{p} = (p - p_\infty)/\mu\omega =$ dimensionless pressure
$N_{\mathrm{Re}} = l^2 \omega\rho/\mu =$ angular Reynolds number
$\tilde{\nabla} = l\nabla$

where p_∞ is the constant pressure at infinity and $\omega = |\boldsymbol{\omega}|$. Upon writing Eq. (2–10.1) in terms of these dimensionless variables, we obtain

$$N_{\mathrm{Re}}\left(\frac{\partial \tilde{\mathbf{v}}}{\partial \tilde{t}} + \tilde{\mathbf{v}} \cdot \tilde{\nabla} \tilde{\mathbf{v}}\right) = -\tilde{\nabla} \tilde{p} + \tilde{\nabla}^2 \tilde{\mathbf{v}} \tag{2–10.3a}$$

$$\tilde{\nabla} \cdot \tilde{\mathbf{v}} = 0 \tag{2–10.3b}$$

Provided that these nondimensional variables remain finite as $N_{\mathrm{Re}} \to 0$, the preceding equations reduce to

$$\mathbf{0} = -\tilde{\nabla} \tilde{p} + \tilde{\nabla}^2 \tilde{\mathbf{v}} \tag{2–10.4a}$$

$$\tilde{\nabla} \cdot \tilde{\mathbf{v}} = 0 \tag{2–10.4b}$$

That the dimensionless variables and derivatives thereof do indeed remain finite as the angular Reynolds number goes to zero may be verified *a posteriori* from the solution of Eq. (2–10.4).

When written in ordinary dimensional form, Eq. (2–10.4) is formally identical to the creeping motion equations for steady flow. It is convenient in these circumstances to refer to Eq. (2–10.2) as the *quasi-static* or *quasi-steady* creeping motion or Stokes equations. The time variable is implicit in these equations.

Other intrinsically nonsteady motions may be examined in a similar vein. In a fairly general case, one has both a translational speed U_0 and frequency factor ω to contend with. This situation arises, for example, in the case of a propeller-like particle settling in a gravitational field. Skew particles of this type normally attain a terminal mode of motion in which they simultaneously translate with velocity $\mathbf{U}_0$ and spin with angular velocity $\boldsymbol{\omega}$. The pendulum-like motion of a sphere undergoing translatory harmonic oscillations at some frequency ω is another example of unsteady motion where both these parameters arise; for the velocity of the sphere at any instant may be described by an equation of the form

$$U = U_o \cos(\omega t + \alpha)$$

where α is a constant.

For these two parameter motions, we define the following dimensionless quantities:

$$\hat{t} = \omega t, \quad \hat{\mathbf{r}} = \frac{\mathbf{r}}{l}, \quad \hat{\mathbf{v}} = \frac{\mathbf{v}}{U_o}$$

$$\hat{p} = (p - p_\infty)\frac{l}{\mu U_o}, \quad \hat{\nabla} = l\nabla$$

$$N_{\mathrm{Re}}^{(t)} = \frac{lU_o\rho}{\mu}, \quad N_{\mathrm{Re}}^{(r)} = \frac{l^2\omega\rho}{\mu}$$

where the latter are, respectively, translational and rotational Reynolds numbers. Upon introducing these into Eq. (2–10.1), the Navier-Stokes and continuity equations adopt the nondimensional forms

$$N_{\mathrm{Re}}^{(r)}\frac{\partial \hat{\mathbf{v}}}{\partial \hat{t}} + N_{\mathrm{Re}}^{(t)}\hat{\mathbf{v}}\cdot\hat{\nabla}\hat{\mathbf{v}} = -\hat{\nabla}\hat{p} + \hat{\nabla}^2\hat{\mathbf{v}} \tag{2–10.5a}$$

$$\hat{\nabla}\cdot\hat{\mathbf{v}} = 0 \tag{2–10.5b}$$

In linearizing these equations by letting the translational Reynolds number go to zero, two possible limiting situations arise, according as the rotational Reynolds number is or is not an independent variable. In gravitational settling of the propeller-like body, ω and U_o are both dependent on the same physical variables and hence are not independent variables. In this case the rotational Reynolds number vanishes with the translational Reynolds number and Eq. (2–10.5) reduces to the quasi-static form of the creeping motion equations. On the other hand, in the forced longitudinal vibration problem, ω can be varied independently of U_0. Here, the "vibrational" Reynolds number $N_{\mathrm{Re}}^{(r)} = l^2\omega\rho/\mu$ need not be small, even though the translational Reynolds number is. In dimensional form Eq. (2–10.5) then becomes

$$\rho\frac{\partial \mathbf{v}}{\partial t} = -\nabla p + \mu\nabla^2\mathbf{v} \tag{2–10.6a}$$

$$\nabla\cdot\mathbf{v} = 0 \tag{2–10.6b}$$

The quasi-static creeping motion equations and the foregoing form of the creeping motion equations differ significantly in that the local acceleration term, $\rho\partial\mathbf{v}/\partial t$, need not be small. Of course if $l^2\omega\rho/\mu$ is also small, the preceding form will be identical to the quasi-static equations. In any event, Eq. (2–10.6) is a *linear* equation and can be solved by relatively straightforward techniques. Laplace transform methods, which remove the time variable, are of great utility in solving the unsteady form of the creeping motion equations.

BIBLIOGRAPHY

1 Andrade, E. N. da C., *Viscosity and Plasticity*. New York: Chemical Publishing Co., 1951.

2 Aoi, T. J. Phys. Soc. Japan **10** (1955), 119.

3 Aris, R., *Vectors, Tensors, and the Basic Equations of Fluid Mechanics*. Englewood Cliffs. N. J.: Prentice-Hall, Inc., 1962.

4 Arnovsky, J. S., J. Appl. Phys. **25** (1954), 48.

5 Bairstow, L., B. M. Cave, and E. D. Lang, Phil. Trans. Roy. Soc. (London) **A223** (1923), 383.

6 Bird, R. B., W. E. Stewart, and E. N. Lightfoot, *Transport Phenomena*. New York: Wiley, 1960.

7 Brenner, H., and J. Happel, J. Fluid Mech. **4** (1958), 195.

8 Brown, G. P., A. Di Nardo, G. K. Cheng, and T. K. Sherwood, J. Appl. Phys. **17** (1946), 802.

9 Caldwell, J., Jour. Royal Tech. College (Glasgow) **2** (1930), 203.

10 Carman, P. C., Proc. Roy. Soc. (London) **A203** (1950), 55.

11 Caswell, B., and W. H. Schwarz, J. Fluid Mech. **13** (1962), 417.

12 Chapman, S., "Progress in International Research on Thermodynamic and Transport Properties," Paper 24 A. S. M. E., New York: Academic Press, 1962.

13 Coleman, B. D., and W. Noll, Ann. N. Y. Acad. Sci. **89** (1961), 672.

14 Drew, T. B., *Handbook of Vector and Polyadic Analysis*. New York: Reinhold, 1961.

15 Emersleben, O., Physik. Z. **26** (1925), 601.

16 Epstein, P. S., Phys. Rev. **23** (1924), 710.

17 Ericksen, J. L., and R. S. Rivlin, J. Rational Mech. Anal. **4** (1955), 323.

18 Faxen, O. H., Nova Acta reg. Sci. Upsaliensis (1927), 1–55.

19 Graetz, L., Z. Math. Physik **25** (1880), 316, 375.

20 Green, A. E., and R. S. Rivlin, Arch. Rational Mech. Anal. **4** (1960), 387.

21 Greenhill, A. G., London Math. Soc. Proc. **13** (1881), 43.

22 Happel, J., Ind. Eng. Chem. **41** (1949), 1161.

23 ———, A. I. Ch. E. Jour. **5** (1959), 174.

24 ——, and R. Pfeffer, A. I. Ch. E. Jour. **6** (1960), 129.

25 Hasimoto, H., J. Fluid Mech. **5** (1959), 317.

26 Hinze, J. O., *Turbulence—An Introduction to Its Mechanism and Theory*. New York: McGraw-Hill, 1959.

27 Hirsch, E. H., J. Appl. Phys. **32** (1961), 977.

28 Hirschfelder, J. O., C. F. Curtiss, and R. B. Bird, *Molecular Theory of Gases and Liquids*. New York: Wiley, 1954.

29 Jenson, V. G., Proc. Roy. Soc. **A249** (1959), 346.

30 Kennard, E. H., *Kinetic Theory of Gases*. New York: McGraw-Hill, 1938.

31 Knudsen, M., Ann. Physik **28** (1909), 75, 999.

32 ——, and S. Weber, Ann. Physik **30** (1911), 981.

33 Krakowski, M., and A. Charnes, Carnegie Institute of Technology Dept. Mathematics, *Tech. Rep.* No. 37, 1953.

34 Lamb, H., *Hydrodynamics* 6th ed. Cambridge: Cambridge Univ. Press, 1932.

35 Landau, L. D., and E. M. Lifshitz, *Fluid Mechanics*. New York: Pergamon Press, 1959.

36 Leslie, F. M., Quart. J. Mech. Appl. Math. **14** (1961), 36.

37 McNown, J. S., N. M. Lee, M. B. McPherson, and S. M. Engez., Proc. VII Intern. Cong. Appl. Mech. **2** (1948), 17.

38 Meksyn, D., *New Methods in Laminar Boundary Layer Theory*. New York: Pergamon, 1961.

39 Milne-Thomson, L. M., *Theoretical Hydrodynamics*. 4th ed., New York: Macmillan, 1960.

40 Navier, L. M. H., Mem. Acad. Sci. **6** (1827), 389.

41 Noll, W., Arch. Rational Mech. Anal. **2** (1958), 197.

42 Oldroyd, J. G., Proc. Roy. Soc. **A245** (1958), 278.

43 Oseen, C. W., *Neuere Methoden und Ergebnisse in der Hydrodynamik*. Leipzig: Akademische Verlag, 1927.

44 Patterson, G. N., *Molecular Flow of Gases*. New York: Wiley, 1956.

45 Pai, S., *Viscous Flow Theory: I—Laminar Flow*. Princeton, N. J.: D. Van Nostrand, 1956.

46 Perry, J. W., ed. *Chemical Engineers' Handbook*, 3rd ed. New York: McGraw-Hill, 1950, p. 1013.

47 Piercy, N. A. V., M. S. Hooper, and H. F. Winny, Phil. Mag. (7) **15** (1933), 647.

48 Prandtl, L., and O. G. Tietjens, *Fundamentals of Hydro- and Aeromechanics*. New York: McGraw-Hill, 1934.

49 Proudman, I., and J. R. A. Pearson, J. Fluid Mech. **2** (1957), 237.

50 Redberger, P. J., and M. E. Charles, Can. J. Chem. Eng. **40** (1962), 148.

51 Reiner, M., *Deformation and Flow*. London: H. K. Lewis & Co., 1949.

52 Rouse, H., *Advanced Mechanics of Fluids*. New York: Wiley, 1959.

53 Saffman, P. G., J. Fluid Mech. **1** (1956), 540.

54 Schlichting, H., *Boundary Layer Theory*. New York: McGraw-Hill, 1961.

55 Schmitt, K., Staub **19** (1959), 416.

56 Schnell, E., J. Appl. Phys. **27** (1956), 1149.

57 Slattery, J. C., and R. B. Bird, Chem. Eng. Sci. **16** (1961), 231.

58 Slezkin, N. A., *Dynamics of Viscous Incompressible Fluids* (Russian). Moscow: Gos. Izdat. Tekh.-Teor. Lit., 1955.

59 Smoluchowski, M., Bull. Acad. Sci. (Cracow) **1a** (1911), 28.

60 Stimson, M., and G. B. Jeffery, Proc. Roy. Soc. (London) **A111** (1926), 110.

61 Tamada, K., and H. Fujikawa, Quart. J. Mech. Appl. Math. **10** (1957), 425; J. Phys. Soc. Japan **14** (1959), 202.

62 Westberg, R., Kungl. Fysiografiska Sallskapets, J. Lund Forhandlingar **14**, No. 1 (1944).

63 Whitehead, A. N., Quart. J Math. **23** (1889), 143.

Some General Solutions and Theorems Pertaining to the Creeping Motion Equations

3

3-1 Introduction

In what follows, we shall restrict our discussion mainly to the steady, incompressible motion of fluids satisfying the creeping motion and continuity equations (2–6.1) and (2–6.2). We have already noted that certain one-dimensional flows (translation in a duct in straight parallel lines) can be reduced to the solution of Laplace's equation (2–5.12), and that solutions are available for a number of cases involving flow of this type.

We should also mention the application of the creeping motion equations to the hydrodynamic theory of lubrication. Reynolds initiated the study[25] of the relative motion of two nearly parallel surfaces. His methods have since been applied to a variety of lubrication problems[14]. In addition to the neglect of inertia, it is assumed that the fluid motion essentially is unidirectional. These same simplifications have also been employed, for example, to investigate the axial motion of a sphere in a circular cylinder containing a viscous fluid, where the diameter of the cylinder only slightly exceeds that of the sphere[8], and to the viscous flow of a fluid through a grating of parallel circular cylinders when the separation between them is small compared with their diameter[17]. Good agreement of experiment with theory is obtained in the former case. Many other similar applications have been made.

In two-dimensional flows, the creeping motion equations do not give finite results for various types of unbounded media, as outlined in Section 2–7. In bounded systems, where a solution is obtainable, the flow pattern is the same in all planes, say xy planes. Thus, we may write

$$\mathbf{v} = \mathbf{i}u(x, y) + \mathbf{j}v(x, y) \tag{3–1.1}$$

Since the vorticity vector is, in general,

$$\boldsymbol{\omega} = \tfrac{1}{2}\nabla \times \mathbf{v} \tag{3–1.2}$$

we find that, for two-dimensional flows, $\boldsymbol{\omega}$ lies perpendicular to the xy plane. If we write $\boldsymbol{\omega} = \mathbf{k}\omega$, then the scalar vorticity is

$$\omega = \frac{1}{2}\left(\frac{\partial v}{\partial x} - \frac{\partial u}{\partial y}\right) \tag{3–1.3}$$

As the flow is incompressible, the creeping motion equations may be written in the form

$$\nabla p = -2\mu\nabla \times \boldsymbol{\omega} \tag{3–1.4}$$

Thus, the equations relating pressure to vorticity are equivalent to the Cauchy-Riemann equations[22] relating the real and imaginary parts of a complex variable:

$$\frac{1}{2\mu}\frac{\partial p}{\partial x} = -\frac{\partial \omega}{\partial y}; \qquad \frac{1}{2\mu}\frac{\partial p}{\partial y} = \frac{\partial \omega}{\partial x} \tag{3–1.5}$$

Two-dimensional incompressible creeping flows may therefore be represented by a suitable analytic function $W(z)$ of the complex variable $z = x + iy$. The real part of $W(z)$ is $p/2\mu$; its imaginary part is $-\omega$.

Incompressible two-dimensional flows may also be derived from a stream function, $\psi(x, y)$. The component velocities are related to the stream function by the expressions

$$\frac{\partial \psi}{\partial y} = u; \qquad \frac{\partial \psi}{\partial x} = -v \tag{3–1.6}$$

The continuity equation

$$\frac{\partial u}{\partial x} + \frac{\partial v}{\partial y} = 0$$

is automatically satisfied by any choice of ψ. In general, ψ is related to the vorticity by the equation

$$\nabla^2 \psi = 2\omega \tag{3–1.7}$$

For creeping flows it is readily shown that ψ satisfies the two-dimensional biharmonic equation

$$\nabla^4 \psi = 0 \tag{3–1.8}$$

where $\nabla^4 \psi = \nabla^2(\nabla^2 \psi)$. Use of a stream function greatly simplifies the solution of all two-dimensional problems.

Flow at small, but nonzero Reynolds numbers may be treated by appropriate perturbation techniques. Thus Dean[9] treats the two-dimensional shearing motion of an incompressible viscous fluid past a projection in a plane wall, the motion being one of uniform shear apart from the disturbance caused by the projection. The biharmonic equation (3–1.8) is solved for the stream function to obtain the zeroth order approximation in Reynolds number. A first order correction is then obtained by an appropriate iterative scheme based essentially on approximating the inertial terms by the zeroth order field.

Three-dimensional solutions of the creeping motion equations are of the greatest value in problems involving particulate media. Since these problems usually involve multiple boundaries it is desirable to have fairly general solutions available, thereby permitting a number of simultaneous boundary conditions to be satisfied. The technique of most general applicability in obtaining such solutions is the classic method of separation of variables. The separability of the creeping motion equations has not been investigated in detail but three-dimensional problems in spherical and cylindrical coordinate systems can be handled effectively in this way. Useful solutions in the form of convenient series representations are thus obtained. *Axisymmetric flow*, or flow past a body of revolution parallel to the axis of revolution, represents a type of three-dimensional flow which can be characterized by a single scalar stream function, as in the case of two-dimensional motion. Separation is then possible in a wider variety of orthogonal coordinate systems, as discussed in Chapter 4. Another general method for dealing with solutions of the linearized equations of motion employs generalized Green's functions. Since the solutions obtained involve integrals, they are in many cases not as convenient as the series representations. Other more specialized methods involve the use of "mirror" images and variational techniques. In the following sections of this chapter, some of these methods will be detailed, with special reference to those of greatest utility for problems of interest in this book.

As is well known[22], a well-posed hydrodynamic problem must state not only the differential equations to be solved but also the boundary conditions. In most cases the shape of the boundary must be specified. A boundary is said to be *closed* if it completely surrounds the fluid involved. A "closed boundary" may go to infinity, but in that case boundary conditions must be specified at infinity. Thus the boundary is closed for a sphere settling axially in a fluid contained in a cylinder of infinite length, if the velocities are specified on the sphere and cylinder wall as well as at $\pm\infty$ along the cylinder axis. A boundary is *open* if it goes to infinity *and* no boundary conditions are imposed on the fluid at infinity. The creeping motion equations belong to a class of partial differential equations known as *elliptic* equations. For these equations it is preferable to specify boundary value problems involving closed boundaries. The boundary conditions usually employed involve specifying either the value of the vector field itself at the boundary or values of the first derivatives of its components along or normal to the boundary.

Establishment of the necessary and sufficient conditions required for a solution of the creeping motion equations to be unique is a mathematical problem which has not been completely solved*. Considerable attention has been given to the most commonly encountered problem; namely, that of existence and uniqueness for the case of a closed boundary on which the velocity components are given assigned values. Thus Lichtenstein[20] and Odquist[23] have demonstrated the existence of a solution for the general case of an incompressible, viscous fluid in a closed domain which consists of a finite number of particles of finite dimensions. In the case of the creeping motion equations the solution is also unique, but at higher Reynolds numbers this is not the case. For example, Taylor's[29] treatment of the flow between two concentric rotating cylinders shows that, when the Reynolds number exceeds a certain value with the inner cylinder rotating relative to the outer cylinder, an instability of flow results which leads to the appearance of another flow which itself is steady. With increasing Reynolds numbers, an unsteady flow of definite periodicity occurs.

For boundary value problems involving derivatives of the vector components or combinations of prescribed velocities and derivatives at the boundaries it is not possible to generalize on the requisite conditions. Usually the physical nature of the problem itself will be intuitively useful in prescribing appropriate boundary conditions leading to a unique solution which exists.

For multiple boundary problems, it is usually necessary either to develop general solutions for each boundary simultaneously and solve for the unknown functions appearing in them by means of an infinite series of simultaneous linear equations, or alternatively, to satisfy the boundary conditions successively (method of reflections) on each internal boundary, together with

*For a rather complete discussion of progress in this area see O. A. Ladyzhenskaya, *The Mathematical Theory of Viscous Incompressible Flow* (New York: Gordon and Breach, 1963).

the common external boundary (if any), separately. The details of these methods will be discussed in connection with particular boundary value problems solved in later chapters of the book. The Lorentz reciprocal theorem, which is generally useful in multiple boundary value problems, will, however, be discussed in this chapter along with several applications.

3-2 Spherical Coordinates

Lamb[19] outlines a general solution of the creeping motion equations in spherical coordinates. Upon taking the divergence of the vector equation (2–6.1) and utilizing Eq. (2–6.2), we find that the pressure field is harmonic, thus satisfying Laplace's equation

$$\nabla^2 p = 0 \tag{3–2.1}$$

This suggests expanding the pressure field in a series of solid spherical harmonics:

$$p = \sum_{n=-\infty}^{\infty} p_n \tag{3–2.2}$$

where p_n is a solid spherical harmonic of order n. This expansion forms the basis of Lamb's general solution

$$\mathbf{v} = \sum_{n=-\infty}^{\infty} \left[\nabla \times (\mathbf{r}\chi_n) + \nabla\Phi_n + \frac{(n+3)}{2\mu(n+1)(2n+3)} r^2 \nabla p_n - \frac{n}{\mu(n+1)(2n+3)} \mathbf{r} p_n \right] \tag{3–2.3}$$

χ_n and Φ_n are each solid spherical harmonics, arising from the solution of the associated homogeneous equations,

$$\nabla^2 \mathbf{v}' = \mathbf{0} \tag{3–2.4}$$

and

$$\nabla \cdot \mathbf{v}' = 0 \tag{3–2.5}$$

The solution of a boundary value problem is effected when these three harmonic functions are determined, for each n, from the prescribed boundary conditions.

We shall outline a very general method for obtaining these harmonic functions when the velocity field is prescribed on a spherical surface. For cases in which the stress field, rather than the velocity field, is partly or wholly prescribed on the sphere, see the original paper* for details; that paper also gives details for applying Lamb's general solution to nonspherical bodies, particularly to slightly deformed spheres.

For the radial component of velocity we have, from Eq. (3–2.3),

*H. Brenner, Chem. Eng. Sci. **19** (1964), 519.

$$v_r = \sum_{n=-\infty}^{\infty} \left[\frac{\partial \Phi_n}{\partial r} + \frac{n+3}{2\mu(n+1)(2n+3)} r^2 \frac{\partial p_n}{\partial r} - \frac{n}{\mu(n+1)(2n+3)} r\, p_n \right] \tag{3–2.6}$$

Further simplification of this expression can be made by the aid of Euler's theorem for homogeneous polynomials,

$$r \frac{\partial h_n}{\partial r} = n h_n \tag{3–2.7}$$

where h_n is any solid spherical harmonic of order n. Application of this relationship to Eq. (3–2.6) yields

$$v_r = \sum_{n=-\infty}^{\infty} \left[\frac{n}{2\mu(2n+3)} r\, p_n + \frac{n}{r} \Phi_n \right] \tag{3–2.8}$$

Differentiation of the foregoing with respect to r and again applying Euler's theorem gives

$$r \frac{\partial v_r}{\partial r} = \sum_{n=-\infty}^{\infty} \left[\frac{n(n+1)}{2\mu(2n+3)} r\, p_n + \frac{n(n-1)}{r} \Phi_n \right] \tag{3–2.9}$$

In addition to these relationships one may also obtain from Eq. (3–2.3) the equation

$$\mathbf{r} \cdot \nabla \times \mathbf{v} = \sum_{n=-\infty}^{\infty} n(n+1)\chi_n \tag{3–2.10}$$

previously given by Lamb. Thus, at the surface of a sphere of radius a we obtain

$$[v_r]_a = \sum_{n=-\infty}^{\infty} \left(\frac{na}{2\mu(2n+3)} [p_n]_a + \frac{n}{a} [\Phi_n]_a \right) \tag{3–2.11}$$

$$\left[r \frac{\partial v_r}{\partial r} \right]_a = \sum_{n=-\infty}^{\infty} \left\{ \frac{n(n+1)a}{2\mu(2n+3)} [p_n]_a + \frac{n(n-1)}{a} [\Phi_n]_a \right\} \tag{3–2.12}$$

and

$$[\mathbf{r} \cdot \nabla \times \mathbf{v}]_a = \sum_{n=-\infty}^{\infty} n(n+1)[\chi_n]_a \tag{3–2.13}$$

where the subscript a denotes evaluation of the function at $r = a$.

If $\mathbf{V} = \mathbf{V}(\theta, \phi)$ denotes the value of the velocity field on the sphere surface, that is,

$$\mathbf{V} = [\mathbf{v}]_a$$

then it may be shown that

$$[v_r]_a = \frac{\mathbf{r}}{r} \cdot \mathbf{V} \tag{3–2.14}$$

$$\left[r \frac{\partial v_r}{\partial r} \right]_a = -r \nabla \cdot \mathbf{V} \tag{3–2.15}$$

$$[\mathbf{r} \cdot \nabla \times \mathbf{v}]_a = \mathbf{r} \cdot \nabla \times \mathbf{V} \tag{3–2.16}$$

where ∇ is the usual three-dimensional nabla operator. The second of these three relationships depends upon the validity of the continuity equation $\nabla \cdot \mathbf{v} = 0$ whereas the remaining two equations are true for an arbitrary vector function $\mathbf{v}$. Because of the properties of solid spherical harmonics, we can write

$$[p_n]_a = \left(\frac{a}{r}\right)^n p_n \tag{3–2.17}$$

with two similar identities for the functions χ_n and Φ_n. These are equally true whether n is positive or negative. Introduction of these relations into Eqs. (3–2.11)–(3–2.13) gives

$$\sum_{n=-\infty}^{\infty} \left[\frac{na}{2\mu(2n+3)}\left(\frac{a}{r}\right)^n p_n + \frac{n}{a}\left(\frac{a}{r}\right)^n \Phi_n\right] = \frac{\mathbf{r}}{r} \cdot \mathbf{V} \tag{3–2.18}$$

$$\sum_{n=-\infty}^{\infty} \left[\frac{n(n+1)a}{2\mu(2n+3)}\left(\frac{a}{r}\right)^n p_n + \frac{n(n-1)}{a}\left(\frac{a}{r}\right)^n \Phi_n\right] = -r\nabla \cdot \mathbf{V} \tag{3–2.19}$$

and

$$\sum_{n=-\infty}^{\infty} \left[n(n+1)\left(\frac{a}{r}\right)^n \chi_n\right] = \mathbf{r} \cdot \nabla \times \mathbf{V} \tag{3–2.20}$$

We shall now specialize these relationships to separate domains involving the regions interior and external to a sphere, respectively.

Region interior to a sphere

When the fluid occupies the interior of a sphere, the condition that the velocity remain finite at the origin limits us to positive harmonics; thus, we put

$$p_n = \Phi_n = \chi_n = 0, \qquad n \leqq 0 \tag{3–2.21}$$

and retain only those terms for which $n \geqq 1$. Now, it is well known that an arbitrary scalar function of the angular spherical coordinates (θ, ϕ) can be uniquely expanded in a series of surface spherical harmonics. This implies that there exist representations of the form

$$\frac{\mathbf{r}}{r} \cdot \mathbf{V} = \sum_{n=1}^{\infty} X_n \tag{3–2.22}$$

$$-r\nabla \cdot \mathbf{V} = \sum_{n=1}^{\infty} Y_n \tag{3–2.23}$$

and

$$\mathbf{r} \cdot \nabla \times \mathbf{V} = \sum_{n=1}^{\infty} Z_n \tag{3–2.24}$$

from which the surface harmonics $X_n(\theta, \phi)$, $Y_n(\theta, \phi)$, and $Z_n(\theta, \phi)$ can be determined from the prescribed surface velocity $\mathbf{V}$. Considered in conjunction with Eqs. (3–2.18)–(3–2.20), these result in the relations

$$\sum_{n=1}^{\infty} \left[\frac{na}{2\mu(2n+3)}\left(\frac{a}{r}\right)^n p_n + \frac{n}{a}\left(\frac{a}{r}\right)^n \Phi_n\right] = \sum_{n=1}^{\infty} X_n \tag{3–2.25}$$

$$\sum_{n=1}^{\infty} \left[\frac{n(n+1)a}{2\mu(2n+3)}\left(\frac{a}{r}\right)^n p_n + \frac{n(n-1)}{a}\left(\frac{a}{r}\right)^n \Phi_n\right] = \sum_{n=1}^{\infty} Y_n \tag{3–2.26}$$

and
$$\sum_{n=1}^{\infty}\left[n(n+1)\left(\frac{a}{r}\right)^{n}\chi_n\right]=\sum_{n=1}^{\infty}Z_n \tag{3–2.27}$$

Equating term-by-term under the summation sign, and solving the three resultant equations for the harmonic functions p_n, Φ_n, and χ_n, we obtain for $n > 0$,

$$p_n=\frac{\mu(2n+3)}{n}\frac{1}{a}\left(\frac{r}{a}\right)^{n}[Y_n-(n-1)X_n] \tag{3–2.28}$$

$$\Phi_n=\frac{1}{2n}a\left(\frac{r}{a}\right)^{n}[(n+1)X_n-Y_n] \tag{3–2.29}$$

$$\chi_n=\frac{1}{n(n+1)}\left(\frac{r}{a}\right)^{n}Z_n \tag{3–2.30}$$

Since the functions X_n, Y_n, and Z_n are known in principle from the prescribed velocity field on the sphere surface, any boundary value problem interior to a sphere may therefore be considered solved in principle.

Region exterior to a sphere

For the situation in which the velocity is required to vanish at infinity, we are restricted to harmonic functions of negative order. The problem of a velocity field which does not vanish at infinity, but is there required to assume the value $\mathbf{v}_\infty=\mathbf{v}_\infty(r,\theta,\phi)$, can always be reduced to the former case of zero velocity, provided that this prescribed field satisfies the creeping motion and continuity equations.

As before, one can now employ a solution obtained by discarding all positive harmonics from Eq. (3–2.3)

$$\mathbf{v}-\mathbf{v}_\infty=\sum_{n-1}^{\infty}\left[\nabla\times(\mathbf{r}\chi_{-(n+1)})+\nabla\Phi_{-(n+1)}-\frac{(n-2)}{\mu 2n(2n-1)}r^2\nabla p_{-(n+1)}+\frac{(n+1)}{\mu n(2n-1)}\mathbf{r}p_{-(n+1)}\right] \tag{3–2.31}$$

Here n has been replaced by $-(n+1)$. By analogy to our previous nomenclature, let $\mathbf{V}_\infty$ denote the field $\mathbf{v}_\infty$ evaluated at the surface of the sphere of radius a; that is,

$$\mathbf{V}_\infty=[\mathbf{v}_\infty]_a \tag{3–2.32}$$

By employing the prescribed boundary conditions, we ultimately obtain the solution

$$p_{-(n+1)}=\frac{\mu(2n-1)}{n+1}\frac{1}{a}\left(\frac{a}{r}\right)^{n+1}[(n+2)X_n+Y_n] \tag{3–2.33}$$

$$\Phi_{-(n+1)}=\frac{1}{2(n+1)}a\left(\frac{a}{r}\right)^{n+1}(nX_n+Y_n) \tag{3–2.34}$$

$$\chi_{-(n+1)}=\frac{1}{n(n+1)}\left(\frac{a}{r}\right)^{n+1}Z_n \tag{3–2.35}$$

The functions X_n, Y_n, Z_n are to be determined from Eqs. (3–2.22)–(3–2.24) by there replacing $\mathbf{V}$ by $\mathbf{V} - \mathbf{V}_\infty$.

Region between concentric spheres

When the fluid is bounded between concentric spheres, there are no conditions to be satisfied at either the origin or at infinity; hence one may resort to harmonic functions of both positive and negative orders. As in the preceding, we can obtain surface harmonic expansions of the type corresponding to arbitrarily prescribed velocity fields on both surfaces. Simultaneous solution of the six equations involved leads to determination of the harmonics which appear in the general solution, Eq. (3–2.3).

Frictional force and torque on a sphere

Once the boundary value problem for flow external to a sphere has been solved, it is possible to calculate the frictional force and torque (about the sphere center) experienced by a sphere in a given field. The stress vector acting across the surface of a sphere of radius r is easily shown to be[19]

$$\mathbf{\Pi}_r = \frac{\mathbf{r}}{r}\cdot\mathbf{\Pi} = -\frac{\mathbf{r}}{r}p + \mu\left(\frac{\partial \mathbf{v}}{\partial r} - \frac{\mathbf{v}}{r}\right) + \frac{\mu}{r}\nabla(\mathbf{r}\cdot\mathbf{v}) \tag{3–2.36}$$

for an incompressible newtonian fluid. By means of Eqs. (3–2.2) and (3–2.3) this can ultimately be expressed in the form,

$$\mathbf{\Pi}_r = \frac{\mu}{r}\sum_{n=-\infty}^{\infty}\Big[(n-1)\nabla\times(\mathbf{r}\chi_n) + 2(n-1)\nabla\Phi_n - \frac{(2n^2+4n+3)}{\mu(n+1)(2n+3)}\mathbf{r}p_n + \frac{n(n+2)}{\mu(n+1)(2n+3)}r^2\nabla p_n\Big] \tag{3–2.37}$$

The frictional force $\mathbf{F}$ exerted by the fluid on the sphere is obtained from Eq. (2–3.1) by integrating this stress around the sphere surface. In this case $dS = r^2 \sin\theta\, d\theta\, d\phi$ is an element of surface area on a sphere. This expression can be evaluated in a very general way by resorting to the following surface integral theorems: In these relations H_n is an arbitrary solid spherical harmonic function of positive or negative order.

$$\int_s H_n\, dS = \begin{cases} 4\pi r^2 H_0 & \text{for } n = 0 \\ 4\pi r^2 H_{-1} & \text{for } n = -1 \\ 0 & \text{for all other } n \end{cases} \tag{3–2.38}$$

$$\int_s \mathbf{r} H_n\, dS = \begin{cases} \dfrac{4\pi r^4}{3}\nabla H_1 & \text{for } n = 1 \\ \dfrac{4\pi r}{3}\nabla(r^3 H_{-2}) & \text{for } n = -2 \\ \mathbf{0} & \text{for all other } n \end{cases} \tag{3–2.39}$$

$$\int_s \nabla H_n \, dS = \begin{cases} 4\pi r^2 \nabla H_1 & \text{for } n = 1 \\ \mathbf{0} & \text{for all other } n \end{cases} \tag{3-2.40}$$

and

$$\int_s \nabla \times (\mathbf{r} H_n) \, dS = \mathbf{0} \quad \text{for all } n \tag{3-2.41}$$

Since $\mathbf{F} = \int_s \mathbf{\Pi}_r \, dS$ we eventually obtain the very simple result,

$$\mathbf{F} = -4\pi \nabla(r^3 p_{-2}) \tag{3-2.42}$$

Note that this formula is not affected by the additional term $\mathbf{v}_\infty$ in Eq. (3-2.31), since there is no contribution by the p_{-2} harmonic to the velocity field at points infinitely distant from the sphere. It is shown in the original paper* that this relation holds for nonspherical particles as well; that is, it gives the force on a particle of *any* shape.

A formula, analogous to Eq. (3-2.42), can be obtained for the torque experienced by a spherical particle about its origin. The force acting on an elementary area of surface on the sphere is $\mathbf{\Pi}_r \, dS$. The moment arm of this force is $\mathbf{r}$. Hence, the torque on the sphere is

$$\mathbf{T}_O = \int_s \mathbf{r} \times \mathbf{\Pi}_r \, dS \tag{3-2.43}$$

Now, from Eq. (3-2.37) we obtain

$$\mathbf{r} \times \mathbf{\Pi}_r = \frac{\mu}{r} \sum_{n=-\infty}^{\infty} \Big[(n-1) r^2 \nabla \chi_n - n(n-1) \mathbf{r} \chi_n - 2(n-1) \nabla \times (\mathbf{r} \Phi_n) - \frac{n(n+2)}{\mu(n+1)(2n+3)} r^2 \nabla \times (\mathbf{r} p_n) \Big] \tag{3-2.44}$$

Application of the integral theorems cited in Eqs. (3-2.38)–(3-2.41) ultimately yields

$$\mathbf{T}_O = -8\pi \mu \nabla(r^3 \chi_{-2}) \tag{3-2.45}$$

Again, it is shown in the original derivation* that this formula is not limited to spherical particles, but gives the torque (about the origin, $r = 0$) experienced by a particle of any shape.

If one wishes only to calculate the hydrodynamic force and torque on a rigid spherical particle, and not the velocity field itself, it may be possible to do so by utilizing Faxen's laws[11]. According to these, if a sphere to which fluid adheres is immersed in an unbounded fluid in motion at infinity with velocity $\mathbf{v}_\infty$, and if the sphere center translates with velocity $\mathbf{U}$ while the sphere spins with angular velocity $\boldsymbol{\omega}$, then the force and torque on the sphere are

$$\mathbf{F} = 6\pi \mu a([\mathbf{v}_\infty]_O - \mathbf{U}) + \mu \pi a^3 (\nabla^2 \mathbf{v}_\infty)_O \tag{3-2.46}$$

*Brenner, *op. cit.*

$$\mathbf{T}_O = 8\pi\mu a^3(\tfrac{1}{2}[\nabla \times \mathbf{v}_\infty]_O - \boldsymbol{\omega}) \tag{3-2.47}$$

where a is the sphere radius. The subscript O implies evaluation at the location of the sphere center.

As a simple illustration, consider the problem of a rigid spherical particle moving axially without spinning in a circular cylindrical tube through which a viscous fluid is moving. We suppose that the cylinder radius is much larger than that of the sphere. The cylinder axis is taken to be the $z = Z$ axis. The spherical particle moves with a constant velocity $\mathbf{U} = \mathbf{k}U$ parallel to the axis, whereas the superficial flow of fluid occurs in the same direction with a mean velocity of $\mathbf{U}_m = \mathbf{k}U_m = \mathbf{k}\frac{1}{2}U_o$. $\mathbf{k}$ is a unit vector in the z direction and U_o is the unperturbed velocity at the tube axis. The cylinder radius is R_o; the perpendicular distance from the longitudinal axis of the cylinder to a point in the fluid is R, and the sphere center is situated at a distance $R = b$ from this axis.

At large distances from the sphere the undisturbed flow is Poiseuillian. Thus we set

$$\mathbf{v}_\infty = 2\,\mathbf{U}_m\left[1 - \left(\frac{R}{R_o}\right)^2\right] \tag{3-2.48}$$

According to Poiseuille's law, the dynamic pressure is

$$p_\infty = \text{constant} - \frac{8\mu U_m}{R_o^2} z \tag{3-2.49}$$

If (r, θ, ϕ) refer to spherical coordinates having their origin at the center of the sphere, as shown in Fig. 7-3.1, it is easy to verify that

$$R^2 = r^2 \sin^2\theta + b^2 + 2br\sin\theta\cos\phi \tag{3-2.50}$$

This makes

$$\mathbf{v}_\infty = \mathbf{k}U_o\left(1 - \frac{r^2}{R_o^2}\sin^2\theta - \frac{b^2}{R_o^2} - \frac{2rb}{R_o^2}\sin\theta\cos\phi\right) \tag{3-2.51}$$

Therefore, since $\mathbf{V} = \mathbf{k}U$,

$$\mathbf{V} - \mathbf{V}_\infty = \mathbf{k}\left[U - U_o\left(1 - \frac{b^2}{R_o^2}\right) + U_o\frac{a^2}{R_o^2}\sin^2\theta + U_o\frac{2ab}{R_o^2}\sin\theta\cos\phi\right] \tag{3-2.52}$$

With the assistance of the transformation,

$$\frac{\mathbf{r}}{r} = \mathbf{i}\sin\theta\cos\phi + \mathbf{j}\sin\theta\sin\phi + \mathbf{k}\cos\theta \tag{3-2.53}$$

we find that

$$\frac{\mathbf{r}}{r}\cdot(\mathbf{V} - \mathbf{V}_\infty) = \left[U - U_o\left(1 - \frac{b^2}{R_o^2}\right)\right]\cos\theta + U_o\frac{a^2}{R_o^2}\sin^2\theta\cos\theta + U_o\frac{2ab}{R_o^2}\sin\theta\cos\theta\cos\phi \tag{3-2.54}$$

In terms of Legendre polynomials and associated Legendre polynomials of the first kind defined, respectively, by the relations

$$P_n(\mu) = \frac{1}{2^n n!} \frac{d^n}{d\mu^n} (\mu^2 - 1)^n \tag{3–2.55}$$

and

$$P_n^m(\mu) = (1 - \mu^2)^{(1/2)m} \frac{d^m}{d\mu^m} P_n(\mu) \tag{3–2.56}$$

we have

$$\cos\theta = P_1(\cos\theta) \tag{3–2.57}$$

$$\sin^2\theta \cos\theta = \tfrac{2}{5}[P_1(\cos\theta) - P_3(\cos\theta)] \tag{3–2.58}$$

and

$$\sin\theta \cos\theta \cos\phi = \tfrac{1}{3} \cos\phi P_2^1(\cos\theta) \tag{3–2.59}$$

If these relationships are introduced into Eq. (3–2.54), we obtain the surface harmonic expansion,

$$\frac{\mathbf{r}}{r} \cdot (\mathbf{V} - \mathbf{V}_\infty) = X_1 + X_2 + X_3 \tag{3–2.60}$$

where

$$X_1 = \left[U - U_o\left(1 - \frac{b^2}{R_o^2}\right) + \frac{2}{5} U_o \left(\frac{a}{R_o}\right)^2\right] P_1(\cos\theta) \tag{3–2.61}$$

$$X_2 = \frac{2}{3} U_o \left(\frac{a}{R_o}\right)\left(\frac{b}{R_o}\right) \cos\phi P_2^1(\cos\theta) \tag{3–2.62}$$

$$X_3 = -\tfrac{2}{5} U_o \left(\frac{a}{R_o}\right)^2 P_3(\cos\theta) \tag{3–2.63}$$

An entirely analogous treatment yields the expansion,

$$-r\nabla \cdot (\mathbf{V} - \mathbf{V}_\infty) = Y_1 + Y_2 + Y_3 \tag{3–2.64}$$

where

$$Y_1 - \frac{4}{5} U_o \left(\frac{a}{R_o}\right)^2 P_1(\cos\theta) \tag{3–2.65}$$

$$Y_2 = \frac{2}{3} U_o \left(\frac{a}{R_o}\right)\left(\frac{b}{R_o}\right) \cos\phi\, P_2^1(\cos\theta) \tag{3–2.66}$$

$$Y_3 = -\frac{4}{5} U_o \left(\frac{a}{R_o}\right)^2 P_3(\cos\theta) \tag{3–2.67}$$

and the expansion

$$\mathbf{r} \cdot \nabla \times (\mathbf{V} - \mathbf{V}_\infty) = Z_1 \tag{3–2.68}$$

where

$$Z_1 = -2 U_o \left(\frac{a}{R_o}\right)\left(\frac{b}{R_o}\right) \sin\phi\, P_1^1(\cos\theta) \tag{3–2.69}$$

We may now use these values in Eqs. (3–2.33)–(3–2.35) to obtain the harmonic functions in the velocity field outside the sphere, in accordance with Eq. (3–2.31). This velocity field $\mathbf{v}$ satisfies the boundary conditions $\mathbf{v} = \mathbf{U}$ at $r = a$ and $\mathbf{v} \to \mathbf{v}_\infty$ at infinity. The appropriate values of the solid spherical harmonics are as follows:

$$p_{-2} = \frac{3}{2}\mu a\left[U - U_0\left(1 - \frac{b^2}{R_0^2}\right) + \frac{2}{3}U_0\left(\frac{a}{R_0}\right)^2\right]r^{-2}P_1(\cos\theta) \quad (3\text{–}2.70)$$

$$p_{-3} = \frac{10}{3}\mu a^2 U_0\left(\frac{b}{R_0}\right)\left(\frac{a}{R_0}\right)r^{-3}\cos\phi P_2^1(\cos\theta) \quad (3\text{–}2.71)$$

$$p_{-4} = -\frac{7}{2}\mu a^3 U_0\left(\frac{a}{R_0}\right)^2 r^{-4}P_3(\cos\theta) \quad (3\text{–}2.72)$$

$$\Phi_{-2} = \frac{1}{4}a^3\left[U - U_0\left(1 - \frac{b^2}{R_0^2}\right) + \frac{6}{5}U_0\left(\frac{a}{R_0}\right)^2\right]r^{-2}P_1(\cos\theta) \quad (3\text{–}2.73)$$

$$\Phi_{-3} = \frac{1}{3}a^4 U_0\left(\frac{b}{R_0}\right)\left(\frac{a}{R_0}\right)r^{-3}\cos\phi\, P_2^1(\cos\theta) \quad (3\text{–}2.74)$$

$$\Phi_{-4} = -\frac{1}{4}a^5 U_0\left(\frac{a}{R_0}\right)^2 r^{-4}P_3(\cos\theta) \quad (3\text{–}2.75)$$

and

$$\chi_{-2} = -a^2 U_0\left(\frac{b}{R_0}\right)\left(\frac{a}{R_0}\right)r^{-2}\sin\phi P_1^1(\cos\theta) \quad (3\text{–}2.76)$$

For reference, we tabulate the values

$$P_1(\cos\theta) = \cos\theta$$
$$P_1^1(\cos\theta) = \sin\theta$$
$$P_2^1(\cos\theta) = 3\sin\theta\cos\theta$$
$$P_3(\cos\theta) = \tfrac{1}{2}(5\cos^3\theta - 3\cos\theta)$$

In accordance with Eq. (3–2.31), the solution of the problem is

$$\begin{aligned}\mathbf{v} - \mathbf{k}U_0\left(1 - \frac{R^2}{R_0^2}\right) &= \nabla \times (\mathbf{r}\chi_{-2}) \\ &\quad + \nabla\left(\Phi_{-2} + \Phi_{-3} + \Phi_{-4} + \frac{1}{2}r^2\frac{p_{-2}}{\mu} - \frac{1}{30}r^2\frac{p_{-4}}{\mu}\right) \quad (3\text{–}2.77) \\ &\quad + \mathbf{r}\left(\frac{p_{-2}}{\mu} + \frac{1}{2}\frac{p_{-3}}{\mu} + \frac{1}{3}\frac{p_{-4}}{\mu}\right)\end{aligned}$$

and

$$p - p_\infty = p_{-2} + p_{-3} + p_{-4}$$

where p_∞ is given in Eq. (3–2.49).

The hydrodynamic force on the sphere may be calculated from Eq. (3–2.42). For this purpose we take the value of p_{-2} given in Eq. (3–2.70) and note that

$$\nabla[rP_1(\cos\theta)] = \nabla z = \mathbf{k} \quad (3\text{–}2.78)$$

The force on the sphere is, therefore,

$$\mathbf{F} = -\mathbf{k}6\pi\mu a\left[U - U_0\left(1 - \frac{b^2}{R_0^2}\right) + \frac{2}{3}U_0\left(\frac{a}{R_0}\right)^2\right] \quad (3\text{–}2.79)$$

To calculate the torque we use Eq. (3–2.45), with the value of χ_{-2} given by Eq. (3–2.76), together with the relationship

$$\nabla[r \sin \phi P_1^1(\cos \theta)] = \nabla y = \mathbf{j} \tag{3–2.80}$$

Hence, the torque about the sphere center is

$$\mathbf{T}_O = \mathbf{j} 8\pi\mu a^2 U_0 \left(\frac{a}{R_0}\right)\left(\frac{b}{R_0}\right) \tag{3–2.81}$$

corresponding to a rotation of the sphere about an axis which is perpendicular to the line joining the origin of the sphere to the longitudinal axis of the cylinder and to the direction of flow. These results check those previously obtained by Simha[26].

We may obtain the latter results directly by employing Faxen's equations (3–2.46) and (3–2.47). In the present instance, we have $\boldsymbol{\omega} = \mathbf{0}$. Furthermore, since the sphere center is located at the point $R = b$, we find from Eq. (3–2.48) that

$$[\mathbf{v}_\infty]_O = \mathbf{k} 2 U_m \left[1 - \left(\frac{b}{R_0}\right)^2\right] \tag{3–2.82}$$

$$[\nabla^2 \mathbf{v}_\infty]_O = \frac{1}{\mu}[\nabla p_\infty]_O = -\mathbf{k}\frac{8U_m}{R_0^2} \tag{3–2.83}$$

$$[\nabla \times \mathbf{v}_\infty]_O = \mathbf{j}\frac{4bU_m}{R_0^2} \tag{3–2.84}$$

Upon setting $U_m = U_0/2$ and substituting these results into Eqs. (3–2.46) and (3–2.47) we obtain results for the force and torque identical with Eqs. (3–2.79) and (3–2.81).

It should be clearly understood that the expressions we have obtained for the hydrodynamic force and torque on a sphere in a circular cylinder are asymptotically correct only in the limit of small a/R_0. Methods for obtaining higher approximations are discussed in Section 7–3.

3-3 Cylindrical Coordinates

In many practical situations involving flow past particles, the fluid is bounded externally by the walls of a long circular cylinder. In this section we develop a general three-dimensional solution of the creeping motion and continuity equations suitable for satisfying arbitrarily prescribed boundary conditions on the surface of an infinitely long circular cylinder.

In terms of cylindrical coordinates (ρ, ϕ, z), the component velocities are given by

$$\mathbf{v} = \mathbf{i}_\rho v_\rho + \mathbf{i}_\phi v_\phi + \mathbf{i}_z v_z \tag{3–3.1}$$

where $\mathbf{i}_\rho$, $\mathbf{i}_\phi$, and $\mathbf{i}_z$ are unit vectors in the direction of each of the three coordinate axes. The creeping motion equations in these coordinates are then

$$\nabla^2 v_z = \frac{1}{\mu}\frac{\partial p}{\partial z} \tag{3-3.2}$$

$$\nabla^2 v_\rho - \frac{v_\rho}{\rho^2} - \frac{2}{\rho^2}\frac{\partial v_\phi}{\partial \phi} = \frac{1}{\mu}\frac{\partial p}{\partial \rho} \tag{3-3.3}$$

$$\nabla^2 v_\phi - \frac{v_\phi}{\rho^2} + \frac{2}{\rho^2}\frac{\partial v_\rho}{\partial \phi} = \frac{1}{\mu}\frac{1}{\rho}\frac{\partial p}{\partial \phi} \tag{3-3.4}$$

where the Laplace operator has the form

$$\nabla^2 = \frac{\partial^2}{\partial \rho^2} + \frac{\partial}{\rho \partial \rho} + \frac{\partial^2}{\rho^2 \partial \phi^2} + \frac{\partial^2}{\partial z^2} \tag{3-3.5}$$

In addition, we must satisfy the continuity equation,

$$\frac{\partial v_\rho}{\partial \rho} + \frac{v_\rho}{\rho} + \frac{\partial v_\phi}{\rho \partial \phi} + \frac{\partial v_z}{\partial z} = 0 \tag{3-3.6}$$

To the particular solution of these equations must be added the general solution of the homogeneous equations,

$$\nabla^2 \mathbf{v}' = \mathbf{0} \tag{3-3.7}$$

$$\nabla \cdot \mathbf{v}' = 0 \tag{3-3.8}$$

Particular solution of the inhomogeneous equations

The method of solution employed here is a generalization[3] of that originally developed by Ladenburg[18] for problems involving axisymmetric motion of a sphere in a cylinder. In that case, however, Haberman and Sayre later used the stream function to provide a more general and sophisticated method of approach. (References to their work are given in Chapter 7.)

As in the case of spherical coordinates, we note that the pressure field satisfies Laplace's equation (3-2.1). Solutions of this equation which are useful for satisfying arbitrary conditions on the surface $\rho =$ constant are of the form

$$p = \sum_{n=-\infty}^{\infty} {\cos \atop \sin} n\phi \int_0^\infty \eta_n(\lambda) {I_n \atop K_n}(\lambda\rho) {\cos \atop \sin}(\lambda z)\, d\lambda \tag{3-3.9}$$

where λ is an arbitrary parameter and $\eta_n(\lambda)$ an arbitrary function, ultimately to be determined from the boundary conditions. The functions $I_n(\lambda\rho)$ and $K_n(\lambda\rho)$ are modified Bessel functions of order n of the first and second kinds, respectively. They satisfy the differential equation

$$I_n''(\lambda\rho) + \frac{1}{\rho} I_n'(\lambda\rho) - \left(\lambda^2 + \frac{n^2}{\rho^2}\right) I_n(\lambda\rho) = 0 \tag{3-3.10}$$

with a similar expression for $K_n(\lambda\rho)$. In this and subsequent equations in this section, the differentiations denoted by primes *are with respect to* ρ. In developing a suitable solution of the inhomogeneous equations, we shall first assume a very simple form of the pressure field. This restriction is later removed. As a particular solution we choose

$$p = \mu A_n(\lambda) \cos n\phi I_n(\lambda\rho) \sin \lambda z \tag{3–3.11}$$

where $A_n(\lambda)$ is an arbitrary function of n and λ.

The z component of the velocity field is assumed to possess a solution of the form

$$v_z = F_n(\lambda, \rho) \cos n\phi \cos \lambda z \tag{3–3.12}$$

Introduction of these expressions into Eq. (3–3.2) shows that F_n is a solution of the equation

$$F_n'' + \frac{1}{\rho} F_n' - \left(\lambda^2 + \frac{n^2}{\rho^2}\right) F_n = A_n(\lambda)\lambda I_n(\lambda\rho) \tag{3–3.13}$$

As evidenced by Eq. (3–3.10), the homogeneous part of the foregoing equation is Bessel's modified equation. Since the general solution of the homogeneous equation is known, it is possible to effect the solution of the inhomogeneous equation (3–3.13) by the method of variation of parameters. This ultimately yields

$$F_n(\lambda, \rho) = B_n(\lambda) I_n(\lambda\rho) + A_n(\lambda) \frac{\rho I_n'(\lambda\rho)}{2\lambda} \tag{3–3.14}$$

where we have rejected that part of the solution containing functions of the second kind. We now have

$$v_z = \left[B_n(\lambda) I_n(\lambda\rho) + A_n(\lambda) \frac{\rho I_n'(\lambda\rho)}{2\lambda} \right] \cos n\phi \cos \lambda z \tag{3–3.15}$$

The element of arbitrariness in the function $B_n(\lambda)$ is eventually removed by simultaneously satisfying the continuity equation, wherein $B_n(\lambda)$ is expressed in terms of $A_n(\lambda)$.

For the remaining velocity components, we assume solutions of the form

$$v_\rho - G_n(\lambda, \rho) \cos n\phi \sin \lambda z \tag{3–3.16}$$

and

$$v_\phi = H_n(\lambda, \rho) \sin n\phi \sin \lambda z \tag{3–3.17}$$

On the basis of Eqs. (3–3.3), (3–3.4), and (3–3.11), we then obtain the simultaneous equations

$$G_n'' + \frac{1}{\rho} G_n' - \left(\lambda^2 + \frac{n^2 + 1}{\rho^2}\right) G_n - \frac{2n}{\rho^2} H_n = A_n(\lambda) I_n'(\lambda\rho) \tag{3–3.18}$$

and

$$H_n'' + \frac{1}{\rho} H_n' - \left(\lambda^2 + \frac{n^2 + 1}{\rho^2}\right) H_n - \frac{2n}{\rho^2} G_n = -A_n(\lambda) \frac{n I_n(\lambda\rho)}{\rho} \tag{3–3.19}$$

Addition of these two equations yields

$$(G_n + H_n)'' + \frac{1}{\rho}(G_n + H_n)' - \left[\lambda^2 + \frac{(n+1)^2}{\rho^2}\right](G_n + H_n) = A_n(\lambda)\lambda I_{n+1}(\lambda\rho) \tag{3–3.20}$$

where we have employed the recurrence relation

$$I'_n(\lambda\rho) - \frac{nI_n(\lambda\rho)}{\rho} = \lambda I_{n+1}(\lambda\rho) \tag{3–3.21}$$

But this equation is of precisely the same form as Eq. (3–3.13) with $n+1$ written in place of n. This immediately leads to the solution

$$G_n + H_n = C_n(\lambda)I_{n+1}(\lambda\rho) + A_n(\lambda)\frac{\rho I'_{n+1}(\lambda\rho)}{2\lambda} \tag{3–3.22}$$

A second relation is obtained between G_n and H_n by subtracting Eq. (3–3.19) from Eq. (3–3.18); namely,

$$(G_n - H_n)'' + \frac{1}{\rho}(G_n - H_n)' - \left[\lambda^2 + \frac{(n-1)^2}{\rho^2}\right](G_n - H_n) = A_n(\lambda)\lambda I_{n-1}(\lambda\rho) \tag{3–3.23}$$

where we have noted that

$$I'_n(\lambda\rho) + \frac{nI_n(\lambda\rho)}{\rho} = \lambda I_{n-1}(\lambda\rho) \tag{3–3.24}$$

Again, this is of the same form as Eq. (3–3.13) except that $n-1$ now appears in place of n. The solution is thus

$$G_n - H_n = D_n(\lambda)I_{n-1}(\lambda\rho) + A_n(\lambda)\frac{\rho I'_{n-1}(\lambda\rho)}{2\lambda} \tag{3–3.25}$$

Equations (3–3.22) and (3–3.25) combine to give

$$G_n(\lambda, \rho) = A_n(\lambda)\frac{\rho}{4\lambda}[I'_{n+1}(\lambda\rho) + I'_{n-1}(\lambda\rho)] \tag{3–3.26}$$

and

$$H_n(\lambda, \rho) = A_n(\lambda)\frac{\rho}{4\lambda}[I'_{n+1}(\lambda\rho) - I'_{n-1}(\lambda\rho)] \tag{3–3.27}$$

The arbitrary functions $C_n(\lambda)$ and $D_n(\lambda)$ have been set equal to zero because similar solutions of the homogeneous equations (3–3.7) and (3–3.8) arise later, and their retention at this point would be redundant.

Differentiation of the recurrence formulas,

$$I_{n+1}(X) + I_{n-1}(X) = 2I'_n(X)$$

and

$$I_{n+1}(X) - I_{n-1}(X) = -\frac{2nI_n(X)}{X}$$

eventually leads to alternate expressions for G_n and H_n.

Thus, we obtain,

$$G_n(\lambda, \rho) = A_n(\lambda)\frac{I''_n(\lambda\rho)}{2\lambda^2} \tag{3–3.28}$$

and

$$H_n(\lambda, \rho) = A_n(\lambda)\frac{n}{2\lambda^2}\left[\frac{I_n(\lambda\rho)}{\rho} - I'_n(\lambda\rho)\right] \tag{3–3.29}$$

The expressions for the ρ and ϕ components of the velocity field are, finally,

$$v_\rho = \frac{A_n(\lambda)}{2\lambda^2} \rho I_n''(\lambda\rho) \cos n\phi \sin \lambda z \tag{3-3.30}$$

and

$$v_\phi = \frac{n A_n(\lambda)}{2\lambda^2}\left[\frac{I_n(\lambda\rho)}{\rho} - I_n'(\lambda\rho)\right] \sin n\phi \sin \lambda z \tag{3-3.31}$$

These two relations in conjunction with the corresponding expressions for v_z and p, Eqs. (3–3.15) and (3–3.11), constitute a solution of Eqs. (3–3.2)–(3–3.4). It still remains to satisfy the equation of continuity. Upon introducing these expressions for the component velocities into the continuity equation (3–3.6), and performing the indicated operations, it is found that this equation is satisfied, provided that

$$\frac{A_n(\lambda)}{2\lambda^2}\left[\rho I_n'''(\lambda\rho) + 2I_n''(\lambda\rho) - \lambda^2\rho I_n'(\lambda\rho) - \frac{n^2 I_n'(\lambda\rho)}{\rho} + \frac{n^2 I_n(\lambda\rho)}{\rho^2}\right] - B_n(\lambda)\lambda I_n(\lambda\rho) = 0 \tag{3-3.32}$$

Multiplication of Eq. (3–3.10) by ρ and subsequent differentiation with respect to ρ shows that the bracketed term in the preceding equation is simply equal to $\lambda^2 I_n(\lambda\rho)$. Hence, the continuity equation is satisfied by choosing

$$B_n(\lambda) = \frac{A_n(\lambda)}{2\lambda^2} \tag{3-3.33}$$

Thus, the expression for v_z takes the form

$$v_z = \frac{A_n(\lambda)}{2\lambda^2}[I_n(\lambda\rho) + \lambda\rho I_n'(\lambda\rho)] \cos n\phi \cos \lambda z \tag{3-3.34}$$

If we now define a harmonic function, Π, by the relation

$$\Pi = \frac{A_n(\lambda)}{2\lambda^2} I_n(\lambda\rho) \cos n\phi \sin \lambda z \tag{3-3.35}$$

it can be shown that

$$v_\rho = \rho \frac{\partial}{\partial\rho}\left(\frac{\partial \Pi}{\partial \rho}\right) \tag{3-3.36}$$

$$v_\phi = \rho \frac{\partial}{\partial\rho}\left(\frac{1}{\rho}\frac{\partial \Pi}{\partial \phi}\right) \tag{3-3.37}$$

$$v_z = \rho \frac{\partial}{\partial\rho}\left(\frac{\partial \Pi}{\partial z}\right) + \frac{\partial \Pi}{\partial z} \tag{3-3.38}$$

and

$$p = -2\mu \frac{\partial^2 \Pi}{\partial z^2} \tag{3-3.39}$$

In vector notation,

$$\mathbf{v} = \rho \frac{\partial}{\partial \rho}(\nabla \Pi) + \mathbf{i}_z \frac{\partial \Pi}{\partial z} \tag{3-3.40}$$

Although this solution has been obtained by considering a special form for the pressure field, the solution given by Eqs. (3–3.39) and (3–3.40) is, in fact, a simultaneous solution of the creeping motion and continuity equations for *any* harmonic function Π, as can be readily verified; that is, for any Π satisfying Laplace's equation

$$\nabla^2 \Pi = 0 \tag{3-3.41}$$

Solution of the homogeneous equations

At this point we turn our attention to the solution of the homogeneous equations (3–3.7) and (3–3.8). As before, we assume that the equations may be solved by simple solutions of the form

$$v'_z = f_n(\lambda, \rho) \cos n\phi \cos \lambda z \tag{3-3.42}$$

$$v'_\rho = g_n(\lambda, \rho) \cos n\phi \sin \lambda z \tag{3-3.43}$$

$$v'_\phi = h_n(\lambda, \rho) \sin n\phi \sin \lambda z \tag{3-3.44}$$

These are substantially identical in appearance to the forms used in the solution of the corresponding inhomogeneous equations. On this basis, if we put $A_n(\lambda) = 0$ in Eqs. (3–3.14), (3–3.22), and (3–3.25), we can immediately infer that the equations $\nabla^2 \mathbf{v}' = \mathbf{0}$ are solved by choosing

$$f_n(\lambda, \rho) = b_n(\lambda) I_n(\lambda\rho) \tag{3-3.45}$$

$$g_n(\lambda, \rho) + h_n(\lambda, \rho) = c_n(\lambda) I_{n+1}(\lambda\rho) \tag{3-3.46}$$

and

$$g_n(\lambda, \rho) - h_n(\lambda, \rho) = d_n(\lambda) I_{n-1}(\lambda\rho) \tag{3-3.47}$$

The latter equations can be solved simultaneously for g_n and h_n. If this is done, and arbitrary functions $e_n(\lambda)$ and $f_n(\lambda)$ defined as follows,

$$e_n(\lambda) = \frac{c_n(\lambda) + d_n(\lambda)}{2\lambda} \tag{3-3.48}$$

$$f_n(\lambda) = \frac{c_n(\lambda) - d_n(\lambda)}{2\lambda} \tag{3-3.49}$$

then, with the aid of the recurrence formulas noted in Eqs. (3–3.21) and (3–3.24), we find

$$v'_\rho = \left[e_n(\lambda) I'_n(\lambda\rho) - f_n(\lambda) \frac{n I_n(\lambda\rho)}{\rho} \right] \cos n\phi \sin \lambda z \tag{3-3.50}$$

$$v'_\phi = \left[f_n(\lambda) I'_n(\lambda\rho) - e_n(\lambda) \frac{n I_n(\lambda\rho)}{\rho} \right] \sin n\phi \sin \lambda z \tag{3-3.51}$$

$$v'_z = b_n(\lambda) I_n(\lambda\rho) \cos n\phi \cos \lambda z \tag{3-3.52}$$

Upon introducing these results into the continuity equation (3–3.8), expressed in cylindrical coordinates [cf. Eq. (3–3.6)], we find that satisfaction of this relationship requires that

$$e_n(\lambda) \left[I''_n(\lambda\rho) + \frac{1}{\rho} I'_n(\lambda\rho) - \frac{n^2 I'_n(\lambda\rho)}{\rho^2} \right] - b_n(\lambda) \lambda I_n(\lambda\rho) = 0 \tag{3-3.53}$$

the coefficient of $f_n(\lambda)$ being zero. Equation (3–3.10) shows that the multiplier of $e_n(\lambda)$ in the preceding expression is equal to $\lambda^2 I_n(\lambda\rho)$. Satisfaction of the continuity equation is thereby assured by choosing

$$b_n(\lambda) = \lambda e_n(\lambda)$$

whence, in place of Eq. (3–3.52), we now write

$$v_z' = e_n(\lambda)\lambda I_n(\lambda\rho)\cos n\phi \cos \lambda z \tag{3–3.54}$$

If we define the harmonic functions

$$\Psi = e_n(\lambda) I_n(\lambda\rho) \cos n\phi \sin \lambda z \tag{3–3.55}$$

and

$$\Omega = -f_n(\lambda) I_n(\lambda\rho) \sin n\phi \sin \lambda z \tag{3–3.56}$$

the component velocities of $\mathbf{v}'$ may then be written in the form

$$v_\rho' = \frac{\partial \Psi}{\partial \rho} + \frac{\partial \Omega}{\rho \partial \phi} \tag{3–3.57}$$

$$v_\phi' = \frac{\partial \Psi}{\rho \partial \phi} - \frac{\partial \Omega}{\partial \rho} \tag{3–3.58}$$

and

$$v_z' = \frac{\partial \Psi}{\partial z} \tag{3–3.59}$$

These relationships can be expressed more concisely in vector notation,

$$\mathbf{v}' = \nabla \Psi + \nabla \times (\mathbf{i}_z \Omega) \tag{3–3.60}$$

The latter constitutes a solution of the homogeneous equations (3–3.7) and (3–3.8). Here again, the final result is true for any harmonic functions,

$$\nabla^2 \Psi = 0 \tag{3–3.61}$$

and

$$\nabla^2 \Omega = 0 \tag{3–3.62}$$

Solution of boundary value problems involving circular cylinders

A complete solution of the creeping motion and continuity equations is obtained by adding together the respective solutions of the inhomogeneous and homogeneous equations, whence

$$\mathbf{v} = \nabla \Psi + \nabla \times (\mathbf{i}_z \Omega) + \rho \frac{\partial}{\partial \rho}(\nabla \Pi) + \mathbf{i}_z \frac{\partial \Pi}{\partial z} \tag{3–3.63}$$

and

$$p = -2\mu \frac{\partial^2 \Pi}{\partial z^2} \tag{3–3.64}$$

provided that

$$\nabla^2\{\Pi, \Psi, \Omega\} = \{0, 0, 0\} \tag{3–3.65}$$

It remains now to show how to apply these results to the solution of boundary-value problems involving circular cylindrical surfaces. By way of illustration, we shall restrict ourselves to situations in which the fluid domain lies entirely within the interior of an infinitely long cylinder on whose surface the velocity field is required to assume arbitrarily prescribed values. Extension of the method to other situations simply involves the use of alternative solutions of Laplace's equation appropriate to the domain under consideration, for example, the region exterior to a cylinder in an otherwise unbounded medium, or the region bounded between two concentric cylinders. Note that no solution exists for *two-dimensional* streaming flow past a circular cylinder in an infinite medium.

We assume that the harmonic functions Π, Ψ, Ω can be expressed as the sum of a series of *cylindrical* harmonic functions of order n:

$$\{\Pi, \Psi, \Omega\} = \sum_{n=-\infty}^{\infty} \{\Pi_n, \Psi_n, \Omega_n\} \tag{3–3.66}$$

where

$$\nabla^2\{\Pi_n, \Psi_n, \Omega_n\} = \{0, 0, 0\} \tag{3–3.67}$$

The method of solution consists in the determination of Π_n, Ψ_n, and Ω_n, for each n, from the conditions prescribed at the boundary $\rho = \rho_o =$ constant.

Suitable solutions of Laplace's equation are of the form

$$\{\Pi_n, \Psi_n, \Omega_n\} = \cos(n\phi + \{\alpha_n, \beta_n, \gamma_n\}) \times \\ \times \int_0^\infty \{\pi_n(\lambda), \psi_n(\lambda), \omega_n(\lambda)\} I_n(\lambda\rho) \cos(\lambda z + \{\delta_\lambda, \xi_\lambda, \zeta_\lambda\}) d\lambda \tag{3–3.68}$$

Thus, we are now required to establish the form taken by the functions $\pi_n(\lambda)$, $\psi_n(\lambda)$, $\omega_n(\lambda)$, . . . , etc., from the boundary conditions, whereupon the problem is solved. This is done by introducing the relations given by Eq. (3–3.68) into Eq. (3–3.66). The resulting expressions, in turn, when substituted into Eq. (3–3.63) yield, for the component velocities on the cylinder wall, expressions of the form

$$[v_\rho]_{\rho=\rho_0} = \sum_{n=-\infty}^{\infty} \cos(n\phi + \sigma_n) \int_0^\infty \xi_n(\lambda) \cos(\lambda z + v_\lambda) d\lambda \tag{3–3.69}$$

with analogous expressions for the remaining two velocity components at the wall.

Here,

$$\sigma_n = \text{function } (\alpha_n, \beta_n, \gamma_n) \tag{3–3.70}$$

$$\xi_n(\lambda) = \text{function}[\pi_n(\lambda), \psi_n(\lambda), \omega_n(\lambda)] \tag{3–3.71}$$

and

$$v_\lambda = \text{function } (\delta_\lambda, \xi_\lambda, \zeta_\lambda) \tag{3–3.72}$$

According to Fourier's theorem any reasonably well-behaved function of the coordinates (ϕ, z) admits of a development having exactly the form indicated by Eq. (3–3.69). Thus, if the prescribed velocity components on the cylinder wall are expanded in this form, the functions σ_n, $\xi_n(\lambda)$, v_λ . . . , etc., are immediately known. With these values, the functions α_n, $\pi_n(\lambda)$, δ_λ, . . . , etc., can be determined via the simultaneous solution of equations such as (3–3.70)–(3–3.72).

This method of solution has been successfully employed[3] to study the behavior of a spherical particle in a circular cylindrical duct through which a viscous fluid is flowing. (Details are discussed in Section 7–3.)

3-4 Integral Representations

For some aspects of hydrodynamic problems and for the proof of existence of solutions, it is desirable to have the solution represented in a closed form, even if it involves an integral form. The Green's function technique can be applied to such an approach, and forms the basis for Oseen's[24] classic book on low Reynolds number hydrodynamics. In this section we shall briefly outline his approach and cite a few applications.

Following Oseen, we employ Cartesian tensors. Unless the contrary is explicitly indicated, the Einstein summation convention will be utilized. Thus, the creeping motion and continuity equations (2–6.1) and (2–6.2), take the form

$$\mu \nabla^2 u_j - \frac{\partial p}{\partial x_j} = 0 \tag{3–4.1}$$

$$\frac{\partial u_j}{\partial x_j} = 0 \tag{3–4.2}$$

where x_1, x_2, x_3 denote a Cartesian coordinate system. Consider a domain B and let S be its bounding surface. Assume that the creeping motion equations have a regular solution in B. Designate by $P^{(0)} \equiv (x_1^{(0)}, x_2^{(0)}, x_3^{(0)})$ any specified point in the interior of B at which we seek a solution of Eqs. (3–4.1) and (3–4.2). To obtain the field caused by a distributed source at a variable point $P \equiv (x_1, x_2, x_3)$, we calculate the contribution of each elementary point P at $P^{(0)}$ and add them all. The legitimacy of this procedure depends, of course, upon the linearity of the basic equations. In order to obtain a solution which depends on any two points P and $P^{(0)}$, we define a second-rank tensor $t_{jk} = t_{jk}(P, P^{(0)})$ as follows:

$$t_{jk} = \delta_{jk} \nabla^2 \psi(r) - \frac{\partial^2 \psi(r)}{\partial x_j \partial x_k}, \qquad r^2 = (x_j - x_j^{(0)})^2, \qquad r > 0 \tag{3–4.3}$$

where $\psi(r)$ is a function only of r, and δ_{jk} is the Kronecker delta:

$$\delta_{jk} = \begin{cases} 1 & \text{for } j = k \\ 0 & \text{for } j \neq k \end{cases}$$

By differentiating Eq. (3–4.3) we find that, for any choice of ψ,

$$\frac{\partial t_{jk}}{\partial x_j} = 0 \tag{3–4.4}$$

If we now assume that ψ satisfies the biharmonic equation

$$\nabla^4 \psi = 0; \qquad \nabla^2 = \frac{\partial^2}{\partial x_1^2} + \frac{\partial^2}{\partial x_2^2} + \frac{\partial^2}{\partial x_3^2} \equiv \frac{\partial}{\partial x_l}\frac{\partial}{\partial x_l} \tag{3–4.5}$$

and we set

$$-\mu \frac{\partial}{\partial x_k} \nabla^2 \psi = p_k \tag{3–4.6}$$

we then obtain

$$\mu\nabla^2 t_{jk} - \frac{\partial p_k}{\partial x_j} = 0 \tag{3-4.7}$$

Thus for each k value (1, 2, or 3) the vector whose three components are t_{1k}, t_{2k}, t_{3k} and the scalar p_k constitute solutions of the creeping motion and continuity equations.

Since ψ depends only on r, the biharmonic equation (3-4.5) written in spherical coordinates becomes

$$\nabla^2\nabla^2\psi = 0, \qquad \nabla^2 = \frac{1}{r^2}\frac{d}{dr}\left(r^2\frac{d}{dr}\right) \tag{3-4.8}$$

Its general solution is $\psi = ar^2 + br + c + (d/r)$, where a, b, c, and d are constants. For reasons soon apparent, we arbitrarily choose $\psi(r) = r$. This yields

$$\nabla^2\psi = \frac{2}{r}, \qquad t_{jk} = \frac{\delta_{jk}}{r} + \frac{(x_j - x_j^{(0)})(x_k - x_k^{(0)})}{r^3} \tag{3-4.9}$$

$$p_k = -2\mu\frac{\partial}{\partial x_k}\left(\frac{1}{r}\right) = 2\mu\frac{(x_k - x_k^{(0)})}{r^3} \tag{3-4.10}$$

Surround $P^{(0)}$ with a sphere of radius $r = \epsilon$ and choose ϵ so small that this sphere lies entirely within S, the bounding surface. $B(\epsilon)$ is the part of the domain of B which falls outside the sphere $r = \epsilon$. By a generalization of Green's formula[22] we have

$$\int_S \left\{t_{jk}\left(\mu\frac{du_j}{dn} - pn_j\right) - u_j\left(\mu\frac{dt_{jk}}{dn} - p_k n_j\right)\right\} dS = 0 \tag{3-4.11}$$

where n_j represents the component of the unit vector drawn outward and normal to the surface S. A proof of this relation is given in the Appendix to this chapter. The boundary surface now consists of two parts, S and the sphere $r = \epsilon$. We have, because the values rt_{jk} are of $O(1)$ throughout, including the point $P^{(0)}$, and because the area of the sphere $r = \epsilon$ is proportional to ϵ^2:

$$\lim_{\epsilon\to 0}\int_{r=\epsilon} t_{jk}\left(\mu\frac{du_j}{dn} - pn_j\right) dS = 0 \tag{3-4.12}$$

Furthermore,

$$\lim_{\epsilon\to 0}\int_{r=\epsilon} u_j\left(\mu\frac{dt_{jk}}{dn} - p_k n_j\right) dS = \lim_{\epsilon\to 0}\mu\int_{r=\epsilon}\left\{u_k + \frac{3u_j(x_j - x_j^{(0)})(x_k - x_k^{(0)})}{r^2}\right\}\frac{dS}{r^2} \tag{3-4.13}$$

By Taylor series expansion,

$$u_j = u_j^{(0)} + r\phi \tag{3-4.14}$$

where $u_j^{(0)} = u_j(P^{(0)})$ and where ϕ denotes a function of P which is finite in the vicinity of $P^{(0)}$. Since

$$\int_{r=\epsilon} \frac{dS}{r^2} = 4\pi, \qquad \int_{r=\epsilon} \frac{(x_j - x_j^{(0)})(x_k - x_k^{(0)})\,dS}{r^2} \frac{dS}{r^2} = \frac{4\pi}{3}\delta_{jk} \tag{3–4.15}$$

we have

$$\lim_{\epsilon\to 0} \int_{r=\epsilon} u_j\left(\mu \frac{dt_{jk}}{dn} - p_k n_j\right) dS = 8\pi\mu u_k(P^{(0)}) \tag{3–4.16}$$

Finally, upon substitution into Eq. (3–4.11), we obtain

$$u_k(P^{(0)}) = \frac{1}{8\pi\mu}\int_S \left\{t_{jk}\left(\mu \frac{du_j}{dn} - pn_j\right) - u_j\left(\mu \frac{dt_{jk}}{dn} - p_k n_j\right)\right\} dS \tag{3–4.17}$$

If we can find tensors T_{jk} and P_k, which inside a closed surface S can be represented in the form $T_{jk} = t_{jk} + \tau_{jk}$, $P_k = p_k + \pi_k$, where τ_{jk}, π_k constitute a regular solution of the creeping motion equations inside S, then one can express the result given in Eq. (3–4.17) in terms of T_{jk} and P_k. In particular if the tensor T_{jk} is chosen so as to vanish when the point P lies on the boundary S, we then obtain

$$u_k(P^{(0)}) = -\frac{1}{8\pi\mu}\int_S u_j\left(\mu \frac{dT_{jk}}{dn} - P_k n_j\right) dS \tag{3–4.18}$$

Thus, T_{jk} represents a Green's dyadic employed in a generalized surface Green's function[22].

In a similiar fashion, one can express the pressure in terms of the same functions which appear in Eq. (3–4.17)

$$p(P^{(0)}) = \frac{1}{4\pi}\int_S \left\{\left(\mu \frac{du_j}{dn} - pn_j\right)\frac{\partial}{\partial x_j}\left(\frac{1}{r}\right) - \mu u_j \frac{d}{dn}\frac{\partial}{\partial x_j}\left(\frac{1}{r}\right)\right\} dS \tag{3–4.19}$$

Oseen has extended solutions of this type to more complicated situations. For example, for the Oseen equation (2–6.4), including a body force exerted on each unit volume of fluid, we have

$$\mu\nabla^2 u_j - \rho U \frac{\partial u_j}{\partial x_1} - \frac{\partial p}{\partial x_j} + \rho X_j = 0 \tag{3–4.20a}$$

and

$$\frac{\partial u_j}{\partial x_j} = 0 \tag{3–4.20b}$$

where U is the uniform stream velocity in the positive x_1 direction, ρ is the fluid density, and X_j is the body force per unit volume. These equations have the fundamental solution

$$\begin{aligned} u_k(P^{(0)}) = \frac{1}{8\pi\mu}\int_S \Big\{ & t_{jk}\left(\mu \frac{du_j}{dn} - pn_j\right) - u_j\left(\mu \frac{dt_{jk}}{dn} - p_k n_j\right) \\ & - \rho U u_j t_{jk} n_1 \Big\} dS + \frac{\rho}{8\pi\mu}\int_B X_j t_{jk}\, dV \end{aligned} \tag{3–4.21}$$

and

$$p(P^{(0)}) = \frac{1}{4\pi}\int_S \left\{\left(\mu\frac{du_j}{dn} - pn_j\right)\frac{\partial}{\partial x_j}\left(\frac{1}{r}\right) - \mu u_j \frac{d}{dn}\frac{\partial}{\partial x_j}\left(\frac{1}{r}\right)\right.$$
$$\left. - \rho U u_j\left[n_1\frac{\partial}{\partial x_j}\left(\frac{1}{r}\right) - n_j\frac{\partial}{\partial x_1}\left(\frac{1}{r}\right)\right]\right\} dS + \frac{\rho}{4\pi}\int_B X_j \frac{\partial}{\partial x_j}\left(\frac{1}{r}\right) dV \tag{3-4.22}$$

Here dV denotes an element of volume. The tensor t_{jk} is defined by the relation

$$t_{jk} = \delta_{jk}\nabla^2\psi - \frac{\partial^2\psi}{\partial x_j x_k} \tag{3-4.23}$$

as previously, but now

$$\psi = \frac{2\mu}{\rho U}\int_0^{\rho U(r-x_1)/2\mu} \frac{1-e^{-\alpha}}{\alpha}\, d\alpha \tag{3-4.24}$$

The tensor p_k is defined as

$$p_k = -2\mu\frac{\partial}{\partial x_k}\frac{1}{(r)} \tag{3-4.25}$$

For most applications involving boundary conditions for spheres and cylinders, it is easier to use direct expansions in harmonic functions than to apply the foregoing integral relationships. In the case of bodies whose dimensions are small with respect to distances from each other and from boundaries, or symmetrical bodies in infinite media, it is possible to represent the disturbance caused by the body in terms of a point force. In such cases one can apply the integral relationships in a simplified form as discussed in the following paragraphs.

Point force applications of Oseen technique

The characteristic feature of the method employed here is that attention is directed to the motion of the fluid produced by the action of exterior forces applied to the individual elements of volume of the fluid. The method was developed extensively by Burgers[7] and is outlined in the following discussion.

Consider first the action of a single force of a given constant magnitude (independent of time) acting at a point P, which may be taken as the origin of a system of x, y, z coordinates. If we wish to use the Oseen equation approximation, we suppose that the fluid originally has a uniform velocity U directed parallel to the x axis and that it assumes this value at a sufficient distance from the point force. The force then produces a disturbance of this uniform motion by superposing upon it a field of motion with the components u, v, w. In the case of a force of sufficiently small intensity these components are given by the following expressions, obtained from Eq. (3-4.21) by omitting the terms relating to the boundary S and designating ρX_j by F_j:

$$u = \frac{1}{8\pi\mu}\left[F_x\nabla^2\psi - F_x\frac{\partial^2\psi}{\partial x^2} - F_y\frac{\partial^2\psi}{\partial x\,\partial y} - F_z\frac{\partial^2\psi}{\partial x\,\partial z}\right] \tag{3–4.26}$$

$$v = \frac{1}{8\pi\mu}\left[F_y\nabla^2\psi - F_x\frac{\partial^2\psi}{\partial x\,\partial y} - F_y\frac{\partial^2\psi}{\partial y^2} - F_z\frac{\partial^2\psi}{\partial y\,\partial z}\right] \tag{3–4.27}$$

$$w = \frac{1}{8\pi\mu}\left[F_z\nabla^2\psi - F_x\frac{\partial^2\psi}{\partial x\,\partial z} - F_y\frac{\partial^2\psi}{\partial y\,\partial z} - F_z\frac{\partial^2\psi}{\partial z^2}\right] \tag{3–4.28}$$

Here F_x, F_y, F_z are the components of the applied force, and ψ is given by Eq. (3–4.24). For small values of $\rho Ur/\mu$ the function ψ can be approximated by

$$\psi = (r - x) - \frac{\rho U}{8\mu}(r^2 - 2rx + x^2) \tag{3–4.29}$$

or, if we wish to consider the solution corresponding to the creeping motion equations, we may take [cf. Eq. (3–4.9)]

$$\psi = r \tag{3–4.30}$$

Since the form given by Eq. (3–4.30) corresponds to the creeping motion equations, the effect of the uniform original motion U and its direction are not taken into account.

Thus, for example, if we take the case of a force acting in the direction of the x axis, we have $F_y = F_z = 0$, and the following expressions are obtained for u, v, w:

$$u = \frac{F_x}{8\pi\mu}\left(\frac{1}{r} + \frac{x^2}{r^3}\right) \tag{3–4.31}$$

$$v = \frac{F_x}{8\pi\mu}\frac{xy}{r^3} \tag{3–4.32}$$

$$w - \frac{F_x}{8\pi\mu}\frac{xz}{r^3} \tag{3–4.33}$$

Lamb[19] gives similar expressions for a point force in the creeping motion approximation.

In the case where forces depend also on time, Oseen has developed appropriate equations for the unsteady state form of the *creeping motion* equations,

$$\mu\nabla^2 u_j - \frac{\partial p}{\partial x_j} - \rho\frac{\partial u_j}{\partial t} + \rho X_j = 0; \qquad \frac{\partial u_k}{\partial x_k} = 0 \qquad (j = 1, 2, 3) \tag{3–4.34}$$

Making the same simplifications as before, Burgers obtained the following approximations for a force applied at the origin of the coordinate system:

$$u = \frac{1}{8\pi\mu}\int_0^t d\tau\left[F_x\nabla^2\psi_1 - F_x\frac{\partial^2\psi_1}{\partial x^2} - F_y\frac{\partial^2\psi_1}{\partial x\,\partial y} - F_z\frac{\partial^2\psi_1}{\partial x\,\partial z}\right] \tag{3–4.35}$$

$$v = \frac{1}{8\pi\mu}\int_0^t d\tau\left[F_y\nabla^2\psi_1 - F_x\frac{\partial^2\psi_1}{\partial x\,\partial y} - F_y\frac{\partial^2\psi_1}{\partial y^2} - F_z\frac{\partial^2\psi_1}{\partial y\,\partial z}\right] \tag{3–4.36}$$

$$w = \frac{1}{8\pi\mu}\int_0^t d\tau \left[F_z \nabla^2 \psi_1 - F_x \frac{\partial^2 \psi_1}{\partial x\, \partial z} - F_y \frac{\partial^2 \psi_1}{\partial y\, \partial z} - F_z \frac{\partial^2 \psi_1}{\partial z^2}\right] \quad (3\text{–}4.37)$$

In these formulas τ is an auxiliary variable, denoting the time between $t = 0$ and the observed time t, whereas the function ψ_1 is defined by the integral:

$$\psi_1 = 2\sqrt{\frac{\mu}{\pi\rho}}\,\frac{1}{r}\int_0^r d\alpha\, \frac{1 - e^{-\rho\alpha^2/4\mu(t-\tau)}}{\sqrt{t-\tau}} \quad (3\text{–}4.38)$$

Here it is assumed that the force is applied at the instant $t = 0$ and that the fluid prior to that instant was everywhere at rest.

If we simplify the situation still further to the case where F_x, F_y, and F_z are kept constant from the instant $t = 0$ onward, then the integration with respect to τ operates only on the function ψ. In that case we may again use Eqs. (3–4.26)–(3–4.28) to describe the motion, provided that we now take

$$\psi = \int_0^t d\tau\, \psi_1 = 2\sqrt{\frac{\mu}{\pi\rho}}\,\frac{1}{r}\int_0^r d\alpha \int_0^t d\tau\, \frac{1 - e^{-\rho\alpha^2/4\mu(t-\tau)}}{\sqrt{t-\tau}} \quad (3\text{–}4.39)$$

Burgers[7] has shown how this integral can be evaluated for sufficiently large values of t by means of a series expansion. He gives the final result for ψ as

$$\psi = r\left(1 - \frac{R}{3\sqrt{\pi}} + \frac{R^3}{240\sqrt{\pi}} - \cdots\right) \quad (3\text{–}4.40)$$

where $R = \sqrt{\rho r^2/\mu t}$. For large values of t (and therefore $R \to 0$), the formula (3–4.30) is obtained, corresponding to the steady state form of the creeping motion equations. Thus, Eq. (3–4.40) may be employed to estimate the error involved in assuming steady state.

Burgers[7] also developed an expression for ψ for the case of very small values of t. In that case

$$\psi \approx \frac{2\mu t}{\rho r} \quad (3\text{–}4.41)$$

Thus, it follows that at small values of t, the velocity components u, v, and w can be written as the derivatives of a potential ϕ,

$$\phi = -\frac{t}{4\pi\rho}\left(F_x \frac{x}{r^3} + F_y \frac{y}{r^3} + F_z \frac{z}{r^3}\right) \quad (3\text{–}4.42)$$

Consequently, under these conditions (t small; r large) we have an irrotational motion with an intensity increasing linearly with time.

As a simple example of the application of these relationships we will show how Stokes' law, Eq. (2–6.3), may be derived, though of course their real value will be in more complicated situations where closed solutions are not possible. Consider a sphere with its center at the origin which is held in a flow with a constant velocity U along the x axis. A force must be exerted in the $-x$ direction to keep it from moving. The resultant of the disturbance produced by holding the sphere stationary will influence the original motion, and Burgers *assumes* that the form of this motion will not differ much from that produced by a point force acting at the origin. Thus, F_x, which is then negative, produces the velocity field given by Eqs. (3–4.31)–(3–4.33). If we wish to consider a sphere of arbitrary radius, a, as producing the force, we may require that the *average* value of the velocity ($U + u$, v, w) must vanish on the surface. Because of symmetry the mean values of v and w will automatically satisfy this condition. As regards u, we write

$$u_{\text{mean}} = \frac{F_x}{8\pi\mu} \frac{\int_0^{\pi}\int_0^{2\pi}\left(\frac{1}{a} + \frac{x^2}{a^3}\right) a^2 \sin\theta \, d\theta \, d\phi}{4\pi a^2} = \frac{F_x}{6\pi\mu a} \tag{3–4.43}$$

The velocity vanishes provided that

$$U + \frac{F_x}{6\pi\mu a} = 0 \tag{3–4.44}$$

or

$$F_x = -6\pi\mu a U \tag{3–4.45}$$

3-5 Generalized Reciprocal Theorem

Useful general relationships regarding the resistance of particles and pressure drops due to fluid moving with respect to particles can be developed from further amplification of the point force approach. These relationships can also be applied in special situations to obtain solutions of problems involving the creeping motion equations.

Many of these developments stem from the work of Lorentz[21] who proved the following relationship for the case of steady, incompressible creeping flows. Let ($\mathbf{v}'$, $\mathbf{\Pi}'$) and ($\mathbf{v}''$, $\mathbf{\Pi}''$) be the velocity and stress fields corresponding to any two motions of the same fluid which conform to Eqs. (2–6.1) and (2–6.2). Then it can be shown that

$$\int_S d\mathbf{S}\cdot\mathbf{\Pi}'\cdot\mathbf{v}'' = \int_S d\mathbf{S}\cdot\mathbf{\Pi}''\cdot\mathbf{v}' \tag{3–5.1}$$

where S is a closed surface bounding any fluid volume V; S may consist of a number of distinct surfaces separated from each other.

Brenner[4] has developed the same theorem in a more general form to include the presence of two different fluids, maintaining only the same geometry of boundaries involved. We present his treatment.

The basic momentum equation is given by Eq. (2–1.5). Thus if external forces, inertial, and unsteady state terms in the preceding are neglected, one obtains $\nabla\cdot\mathbf{\Pi} = \mathbf{0}$; or, expressed in terms of Cartesian tensors,

$$\frac{\partial \Pi_{ij}}{\partial x_j} = 0 \tag{3–5.2}$$

where we have adopted the usual summation convention.

From Eqs. (2–1.8) and (2–1.12), one has for an incompressible fluid

$$\Pi_{ij} = -\delta_{ij}p + 2\mu\Delta_{ij} \tag{3–5.3}$$

where δ_{ij} is the Kronecker delta, and

$$\Delta_{ij} = \frac{1}{2}\left(\frac{\partial v_i}{\partial x_j} + \frac{\partial v_j}{\partial x_i}\right)$$

is the rate of shear tensor for an incompressible fluid. Now consider two possible fluid motions, (v'_i, Π'_{ij}) and (v''_i, Π''_{ij}). Upon writing Eq. (3–5.3) for the primed motion and multiplying the result by Δ''_{ij}, we obtain

$$\Pi'_{ij}\Delta''_{ij} = -p'\,\delta_{ij}\,\Delta''_{ij} + 2\mu'\Delta'_{ij}\Delta''_{ij} \tag{3–5.4}$$

where, by writing μ', we have allowed for the possibility that the primed and double-primed motions occur in two different fluids of unequal viscosities. But, for an incompressible fluid,

$$\delta_{ij}\Delta''_{ij} = \frac{\partial v''_j}{\partial x_j} = \nabla\cdot\mathbf{v}'' = 0 \tag{3–5.5}$$

Thus,

$$\Pi'_{ij}\Delta''_{ij} = 2\mu'\Delta'_{ij}\Delta''_{ij} \tag{3–5.6}$$

Likewise, interchanging the primed and double-primed symbols in the foregoing, one obtains

$$\Pi''_{ij}\Delta'_{ij} = 2\mu''\Delta''_{ij}\Delta'_{ij} \tag{3–5.7}$$

But

$$\Delta'_{ij}\Delta''_{ij} = \Delta''_{ij}\Delta'_{ij} \tag{3–5.8}$$

It follows that

$$\mu''\Pi'_{ij}\Delta''_{ij} = \mu'\Pi''_{ij}\Delta'_{ij} \tag{3–5.9}$$

or, in dyadic notation,

$$\mu''\mathbf{\Pi}' : \mathbf{\Delta}'' = \mu'\mathbf{\Pi}'' : \mathbf{\Delta}' \tag{3–5.10}$$

where the double-dot nomenclature is that of Gibbs[10,12].

Consider

$$\mathbf{\Pi}' : \mathbf{\Delta}'' = \Pi'_{ij}\Delta''_{ij} = \frac{1}{2}\Pi'_{ij}\frac{\partial v''_i}{\partial x_j} + \frac{1}{2}\Pi'_{ij}\frac{\partial v''_j}{\partial x_i} \tag{3–5.11}$$

In consequence of the symmetry of the pressure tensor, $\Pi_{ij} = \Pi_{ji}$, the last term in the preceding can be written as

$$\Pi'_{ij}\frac{\partial v''_j}{\partial x_i} = \Pi'_{ji}\frac{\partial v''_j}{\partial x_i} = \Pi'_{ij}\frac{\partial v''_i}{\partial x_j} \tag{3–5.12}$$

where the dummy summation indices i and j have been interchanged to obtain the last expression. Thus,

$$\mathbf{\Pi}' : \mathbf{\Delta}'' = \Pi'_{ij}\frac{\partial v''_i}{\partial x_j} = \frac{\partial}{\partial x_j}(\Pi'_{ij}v''_i) - v''_i\frac{\partial \Pi'_{ij}}{\partial x_j} \tag{3–5.13}$$

The last term vanishes because of Eq. (3–5.2). Hence, reverting to dyadic notation and multiplying by μ'', one obtains

$$\mu''\mathbf{\Pi}' : \mathbf{\Delta}'' = \mu''\nabla\cdot(\mathbf{\Pi}'\cdot\mathbf{v}'') \tag{3–5.14}$$

An identical expression applies if the primed and double-primed superscripts are interchanged. In accordance with Eq. (3–5.10) the expressions thus obtained are equal. Hence,

$$\mu''\nabla\cdot(\mathbf{\Pi}'\cdot\mathbf{v}'') = \mu'\nabla\cdot(\mathbf{\Pi}''\cdot\mathbf{v}') \tag{3–5.15}$$

If both sides of the foregoing are multiplied by an element of fluid volume dV and integrated over an arbitrary fluid volume, V, the resultant volume integrals may be converted to surface integrals by application of Gauss' divergence theorem. In this way, one obtains

$$\mu''\int_S d\mathbf{S}\cdot\mathbf{\Pi}'\cdot\mathbf{v}'' = \mu'\int_S d\mathbf{S}\cdot\mathbf{\Pi}''\cdot\mathbf{v}' \tag{3–5.16}$$

in which S is any closed surface drawn in the fluid. If $\mu' = \mu''$, Eq. (3–5.1) results.

A mirror image technique

We discuss here problems involving objects in the vicinity of a plane wall. The problem of a sphere approaching a plane wall can be solved exactly by use of bipolar coordinates[5]. A somewhat simpler first approximation to problems involving a single plane wall was developed by Lorentz[21] as one of the applications of the reciprocal theorem.

Assume that we have a solution $(\mathbf{v}', p')$ of the creeping motion and continuity equations. A new solution is then given by the following formulas, as is easily verified:

$$u'' = u' - 2x\frac{\partial u'}{\partial x} + \frac{x^2}{\mu}\frac{\partial p'}{\partial x} \tag{3–5.17}$$

$$v'' = -v' - 2x\frac{\partial u'}{\partial y} + \frac{x^2}{\mu}\frac{\partial p'}{\partial y} \tag{3–5.18}$$

$$w'' = -w' - 2x\frac{\partial u'}{\partial z} + \frac{x^2}{\mu}\frac{\partial p'}{\partial z} \tag{3–5.19}$$

$$p'' = p' + 2x\frac{\partial p'}{\partial x} - 4\mu\frac{\partial u'}{\partial x} \tag{3–5.20}$$

The solution given by Eqs. (3–5.17)–(3–5.20) satisfies the following boundary conditions on the yz plane (that is, at $x = 0$):

$$u'' = u', \quad v'' = -v', \quad w'' = -w' \tag{3–5.21}$$

The motion described by $(\mathbf{v}'', p'')$ corresponds exactly to a reflection from a solid wall of the motion u'. Thus at any point on the positive side of the wall the motion $\mathbf{v}''$ is the mirror image of the motion $\mathbf{v}'$ which would exist at a point an equal distance on the negative side of the wall if no wall were present. The field obtained by adding $\mathbf{v}'$ and $\mathbf{v}''$ corresponds therefore to a motion with no slip at the plane $x = 0$.

For example, we can take a sphere of radius a moving in the direction of the x axis with a velocity $-U$, whose midpoint is located at the coordinates $(-l, 0, 0)$. For a point force, the field produced by this falling sphere moving perpendicularly toward the wall $x = 0$ is given by applying Eqs. (3–4.31)–(3–4.33), noting that $r^2 = (x + l)^2 + y^2 + z^2$. Hence,

$$u' = -\frac{3}{4}aU\left\{\frac{(x + l)^2}{r^3} + \frac{1}{r}\right\} \tag{3–5.22}$$

$$v' = -\frac{3}{4}aU\frac{(x + l)y}{r^3} \tag{3–5.23}$$

$$w' = -\frac{3}{4}aU\frac{(x + l)z}{r^3} \tag{3–5.24}$$

$$p' = -\frac{3}{2}\mu a U\frac{(x + l)}{r^3} + \text{constant} \tag{3–5.25}$$

Substitution of these values into Eqs. (3–5.17)–(3–5.20), gives, for the motion reflected from the wall,

$$u'' = -\frac{3}{4}aU\left\{\frac{1}{r} + \frac{x^2 + l^2}{r^3} + \frac{6lx(x + l)^2}{r^5}\right\} \tag{3–5.26}$$

$$v'' = -\frac{3}{4}aU\left\{\frac{(x - l)y}{r^3} + \frac{6lxy(x + l)}{r^5}\right\} \tag{3–5.27}$$

$$w'' = -\frac{3}{4}aU\left\{\frac{(x - l)z}{r^3} + \frac{6lxz(x + l)}{r^5}\right\} \tag{3–5.28}$$

Without investigating further the reflection of this motion from the sphere, one can obtain the resistance that this body experiences for small values of a/l. According to the preceding formulas, the component velocities at the location of the center of the sphere are

$$u'' = -\frac{9}{8}\frac{a}{l}U, \qquad v'' = w'' = 0 \tag{3–5.29}$$

Hence, the sphere has a velocity relative to the surrounding fluid equal to $[1 + (9/8)\, a/l]\, U$. Thus, the drag on the sphere is

$$F = 6\pi\mu aU\left(1 + \frac{9}{8}\frac{a}{l}\right) \tag{3–5.30}$$

It can be shown in similar fashion that the resistance is increased for a sphere moving *parallel* to the wall by a factor of $1 + (9/16)\, a/l$. As long as the creeping motion equations are applicable there is no force tending to move the sphere toward or away from the wall.

3-6 Energy Dissipation

Knowledge of the rate at which energy is dissipated in a moving fluid often furnishes useful results without requiring detailed solutions of the equations of motion[6]. As a simple example we shall show how to obtain estimates of the pressure drop experienced in creeping flow through systems containing suspended particles.

Consider a given unperturbed motion ($\mathbf{v}^{(0)}$, $p^{(0)}$) which satisfies the creeping motion equations. The rate of mechanical energy dissipation per unit time in a volume Q_o of fluid is, from Eq. (2–2.4),

$$E^{(0)} = \int_{Q_o} \Phi^{(0)}\, dQ \tag{3–6.1}$$

$\Phi^{(0)}$ is the local rate of energy dissipation per unit time per unit volume. For the creeping motion equations, with no body forces involved, we have from Eq. (3–5.2) that $\nabla \cdot \mathbf{\Pi} = \mathbf{0}$. Furthermore, for an incompressible fluid, $\nabla \cdot \mathbf{v} = 0$. Thus, from the mechanical energy balance, Eq. (2–2.2), we obtain

$$\Phi^{(0)} = \nabla \cdot (\mathbf{v}^{(0)} \cdot \mathbf{\Pi}^{(0)})$$

Substituting this result into Eq. (3–6.1) and applying Gauss' divergence theorem, we obtain

$$E^{(0)} = \int_{S_o} \mathbf{v}^{(0)} \cdot \mathbf{\Pi}^{(0)} \cdot d\mathbf{S} \tag{3–6.2}$$

where S_o is a closed surface bounding the volume Q_o. This states that the rate at which energy is being dissipated is equal to the rate at which the stresses acting over the surface are doing work.

Now, suppose that a number of rigid particles of any shape are immersed in the flowing fluid, and let ($\mathbf{v}, p$) denote the new state of motion. The surface of the ith particle is denoted by S_i and its volume by Q_i. The totality of all surfaces bounding the fluid is now $S_o + \sum_i S_i$ while the total volume occupied by fluid is $Q_o - \sum_i Q_i$. It is assumed that the new, disturbed motion satisfies exactly the same boundary conditions on S_o as does the original, undisturbed motion, that is, $\mathbf{v} = \mathbf{v}^{(0)}$ on S_o, and that ($\mathbf{v}, p$) satisfies the creeping motion equations. By application of the reciprocal theorem (3–5.1) to the surface $S_o + \sum_i S_i$ one can show that[6]

$$E^* = \int_{S_o} \mathbf{v} \cdot (\mathbf{\Pi} - \mathbf{\Pi}^{(0)}) \cdot d\mathbf{S} + \sum_i \int_{S_i} \mathbf{v} \cdot (\mathbf{\Pi} - \mathbf{\Pi}^{(0)}) \cdot d\mathbf{S} \tag{3–6.3}$$

where E^* is the *additional* rate at which dissipation is occurring as a result of the presence of suspended particles. That is,

$$E = E^* + E^{(0)} \tag{3–6.4}$$

where E is the energy dissipation arising from the disturbed motion ($\mathbf{v}, p$).

As an example of the application of Eq. (3–6.3) consider an infinitely long cylindrical tube whose surface S_o consists of the vertical tube walls and, at either end of the cylinder, the horizontal planes $z = \pm\infty$ on which perturbing effects due to the presence of the particles in the cylinder are no longer felt[6]. The particles are confined to a finite region within the duct.

The cylinder need not be circular in cross section. Thus, the boundary conditions for the perturbed and unperturbed flows are

(a) $\mathbf{v} = \mathbf{v}^{(0)} = \mathbf{0}$ on the container walls

(b) $\mathbf{v} = \mathbf{v}^{(0)}$ at $z = \pm\infty$

(c) $\mathbf{v} = \mathbf{0}$ on each S_i

where, for simplicity of illustration, we have assumed that the particles are restrained from moving, as evidenced by the last boundary condition.

For the first term in Eq. (3–6.3) we obtain from these boundary conditions[6]

$$\int_{S_0} \mathbf{v}\cdot(\mathbf{\Pi} - \mathbf{\Pi}^{(0)})\cdot d\mathbf{S} = \Delta p^* A V_m$$

where A is the superficial cross-sectional area of the duct, V_m is the superficial velocity of flow through the duct, and

$$\Delta p^* = \Delta p - \Delta p^{(0)} \tag{3–6.5}$$

is the *additional* pressure drop caused by the presence of the particles. Here, Δp and $\Delta p^{(0)}$ refer to the actual pressure drops with and without the particles present, with fluid flowing at the same superficial velocity.

For the second class of integrals in Eq. (3–6.3) the boundary conditions yield

$$\int_{S_i} \mathbf{v}\cdot(\mathbf{\Pi} - \mathbf{\Pi}^{(0)})\cdot d\mathbf{S} = 0$$

since $\mathbf{v} = \mathbf{0}$ on S_i. Hence we have that

$$E^* = \Delta p^* A V_m \tag{3–6.6}$$

But it can also be shown[6] for these circumstances that

$$E^* = \sum_i \mathbf{v}_{[i]}^{(0)} \cdot \mathbf{F}_i \tag{3–6.7}$$

where $\mathbf{v}_{[i]}^{(0)}$ is the unperturbed velocity field evaluated at the center of the ith particle, and $\mathbf{F}_i$ is the hydrodynamic force on the ith particle. Equating the latter two expressions yields

$$\Delta p^* = \frac{\sum_i \mathbf{v}_{[i]}^{(0)} \cdot \mathbf{F}_i}{V_m A} \tag{3–6.8}$$

This result is valid when the particle dimensions are small with respect to that of the apparatus.

Suppose, for illustrative purposes, that only one stationary spherical particle is present in a circular cylindrical tube of radius R_0 through which viscous fluid flows. If the center of the sphere is located at a distance b from the cylinder axis, the unperturbed parabolic field evaluated at the center of the particle is

$$\mathbf{v}_{[i]}^{(0)} = \mathbf{k}2V_m\left(1 - \frac{b^2}{R_o^2}\right) \tag{3–6.9}$$

where $\mathbf{k}$ is a unit vector in the flow direction. Also, from Stokes' law

$$\mathbf{F}_i = 6\pi\mu a\mathbf{v}_{[i]}^{(0)} \tag{3–6.10}$$

Thus,

$$\mathbf{v}_{[i]}^{(0)} \cdot \mathbf{F}_i = 24\pi\mu a V_m^2\left(1 - \frac{b^2}{R_o^2}\right)^2 \tag{3–6.11}$$

These combine to give

$$\Delta p^* = \frac{24\mu a V_m}{R_o^2}\left(1 - \frac{b^2}{R_o^2}\right)^2 \tag{3–6.12}$$

for the additional pressure drop caused by the presence of the sphere. This result, which is valid only for small a/R_o, was obtained previously by a considerably more complicated analysis which required the detailed calculation of the field $\mathbf{v}$ with the cylinder[3]. As shown in the original derivation[6], Eq. (3–6.8) may be applied to other situations involving *moving* particles. The same general treatment involving complete energy balances, allowing for rotational and dilational effects produced by small particles in a shearing field, can also be applied to the study of viscosity of dilute suspensions in creeping motion. These applications are discussed in Sections 9–1 and 9–2.

It is also possible to use energy relationships directly, without solving the particular boundary value problem, by exploiting variational techniques. A brief account of this approach is given here, though it has not yet generally proved of great value in resolving significant problems encountered in flow through particulate systems.

Variational principles for creeping motion

A well-known theorem, due to Helmholtz[15,19] may be stated as follows: The dissipation of energy in creeping flow is less than in any other incompressible, continuous motion consistent with the same boundary conditions. That is, if we consider the class of continuous vector fields $\mathbf{v}$ satisfying prescribed velocity boundary conditions on some closed surface S, and satisfying the auxiliary condition $\nabla \cdot \mathbf{v} = 0$, then the particular member of this class which minimizes the functional $E(\mathbf{v})$ defined by

$$E = \int_V \Phi \, dV \tag{3–6.13}$$

(in which $\Phi = 2\mu\mathbf{\Delta}:\mathbf{\Delta}$, where $\mathbf{\Delta} = \frac{1}{2}[\nabla\mathbf{v} + (\nabla\mathbf{v})^{+}]$) will be such that $\mathbf{v}$ satisfies, at each point of the fluid, the differential equation

$$\nabla^2\mathbf{v} = \frac{1}{\mu}\nabla q \tag{3–6.14}$$

where q is some scalar function. Here V is the volume contained within S. This variational principle may be employed to determine the fluid motion in a given domain when the velocity is prescribed at all points of the bounding surface, including those portions of the surface through which fluid is streaming in and out. In applying this principle, one would normally select a trial velocity distribution such that the equation of continuity and boundary conditions is satisfied. Then, of all the possible fluid motions satisfying these conditions, that motion which minimizes Eq. (3–6.13) will be the true steady state creeping motion. Bird[1] has given a similar variational principle for the steady laminar motion of incompressible non-newtonian fluids for the case where inertial terms may be neglected in the equations of motion. He has also called attention to other similar treatments[2]. Another generalization which refers to the slow movement of a viscous incompressible fluid at nonuniform temperature has been presented by Glansdorff, Prigogine, and Hays[13].

The variational approach using the creeping motion equations has been applied to the theory of lubrication, where it is assumed that a bearing may be idealized to the two-dimensional problem of two nearly parallel surfaces sliding over each other with a film of lubricant between them. The non-homogeneous second-order partial differential equation basic to these treatments was originally derived by Reynolds[25] and solved approximately by him, as mentioned earlier. Several authors have reported numerical and analog solutions of the two-dimensional Reynolds equations and Hays[14] has presented a general method using the variational approach.

Christopherson and Dowson[8] developed an approximate theory to describe the behavior of a heavy ball passing slowly down a vertical tube having a diameter only slightly in excess of that of the ball, and filled with a viscous fluid. It is shown that, according to this theory, the equations of motion can be satisfied when the ball takes any degree of eccentricity in the tube and furthermore that any given eccentricity requires a particular velocity of rotation about a horizontal axis. It is found that the eccentricity ratio corresponding to the minimum dissipation of energy for a given velocity of descent (that is, to the maximum rate of fall for a given weight of ball) is about 0.98, and that this velocity is somewhat more than twice the velocity corresponding to zero eccentricity. Experiments were made in an attempt to establish whether the falling ball would in fact assume the position corresponding to minimum dissipation of energy. But exact verification was not possible. The principle one would like to prove is that if a region is selected in which one or more solid objects are moving and are free to take up a variety of positions and velocities (that is, velocities and boundaries not exactly specified as in the Helmholtz theorem), the position and velocity actually taken will be that corresponding to minimum dissipation.

Based on an analogy between the field equations for an elastic solid in equilibrium and a viscous newtonian fluid in steady creeping flow, Hill and Power[16] derived a pair of extremum principles. Stewart[28] discussed these complementary variational principles and applied them to the problem of laminar flow in uniform ducts. These theorems bracket the energy dissipation in a given boundary value problem between upper and lower bounds corresponding to arbitrarily chosen admissible functions. The one function which affords an upper limit is given by Helmholtz's theorem. For the lower bound, stresses must be assumed which will result in a finite force and/or couple on the body in question. A number of applications are cited[16], including the case of the translation of a sphere in an unbounded fluid, and it is shown by way of illustration that

$$\frac{40}{7} \leq \frac{F}{\pi\mu a U} \leq \frac{56}{9} \tag{3–6.15}$$

for the force on a spherical particle of radius a. Thus $F/\pi\mu a U \approx 5.97$ with a maximum error of 0.25 as compared with the actual value of 6 given by Stokes' law. Use of these principles, however, leads the authors to an apparently incorrect conclusion, namely that the drag on a body S tends to be increased by the presence of other bodies $S_1, \ldots, S_n$, which are either in fixed position or free to move without restraint. Smoluchowski[27] showed, in fact, if the bodies move in an unbounded medium, that the reverse is true.

In general it must be concluded that variational techniques, although interesting, have not yet reached the stage of development where they may be applied to significant three-dimensional problems, such as are involved in particulate systems.

Appendix to Chapter 3

In order to prove Eq. (3–4.11) we proceed as follows: According to Eqs. (3–4.1), (3–4.2), (3–4.4), and (3–4.7), we have that

$$\mu\nabla^2 u_j - \frac{\partial p}{\partial x_j} = 0 \tag{3–A.1}$$

$$\frac{\partial u_j}{\partial x_j} = 0 \tag{3–A.2}$$

and

$$\mu\nabla^2 t_{jk} - \frac{\partial p_k}{\partial x_j} = 0 \tag{3–A.3}$$

$$\frac{\partial t_{jk}}{\partial x_j} = 0 \tag{3–A.4}$$

By Green's second identity we have, for any functions f and g,

$$\int_V (f\nabla^2 g - g\nabla^2 f)\, dV = \int_S \left(f\frac{dg}{dn} - g\,\frac{df}{dn}\right) dS \tag{3–A.5}$$

where S is a closed surface bounding the volume V. Now choose

$$f = u_j \quad \text{and} \quad g = t_{jk}$$

Substituting into the foregoing and utilizing Eqs. (3–A.1) and (3–A.3), we obtain

$$\frac{1}{\mu}\int_V\left(u_j\frac{\partial p_k}{\partial x_j}-t_{jk}\frac{\partial p}{\partial x_j}\right)dV=\int_S\left(u_j\frac{dt_{jk}}{dn}-t_{jk}\frac{du_j}{dn}\right)dS \tag{3–A.6}$$

But
$$u_j\frac{\partial p_k}{\partial x_j}=\frac{\partial}{\partial x_j}(u_jp_k)-p_k\frac{\partial u_j}{\partial x_j}=\frac{\partial}{\partial x_j}(u_jp_k)$$

and
$$t_{jk}\frac{\partial p}{\partial x_j}=\frac{\partial}{\partial x_j}(t_{jk}p)-p\frac{\partial t_{jk}}{\partial x_j}=\frac{\partial}{\partial x_j}(t_{jk}p)$$

By Gauss' divergence theorem,

$$\int_V\frac{\partial}{\partial x_j}(u_jp_k)\,dV=\int_S dS_ju_jp_k$$

and
$$\int_V\frac{\partial}{\partial x_j}(t_{jk}p)\,dV=\int_S dS_jt_{jk}p$$

Since $dS_j = n_j\,dS$, we find upon substituting these into Eq. (3–A.6) that Eq. (3–4.11) is indeed valid.

BIBLIOGRAPHY

1. Bird, R. B., Phys. Fluids **3** (1960), 539.
2. ———, Phys. Fluids **5** (1962), 502.
3. Brenner, H., and J. Happel, J. Fluid Mech. **4** (1958), 195.
4. Brenner, H., Chem. Eng. Sci. **18** (1963), 1.
5. ———, Chem. Eng. Sci. **16** (1961), 242.
6. ———, Phys. Fluids **1** (1958), 338.
7. Burgers, J. M., "Second Report on Viscosity and Plasticity," *Kon. Ned. Akad. Wet., Verhand.* (Eerste Sectie), Dl. XVI, No. 4 (1938) pp. 1–287. Amsterdam: N. V. Noord-Hollandsche, Uitgeversmaatschappi, 1938.
8. Christopherson, D. G., and D. Dowson, Proc. Roy. Soc. (London) **A251** (1959), 550.
9. Dean, W. R., Proc. Camb. Phil. Soc. **40** (1944), 19.
10. Drew, T. B., *Handbook of Vector and Polyadic Analysis.* New York: Reinhold, 1961.
11. Faxen, H., Arkiv. Mat. Astron. och Fys. **17,** No. 1 (1922); **20,** No. 8 (1927). See also his dissertation, Uppsala Univ., Uppsala, Sweden (1921).
12. Gibbs, J. W., and E. B. Wilson, *Vector Analysis.* New York: Dover, 1960.
13. Glansdorff, P., I. Prigogine, and D. F. Hays, Phys. Fluids **5** (1962), 144.
14. Hays, D. F., J. Basic Eng. **81** (1959), 13.

15. Helmholtz, H. L. F., Verhandl. Naturhist. med., ser. 5, October 30, 1868; or Wiss. Abhandl. **1** (1882), 223.

16. Hill, R., and G. Power, Quart. J. Mech. Appl. Math. **9** (1956), 313.

17. Keller, J. B., J. Fluid Mech. **18** (1964), 94.

18. Ladenburg, R., Ann. der Physik **23** (1907), 447; see also his dissertation, Leipzig (1906).

19. Lamb, H., *Hydrodynamics*, 6th ed. London: Cambridge Univ. Press, 1932; New York: Dover, 1945.

20. Lichtenstein, L., Math. Z. **28** (1928), 387.

21. Lorentz, H. A., Abhandl. theoret. Phys. **1** (1906), 23.

22. Morse, P. M., and H. Feshbach, *Methods of Theoretical Physics*. New York: McGraw-Hill, 1953.

23. Odquist, F. K. G., Math. Z. **32** (1930), 329; also Arkiv. Mat. Astron. och Fys. **22A**, No. 28 (1931).

24. Oseen, C. W., *Neuere Methoden und Ergebnisse in der Hydrodynamik*. Leipzig: Akademische Verlagsgesellschaft, 1927.

25. Reynolds, O., Phil. Trans. Roy. Soc. (London) **177** (1887), 157.

26. Simha, R., Kolloid-Z., **76** (1936), 16.

27. Smoluchowski, M., Proc. 5th Intern. Cong. Math., Cambridge 1912, Vol. II, p. 192.

28. Stewart, W. E., A. I. Ch. E. Jour. **8** (1962), 425.

29. Taylor, G. I., Phil. Trans. Roy. Soc. (London) **A223** (1923), 289.

Axisymmetrical Flow 4

4-1 Introduction

The introduction of a stream function serves to unify the method of attack on all two-dimensional incompressible fluid motions. For those situations the solution of the equations of motion is reduced to the search for a *single scalar* function. Unfortunately, in the general case of three-dimensional motions this unified method of approach is denied us. Specific solutions of the equations of motion must be developed for each different boundary geometry. There exist, however, a number of classes of three-dimensional flows which can still be uniquely characterized by means of a single scalar function. Each of these involves a certain mode of symmetry in the flow pattern.

The most important from the standpoint of flow in particulate systems is typified by streaming flow past a body of revolution, parallel to its symmetry axis. Such motions are termed *axisymmetric* (or, sometimes, axially symmetric). They are characterized by the existence of a stream function. In this chapter we shall develop a number of exact solutions which are obtainable for this type of motion.

4-2 Stream Function

Let ϕ denote the azimuthal angle of a point lying in a plane perpendicular to that of the axis of revolution of the body. An axisymmetrical fluid motion is then one for which: (a) the velocity is independent of this angle, that is,

$$\frac{\partial \mathbf{v}}{\partial \phi} = \mathbf{0} \tag{4–2.1}$$

and (b) the azimuthal component of velocity is everywhere zero,

$$\mathbf{i}_\phi \cdot \mathbf{v} = 0 \tag{4–2.2}$$

The fluid motion is therefore the same in every meridian plane, $\phi =$ constant. On the basis of Eq. (4–2.2), it follows that the streamlines lie in meridian planes. In conjunction with Eq. (4–2.1), this further implies that the stream surfaces are coaxial surfaces of revolution.

Here and in the sequel, we shall consistently let the z axis coincide with the axis of revolution of the body past or through which the fluid streams. Imagine a curve, a, lying in a meridian plane, which connects some point, $\mathbf{R}(\varpi, z)$, in that plane with any point $\mathbf{R}_o(0, z_o)$ lying along the axis of symmetry, as in Fig. 4–2.1(a). Upon rotation of this meridian curve through an angle 2π, the diaphragm-like surface, A, shown in Fig. 4–2.1(b), is generated.

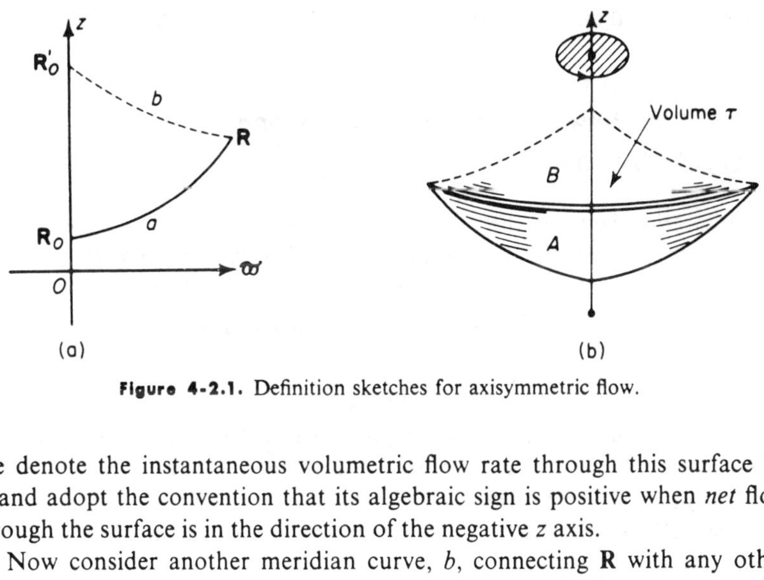

Figure 4-2.1. Definition sketches for axisymmetric flow.

We denote the instantaneous volumetric flow rate through this surface by Q and adopt the convention that its algebraic sign is positive when *net* flow through the surface is in the direction of the negative z axis.

Now consider another meridian curve, b, connecting $\mathbf{R}$ with any other point, $\mathbf{R}_o'$, lying on the axis of revolution. Rotation of this curve generates the surface B. For an *incompressible* fluid the equation of continuity is

$$\nabla \cdot \mathbf{v} = 0 \tag{4–2.3}$$

It is apparent that the instantaneous volumetric flow rate must be the same through both A and B, as there can be no accumulation or depletion of fluid within the volume, τ, bounded externally by these surfaces. The value of Q, at any instant, is therefore uniquely determined by the position vector, $\mathbf{R}$, alone. In particular, it is independent of the choice of a meridian curve joining **R** to the axis.

It is convenient to define a quantity called the *stream function*, at the point **R**, by the relation

$$\psi = \psi(\mathbf{R}\,;t) = \frac{Q}{2\pi} \tag{4–2.4}$$

The foregoing considerations insure that the instantaneous value of ψ is a unique scalar point function. According to its definition, $\psi = 0$ along the axis of revolution.

This quantity was first introduced into hydrodynamics by Stokes and is often referred to as Stokes' stream function or Stokes' current function. It is to be distinguished from the two-dimensional Lagrangian stream function mentioned at the beginning of Chapter 3. Extended discussions of Stokes' stream function may be found in the works of Sampson[32], Kneale[18], and Yih[39]. The existence of a stream function is predicated solely on certain hypotheses of symmetry and incompressibility. It is not surprising, therefore, that the stream function plays an important role in branches of hydrodynamics other than those considered here.

4-3 Relation Between Stream Function and Local Velocity

The connection between the stream function at a point and the local fluid velocity can be established by application of the various definitions. Consider Fig. 4–3.1. The instantaneous volumetric flow rate through the surface of revolution S, generated by rotation of the meridian curve $\widehat{\mathbf{R}_o\mathbf{R}}$, is

$$Q = \int_S \mathbf{v}\cdot\mathbf{n}\, dS = \int_{\mathbf{R}_o}^{\mathbf{R}} \int_{\phi=0}^{2\pi} \mathbf{v}\cdot\mathbf{n}\varpi\, d\phi\,|d\mathbf{R}| \tag{4–3.1}$$

where $|d\mathbf{R}|$ is an element of arc length in a meridian plane and $\mathbf{n}$ is a unit outer normal to the surface. Because the surface is one of revolution, $\mathbf{n}$ necessarily lies in a meridian plane. Integration with respect to ϕ, followed by subsequent application of the definition of the stream function, Eq. (4–2.4), gives

$$\psi = \int_{\mathbf{R}_o}^{\mathbf{R}} \mathbf{v}\cdot\mathbf{n}\varpi\,|d\mathbf{R}| \tag{4–3.2}$$

Since the stream function vanishes along the symmetry axis we have, successively,

$$\psi = \int_{\mathbf{R}_o}^{\mathbf{R}} d\psi = \int_{\mathbf{R}_o}^{\mathbf{R}} d\mathbf{R}\cdot\nabla\psi = \int_{\mathbf{R}_o}^{\mathbf{R}} \mathbf{t}\cdot\nabla\psi\,|d\mathbf{R}| \tag{4–3.3}$$

where $\mathbf{t}$ is a unit tangent vector to the meridian curve, in the sense indicated in the sketch. Upon equating these alternative expressions for ψ, we obtain

$$\int_{\mathbf{R}_0}^{\mathbf{R}} (\mathbf{v}\cdot\mathbf{n}\varpi - \mathbf{t}\cdot\nabla\psi)\,|d\mathbf{R}| = 0 \qquad (4\text{–}3.4)$$

As $\mathbf{R}$ is arbitrary, this condition requires that the integrand vanish at every point; that is,

$$\mathbf{n}\cdot\mathbf{v}\varpi - \mathbf{t}\cdot\nabla\psi = 0 \qquad (4\text{–}3.5)$$

Now, the system of unit vectors ($\mathbf{t}$, $\mathbf{n}$, $\mathbf{i}_\phi$), in that order, forms a right-handed system; hence,

$$\mathbf{t} = \mathbf{n} \times \mathbf{i}_\phi \qquad (4\text{–}3.6)$$

This leads to the relation

$$\mathbf{n}\cdot(\mathbf{v}\varpi - \mathbf{i}_\phi \times \nabla\psi) = 0 \qquad (4\text{–}3.7)$$

But, at a given point of the fluid, the direction of $\mathbf{n}$ is arbitrary. Consequently, the expression in parentheses must vanish, and thus

$$\mathbf{v} = \frac{1}{\varpi}\mathbf{i}_\phi \times \nabla\psi = \nabla\phi \times \nabla\psi \qquad (4\text{–}3.8)$$

or, alternatively,

$$\mathbf{v} = -\nabla \times \left(\mathbf{i}_\phi \frac{\psi}{\varpi}\right) = -\nabla \times (\psi\nabla\phi) \qquad (4\text{–}3.9)$$

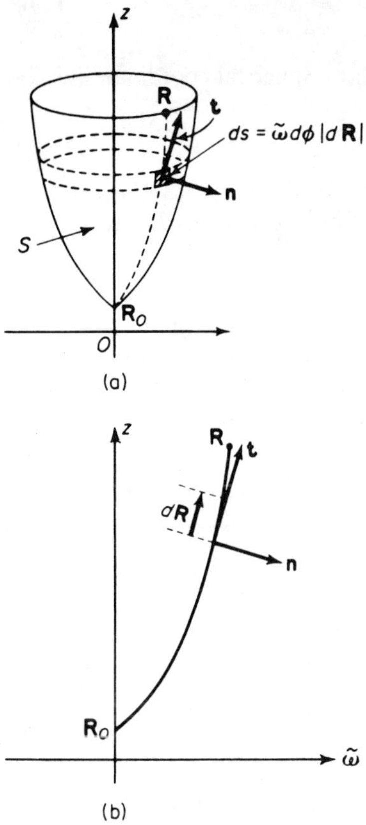

Figure 4-3.1. Unit vectors for axisymmetric flows.

The latter form shows clearly that the condition of incompressibility, $\nabla\cdot\mathbf{v} = 0$, is automatically fulfilled, regardless of the choice of ψ; for the divergence operator annihilates the curl of any vector function. It is equally obvious that the symmetry conditions Eqs. (4–2.1) and (4–2.2) are satisfied, as the stream function is independent of the azimuthal angle.

4-4 Stream Function in Various Coordinate Systems

The components of velocity in any system of coordinates can be obtained, most readily from Eq. (4–3.8) by expressing the ∇–operator in that system of coordinates and performing the indicated operations. As particular examples we have:

(i) Cylindrical coordinates, $\psi = \psi(\varpi, z; t)$

$$v_\varpi = \frac{1}{\varpi}\frac{\partial\psi}{\partial z}; \qquad v_z = -\frac{1}{\varpi}\frac{\partial\psi}{\partial\varpi} \tag{4–4.1}$$

(ii) Spherical coordinates, $\psi = \psi(r, \theta; t)$

$$v_r = -\frac{1}{r^2 \sin\theta}\frac{\partial\psi}{\partial\theta}; \qquad v_\theta = \frac{1}{r\sin\theta}\frac{\partial\psi}{\partial r} \tag{4–4.2}$$

More generally, if (q_1, q_2, ϕ), in that order, constitute a right-handed system of orthogonal curvilinear coordinates of revolution (see Section A–14),

$$v_1 = -\frac{h_2}{\varpi}\frac{\partial\psi}{\partial q_2}; \qquad v_2 = \frac{h_1}{\varpi}\frac{\partial\psi}{\partial q_1} \tag{4–4.3}$$

with $\varpi = \varpi(q_1, q_2)$.

The stream function enables us to transform velocity components rapidly from one system of coordinates to another. By the chain rule for partial differentiation

$$\frac{\partial\psi}{\partial q_i} = \frac{\partial\psi}{\partial z}\frac{\partial z}{\partial q_i} + \frac{\partial\psi}{\partial\varpi}\frac{\partial\varpi}{\partial q_i} \qquad (i = 1, 2)$$

so that, for example, from Eqs. (4–4.1) and (4–4.3)

$$v_1 = h_2\left(v_z\frac{\partial\varpi}{\partial q_2} - v_\varpi\frac{\partial z}{\partial q_2}\right); \qquad v_2 = h_1\left(v_\varpi\frac{\partial z}{\partial q_1} - v_z\frac{\partial\varpi}{\partial q_1}\right) \tag{4–4.4}$$

Conversely,

$$\frac{\partial\psi}{\partial z} = \frac{\partial\psi}{\partial q_1}\frac{\partial q_1}{\partial z} + \frac{\partial\psi}{\partial q_2}\frac{\partial q_2}{\partial z}$$

whence

$$v_\varpi = \frac{v_2}{h_1}\frac{\partial q_1}{\partial z} - \frac{v_1}{h_2}\frac{\partial q_2}{\partial z} \tag{4–4.5}$$

the analogous expression for v_z following in similar fashion.

4-5 Intrinsic Coordinates

It is sometimes useful to describe the velocities at the surface of a body of revolution in terms of a system of coordinates which are intrinsic to the surface itself. Specifically, denote by **n** the unit *outer* normal at a point on the surface of the body, and by **s** a unit tangent vector to a meridian curve on the surface, as in Fig. 4–5.1. On account of symmetry, these vectors necessarily lie in a meridian plane. The sense of **s** is to be such that the system of unit vectors ($\mathbf{n}$, $\mathbf{s}$, $\mathbf{i}_\phi$) are right-handed in that order. By the symbols δn and δs we shall mean elements of arc length in the directions in which n and s, respectively, are pointed. (The symbols δl_n and δl_s for these quantities would be more consistent with the nomenclature outlined in Appendix A, but they are somewhat unwieldy.)

We may regard the foregoing as a system of local orthogonal curvilinear coordinates by assigning them the following values:

$$\delta q_1 = \delta n, \quad \delta q_2 = \delta s, \quad \delta q_3 = \delta\phi$$

$$\mathbf{i}_1 = \mathbf{n}, \quad \mathbf{i}_2 = \mathbf{s}, \quad \mathbf{i}_3 = \mathbf{i}_\phi \qquad (4\text{–}5.1)$$

$$h_1 = 1, \quad h_2 = 1, \quad h_3 = \frac{1}{\varpi}$$

In particular, the nabla operator is of the form

$$\nabla = \mathbf{n}\frac{\partial}{\partial n} + \mathbf{s}\frac{\partial}{\partial s} + \mathbf{i}_\phi \frac{1}{\varpi}\frac{\partial}{\partial \phi} \qquad (4\text{–}5.2)$$

so that

$$v_n = -\frac{1}{\varpi}\frac{\partial \psi}{\partial s}; \quad v_s = \frac{1}{\varpi}\frac{\partial \psi}{\partial n} \qquad (4\text{–}5.3)$$

These give the normal and tangential velocities, respectively, at a point on the surface.

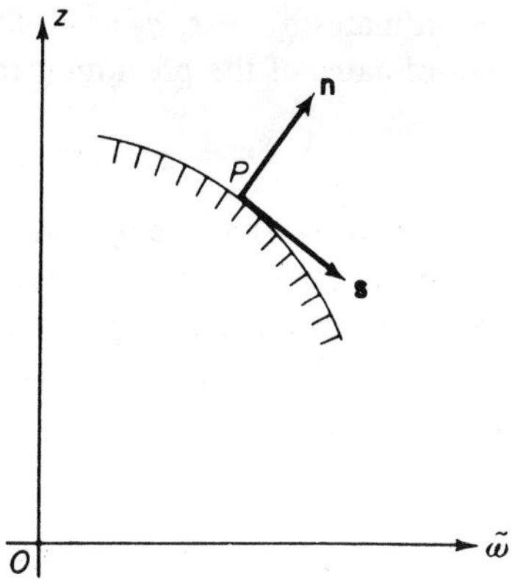

Figure 4-5.1. Normal and tangential unit vectors.

No significance attaches to these coordinates at points removed from the surface of the body. We can, however, extend their meaning by applying them to *any* surface of revolution, coaxial with the z axis, bearing in mind their altered significance. In particular, if the surface is one of the instantaneous stream surfaces of the flow, it follows at once that $v_n = 0$, whereas the absolute value of v_s gives the speed along the instantaneous streamline passing through the point in question.

For subsequent reference we tabulate here the following useful formulas relating to these coordinates:

$$\mathbf{s}\cdot\mathbf{i}_\varpi = \frac{\partial \varpi}{\partial s} = \frac{\partial z}{\partial n} = \mathbf{n}\cdot\mathbf{i}_z$$

and (4–5.4)

$$\mathbf{n}\cdot\mathbf{i}_\varpi = \frac{\partial \varpi}{\partial n} = -\frac{\partial z}{\partial s} = -\mathbf{s}\cdot\mathbf{i}_z$$

To prove the first set, we have that

$$\begin{aligned}\frac{\partial \varpi}{\partial s} &= \mathbf{s}\cdot\nabla\varpi = \mathbf{s}\cdot\mathbf{i}_\varpi = \mathbf{s}\cdot\mathbf{i}_\phi \times \mathbf{i}_z \\ &= \mathbf{s}\times\mathbf{i}_\phi\cdot\mathbf{i}_z = \mathbf{n}\cdot\mathbf{i}_z = \mathbf{n}\cdot\nabla z = \frac{\partial z}{\partial n}\end{aligned} \qquad (4\text{–}5.5)$$

A proof of the second set follows along similar lines.

In the solution of problems relating to a boundary of particular shape, it is often possible to choose an appropriate system of curvilinear coordinates of revolution such that the surface of the body is a member of the family of coordinate surfaces q_N = constant. Orthogonal to these, in a meridian plane, are the curvilinear coordinate curves q_S = constant. For example, in problems dealing with spherical boundaries it is natural to choose spherical coordinates $q_N = r$, $q_S = \theta$. These are closely connected with the intrinsic coordinates of the previous paragraph through the relations

$$\delta n = \pm\, \delta l_N = \pm\, \frac{\delta q_N}{h_N}, \qquad \delta s = \pm\, \delta l_S = \pm\, \frac{\delta q_S}{h_S} \tag{4–5.6}$$

the algebraic signs being positive or negative according as the unit vectors $\mathbf{i}_N$ and $\mathbf{i}_S$ coincide with, or are oppositely directed from, the corresponding unit vectors **n** and **s**, respectively. Thus, in the case of spherical coordinates, $\mathbf{i}_r = \mathbf{n}$ and $\mathbf{i}_\theta = \mathbf{s}$, in which instance the signs are positive. [See Fig. A–15.1(b).]

4-6 Properties of the Stream Function

The instantaneous stream surfaces within a fluid in axisymmetrical motion are characterized by the values $\psi =$ constant, the streamlines being given by the intersection of these with the meridian planes. To prove this we have, for an arbitrary displacement in space,

$$d\psi = d\mathbf{R}\cdot\nabla\psi \tag{4–6.1}$$

But, from Eq. (4–3.8)

$$\nabla\psi = -\mathbf{i}_\phi \times \varpi\mathbf{v} \tag{4–6.2}$$

whence,

$$d\psi = -\mathbf{i}_\phi \times \varpi\mathbf{v}\cdot d\mathbf{R} = -\mathbf{i}_\phi\varpi\cdot(\mathbf{v} \times d\mathbf{R}) \tag{4–6.3}$$

This gives the change in the stream function for an arbitrary displacement, $d\mathbf{R}$. If we now let $d\mathbf{R}$ refer to a displacement *along* a streamline or stream surface then, by definition,

$$\mathbf{v} \times d\mathbf{R} = \mathbf{0} \tag{4–6.4}$$

as **v** is parallel to $d\mathbf{R}$. Hence,

$$d\psi = 0 \tag{4–6.5}$$

from which it follows that $\psi =$ constant along the streamlines and surfaces.

It follows without difficulty from Eq. (4–3.8) that the speed along a streamline is given by

$$|\mathbf{v}| = \frac{1}{\varpi}|\nabla\psi| \tag{4–6.6}$$

Alternatively, if δn refers to a distance measured normal to a stream surface and **n** is a unit vector pointing in the direction in which δn is measured, we have that

$$\nabla\psi = \mathbf{n}\frac{\partial\psi}{\partial n} \tag{4–6.7}$$

and thus an equivalent expression for the speed is

$$|\mathbf{v}| = \frac{1}{\varpi}\left|\frac{\partial\psi}{\partial n}\right| \tag{4–6.8}$$

The stream function has been defined in such a way as to vanish everywhere on the axis of revolution. This results in the assignment of absolute values to the stream function at each point of the fluid. Absolute values for this function are, however, without physical significance since only the derivatives of ψ appear in the expressions for the velocity components. Thus, the stream function should be regarded as defined only to within an arbitrary additive constant.

That a stream function exists for axisymmetrical motions depends entirely on the *kinematical* assumption of incompressibility. Thus, the stream function arises not only in viscous fluid motions but, for example, in ideal fluid motions as well; for these two flows differ only in their *dynamical* assumptions. Furthermore, the existence of a stream function is not limited to steady motions.

We note further that whereas a stream function exists in *all* cases of incompressible flow in *two dimensions*, it exists in the case of three-dimensional motions only when the latter are axially symmetric.

4-7 Dynamic Equation Satisfied by the Stream Function

To obtain the dynamical equation satisfied by Stokes' stream function we have for the vorticity vector* in cylindrical coordinates, from Eq. (A–9.19),

$$\boldsymbol{\zeta} = \nabla \times \mathbf{v} = \mathbf{i}_\phi \left(\frac{\partial v_\varpi}{\partial z} - \frac{\partial v_z}{\partial \varpi}\right) = \mathbf{i}_\phi \left\{\frac{\partial}{\partial z}\left(\frac{1}{\varpi}\frac{\partial \psi}{\partial z}\right) + \frac{\partial}{\partial \varpi}\left(\frac{1}{\varpi}\frac{\partial \psi}{\partial \varpi}\right)\right\} = \frac{\mathbf{i}_\phi}{\varpi} E^2 \psi \tag{4–7.1}$$

where the operator E^2 is given by

$$E^2 = \varpi \frac{\partial}{\partial \varpi}\left(\frac{1}{\varpi}\frac{\partial}{\partial \varpi}\right) + \frac{\partial^2}{\partial z^2} = \nabla^2 - \frac{2}{\varpi}\frac{\partial}{\partial \varpi} \tag{4–7.2}$$

Repeated application of the operator $\nabla\times$ to Eq. (4–7.1) gives, successively, having recourse to Eq. (A–9.19),

$$\nabla \times \boldsymbol{\zeta} = -\frac{\mathbf{i}_\varpi}{\varpi}\frac{\partial}{\partial z}(E^2\psi) + \frac{\mathbf{i}_z}{\varpi}\frac{\partial}{\partial \varpi}(E^2\psi) \tag{4–7.3}$$

and

$$\nabla \times (\nabla \times \boldsymbol{\zeta}) = -\frac{\mathbf{i}_\phi}{\varpi}\left\{\varpi \frac{\partial}{\partial \varpi}\left(\frac{1}{\varpi}\frac{\partial}{\partial \varpi}\right) + \frac{\partial^2}{\partial z^2}\right\} E^2\psi = -\frac{\mathbf{i}_\phi}{\varpi} E^2(E^2\psi)$$

$$= -\frac{\mathbf{i}_\phi}{\varpi} E^4\psi \tag{4–7.4}$$

*Previously, we called the quantity $\boldsymbol{\omega} = \frac{1}{2}\nabla \times \mathbf{v}$ the vorticity. We alter our definition here to avoid having to carry along the extraneous factor of one-half.

In the case of an incompressible fluid the Navier-Stokes equations are of the form given by Eq. (2–1.15),

$$\rho\left(\frac{\partial \mathbf{v}}{\partial t} + \mathbf{v}\cdot\nabla\mathbf{v}\right) - \mu\nabla^2\mathbf{v} = -\nabla p \tag{4–7.5}$$

This can be expressed in an alternative form by introducing the vector identities

$$\mathbf{v}\cdot\nabla\mathbf{v} = \tfrac{1}{2}\nabla\mathbf{v}^2 - \mathbf{v}\times(\nabla\times\mathbf{v}) \tag{4–7.6}$$

and

$$\nabla^2\mathbf{v} = \nabla(\nabla\cdot\mathbf{v}) - \nabla\times(\nabla\times\mathbf{v}) \tag{4–7.7}$$

The term $\nabla\cdot\mathbf{v}$ vanishes in the last equation on account of the assumption of incompressibility. Upon substituting these into Eq. (4–7.5) and taking the curl of both sides of the resultant equation the pressure is eliminated, since this operator annihilates the gradient of any scalar. With the substitution $\boldsymbol{\zeta} = \nabla\times\mathbf{v}$, we then obtain

$$\frac{\partial\boldsymbol{\zeta}}{\partial t} - \nabla\times(\mathbf{v}\times\boldsymbol{\zeta}) + \upsilon\nabla\times(\nabla\times\boldsymbol{\zeta}) = \mathbf{0} \tag{4–7.8}$$

where υ is the kinematic viscosity.

Equation (4–3.8), in conjunction with Eqs. (4–7.1) and (4–7.4), enables us to express the foregoing in terms of the stream function as follows:

$$\frac{\mathbf{i}_\phi}{\varpi}\frac{\partial}{\partial t}(E^2\psi) - \nabla\times\left(\frac{E^2\psi}{\varpi^2}\nabla\psi\right) - \mathbf{i}_\phi\frac{\upsilon}{\varpi}E^4\psi = \mathbf{0} \tag{4–7.9}$$

The middle term is equal to $-\nabla\psi\times\nabla(E^2\psi/\varpi^2)$, which has a component only in the ϕ direction. Therefore, the differential equation satisfied by the stream function is

$$\upsilon E^4\psi - \mathbf{i}_\phi\cdot\varpi\nabla\psi\times\nabla\left(\frac{E^2\psi}{\varpi^2}\right) - \frac{\partial}{\partial t}(E^2\psi) = 0 \tag{4–7.10}$$

or

$$\upsilon E^4\psi - \varpi\left(\frac{\partial\psi}{\partial z}\frac{\partial}{\partial\varpi} - \frac{\partial\psi}{\partial\varpi}\frac{\partial}{\partial z}\right)\frac{E^2\psi}{\varpi^2} - \frac{\partial}{\partial t}(E^2\psi) = 0 \tag{4–7.11}$$

This scalar equation is of the fourth order in ψ. It is not linear because of the presence of the central term contributed by the inertial terms, $\mathbf{v}\cdot\nabla\mathbf{v}$. For the case of creeping motions the nonlinear term is omitted; hence,

$$E^4\psi - \frac{1}{\upsilon}\frac{\partial}{\partial t}(E^2\psi) = 0 \tag{4–7.12}$$

which is of the same order as the original equation. If the latter motion is steady we obtain

$$E^4\psi = 0 \tag{4–7.13}$$

as the equation of motion.

It is of interest to note from Eq. (4–7.1) that in the case of irrotational motion, $\boldsymbol{\zeta} = \mathbf{0}$, the equation of motion is given by

$$E^2\psi = 0 \tag{4–7.14}$$

An expression for the operator E^2 in curvilinear coordinates can be obtained by applying the definition given in Eq. (4–7.1). If (q_1, q_2, ϕ)—in that order—are a set of orthogonal curvilinear coordinates of revolution, Eq. (A–5.4) yields, for the case of axisymmetrical motion,

$$\boldsymbol{\zeta} = \nabla \times \mathbf{v} = \mathbf{i}_\phi h_1 h_2 \left[\frac{\partial}{\partial q_1}\left(\frac{v_2}{h_2}\right) - \frac{\partial}{\partial q_2}\left(\frac{v_1}{h_1}\right)\right] \tag{4–7.15}$$

Expressions for the curvilinear velocity components v_1 and v_2, in terms of the stream function, are given in Eq. (4–4.3). Hence, upon substituting into Eq. (4–7.1), we obtain

$$E^2\psi = \varpi h_1 h_2 \left[\frac{\partial}{\partial q_1}\left(\frac{1}{\varpi}\frac{h_1}{h_2}\frac{\partial\psi}{\partial q_1}\right) + \frac{\partial}{\partial q_2}\left(\frac{1}{\varpi}\frac{h_2}{h_1}\frac{\partial\psi}{\partial q_2}\right)\right] \tag{4–7.16}$$

or, simply,

$$E^2 = \varpi h_1 h_2 \left[\frac{\partial}{\partial q_1}\left(\frac{1}{\varpi}\frac{h_1}{h_2}\frac{\partial}{\partial q_1}\right) + \frac{\partial}{\partial q_2}\left(\frac{1}{\varpi}\frac{h_2}{h_1}\frac{\partial}{\partial q_2}\right)\right] \tag{4–7.17}$$

In circular cylindrical coordinates (z, ϖ, ϕ), $h_z = h_\varpi = 1$, and we reproduce Eq. (4–7.2). Likewise, in spherical coordinates

$$E^2 = \frac{\partial^2}{\partial r^2} + \frac{\sin\theta}{r^2}\frac{\partial}{\partial\theta}\left(\frac{1}{\sin\theta}\frac{\partial}{\partial\theta}\right) \tag{4–7.18}$$

In the important case of a *conjugate** system of revolution (see Section A–16), defined by a conformal transformation of the type

$$z + i\varpi = f(\xi + i\eta) \tag{4–7.19}$$

the metrical coefficients are equal, and in place of Eq. (4–7.17) we obtain the simple expression

$$E^2 = \varpi h^2 \left[\frac{\partial}{\partial\xi}\left(\frac{1}{\varpi}\frac{\partial}{\partial\xi}\right) + \frac{\partial}{\partial\eta}\left(\frac{1}{\varpi}\frac{\partial}{\partial\eta}\right)\right] \tag{4–7.20}$$

where $h = h_\xi = h_\eta$. This should be compared with the ϕ–independent Laplace operator in these coordinates,

$$\nabla^2 = \frac{h^2}{\varpi}\left[\frac{\partial}{\partial\xi}\left(\varpi\frac{\partial}{\partial\xi}\right) + \frac{\partial}{\partial\eta}\left(\varpi\frac{\partial}{\partial\eta}\right)\right] \tag{4–7.21}$$

As will appear in subsequent sections, solutions of the equation $E^4\psi = 0$ may be derived by various means from solutions of the second-order equation $E^2\psi = 0$. The operator E^2 is closely related to the Laplace operator ∇^2 as can be seen, for example, by comparing them in cylindrical coordinates:

$$E^2 = \varpi\frac{\partial}{\partial\varpi}\left(\frac{1}{\varpi}\frac{\partial}{\partial\varpi}\right) + \frac{\partial^2}{\partial z^2} = \frac{\partial^2}{\partial\varpi^2} - \frac{1}{\varpi}\frac{\partial}{\partial\varpi} + \frac{\partial^2}{\partial z^2} \tag{4–7.22}$$

*The only nonconjugate system of revolution discussed in Appendix A is spherical coordinates.

$$\nabla^2 = \frac{1}{\varpi}\frac{\partial}{\partial\varpi}\left(\varpi\frac{\partial}{\partial\varpi}\right) + \frac{\partial^2}{\partial z^2} = \frac{\partial^2}{\partial\varpi^2} + \frac{1}{\varpi}\frac{\partial}{\partial\varpi} + \frac{\partial^2}{\partial z^2} \tag{4–7.23}$$

Because of this similarity the well-known theory of solutions of Laplace's equation, $\nabla^2 V = 0$, in various systems of coordinates is of inestimable value in guiding one to the corresponding solutions of the equation $E^2\psi = 0$ and, ultimately, to the more pertinent equation $E^4\psi = 0$.

Even more to the point is the fact that irrotational flows satisfy the equation $E^2\psi = 0$. These motions have been extensively studied in connection with their applications in ideal fluid theory. Numerous examples are cited in the treatises of Lamb[19] and Milne-Thomson[24].

4-8 Uniform Flow

When a fluid streams everywhere with uniform velocity $\mathbf{v} = \mathbf{i}_z U$, we have at each point of the fluid

$$v_z = -\frac{1}{\varpi}\frac{\partial\psi}{\partial\varpi} = U, \qquad v_\varpi = \frac{1}{\varpi}\frac{\partial\psi}{\partial z} = 0$$

Upon integration this yields

$$\psi = -\tfrac{1}{2}\varpi^2 U = -\tfrac{1}{2}Ur^2\sin^2\theta \tag{4–8.1}$$

which is the stream function for uniform flow with velocity U in the direction of the positive z axis.

This result also follows at once from the fundamental definition of the stream function, Eq. (4–2.4), by noting that the volumetric flow rate Q through the circular section shown in Fig. 4–8.1 is

$$Q = -\pi\varpi^2 U = 2\pi\psi \tag{4–8.2}$$

the negative sign being in accord with the convention of Section 4–2 regarding the direction of net flow.

As is immediately apparent, the stream surfaces $\psi = \text{constant}$ are concentric cylinders, $\varpi = \text{constant}$. The streamlines, formed by the intersection of these surfaces with the meridian planes, are straight lines parallel to the z axis.

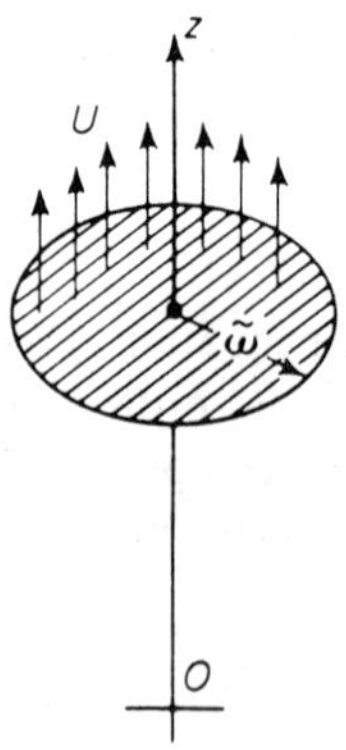

Figure 4-8.1. Uniform streaming flow.

4-9 Point Source or Sink

A *point source* or *sink* is an infinitesimal region of space from which fluid issues radially in all directions. When the flow is directed inwardly, toward the point, it is termed a *sink*. Point sources and sinks are abstractions which cannot be realized in nature, although they can be simulated more or less closely.

Suppose that an incompressible fluid radiates from the source into an unbounded region at a volumetric flow rate q. The flow is purely radial. If we imagine a spherical envelope of fluid of radius r, having the source at its center, the radial component of velocity on its surface is given by

$$v_r = \frac{q}{4\pi r^2} \tag{4–9.1}$$

while the tangential component vanishes identically. The term $4\pi r^2$ is the surface area of the sphere. Therefore,

$$v_r = -\frac{1}{r^2 \sin\theta}\frac{\partial\psi}{\partial\theta} = \frac{q}{4\pi r^2}, \qquad v_\theta = \frac{1}{r\sin\theta}\frac{\partial\psi}{\partial r} = 0 \tag{4–9.2}$$

from which we obtain by integration

$$\psi = -\frac{q}{4\pi}(1 - \cos\theta) \tag{4–9.3}$$

the integration constant having been chosen so as to make ψ vanish on the axis, $\theta = 0$. The stream function for a point sink corresponds to a negative value of q.

As would be expected on physical grounds alone, the stream surfaces are cones, $\theta =$ constant, having their apex at the source. The corresponding streamlines are rays emanating from this point. These results and conclusions hold whether or not the flow is steady provided that q is interpreted as the *instantaneous* rate of flow.

Equation (4–9.3) is of a purely kinematical nature, having been obtained without the introduction of any *dynamic* assumptions. As such it is applicable to any class of incompressible flows for which the motion represented by it is dynamically possible. This question can always be resolved by direct substitution into the dynamic equations of motion adopted. In particular we note that Eq. (4–9.3) satisfies the equations of irrotational motion, $E^2\psi = 0$. In accord with Eqs. (4–7.11), (4–7.13), and (4–7.14) the motion is therefore, at the same time, an exact solution of the Navier-Stokes, creeping motion, and potential flow equations. Their different dynamical assumptions lead, however, to different expressions for the pressures and stresses

4-10 Source and Sink of Equal Strength

Some interesting flow patterns can be generated by distributing sources and sinks of various strengths throughout the fluid. The distributions may be discrete or continuous. We restrict ourselves, here, to those combinations

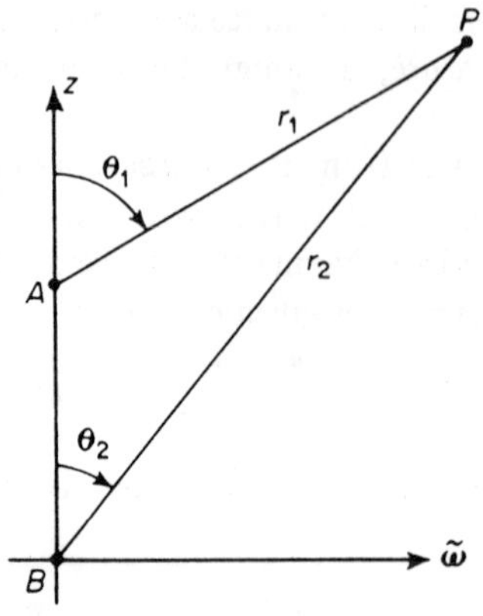

Figure 4-10.1. Coordinates for source and sink.

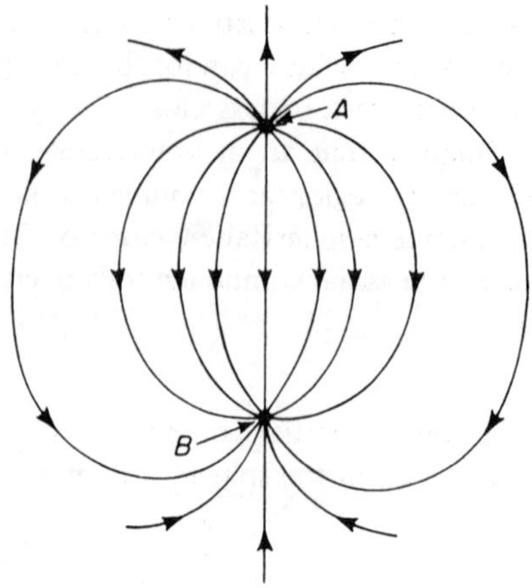

Figure 4-10.2. Streamlines for source and sink of equal strength.

which give rise to axisymmetrical flows. As the individual contributions each satisfy the differential equation $E^2\psi = 0$, it follows from the linearity of the operator E^2 that the motion obtained by superposing these in arbitrary fashion is itself a solution of this equation. They are therefore, automatically, solutions of the creeping motion equation, $E^4\psi = 0$.

As a simple example, consider a source and sink of equal strength situated on either side of the z axis at points equidistant from the origin (Fig. 4–10.1). We shall suppose that fluid issues from the source, A, at a volumetric flow rate q and into the sink, B, at the same rate. The value of the stream function at P due to the source at A is given by Eq. (4–9.3) with $\theta = \theta_1$, whereas the value at P due to the sink at B is

$$\psi_{\text{sink}} = \frac{q}{4\pi}(1 - \cos\theta_2) \tag{4–10.1}$$

Hence, the value of the stream function at the point P due to the combination is

$$\psi(P) = \frac{q}{4\pi}(\cos\theta_1 - \cos\theta_2) \tag{4–10.2}$$

The streamlines are thus given by the equation $\cos\theta_1 - \cos\theta_2 = \text{constant}$, and are depicted in Fig. 4–10.2.

4-11 Finite Line Source

As an example of a continuous distribution of sources, consider a finite line source extending along the z axis from the origin to point $A(0, a)$, as shown in Fig. 4–11.1. The distribution of sources along this line segment can be characterized by prescribing the local volumetric flow rate per unit length, $Q = Q(\xi)$, at a given point on the line; that is, if δq is the volumetric flow rate from a length $\delta\xi$ of the line we have that

$$\delta q = Q(\xi)\delta\xi \qquad (4\text{–}11.1)$$

The value of the stream function $\delta\psi$ at the point $P(\varpi, z)$ due to this source at ξ is, from Eq. (4–9.3),

$$\delta\psi(P) = \frac{\delta q}{4\pi}\cos\theta = \frac{Q(\xi)}{4\pi}\cos\theta\, d\xi \qquad (4\text{–}11.2)$$

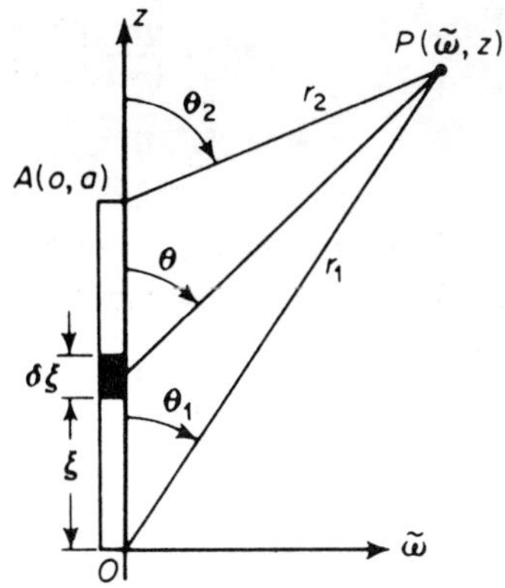

Figure 4-11.1. Coordinates for a finite line source.

We have for simplicity discarded the arbitrary convention that ψ vanish along the symmetry axis.

Upon superposing the stream functions, we obtain for the continuous source distribution

$$\psi(\varpi, z) = \frac{1}{4\pi}\int_0^a Q(\xi)\cos\theta\, d\xi \qquad (4\text{–}11.3)$$

This gives the stream function at the point (ϖ, z) due to the distribution of sources lying along the line segment of the z axis from O to A. To express θ in terms of ξ we note from Fig. 4–11.1 that

$$\cot\theta = \frac{z - \xi}{\varpi} \qquad (4\text{–}11.4)$$

Upon solving this for $\cos\theta$ and substituting into Eq. (4–11.3) one obtains

$$\psi(\varpi, z) = \frac{1}{4\pi}\int_0^a Q(\xi)\frac{z - \xi}{\sqrt{\varpi^2 + (z - \xi)^2}}\, d\xi \qquad (4\text{–}11.5)$$

In performing the integration, ϖ and z are to be regarded as constants.

The total volumetric flow rate from the line source is

$$q = \int_0^a Q(\xi)\, d\xi \qquad (4\text{–}11.6)$$

Equation (4–11.5) may be employed to calculate the stream function for an arbitrary distribution of sources of various strengths along the line segment. When Q is constant, the integration is effected most easily by resorting directly to Eq. (4–11.3). We find upon differentiating Eq. (4–11.4) that

$$d\xi = \varpi\csc^2\theta\, d\theta \qquad (4\text{–}11.7)$$

and thus

$$\psi = \frac{q\varpi}{4\pi a}\int_{\theta_1}^{\theta_2}\frac{\cos\theta}{\sin^2\theta}\,d\theta = \frac{q}{4\pi a}\left(\frac{\varpi}{\sin\theta_1} - \frac{\varpi}{\sin\theta_2}\right)$$

$$= \frac{q}{4\pi a}(r_1 - r_2) \tag{4–11.8}$$

The streamlines, given by $r_1 - r_2 =$ constant, are hyperbolas with foci at O and A, as shown in Fig. 4–11.2.

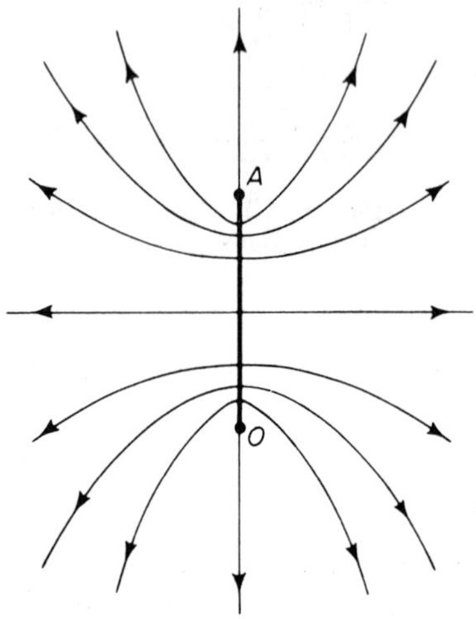

Figure 4-11.2. Streamlines for flow from a uniform finite line source.

For those points on the positive z axis lying *above* the line source, we have $r_1 - r_2 = a$, and therefore $\psi = q/4\pi$. Conversely, *below* the source, $\psi = -q/4\pi$.

The results of this section can also be concisely expressed in prolate spheroidal coordinates (see Section A–17).

4-12 Point Force

Consider an isolated force acting *on* the fluid and having a component D in the direction of the *negative* z axis, the surrounding fluid being unbounded.

The resulting fluid motion is axisymmetrical. For a force at the origin of coordinates, we have in creeping motion—cf. Eqs. (3–4.31) to (3–4.33),

$$v_\varpi = -\frac{D}{8\pi\mu}\frac{\varpi z}{r^3} = \frac{1}{\varpi}\frac{\partial\psi}{\partial z} \tag{4–12.1}$$

and

$$v_z = \frac{D}{8\pi\mu}\left(\frac{\varpi^2}{r^3} - \frac{2}{r}\right) = -\frac{1}{\varpi}\frac{\partial\psi}{\partial\varpi} \tag{4–12.2}$$

where $r^2 = \varpi^2 + z^2$. The stream function is, therefore,

$$\psi = \frac{D}{8\pi\mu}\frac{\varpi^2}{r} \tag{4–12.3}$$

As before, the pressure is given by

$$p = -\frac{D}{4\pi}\frac{z}{r^3} \tag{4–12.4}$$

It can easily be verified that Eq. (4–12.3) satisfies the differential equation of creeping flow, $E^4\psi = 0$.

The solution Eq. (4–12.3) may be regarded as fundamental to axisymmetrical creeping flows. Discrete or continuous distributions of point forces may be superposed to synthesize various axisymmetrical motions, in much the same manner as done for the sources and sinks of the previous sections.

4-13 Boundary Conditions Satisfied by the Stream Function

The most common boundary-value problem encountered in axisymmetrical motion is that of a body of revolution moving with constant velocity, $\mathbf{U} = \mathbf{i}_z U$, in an infinite fluid, parallel to its axis of rotation. The boundary conditions on the body are of two types, kinematical and dynamical.

If we refer the motion to an observer at rest with respect to the fluid, the kinematical condition describing the impenetrability of the surface is

$$(\mathbf{v} - \mathbf{U})\cdot\mathbf{n} = 0 \tag{4–13.1}$$

This holds whether or not the surface is solid, unless it be porous.

For a *solid* body, the dynamical condition of no relative tangential velocity at the interface requires that

$$(\mathbf{v} - \mathbf{U})\cdot\mathbf{s} = 0 \tag{4–13.2}$$

The vectors **n** and **s** are those discussed in Section 4–5.

These conditions may be expressed in terms of the stream function. Application of Eqs. (4–5.3) and (4–5.4) enables us to write the kinematical condition in the form

$$-\frac{1}{\varpi}\frac{\partial\psi}{\partial s} - \frac{\partial\varpi}{\partial s}U = 0 \tag{4–13.3}$$

or, alternatively,

$$\frac{\partial}{\partial s}(\psi + \tfrac{1}{2}\varpi^2 U) = 0 \tag{4–13.4}$$

Upon integrating this around the body one obtains

$$\psi + \tfrac{1}{2}\varpi^2 U = 0 \tag{4–13.5}$$

where we have set the integration constant equal to zero to be consistent with the convention that ψ vanishes along the axis, $\varpi = 0$.

A similar analysis gives, for the dynamic condition at the surface of the solid,

$$\frac{\partial}{\partial n}(\psi + \tfrac{1}{2}\varpi^2 U) = 0 \tag{4–13.6}$$

If the surface is a member of the family of curvilinear coordinate surfaces $q_N =$ constant, we have

$$\delta n = \left|\frac{\delta q_N}{h_N}\right|$$

and thus, in place of Eq. (4–13.6), we may write

$$\frac{\partial}{\partial q_N}(\psi + \tfrac{1}{2}\varpi^2 U) = 0 \tag{4–13.7}$$

If the fluid extends to infinity and is at rest there, we must have

$$\frac{\psi}{r^2} \to 0 \quad \text{as } r \to \infty \tag{4–13.8}$$

in order that the velocity vanish.

It is sometimes simpler to regard the solid as being at rest with fluid streaming past it, the latter attaining a uniform velocity $\mathbf{U} = -\mathbf{i}_z U$ at infinity, in the direction of the *negative* z axis. As earlier, it follows from Eq. (4–5.3), that the vanishing of the normal velocity at the surface requires $\partial\psi/\partial s = 0$, which leads ultimately to the condition

$$\psi = 0 \tag{4–13.9}$$

at the surface. Likewise, in the case of no tangential velocity we require

$$\frac{\partial \psi}{\partial n} = 0 \quad \text{or} \quad \frac{\partial \psi}{\partial q_N} = 0 \tag{4–13.10}$$

at the surface of the solid.

If the fluid extends indefinitely in all directions, the condition to be satisfied at points far distant from the surface is one of uniform motion in the *negative* z direction. Therefore, from Eq. (4–8.1), bearing in mind the altered direction of streaming, we must have

$$\psi \to \tfrac{1}{2}\varpi^2 U \quad \text{as } r \to \infty \tag{4–13.11}$$

The stream functions for the two classes of motion just described differ at each point of the fluid solely by the term $\frac{1}{2}\varpi^2 U$, corresponding to a constant velocity difference U. Specifically, if $\dot{\psi}$ is the stream function for a body translating in the positive z direction with velocity U through a fluid at rest at infinity, and $\ddot{\psi}$ is the stream function for streaming flow in the negative z direction with velocity U past the stationary body, then

$$\dot{\psi} + \tfrac{1}{2}\varpi^2 U = \ddot{\psi} \tag{4–13.12}$$

In the case of streaming flow past a stationary body, Eq. (4–13.9) shows that the body is itself a stream surface. That this should be so is at once evident by considering the paths followed by the fluid particles traversing the obstacle. On the other hand, Eq. (4–13.5) shows that this is not the case for a moving body. Rather, the fluid particles are pushed or dragged along by the body, and the streamlines therefore intersect the body. In the case of a rigid body to whose surface fluid adheres, it is apparent that the paths of the particles at the surface must be in the same direction as that in which the fluid is moving. It follows that the streamlines must be parallel to the z axis at the surface of the solid; that is,

$$\frac{\partial \psi}{\partial z} = 0 \tag{4–13.13}$$

at the solid surface. This may be demonstrated analytically by noting that $v_z = U$ and $v_\varpi = 0$ at the surface, and resorting to the relation (4–4.1)

$$\frac{\partial \psi}{\partial z} = \varpi v_\varpi \tag{4–13.14}$$

Equation (4–13.13) ceases to apply in the event of nonadherence at the surface, as, for example, in the case of a liquid droplet.

Streamlines for these two alternative cases of relative motion are shown in Figs. 4–17.2 and 4–18.2 for a solid spherical particle.

4-14 Drag on a Body

The stress dyadic for an incompressible viscous fluid is given by Eq. (2–1.8),

$$\mathbf{\Pi} = -p\mathbf{I} + 2\mu\mathbf{\Delta} \tag{4–14.1}$$

where from Eq. (2–1.12)

$$\mathbf{\Delta} = \tfrac{1}{2}[\nabla\mathbf{v} + (\nabla\mathbf{v})^{\dagger}] \tag{4–14.2}$$

We propose to express the stress at the surface of the body in the intrinsic coordinates of Section 4–5. It follows at once from Eq. (4–5.1) that the idemfactor is

$$\mathbf{I} = \mathbf{nn} + \mathbf{ss} + \mathbf{i}_\phi \mathbf{i}_\phi \tag{4–14.3}$$

Setting,

$$\mathbf{v} = \mathbf{n}v_n + \mathbf{s}v_s \tag{4–14.4}$$

the rate of deformation tensor, obtained from Eq. (A–7.7), takes the form

$$\begin{aligned}\mathbf{\Delta} = \mathbf{nn}\frac{\partial v_n}{\partial n} + \mathbf{ns}\frac{1}{2}\left(\frac{\partial v_n}{\partial s} + \frac{\partial v_s}{\partial n}\right) + \mathbf{sn}\frac{1}{2}\left(\frac{\partial v_n}{\partial s} + \frac{\partial v_s}{\partial n}\right) + \mathbf{ss}\frac{\partial v_s}{\partial s} \\ + \mathbf{i}_\phi \mathbf{i}_\phi\left\{\varpi v_n \frac{\partial}{\partial n}\left(\frac{1}{\varpi}\right) + \varpi v_s \frac{\partial}{\partial s}\left(\frac{1}{\varpi}\right)\right\}\end{aligned} \tag{4–14.5}$$

The stress vector acting across an element of surface area whose outer normal is $\mathbf{n}$ is then of the form

$$\mathbf{\Pi}_n = \mathbf{\Pi}\cdot\mathbf{n} = \mathbf{n}\left(-p + 2\mu\frac{\partial v_n}{\partial n}\right) + \mathbf{s}\mu\left(\frac{\partial v_n}{\partial s} + \frac{\partial v_s}{\partial n}\right) \tag{4–14.6}$$

The first and second terms correspond, respectively, to the normal and tangential stresses. Equation (4–14.6) may be expressed alternatively as follows:

$$\mathbf{\Pi}_n = -\mathbf{n}p + 2\mu\left(\mathbf{n}\frac{\partial v_n}{\partial n} + \mathbf{s}\frac{\partial v_n}{\partial s}\right) + \mathbf{s}\mu\left(\frac{\partial v_s}{\partial n} - \frac{\partial v_n}{\partial s}\right) \tag{4–14.7}$$

Upon availing ourselves of Eq. (A–5.4), we find that

$$\boldsymbol{\zeta} = \nabla\times\mathbf{v} = \mathbf{i}_\phi\left(\frac{\partial v_s}{\partial n} - \frac{\partial v_n}{\partial s}\right) \tag{4–14.8}$$

In conjunction with the expression for the ∇-operator given in Eq. (4–5.2), these enable us to write

$$\mathbf{\Pi}_n = -\mathbf{n}p + 2\mu\nabla v_n + \mathbf{s}\mu(\mathbf{i}_\phi\cdot\boldsymbol{\zeta}) \tag{4–14.9}$$

Finally, the relations (4–5.3) and (4–7.1), connecting the normal velocity and vorticity, respectively, to the stream function, lead to

$$\mathbf{\Pi}_n = -\mathbf{n}p - 2\mu\nabla\left(\frac{1}{\varpi}\frac{\partial\psi}{\partial s}\right) + \mathbf{s}\mu\frac{1}{\varpi}E^2\psi \tag{4–14.10}$$

Owing to the symmetry of flow, the integral effect of the stresses acting over the entire surface of the body give rise to only one significant dynamic parameter—a force acting parallel to the axis of revolution. According to our various conventions, the force exerted *on* the body *by* the fluid in the positive z direction will be

$$\begin{aligned} F_z &= \int_S \mathbf{\Pi}_z\cdot d\mathbf{S} = \int_S \mathbf{\Pi}_z\cdot\mathbf{n}\,dS = \int_S \Pi_{zn}\,dS \\ &= \int_S \Pi_{nz}\,dS \end{aligned} \tag{4–14.11}$$

where, in our present coordinates, the element of surface area is

$$\delta S = 2\pi\varpi\,\delta s \tag{4–14.12}$$

For the component of the stress vector in the positive z direction, we find

$$\Pi_{nz} = \mathbf{\Pi}_n\cdot\mathbf{i}_z = -(\mathbf{n}\cdot\mathbf{i}_z)p - 2\mu(\mathbf{i}_z\cdot\nabla)\left(\frac{1}{\varpi}\frac{\partial\psi}{\partial s}\right) + (\mathbf{s}\cdot\mathbf{i}_z)\mu\frac{1}{\varpi}E^2\psi \tag{4–14.13}$$

which, with the aid of Eq. (4–5.4), can be put in the form

$$\begin{aligned} \Pi_{nz} &= -p\frac{\partial\varpi}{\partial s} - 2\mu\frac{\partial}{\partial z}\left(\frac{1}{\varpi}\frac{\partial\psi}{\partial s}\right) - \mu\frac{1}{\varpi}\frac{\partial\varpi}{\partial n}E^2\psi \\ &= -\frac{1}{2\varpi}\left\{\frac{\partial}{\partial s}(\varpi^2 p) - \varpi^2\frac{\partial p}{\partial s}\right\} - 2\mu\frac{1}{\varpi}\frac{\partial}{\partial z}\left(\frac{\partial\psi}{\partial s}\right) - \mu\frac{1}{\varpi}\frac{\partial\varpi}{\partial n}E^2\psi \end{aligned} \tag{4–14.14}$$

The force on the body is then

$$\begin{aligned} F_z = &-\pi\int\frac{\partial}{\partial s}(\varpi^2 p)\,\delta s + \pi\int\varpi^2\frac{\partial p}{\partial s}\,\delta s \\ &- 4\pi\mu\int\frac{\partial}{\partial s}(\varpi v_\varpi)\,\delta s - 2\pi\mu\int\frac{\partial\varpi}{\partial n}E^2\psi\,\delta s \end{aligned} \tag{4–14.15}$$

where the integrals are to be taken around a meridian section of the body in a sense making a positive right-angle with **n** (see Fig. 4–5.1), the end points of the path of integration being the upper and lower points at which the boundary surface intersects the axis of symmetry. Since $\varpi = 0$ at these terminal points, the first and third integrals vanish identically, leaving

$$F_z = \pi\int\varpi^2\frac{\partial p}{\partial s}\,\delta s - 2\pi\mu\int\frac{\partial\varpi}{\partial n}E^2\psi\,\delta s \tag{4–14.16}$$

To rid ourselves of the pressure term, the results of Section 4–15 may be applied. These show that in steady creeping flow,

$$\frac{\partial p}{\partial s} = \frac{\mu}{\varpi}\frac{\partial}{\partial n}(E^2\psi) \tag{4-14.17}$$

Substitution into the previous expression yields the following simple expression for the force exerted on the body:

$$F_z = \mu\pi \int \varpi^3 \frac{\partial}{\partial n}\left(\frac{E^2\psi}{\varpi^2}\right) \delta s \tag{4-14.18}$$

This is identical to a corresponding expression given by Stimson and Jeffery[36]. It applies quite generally to steady creeping flows regardless of the conditions prevailing at the boundaries and, in particular, is not limited to unbounded fluids. It is noteworthy that it also applies to the complete Navier-Stokes equations in the special case where the velocity vanishes at the surface; for the inertial term, $\mathbf{v}\cdot\nabla\mathbf{v}$, which would normally contribute to the expression for the pressure change $\partial p/\partial s$ along the boundary vanishes at each point of the surface. In this sense it agrees with a formula due to Walton[38].

Evaluation of the resistance by application of Eq. (4–14.18) is often a troublesome problem. If the medium is unbounded we may proceed otherwise by availing ourselves of the results of Section 4–12, for an axisymmetric point force. At a sufficiently large distance from the obstacle the stream function must become identical to that which would be generated by the action of a point force equal in magnitude to the drag on the obstacle, *provided that the fluid is at rest at infinity.* Thus it follows from Eq. (4–12.3), by there replacing D by F_z, that the force exerted by the fluid on the body in the positive z direction is given by the relation

$$F_z = 8\pi\mu \lim_{r\to\infty} \frac{r\psi}{\varpi^2} \tag{4-14.19*}$$

where ψ is the stream function for the motion under consideration. A more rigorous demonstration is given by Payne and Pell[27], to whom this relation is due.

If the fluid is not at rest at infinity, Eq. (4–14.19) is inapplicable; if, however, ψ_∞ denotes the stream function corresponding to the fluid motion at infinity, then the stream function $\psi - \psi_\infty$ gives a state of rest at infinity and in place of Eq. (4–14.19) we have

$$F_z = 8\pi\mu \lim_{r\to\infty} \frac{r(\psi - \psi_\infty)}{\varpi^2} \tag{4-14.20}$$

For example, if the fluid motion at infinity is one of uniform flow with velocity U in the negative z direction, then

*Recall that in the analysis of Section 4–12, D is the force exerted by the particle on the fluid in the negative z direction. By Newton's law of action and reaction this is the same as the force exerted by fluid on particle in the *positive* z direction.

$$\psi_\infty = \tfrac{1}{2}\varpi^2 U \tag{4–14.21}$$

In general, ψ_∞ is the fluid motion which would prevail in the absence of the particle.

Another method for computing the force on a body translating with uniform velocity U in an unbounded fluid depends on the fact that $|F_z U|$ gives the work done by the stresses acting over the surface of the body which, in turn, is equal to the rate at which energy is being dissipated within the fluid. A calculation of the latter, therefore, immediately gives the force on the body.

4-15 Pressure

The relationship between the pressure in steady creeping flow and the stream function may be established as follows:

$$\frac{1}{\mu}\nabla p = \nabla^2 \mathbf{v} = -\nabla \times \boldsymbol{\zeta} = -\nabla \times \left(\frac{\mathbf{i}_\phi}{\varpi} E^2\psi\right)$$
$$= -\frac{\mathbf{i}_\phi}{\varpi} \times \nabla(E^2\psi) \tag{4–15.1}$$

For the ∇-operator in a curvilinear coordinate system of rotation, (q_1, q_2, ϕ), right-handed in that order, we have

$$\nabla = \mathbf{i}_1 h_1 \frac{\partial}{\partial q_1} + \mathbf{i}_2 h_2 \frac{\partial}{\partial q_2} \tag{4–15.2}$$

Upon substituting into both sides of the first expression and equating components we obtain

$$\frac{\partial p}{\partial q_1} = -\frac{\mu}{\varpi}\frac{h_2}{h_1}\frac{\partial}{\partial q_2}(E^2\psi)$$

and

$$\frac{\partial p}{\partial q_2} = \frac{\mu}{\varpi}\frac{h_1}{h_2}\frac{\partial}{\partial q_1}(E^2\psi) \tag{4–15.3}$$

The pressure may be obtained by integration of these relations.

When the coordinate system of revolution is of the *conjugate* type (Section A–16), that is, defined by a conformal transformation of the type

$$z + i\varpi = f(\xi + i\eta) \tag{4–15.4}$$

the metrical coefficients are equal, whence Eq. (4–15.3) reduces to

$$\frac{\partial p}{\partial \xi} = -\frac{\mu}{\varpi}\frac{\partial}{\partial \eta}(E^2\psi)$$
$$\frac{\partial p}{\partial \eta} = \frac{\mu}{\varpi}\frac{\partial}{\partial \xi}(E^2\psi) \tag{4–15.5}$$

where (ξ, η, ϕ) are right-handed in that order.

4-16 Separable Coordinate Systems

Before undertaking the solution of the equations of motion, $E^4\psi = 0$, in specific coordinate systems, it is well to give a brief account of methods employed in the solution of simpler, but related, partial differential equations, which may be of value in the sequel. As will be put in evidence later, general methods for solving axially symmetric flow problems in *any* orthogonal conjugate coordinate system of revolution can be made to depend upon solutions of the second-order equation*

$$L_k(\chi) = 0 \tag{4–16.1}$$

where L_k is the operator

$$L_k = \frac{\partial^2}{\partial z^2} + \frac{\partial^2}{\partial \varpi^2} + \frac{k}{\varpi}\frac{\partial}{\partial \varpi} \qquad (-\infty < k < \infty) \tag{4–16.2}$$

The solutions, $\chi^{(k)}(z, \varpi)$, of the preceding equation are referred to as *generalized axisymmetrical potential functions.* It follows from the results of Section 4–7 that special cases of the foregoing are

$$L_1 = \nabla^2, \qquad L_{-1} = E^2 \tag{4–16.3}$$

For a conjugate coordinate system of revolution

$$z + i\varpi = f(\xi + i\eta) \tag{4–16.4}$$

we have that

$$\frac{\partial^2}{\partial z^2} + \frac{\partial^2}{\partial \varpi^2} = h^2\left(\frac{\partial^2}{\partial \xi^2} + \frac{\partial^2}{\partial \eta^2}\right) \tag{4–16.5}$$

where $h = h_\xi = h_\eta$ is the metrical coefficient. Also, from Eq. (A–14.8),

$$\frac{\partial}{\partial \varpi} = h^2\left(\frac{\partial \varpi}{\partial \xi}\frac{\partial}{\partial \xi} + \frac{\partial \varpi}{\partial \eta}\frac{\partial}{\partial \eta}\right)$$

The generalized operator is therefore given in these coordinates by

$$L_k = \varpi^{-k}h^2\left\{\frac{\partial}{\partial \xi}\left(\varpi^k\frac{\partial}{\partial \xi}\right) + \frac{\partial}{\partial \eta}\left(\varpi^k\frac{\partial}{\partial \eta}\right)\right\} \tag{4–16.6}$$

We propose to examine the conditions under which Eq. (4–16.1) may be solved by separation of variables via a substitution of the form

$$\chi^{(k)} = \varpi^{-k/2}U(\xi)V(\eta) \tag{4–16.7}$$

Upon performing the indicated operations it is found that

*For a further discussion of separable coordinates, see the text by Morse and Feshbach[25].

$$L_k(\chi^{(k)}) = \varpi^{-k} h^2 \left[-\frac{k}{2}\left(\frac{k}{2} - 1\right) \varpi^{(k/2)-2}\, UV \left\{ \left(\frac{\partial \varpi}{\partial \xi}\right)^2 + \left(\frac{\partial \varpi}{\partial \eta}\right)^2 \right\} - \frac{k}{2} \varpi^{(k/2)-1}\, UV \left(\frac{\partial^2 \varpi}{\partial \xi^2} + \frac{\partial^2 \varpi}{\partial \eta^2} \right) + \varpi^{k/2}(U''V + V''U) \right] \tag{4–16.8}$$

Now it follows from Eq. (4–16.5) that

$$\frac{\partial^2 \varpi}{\partial \xi^2} + \frac{\partial^2 \varpi}{\partial \eta^2} = 0 \tag{4–16.9}$$

Furthermore, from the results of Section A–16, it is easily shown that

$$\left(\frac{\partial \varpi}{\partial \xi}\right)^2 + \left(\frac{\partial \varpi}{\partial \eta}\right)^2 = \frac{1}{h^2} \tag{4–16.10}$$

Hence the equation

$$L_k(\chi^{(k)}) = 0 \tag{4–16.11}$$

is satisfied, provided that

$$-\frac{k}{2}\left(\frac{k}{2} - 1\right)\left(\frac{1}{h^2 \varpi^2}\right) UV + U''V + V''U = 0 \tag{4–16.12}$$

or, dividing by UV and rearranging,

$$\frac{U''}{U} + \frac{V''}{V} - \frac{k}{2}\left(\frac{k}{2} - 1\right)\frac{1}{h^2 \varpi^2} = 0 \tag{4–16.13}$$

It follows, therefore, that a sufficient condition for separability of the coordinates is

$$\frac{1}{h^2 \varpi^2} = P(\xi) + Q(\eta) \tag{4–16.14}$$

which may also be written as

$$\frac{\partial^2}{\partial \xi\, \partial \eta}\left(\frac{1}{h^2 \varpi^2}\right) = 0 \tag{4–16.15}$$

All the conjugate coordinate systems of revolution in Appendix A meet this condition. For example, in oblate spheroidal coordinates, Section A–18, we have

$$\varpi = c \cosh \xi \sin \eta, \qquad h = \frac{1}{c(\cosh^2 \xi - \sin^2 \eta)^{1/2}}$$

and thus,

$$\frac{1}{h^2 \varpi^2} = \frac{1}{\sin^2 \eta} - \frac{1}{\cosh^2 \xi}$$

which is of the type required by Eq. (4–16.14) for separability.

Combining Eqs. (4–16.13) and (4–16.14) results in

$$\frac{U''(\xi)}{U(\xi)} - \frac{k}{2}\left(\frac{k}{2} - 1\right) P(\xi) = -\frac{V''(\eta)}{V(\eta)} + \frac{k}{2}\left(\frac{k}{2} - 1\right) Q(\eta)$$

As the left side is independent of η, and the right side independent of ξ, each must be equal to a constant, say, λ. Coordinates separable in the sense of

Eq. (4–16.7) therefore lead to the following total differential equations of the second order:

$$U''(\xi) - \left[\frac{k}{2}\left(\frac{k}{2} - 1\right) P(\xi) + \lambda\right] U(\xi) = 0$$

and

$$V''(\eta) - \left[\frac{k}{2}\left(\frac{k}{2} - 1\right) Q(\eta) - \lambda\right] V(\eta) = 0 \tag{4–16.16}$$

The choice of the separation constant λ depends entirely upon the form of the functions P and Q and, hence, ultimately on the system of coordinates. Solutions of Eq. (4–16.7) for different values of the arbitrary parameter λ may be combined by superposition (either summation or integration) to obtain more general solutions of Eq. (4–16.1).

It is immediately evident from Eq. (4–16.7) that in those coordinate systems where ϖ has the form of a product,

$$\varpi = g(\xi)h(\eta) \tag{4–16.17}$$

the substitution

$$\chi = u(\xi)v(\eta) \tag{4–16.18}$$

will itself provide a separable solution, without recourse to the extraneous factor of $\varpi^{-k/2}$. This is the case for the systems discussed in Appendix A, except for bipolar and toroidal coordinates.

Applications of the general theory discussed here will be found interspersed throughout the sequel.

4-17 Translation of a Sphere

The most important example of an axisymmetrical flow is that arising from the motion of a solid sphere moving with constant velocity through an unbounded fluid otherwise at rest. This problem was first treated by Stokes and solved by him in terms of the stream function, which he created specifically for this purpose.

We shall suppose that the radius of the sphere is a and that it moves in the positive z direction with velocity U (Fig. 4–17.1). The problem is most amenable to treatment in spherical coordinates. Upon putting $\varpi = r \sin \theta$ in Eqs. (4–13.5) and (4–13.7), the boundary conditions to be satisfied on the surface of the sphere, $r = a$, become

$$\psi\big|_{r=a} = -\tfrac{1}{2} U a^2 \sin^2\theta$$

$$\left.\frac{\partial \psi}{\partial r}\right|_{r=a} = -Ua \sin^2\theta \tag{4–17.1}$$

Figure 4-17.1. Translating sphere.

As the fluid is at rest at infinity we must also take account of the condition

$$\frac{\psi}{r^2} \to 0 \quad \text{as } r \to \infty \tag{4–17.2}$$

The stream function is to satisfy the differential equation

$$E^4\psi = 0 \tag{4–17.3}$$

where, in spherical coordinates,

$$E^2 = \frac{\partial}{\partial r^2} + \frac{\sin\theta}{r^2}\frac{\partial}{\partial\theta}\left(\frac{1}{\sin\theta}\frac{\partial}{\partial\theta}\right) \tag{4–17.4}$$

A trial solution of the form

$$\psi = \sin^2\theta\, F(r) \tag{4–17.5}$$

is suggested by the boundary conditions Eq. (4–17.1). In conjunction with Eq. (4–17.4), we have, successively,

$$E^2\psi = \sin^2\theta\left(F'' - \frac{2}{r^2}F\right) = \sin^2\theta\, f(r) \tag{4–17.6}$$

and

$$E^4\psi = \sin^2\theta\left(f'' - \frac{2}{r^2}f\right) = 0 \tag{4–17.7}$$

This requires that

$$f'' - \frac{2}{r^2}f = 0 \tag{4–17.8}$$

the solution of which is

$$f(r) = Ar^2 + \frac{B}{r} \tag{4–17.9}$$

Therefore, from Eq. (4–17.6),

$$F'' - \frac{2}{r^2}F = Ar^2 + \frac{B}{r} \tag{4–17.10}$$

whence,

$$F(r) = \frac{A}{10}r^4 - \frac{1}{2}Br + Cr^2 + \frac{D}{r} \tag{4–17.11}$$

This makes

$$\psi = \sin^2\theta\left(\frac{1}{10}Ar^4 - \frac{1}{2}Br + Cr^2 + \frac{D}{r}\right) \tag{4–17.12}$$

In accordance with Eq. (4–17.2) we must set $A = 0$, $C = 0$. The remaining boundary conditions, Eq. (4–17.1), when applied to Eq. (4–17.12), result in the following simultaneous equations to determine the remaining constants, B and D:

$$\begin{aligned} -\frac{1}{2}Ua^2 &= -\frac{1}{2}Ba + \frac{D}{a} \\ -Ua &= -\frac{1}{2}B - \frac{D}{a^2} \end{aligned} \tag{4–17.13}$$

These have as their solution

$$B = \tfrac{3}{2}Ua, \qquad D = \tfrac{1}{4}Ua^3 \tag{4–17.14}$$

The stream function is then

$$\psi = \frac{1}{4}\,Ua^2 \sin^2\theta \left(\frac{a}{r} - 3\frac{r}{a}\right) \tag{4–17.15}$$

or

$$\psi = \frac{1}{4}\,Ur^2 \sin^2\theta \left[\left(\frac{a}{r}\right)^3 - 3\left(\frac{a}{r}\right)\right] \tag{4–17.16}$$

whereas the corresponding velocity components, obtained from Eq. (4–4.2), are

$$v_r = -\frac{1}{2}\,U\cos\theta \left(\frac{a}{r}\right)^2 \left(\frac{a}{r} - 3\frac{r}{a}\right) \tag{4–17.17}$$

and

$$v_\theta = -\frac{1}{4}\,U\sin\theta \left(\frac{a}{r}\right)\left[\left(\frac{a}{r}\right)^2 + 3\right] \tag{4–17.18}$$

A sketch of a few streamlines, $\psi =$ constant, is given in Fig. 4–17.2.

The pressure may be obtained by utilizing Eq. (4–15.3) in spherical coordinates,

$$\frac{\partial p}{\partial r} = -\frac{\mu}{r^2 \sin\theta}\frac{\partial}{\partial\theta}(E^2\psi)$$
$$\frac{\partial p}{\partial\theta} = \frac{\mu}{\sin\theta}\frac{\partial}{\partial r}(E^2\psi) \tag{4–17.19}$$

Upon putting $A = 0$ in Eq. (4–17.9) and substituting into Eq. (4–17.6), we obtain

$$E^2\psi = B\,\frac{\sin^2\theta}{r} \tag{4–17.20}$$

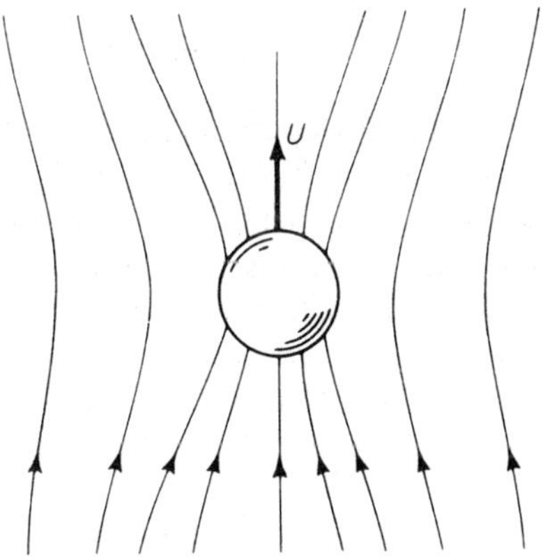

Figure 4-17.2. Streamlines for a moving sphere.

and thus

$$dp = \frac{\partial p}{\partial r}\,dr + \frac{\partial p}{\partial\theta}\,d\theta$$
$$= -\mu B\left(\frac{2\cos\theta\,dr}{r^3} + \frac{\sin\theta\,d\theta}{r^2}\right) = \mu B d\left(\frac{\cos\theta}{r^2}\right) \tag{4–17.21}$$

integration of which results in

$$p = p_\infty + \frac{3}{2}\,\mu a U\,\frac{\cos\theta}{r^2} \tag{4–17.22}$$

where p_∞ is the uniform pressure at infinity.

To obtain the force exerted by the fluid on the sphere set $\delta n = \delta r$ and $\delta s = r\delta\theta$ in Eq. (4–14.18) and observe that the lower and upper integration limits are $\theta = 0$ and π, respectively. This gives

$$F_z = \mu\pi \int_0^\pi r^3 \sin^3\theta \frac{\partial}{\partial r}\left(\frac{B \sin^2\theta/r}{r^2 \sin^2\theta}\right) r\, d\theta = -3B\mu\pi \int_0^\pi \sin^3\theta\, d\theta$$

$$= -4\pi\mu B \tag{4–17.23}$$

which holds independently of the values of the constants A, B, C, D in Eq. (4–17.12). In the present instance, this leads to

$$F_z = -6\pi\mu a U \tag{4–17.24}$$

This well-known result is known as Stokes' law of resistance. The negative sign implies that the force exerted by the fluid on the sphere is opposite in direction to the motion of the latter; the fluid thereby opposes the motion of the particle through it. In order to maintain the steady motion it is necessary to continuously apply a force of the same magnitude to the sphere in the direction of its motion. In practice, this normally results from the action of gravity on the mass of the sphere.

The central result of the analysis, Eq. (4–17.24), might have been obtained more simply through application of the formula (4–14.19) to the stream function Eq. (4–17.15) as follows:

$$F_z = 8\pi\mu \lim_{r\to\infty} \frac{Ua^2}{4r}\left(\frac{a}{r} - 3\frac{r}{a}\right) = -6\pi\mu a U \tag{4–17.25}$$

Some investigators attach separate significance to those parts of the total resistance which arise from the integrated effect of the *normal* and *tangential* stresses, respectively. The individual resistances which correspond to these two types of stress are termed the *form* (or profile) drag and *skin* drag, respectively. These can be obtained without difficulty from the decomposition of the stress vector in Eq. (4–14.6) into normal and shear contributions. As $\partial v_n/\partial n = 0$ on the sphere surface, the normal stress is simply—$\mathbf{n}p$. The component of this stress in the z direction is, with the aid of Eq. (4–5.4),

$$(\Pi_{nz})_{\text{normal}} = -\mathbf{i}_z \cdot \mathbf{n}p = -\frac{\partial \varpi}{\partial s} p$$

$$= -\frac{\partial (r \sin\theta)}{r\, \partial\theta} p = -p\cos\theta \tag{4–17.26}$$

Integrating this over the sphere surface gives the form drag,

$$F_z' = -\int_{\theta=0}^{\pi} \int_{\phi=0}^{2\pi} [p]_{r=a} \cos\theta\, dS \tag{4–17.27}$$

where $dS = a^2 \sin\theta\, d\theta\, d\phi$ is an element of surface area. This is readily integrated after substitution of the expression for the pressure, given in Eq. (4–17.22). The result is

$$F_z' = -2\pi\mu a U \tag{4–17.28}$$

To account for the total resistance, the skin (or viscous) drag must be

$$F_z'' = -4\pi\mu a U \tag{4–17.29}$$

Thus, the ratio of skin to form drag is 2:1. These results might also have been obtained by integrating the normal and tangential stresses, Π_{rr} and $\Pi_{r\theta}$, respectively, around the surface of the sphere.

4-18 Flow Past a Sphere

A problem closely related to that of the previous section is that of streaming flow past a stationary sphere, depicted in Fig. 4–18.1. If we regard the fluid

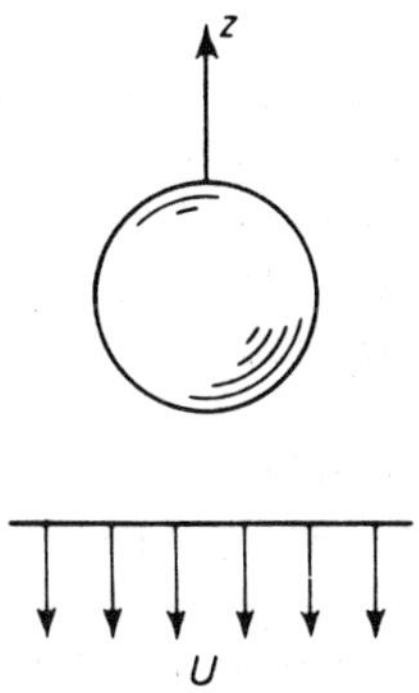

Figure 4-18.1. Streaming flow past a sphere.

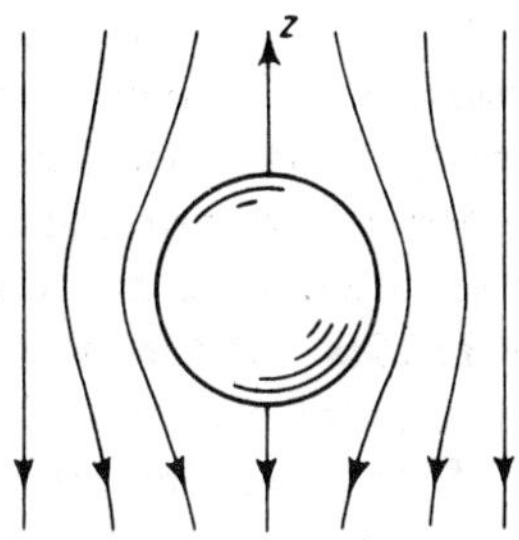

Figure 4-18.2. Streamlines for streaming flow past a sphere.

as streaming in the *negative* z direction with uniform velocity U at points infinitely distant from the sphere, the condition of uniform flow at infinity, Eq. (4–13.11), requires that

$$\psi \rightarrow \tfrac{1}{2} U r^2 \sin^2\theta \quad \text{as } r \rightarrow \infty \tag{4–18.1}$$

As the boundary conditions to be satisfied on the sphere we have, from Eqs. (4–13.9) and (4–13.10),

$$\psi = 0 \quad \text{and} \quad \frac{\partial \psi}{\partial r} = 0 \quad \text{at } r = a \tag{4–18.2}$$

The problem may be solved anew by redetermining the constants in Eq. (4–17.12). Alternatively, we have merely to superpose a uniform flow in the direction of z negative on the solution of the previous section, thereby obtaining

$$\psi = \frac{1}{4} U a^2 \sin^2\theta \left[\left(\frac{a}{r}\right) - 3\left(\frac{r}{a}\right) + 2\left(\frac{r}{a}\right)^2\right] \tag{4–18.3}$$

or

$$\psi = \frac{1}{2} U r^2 \sin^2\theta \left[\frac{1}{2}\left(\frac{a}{r}\right)^3 - \frac{3}{2}\left(\frac{a}{r}\right) + 1\right] \tag{4–18.4}$$

The component velocities may be obtained at once by adding to those of the previous section the values

$$v_r' = -U\cos\theta, \qquad v_\theta' = U\sin\theta \tag{4–18.5}$$

for uniform flow in the negative z direction. The results of this calculation are noted here for reference:

$$v_r = -\frac{1}{2}U\cos\theta\left[\left(\frac{a}{r}\right)^3 - 3\left(\frac{a}{r}\right) + 2\right] \tag{4–18.6}$$

and

$$v_\theta = -\frac{1}{4}U\sin\theta\left[\left(\frac{a}{r}\right)^3 + 3\left(\frac{a}{r}\right) - 4\right] \tag{4–18.7}$$

These fulfill the condition of vanishing velocity at the sphere surface, $r = a$.

The pressure and viscous stresses, depending as they do on first derivatives of velocity, have the same values as before; the constant velocity difference between the two solutions cannot affect them. In particular, the force on the sphere continues to be given by Eq. (4–17.24). The negative sign there now implies that the sphere tends to be dragged along by the moving fluid; that is, in the negative z direction.

Some of the streamlines derived from Eq. (4–18.3) are plotted in Fig. 4–18.2. In contrast to the results of the previous section, the sphere is now itself a stream surface and corresponds to the value $\psi = 0$. That the preceding configuration differs from that of Fig. 4–17.2 for a translating sphere in an otherwise quiescent fluid is to be expected on the basis of the remarks made at the close of Section 4–13.

In either case, the most striking aspect of the diagrams is the complete absence of asymmetry about the midplane, $z = 0$. This is possible only at low Reynolds numbers where the fluid is sensibly devoid of inertia. Experimental observation at higher particle Reynolds numbers reveals the presence of asymmetries, manifesting themselves in the form of vortices and in the existence of a wake to the rear of the sphere.

4-19 Terminal Settling Velocity

A spherical particle falling under the influence of gravity in a viscous fluid ultimately attains a uniform velocity in which the gravitational forces experienced by it are counterbalanced by the hydrodynamic forces. It henceforth falls at this constant speed, called its *terminal settling velocity*, U_∞. This is true, of course, whether or not the motion is sufficiently slow to come within the province of the creeping motion equations, though we shall confine our attention solely to this case. The rate of attainment of this uniform motion from any other state of motion, for example, rest, is an unsteady state problem.

The gravitational force acting on the particle, allowing for the buoyancy of the surrounding fluid, is

$$F = (\rho' - \rho) g \tfrac{4}{3}\pi a^3 \tag{4–19.1}$$

where ρ' is the (mean) density of the sphere and g is the local acceleration of gravity. On the other hand, Stokes' law gives for the frictional resistance

$$F = 6\pi\mu a U_\infty \tag{4–19.2}$$

Equating these, we find

$$U_\infty = \frac{2}{9} a^2 \frac{\rho' - \rho}{\mu} g \tag{4–19.3}$$

In the case where the particle is less dense than the surrounding fluid, $\rho' - \rho < 0$, this gives the constant velocity with which the particle would rise through the surrounding fluid.

If the fluid is gaseous, it is often satisfactory to use the approximation $\rho' \gg \rho$ in which event Eq. (4–19.3) reduces to

$$U_\infty \approx \frac{2a^2 \rho' g}{9\mu} \tag{4–19.4}$$

The validity of Eq. (4–19.3) has been much discussed in connection with the important applications to which it lends itself.

4-20 Slip at the Surface of a Sphere

It is of some interest to examine the possibility that fluid may slip at the surface of the sphere. This problem was first solved by Basset[2]. The most plausible hypothesis one can frame under these circumstances is that the tangential velocity* of fluid relative to the solid at a point on its surface is proportional to the tangential stress prevailing at that point. The constant of proportionality, β, between these two quantities may be termed a *coefficient of sliding friction*. Granting its existence, it is assumed to depend only on the nature of the fluid and solid surface.

When the sphere is at rest and fluid streams past it, the hypothesis takes the form

$$\beta v_\theta = \Pi_{r\theta} \quad \text{at } r = a \tag{4–20.1}$$

Now, for axisymmetric flows,

$$\Pi_{r\theta} = \mu \left[\frac{1}{r} \frac{\partial v_r}{\partial \theta} + r \frac{\partial}{\partial r} \left(\frac{v_\theta}{r} \right) \right] \tag{4–20.2}$$

Taking account of the condition that the normal velocity v_r vanishes on the

*There can be no question of the validity of the condition that the relative *normal* velocity vanishes at the surface, as this is a purely kinematical condition.

surface, we can reformulate Eq. (4–20.1) in terms of the stream function as follows:

$$\beta \frac{1}{r}\frac{\partial \psi}{\partial r} = \mu r \frac{\partial}{\partial r}\left(\frac{1}{r^2}\frac{\partial \psi}{\partial r}\right) \quad \text{at } r = a \tag{4–20.3}$$

The vanishing of v_r on the sphere is equivalent to the condition

$$\psi = 0 \quad \text{at} \quad r = a \tag{4–20.4}$$

In addition, the condition of uniform flow at infinity in the direction of the negative z axis requires that

$$\psi \to \tfrac{1}{2} U r^2 \sin^2 \theta \quad \text{as} \quad r \to \infty \tag{4–20.5}$$

The problem is uniquely determined by equations (4–20.3)–(4–20.5).

The general solution given by Eq. (4–17.12) is sufficient for our purposes. It is an easy matter to show that the constants are given by

$$\begin{aligned} A &= 0, \qquad B = \frac{3}{2} Ua \frac{\beta a + 2\mu}{\beta a + 3\mu} \\ C &= \frac{1}{2} U, \qquad D = \frac{1}{4} Ua^3 \frac{\beta a}{\beta a + 3\mu} \end{aligned} \tag{4–20.6}$$

The force acting on the sphere can be obtained from Eq. (4–17.23) by there inserting the value of B just given. Hence,

$$F_z = -6\pi\mu a U \frac{\beta a + 2\mu}{\beta a + 3\mu} \tag{4–20.7}$$

As a particular case of this we have for no slip, $\beta = \infty$,

$$F_z = -6\pi\mu a U \tag{4–20.8}$$

which reproduces Stokes' law. In the case of perfect slip, $\beta = 0$,

$$F_z = -4\pi\mu a U \tag{4–20.9}$$

and the resistance is less than in Stokes' theory by a factor of $\frac{2}{3}$.

Epstein[9] noted in a development from the kinetic point of view that Basset's phenomenological assumption, Eq. (4–20.1), is valid only for the case where the square of the ratio l/a can be neglected (l = mean free path of a gas molecule). He derived an expression for the resistance F_z directly, rather than using hydrodynamic theory for making the computation, and arrived at the simpler expression—see Eq. (2–8.1)

$$F_z = -\, 6\pi\mu a U\left(1 - \frac{\mu}{\beta a}\right) \approx -\frac{6\pi\mu a U}{1 + (\mu/\beta a)} \tag{4–20.10}$$

This formula seems to have more experimental support than Basset's for gases, where l/a is large enough to affect the correction factor but where higher power terms in l/a may be neglected. (See Section 2–8.)

4-21 Fluid Sphere

The translation of a liquid sphere was first treated, independently, by Rybczynski[31] and Hadamard[13]. Interfacial tension acting at the junction between the two immiscible fluids tends to maintain a spherical shape against the shearing stresses which tend to deform it. If the motion is sufficiently slow or the particle sufficiently small, the droplet will be spherical, at least in the first approximation*.

We distinguish between the separate fluid motions occurring inside and outside of the globule by appending the superscripts i and o, respectively, where ambiguity would otherwise result. It is assumed that the sphere is macroscopically at rest while the external fluid streams past it with velocity U in the direction of z negative. This leads to the boundary condition

$$\psi^{(o)} \to \tfrac{1}{2} U r^2 \sin^2\theta \quad \text{as } r \to \infty \tag{4–21.1}$$

The kinematical condition of mutual impenetrability at the interface, $v_r = 0$, requires that we set

$$\psi^{(i)} = 0 \quad \text{at } r = a \tag{4–21.2}$$

and

$$\psi^{(o)} = 0 \quad \text{at } r = a \tag{4–21.3}$$

It is further assumed that the tangential velocity v_θ is continuous across the interface, whereupon

$$\frac{\partial \psi^{(i)}}{\partial r} = \frac{\partial \psi^{(o)}}{\partial r} \quad \text{at } r = a \tag{4–21.4}$$

If we assume that the usual *equilibrium* theory of interfacial tension is applicable to the phenomenon at hand, the sole effect of interfacial tension, γ, is to bring about a discontinuity in the *normal* stress, Π_{rr}, across the interface. This manifests itself by the pressure difference

$$p^{(i)} = p^{(o)} + \frac{2\gamma}{a} \tag{4–21.5}$$

at each point on the sphere surface. The greater pressure in the interior arises from a stretching of the interfacial barrier, in much the same way as the flexible skin of a balloon requires a larger pressure inside than outside. The existence of an interfacial tension does not, therefore, contribute to the *tangential* stress, $\Pi_{r\theta}$, at the interface; hence, the latter must be continuous across the boundary. Employing the methods of the previous section, this condition is expressible, after some simplification, by the relation

*For a calculation of the distortion which occurs when inertial effects are no longer negligible see the treatment of Taylor, T. D., and A. Acrivos, J. Fluid Mech. **18** (1964), 466.

$$\mu_i \frac{\partial}{\partial r}\left(\frac{1}{r^2}\frac{\partial \psi^{(i)}}{\partial r}\right) = \mu_o \frac{\partial}{\partial r}\left(\frac{1}{r^2}\frac{\partial \psi^{(o)}}{\partial r}\right) \quad \text{at } r = a \tag{4–21.6}$$

where μ_i and μ_o are, respectively, the viscosities of the internal and external media.

These six conditions are sufficient to yield a unique solution. For the stream functions appropriate to the external and internal motions, we again avail ourselves of the general solution given by Eq. (4–17.12),

$$\psi^{(o)} = \sin^2\theta\left(\frac{1}{10}Ar^4 - \frac{1}{2}Br + Cr^2 + \frac{D}{r}\right), \quad r \geq a \tag{4–21.7}$$

and

$$\psi^{(i)} = \sin^2\theta\left(\frac{1}{10}Er^4 - \frac{1}{2}Fr + Gr^2 + \frac{H}{r}\right), \quad r \leq a \tag{4–21.8}$$

The condition at infinity immediately requires that we put

$$A = 0, \qquad C = \tfrac{1}{2}U \tag{4–21.9}$$

whereas the finiteness of the component velocities at the origin, $r = 0$, demands that

$$F = 0, \qquad H = 0 \tag{4–21.10}$$

The four arbitrary constants yet to be evaluated can be determined in straightforward fashion by application of the remaining boundary conditions, Eqs. (4–21.2)–(4–21.4), and (4–21.6). Solution of the resultant set of four simultaneous equations yields

$$\begin{aligned} B &= \frac{3}{2}Ua\frac{1 + \frac{2}{3}\sigma}{1 + \sigma}, \qquad & D &= \frac{1}{4}Ua^3\frac{1}{1 + \sigma} \\ E &= \frac{5}{2}\frac{U}{a^2}\frac{\sigma}{1 + \sigma}, \qquad & G &= -\frac{1}{4}U\frac{\sigma}{1 + \sigma} \end{aligned} \tag{4–21.11}$$

where σ is the viscosity ratio μ_o/μ_i.

Typical streamlines for both the internal and external motions are depicted in Fig. 4–21.1. Circulation within the droplet is clearly evident.

The drag on the sphere may be calculated at once from the general expression Eq. (4–17.23), applicable to the external motion. Substituting the foregoing value of B, we find

$$F_z = -6\pi\mu_o aU\frac{1 + \frac{2}{3}\sigma}{1 + \sigma} \tag{4–21.12}$$

It should come as no surprise that the interfacial tension γ does not figure explicitly in any of these results, as it merely alters the pressure at an internal point by the constant amount $2\gamma/a$. Since the dynamic pressure is itself determinate only to within an extraneous additive constant, depending upon the uniform pressure at infinity, it is clear that this additive interfacial factor cannot contribute to the final result.

If in Eq. (4–21.12) we put $\mu_i = \infty$, we reproduce Stokes' law.

For the case of a *gaseous* bubble rising slowly through a liquid we have $\mu_o \gg \mu_i$ and thus

$$F_z \approx -4\pi \mu_o a U \qquad (4\text{–}21.13)$$

This result is identical to that previously given for a sphere at whose surface perfect slip occurs.

It is an experimental fact that though large gaseous bubbles circulate internally, small bubbles behave like solid spheres, displaying terminal rise velocities closely approaching Stokes' law. Levich[21] has shown that this most likely is attributable to surface active agents, present as trace impurities, which accumulate at the interface. These tend to be swept to the rear of the bubble, creating a nonuniform concentration of surface active agent over the bubble surface. This, in turn, results in a gradient in interfacial tension and leads to a system of interfacial stresses. These stresses oppose interfacial motion and thus inhibit internal circulation.

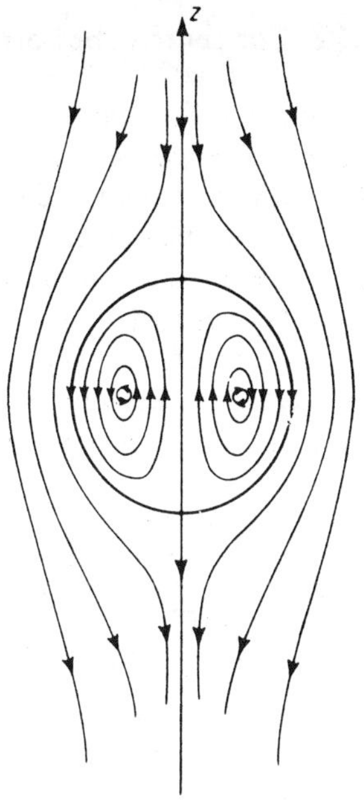

Figure 4-21.1. Streamlines for liquid droplet showing internal circulation.

Earlier, Boussinesq[5] had criticized* the theories advanced by Rybczynski and Hadamard. He argued that, as the interface was undergoing continuous deformation, it was fundamentally wrong to use a theory of capillarity based on principles gleaned from *static* experiments. Rather, the state of stress at the interface should in general depend upon its rate of deformation, in much the same way as the concept of pressure in a static fluid gives way to a more general stress system for moving fluids. Pursuing this line of reasoning, Boussinesq ultimately obtained the following formula for the resistance of a spherical liquid droplet:

$$F_z = -6\pi \mu_o a U \frac{(\epsilon/a) + 2\mu_o + 3\mu_i}{(\epsilon/a) + 3\mu_o + 3\mu_i} \qquad (4\text{–}21.14)$$

where ϵ is in the nature of a "surface viscosity." When ϵ is put equal to zero the results cited previously are obtained.

A general mathematical formalism, appropriate to the dynamics of newtonian fluid-fluid interfaces, has been developed by Scriven[34]. It is reproduced in Aris' book, cited in Chapter 2.

*The analysis is reproduced in the treatise of Dryden, Murnaghan, and Bateman[8].

4-22 Concentric Spheres

The motion of a sphere at the instant it passes the center of a spherical container is of interest as a guide to the importance of wall effects in the motion of a single particle, in addition to providing a model of interactions among particles in unbounded multiparticle systems. A number of authors have given appropriate solutions of the creeping motion equations (neglecting unsteady-state terms) for a variety of boundary conditions.

The radii of the inner and outer spheres are a and b, respectively, and the inner sphere is assumed to remain at rest while the outer sphere moves in the negative z direction with constant velocity U, as in Fig. 4–22.1. Apart from a constant velocity U, the problem is identical to that of a sphere moving with uniform velocity U, in the positive z direction, at the center of an outer sphere which is at rest.

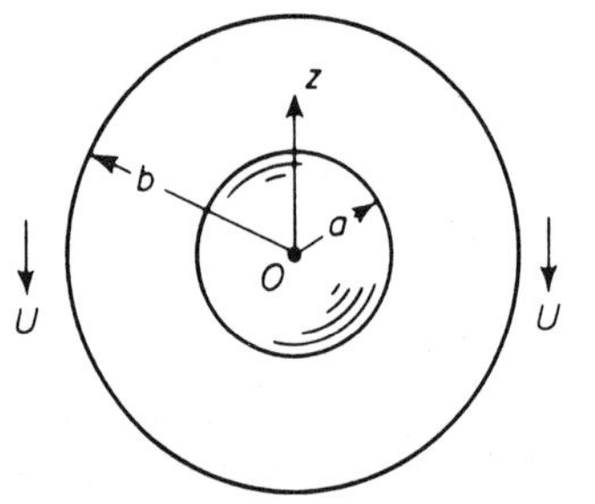

Figure 4-22.1. Concentric spheres in relative motion.

Cunningham[7], Williams[37], and Lee[20] have, independently, presented solutions for the case of a *solid* inner sphere. Haberman and Sayre[12] give the analogous solution for a *liquid* inner sphere. In both cases the outer container is assumed to be rigid, so that fluid adheres to it. We shall follow the development of the liquid sphere, as the corresponding relations for a solid sphere constitute a special case, obtained by letting the viscosity of the liquid sphere become infinite.

As before, we must distinguish between the internal and external motions. The boundary conditions to be satisfied at the surface of the liquid sphere, $r = a$, are identical to Eqs. (4–21.2; 3; 4; 6). To these we must add the conditions

$$\psi^{(o)} - \tfrac{1}{2}Ur^2\sin^2\theta = 0 \tag{4–22.1}$$

and

$$\frac{\partial}{\partial r}\left(\psi^{(o)} - \frac{1}{2}Ur^2\sin^2\theta\right) = 0 \tag{4–22.2}$$

to be satisfied at the outer surface, $r = b$. In accord with Eqs. (4–13.5) and (4–13.7), these describe a uniform motion in the direction of the *negative* z axis.

The stream functions given in Eqs. (4–21.7) and (4–21.8) are suitable for describing the motion under consideration. In order that the internal motion be finite at the origin, we require

$$F = 0, \qquad H = 0 \tag{4–22.3}$$

The remaining constants may be obtained by solving the set of simultaneous equations obtained by satisfaction of the six boundary conditions. After much reduction we find

$$
\begin{aligned}
A &= -\frac{5}{2}\frac{U}{a^2}\left[\left(\frac{3}{2}+\sigma\right)\lambda^3 - \frac{3}{2}\lambda^5\right]\alpha \\
B &= \frac{3}{2}\,Ua\left[1+\frac{2}{3}\sigma-(1-\sigma)\lambda^5\right]\alpha \\
C &= \frac{1}{2}\,U\left[1+\sigma+\frac{5}{4}\lambda^3-\frac{3}{2}\left(\frac{3}{2}-\sigma\right)\lambda^5\right]\alpha \\
D &= \frac{1}{4}\,Ua^3[1-(1-\sigma)\lambda^3]\,\alpha \\
E &= \frac{5}{2}\frac{U}{a^2}\sigma\left(1-\frac{5}{2}\lambda^3+\frac{3}{2}\lambda^5\right)\alpha \\
G &= -\frac{1}{4}\,U\sigma\left(1-\frac{5}{2}\lambda^3+\frac{3}{2}\lambda^5\right)\alpha
\end{aligned}
\tag{4–22.4}
$$

where

$$
\frac{1}{\alpha} = 1+\sigma-\frac{3}{2}\left(\frac{3}{2}+\sigma\right)\lambda+\frac{5}{2}\lambda^3-\frac{3}{2}\left(\frac{3}{2}-\sigma\right)\lambda^5+(1-\sigma)\lambda^6 \tag{4–22.5}
$$

and

$$
\sigma = \frac{\mu_o}{\mu_i} = \frac{\text{external viscosity}}{\text{internal viscosity}} \tag{4–22.6}
$$

in addition to

$$
\lambda = \frac{a}{b} = \frac{\text{inner radius}}{\text{outer radius}} \tag{4–22.7}
$$

These values reduce properly to those of Section 4–21 when the medium is unbounded, corresponding to a diameter ratio $\lambda = 0$.

The drag may again be calculated from Eq. (4–17.23). Having regard to Eq. (4–21.12) we write it in the form:

$$
F_z = -6\pi\mu_o aU\,\frac{1+\frac{2}{3}\sigma}{1+\sigma}\,K \tag{4–22.8}
$$

where the wall correction factor K is defined by

$$
K = \frac{\text{drag in the presence of outer sphere}}{\text{drag in infinite medium}} \tag{4–22.9}
$$

In the present instance,

$$
K = \frac{1-\left(\dfrac{1-\sigma}{1+\frac{2}{3}\sigma}\right)\lambda^5}{1-\dfrac{9}{4}\left(\dfrac{1+\frac{2}{3}\sigma}{1+\sigma}\right)\lambda+\dfrac{5}{2}\left(\dfrac{1}{1+\sigma}\right)\lambda^3-\dfrac{9}{4}\left(\dfrac{1-\frac{2}{3}\sigma}{1+\sigma}\right)\lambda^5+\left(\dfrac{1-\sigma}{1+\sigma}\right)\lambda^6} \tag{4–22.10}
$$

Some special cases of the foregoing are

(1) *Rigid sphere** ($\sigma = 0$):

$$K = \frac{1 - \lambda^5}{1 - (9/4)\lambda + (5/2)\lambda^3 - (9/4)\lambda^5 + \lambda^6} \tag{4–22.11}$$

(2) *Fluid sphere of vanishing viscosity* ($\sigma = \infty$):

$$K = \frac{1 + (3/2)\lambda^5}{1 - (3/2)\lambda + (3/2)\lambda^5 - \lambda^6} \tag{4–22.12}$$

(3) *Fluid sphere with viscosity equal to that of the external medium* ($\sigma = 1$):

$$K = \frac{1}{1 - (15/8)\lambda + (5/4)\lambda^3 - (3/8)\lambda^5} \tag{4–22.13}$$

Values of the wall correction factors for these three cases are shown as a function of λ in Table 4–22.1 and Fig. 4–22.2.

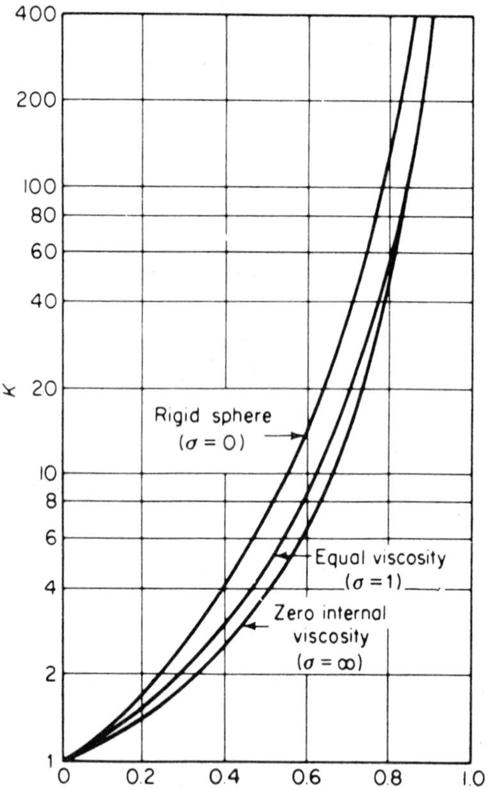

Figure 4-22.2. Wall correction factors for spheres moving in a rigid spherical container.

*This equation agrees with that of Cunningham[7] and Lee[20] except that Cunningham shows a *minus* λ^6 term because of a printing error.

Streamlines for a solid sphere in an enclosure are shown by Williams[37] and compared as far as possible with experiment.

TABLE 4–22.1
WALL CORRECTION FACTORS FOR SPHERES MOVING IN A RIGID SPHERICAL CONTAINER

λ	Solid Spheres	Fluid Spheres	
		$\sigma = 1$	$\sigma = \infty$
		K	
0.0	1.000	1.000	1.000
0.1	1.286	1.229	1.176
0.2	1.756	1.575	1.428
0.3	2.573	2.126	1.815
0.4	4.106	3.066	2.469
0.5	7.294	4.831	3.722
0.6	14.948	8.636	6.569
0.7	37.830	18.762	14.826
0.8	138.224	58.480	50.773
0.9	1209.778	431.779	439.156
1.0	∞	∞	∞

Closely related to the problems previously discussed is Happel's[15] solution of the creeping motion equations for a solid sphere at the center of an outer *frictionless* spherical envelope. When the inner sphere, $r = a$, falls with velocity U, the boundary conditions on the outer envelope, $r = b$, are such that the normal velocity, v_r, and tangential stress, $\Pi_{r\theta}$, vanish. Although not originally formulated in terms of the stream function, the problem is one of axisymmetrical flow to which the general solution given by Eq. (4–17.12) would be applicable. The central result of the calculation is that the drag on the sphere is given by

$$F_z = -6\pi\mu a U K \tag{4–22.14}$$

where the correction to Stokes' law is

$$K = \frac{1 + (2/3)\lambda^5}{1 - (3/2)\lambda + (3/2)\lambda^5 - \lambda^6} \tag{4–22.15}$$

Except for the numerator this is identical to Eq. (4–22.12), where the boundary conditions on the inner and outer spheres are exactly the opposite of those considered by Happel.

4-23 General Solution in Spherical Coordinates

In this section we provide a systematic method of approach to the problem of solving the differential equation of axisymmetric creeping flow in spherical coordinates. The solution is due to Sampson[32]. It is also given, independently, by Savic[33] and Haberman and Sayre[12]. The development outlined here follows that of the latter quite closely. The pertinent equation satisfied by the stream function is

$$E^2(E^2\psi) = 0 \tag{4–23.1}$$

where, in spherical coordinates,

$$E^2 = \frac{\partial^2}{\partial r^2} + \frac{1-\zeta^2}{r^2}\frac{\partial^2}{\partial \zeta^2} \tag{4–23.2}$$

For brevity we have put

$$\zeta = \cos\theta \tag{4–23.3}$$

Solutions of Eq. (4–23.1) may be obtained by setting

$$\psi = \psi^{(1)} + \psi^{(2)} \tag{4–23.4}$$

in which

$$E^2\psi^{(1)} = 0 \tag{4–23.5}$$

and

$$E^2\psi^{(2)} = W \tag{4–23.6}$$

where $W(r, \zeta)$ is itself a solution of the equation

$$E^2 W = 0 \tag{4–23.7}$$

The homogeneous equation (4–23.5) may be solved by separation of variables as follows:

$$\psi^{(1)} = R(r)\, Z(\zeta) \tag{4–23.8}$$

which, upon substitution in Eq. (4–23.5), yields

$$\frac{r^2}{R}\frac{d^2R}{dr^2} + \frac{1-\zeta^2}{Z}\frac{d^2Z}{d\zeta^2} = 0 \tag{4–23.9}$$

As the last term is independent of r, the same must be true of the first term. This can be true only if the lead term is constant, say $n(n-1)$. This leads to the two second-order equations

$$r^2\frac{d^2R}{dr^2} - n(n-1)R = 0 \tag{4–23.10}$$

and

$$(1-\zeta^2)\frac{d^2Z}{d\zeta^2} + n(n-1)Z = 0 \tag{4–23.11}$$

Degenerate cases occur for $n = 0, 1$.

The first of these equations has as its solution

$$R(r) = \alpha_n r^n + \beta_n r^{-n+1} \tag{4–23.12}$$

whereas the latter is Gegenbauer's equation of degree $-\frac{1}{2}$. It is also closely related to Legendre's equation,

$$\frac{d}{d\zeta}\left[(1-\zeta^2)\frac{dY}{d\zeta}\right] + n(n+1)Y = 0 \tag{4–23.13}$$

which has the Legendre functions of the first and second kind, $P_n(\zeta)$ and $Q_n(\zeta)$, as its independent solutions.

As in the case of Legendre's equation, we need only consider integral values of n. Moreover, if we put $n = -v + 1$, then $n(n-1) = v(v-1)$. It follows that we need focus attention on the positive integers, $n \geq 0$. In the present range of interest, $-1 \leq \zeta \leq 1$, we write the independent solutions of Eq. (4–23.11) in the form

$$Z(\zeta) = \nu_n \mathscr{I}_n(\zeta) + \delta_n \mathscr{H}_n(\zeta) \tag{4–23.14}$$

where $\mathscr{I}_n(\zeta)$ and $\mathscr{H}_n(\zeta)$ are, for want of a simple name, termed Gegenbauer functions of the first and second kinds, respectively*.

The properties of these functions have been exhaustively investigated by Sampson[32] in connection with hydrodynamic applications. For our present purposes, their properties can be deduced most readily from their connection with the corresponding Legendre functions of the first and second kinds,†

$$\mathscr{I}_n(\zeta) = \frac{P_{n-2}(\zeta) - P_n(\zeta)}{2n-1}, \qquad \mathscr{H}_n(\zeta) = \frac{Q_{n-2}(\zeta) - Q_n(\zeta)}{2n-1} \quad \text{for } n \geq 2 \tag{4–23.15}$$

the validity of which is easily demonstrated from Eqs. (4–23.11) and (4–23.13). The extraneous factor of $(2n-1)$ is introduced to normalize the functions. The definitions adopted here differ in algebraic sign from those of Sampson, but are identical in all other respects.

In the degenerate cases $n = 0$ and 1, we *define*

$$\begin{aligned} \mathscr{I}_0(\zeta) &= 1, & \mathscr{I}_1(\zeta) &= -\zeta \\ \mathscr{H}_0(\zeta) &= -\zeta, & \mathscr{H}_1(\zeta) &= -1 \end{aligned} \tag{4–23.16}$$

which are obvious solutions of Eq. (4–23.11). For values of $n \geq 2$, we find from Eq. (4–23.15) that the functions of the first kind are given by a finite series of polynomials as follows:

$$\begin{aligned} \mathscr{I}_n(\zeta) = \frac{1\cdot 3 \ldots (2n-3)}{1\cdot 2 \ldots n} \bigg[\zeta^n - \frac{n(n-1)}{2(2n-3)} \zeta^{n-2} \\ + \frac{n(n-1)(n-2)(n-3)}{2\cdot 4\cdot (2n-3)(2n-5)} \zeta^{n-4} - \cdots \bigg] \end{aligned} \tag{4–23.17}$$

$$= -\frac{1}{(n-1)!} \left(\frac{d}{d\zeta}\right)^{n-2} \left(\frac{\zeta^2 - 1}{2}\right)^{n-1} \tag{4–23.18}$$

Some values are

$$\begin{aligned} &\mathscr{I}_0(\zeta) = 1, \quad \mathscr{I}_1(\zeta) = -\zeta, \quad \mathscr{I}_2(\zeta) = \tfrac{1}{2}(1-\zeta^2) \\ &\mathscr{I}_3(\zeta) = \tfrac{1}{2}(1-\zeta^2)\zeta, \quad \mathscr{I}_4(\zeta) = \tfrac{1}{8}(1-\zeta^2)(5\zeta^2 - 1) \\ &\mathscr{I}_5(\zeta) = \tfrac{1}{8}(1-\zeta^2)(7\zeta^2 - 3)\zeta, \ldots \end{aligned} \tag{4–23.19}$$

*More precisely they are Gegenbauer functions of order n and degree $-\frac{1}{2}$.
†For a discussion of these Legendre functions see MacRobert[22].

For the functions of the second kind, we find from Eq. (4–23.15) that, when $n \geq 2$ and $-1 \leq \zeta \leq 1$,

$$\mathscr{H}_n(\zeta) = \frac{1}{2}\mathscr{I}_n(\zeta) \ln \frac{1+\zeta}{1-\zeta} + \mathscr{K}_n(\zeta) \tag{4–23.20}$$

where

$$\mathscr{K}_n(\zeta) = -\sum_{k}^{\frac{1}{2}n \leq k < \frac{1}{2}(n+1)} \frac{(2n-4k+1)}{(2k-1)(n-k)}\left[1 - \frac{(2k-1)(n-k)}{n(n-1)}\right]\mathscr{I}_{n-2k+1}(\zeta) \tag{4–23.21}$$

the series beginning with either $\mathscr{I}_0$ or $\mathscr{I}_1$, according as n is odd or even. Typical values of the latter are

$$\begin{aligned} &\mathscr{K}_2(\zeta) = \tfrac{1}{2}\zeta, \quad \mathscr{K}_3(\zeta) = \tfrac{1}{6}(3\zeta^2 - 2) \\ &\mathscr{K}_4(\zeta) = \tfrac{1}{24}(15\zeta^2 - 13)\zeta, \quad \mathscr{K}_5(\zeta) = \tfrac{1}{120}(105\zeta^4 - 115\zeta^2 + 16) \end{aligned} \tag{4–23.22}$$

The functions of the second kind become infinite along the axis, $\zeta = \pm 1$.

Upon collecting results we find as possible solutions of Eq. (4–23.5)

$$\psi^{(1)} = \sum_{n=0}^{\infty} (A_n r^n + B_n r^{-n+1}) \frac{\mathscr{I}_n}{\mathscr{H}_n}(\zeta) \tag{4–23.23}$$

The same expression, but with different constants, say a_n and b_n, provides a solution to Eq. (4–23.7). The inhomogeneous equation (4–23.6) may also be solved by separation of variables through a substitution of the form

$$\psi^{(2)} = \sum_{n=0}^{\infty} \pi_n(r) \frac{\mathscr{I}_n}{\mathscr{H}_n}(\zeta) \tag{4–23.24}$$

Applying the operator E^2 to the foregoing, taking into account that $\mathscr{I}_n(\zeta)$ and $\mathscr{H}_n(\zeta)$ satisfy the differential equation (4–23.11) gives, as the equation satisfied by $\pi_n(r)$,

$$\frac{d^2\pi_n}{dr^2} - n(n-1)\frac{\pi_n}{r^2} = a_n r^n + b_n r^{-n+1} \tag{4–23.25}$$

of which a particular solution is

$$\pi_n(r) = \frac{a_n r^{n+2}}{2(2n+1)} - \frac{b_n r^{-n+3}}{2(2n-3)} = C_n r^{n+2} + D_n r^{-n+3} \tag{4–23.26}$$

Upon substitution in Eq. (4–23.24) we obtain a particular integral of Eq. (4–23.6).

In accord with Eq. (4–23.4), a complete solution for the stream function in spherical coordinates is of the form

$$\begin{aligned} \psi(r,\theta) &= \sum_{n=0}^{\infty} (A_n r^n + B_n r^{-n+1} + C_n r^{n+2} + D_n r^{-n+3})\mathscr{I}_n(\zeta) \\ &+ \sum_{n=2}^{\infty} (A'_n r^n + B'_n r^{-n+1} + C'_n r^{n+2} + D'_n r^{-n+3})\mathscr{H}_n(\zeta) \end{aligned} \tag{4–23.27}$$

Beginning the last summation at $n = 2$ avoids redundancy, as is apparent from Eq. (4–23.16).

As a particular case of this we have, for $n = 2$,

$$\mathscr{I}_2(\cos\theta) = \tfrac{1}{2}\sin^2\theta \tag{4–23.28}$$

and thus

$$\psi = \frac{1}{2}\sin^2\theta\left(A_2 r^2 + \frac{B_2}{r} + C_2 r^4 + D_2 r\right) \tag{4–23.29}$$

which is identical to Eq. (4–17.12).

The component velocities, obtained from Eq. (4–4.2) and the relations

$$\frac{d\mathscr{I}_n(\zeta)}{d\zeta} = -P_{n-1}(\zeta), \qquad \frac{d\mathscr{H}_n(\zeta)}{d\zeta} = -Q_{n-1}(\zeta) \tag{4–23.30}$$

valid for $n \geq 1$*, are

$$\begin{aligned} v_r = &-\sum_{n=1}^{\infty}(A_n r^{n-2} + B_n r^{-n-1} + C_n r^n + D_n r^{-n+1})\,P_{n-1}(\zeta) \\ &-\sum_{n=2}^{\infty}(A'_n r^{n-2} + B'_n r^{-n-1} + C'_n r^n + D'_n r^{-n+1})\,Q_{n-1}(\zeta) \end{aligned} \tag{4–23.31}$$

and

$$\begin{aligned} v_\theta = &\sum_{n=0}^{\infty}[nA_n r^{n-2} - (n-1)B_n r^{-n-1} + (n+2)C_n r^n \\ &\qquad - (n-3)D_n r^{-n+1}]\frac{\mathscr{I}_n(\zeta)}{\sin\theta} \\ &+\sum_{n=2}^{\infty}[nA'_n r^{n-2} - (n-1)B'_n r^{-n-1} + (n+2)C'_n r^n \\ &\qquad - (n-3)D'_n r^{-n+1}]\frac{\mathscr{H}_n(\zeta)}{\sin\theta} \end{aligned} \tag{4–23.32}$$

The pressure distribution may be obtained by integration of Eq. (4–17.19) for the radial and tangential pressure gradients. Apart from an arbitrary additive constant this results in

$$\begin{aligned} \frac{1}{\mu}p = &-6D_0 Q_0(\zeta) - 6C_1\ln(r\sin\theta) + 2D_1\frac{1}{r} \\ &-\sum_{n=2}^{\infty}\left[\frac{2(2n+1)}{(n-1)}C_n r^{n-1} + \frac{2(2n-3)}{n}D_n r^{-n}\right]P_{n-1}(\zeta) \\ &-\sum_{n=2}^{\infty}\left[\frac{2(2n+1)}{(n-1)}C'_n r^{n-1} + \frac{2(2n-3)}{n}D'_n r^{-n}\right]Q_{n-1}(\zeta) \end{aligned} \tag{4–23.33}$$

As mentioned earlier, the functions of the second kind become infinite along the axis, $\zeta = \pm 1$. In circumstances where a singularity of this type would be prohibited by the physical nature of the problem, the primed constants in Eq. (4–23.27) must vanish for all n. Furthermore, in regard to the unprimed constants, the values $n = 0$ and 1 lead to infinite tangential velocities, v_θ, at $\theta = 0$ and π. In most, but not all, applications to spherical coordinates the stream function will therefore be of the general form

$$\psi(r,\theta) = \sum_{n=2}^{\infty}(A_n r^n + B_n r^{-n+1} + C_n r^{n+2} + D_n r^{-n+3})\mathscr{I}_n(\zeta) \tag{4–23.34}$$

*The term $n = 0$ contributes nothing to v_r since the derivative of $\mathscr{I}_0(\zeta)$ vanishes.

When such is the case, the force exerted by the fluid external to any spherical boundary, $r =$ constant, on the boundary, is given quite generally by the expression

$$F_z = 4\pi\mu D_2 \tag{4–23.35}$$

This follows from an integration of Eq. (4–14.18) over any spherical boundary, and requires the use of the relations

$$\int_{-1}^{+1} \mathscr{I}_n(\zeta)\, d\zeta = \begin{cases} 2 & \text{for } n = 0 \\ \frac{2}{3} & \text{for } n = 2 \\ 0 & \text{for } n \neq 0 \text{ or } 2 \end{cases} \tag{4–23.36}$$

Solutions of Eq. (4–23.34), satisfying *arbitrary* boundary conditions on the surface of a sphere, $r =$ constant, may be obtained by the usual methods of potential theory through the use of orthogonality relations of the type[32]

$$\int_{-1}^{+1} \frac{\mathscr{I}_m(\zeta)\, \mathscr{I}_n(\zeta)}{1 - \zeta^2}\, d\zeta = \begin{cases} 0 & \text{for } m \neq n \\ \dfrac{2}{n(n-1)(2n-1)} & \text{for } m = n \end{cases} \tag{4–23.37}$$

valid when neither m nor n is 0 or 1, and

$$\int_{-1}^{+1} P_n(\zeta) P_m(\zeta)\, d\zeta = \begin{cases} 0 & \text{for } m \neq n \\ \dfrac{2}{2n+1} & \text{for } m = n \end{cases} \tag{4–23.38}$$

These permit the expansion of an arbitrary function of ζ in a series of characteristic functions. For the Gegenbauer functions we have

$$f(\zeta) = \sum_{n=0}^{\infty} a_n \mathscr{I}_n(\zeta) \tag{4–23.39}$$

in which case the orthogonality condition leads to

$$a_n = \frac{1}{2} n(n-1)(2n-1) \int_{-1}^{+1} \frac{f(\zeta)\, \mathscr{I}_n(\zeta)}{1 - \zeta^2}\, d\zeta \tag{4–23.40}$$

The corresponding expansion for the Legendre polynomials is

$$g(\zeta) = \sum_{n=0}^{\infty} b_n P_n(\zeta) \tag{4–23.41}$$

where

$$b_n = \tfrac{1}{2}(2n+1) \int_{-1}^{+1} g(\zeta)\, P_n(\zeta)\, d\zeta \tag{4–23.42}$$

In Section 4–25 flow past an approximate sphere is treated by this method.

4-24 Flow Through a Conical Diffuser

The problem of slow steady flow through a right circular cone may be treated in spherical coordinates.

We shall suppose that the vertex of the cone coincides with the origin, O, and that the axis of the cone is oriented along the z axis. The surface of the cone is taken as the surface of revolution $\theta = \theta_o =$ constant (Fig. 4–24.1). The streamlines are obviously straight lines radiating from the origin and therefore given by the values $\theta =$ constant. From this observation, it follows at once that the stream function must be independent of r, that is,

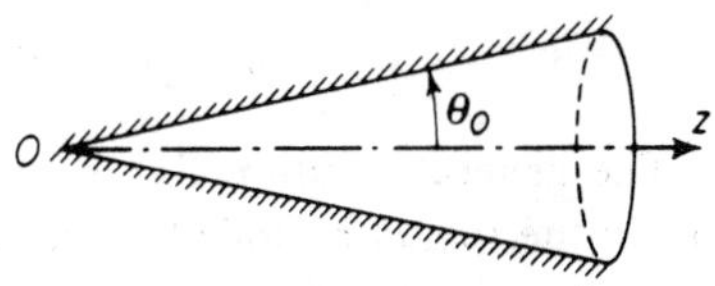

Figure 4-24.1. Flow through a cone.

$$\psi = \psi(\theta) \tag{4–24.1}$$

This is equivalent, of course, to $v_\theta = 0$. We therefore select those solutions of Eq. (4–23.27) which are independent of r. Thus,

$$\begin{aligned}\psi(\zeta) &= A_0 \mathscr{I}_0(\zeta) + B_1 \mathscr{I}_1(\zeta) + D_3 \mathscr{I}_3(\zeta) + D_3' \mathscr{H}_3(\zeta) \\ &= A_0 - B_1\zeta + \frac{1}{2} D_3(1 - \zeta^2)\zeta \\ &\quad + D_3'\left[\frac{1}{4}\zeta(1 - \zeta^2)\ln\left(\frac{1+\zeta}{1-\zeta}\right) + \frac{1}{6}(3\zeta^2 - 2)\right]\end{aligned} \tag{4–24.2}$$

where, as before, $\zeta = \cos\theta$ and $\zeta_o = \cos\theta_o$.

The finiteness of the velocity along the axis requires that we put $D_3' = 0$. The remaining constants are determined from the conditions that the stream function vanish along the axis,

$$\psi = 0 \quad \text{at } \zeta = 1 \tag{4–24.3}$$

and that the velocity v_r vanish at the wall,

$$\frac{d\psi}{d\zeta} = 0 \quad \text{at } \zeta = \zeta_o \tag{4–24.4}$$

In addition to these, the volumetric rate of flow of fluid through the duct in the positive z direction, q, must be prescribed. It follows at once from the definition of the stream function, Eq. (4–2.4), that this requires

$$\psi(\zeta_o) = -\frac{q}{2\pi} \tag{4–24.5}$$

Equations (4–24.3)–(4–24.5) lead to the values

$$A_0 = B_1 = -\frac{q}{2\pi}\frac{1 - 3\zeta_o^2}{(1 + 2\zeta_o)(1 - \zeta_o)^2} \tag{4–24.6}$$

$$D_3 = -\frac{q}{\pi}\frac{1}{(1 + 2\zeta_o)(1 - \zeta_o)^2} \tag{4–24.7}$$

The radial velocity may be obtained from Eq. (4–23.31) with the aid of the relations

$$P_0(\zeta) = 1, \quad P_2(\zeta) = \tfrac{1}{2}(3\zeta^2 - 1)$$

whence
$$v_r = \frac{3q}{2\pi r^2} \frac{\zeta^2 - \zeta_o^2}{(1 + 2\zeta_o)(1 - \zeta_o)^2} \tag{4–24.8}^*$$

For the pressure distribution we obtain from Eq. (4–23.33)

$$p = p_\infty - \frac{\mu q}{\pi r^3} \frac{1 - 3\zeta^2}{(1 + 2\zeta_o)(1 - \zeta_o)^2} \tag{4–24.9}^*$$

The general solution Eq. (4–24.2) is suitable for treating the problem of flow in the region bounded on either side by concentric cones.

A special case of interest occurs for $\theta_o = \frac{1}{2}\pi$, that is, $\zeta_o = 0$. This corresponds to a point source situated in a solid plane wall from which fluid issues at a rate q into the semi-infinite region $z > 0$, as depicted in Fig. 4–24.2. Here,

$$\psi = -\frac{q}{2\pi}(1 - \cos^3\theta) \tag{4–24.10}$$

which may be compared with the expression for a point source in the absence of the wall, Eq. (4–9.3). The present solution gives the streamlines which would exist at *large distances* from a finite-sized hole in a wall through which fluid flowed at rate q.

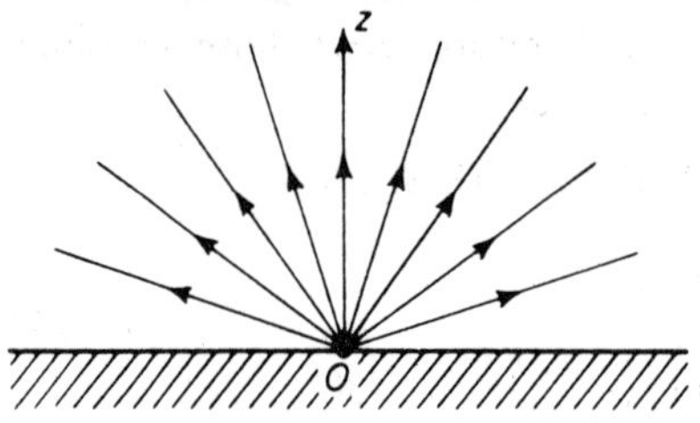

Figure 4-24.2. Point source in a plane wall.

As $\theta_o \to 0$, the cone approximates a circular cylinder and the present results reproduce those for Poiseuille flow in a circular cylinder. We have for small θ_o, since $\theta_o \geq \theta \geq 0$,

$$\zeta = \cos\theta = 1 - \tfrac{1}{2}\theta^2 + \tfrac{1}{24}\theta^4 - \cdots \tag{4–24.11}$$

whence
$$\zeta^2 - \zeta_o^2 \approx \theta_o^2 - \theta^2 \tag{4–24.12}$$

$$(1 + 2\zeta_o)(1 - \zeta_o)^2 \approx \tfrac{3}{4}\theta_o^4 \tag{4–24.13}$$

$$1 - 3\zeta^2 \approx -2 \tag{4–24.14}$$

With these approximations

$$v_r \approx \frac{2q}{\pi(r\theta_o)^4}[(r\theta_o)^2 - (r\theta)^2] \tag{4–24.15}$$

and
$$\frac{\partial p}{\partial r} \approx -\frac{8\mu q}{\pi(r\theta_o)^4} \tag{4–24.16}$$

Putting

$$a = r\sin\theta_o \approx r\theta_o; \qquad \varpi = r\sin\theta \approx r\theta; \qquad z = r\cos\theta \approx r$$

these become

$$v_z = \frac{2q}{\pi a^4}(a^2 - \varpi^2) \tag{4–24.17}$$

and
$$\frac{dp}{dz} = -\frac{8\mu q}{\pi a^4} \tag{4–24.18}$$

in agreement with Eq. (2–5.8).

*These results are in accord with Slezkin's[35].

The most striking aspect of the solution for Stokes flow in a cone is the lack of distinction between converging and diverging flows, $q < 0$ and $q > 0$, respectively. This is not true when inertial effects are taken into account*; for if we replace $\mathbf{v}$ by $-\mathbf{v}$ in the complete Navier-Stokes equations, the inertial terms $\rho\mathbf{v}\cdot\nabla\mathbf{v}$ remain unaltered, whereas the viscous terms $\mu\nabla^2\mathbf{v}$ undergo a change in algebraic sign. Hence, the solutions of the Navier-Stokes equations are not, in general, invariant to a reversal of the direction of net flow.

Exact solutions of the Navier-Stokes equations and of Prandtl's boundary layer equations are available for the analogous two-dimensional problem of plane flow between nonparallel walls[3, 10, 14, 16, 23].

4-25 Flow Past an Approximate Sphere

The problem of symmetrical flow past a spheroid which departs but little in shape from a sphere may be treated, as by Sampson[32], through the general methods of Section 4–23. Further generalization to asymmetrical flows is discussed in Section 5–9. As the equation of the surface we take the polar form $r = c\{1 + f(\theta)\}$. The orthogonality relations, Eqs. (4–23.37) and (4–23.38), assure us that $f(\theta)$ can be developed in the form $\sum_{m=2}^{\infty} \beta_m \mathscr{I}_m(\cos\theta)$ under fairly general circumstances. We therefore consider the surface

$$r = c\{1 + \beta_m \mathscr{I}_m(\zeta)\} \tag{4–25.1}$$

and assume that the coefficients β_m are sufficiently small that squares and higher powers may be neglected; that is,

$$\left(\frac{r}{c}\right)^k \approx 1 + k\beta_m \mathscr{I}_m(\zeta) \tag{4–25.2}$$

where k may be positive or negative.

For streaming flow with velocity U in the negative z direction a suitable form of the stream function is obtained by setting $A_2 = U$, $C_2 = 0$, and $A_n = C_n = 0$ $(n \geq 3)$ in Eq. (4–23.34), and writing it in a form where the remaining coefficients are dimensionless:

$$\begin{aligned} \frac{\psi}{Uc^2} = {} & \left[\left(\frac{r}{c}\right)^2 + B_2\left(\frac{c}{r}\right) + D_2\left(\frac{r}{c}\right)\right]\mathscr{I}_2(\zeta) \\ & + \sum_{n=3}^{\infty}\left[B_n\left(\frac{c}{r}\right)^{n-1} + D_n\left(\frac{c}{r}\right)^{n-3}\right]\mathscr{I}_n(\zeta) \end{aligned} \tag{4–25.3}$$

*For a treatment of the problem of converging flow in an infinite right circular cone at *non-zero* Reynolds numbers see R. C. Ackerberg, J. Fluid Mech. **21** (1965), 47. Also pertinent in this regard are the remarks of S. Goldstein, J. Fluid. Mech. **21** (1965), 33.

This satisfies the condition at infinity. The only coefficients which contribute to the solution of flow past a sphere, treated in Section 4–18, are B_2 and D_2. Since the motion will not be far different from that which occurs when the surface is a perfect sphere, all the coefficients in Eq. (4–25.3) will be of the order of β_m, excepting the aforementioned. Therefore, except when these coefficients enter, we may disregard the departure from a spherical form and set $r = c$ in place of Eq. (4–25.1).

The coefficients B_n, D_n $(n \geq 2)$ are to be determined from the boundary conditions $\psi = 0$ and $\partial\psi/\partial r = 0$ on the surface, Eq. (4–25.1). The first of these leads to

$$0 = (1 + B_2 + D_2)\mathscr{I}_2(\zeta) + (2 - B_2 + D_2)\beta_m \mathscr{I}_m(\zeta)\mathscr{I}_2(\zeta) + \sum_{n=3}^{\infty}(B_n + D_n)\mathscr{I}_n(\zeta) \tag{4–25.4}$$

whereas the second results in

$$0 = (2 - B_2 + D_2)\mathscr{I}_2(\zeta) + (2 + 2B_2)\beta_m \mathscr{I}_m(\zeta)\mathscr{I}_2(\zeta) - \sum_{n=3}^{\infty}[(n-1)B_n + (n-3)D_n]\mathscr{I}_n(\zeta) \tag{4–25.5}$$

The leading terms of Eqs. (4–25.4) and (4–25.5) require that

$$1 + B_2 + D_2 = 0 \quad \text{and} \quad 2 - B_2 + D_2 = 0 \tag{4–25.6}$$

Solving these simultaneously gives the values $B_2 = \frac{1}{2}$, $D_2 = -\frac{3}{2}$ which are, therefore, the same as for a sphere. Upon substituting in Eqs. (4–25.4) and (4–25.5), we now require that

$$0 = \sum_{n=3}^{\infty}(B_n + D_n)\mathscr{I}_n(\zeta) \tag{4–25.7}$$

and

$$0 = 3\beta_m \mathscr{I}_m(\zeta)\mathscr{I}_2(\zeta) - \sum_{n=3}^{\infty}[(n-1)B_n + (n-3)D_n]\mathscr{I}_n(\zeta) \tag{4–25.8}$$

As the $\mathscr{I}_n$ are independent, the first of these can be satisfied only if

$$B_n = -D_n \quad (n \geq 3) \tag{4–25.9}$$

whence we must also have

$$0 = 3\beta_m \mathscr{I}_m(\zeta)\mathscr{I}_2(\zeta) + 2\sum_{n=3}^{\infty} D_n \mathscr{I}_n(\zeta) \tag{4–25.10}$$

To solve the foregoing for D_n we require the identity

$$\mathscr{I}_m \mathscr{I}_2 = -\frac{(m-2)(m-3)}{2(2m-1)(2m-3)}\mathscr{I}_{m-2} + \frac{m(m-1)}{(2m+1)(2m-3)}\mathscr{I}_m - \frac{(m+1)(m+2)}{2(2m-1)(2m+1)}\mathscr{I}_{m+2} \tag{4–25.11}$$

valid for $m \geq 2$. This identity may be established by first demonstrating that

$$(m+1)\mathscr{I}_{m+1}(\zeta) - (2m-1)\zeta\mathscr{I}_m(\zeta) + (m-2)\mathscr{I}_{m-1}(\zeta) = 0 \tag{4–25.12}$$

which follows directly from Eq. (4–23.5) and the recurrence formula[22]

$$(n+1)P_{n+1}(\zeta) - (2n+1)\zeta P_n(\zeta) + nP_{n-1}(\zeta) = 0$$

Upon multiplying Eq. (4–25.12) by ζ and reapplying it to the terms $\zeta \mathscr{I}_{m+1}$ and $\zeta \mathscr{I}_{m-1}$ we find, for $m \neq 0$ or 1,

$$\zeta^2 \mathscr{I}_m(\zeta) = \frac{(m-2)(m-3)}{(2m-1)(2m-3)} \mathscr{I}_{m-2}(\zeta) + \frac{2m^2-2m-3}{(2m+1)(2m-3)} \mathscr{I}_m(\zeta) + \frac{(m+1)(m+2)}{(2m-1)(2m+1)} \mathscr{I}_{m+2}(\zeta) \qquad (4\text{–}25.13)$$

Combining this with the identity $\zeta^2 = 1 - 2\mathscr{I}_2(\zeta)$ proves the relation.

It now follows, upon comparing corresponding terms in Eq. (4–25.10), that all the coefficients D_n are zero except when n has the values $m-2$, m, and $m+2$, in which cases the coefficients are

$$D_{m-2} = \frac{3(m-2)(m-3)}{4(2m-1)(2m-3)} \beta_m \qquad (4\text{–}25.14)$$

$$D_m = -\frac{3m(m-1)}{2(2m+1)(2m-3)} \beta_m \qquad (4\text{–}25.15)$$

$$D_{m+2} = \frac{3(m-2)(m-3)}{4(2m-1)(2m-3)} \beta_m \qquad (4\text{–}25.16)$$

The coefficients B_m may be obtained from these through applications of Eq. (4–25.9).

Collecting results and substituting in Eq. (4–25.3) gives the stream function

$$\begin{aligned}\psi = Uc^2 &\left[\left(\frac{r}{c}\right)^2 + \frac{1}{2}\left(\frac{c}{r}\right) - \frac{3}{2}\left(\frac{r}{c}\right)\right] \mathscr{I}_2(\zeta) \\ &- \frac{3}{4} Uc^2 \frac{(m-2)(m-3)}{(2m-1)(2m-3)} \beta_m \left[\left(\frac{c}{r}\right)^{m-3} - \left(\frac{c}{r}\right)^{m-5}\right] \mathscr{I}_{m-2}(\zeta) \\ &+ \frac{3}{2} Uc^2 \frac{m(m-1)}{(2m+1)(2m-3)} \beta_m \left[\left(\frac{c}{r}\right)^{m-1} - \left(\frac{c}{r}\right)^{m-3}\right] \mathscr{I}_m(\zeta) \\ &- \frac{3}{4} Uc^2 \frac{(m+1)(m+2)}{(2m-1)(2m+1)} \beta_m \left[\left(\frac{c}{r}\right)^{m+1} - \left(\frac{c}{r}\right)^{m-1}\right] \mathscr{I}_{m+2}(\zeta)\end{aligned} \qquad (4\text{–}25.17)$$

From this formula the solution for the case $r = c\,[1 + \sum_m \beta_m \mathscr{I}_m(\cos\theta)]$ can be built up.

As a particular example of this relation consider the *oblate* spheroid

$$\frac{x^2+y^2}{a^2} + \frac{z^2}{a^2(1-\epsilon)^2} = 1 \qquad (4\text{–}25.18)$$

whose equatorial radius is a, in which ϵ is so small that squares and higher powers of it may be neglected. Its polar equation is

$$r = a(1 - \epsilon \cos^2\theta)$$
$$= a\{1 - \epsilon + 2\epsilon \mathscr{I}_2(\zeta)\} \tag{4–25.19}$$

or, if we put

$$c = a(1 - \epsilon) \tag{4–25.20}$$
$$r = c\{1 + 2\epsilon \mathscr{I}_2(\zeta)\} \tag{4–25.21}$$

Upon comparison with Eq. (4–25.1) one is led to the value $m = 2$ and $\beta_2 = 2\epsilon$. Availing ourselves of Eq. (4–25.20), we find from Eq. (4–25.17) that the stream function is given by

$$\psi = Ua^2\left[\left(\frac{r}{a}\right)^2 + \frac{1}{2}\left(1 - \frac{3}{5}\epsilon\right)\left(\frac{a}{r}\right) - \frac{3}{2}\left(1 - \frac{1}{5}\epsilon\right)\left(\frac{r}{a}\right)\right]\mathscr{I}_2(\zeta)$$
$$- \frac{6}{5}Ua^2\epsilon\left[\left(\frac{a}{r}\right)^3 - \left(\frac{a}{r}\right)\right]\mathscr{I}_4(\zeta) \tag{4–25.22}$$

The force on the spheroid may be obtained without difficulty from Eq. (4–14.20) by there setting $\psi_\infty = Ur^2\mathscr{I}_2(\zeta)$. This yields

$$F_z = -6\pi\mu aU(1 - \tfrac{1}{5}\epsilon) \tag{4–25.23}$$

This force is less than would be exerted on a sphere of radius equal to the equatorial radius of the spheroid. That the force here is smaller appears quite reasonable since both the surface area and volume of the oblate spheroid are less than that of the sphere. The relative smallness of this reduction in resistance is not surprising since the polar regions of the sphere contribute least to its resistance; hence, their removal does not have a profound effect on its resistance.

A more appropriate comparison might be made between Eq. (4–25.23) and the resistance of a sphere of equal volume or surface area. The volume of the spheroid Eq. (4–25.18) is $\frac{4}{3}\pi a^3(1 - \epsilon)$. Hence, a sphere of equal volume would have a radius of $a(1 - \frac{1}{3}\epsilon)$ and its resistance would be

$$F_z = -6\pi\mu aU(1 - \tfrac{1}{3}\epsilon) \tag{4–25.24}$$

A sphere of equivalent volume therefore has a smaller resistance than the spheroid. Likewise, since the surface area of the spheroid is $4\pi a^2(1 - \frac{2}{3}\epsilon)$ a sphere of equal area has a radius of $a(1 - \frac{1}{3}\epsilon)$. Thus, the smaller resistance of the sphere holds also on the basis of equal surface area.

Equation (4–25.23) may also be deduced from the *exact* expression for the force on an oblate spheroid, given in Section 4–26 for flow parallel to its axis of revolution. Using this result, we have

$$F_z = -\frac{6\pi\mu aU}{\frac{3}{4}\sqrt{\lambda_o^2 + 1}\,[\lambda_o - (\lambda_o^2 - 1)\cot^{-1}\lambda_o]} \tag{4–25.25}$$

where

$$\lambda_o = \frac{b}{\sqrt{a^2 - b^2}} \tag{4–25.26}$$

As before, a is the equatorial radius and b the polar radius, which in the present instance is $a(1 - \epsilon)$. For large λ_o the expansions,

$$\cot^{-1}\lambda_o = \frac{1}{\lambda_o} - \frac{1}{3\lambda_o^3} + \frac{1}{5\lambda_o^5} - \frac{1}{7\lambda_o^7} + \cdots \tag{4–25.27}$$

and

$$\sqrt{\lambda_o^2 + 1} = \lambda_o\left(1 + \frac{1}{2\lambda_o^2} - \frac{1}{8\lambda_o^4} + \cdots\right) \tag{4–25.28}$$

each valid for $\lambda_o > 1$, ultimately lead to

$$F_z = -6\pi\mu aU(1 - \tfrac{1}{5}\epsilon - \tfrac{33}{175}\epsilon^2 - \cdots) \tag{4–25.29}$$

in accord with Eq. (4–25.23) for small ϵ.

4-26 Oblate Spheroid

We take up here the problem of streaming flow past an oblate spheroid, parallel to its axis of revolution as in Fig. 4–26.1. The spheroid is assumed to remain at rest while fluid streams past it with velocity U in the direction of the negative z axis. On account of the prevailing symmetry, the flow is axisymmetric. The results of this section may also be deduced from Oberbeck's[26] general treatment of the translation of any ellipsoid parallel to a principal axis. A discussion of the latter is presented in Section 5–11. Alternative treatments of axisymmetrical creeping flows past spheroids are given by Sampson[32] and Payne and Pell[27]. Aoi[1] treats the flow of viscous fluids past oblate and prolate spheroids on the basis of the Oseen equations.

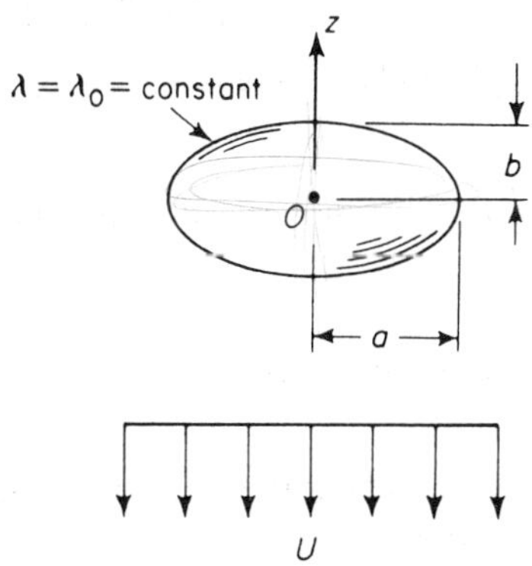

Figure 4-26.1. Flow past an oblate spheroid.

The system of coordinates appropriate to the present problem are oblate spheroidal coordinates, (ξ, η, ϕ), discussed at length in Section A–18. For brevity, we put

$$\lambda = \sinh\xi, \qquad \zeta = \cos\eta \tag{4–26.1}$$

The coordinates surfaces λ (or ξ) = constant are a family of confocal oblate spheroids. We designate the particular spheroid of interest by the value λ_o.

With these substitutions we have, from Eq. (A–18.2),

$$\varpi = c\sqrt{\lambda^2 + 1}\sqrt{1 - \zeta^2}, \qquad z = c\lambda\zeta \tag{4–26.2}$$

The parameters λ and ζ range over the values

$$\infty > \lambda \geq 0, \qquad 1 \geq \zeta \geq -1$$

The following relations obtained from Eq. (4–26.1) prove useful in the sequel:

$$\frac{\partial}{\partial \xi} = \sqrt{\lambda^2 + 1}\frac{\partial}{\partial \lambda}, \quad \frac{\partial}{\partial \eta} = -\sqrt{1 - \zeta^2}\frac{\partial}{\partial \zeta} \tag{4–26.3}$$

It follows from Eqs. (4–13.9) and (4–13.10) that the vanishing of the normal and tangential velocities on the spheroid, $\lambda = \lambda_o$, requires that

$$\psi = 0 \quad \text{for } \lambda = \lambda_o \tag{4–26.4}$$

and

$$\frac{\partial \psi}{\partial \lambda} = 0 \quad \text{for } \lambda = \lambda_o \tag{4–26.5}$$

respectively. In addition to these, the condition of streaming flow at infinity, Eq. (4–13.11), requires that

$$\psi \to \tfrac{1}{2} U c^2 (\lambda^2 + 1)(1 - \zeta^2) \quad \text{as } \psi \to \infty \tag{4–26.6}$$

With the aid of Eq. (4–26.3), the expression for the differential operator E^2, given by Eq. (4–7.20), here takes the form

$$E^2 = \frac{1}{c^2(\lambda^2 + \zeta^2)}\left[(\lambda^2 + 1)\frac{\partial^2}{\partial \lambda^2} + (1 - \zeta^2)\frac{\partial^2}{\partial \zeta^2}\right] \tag{4–26.7}$$

The boundary conditions suggest a trial solution of the form

$$\psi = (1 - \zeta^2)\, g(\lambda) \tag{4–26.8}$$

Upon substitution in the preceding equations, this yields

$$E^2 \psi = \frac{(1 - \zeta^2)}{c^2(\lambda^2 + \zeta^2)} G(\lambda) \tag{4–26.9}$$

where we have set

$$G(\lambda) = (\lambda^2 + 1) g''(\lambda) - 2 g(\lambda) \tag{4–26.10}$$

A second application of the operator E^2 to Eq. (4–26.9) gives, after a lengthy reduction,

$$E^4 \psi = \frac{(\lambda^2 + 1)(1 - \zeta^2)}{c^4(\lambda^2 + \zeta^2)}[4(G - \lambda G') + (\lambda^2 + \zeta^2) G''] \tag{4–26.11}$$

To satisfy the equation of creeping motion,

$$E^4 \psi = 0 \tag{4–26.12}$$

the term in brackets must vanish. But, as G is a function solely of λ, this can occur only if the following relations are *simultaneously* satisfied:

$$G'' = 0$$

$$G - \lambda G' = 0 \tag{4–26.13}$$

These have as their only solution

$$G = C_1 \lambda \tag{4–26.14}$$

Upon substituting in Eq. (4–26.10) we obtain

$$(\lambda^2 + 1)g'' - 2g = C_1\lambda \tag{4–26.15}$$

of which a particular integral is evidently

$$g = -\tfrac{1}{2}C_1\lambda \tag{4–26.16}$$

The remaining part of the solution of Eq. (4–26.15) corresponds to the solution of the *homogeneous* equation

$$(\lambda^2 + 1)g'' - 2g = 0 \tag{4–26.17}$$

which can be written in the form

$$\frac{d}{d\lambda}[(\lambda^2 + 1)g' - 2\lambda g] = 0 \tag{4–26.18}$$

As a first integral of Eq. (4–26.17) we therefore have

$$(\lambda^2 + 1)g' - 2\lambda g = C_2 \tag{4–26.19}$$

or, alternatively,

$$\frac{d}{d\lambda}\left[\frac{g}{\lambda^2 + 1}\right] = \frac{C_2}{(\lambda^2 + 1)^2} \tag{4–26.20}$$

Integration of the foregoing gives

$$g(\lambda) = C_2(\lambda^2 + 1)\int \frac{d\lambda}{(\lambda^2 + 1)^2} + C_3(\lambda^2 + 1) \tag{4–26.21}$$

Upon integrating by parts, we find

$$\int \frac{d\lambda}{(\lambda^2 + 1)^2} = \frac{1}{2}\left(\frac{\lambda}{\lambda^2 + 1} - \cot^{-1}\lambda\right) \tag{4–26.22}$$

The solution of the homogeneous equation is therefore

$$g(\lambda) = \tfrac{1}{2}C_2[\lambda - (\lambda^2 + 1)\cot^{-1}\lambda] + C_3(\lambda^2 + 1) \tag{4–26.23}$$

The general solution of Eq. (4–26.15) may be obtained by summing Eqs. (4–26.16) and (4–26.23). It follows from Eq. (4–26.8) that a possible form for the stream function is

$$\psi = (1 - \zeta^2)\{-\tfrac{1}{2}C_1\lambda + \tfrac{1}{2}C_2[\lambda - (\lambda^2 + 1)\cot^{-1}\lambda] + C_3(\lambda^2 + 1)\} \tag{4–26.24}$$

That part of the preceding solution involving the arbitrary constant C_1 is a solution of $E^4\psi = 0$. The remainder of the solution involving the constants C_2 and C_3 satisfies $E^2\psi = 0$.

The boundary condition at infinity, Eq. (4–26.6), obviously requires that

$$C_3 = \tfrac{1}{2}Uc^2 \tag{4–26.25}$$

To evaluate C_1 and C_2 we utilize the boundary conditions on the spheroid, Eqs. (4–26.4) and (4–26.5). The two simultaneous equations which result have as their solution

$$C_1 = \frac{2Uc^2}{\lambda_o - (\lambda_o^2 - 1)\cot^{-1}\lambda_o} \tag{4–26.26}$$

and

$$C_2 = -\frac{Uc^2(\lambda_o^2 - 1)}{\lambda_o - (\lambda_o^2 - 1)\cot^{-1}\lambda_o} \tag{4–26.27}$$

Upon substituting these constants into Eq. (4–26.24) we finally obtain

$$\psi = \frac{1}{2} U\varpi^2 \left\{1 - \frac{[\lambda/(\lambda^2+1)] - [(\lambda_o^2-1)/(\lambda_o^2+1)]\cot^{-1}\lambda}{[\lambda_o/(\lambda_o^2+1)] - [(\lambda_o^2-1)/(\lambda_o^2+1)]\cot^{-1}\lambda_o}\right\} \tag{4–26.28}$$

where

$$\lambda_o = \sinh \xi_o \tag{4–26.29}$$

The corresponding expression for an oblate spheroid translating with velocity U in the positive z direction may be obtained by subtracting $\frac{1}{2} U\varpi^2$ from the foregoing, in accord with Eq. (4–13.12). The result is

$$\psi = -\frac{1}{2} U\varpi^2 \frac{[\lambda/(\lambda^2+1)] - [(\lambda_o^2-1)/(\lambda_o^2+1)]\cot^{-1}\lambda}{[\lambda_o/(\lambda_o^2+1)] - [(\lambda_o^2-1)/(\lambda_o^2+1)]\cot^{-1}\lambda_o} \tag{4–26.30}$$

The force exerted by the fluid on the spheroid can be obtained from Eq. (4–14.19). In the present instance, since $r \to c\lambda$ at large distances from the spheroid, we have

$$F_z = 8\pi\mu c \lim_{\lambda\to\infty} \frac{\lambda\psi}{\varpi^2} \tag{4–26.31}$$

This limit is easily obtained from Eq. (4–26.30) by availing ourselves of the limiting relations

$$\lim_{\lambda\to\infty} \lambda \cot^{-1}\lambda = 1 \tag{4–26.32}$$

and

$$\lim_{\lambda\to\infty} \frac{\lambda^2}{\lambda^2+1} = 1 \tag{4–26.33}$$

Thus we find

$$F_z = -\frac{8\pi\mu c U}{\lambda_o - (\lambda_o^2-1)\cot^{-1}\lambda_o} \tag{4–26.34}$$

This formula may be expressed in terms of the basic dimensions of the spheroid, a and b, by observing that

$$c = \sqrt{a^2 - b^2} \tag{4–26.35}$$

and

$$\lambda_o = \frac{b}{c} = \left[\left(\frac{a}{b}\right)^2 - 1\right]^{-1/2} \tag{4–26.36}$$

To compare the resistance of the oblate spheroid with a sphere of radius a we write Eq. (4–26.34) in the form

$$F_z = -6\pi\mu a U K \tag{4–26.37}$$

where $K = K(b/a)$ is the correction to Stokes' law given by

$$K = \frac{1}{\frac{3}{4}\sqrt{\lambda_o^2+1}\,[\lambda_o - (\lambda_o^2-1)\cot^{-1}\lambda_o]} \tag{4–26.38}$$

Some values of K are tabulated in Table 4–26.1. These results show the very minor effect of spheroid "thickness" (that is, polar radius).

TABLE 4–26.1
RESISTANCE OF OBLATE AND PROLATE SPHEROIDS EXPRESSED IN TERMS OF THE STOKES' LAW CORRECTION FACTOR FOR A SPHERE HAVING THE SAME EQUATORIAL RADIUS

b/a	K (for oblate spheroid)	K (for prolate spheroid)
0.0	0.84882639	∞
0.1	0.85245060	2.6471358
0.2	0.86145221	1.7848095
0.3	0.87394886	1.4697413
0.4	0.88880656	1.3050489
0.5	0.90530533	1.2039411
0.6	0.92296815	1.1358194
0.7	0.94146887	1.0870324
0.8	0.96057733	1.0505422
0.9	0.98012819	1.0223468
1.0	1	1

4-27 Circular Disk

As $\lambda_0 \to 0$, the oblate spheroid considered in the previous section degenerates to a flat circular disk, infinitesimally thin. In particular, for a disk of radius a moving broadside on, as in Fig. 4–27.1, we find from Eq. (4–26.30) that the stream function is given by

$$\psi = -\frac{U\varpi^2}{\pi}\left(\frac{\lambda}{\lambda^2+1} + \cot^{-1}\lambda\right) \qquad (4\text{–}27.1)$$

For the force on the disk, Eq. (4–26.34) leads to

$$F_z = -16\mu a U \qquad (4\text{–}27.2)$$

Alternative methods for arriving at these results are given by Sampson[32], Ray[28], Roscoe[30], and Gupta[11]. The first of these authors provides detailed drawings of the streamlines and lines of constant pressure.

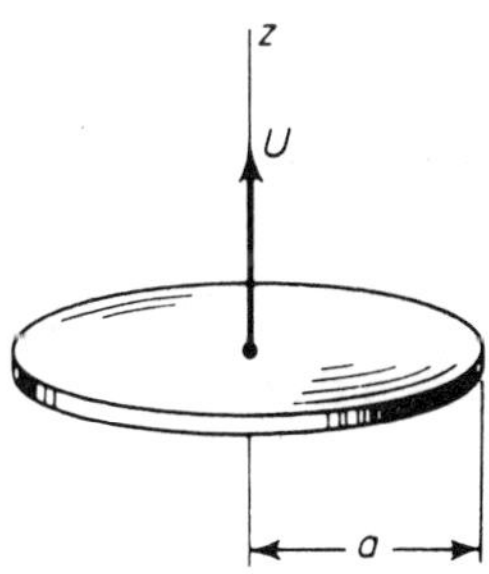

Figure 4-27.1. Translation of a circular disk.

Experiments performed at low Reynolds numbers confirm Eq. (4–27.2). These are discussed in Section 5–11, in the portion dealing with ellipsoids. In practice one cannot, of course, experiment with disks of infinitesimal thickness. The results of the previous section show, however, that even when the "thickness" of the spheroid is as much as one-tenth of its diameter, the coefficient in Eq. (4–27.2) is only increased to 16.07.

4-28 Flow in a Venturi Tube

Oblate spheroidal coordinates are useful in discussing the slow viscous flow of a fluid through a convergent-divergent nozzle having the form of a one-sheeted hyperboloid of revolution—see Fig. A–18.1(c). In particular, the coordinate surfaces η = constant are hyperboloids of this type.

As in Section 4–26, we define

$$\lambda = \sinh \xi, \qquad \zeta = \cos \eta \tag{4–28.1}$$

The surface of the hyperboloid which coincides with the tube is denoted by ζ_o, as in Fig. 4–28.1. We assume, subject to verification, that the stream surfaces coincide with the surfaces of the confocal hyperboloids, ζ = constant; thus

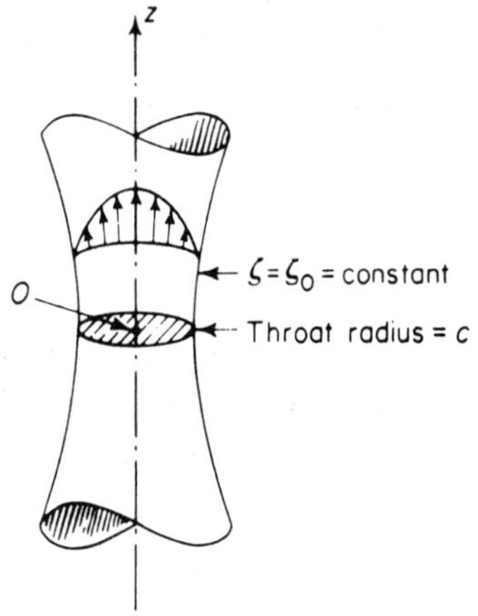

Figure 4-28.1. Flow through a "Venturi tube."

$$\psi = \psi(\zeta) \tag{4–28.2}$$

With this assumption, the normal velocity vanishes identically on the surface of the tube. The boundary condition of no tangential velocity at the tube surface requires that

$$\frac{d\psi}{d\zeta} = 0 \quad \text{for } \zeta = \zeta_o \tag{4–28.3}$$

The normalizing condition that the stream function vanish along the tube axis requires that

$$\psi = 0 \quad \text{for } \zeta = 1 \tag{4–28.4}$$

If q denotes the volumetric flow rate in the positive z direction it then follows from the definition of the stream function that

$$\psi(\zeta_o) = -\frac{q}{2\pi} \tag{4–28.5}$$

where $\psi(\zeta_o)$ denotes the value of the stream function at the tube wall.

One should note here the similarity between the present problem and the analogous problem of flow in a cone, discussed in Section 4–24.

In accord with Eq. (4–28.2) we seek those solutions of the fundamental differential equation

$$E^4\psi = 0 \tag{4–28.6}$$

which are independent of λ. An expression for the operator E^2 in oblate spheroidal coordinates is given by Eq. (4–26.7). Thus, we find

$$E^2\psi = \frac{F(\zeta)}{c^2(\lambda^2 + \zeta^2)} \tag{4–28.7}$$

where we have set

$$F(\zeta) = (1 - \zeta^2)\frac{d^2\psi}{d\zeta^2} \tag{4–28.8}$$

A second application of the operator E^2 leads to

$$\begin{aligned} E^4\psi &= \frac{1}{c^2(\lambda^2 + \zeta^2)}\left[(\lambda^2 + 1)\frac{\partial^2}{\partial\lambda^2} + (1 - \zeta^2)\frac{\partial^2}{\partial\zeta^2}\right]\frac{F(\zeta)}{c^2(\lambda^2 + \zeta^2)} \\ &= \frac{1}{c^4(\lambda^2 + \zeta^2)^4}\Big[\lambda^4\{6F + (1 - \zeta^2)F''\} \\ &\quad + 2\lambda^2\{2F - 2\zeta(1 - \zeta^2)F' + \zeta^2(1 - \zeta^2)F''\} \\ &\quad + \{2\zeta^2(2 - 3\zeta^2)F - 4\zeta^3(1 - \zeta^2)F' + \zeta^4(1 - \zeta^2)F''\}\Big] \end{aligned} \tag{4–28.9}$$

As F is a function of only ζ, satisfaction of Eq. (4–28.6) obviously requires that F simultaneously satisfy the following three differential equations:

$$\begin{gathered} (1 - \zeta^2)F'' + 6F = 0 \\ \zeta^2(1 - \zeta^2)F'' - 2\zeta(1 - \zeta^2)F' + 2F = 0 \\ \zeta^2(1 - \zeta^2)F'' - 4\zeta(1 - \zeta^2)F' + 2(2 - 3\zeta^2)F = 0 \end{gathered} \tag{4–28.10}$$

If the value of F'' from the first equation is substituted into the two remaining equations they yield the same first-order differential equation. Hence, the foregoing relations will be satisfied if the following two equations are *simultaneously* satisfied:

$$\begin{gathered} (1 - \zeta^2)F'' + 6F = 0 \\ \zeta(1 - \zeta^2)F' + (3\zeta^2 - 1)F = 0 \end{gathered} \tag{4–28.11}$$

Upon comparison with Eq. (4–23.11) with $n = 3$, it appears that the first of these equations has the solution

$$F = C_1\mathscr{I}_3(\zeta) + C_2\mathscr{H}_3(\zeta) \tag{4–28.12}$$

The second equation may be integrated directly with the result

$$F = C_3\zeta(1 - \zeta^2) = 2C_3\mathscr{I}_3(\zeta) \tag{4–28.13}$$

The solution which remains finite at $\zeta = 1$ is, therefore,

$$F(\zeta) = \tfrac{1}{2}C_1\zeta(1 - \zeta^2) \tag{4–28.14}$$

Combining this with Eq. (4–28.8) we find as the differential equation satisfied by the stream function

$$\frac{d^2\psi}{d\zeta^2} = \frac{1}{2}C_1\zeta \tag{4–28.15}$$

This has the obvious solution

$$\psi = A + B\zeta + C\zeta^3 \tag{4–28.16}$$

where we have set $C = \frac{1}{12}C_1$.

The three foregoing constants are easily evaluated from the boundary conditions Eqs. (4–28.3)–(4–28.5), the result being

$$A = -\frac{q}{2\pi}\frac{1 - 3\zeta_o^2}{(1 + 2\zeta_o)(1 - \zeta_o)^2}$$

$$B = -\frac{q}{2\pi}\frac{3\zeta_o^2}{(1 + 2\zeta_o)(1 - \zeta_o)^2} \tag{4–28.17}$$

$$C = \frac{q}{2\pi}\frac{1}{(1 + 2\zeta_o)(1 - \zeta_o)^2}$$

This leads to the following expression for the stream function:

$$\psi = \frac{q}{2\pi}\frac{\zeta(\zeta^2 - 3\zeta_o^2) - (1 - 3\zeta_o^2)}{(1 + 2\zeta_o)(1 - \zeta_o)^2} \tag{4–28.18}$$

It is of some interest to calculate the pressure distribution within the tube. The pressure gradients are, from Eqs. (4–15.5) and (4–26.2)–(4–26.3),

$$\frac{\partial p}{\partial \lambda} = \frac{\mu}{c(\lambda^2 + 1)}\frac{\partial}{\partial \zeta}(E^2\psi) \tag{4–28.19}$$

and

$$\frac{\partial p}{\partial \zeta} = -\frac{\mu}{c(1 - \zeta^2)}\frac{\partial}{\partial \lambda}(E^2\psi) \tag{4–28.20}$$

But

$$E^2\psi = \frac{(1 - \zeta^2)}{c^2(\lambda^2 + \zeta^2)}\frac{d^2\psi}{d\zeta^2} = \frac{6C}{c^2}\frac{\zeta(1 - \zeta^2)}{(\lambda^2 + \zeta^2)} \tag{4–28.21}$$

Thus

$$\frac{\partial p}{\partial \zeta} = -\frac{6C\mu}{c^3}\frac{\partial}{\partial \zeta}\left(\frac{\lambda}{\lambda^2 + \zeta^2}\right) \tag{4–28.22}$$

$$\frac{\partial p}{\partial \lambda} = -\frac{6C\mu}{c^3}\frac{\partial}{\partial \lambda}\left(\frac{\lambda}{\lambda^2 + \zeta^2} + \tan^{-1}\lambda\right) \tag{4–28.23}$$

Since,

$$dp = \frac{\partial p}{\partial \lambda}d\lambda + \frac{\partial p}{\partial \zeta}d\zeta \tag{4–28.24}$$

one obtains

$$dp = -\frac{6C\mu}{c^3}d\left(\frac{\lambda}{\lambda^2 + \zeta^2} + \tan^{-1}\lambda\right) \tag{4–28.25}$$

Integration of this expression gives

$$p = p_0 - \frac{3q\mu}{\pi c^3(1 + 2\zeta_o)(1 - \zeta_o)^2}\left(\frac{\lambda}{\lambda^2 + \zeta^2} + \tan^{-1}\lambda\right) \tag{4–28.26}$$

where p_0 is the uniform pressure at the throat of the tube, $\lambda = 0$.

To obtain the pressure drop experienced by a fluid traversing the duct, we note that the pressure at (positive) infinity (corresponding to $\lambda \to \infty$) is

$$p_\infty = p_0 - \frac{3q\mu}{2c^3(1 + 2\zeta_o)(1 - \zeta_o)^2} \tag{4–28.27}$$

On account of symmetry, the pressure drop from $-\infty$ to $+\infty$ is

$$\Delta P = 2(p_0 - p_\infty)$$

whence

$$\Delta P = \frac{3q\mu}{c^3(1 + 2\zeta_0)(1 - \zeta_0)^2} \tag{4–28.28}$$

4-29 Flow Through a Circular Aperture

The problem of flow through a circular hole in a plane wall is solved by setting $\zeta_0 = 0$ in the results of the previous section. The radius of the hole is c, as in Fig. 4–29.1.

For the stream function we obtain

$$\psi = -\frac{q}{2\pi}(1 - \zeta^3)$$

$$= -\frac{q}{2\pi}(1 - \cos^3 \eta) \tag{4–29.1}$$

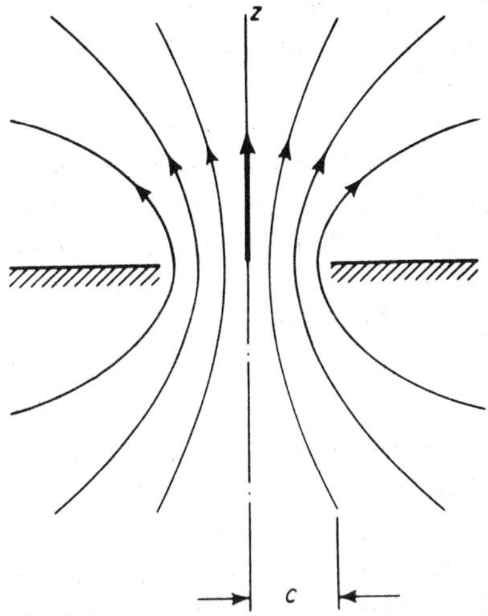

Figure 4-29.1. Flow through a circular orifice.

As before, the streamlines are hyperbolas. At large distances from the opening the solution becomes identical to that for flow from a point source situated in a plane wall, for which the stream function is given by Eq. (4–24.10).

Of particular interest is the pressure drop experienced by the fluid flowing through the aperture, given by

$$\Delta P = \frac{3q\mu}{c^3} \tag{4–29.2}$$

This result was first given by Sampson[32] although a typographical error appears in his final result. An alternative method of solution is given by Roscoe[30], who employs an electrostatic analogy.

The relation between pressure difference and rate of flow has been determined experimentally by different investigators who make no reference to Sampson's theoretical work. Thus, Bond[4] observed the rate of flow of glycerine-water mixtures through a hole 0.1469 cm in diameter drilled in a plate 0.0075 cm thick. He found the pressure difference could be expressed by the formula $(16k/\pi)(q\mu/c^3)$ with $k = 0.631 \pm 0.01$, which is equivalent to Eq. (4–29.2) with the constant equal to 3.21 ± 0.05 instead of the theoretical value of 3. He attempted to correct his result so as to find k for an infinitely thin plane (obtaining a value of 0.580 ± 0.01, which is in good agreement with the present theory), but the correction he employed requires further justification. Johansen[17] studied the flow of castor oil through circular orifices situated within a pipe, the orifices having sharp edges, being beveled at 45° on the low pressure side. The smallest orifice used by him had a diameter of 0.090 times the diameter of the pipe, and it is evident from his results with different orifice diameters that the finite width of the pipe does affect the result in this case. By measurement of the graphical presentation of his results, a figure of 2.99 is obtained versus the theoretical value of 3 in Eq. (4–29.2). The accuracy of his observations is probably about the same as Bond's.

In the results of both writers a departure from the direct proportionality between ΔP and q is observable at high rates of flow, this departure becoming more noticeable in the observations when the dimensionless quantity $q\rho/c\mu$ exceeds 10. Alternatively, if U_m denotes the mean velocity of flow through the hole, we have $q = U_m \pi c^2$ and thus the departure from linearity becomes appreciable when

$$\frac{cU_m\rho}{\mu} > 3.2 \tag{4–29.3}$$

At Reynolds numbers above this value, inertial effects become more pronounced.

The type of behavior predicted by Eq. (4–29.2) differs radically from that observed at higher Reynolds numbers. In particular, for $cU_m\rho/\mu$ greater than about 30,000 the volumetric flow rate is given by an expression of the type[20]

$$q = C_o \pi c^2 \sqrt{\frac{2\,\Delta P}{\rho}} \tag{4–29.4}$$

where C_o is an orifice coefficient, having a value of about 0.61.

4-30 Prolate Spheroid

The motion of a rigid prolate spheroid parallel to its axis of revolution (Fig. 4–30.1) may be calculated by methods very similar to those employed in Section 4–26 for an oblate spheroid. The coordinates appropriate to this problem are prolate spheroidal coordinates, (ξ, η, ϕ), defined by

$$z + i\varpi = c \cosh(\xi + i\eta) \tag{4–30.1}$$

discussed in Section A–17. For conciseness, we set

$$\tau = \cosh \xi, \qquad \zeta = \cos \eta \tag{4–30.2}$$

The surfaces τ = constant are then prolate spheroids. We designate the particular spheroid of interest by τ_0.

Rather than solving the problem anew, we can proceed more directly by availing ourselves of the analogous results for an *oblate* spheroid. In the case of the oblate spheroid we have from Eqs. (A–18.2) and (4–26.1) that

$$z + i\varpi = c(\lambda\zeta + i\sqrt{\lambda^2+1}\sqrt{1-\zeta^2}) \tag{4–30.3}$$

On the other hand it follows from Eqs. (A–17.2) and (4–30.2) that prolate spheroidal coordinates are defined by

$$z + i\varpi = c(\tau\zeta + i\sqrt{\tau^2-1}\sqrt{1-\zeta^2}) \tag{4–30.4}$$

As may now be easily verified, if in Eq. (4–30.3) we replace λ by $i\tau$ and c by $-ic$ we obtain Eq. (4–30.4). It follows then that if we make these two substitutions in the various relations already obtained for an oblate spheroid, λ_o, we shall obtain the solution of the analogous problem for the prolate spheroid, τ_o.

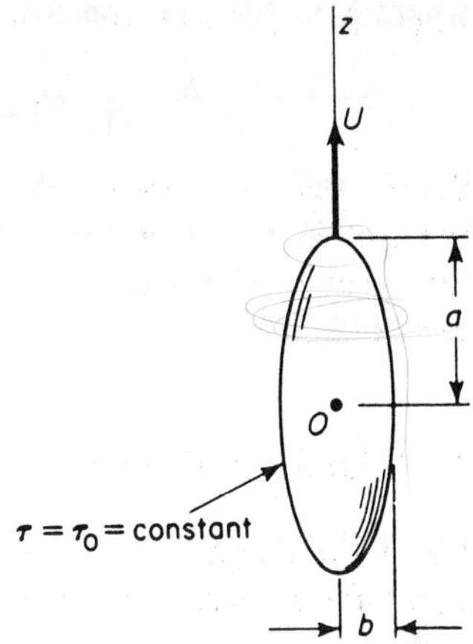

Figure 4-30.1. Translation of a prolate spheroid.

Thus, the stream function for a prolate spheroid translating with velocity U in the positive z direction parallel to its axis of revolution is, from Eq. (4–26.30),

$$\psi = -\frac{1}{2}U\varpi^2 \frac{[(\tau_o^2+1)/(\tau_o^2-1)]\coth^{-1}\tau - [\tau/(\tau^2-1)]}{[(\tau_o^2+1)/(\tau_o^2-1)]\coth^{-1}\tau_o - [\tau_o/(\tau_o^2-1)]} \tag{4–30.5}$$

where we have utilized the relation

$$\cot^{-1}(i\tau) = -i\coth^{-1}\tau \tag{4–30.6}$$

In view of the relation

$$\coth^{-1}\tau = \frac{1}{2}\ln\left(\frac{\tau+1}{\tau-1}\right) \tag{4–30.7}$$

Eq. (4–30.5) may be expressed in alternative form.

In a similar vein, the force on the prolate spheroid is, from Eq. (4–26.34),

$$F_z = -\frac{8\pi\mu c U}{(\tau_o^2+1)\coth^{-1}\tau_o - \tau_o} \tag{4–30.8}$$

where

$$c = \sqrt{a^2-b^2} \tag{4–30.9}$$

a and b being the polar and equatorial radii, respectively, as in Fig. 4–30.1. In terms of the basic dimensions of the spheroid we have

$$\tau_o = \cosh\xi_o = \frac{a}{c} = \left[1-\left(\frac{b}{a}\right)^2\right]^{-1/2} \tag{4–30.10}$$

To compare the resistance of the prolate spheroid with a sphere having the same equatorial radius b, write Eq. (4–30.8) in the form

$$F_z = -6\pi\mu bUK \tag{4–30.11}$$

where $K = K(b/a)$ is the correction to Stokes' law given by

$$K = \frac{1}{\frac{3}{4}\sqrt{\tau_0^2 - 1}\,[(\tau_0^2 + 1)\coth^{-1}\tau_0 - \tau_0]} \tag{4–30.12}$$

Some values of K are given in Table 4–26.1. In comparing these results with those of Section 4–26 for an oblate spheroid, it must be borne in mind that the meaning of a and b is interchanged. In both cases a is the longest of the two semiaxes.

4-31 Elongated Rod

When the major axis, a, of the prolate spheroid is much greater than its equatorial radius, b, the spheroid resembles a long thin rod. For this limiting case

$$\tau_0 = \left[1 - \left(\frac{b}{a}\right)^2\right]^{-1/2} \approx 1 + \frac{1}{2}\left(\frac{b}{a}\right)^2 \tag{4–31.1}$$

and

$$\coth^{-1}\tau_0 = \frac{1}{2}\ln\frac{\tau_0 + 1}{\tau_0 - 1} \approx \ln 2 + \ln\left(\frac{a}{b}\right) \tag{4–31.2}$$

and

$$c \approx a \tag{4–31.3}$$

The force on the rod is then, from Eq. (4–30.8),

$$\begin{aligned} F_z &= -\frac{4\pi\mu a U}{\ln(a/b) + \ln 2 - 1/2} \\ &= -\frac{4\pi\mu a U}{\ln(a/b) + 0.19315} \end{aligned} \tag{4–31.4}$$

Because of the presence of the logarithmic term, the resistance changes but slowly with the ratio a/b.

4-32 Axisymmetric Flow Past a Spherical Cap

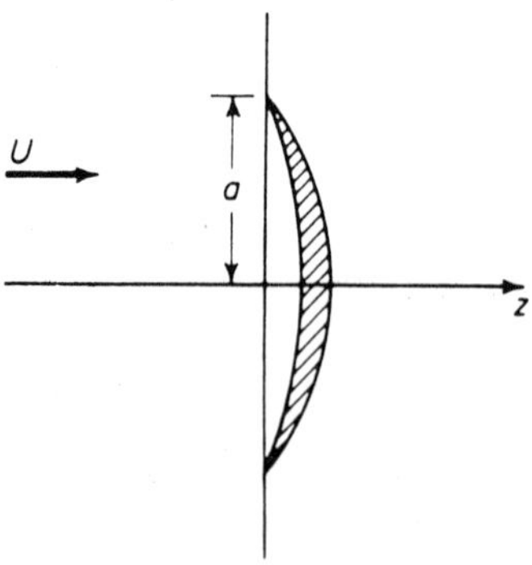

Figure 4-32.1. Flow past a lens-shaped body.

Payne and Pell[27] have obtained general solutions for the case of axisymmetric flow relative to lens-shaped bodies, Fig. 4–32.1. By a suitable limiting process, they derived the stream function and drag coefficient for the flow past a spherical cap from the stream function for flow past a lens-shaped body. Their method involves the use of peripolar coordinates. Numerical evaluation is complicated so that the authors completed calculations only for the case of a spherical cap whose semiangle is $\pi/2$.

A later study by Collins[6] indicates that the value given by Payne and Pell for the drag on a hemispherical cap is incorrect.

Collins employed spherical polar coordinates and found the stream function of the flow by solving two pairs of dual series equations in associated Legendre functions. His results are applicable in general for calculation of the stream function and drag due to a rigid spherical cap, $r = a\,(0 \leq \theta \leq \alpha)$, in a uniform flow of velocity U:

$$F_z = -Ua\mu(6\alpha + 8 \sin \alpha + \sin 2\alpha) \tag{4–32.1}$$

Thus the drag on a sphere ($\alpha = \pi$) reduces to $-6\pi\mu a U$. For a circular disk ($\alpha \to 0$, $a \to \infty$, $a\alpha \to b$, b being the radius of the disk) it is $-16\mu Ub$. The drag on a cap of semiangle $\pi/2$ is $-(3\pi + 8)\,\mu Ua = -17.425\mu Ua$, instead of $-17.52\mu Ua$ obtained by Payne and Pell.

BIBLIOGRAPHY

1. Aoi, T., J. Phys. Soc. Japan **10** (1955), 119; *Ibid.* **10** (1955), 130.
2. Basset, A. B., *A Treatise on Hydrodynamics*, Vol. 2. New York: Dover, 1961.
3. Bateman, H., *Hydrodynamics*, Pt. II. *Nat. Res. Council Bull.* No. 84, 1932.
4. Bond, W. N., Proc. Phys. Soc. **33** (1921), 225; **34** (1922), 139.
5. Boussinesq, J., Compt. Rend. Acad. Sci. (Paris) **156** (1913), 124.
6. Collins, W. D., Mathematika **10** (1963), 72.
7. Cunningham, E., Proc. Royal Soc. (London) **A83** (1910), 357.
8. Dryden, H. L., F. D. Murnaghan, and H. Bateman, *Hydrodynamics*. New York: Dover, 1956.
9. Epstein, P. S., Phys. Rev. **23** (1924), 710.
10. Goldstein, S., *Modern Developments in Fluid Dynamics*, Vol. I. Oxford: Oxford Univ. Press, 1938.
11. Gupta, S. C., ZAMP **8** (1957), 257.
12. Haberman, W. L., and R. M. Sayre, *David W. Taylor Model Basin Report* no. 1143, Washington, D. C., 1958.
13. Hadamard, J. S., Compt. Rend. Acad. Sci. (Paris) **152** (1911), 1735; **154** (1912), 109.
14. Hamel, G., Jahresber. Dt. Math. Ver. **25** (1917), 34; also NACA, Tech. Memo No. 1342.
15. Happel, J., A. I. Ch. E. Jour. **4** (1958), 197.
16. Jeffery, G. B., Proc. Lond. Math. Soc. **14** (1915), 327.
17. Johansen, F. C., Proc. Roy. Soc. **A126** (1930), 231.
18. Kneale, S. G., Ph. D. Thesis, Harvard Univ., Cambridge, Mass., 1953.

19. Lamb, H., *Hydrodynamics*, 6th ed. Cambridge: Cambridge Univ. Press, 1932.
20. Lee, H. M., M. S. Thesis, Univ. of Iowa, Iowa City, Iowa, 1947.
21. Levich, V. G., *Physicochemical Hydrodynamics*. Englewood Cliffs, N. J.: Prentice-Hall, Inc., 1962.
22. MacRobert, T. M., *Spherical Harmonics*, 2nd ed. New York: Dover, 1947.
23. Millsaps, K., and K. Pohlhausen, J. Aero. Sci. **20** (1953), 187.
24. Milne-Thomson, L. M., *Theoretical Hydrodynamics*, 4th ed. New York: Macmillan, 1960.
25. Morse, P. M., and H. Feshbach, *Methods of Theoretical Physics*, Vol. I New York: McGraw-Hill, 1953.
26. Oberbeck, A., J. reine angew. Math. **81** (1876), 62.
27. Payne, L. E., and W. H. Pell, J. Fluid Mech. **7** (1960), 529.
28. Ray, M., Phil. Mag. (ser. 7), **21** (1936), 546.
29. Rouse, H., *Engineering Hydraulics*. New York: Wiley, 1950, p. 35.
30. Roscoe, R., Phil. Mag. **40** (1949), 338.
31. Rybczynski, W., Bull. Acad. Sci. Cracovie (ser. A) (1911), 40.
32. Sampson, R. A., Phil. Trans. Roy. Soc. **A182** (1891), 449.
33. Savic, P., Rept. No. MT-22, Nat. Res. Council Canada (Ottawa), July 31, 1953.
34. Scriven, L. E., Chem. Eng. Sci. **12** (1960), 98.
35. Slezkin, N. A., *Dynamics of Viscous Incompressible Fluids*, (in Russian). Moscow: Gos. Izdat. Tikhi-Teor. Lit., 1955.
36. Stimson, M., and G. B. Jeffery, Proc. Roy. Soc. (London), **A111** (1926), 110.
37. Williams, W. E., Phil. Mag. (6th ser.), **29** (1915), 526.
38. Walton, T. S., "Equations for the numerical determination of the steady flow of a viscous liquid past a solid of revolution," Naval Ordnance Lab., White Oak, Silver Springs, Md. October 19, 1949.
39. Yih, C. S., La Houille Blanche **12**, No. 3, 1957.

The Motion of a Rigid Particle of Arbitrary Shape in an Unbounded Fluid

5

5-1 Introduction

In the previous chapter, formulas were developed for the resistance of axisymmetrical particles moving parallel to their symmetry axes. In such cases, symmetry requires that the hydrodynamic force on the particle be parallel to its direction of motion. Before proceeding to other cases, where it is impossible to take advantage of such symmetry, it is appropriate to investigate the effect of particle orientation on resistance in a general way. This will provide a background for using both theoretical and experimental information on the resistance of particles of more complicated shape. The development of this chapter is along the lines presented by Brenner[6–13].

At sufficiently small Reynolds numbers, it is generally agreed that the magnitude of the force on a rigid particle of arbitrary shape in a streaming flow of viscous fluid is directly proportional to both the fluid viscosity and the magnitude of the free stream velocity. This result follows from an elementary dimensional analysis of the equations of motion and boundary conditions. But dimensional arguments furnish no information as to the relation between the *directions* of the stream velocity vector, **U**, and hydrodynamic force vector, **F**. These vectors are not generally parallel because the body experiences not only drag forces, parallel to the stream velocity, but lateral (lift) forces as well, at right angles to the stream. In the case of a particle settling in a gravitational field, the effect may cause the particle to drift off to one side.

Any general relationship connecting the vector velocity and vector force must take account of the *orientation* of the body with respect to either the stream velocity vector in a streaming flow or the gravity force vector in the case of a settling particle. Gans[21] has discussed the effects of orientation on settling velocity for the particular case of an ellipsoidal particle. More recently, Landau and Lifshitz[33] have presented a general formula for the effect of orientation on the force experienced by an arbitrary body in a streaming flow.

The rotational behavior of a particle is more complicated. If the body has certain well-defined symmetry properties, it may possess a center of hydrodynamic stress. In the absence of external torques, such a body will attain a steady terminal motion in which it merely translates, without rotating, as it settles. Certain particles of skew shape, typically propeller-like bodies, fail to possess such a point and may rotate as they settle in a gravitational field. If such particles experience lateral forces as well, a spiraling downward motion may result. When the spinning particle can change its orientation relative to the direction of gravity, a wobbling motion may also occur.

For cases where the quasi-static form of the creeping motion equations are applicable, the hydrodynamic force and torque (about any origin O) exerted on a rigid particle of *arbitrary shape*, translating and spinning in a fluid at rest at infinity, are dependent upon three fundamental second-rank tensors (dyadics), which are intrinsic geometrical properties of the body. These are

(a) A translation tensor $\mathbf{K}$ (termed resistance tensor in Brenner's[6] first paper; but later, more appropriately, the term *resistance* is reserved for all modes of hydrodynamic interaction). This dyadic is symmetric.

(b) A rotation tensor $\mathbf{\Omega}_O$ which depends on the location of the origin O. This dyadic is symmetric.

(c) A coupling tensor $\mathbf{C}_O$ which depends on the location of the origin O, and represents a coupling of the translational and rotational motions. In general, this dyadic is not symmetric.

A unique point exists for all bodies at which the coupling tensor is symmetric. This point is termed the *center of reaction*, R. If $\mathbf{C}_R = \mathbf{0}$ the point R may be designated a *center of hydrodynamic stress*. If such a center exists, a body falling under the influence of gravity will reach a steady state of motion in which it translates without rotation.

When a rigid particle of arbitrary shape translates and rotates in a fluid the motion is inherently unsteady. However, as we have discussed in Section 2–10, when both the translational and rotational Reynolds numbers

$$N_{\text{Re}}^{(t)} = \frac{cU_o\rho}{\mu}, \quad N_{\text{Re}}^{(r)} = \frac{c^2\omega\rho}{\mu} \tag{5–1.1}$$

are small, it is permissible to adopt the *quasi-static* form of the creeping motion equations,

$$\nabla^2\mathbf{v} = \frac{1}{\mu}\nabla p \tag{5–1.2}$$

$$\nabla\cdot\mathbf{v} = 0 \tag{5–1.3}$$

In Eq. (5–1.1), c is a characteristic particle dimension; U_o is, say, the maximum speed with which any point in the particle moves; and ω is the magnitude of the spin.

As we are primarily interested in the forces and moments resulting from the fluid motion alone, the pressure p in Eq. (5–1.2) will be interpreted here as the *dynamic*, rather than total pressure. The effects of hydrostatic pressure are thereby temporarily ignored.

Let O be *any* point rigidly affixed to the particle and denote by $\mathbf{U}_o$ the instantaneous velocity of this point, and by $\boldsymbol{\omega}$ the instantaneous angular velocity of the particle. Since $\boldsymbol{\omega}$ is a *free* vector which may be displaced parallel to itself, it is not necessary that O lie on the instantaneous axis of rotation of the particle. In view of the no-slip condition at the surface S_p of the particle, the instantaneous boundary condition at the particle surface is

$$\mathbf{v} = \mathbf{U}_o + \boldsymbol{\omega}\times\mathbf{r}_o \quad \text{on } S_p \tag{5–1.4}$$

where $\mathbf{r}_o$ is the position vector of a point relative to an origin at O. As the fluid is at rest at infinity the boundary condition there is

$$\mathbf{v}\to\mathbf{0} \quad \text{as } r_o\to\infty \tag{5–1.5}$$

The linearity of the equations of motion and boundary conditions enables us to write

$$\mathbf{v} = \bar{\mathbf{v}}_o + \bar{\bar{\mathbf{v}}}_o \tag{5–1.6a}$$

$$p = \bar{p}_o + \bar{\bar{p}}_o \tag{5–1.6b}$$

where the single and double overbars refer, respectively, to translational and rotational fields satisfying the following differential equations and boundary conditions:

$$\nabla^2\bar{\mathbf{v}}_o = \frac{1}{\mu}\nabla\bar{p}_o \tag{5–1.7a}$$

$$\nabla\cdot\bar{\mathbf{v}}_o = 0 \tag{5–1.7b}$$

$$\bar{\mathbf{v}}_o = \mathbf{U}_o \quad \text{on } S_p \tag{5–1.8}$$

$$\bar{\mathbf{v}}_o\to\mathbf{0} \quad \text{as } r_o\to\infty \tag{5–1.9}$$

and
$$\nabla^2 \bar{\bar{\mathbf{v}}}_O = \frac{1}{\mu} \nabla \bar{\bar{p}}_O \tag{5–1.10a}$$

$$\nabla \cdot \bar{\bar{\mathbf{v}}}_O = 0 \tag{5–1.10b}$$

$$\bar{\bar{\mathbf{v}}}_O = \boldsymbol{\omega} \times \mathbf{r}_O \quad \text{on } S_p \tag{5–1.11}$$

$$\bar{\bar{\mathbf{v}}}_O \rightarrow \mathbf{0} \quad \text{as } r_O \rightarrow \infty \tag{5–1.12}$$

The subscripts O on the velocity and pressure fields serve to remind that the separate translational and rotational fields, of which $\mathbf{v}$ and p in Eq. (5–1.6) are comprised, both depend on the location of O. The translational field $(\bar{\mathbf{v}}_O, \bar{p}_O)$ depends on the location of O through the dependence of $\mathbf{U}_O$ on O. On the other hand, the rotational field $(\bar{\bar{\mathbf{v}}}_O, \bar{\bar{p}}_O)$ depends on the choice of O via the dependence of $\mathbf{r}_O$ on O.

The hydrodynamic forces exerted by the fluid on the particle by virtue of the individual translational and rotational motions are, respectively,

$$\bar{\mathbf{F}}_O = \int_{S_p} d\mathbf{S} \cdot \bar{\mathbf{\Pi}}_O \tag{5–1.13}$$

and
$$\bar{\bar{\mathbf{F}}}_O = \int_{S_p} d\mathbf{S} \cdot \bar{\bar{\mathbf{\Pi}}}_O \tag{5–1.14}$$

where $d\mathbf{S}$ is directed into the fluid and $\mathbf{\Pi}$ is the pressure tensor. For an incompressible newtonian fluid, the latter is given by the relation [see Eqs. (2–1.8) and (2–1.12)]:

$$\mathbf{\Pi} = -\mathbf{I}p + \mu[\nabla \mathbf{v} + (\nabla \mathbf{v})^{+}] \tag{5–1.15}$$

in which $\mathbf{I}$ is the dyadic idemfactor and $(\nabla \mathbf{v})^{+}$ is the transpose of $\nabla \mathbf{v}$. The *total* force experienced by the particle in consequence of its combined translational and rotational motion is

$$\mathbf{F} = \bar{\mathbf{F}}_O + \bar{\bar{\mathbf{F}}}_O \tag{5–1.16}$$

For a given rigid body motion this force must be independent of the choice of O.

In similar fashion, the hydrodynamic torques about O, which the particle experiences because of its individual translational and rotational motions, are, respectively,

$$\bar{\mathbf{T}}_O = \int_{S_p} \mathbf{r}_O \times (d\mathbf{S} \cdot \bar{\mathbf{\Pi}}_O) \tag{5–1.17}$$

and
$$\bar{\bar{\mathbf{T}}}_O = \int_{S_p} \mathbf{r}_O \times (d\mathbf{S} \cdot \bar{\bar{\mathbf{\Pi}}}_O) \tag{5–1.18}$$

whereas the *total* torque about O is

$$\mathbf{T}_O = \bar{\mathbf{T}}_O + \bar{\bar{\mathbf{T}}}_O \tag{5–1.19}$$

In contrast to $\mathbf{F}$, this torque obviously depends on the choice of O.

5-2 Translational Motions

In consequence of the linearity of the translational equations of motion (5–1.7)–(5–1.9) there exists a dyadic "velocity" field $\bar{\mathbf{V}} = \bar{\mathbf{V}}(\mathbf{r})$ (not symmetric in general) and a vector "pressure" field $\bar{\mathbf{P}} = \bar{\mathbf{P}}(\mathbf{r})$, such that, with respect to translational motions, we may write

$$\bar{\mathbf{v}}_O = \bar{\mathbf{V}}\cdot\mathbf{U}_O \tag{5–2.1a}$$

$$\bar{p}_O = \mu\bar{\mathbf{P}}\cdot\mathbf{U}_O \tag{5–2.1b}$$

At any point in the fluid, $(\bar{\mathbf{V}}, \bar{\mathbf{P}})$ are independent of the viscosity of the fluid, the location of O, and of the magnitude and direction of $\mathbf{U}_O$. Rather, they depend only on the surface geometry of the particle and on the position vector of the fluid point under discussion relative to *any* origin fixed in the particle.

To prove the existence of these higher-order fields we note that, in terms of them, Eqs. (5–1.7)–(5–1.9) may be written in the form:

$$(\nabla^2\bar{\mathbf{V}} - \nabla\bar{\mathbf{P}})\cdot\mathbf{U}_O = \mathbf{0}$$

$$(\nabla\cdot\bar{\mathbf{V}})\cdot\mathbf{U}_O = 0$$

$$(\bar{\mathbf{V}} - \mathbf{I})\cdot\mathbf{U}_O = \mathbf{0} \quad \text{on } S_p$$

$$\bar{\mathbf{V}}\cdot\mathbf{U}_O \to \mathbf{0} \quad \text{as } r \to \infty$$

As the vector $\mathbf{U}_O$ is *arbitrary*, it follows from the definition of equality of vectors and dyadics that these relations imply

$$\nabla^2\,\bar{\mathbf{V}} = \nabla\,\bar{\mathbf{P}}$$

$$\nabla\cdot\bar{\mathbf{V}} = \mathbf{0}$$

$$\bar{\mathbf{V}} = \mathbf{I} \quad \text{on } S_p$$

$$\bar{\mathbf{V}} \to \mathbf{O} \text{ as } r \to \infty$$

That this system of equations is independent of μ, $\mathbf{U}_O$, and O proves the contention.

As a simple example, we note that for a sphere of radius a, the Stokes' "velocity" field is[8] (see also Section 4–17)

$$\bar{\mathbf{V}} = \frac{3}{4}a\left(\frac{\mathbf{rr}}{r^3} + \frac{\mathbf{I}}{r}\right) - \frac{1}{4}a^3\left(\frac{3\mathbf{rr}}{r^5} - \frac{\mathbf{I}}{r^3}\right)$$

and
$$\bar{\mathbf{P}} = \frac{3}{2}a\frac{\mathbf{r}}{r^3}$$

where $\mathbf{r}$ is measured from the sphere center.

From Eqs. (5–1.15) and (5–2.1) we obtain*

$$\bar{\mathbf{\Pi}}_o = \mu[-\mathbf{I}\bar{\mathbf{P}} + \nabla\bar{\mathbf{V}} + {}^{+}(\nabla\bar{\mathbf{V}})]\cdot\mathbf{U}_o$$

Thus, there exists a triadic "stress" field, $\bar{\mathscr{P}} = \bar{\mathscr{P}}(\mathbf{r})$, defined by the relation

$$\bar{\mathbf{\Pi}}_o = \mu\bar{\mathscr{P}}\cdot\mathbf{U}_o \tag{5–2.2}$$

where

$$\bar{\mathscr{P}} = -\mathbf{I}\bar{\mathbf{P}} + \nabla\bar{\mathbf{V}} + {}^{+}(\nabla\bar{\mathbf{V}})$$

Like $\bar{\mathbf{V}}$ and $\bar{\mathbf{P}}$, this triadic field depends solely on the geometry of the particle. Though it is not essential to our arguments, we note as a matter of interest that $\bar{\mathscr{P}} = {}^{+}\bar{\mathscr{P}}$, which is a consequence of the symmetry of the pressure tensor.

In consequence of Eq. (5–2.2), Eqs. (5–1.13) and (5–1.17) may be written in the forms

$$\bar{\mathbf{F}}_o = \mu\int_{S_p} d\mathbf{S}\cdot\bar{\mathscr{P}}\cdot\mathbf{U}_o \tag{5–2.3}$$

and

$$\bar{\mathbf{T}}_o = \mu\int_{S_p} \mathbf{r}_o \times (d\mathbf{S}\cdot\bar{\mathscr{P}})\cdot\mathbf{U}_o \tag{5–2.4}$$

Thus, if we define the *constant* dyadics

$$\mathbf{K} = -\int_{S_p} d\mathbf{S}\cdot\bar{\mathscr{P}} \tag{5–2.5}$$

and

$$\mathbf{C}_o = -\int_{S_p} \mathbf{r}_o \times (d\mathbf{S}\cdot\bar{\mathscr{P}}) \tag{5–2.6}$$

equations (5–2.3) and (5–2.4) become

$$\bar{\mathbf{F}}_o = -\mu\mathbf{K}\cdot\mathbf{U}_o \tag{5–2.7}$$

$$\bar{\mathbf{T}}_o = -\mu\mathbf{C}_o\cdot\mathbf{U}_o \tag{5–2.8}$$

It is reasonable to term $\mathbf{K}$ the translation tensor. It is an intrinsic geometric property of the body, dependent solely on its size and shape. In particular, it is independent of the orientation and velocity of the body and of the fluid properties. It has the dimensions of length, and it uniquely characterizes the resistance of the body to translational motions at small Reynolds numbers.

As we shall now demonstrate, the translation tensor is symmetric. Consider a particle translating with either the velocity $\mathbf{U}_o'$ or $\mathbf{U}_o''$. Both velocities are arbitrary as to magnitude and direction. The velocity fields arising from these motions satisfy the boundary conditions [see Eqs. (5–1.8) and (5–1.9)]

*With regard to transposes of polyadics we adopt the convention that the transposition operator $^{+}$ has the following properties:

$$^{+}(\mathbf{a\,b\,c\,d}\ldots) = \mathbf{b\,a\,c\,d}\ldots$$

and

$$(\ldots\mathbf{a\,b\,c\,d})^{+} = \ldots\mathbf{a\,b\,d\,c}$$

where $\mathbf{a}$, $\mathbf{b}$, ... may be any vectors.

$$\bar{\mathbf{v}}'_O = \begin{cases} \mathbf{U}'_O & \text{on } S_p \\ \mathbf{0} & \text{at } \infty \end{cases}$$

$$\bar{\mathbf{v}}''_O = \begin{cases} \mathbf{U}''_O & \text{on } S_p \\ \mathbf{0} & \text{at } \infty \end{cases}$$

Let $\bar{\mathbf{\Pi}}'_O$ and $\bar{\mathbf{\Pi}}''_O$, respectively, be the pressure tensors arising from these two motions. From the reciprocal theorem Eq. (3–5.1) we then have

$$\int_S \bar{\mathbf{v}}''_O \cdot \bar{\mathbf{\Pi}}'_O \cdot d\mathbf{S} = \int_S \bar{\mathbf{v}}'_O \cdot \bar{\mathbf{\Pi}}''_O \cdot d\mathbf{S}$$

Choose S to be the closed surface consisting of the surface S_p of the particle and a spherical surface S_∞ of indefinitely large radius containing the particle near its center. Now, any Stokesian velocity field arising from the translational motion of a particle through a fluid at rest at infinity can be shown to decay like r^{-1} at sufficiently large distances r from the particle. The creeping motion equations (5–1.2) show that the pressure field must then decay like r^{-2}. Hence, the newtonian stress relation (5–1.15) shows that the pressure tensor decays like r^{-2}. But $d\mathbf{S}$ behaves like r^2. Accordingly, the integrals of $\bar{\mathbf{v}}''_O \cdot \bar{\mathbf{\Pi}}'_O \cdot d\mathbf{S}$ and $\bar{\mathbf{v}}'_O \cdot \bar{\mathbf{\Pi}}''_O \cdot d\mathbf{S}$ over S_∞ are of $O(r^{-1})$, and vanish as we let the radius of the outer sphere tend to infinity. The reciprocal theorem therefore adopts the form

$$\int_{S_p} \bar{\mathbf{v}}''_O \cdot \bar{\mathbf{\Pi}}'_O \cdot d\mathbf{S} = \int_{S_p} \bar{\mathbf{v}}'_O \cdot \bar{\mathbf{\Pi}}''_O \cdot d\mathbf{S} \tag{5–2.9}$$

Utilizing the boundary conditions satisfied by these velocity fields on S_p, we obtain

$$\mathbf{U}''_O \cdot \int_{S_p} \bar{\mathbf{\Pi}}'_O \cdot d\mathbf{S} = \mathbf{U}'_O \cdot \int_{S_p} \bar{\mathbf{\Pi}}''_O \cdot d\mathbf{S}$$

From Eq. (5–1.13) and the symmetry of the pressure tensor, it is clear that these integrals are precisely the hydrodynamic forces $\bar{\mathbf{F}}'_O$ and $\bar{\mathbf{F}}''_O$, respectively, arising from the translational motions. Hence,

$$\mathbf{U}''_O \cdot \bar{\mathbf{F}}'_O - \mathbf{U}'_O \cdot \bar{\mathbf{F}}''_O$$

With the aid of Eq. (5–2.7) this may be written as

$$\mathbf{U}''_O \cdot \mathbf{K} \cdot \mathbf{U}'_O = \mathbf{U}'_O \cdot \mathbf{K} \cdot \mathbf{U}''_O$$

But, from the definition of the transpose of a dyadic, one has the identities

$$\mathbf{U}'_O \cdot \mathbf{K} \cdot \mathbf{U}''_O = \mathbf{U}'_O \cdot (\mathbf{U}''_O \cdot \mathbf{K}) = \mathbf{U}''_O \cdot \mathbf{K}^{\dagger} \cdot \mathbf{U}'_O$$

whence $$\mathbf{U}''_O \cdot \mathbf{K} \cdot \mathbf{U}'_O = \mathbf{U}''_O \cdot \mathbf{K}^{\dagger} \cdot \mathbf{U}'_O$$

As the vectors $\mathbf{U}'_O$ and $\mathbf{U}''_O$ have been chosen arbitrarily, it follows from the definition of equality of two dyadics[17, 22] that

$$\mathbf{K} = \mathbf{K}^{\dagger} \tag{5–2.10}$$

This is precisely the condition that $\mathbf{K}$ be symmetric.

Landau and Lifshitz[33,34] offer an alternative proof of the symmetry of the translation tensor, although it must be remarked that they offer no proof of the *existence* of the latter in the first place. Rather, they assume at the outset that the force on an arbitrary body may be expressed as a linear vector function of its velocity. A complex chain of reasoning, deriving from Onsager's reciprocal relations via the thermodynamics of irreversible processes, is then invoked to prove symmetry of the tensor. Their ingenious proof is quite remarkable in the sense that the fluid itself is never explicitly considered in the analysis, except insofar as its momentary thermodynamic state is assumed to be uniquely determined by the instantaneous position and velocity of the particle. In particular, the usual equations of fluid dynamics are never invoked.* For the unsteady motions analyzed by them, the assumption that the momentary thermodynamic state of their fluid-particle system is uniquely determined by the instantaneous position and velocity of the particle is tantamount to neglecting both the convective and local acceleration terms in the equations of fluid motion, and assuming the fluid to be incompressible. Otherwise the state of the fluid at any instant would also depend upon the past history of the particle motion. One may therefore look upon these results as providing indirect confirmation of Onsager's relations.†

We shall call the dyadic $\mathbf{C}_O$ the *coupling tensor* at O. It depends only upon the exterior geometry of the particle and upon the location of O. To determine the manner in which $\mathbf{C}$ varies with origin, we note that since $\bar{\mathscr{P}}$ is independent of the choice of origin, Eq. (5–2.6) may be written as

$$\mathbf{C}_P = -\int_{S_p} \mathbf{r}_P \times (d\mathbf{S}\cdot\bar{\mathscr{P}})$$

where P is any point affixed to the particle. Upon subtracting Eq. (5–2.6) from the foregoing and utilizing Eq. (5–2.5), it easily follows that

$$\mathbf{C}_P = \mathbf{C}_O - \mathbf{r}_{OP} \times \mathbf{K} \tag{5–2.11}$$

where $\mathbf{r}_{OP}$ is the position vector of P relative to an origin at O. Equation (5–2.11) gives the law by which the coupling tensor transforms from point to point.

*Moreover, their analysis does not explicitly take account of the boundary conditions on the particle surface.

†In the first volume of their series, L. D. Landau and E. M. Lifshitz (*Mechanics*, Reading, Mass.: Addison-Wesley, 1960, p. 76) state, in effect, that the symmetry of the translation tensor (or the entire resistance matrix, for that matter) cannot be established by purely mechanical arguments—but, rather, requires the use of statistical physics as embodied in Onsager's principle. This claim is contradicted by the proof offered here, though it must be pointed out that the proof requires that the pressure tensor be symmetric. And symmetry of the latter is not demanded by the general principles of continuum mechanics if allowance is made for the existence of body couples and couple stresses (cf. the first footnote in Section 2–1).

Inasmuch as the dyadic $\mathbf{r}_{OP} \times \mathbf{K}$ is not symmetric, it follows that the coupling tensor cannot, in general, be symmetric. Neither could it be antisymmetric in general. The properties of the coupling tensor are discussed at further length beginning on page 174.

The translation tensor

Let (x_1, x_2, x_3) denote any system of cartesian coordinates fixed in the particle and let $(\mathbf{i}_1, \mathbf{i}_2, \mathbf{i}_3)$ be unit vectors parallel to the coordinate axes. The translation tensor can then be expressed in nonion form,

$$\begin{aligned} \mathbf{K} = {} & \mathbf{i}_1\mathbf{i}_1 K_{11} + \mathbf{i}_1\mathbf{i}_2 K_{12} + \mathbf{i}_1\mathbf{i}_3 K_{13} \\ & + \mathbf{i}_2\mathbf{i}_1 K_{21} + \mathbf{i}_2\mathbf{i}_2 K_{22} + \mathbf{i}_2\mathbf{i}_3 K_{23} \\ & + \mathbf{i}_3\mathbf{i}_1 K_{31} + \mathbf{i}_3\mathbf{i}_2 K_{32} + \mathbf{i}_3\mathbf{i}_3 K_{33} \end{aligned} \tag{5-2.12}$$

Symmetry requires that

$$K_{12} = K_{21}, \qquad K_{23} = K_{32}, \qquad K_{31} = K_{13} \tag{5-2.13}$$

The translation tensor can thus generally be characterized by six independent scalar resistance coefficients. Although the numerical values of these constant coefficients depend upon the particular set of cartesian coordinates selected, the invariant nature of $\mathbf{K}$ transcends any coordinate system.

As an immediate consequence of the properties of symmetric tensors, every *arbitrary* particle must possess at least three mutually perpendicular axes, fixed in it, such that if it be translating without rotation parallel to one of them, it will experience a force only in this direction; that is, there will be no lateral forces. Hereafter, these are referred to as *principal axes of translation*. The K_{ij} coefficients corresponding to this special coordinate system are termed the *principal translational resistances* of the particle.

These principal resistances correspond to the three eigenvalues, K_i $(i = 1, 2, 3)$, of the characteristic equation

$$\det [\mathbf{K} - K\mathbf{I}] = 0 \tag{5-2.14}$$

Thus, if the six scalar resistances, K_{jk} $(j, k = 1, 2, 3)$, be known experimentally in any system of cartesian coordinates, the principal translational resistances may be obtained by solving the third-order determinantal equation,

$$\det \begin{bmatrix} K_{11} - K & K_{12} & K_{13} \\ K_{21} & K_{22} - K & K_{23} \\ K_{31} & K_{32} & K_{33} - K \end{bmatrix} = 0 \tag{5-2.15}$$

for the three roots of the resulting cubic equation in K. Since the characteristic roots of any real symmetric matrix are necessarily real[1, 20, 25], one is assured that the three eigenvalues thus obtained will each be real. Moreover, as shown by the argument which follows, they are essentially positive.

The instantaneous rate, E, at which mechanical energy is dissipated by the translational motion of a settling particle is $-\mathbf{U}\cdot\mathbf{F}$* which, in the present approximation, takes the form $\mu\mathbf{U}\cdot\mathbf{K}\cdot\mathbf{U}$. Thus, if (U_1, U_2, U_3) be the components of $\mathbf{U}$ along the principal axes,

$$E = \mu(K_1U_1^2 + K_2U_2^2 + K_3U_3^2) \tag{5–2.16}$$

Since E and μ are essentially positive, this requires that K_1, K_2, and K_3 each be greater than zero.

From a dynamic, rather than thermodynamic, point of view the significance of this result is that, since

$$\mathbf{U}\cdot\mathbf{F} = |\mathbf{U}|\,|\mathbf{F}|\cos(\mathbf{U}, \mathbf{F}) \tag{5–2.17}$$

the angle between the direction of the hydrodynamic force acting on a translating body and its direction of motion must be obtuse.

The principal axes of translational resistance lie parallel to the three eigenvectors, $\mathbf{e}_i$ $(i = 1, 2, 3)$, of the translation tensor, which are governed by the equation

$$\mathbf{K}\cdot\mathbf{e}_i = K_i\mathbf{e}_i \tag{5–2.18}$$

The solutions of this equation relate each of the eigenvectors to the corresponding unit vectors in Eq. (5–2.12). If, therefore, we write

$$\mathbf{e}_i = \mathbf{i}_1 e_{i1} + \mathbf{i}_2 e_{i2} + \mathbf{i}_3 e_{i3} \tag{5–2.19}$$

the components (e_{i1}, e_{i2}, e_{i3}) of each of the eigenvectors correspond, for each i, to the nontrivial solutions of the following three simultaneous equations:

$$\begin{aligned} e_{i1}(K_{11} - K_i) + e_{i2}K_{12} + e_{i3}K_{13} &= 0 \\ e_{i1}K_{21} + e_{i2}(K_{22} - K_i) + e_{i3}K_{23} &= 0 \\ e_{i1}K_{31} + e_{i2}K_{32} + e_{i3}(K_{33} - K_i) &= 0 \end{aligned} \tag{5–2.20}$$

By these means, the directions of the three principal axes can be determined from a knowledge of the six resistance coefficients, K_{jk}, in any set of cartesian coordinates fixed in the particle.

If the eigenvectors $\mathbf{e}_i$ are *normalized* (that is, are *unit* vectors in the principal axis system), the translation dyadic can be expressed in the symmetric (trinomial) form

$$\mathbf{K} = \mathbf{e}_1\mathbf{e}_1K_1 + \mathbf{e}_2\mathbf{e}_2K_2 + \mathbf{e}_3\mathbf{e}_3K_3 \tag{5–2.21}$$

A spherical particle (radius $= a$), being an isotropic body, represents a degenerate case in which the principal resistances are identical and in which all directions are eigenvectors. From Stokes' law, one has $K_1 = K_2 = K_3 = 6\pi a$; whence,

$$\mathbf{K} = \mathbf{I}6\pi a \tag{5–2.22}$$

*In discussing pure translational motions, we may temporarily drop both the overbar and the subscript O since, in the absence of rotation, all points fixed in the particle move with the same velocity.

Substitution in Eq. (5–2.7) thereby yields

$$\mathbf{F} = -6\pi\mu a\mathbf{U} \tag{5–2.23}$$

which correctly indicates that the force on the body is antiparallel to the velocity vector. Other translationally isotropic bodies are discussed in Section 5–5, Case 4.

In the event that a body is *orthotropic*, that is, possesses three mutually perpendicular symmetry planes (for example, an ellipsoid or right-angled parallelepiped), it is clear from symmetry considerations that the principal axes lie normal to these planes. Translational motion normal to a symmetry plane obviously results in a force parallel to the direction of motion.

5-3 Rotational Motions

The rotational Stokes field, $(\bar{\bar{\mathbf{v}}}_O, \bar{\bar{p}}_O)$, satisfying Eqs. (5–1.10)–(5–1.12), may be expressed in the form

$$\bar{\bar{\mathbf{v}}}_O = \bar{\bar{\mathbf{V}}}_O \cdot \boldsymbol{\omega} \tag{5–3.1a}$$

$$\bar{\bar{p}}_O = \mu \bar{\bar{\mathbf{P}}}_O \cdot \boldsymbol{\omega} \tag{5–3.1b}$$

where the dyadic "velocity" field $\bar{\bar{\mathbf{V}}}_O = \bar{\bar{\mathbf{V}}}_O(\mathbf{r})$, and the vector "pressure" field $\bar{\bar{\mathbf{P}}}_O = \bar{\bar{\mathbf{P}}}_O(\mathbf{r})$ depend, at a given fluid point, only on the size and shape of the particle and the location of O. To prove the existence of these fields we note that, in terms of them, Eqs. (5–1.10)–(5–1.12) can be written in the form

$$\nabla^2 \bar{\bar{\mathbf{V}}}_O - \nabla \bar{\bar{\mathbf{P}}}_O = \mathbf{O}$$

$$\nabla \cdot \bar{\bar{\mathbf{V}}}_O = \mathbf{0}$$

$$\bar{\bar{\mathbf{V}}}_O = \boldsymbol{\epsilon} \cdot \mathbf{r}_O \quad \text{on } S_p$$

$$\bar{\bar{\mathbf{V}}}_O \to \mathbf{O} \quad \text{as } r \to \infty$$

in which $\boldsymbol{\epsilon} = -\mathbf{I} \times \mathbf{I}$ is the isotropic triadic*. That this system of equations is independent of μ and $\boldsymbol{\omega}$ proves the existence of $\bar{\bar{\mathbf{V}}}_O$ and $\bar{\bar{\mathbf{P}}}_O$. The latter depend on the location of O only through the dependence of $\mathbf{r}_O$ on O. As an example, for a sphere of radius a, the velocity and pressure fields satisfying Eqs. (5–1.10)–(5–1.12) are[8]

$$\bar{\bar{\mathbf{v}}}_O = \boldsymbol{\omega} \times \mathbf{r}_O \left(\frac{a}{r}\right)^3 \quad \text{and} \quad \bar{\bar{p}}_O = 0$$

*In terms of any right-handed system of orthogonal unit vectors, $(\mathbf{i}_1, \mathbf{i}_2, \mathbf{i}_3)$, this may be written in the form

$$\boldsymbol{\epsilon} = \sum_{j=1}^{3} \sum_{k=1}^{3} \sum_{l=1}^{3} \mathbf{i}_j \mathbf{i}_k \mathbf{i}_l \epsilon_{jkl}$$

where ϵ_{jkl} is the permutation symbol.

Thus we obtain

$$\bar{\mathbf{v}}_O = \boldsymbol{\epsilon}\cdot\mathbf{r}_O\left(\frac{a}{r}\right)^3$$

$$\bar{\mathbf{P}}_O = \mathbf{0}$$

provided that O is chosen at the center of the sphere.

It follows from Eqs. (5–1.15) and (5–3.1) that the pressure tensor associated with the rotational motion can be expressed in the form

$$\bar{\mathbf{\Pi}}_O = \mu\bar{\mathscr{P}}_O\cdot\boldsymbol{\omega} \tag{5–3.2}$$

where, at a given fluid point, the stress triadic $\bar{\mathscr{P}}_O$ depends only on the geometry of the particle and the location of O. Hence, from Eqs. (5–1.14) and (5–1.18), the force and torque on the particle arising from the rotational motion may be expressed in the forms

$$\bar{\mathbf{F}}_O = -\mu\mathbf{D}_O\cdot\boldsymbol{\omega} \tag{5–3.3}$$

$$\bar{\mathbf{T}}_O = -\mu\mathbf{\Omega}_O\cdot\boldsymbol{\omega} \tag{5–3.4}$$

where

$$\mathbf{D}_O = -\int_{S_p} d\mathbf{S}\cdot\bar{\mathscr{P}}_O \tag{5–3.5}$$

$$\mathbf{\Omega}_O = -\int_{S_p} \mathbf{r}_O \times (d\mathbf{S}\cdot\bar{\mathscr{P}}_O) \tag{5–3.6}$$

The constant dyadics $\mathbf{D}_O$ and $\mathbf{\Omega}_O$ depend only on the size and shape of the particle and the location of O.

The dyadic $\mathbf{\Omega}_O$ is termed the *rotation tensor* at O. We shall, in the next paragraph, discuss its properties. No special name need be assigned to the dyadic $\mathbf{D}_O$; for, as we shall now prove, this dyadic is equal to the transpose of the coupling tensor at O, and so is not an independent resistance parameter. To prove this we again apply the reciprocal theorem—see Eq. (5–2.9)—in the form

$$\int_{S_p} \bar{\mathbf{v}}_O\cdot\bar{\mathbf{\Pi}}_O\cdot d\mathbf{S} = \int_{S_p} \bar{\mathbf{v}}_O\cdot\bar{\mathbf{\Pi}}_O\cdot d\mathbf{S} \tag{5–3.7}$$

As before, we have used order-of-magnitude estimates of the velocity and pressure tensor fields to show that the integrals vanish over the distant surface S_∞. In view of the boundary conditions (5–1.8) and (5–1.11), the symmetry of the pressure tensor, and the vector identity $\mathbf{a}\times\mathbf{b}\cdot\mathbf{c} = \mathbf{a}\cdot\mathbf{b}\times\mathbf{c}$, we may write the foregoing in the form

$$\mathbf{U}_O\cdot\int_{S_p} d\mathbf{S}\cdot\bar{\mathbf{\Pi}}_O = \boldsymbol{\omega}\cdot\int_{S_p} \mathbf{r}_O \times (d\mathbf{S}\cdot\bar{\mathbf{\Pi}}_O)$$

Utilizing Eqs. (5–1.14) and (5–1.17) this becomes

$$\mathbf{U}_O\cdot\bar{\mathbf{F}}_O = \boldsymbol{\omega}\cdot\bar{\mathbf{T}}_O$$

Hence, substituting Eqs. (5–3.3) and (5–2.8) into the preceding, we obtain

$$\mathbf{U}_O\cdot\mathbf{D}_O\cdot\boldsymbol{\omega} = \boldsymbol{\omega}\cdot\mathbf{C}_O\cdot\mathbf{U}_O$$

But, from the definition of the transpose of a dyadic, it follows that

$$\boldsymbol{\omega}\cdot\mathbf{C}_O\cdot\mathbf{U}_O = \mathbf{U}_O\cdot\mathbf{C}_O^{\dagger}\cdot\boldsymbol{\omega}$$

Upon substituting this identity into the preceding expression and invoking the definition of equality of dyadics, it follows at once that

$$\mathbf{D}_O = \mathbf{C}_O^{\dagger} \tag{5–3.8}$$

which proves the contention. Thus, from Eq. (5–3.3), the hydrodynamic force experienced by a particle rotating about any axis through O is

$$\bar{\mathbf{F}}_O = -\mu\mathbf{C}_O^{\dagger}\cdot\boldsymbol{\omega} \tag{5–3.9}$$

The rotation tensor

In contrast to the translation tensor $\mathbf{K}$, the rotation dyadic $\boldsymbol{\Omega}_O$ varies with choice of origin O. In this sense, Eq. (5–3.4) is an analog of the linear relation existing between the angular momentum and angular velocity vectors of a rigid rotating body, the proportionality coefficient there being the moment of inertia tensor[23], which depends upon the choice of origin. The dependence of $\boldsymbol{\Omega}$ on the choice of origin will be established in Section 5–4 — see Eq. (5–4.10).

As we now show, the rotation tensor is symmetric at all points. Consider a particle rotating about any axis through O, which point we regard as being at rest relative to the stationary fluid at infinity. Consider the cases where the particle rotates with angular velocities $\boldsymbol{\omega}'$ or $\boldsymbol{\omega}''$ about axes through O. These velocities are arbitrary with respect to both magnitude and direction. The velocity fields arising from these motions satisfy the boundary conditions—see Eqs. (5-1.11) and (5-1.12):

$$\bar{\mathbf{v}}_O' = \begin{cases} \boldsymbol{\omega}' \times \mathbf{r}_O & \text{on } S_p \\ \mathbf{0} & \text{at } \infty \end{cases}$$

$$\bar{\mathbf{v}}_O'' = \begin{cases} \boldsymbol{\omega}'' \times \mathbf{r}_O & \text{on } S_p \\ \mathbf{0} & \text{at } \infty \end{cases}$$

The corresponding pressure tensors are denoted by $\bar{\boldsymbol{\Pi}}_O'$ and $\bar{\boldsymbol{\Pi}}_O''$, respectively. As in our proof of the symmetry of the translation tensor, we again invoke the reciprocal theorem and argue that the surface integrals over the surface S_∞ at infinity vanish. Thus, we obtain

$$\int_{S_p} \bar{\mathbf{v}}_O''\cdot\bar{\boldsymbol{\Pi}}_O'\cdot d\mathbf{S} = \int_{S_p} \bar{\mathbf{v}}_O'\cdot\bar{\boldsymbol{\Pi}}_O''\cdot d\mathbf{S}$$

Utilizing the boundary conditions satisfied by these velocity fields, the preceding equation adopts the form

$$\boldsymbol{\omega}''\cdot\int_{S_p} \mathbf{r}_O \times \bar{\boldsymbol{\Pi}}_O'\cdot d\mathbf{S} = \boldsymbol{\omega}'\cdot\int_{S_p} \mathbf{r}_O \times \bar{\boldsymbol{\Pi}}_O''\cdot d\mathbf{S}$$

in which we have invoked the vector identity $\mathbf{a}\times\mathbf{b}\cdot\mathbf{c} = \mathbf{a}\cdot\mathbf{b}\times\mathbf{c}$. It is clear from Eq. (5–1.18), coupled with the symmetry of the pressure tensor, that these integrals are precisely the hydrodynamic torques $\bar{\mathbf{T}}'_O$ and $\bar{\mathbf{T}}''_O$, respectively, experienced by the particle in consequence of the two spinning motions characterized by $\boldsymbol{\omega}'$ and $\boldsymbol{\omega}''$. Thus,

$$\boldsymbol{\omega}''\cdot\bar{\mathbf{T}}'_O = \boldsymbol{\omega}'\cdot\bar{\mathbf{T}}''_O$$

With the aid of Eq. (5–3.4) these become

$$\boldsymbol{\omega}''\cdot\boldsymbol{\Omega}_O\cdot\boldsymbol{\omega}' = \boldsymbol{\omega}'\cdot\boldsymbol{\Omega}_O\cdot\boldsymbol{\omega}''$$

Since, by definition of the transpose of a dyadic,

$$\boldsymbol{\omega}'\cdot\boldsymbol{\Omega}_O\cdot\boldsymbol{\omega}'' = \boldsymbol{\omega}''\cdot\boldsymbol{\Omega}_O^{+}\cdot\boldsymbol{\omega}'$$

we obtain

$$\boldsymbol{\omega}''\cdot\boldsymbol{\Omega}_O\cdot\boldsymbol{\omega}' = \boldsymbol{\omega}''\cdot\boldsymbol{\Omega}_O^{+}\cdot\boldsymbol{\omega}'$$

As the vectors $\boldsymbol{\omega}'$ and $\boldsymbol{\omega}''$ are both arbitrary, we conclude from the definition of equality of dyadics that

$$\boldsymbol{\Omega}_O = \boldsymbol{\Omega}_O^{+}$$

This shows that the rotation tensor is symmetric for any choice of origin O.

Since $\boldsymbol{\Omega}_O$ is symmetric, one may speak of the principal axes of rotation at O and the principal resistances to rotation at O. These correspond, of course, to the eigenvectors and eigenvalues of $\boldsymbol{\Omega}_O$.

As an application of the results of the present section consider, say, a cube-shaped body rotating about an axis which passes through its geometric center, G. It is obvious from symmetry considerations that if the cube rotates about one of its three symmetry axes (that is, those axes normal to the faces of the cube which pass through its center) the torque $\mathbf{T}_G$ about its center will be parallel to $\boldsymbol{\omega}$. Hence, these three directions must be eigenvectors of the rotation tensor at G. As these directions are mutually orthogonal, they must obviously be the principal axes of $\boldsymbol{\Omega}$ at the point under discussion. If, therefore, $(\mathbf{e}_1, \mathbf{e}_2, \mathbf{e}_3)$ be unit vectors along each of the symmetry axes (Gx_1, Gx_2, Gx_3) of the cube, one may write

$$\boldsymbol{\Omega}_G = \mathbf{e}_1\mathbf{e}_1\Omega_1 + \mathbf{e}_2\mathbf{e}_2\Omega_2 + \mathbf{e}_3\mathbf{e}_3\Omega_3 \tag{5–3.10}$$

where $(\Omega_1, \Omega_2, \Omega_3)$ are the principal resistances to rotation at G. But rotations about the Gx_1, Gx_2, and Gx_3 axes are clearly indistinguishable. Obviously one must have $\Omega_1 = \Omega_2 = \Omega_3 = \Omega$, say, whereupon it follows that $\boldsymbol{\Omega}_G = \mathbf{I}\,\Omega$. Substitution into Eq. (5–3.4) now yields

$$\mathbf{T}_G = -\mu\Omega\boldsymbol{\omega} \tag{5–3.11}$$

for the couple about the geometric center of the cube.

This relation shows that, in creeping motion, a cube is isotropic with respect to rotation about its center; that is to say, the resistance to rotation is the same about *any* axis which passes through its center. The cube is not isotropic with respect to rotation about any other point.

It will be shown in Section 5–5 that there exist a large variety of symmetric bodies which enjoy the property of being isotropic with respect to rotation about their geometric centers (see Case 4).

A sphere is, of course, the simplest example of a body isotropic with respect to rotation about its center. The couple experienced by a sphere rotating about an axis through its center is[32]

$$\mathbf{T}_G = -8\pi\mu a^3\boldsymbol{\omega} \tag{5–3.12}$$

in which a is the sphere radius. Thus, the rotation tensor at its center is

$$\boldsymbol{\Omega}_G = \mathbf{I}8\pi a^3 \tag{5–3.13}$$

It is interesting to note that a circular disk is isotropic with respect to rotation about its center, whereas it is anisotropic for translational motions. The couple on the disk is the same for broadside rotation about a diameter[28,45] as it is for edgewise rotation about its symmetry axis[21,29], the couple in either case being

$$\mathbf{T}_G = -\tfrac{32}{3}\mu c^3\boldsymbol{\omega} \tag{5–3.14}$$

where c is the disk radius. Hence, the rotation tensor at the disk center is

$$\boldsymbol{\Omega}_G = \mathbf{I}\tfrac{32}{3}c^3 \tag{5–3.15}$$

which shows the isotropy of the disk in this regard.

5-4 Combined Translation and Rotation

Return now to the general case where the particle may simultaneously translate and rotate. In this case the force on the body, given by Eq. (5–1.16), and the torque about O, given by Eq. (5–1.19), may be written as

$$\mathbf{F} = -\mu\mathbf{K}\cdot\mathbf{U}_O - \mu\mathbf{C}_O^{\dagger}\cdot\boldsymbol{\omega} \tag{5–4.1}$$

$$\mathbf{T}_O = -\mu\mathbf{C}_O\cdot\mathbf{U}_O - \mu\boldsymbol{\Omega}_O\cdot\boldsymbol{\omega} \tag{5–4.2}$$

If c is a characteristic particle dimension, then $\mathbf{K}$, $\mathbf{C}_O$, and $\boldsymbol{\Omega}_O$ have the dimensions c, c^2 and c^3, respectively.

The force on the body must, of course, be independent of the choice of O; that is, if in place of O we choose some other reference point P, then $\mathbf{F}$ must have the form

$$\mathbf{F} = -\mu\mathbf{K}\cdot\mathbf{U}_P - \mu\mathbf{C}_P^{\dagger}\cdot\boldsymbol{\omega} \tag{5–4.3}$$

To prove that this indeed follows from Eq. (5–4.1), we note that, for rigid body motions, the velocities of O and P are connected by the relation $\mathbf{U}_O = \mathbf{U}_P - \boldsymbol{\omega}\times\mathbf{r}_{PO}$, or since $\mathbf{r}_{OP} = -\mathbf{r}_{PO}$,

$$\mathbf{U}_O = \mathbf{U}_P + \mathbf{r}_{OP}\times\boldsymbol{\omega} \tag{5–4.4}$$

Furthermore, since $\mathbf{K}$ is symmetric, it follows from Eq. (5–2.11) that

$$\mathbf{C}_O^{\dagger} = \mathbf{C}_P^{\dagger} - \mathbf{K}\times\mathbf{r}_{OP} \tag{5–4.5}$$

Upon substituting Eqs. (5–4.4) and (5–4.5) into Eq. (5–4.1) and utilizing the identity $\mathbf{K}\cdot(\mathbf{r}_{OP} \times \boldsymbol{\omega}) = (\mathbf{K} \times \mathbf{r}_{OP})\cdot\boldsymbol{\omega}$, which is valid for any dyadic $\mathbf{K}$, the correctness of Eq. (5–4.3) follows immediately.

By means of Eqs. (5–4.1) and (5–4.2) we can obtain the law by which the rotation tensor transforms from point to point. If $\mathbf{T}_O$ and $\mathbf{T}_P$ are the respective hydrodynamic torques about O and P when the particle simultaneously translates and rotates, then

$$\mathbf{T}_O = \int_{S_p} \mathbf{r}_O \times (d\mathbf{S}\cdot\mathbf{\Pi}) \tag{5–4.6}$$

and

$$\mathbf{T}_P = \int_{S_p} \mathbf{r}_P \times (d\mathbf{S}\cdot\mathbf{\Pi}) \tag{5–4.7}$$

where $\mathbf{\Pi}$ is the pressure tensor arising from the combined translational and rotational motion. This pressure tensor is independent of the choice of origin. Upon subtracting the latter equations we obtain

$$\mathbf{T}_O - \mathbf{T}_P = \mathbf{r}_{OP} \times \mathbf{F} \tag{5–4.8}$$

in which

$$\mathbf{F} = \int_{S_p} d\mathbf{S}\cdot\mathbf{\Pi} \tag{5–4.9}$$

is the hydrodynamic force on the particle. $\mathbf{F}$ is given by Eq. (5–4.1), $\mathbf{T}_O$ by Eq. (5–4.2), and $\mathbf{T}_P$ by an equation of the form Eq. (5–4.2) in which P appears in place of O. When these are substituted into Eq. (5–4.8) one obtains

$$-\mu(\mathbf{C}_O\cdot\mathbf{U}_O - \mathbf{C}_P\cdot\mathbf{U}_P) - \mu(\mathbf{\Omega}_O - \mathbf{\Omega}_P)\cdot\boldsymbol{\omega} = -\mu\mathbf{r}_{OP} \times (\mathbf{K}\cdot\mathbf{U}_O + \mathbf{C}_O^{\dagger}\cdot\boldsymbol{\omega})$$

$\mathbf{C}_P$ may be eliminated from this equation via Eq. (5–2.11), and $\mathbf{U}_P$ via Eq. (5–4.4). Utilizing some elementary identities to simplify the resulting expression, and taking into account that $\boldsymbol{\omega}$ is an *arbitrary* vector, we ultimately obtain

$$\mathbf{\Omega}_P = \mathbf{\Omega}_O - \mathbf{r}_{OP} \times \mathbf{K} \times \mathbf{r}_{OP} + \mathbf{C}_O \times \mathbf{r}_{OP} - \mathbf{r}_{OP} \times \mathbf{C}_O^{\dagger} \tag{5–4.10}$$

which expresses the dependence of the rotation tensor on position. We note that the dyadics $\mathbf{r}_{OP} \times \mathbf{K} \times \mathbf{r}_{OP}$, and $\mathbf{C}_O \times \mathbf{r}_{OP} - \mathbf{r}_{OP} \times \mathbf{C}_O^{\dagger}$ are each symmetric, as they must be if the rotation tensor is to be symmetric at *all* points.

The coupling tensor

In contrast to the translation and rotation tensors, which are symmetric at all points, the coupling tensor is not generally symmetric. As we shall now prove, however, every particle, irrespective of shape, possesses a unique, intrinsic geometrical point at which the coupling tensor is symmetric. We shall term this point the *center of hydrodynamic reaction* (or, more simply, the *center of reaction*) and denote it by the symbol R.

To prove the existence and uniqueness of this point, let O be any point, as before. In general the coupling tensor will not be symmetric at O, but it can be uniquely decomposed into symmetric and antisymmetric parts as follows:

$$\mathbf{C}_O = \tfrac{1}{2}(\mathbf{C}_O + \mathbf{C}_O^\dagger) + \tfrac{1}{2}(\mathbf{C}_O - \mathbf{C}_O^\dagger) \tag{5-4.11}$$

From the transformation law for the coupling tensor, Eq. (5–2.11), we have at *any* point R

$$\mathbf{C}_R = \mathbf{C}_O - \mathbf{r}_{OR} \times \mathbf{K} \tag{5-4.12}$$

But $\mathbf{r}_{OR} \times \mathbf{K}$ can also be uniquely decomposed into symmetric and antisymmetric parts:

$$\mathbf{r}_{OR} \times \mathbf{K} = \tfrac{1}{2}(\mathbf{r}_{OR} \times \mathbf{K} - \mathbf{K} \times \mathbf{r}_{OR}) + \tfrac{1}{2}(\mathbf{r}_{OR} \times \mathbf{K} + \mathbf{K} \times \mathbf{r}_{OR}) \tag{5-4.13}$$

in which we have taken account of the symmetry of $\mathbf{K}$. The leading term is symmetric and the last term antisymmetric. Hence, if the coupling tensor is to be symmetric at R, we require that $\mathbf{C}_R - \mathbf{C}_R^\dagger = \mathbf{0}$, that is,

$$(\mathbf{C}_O - \mathbf{C}_O^\dagger) - (\mathbf{r}_{OR} \times \mathbf{K} + \mathbf{K} \times \mathbf{r}_{OR}) = \mathbf{0} \tag{5-4.14}$$

Thus, given the nonsymmetric dyadic $\mathbf{C}_O$ and the symmetric dyadic $\mathbf{K}$, we must prove that it is always possible to find a vector $\mathbf{r}_{OR}$ satisfying Eq. (5–4.14).

The solution of Eq. (5–4.14) is readily found to be

$$\mathbf{r}_{OR} = -[(\mathbf{I}:\mathbf{K})\mathbf{I} - \mathbf{K}]^{-1}\cdot\boldsymbol{\epsilon}:\mathbf{C}_O \tag{5-4.15}$$

in which the double-dot notation is that of Gibbs and $\boldsymbol{\epsilon}$ is the alternating, isotropic triadic. As $\mathbf{K}$ is a positive-definite symmetric dyadic, its three eigenvalues, K_1, K_2, K_3, are essentially positive. The determinant

$$\det[(\mathbf{I}:\mathbf{K})\mathbf{I} - \mathbf{K}] = \det\begin{bmatrix} K_2 + K_3 & 0 & 0 \\ 0 & K_3 + K_1 & 0 \\ 0 & 0 & K_1 + K_2 \end{bmatrix}$$
$$= (K_1 + K_2)(K_2 + K_3)(K_3 + K_1)$$

is, therefore, essentially positive and so cannot vanish. Hence, the bracketed dyadic in Eq. (5–4.15) is nonsingular, and we may conclude that $\mathbf{r}_{OR}$ always exists.

For later reference we shall, at this point, resolve $\mathbf{r}_{OR}$ into components in a system of cartesian axes parallel to the principal axes of $\mathbf{K}$. Let $\mathbf{e}_1$, $\mathbf{e}_2$, $\mathbf{e}_3$ be the normalized eigenvectors of $\mathbf{K}$, ordered in such a way that the axes are right-handed in the cyclical order $1 \to 2 \to 3$; that is, $\mathbf{e}_1 \times \mathbf{e}_2 \cdot \mathbf{e}_3 = +1$. Then we may write

$$\mathbf{K} = \mathbf{e}_1\mathbf{e}_1 K_1 + \mathbf{e}_2\mathbf{e}_2 K_2 + \mathbf{e}_3\mathbf{e}_3 K_3$$

and

$$\mathbf{r}_{OR} = \mathbf{e}_1(x_{OR})_1 + \mathbf{e}_2(x_{OR})_2 + \mathbf{e}_3(x_{OR})_3$$

and

$$\begin{aligned}\mathbf{C}_O = {} & \mathbf{e}_1\mathbf{e}_1 C_{11}^{(O)} + \mathbf{e}_1\mathbf{e}_2 C_{12}^{(O)} + \mathbf{e}_1\mathbf{e}_3 C_{13}^{(O)} \\ & + \mathbf{e}_2\mathbf{e}_1 C_{21}^{(O)} + \mathbf{e}_2\mathbf{e}_2 C_{22}^{(O)} + \mathbf{e}_2\mathbf{e}_3 C_{23}^{(O)} \\ & + \mathbf{e}_3\mathbf{e}_1 C_{31}^{(O)} + \mathbf{e}_3\mathbf{e}_2 C_{32}^{(O)} + \mathbf{e}_3\mathbf{e}_3 C_{33}^{(O)}\end{aligned}$$

in which case Eq. (5–4.15) is equivalent to

$$(x_{OR})_1 = -\frac{C_{23}^{(O)} - C_{32}^{(O)}}{K_2 + K_3}, \quad (x_{OR})_2 = -\frac{C_{31}^{(O)} - C_{13}^{(O)}}{K_3 + K_1}, \quad (x_{OR})_3 = -\frac{C_{12}^{(O)} - C_{21}^{(O)}}{K_1 + K_2} \tag{5-4.16}$$

We have demonstrated that all bodies possess at least one point at which the coupling tensor is symmetric. We now prove that there is only one such point. Let P be any point at which the coupling tensor is symmetric. Now, from Eq. (5–2.11),

$$\mathbf{C}_P = \mathbf{C}_R - \mathbf{r}_{RP} \times \mathbf{K}$$

As $\mathbf{C}_R$ is symmetric, in order that $\mathbf{C}_P$ also be symmetric the antisymmetric part of $\mathbf{r}_{RP} \times \mathbf{K}$ must vanish; that is, apart from a factor of one-half, we require that

$$\mathbf{r}_{RP} \times \mathbf{K} - \mathbf{K} \times \mathbf{r}_{RP} = \mathbf{O}$$

or, what is equivalent,

$$[(\mathbf{I} : \mathbf{K})\mathbf{I} - \mathbf{K}] \cdot \mathbf{r}_{RP} = \mathbf{0}$$

Since the bracketed term is a *complete* dyadic, that is,

$$\det[(\mathbf{I} : \mathbf{K})\mathbf{I} - \mathbf{K}] \neq 0$$

then the preceding relation can only be satisfied if

$$\mathbf{r}_{RP} = \mathbf{0}$$

whence P and R must be one and the same point. Thus, for a given body, the center of reaction is a *unique* point.

Since $\mathbf{C}_R = \mathbf{C}_R^\dagger$, the fundamental relations (5–4.1) and (5–4.2) adopt their most symmetric form at R, namely,

$$\mathbf{F} = -\mu(\mathbf{K} \cdot \mathbf{U}_R + \mathbf{C}_R \cdot \boldsymbol{\omega}) \tag{5-4.17a}$$

$$\mathbf{T}_R = -\mu(\mathbf{C}_R \cdot \mathbf{U}_R + \boldsymbol{\Omega}_R \cdot \boldsymbol{\omega}) \tag{5-4.17b}$$

It is convenient to refer to the principal axes of $\mathbf{C}_R$ as the *principal axes of coupling*. These axes are mutually perpendicular and possess the following property: If the body is restrained from either translating or rotating and fluid allowed to stream past it, parallel to a principal axis of coupling, the hydrodynamic torque on the body will be parallel to the stream velocity vector. Conversely, if the body rotates about an axis through R is such a way that $\boldsymbol{\omega}$ is parallel to a principal axis of coupling, and if R is at rest relative to the fluid at infinity, then the hydrodynamic force on the body will be parallel to the angular velocity vector $\boldsymbol{\omega}$.

As we shall discuss in detail in Section 5–5, certain classes of bodies exist whose geometric symmetry is such that $\mathbf{C}_R = \mathbf{O}$. In such cases Eq. (5–4.17) shows that the translational and rotational motions are "uncoupled," and the center of reaction becomes a "center of hydrodynamic stress." The latter then plays a role comparable to that played by the center of mass in rigid body dynamics, in the sense that the hydrodynamic force depends only on the instantaneous translational velocity of R, whereas the hydrodynamic torque (about R) depends only on the instantaneous angular velocity. For such bodies the transformation law for $\mathbf{\Omega}$, Eq. (5–4.10), reduces to

$$\mathbf{\Omega}_P = \mathbf{\Omega}_C - \mathbf{r}_{CP} \times \mathbf{K} \times \mathbf{r}_{CP} \tag{5–4.18}$$

where $C \equiv R$ is the center of hydrodynamic stress. This relationship is similar to the well-known "parallel axis" theorem[23] of mechanics, which relates the moment of inertia tensor at any point to its value at the center of mass of a body.

Energy dissipation

In Stokes flow, where the time rate of change of the kinetic energies of particle and fluid are negligible, and potential energy effects absent, the instantaneous mechanical energy dissipation rate, E, is equal to the instantaneous rate at which the stresses acting over the surfaces bounding the fluid are doing work upon it. Hence,

$$E = -\int_{S_p+S_\infty} d\mathbf{S}\cdot\mathbf{\Pi}\cdot\mathbf{v} \tag{5–4.19}$$

in which the integration extends over the surface S_p of the particle, and over the surface of the fluid at infinity, S_∞. For the translational and rotational motions under consideration, $\mathbf{v}$ is at most of $O(r^{-1})$; hence, $\mathbf{\Pi}$ is at most of $O(r^{-2})$ as $r \to \infty$. Since $d\mathbf{S} = O(r^2)$, the integral over S_∞ vanishes in the limit, and we remain with only the integral over S_p. On this surface, $\mathbf{v}$ is given by Eq. (5–1.4). Hence, since $\mathbf{T}_O$ and $\mathbf{F}$ are given by Eqs. (5–4.6) and (5–4.9), respectively, we obtain

$$E = -\mathbf{U}_O\cdot\mathbf{F} - \boldsymbol{\omega}\cdot\mathbf{T}_O \tag{5–4.20}$$

Substituting Eqs. (5–4.1) and (5–4.2) into the foregoing, we find

$$E = \mu(\mathbf{U}_O\cdot\mathbf{K}\cdot\mathbf{U}_O + 2\boldsymbol{\omega}\cdot\mathbf{C}_O\cdot\mathbf{U}_O + \boldsymbol{\omega}\cdot\mathbf{\Omega}_O\cdot\boldsymbol{\omega}) \tag{5–4.21}$$

Among other things, the invariance of E to a choice of origin provides a simple, alternative method for determining the variation of the coupling and rotation tensors with position. Thus, if P is any point rigidly affixed to the particle, we must have

$$E = \mu(\mathbf{U}_P\cdot\mathbf{K}\cdot\mathbf{U}_P + 2\boldsymbol{\omega}\cdot\mathbf{C}_P\cdot\mathbf{U}_P + \boldsymbol{\omega}\cdot\mathbf{\Omega}_P\cdot\boldsymbol{\omega})$$

If we now substitute Eq. (5–4.4) into Eq. (5–4.21), equate the result to the foregoing, and take into account that $\mathbf{U}_P$ and $\boldsymbol{\omega}$ are *arbitrary* vectors, the transformation laws—Eqs. (5–2.11) and (5–4.10)—easily follow.

As the dissipation rate must be essentially positive, irrespective of the values of $\mathbf{U}_O$ and $\boldsymbol{\omega}$, certain inequalities must exist between the elements of $\mathbf{C}_O$ and those of $\mathbf{K}$ and $\mathbf{\Omega}_O$. To obtain these, consider any system of cartesian coordinates and let $(\mathbf{U}_O)$, $(\boldsymbol{\omega})$, $(\mathbf{F})$, and $(\mathbf{T}_O)$ be column matrices whose scalar elements are the corresponding scalar components of these vectors, and let

$(\mathbf{K})$, $(\mathbf{\Omega}_O)$, and $(\mathbf{C}_O)$ be square 3×3 matrices whose elements are the components of the corresponding dyadics. Then Eqs. (5–4.1) and (5–4.2) may be written in terms of partitioned matrices as follows:

$$\begin{pmatrix} (\mathbf{F}) \\ (\mathbf{T}_O) \end{pmatrix} = -\mu \begin{pmatrix} (\mathbf{K}) & (\mathbf{C}_O)^{\dagger} \\ (\mathbf{C}_O) & (\mathbf{\Omega}_O) \end{pmatrix} \begin{pmatrix} (\mathbf{U}_O) \\ (\boldsymbol{\omega}) \end{pmatrix} \tag{5–4.22}$$

in which $(\mathbf{C}_O)^{\dagger}$ is the transpose of the matrix $(\mathbf{C}_O)$. Since, from Eq. (5–4.20),

$$E = -\{(\mathbf{U}_O)^{\dagger}(\boldsymbol{\omega})^{\dagger}\} \begin{pmatrix} (\mathbf{F}) \\ (\mathbf{T}_O) \end{pmatrix}$$

we obtain

$$E = \mu\{(\mathbf{U}_O)^{\dagger}(\boldsymbol{\omega})^{\dagger}\} \begin{pmatrix} (\mathbf{K}) & (\mathbf{C}_O)^{\dagger} \\ (\mathbf{C}_O) & (\mathbf{\Omega}_O) \end{pmatrix} \begin{pmatrix} (\mathbf{U}_O) \\ (\boldsymbol{\omega}) \end{pmatrix} \tag{5–4.23}$$

It is natural to refer to the preceding 6×6 square matrix as the *resistance matrix*. It is a real *symmetric* matrix, its symmetry being a consequence of the symmetry of the submatrices $(\mathbf{K})$ and $(\mathbf{\Omega}_O)$. Moreover, the condition that the energy dissipation be positive requires that the resistance matrix be positive-definite at all points O. (The submatrices $(\mathbf{K})$ and $(\mathbf{\Omega}_O)$ are, of course, each separately positive-definite). Among other things, this requires that

$$K_{jj}\Omega_{kk}^{(O)} > (C_{kj}^{(O)})^2 \quad (j, k = 1, 2, 3) \tag{5–4.24}$$

as can be shown by supposing that, for particular values of j and k, only the components $U_j^{(O)}$ and ω_k in Eq. (5–4.21) are non-zero.

Of special interest are helicoidally isotropic bodies, discussed further in Section 5–5, Case 10. These bodies are characterized by the fact that $\mathbf{K}$ is isotropic and that $\mathbf{C}_O$ and $\mathbf{\Omega}_O$ are isotropic at the center of reaction; that is,

$$(\mathbf{K}) = K(\delta_{jk}), \qquad (\mathbf{\Omega}_R) = \Omega(\delta_{jk}), \qquad (\mathbf{C}_R) = C(\delta_{jk})$$

where (δ_{jk}) is the unit diagonal matrix. Thus, here the condition that the resistance matrix be positive-definite requires simply that the three scalar coefficients, K, Ω, and C, characterizing the resistance of the body, satisfy the inequality

$$K\Omega > C^2 \tag{5–4.25}$$

Whereas K and Ω are positive, we shall show later that C may be either positive or negative, according as its sense of helicoidal isotropy is either left- or right-handed, respectively.

The function E appearing in Eq. (5–4.21) is a homogeneous, quadratic function of the velocities $\mathbf{U}_O$ and $\boldsymbol{\omega}$. It is therefore intimately related to the *Rayleigh dissipation function*, F, for systems in which the "frictional forces are proportional to the velocities", the relation being $E = 2F$. A large body of literature deals with the properties of this function, especially with regard to applications of Lagrange's equations to the damped vibrations of non-conservative systems.[44]

An Example

As a simple, nontrivial example of a body for which we can explicitly obtain the translation, rotation, and coupling tensors, consider the idealized, two-bladed "screw-propeller" shown in Fig. 5–4.1. It is formed by joining

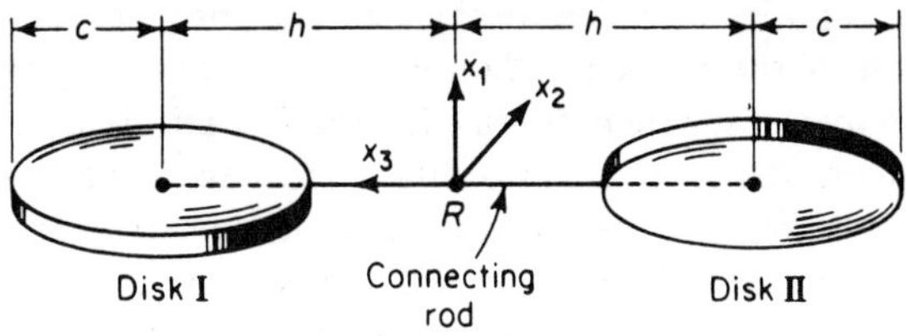

Figure 5-4.1. Two circular disks joined together to form a "screw-propeller."

together two infinitesimally thin circular disks (radii $= c$) by means of a rigid rod (of negligible hydrodynamic resistance) whose extension lies in the planes of both disks and there coincides with a diameter of each of the disks. The center-to-center spacing between disks is $2h$ and the smaller angle between the planes of the disks is $2|\theta|$, whence $\pi/2 \geqq |\theta| \geqq 0$. To define the sense of this angle, imagine that the disks are free to rotate relative to the connecting rod, and that the planes of the two disks are initially coincident. View the rods so that the connecting rod appears as a point to the eye. Designate the disk nearest the observer as disk I and that furthest from the observer as disk II, and suppose that the propeller is formed by rotating the planes of each of the disks through equal, but opposite, angles $|\theta|$ about the axis of the connecting rod. Let θ be the angle measured *clockwise*, relative to the observer, between the initial plane of coincidence of the disks and the plane of disk I as depicted in Fig. 5–4.2. The two propellers shown in Fig.

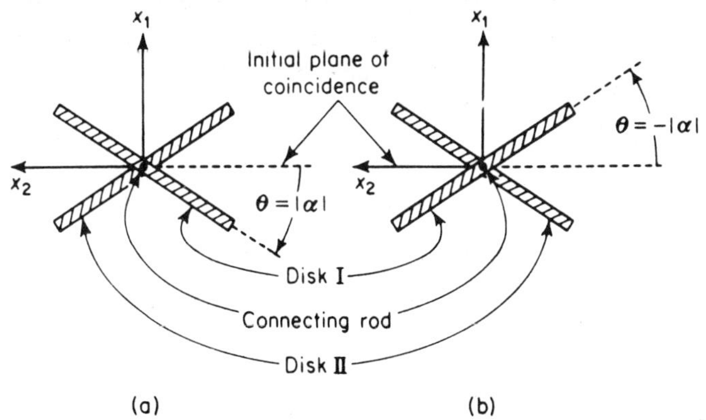

Figure 5-4.2. Side view of circular disks.

5–4.2(a) and 5–4.2(b), differing only in the algebraic sign of θ, are mirror images, one being right-handed and the other left-handed.

It is convenient to characterize the three fundamental tensors in terms of a system of cartesian coordinates fixed in the particle and defined as follows: Let the origin of the system lie at the midpoint of the connecting rod. Denote this point by R, which we shall subsequently prove is indeed the center of reaction of the propeller. Consider the situation *before* the disks were rotated into their ultimate positions. Let the Rx_2 axis lie in the common initial plane of coincidence of the two disks, and have it point toward the left hand of the observer. Let the Rx_1 axis be perpendicular to this plane and directed vertically upward. Finally, let the Rx_3 axis be directed along the connecting rod, pointing from disk II toward disk I, that is, into the eye of the observer. (In Fig. 5–4.2 the Rx_3 axis is directed out of the plane of the page). This system of coordinates is then right-handed in the order $1 \to 2 \to 3$. We shall refer to these as the *natural axes* of the propeller.

It is assumed that the resistance of the connecting rod is nil and that the disks are sufficiently far apart for hydrodynamic interaction between them to be negligible. This occurs for $c/h \ll 1$. Thus, the properties of the propeller may be obtained by simply superposing the hydrodynamic properties of the individual disks. Let $\mathbf{K}^{\mathrm{I}}$ be the translation tensor for disk I and let $\mathbf{\Omega}_R^{\mathrm{I}}$ and $\mathbf{C}_R^{\mathrm{I}}$ be the rotation- and coupling-tensors for this disk at point R. Likewise, let $\mathbf{K}^{\mathrm{II}}$, $\mathbf{\Omega}_R^{\mathrm{II}}$, and $\mathbf{C}_R^{\mathrm{II}}$ be the comparable properties for disk II. Then, if $\mathbf{K}$, $\mathbf{\Omega}_R$, and $\mathbf{C}_R$ be the corresponding tensors for the "screw-propeller," we have for small c/h that

$$\mathbf{K} = \mathbf{K}^{\mathrm{I}} + \mathbf{K}^{\mathrm{II}} \tag{5–4.26}$$

$$\mathbf{\Omega}_R = \mathbf{\Omega}_R^{\mathrm{I}} + \mathbf{\Omega}_R^{\mathrm{II}} \tag{5–4.27}$$

$$\mathbf{C}_R = \mathbf{C}_R^{\mathrm{I}} + \mathbf{C}_R^{\mathrm{II}} \tag{5–4.28}$$

To compute $\mathbf{K}^{\mathrm{I}}$ in terms of the natural propeller axes, let $\bar{x}_1, \bar{x}_2, \bar{x}_3$ be a right-handed system of cartesian coordinates lying in the plane of disk I and normal to it, as in Fig. 5–4.3. The $\bar{x}_3$ and x_3 axes coincide in the plane of the disk and are directed out of the page. As the barred axes are the principal axes of translation of the circular disk, we have, from the solution of Stokes' equations for a translating disk[32, 39],

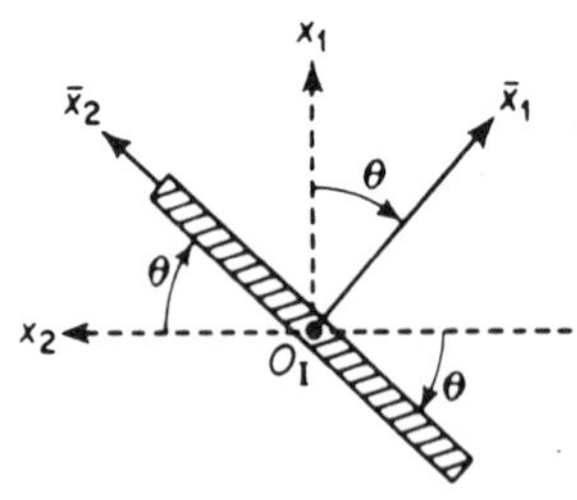

Figure 5-4.3. Intrinsic axes of disk I and natural axes of the "propeller."

$$\mathbf{K}^{\mathrm{I}} = \tfrac{16}{3}c(3\bar{\mathbf{i}}_1\bar{\mathbf{i}}_1 + 2\bar{\mathbf{i}}_2\bar{\mathbf{i}}_2 + 2\bar{\mathbf{i}}_3\bar{\mathbf{i}}_3) \tag{5–4.29}$$

the $\bar{\mathbf{i}}_k$'s being unit vectors. By means of the relations

$$\begin{aligned}\bar{\mathbf{i}}_1 &= \mathbf{i}_1 \cos\theta - \mathbf{i}_2 \sin\theta \\ \bar{\mathbf{i}}_2 &= \mathbf{i}_1 \sin\theta + \mathbf{i}_2 \cos\theta \\ \bar{\mathbf{i}}_3 &= \mathbf{i}_3\end{aligned} \tag{5–4.30}$$

we then obtain

$$\mathbf{K}^{\mathrm{I}} = \tfrac{16}{3}c\,[\mathbf{i}_1\mathbf{i}_1(2+\cos^2\theta) + \mathbf{i}_2\mathbf{i}_2\,(2+\sin^2\theta) + 2\,\mathbf{i}_3\mathbf{i}_3 - (\mathbf{i}_1\mathbf{i}_2 + \mathbf{i}_2\mathbf{i}_1)\sin\theta\cos\theta] \tag{5–4.31}$$

Similarly,

$$\mathbf{K}^{\mathrm{II}} = \tfrac{16}{3}c[\mathbf{i}_1\mathbf{i}_1\,(2+\cos^2\theta) + \mathbf{i}_2\mathbf{i}_2\,(2+\sin^2\theta) + 2\,\mathbf{i}_3\mathbf{i}_3 + (\mathbf{i}_1\mathbf{i}_2 + \mathbf{i}_2\mathbf{i}_1)\sin\theta\cos\theta] \tag{5–4.32}$$

On adding these we obtain, to the zeroth order in c/h,

$$\mathbf{K} = \tfrac{32}{3}c[\mathbf{i}_1\mathbf{i}_1(2+\cos^2\theta) + \mathbf{i}_2\mathbf{i}_2\,(2+\sin^2\theta) + 2\,\mathbf{i}_3\mathbf{i}_3] \tag{5–4.33}$$

It follows from this that the principal axes of translation of the propeller coincide with the natural axes of the propeller. Moreover, as seems intuitively plausible, the translation tensor is the same for a given propeller as it is for its mirror image, as may be seen upon replacing θ by $-\theta$.

In order to compute $\mathbf{\Omega}_R^{\mathrm{I}}$, let $\mathbf{\Omega}_{\mathrm{I}}^{\mathrm{I}}$ be the rotation tensor for disk I at its own geometric center. By symmetry arguments, it is clear that the coupling tensor for disk I vanishes at the center of disk I. Hence, from Eq. (5–4.10), we obtain

$$\mathbf{\Omega}_R^{\mathrm{I}} = \mathbf{\Omega}_{\mathrm{I}}^{\mathrm{I}} - \mathbf{r}_{IR} \times \mathbf{K}^{\mathrm{I}} \times \mathbf{r}_{IR} \tag{5–4.34}$$

where $\mathbf{r}_{IR}$ is the position vector of R relative to an origin at the center of disk I. From Fig. 5–4.1 we see that

$$\mathbf{r}_{IR} = -\mathbf{i}_3 h \tag{5–4.35}$$

Furthermore, for a circular disk,

$$\mathbf{\Omega}_{\mathrm{I}}^{\mathrm{I}} = \mathbf{I}\tfrac{32}{3}c^3 \tag{5–4.36}$$

in which $\mathbf{I}$ is the dyadic idemfactor. Consequently,

$$\mathbf{\Omega}_R^{\mathrm{I}} = \mathbf{I}\tfrac{32}{3}c^3 - \mathbf{i}_3 \times \mathbf{K}^{\mathrm{I}} \times \mathbf{i}_3 h^2 \tag{5–4.37}$$

Similarly,

$$\mathbf{\Omega}_R^{\mathrm{II}} = \mathbf{I}\tfrac{32}{3}c^3 - \mathbf{i}_3 \times \mathbf{K}^{\mathrm{II}} \times \mathbf{i}_3 h^2 \tag{5–4.38}$$

Adding these in the light of Eqs. (5–4.26) and (5–4.27), we obtain

$$\mathbf{\Omega}_R = \mathbf{I}\tfrac{64}{3}c^3 - \mathbf{i}_3 \times \mathbf{K} \times \mathbf{i}_3 h^2 \tag{5–4.39}$$

Now, $\mathbf{I} = \mathbf{i}_1\mathbf{i}_1 + \mathbf{i}_2\mathbf{i}_2 + \mathbf{i}_3\mathbf{i}_3$, whereas $\mathbf{K}$ is given by Eq. (5–4.33). Hence,

$$\mathbf{\Omega}_R = \mathbf{i}_1\mathbf{i}_1\frac{32}{3}ch^2\left(2+\sin^2\theta + 2\frac{c^2}{h^2}\right) + \mathbf{i}_2\mathbf{i}_2\frac{32}{3}ch^2\left(2+\cos^2\theta + 2\frac{c^2}{h^2}\right) + \mathbf{i}_3\mathbf{i}_3\frac{64}{3}c^3$$

Since c/h is small compared with unity we may neglect the terms in parentheses of $O(c/h)^2$. This is tantamount to neglecting the torque arising from the rotation of the disks about their own centers compared with the torque arising from the forces experienced by the disks as they translate through the fluid during rotation of the propeller about an axis through R. Hence, to terms of zeroth order in c/h,

$$\mathbf{\Omega}_R = \tfrac{32}{3}c[\mathbf{i}_1\mathbf{i}_1 h^2(2 + \sin^2\theta) + \mathbf{i}_2\mathbf{i}_2 h^2(2 + \cos^2\theta) + \mathbf{i}_3\mathbf{i}_3 2c^2] \qquad (5\text{–}4.40)$$

As in the case of the translation tensor, the principal axes of rotation at R coincide with the natural axes of the propeller; moreover, $\mathbf{\Omega}_R$ is the same for a given propeller as for its mirror-image.

As is clear from the lack of skew-symmetry of a circular disk,

$$\mathbf{C}_{\mathrm{I}}^{\mathrm{I}} = \mathbf{O} \qquad (5\text{–}4.41)$$

Thus, from Eq. (5–2.11)

$$\mathbf{C}_R^{\mathrm{I}} = -\mathbf{r}_{\mathrm{I}R} \times \mathbf{K}^{\mathrm{I}} \qquad (5\text{–}4.42)$$

$\mathbf{K}^{\mathrm{I}}$ is given by Eq. (5–4.31) and $\mathbf{r}_{\mathrm{I}R}$ by Eq.(5–4.35). Similarly, we may compute

$$\mathbf{C}_R^{\mathrm{II}} = -\mathbf{r}_{\mathrm{II}R} \times \mathbf{K}^{\mathrm{II}}$$

in which $\mathbf{r}_{\mathrm{II}R} = +\mathbf{i}_3 h$ and $\mathbf{K}^{\mathrm{II}}$ is given by Eq. (5–4.32). Upon adding the resulting expressions for $\mathbf{C}_R^{\mathrm{I}}$ and $\mathbf{C}_R^{\mathrm{II}}$, one finds

$$\mathbf{C}_R = \tfrac{32}{3}ch(\mathbf{i}_1\mathbf{i}_1 - \mathbf{i}_2\mathbf{i}_2)\sin\theta\cos\theta \qquad (5\text{–}4.43)$$

We note that $\mathbf{C}_R$ is symmetrical, confirming that the center of reaction of the propeller lies at the midpoint of the connecting rod. Furthermore, the principal axes of coupling coincide with the natural axes of the propeller. Note also that $\mathbf{C}_R$ changes in algebraic sign upon replacing θ by $-\theta$. Finally, we observe that the various inequalities implied by Eq. (5–4.24) are all satisfied by the present example.

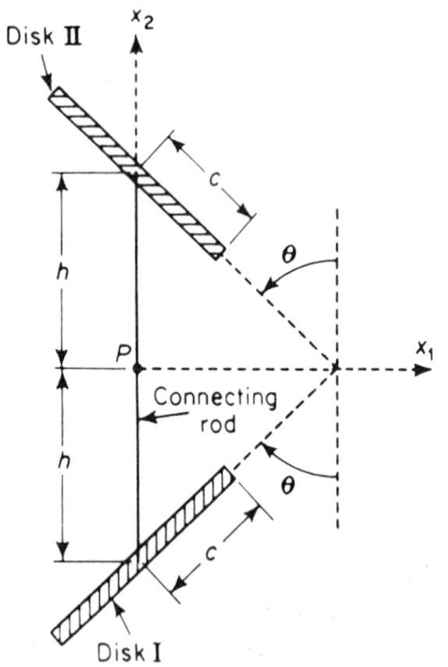

Figure 5-4.4. Side view of an "impeller" formed from two circular disks.

Other enlightening examples consisting of two or more distant circular disks rigidly joined together in other ways may be similarly constructed. Thus, consider the "impeller"-like body shown edge-on in Fig. 5–4.4, the circular disks being joined along their centers. The x_3 axis is directed out of the plane of the page. A calculation similar to the preceding shows that $\mathbf{K}^{\mathrm{I}}$, $\mathbf{K}^{\mathrm{II}}$ and, hence, $\mathbf{K}$ have the same values as in Eqs. (5–4.31), (5–4.32), and (5–4.33), respectively. Though it might be suspected that the center of reaction lies at the centroid of the body, P, a simple calculation shows that this is not so—for $\mathbf{C}_P$ is not symmetric, but rather has the value

$$\mathbf{C}_P = -\mathbf{i}_3\mathbf{i}_2 \tfrac{32}{3} ch \sin\theta \cos\theta \tag{5–4.44}$$

Further calculation reveals that the center of reaction lies at the point

$$\mathbf{r}_{PR} = -\mathbf{i}_1 \frac{h \sin\theta \cos\theta}{4 + \sin^2\theta} \tag{5–4.45}$$

at which point

$$\mathbf{C}_R = -(\mathbf{i}_2\mathbf{i}_3 + \mathbf{i}_3\mathbf{i}_2)\frac{64ch \sin\theta \cos\theta}{3(4 + \sin^2\theta)} \tag{5–4.46}$$

Since

$$\mathbf{i}_2\mathbf{i}_3 + \mathbf{i}_3\mathbf{i}_2 = \mathbf{e}_2\mathbf{e}_2 - \mathbf{e}_3\mathbf{e}_3$$

in which

$$\mathbf{e}_2 = \frac{1}{\sqrt{2}}(\mathbf{i}_2 - \mathbf{i}_3), \qquad \mathbf{e}_3 = \frac{1}{\sqrt{2}}(\mathbf{i}_2 + \mathbf{i}_3)$$

it follows that the principal axes of coupling lie in a plane perpendicular to the x_1 axis, and there make angles of 45° with the x_2 and x_3 coordinate axes.

This example clearly shows: (1) the center of reaction is a unique geometrical point, different in general from the centroid of the body; (2) the principal axes of the various tensors at the center of reaction do not generally coincide.

5-5 Symmetrical Particles

A complete characterization of the hydrodynamic resistance of rigid particles to translational and rotational motions in Stokes flow requires a knowledge of 21 independent scalar resistance coefficients; namely, 6 for the translation tensor, 6 for the rotation tensor, and 9 for the coupling tensor. The number of such coefficients is reduced by any geometric symmetry possessed by the particle. This section presents a systematic investigation of the effects of various types of symmetry on the structure of the three fundamental tensors. These efforts have been strongly influenced by comparable treatments of "hydrokinetic symmetry" by Kelvin[31], Larmor[35], and Lamb[32] in connection with the kinetic energy of *ideal* fluids in irrotational, acyclic motion generated by the impulsive motion of a rigid particle from rest in a fluid initially at rest.[37] In that case, the combined, instantaneous kinetic energy of the particle-fluid system is a homogeneous, quadratic function of the instantaneous linear and angular velocity components of the particle. Since, in Stokes flow the dissipation function is also a homogeneous, quadratic function of these velocities, a perfect analogy exists—at least insofar as symmetry considerations are concerned.

Imagine the body to be fixed in space, and let Ox_i ($i = 1, 2, 3$) be any system of cartesian axes fixed in space relative to an arbitrary origin O affixed to the particle. Denote by $O\bar{x}_i$ another set of axes, initially coincident with Ox_i, which we rotate and/or reflect. Consider the class of bodies whose symmetry is such that the surface of the particle bears the same relation to the Ox_i axes as it does to the $O\bar{x}_i$ axes *after* the latter are rotated about O and/or reflected in a plane containing O; that is, if the surface of the body is described by the equation $f(x_1, x_2, x_3) = 0$, then $f(\bar{x}_1, \bar{x}_2, \bar{x}_3) = 0$ describes the surface of this same body.

In cartesian tensor notation, the rate of energy dissipation in the unbarred and barred system of coordinates is, from Eq. (5–4.21),

$$E = \mu(U_i^{(O)} K_{ij} U_j^{(O)} + 2\omega_i C_{ij}^{(O)} U_j^{(O)} + \omega_i \Omega_{ij}^{(O)} \omega_j) \tag{5–5.1}$$

and

$$\bar{E} = \mu(\bar{U}_k^{(O)} \bar{K}_{kl} \bar{U}_l^{(O)} + 2\bar{\omega}_k \bar{C}_{kl}^{(O)} \bar{U}_l^{(O)} + \bar{\omega}_k \bar{\Omega}_{kl}^{(O)} \bar{\omega}_l) \tag{5–5.2}$$

in which the usual summation convention applies. The superscript (O) refers to the origin O. The general transformation from Ox_i to $O\bar{x}_i$ coordinates is

$$\bar{x}_i = a_{ij} x_j \tag{5–5.3}$$

where $(\mathbf{a})$ is the orthogonal matrix of direction cosines

$$(\mathbf{a}) = \begin{pmatrix} a_{11} & a_{12} & a_{13} \\ a_{21} & a_{22} & a_{23} \\ a_{31} & a_{32} & a_{33} \end{pmatrix} \tag{5–5.4}$$

whose elements are

$$a_{jk} = \bar{\mathbf{i}}_j \cdot \mathbf{i}_k = \cos(O\bar{x}_j, Ox_k) \tag{5–5.5}$$

As the scalar E is a tensor of rank zero, its law of transformation is

$$E = \bar{E} \tag{5–5.6}$$

Similarly, the (polar) vector $\bar{U}_k^{(O)}$ is a true vector and so transforms according to the relation

$$U_m^{(O)} = a_{nm} \bar{U}_n^{(O)} \tag{5–5.7}$$

On the other hand, the angular velocity vector $\bar{\omega}_m$ is a pseudovector (axial vector) and transforms according to the law

$$\omega_m = |\mathbf{a}| a_{nm} \bar{\omega}_n \tag{5–5.8}$$

in which $|\mathbf{a}|$ denotes the determinant of the matrix $(\mathbf{a})$. The value of this determinant is either -1 or $+1$ according as the transformation $Ox_i \rightarrow O\bar{x}_i$ does or does not alter the hand of the axes.

Substituting Eqs. (5–5.6)–(5–5.8) into Eq. (5–5.1) and comparing the resulting expression with Eq. (5–5.2), we find that K_{ij}, $C_{ij}^{(O)}$, and $\Omega_{ij}^{(O)}$ transform as follows:

$$\begin{aligned} \bar{K}_{kl} &= a_{ki} a_{lj} K_{ij} \\ \bar{C}_{kl}^{(O)} &= |\mathbf{a}|\, a_{ki} a_{lj} C_{ij}^{(O)} \\ \bar{\Omega}_{kl}^{(O)} &= a_{ki} a_{lj} \Omega_{ij}^{(O)} \end{aligned} \tag{5–5.9}$$

where it has been noted that $|\mathbf{a}||\mathbf{a}| = +1$. Hence, K_{ij} and $\Omega_{ij}^{(O)}$ are true, second-rank tensors, whereas $C_{ij}^{(O)}$ is a pseudo, second-rank tensor because of the question of algebraic sign in its transformation law.

Equations (5–5.9) are, of course, applicable to *any* cartesian coordinate transformation involving a common origin. However, it also follows from the particular way in which we have defined the Ox_i and $O\bar{x}_i$ systems, and from the invariance of the Stokes equations and boundary conditions to rotation and/or reflection, that

$$\begin{aligned} \bar{K}_{kl} &= K_{kl} \\ \bar{C}_{kl}^{(O)} &= C_{kl}^{(O)} \\ \bar{\Omega}_{kl}^{(O)} &= \Omega_{kl}^{(O)} \end{aligned} \tag{5–5.10}$$

Therefore,

$$\begin{aligned} K_{kl} &= a_{ki}a_{lj}K_{ij} \\ C_{kl}^{(O)} &= |\mathbf{a}|\,a_{ki}a_{lj}C_{ij}^{(O)} \\ \Omega_{kl}^{(O)} &= a_{ki}a_{lj}\Omega_{ij}^{(O)} \end{aligned} \tag{5–5.11}$$

These relations place certain restrictions on the components of the three fundamental tensors which we shall now investigate for a few important specific cases. Since K_{kl} and $\Omega_{kl}^{(O)}$ in Eq. (5–5.11) both transform according to the same law, and since both are symmetric, it suffices to investigate either one.

CASE 1: One plane of symmetry

To illustrate the general application of Eq. (5–5.11) we shall investigate this special case in some detail. Take the Ox_1-Ox_2 plane as the plane of symmetry. The existence of a plane of symmetry implies that the body is identical to itself when reflected in the plane. Symbolically, then, the transformation is

$$1 \to \bar{1}, \quad 2 \to \bar{2}, \quad 3 \to -\bar{3}$$

for which

$$(\mathbf{a}) = \begin{pmatrix} 1 & \cdot & \cdot \\ \cdot & 1 & \cdot \\ \cdot & \cdot & -1 \end{pmatrix}$$

and $|\mathbf{a}| = -1$. Equation (5–5.11) yields various relations, of which the following are typical:

$$K_{11} = K_{11}, \qquad K_{12} = K_{12}, \ldots, \qquad K_{23} = -K_{23}, \ldots$$

and $$C_{11}^{(O)} = -C_{11}^{(O)}, \qquad C_{12}^{(O)} = -C_{12}^{(O)}, \ldots, \qquad C_{23}^{(O)} = C_{23}^{(O)}, \ldots$$

In order that these relations may be satisfied, we require that

$$\begin{aligned} K_{23} = K_{31} = 0, \qquad \Omega_{23}^{(O)} = \Omega_{31}^{(O)} = 0 \\ C_{11}^{(O)} = C_{12}^{(O)} = C_{21}^{(O)} = C_{22}^{(O)} = C_{33}^{(O)} = 0 \end{aligned} \tag{5–5.12}$$

it being understood that, since K_{ij} and $\Omega_{ij}^{(O)}$ are symmetric, these also imply, for example, that $K_{32} = 0$. Consequently,

$$(\mathbf{K}) = \begin{pmatrix} K_{11} & K_{12} & \cdot \\ K_{12} & K_{22} & \cdot \\ \cdot & \cdot & K_{33} \end{pmatrix}, \quad (\boldsymbol{\Omega}_O) = \begin{pmatrix} \Omega_{11}^{(O)} & \Omega_{12}^{(O)} & \cdot \\ \Omega_{12}^{(O)} & \Omega_{22}^{(O)} & \cdot \\ \cdot & \cdot & \Omega_{33}^{(O)} \end{pmatrix},$$

$$(\mathbf{C}_O) = \begin{pmatrix} \cdot & \cdot & C_{13}^{(O)} \\ \cdot & \cdot & C_{23}^{(O)} \\ C_{31}^{(O)} & C_{32}^{(O)} & \cdot \end{pmatrix} \tag{5–5.13}$$

If $\mathbf{K}$ is written in its dyadic form

$$\mathbf{K} = \mathbf{i}_1\mathbf{i}_1 K_{11} + \mathbf{i}_2\mathbf{i}_2 K_{22} + \mathbf{i}_3\mathbf{i}_3 K_{33} + (\mathbf{i}_1\mathbf{i}_2 + \mathbf{i}_2\mathbf{i}_1) K_{12}$$

we see that $\mathbf{K}\cdot\mathbf{i}_3$ is parallel to $\mathbf{i}_3$. Hence, $\mathbf{i}_3$ is an eigenvector of $\mathbf{K}$. Therefore, if a body possesses a plane of symmetry, one of the three principal axes of translation lies normal to the plane, and the remaining two (which are mutually perpendicular to the third) lie parallel to the plane. Similar remarks apply to the principal axes of rotation at any point O lying in the plane.

CASE 2: Two mutually perpendicular symmetry planes

If the particle has a second plane of symmetry at right angles to the first we may take this as the Ox_3-Ox_1 plane, where O lies anywhere along the line of intersection of the planes. Besides the relations in Eq. (5–5.12), the following additional relations must be satisfied:

$$K_{12} = 0, \qquad \Omega_{12}^{(O)} = 0, \quad \text{and} \quad C_{13}^{(O)} = C_{31}^{(O)} = 0 \tag{5–5.14}$$

Hence,

$$(\mathbf{K}) = \begin{pmatrix} K_{11} & \cdot & \cdot \\ \cdot & K_{22} & \cdot \\ \cdot & \cdot & K_{33} \end{pmatrix}, \quad (\boldsymbol{\Omega}_O) = \begin{pmatrix} \Omega_{11}^{(O)} & \cdot & \cdot \\ \cdot & \Omega_{22}^{(O)} & \cdot \\ \cdot & \cdot & \Omega_{33}^{(O)} \end{pmatrix},$$

$$(\mathbf{C}_O) = \begin{pmatrix} \cdot & \cdot & \cdot \\ \cdot & \cdot & C_{23}^{(O)} \\ \cdot & C_{32}^{(O)} & \cdot \end{pmatrix} \tag{5–5.15}$$

Two of the principal axes of $\mathbf{K}$ therefore lie normal to the two symmetry planes, whereas the third principal axis lies parallel to their line of intersection. The rotation tensor at any point along this line has the same principal axes as $\mathbf{K}$.

The "impeller" shown in Fig. 5–4.4 is an example of a body of this class, for if O lies anywhere along the Px_1 axis, the planes Ox_1-Ox_2 and Ox_3-Ox_1 are planes of symmetry. The further symmetry result, $C_{23}^{(P)} = 0$, implied by Eq. (5–4.44) is *not* deducible by our present methods of analysis. Rather, group theory methods, such as are employed in crystallography[46], would be required; for this symmetry restriction is not merely a consequence of the over-all symmetry of the impeller, but rather of the symmetry of the separate elements of which it is comprised. Thus, in general, it is possible for a body to possess two perpendicular symmetry planes without the individual elements of which it is composed possessing any symmetry whatsoever.

Case 3: Three mutually perpendicular symmetry planes (orthotropic bodies)

If the particle has a third plane of symmetry (the Ox_2-Ox_3 plane) at right angles to the two former ones, then, in addition to Eqs. (5–5.12) and (5–5.14), we also require that

$$C_{23}^{(O)} = C_{32}^{(O)} = 0 \tag{5–5.16}$$

Thus, if R denotes the point of intersection of the three planes, we have

$$(\mathbf{K}) = \begin{pmatrix} K_{11} & \cdot & \cdot \\ \cdot & K_{22} & \cdot \\ \cdot & \cdot & K_{33} \end{pmatrix}, \quad (\boldsymbol{\Omega}_R) = \begin{pmatrix} \Omega_{11}^{(R)} & \cdot & \cdot \\ \cdot & \Omega_{22}^{(R)} & \cdot \\ \cdot & \cdot & \Omega_{33}^{(R)} \end{pmatrix}, \quad (\mathbf{C}_R) = (\mathbf{O}) \tag{5–5.17}$$

so that this point constitutes the center of reaction of the body, at which point the coupling tensor is identically zero.

Case 4: Spherically isotropic bodies

If the preceding solid is such that the form of the body be similarly related to each of the three mutually perpendicular coordinate planes (for example, a sphere or cube), the coordinate axes Ox_1, Ox_2, and Ox_3 are indistinguishable; hence,

$$\begin{aligned} K_{11} &= K_{22} = K_{33} = K, \text{ say,} \\ \Omega_{11}^{(R)} &= \Omega_{22}^{(R)} = \Omega_{33}^{(R)} = \Omega, \text{ say.} \end{aligned} \tag{5–5.18}$$

Therefore

$$(\mathbf{K}) = K\begin{pmatrix} 1 & \cdot & \cdot \\ \cdot & 1 & \cdot \\ \cdot & \cdot & 1 \end{pmatrix}, \quad (\boldsymbol{\Omega}_R) = \Omega\begin{pmatrix} 1 & \cdot & \cdot \\ \cdot & 1 & \cdot \\ \cdot & \cdot & 1 \end{pmatrix}, \quad (\mathbf{C}_R) = (\mathbf{O}) \tag{5–5.19}$$

or, in dyadic notation,

$$\mathbf{K} = \mathbf{I}K, \qquad \boldsymbol{\Omega}_R = \mathbf{I}\Omega, \qquad \mathbf{C}_R = \mathbf{O}$$

We shall term such bodies *spherically isotropic*. Characterization of their hydrodynamic resistance requires knowledge of only the two scalar coefficients, K and Ω. All regular polyhedra and bodies derived from them by symmetrically cutting or rounding the corners and/or edges and/or faces are spherically isotropic[8]. Thus, substitution in Eqs. (5–4.1) and (5–4.2) gives

$$\mathbf{F} = -\mu K \mathbf{U}_R \quad \text{and} \quad \mathbf{T}_R = -\mu\Omega\boldsymbol{\omega}$$

Experiments by Pettyjohn and Christiansen[42] on the fall of isometric particles in the Stokes regime confirm the preceding conclusions. They studied the settling of tetrahedra, cubes, octahedra, and cube-octahedra and observed the settling velocities to be independent of orientation to within their experimental uncertainty of 0.5 per cent. The particles, dropped with random orientation into the fluid, showed no tendency either to rotate or set themselves in any particular position with respect to their direction of motion through the fluid. Moreover, they fell vertically downward. These points are further discussed in Section 5–7.

CASE 5: Bodies of revolution

If the body is a solid of revolution (say about the Ox_1 axis) it is symmetrical with respect to *all* planes containing its axis (although not necessarily with respect to any plane perpendicular to this axis, for example, a hemisphere). The results of Case 2 are applicable, where O now refers to any point along the axis. Furthermore, the form of the solid is unaltered if we rotate the Ox_2 and Ox_3 axes through any angle, say, a positive right angle. The transformation appropriate to this rotation is

$$1 \to \bar{1}, \quad 2 \to -\bar{3}, \quad 3 \to \bar{2}$$

whence

$$(\mathbf{a}) = \begin{pmatrix} 1 & \cdot & \cdot \\ \cdot & \cdot & 1 \\ \cdot & -1 & \cdot \end{pmatrix}$$

and $|\mathbf{a}| = +1$. Thus, in addition to those relations set forth in Eq. (5–5.15), we also require that

$$K_{22} = K_{33}, \qquad \Omega_{22}^{(O)} = \Omega_{33}^{(O)}, \qquad C_{23}^{(O)} = -C_{32}^{(O)} \tag{5–5.20}$$

Consequently,

$$(\mathbf{K}) = \begin{pmatrix} K_{11} & \cdot & \cdot \\ \cdot & K_{22} & \cdot \\ \cdot & \cdot & K_{22} \end{pmatrix}, \quad (\boldsymbol{\Omega}_O) = \begin{pmatrix} \Omega_{11}^{(O)} & \cdot & \cdot \\ \cdot & \Omega_{22}^{(O)} & \cdot \\ \cdot & \cdot & \Omega_{22}^{(O)} \end{pmatrix},$$
$$(\mathbf{C}_O) = \begin{pmatrix} \cdot & \cdot & \cdot \\ \cdot & \cdot & C_{23}^{(O)} \\ \cdot & -C_{23}^{(O)} & \cdot \end{pmatrix} \tag{5–5.21}$$

When the preceding values are substituted into Eq. (5–4.16) we obtain $(x_{OR})_2 = (x_{OR})_3 = 0$. The center of reaction therefore lies along the axis of revolution of the body. Now, according to Eq. (5–5.21), $(\mathbf{C}_O)$ is antisymmetric at all points along the axis. But according to the definition of the center of reaction, $(\mathbf{C}_R)$ must be symmetric. These conditions are all satisfied if and only if $\mathbf{C}_R = \mathbf{0}$. If the body lacks a plane of symmetry perpendicular to the axis it is not possible to predict *a priori* the general location of this point. The dumbbell example (Section 5–6), consisting of two unequal spheres, is a simple example of such a body.

CASE 6: Skew-symmetry about an axis

Suppose a body has a sort of skew-symmetry about the Ox_1 axis—for example, that it is identical to itself when turned through 180° about this axis, but lacks a plane of symmetry either normal to, or containing, this axis (for example, the two-bladed screw-propeller in Fig. 5–4.1). We find in this case that

$$(\mathbf{K}) = \begin{pmatrix} K_{11} & \cdot & \cdot \\ \cdot & K_{22} & K_{23} \\ \cdot & K_{23} & K_{33} \end{pmatrix}, \quad (\mathbf{\Omega}_O) = \begin{pmatrix} \Omega_{11}^{(O)} & \cdot & \cdot \\ \cdot & \Omega_{22}^{(O)} & \Omega_{23}^{(O)} \\ \cdot & \Omega_{23}^{(O)} & \Omega_{33}^{(O)} \end{pmatrix},$$
$$(\mathbf{C}_O) = \begin{pmatrix} C_{11}^{(O)} & \cdot & \cdot \\ \cdot & C_{22}^{(O)} & C_{23}^{(O)} \\ \cdot & C_{32}^{(O)} & C_{33}^{(O)} \end{pmatrix} \tag{5–5.22}$$

The screw-propeller in Fig. 5–4.1 possesses this type of skew-symmetry about *both* the Rx_1 and Rx_2 axes. Thus, in this example, it is further required that $K_{23} = \Omega_{23}^{(R)} = C_{23}^{(R)} = C_{32}^{(R)} = 0$. The additional symmetry relations, $C_{33}^{(R)} = 0$ and $C_{22}^{(R)} = -C_{11}^{(R)}$, implied by Eq. (5–4.43) are, again, not deducible by our present techniques.

CASE 7: Helicoidal symmetry about an axis

If, in addition to the conditions of the preceding case, the body is also identical to itself when turned through 90° about the Ox_1 axis (for example, a four-bladed screw-propeller) we find that

$$(\mathbf{K}) = \begin{pmatrix} K_{11} & \cdot & \cdot \\ \cdot & K_{22} & \cdot \\ \cdot & \cdot & K_{22} \end{pmatrix}, \quad (\mathbf{\Omega}_O) = \begin{pmatrix} \Omega_{11}^{(O)} & \cdot & \cdot \\ \cdot & \Omega_{22}^{(O)} & \cdot \\ \cdot & \cdot & \Omega_{22}^{(O)} \end{pmatrix},$$
$$(\mathbf{C}_O) = \begin{pmatrix} C_{11}^{(O)} & \cdot & \cdot \\ \cdot & C_{22}^{(O)} & C_{23}^{(O)} \\ \cdot & -C_{23}^{(O)} & C_{22}^{(O)} \end{pmatrix} \tag{5–5.23}$$

where O lies along the axis. Equation (5–4.16) then shows that the center of reaction must lie along the axis. Moreover, since $(\mathbf{C}_R)$ is symmetrical, Eq. (5–5.23) requires that $C_{23}^{(R)} = 0$. Consequently,

$$(\mathbf{C}_R) = \begin{pmatrix} C_{11}^{(R)} & \cdot & \cdot \\ \cdot & C_{22}^{(R)} & \cdot \\ \cdot & \cdot & C_{22}^{(R)} \end{pmatrix} \tag{5–5.24}$$

The form of the foregoing expressions is unaltered when the axes of Ox_2 and Ox_3 are simultaneously rotated in their own plane through *any* angle.

The body is therefore said to possess *helicoidal symmetry*[31,32,35] about the Ox_1 axis. For later reference we note that the axis of helicoidal symmetry of the body is a principal axis of coupling, and that all directions in the plane perpendicular to it are eigenvectors for the coupling tensor at R.

CASE 8: Helicoidal symmetry under more general circumstances

The results of the preceding section may be generalized so as to apply to bodies of more general shape. Suppose the form of the solid retains the same relations to the Ox_2 and Ox_3 axes when turned about the Ox_1 axis through any given angle θ in *either* direction, except $\theta = \pi$. The case $\theta = 0$ is, of course, trivial. The general transformation from Ox_i to $O\bar{x}_i$ axes, corresponding to a positive rotation through any angle θ about the Ox_1 axis, is

$$(\mathbf{a}) = \begin{pmatrix} 1 & \cdot & \cdot \\ \cdot & \cos\theta & \sin\theta \\ \cdot & -\sin\theta & \cos\theta \end{pmatrix}$$

whence $|\mathbf{a}| = +1$. From Eq. (5–5.9) we then obtain

$$\begin{aligned}
\bar{C}_{11}^{(O)} &= C_{11}^{(O)} \\
\bar{C}_{12}^{(O)} &= C_{12}^{(O)} \cos\theta + C_{13}^{(O)} \sin\theta \\
\bar{C}_{13}^{(O)} &= -C_{12}^{(O)} \sin\theta + C_{13}^{(O)} \cos\theta \\
\bar{C}_{21}^{(O)} &= C_{21}^{(O)} \cos\theta + C_{31}^{(O)} \sin\theta \\
\bar{C}_{22}^{(O)} &= C_{22}^{(O)} \cos^2\theta + (C_{23}^{(O)} + C_{32}^{(O)}) \sin\theta\cos\theta + C_{33}^{(O)} \sin^2\theta \\
\bar{C}_{23}^{(O)} &= C_{23}^{(O)} \cos^2\theta - C_{32}^{(O)} \sin^2\theta + (C_{33}^{(O)} - C_{22}^{(O)}) \sin\theta\cos\theta \\
\bar{C}_{31}^{(O)} &= -C_{21}^{(O)} \sin\theta + C_{31}^{(O)} \cos\theta \\
\bar{C}_{32}^{(O)} &= (C_{33}^{(O)} - C_{22}^{(O)}) \sin\theta\cos\theta - C_{23}^{(O)} \sin^2\theta + C_{32}^{(O)} \cos^2\theta \\
\bar{C}_{33}^{(O)} &= C_{22}^{(O)} \sin^2\theta - (C_{23}^{(O)} + C_{32}^{(O)}) \sin\theta\cos\theta + C_{33}^{(O)} \cos^2\theta
\end{aligned}$$

with similar expressions for $\bar{K}_{kl}$ and $\bar{\Omega}_{kl}^{(O)}$. The condition that these expressions be invariant to a reversal of the algebraic sign of θ requires that

$$\begin{aligned}
C_{12}^{(O)} = C_{21}^{(O)} = C_{31}^{(O)} &= C_{13}^{(O)} = 0 \\
C_{23}^{(O)} + C_{32}^{(O)} &= 0 \\
C_{33}^{(O)} - C_{22}^{(O)} &= 0
\end{aligned} \tag{5–5.25}$$

Hence $(\mathbf{K})$, $(\mathbf{\Omega}_O)$, and $(\mathbf{C}_O)$ must be of exactly the form indicated in Eq. (5–5.23). Such bodies therefore possess helicoidal symmetry.

CASE 9: Helicoidal symmetry about an axis plus reflection symmetry in a plane containing this axis

If we take Ox_1 to be the axis of helicoidal symmetry and Ox_1-Ox_2 to be the plane of symmetry containing this axis, then Eqs. (5–5.13) and (5–5.23) must be *simultaneously* satisfied. Equation (5–5.21) is therefore applicable to the present case, and we conclude that the hydrodynamic behavior of bodies of the type under discussion is identical to that of bodies of revolution.

This case arises, for example, when all the cross sections are regular polygons (it may be with rounded faces and corners), derived from the same fundamental form, and having one common axial plane of symmetry. It applies, for example, to a right prism or pyramid standing on a base having the form of a regular polygon. It also applies to a rhombohedron.

CASE 10: Helicoidal isotropy

Returning to the discussion of Cases 7 and 8, suppose the body has a second axis of helicoidal symmetry situated anywhere (say the $O'x_1'$ axis). Then, as in Eq. (5–5.24), there is one unique point (say, R') lying along this axis at which the coupling tensor assumes the *symmetric* form

$$(\mathbf{C}_R') = \begin{pmatrix} C_{1'1'}^{(R')} & \cdot & \cdot \\ \cdot & C_{2'2'}^{(R')} & \cdot \\ \cdot & \cdot & C_{2'2'}^{(R')} \end{pmatrix} \tag{5–5.26}$$

But, as discussed in Section 5–4, a body has one and only one point at which the coupling tensor is symmetric. Hence—see Eq. (5–5.24)—we conclude that R and R' must be one and the same point. If a body has two axes of helicoidal symmetry they must therefore intersect at the center of reaction of the body.

As the axes Rx_1' and Rx_1 must *both* be principal axes of coupling, and since the principal axes of coupling are mutually perpendicular, the Rx_1' axis must lie in a plane perpendicular to Rx_1. But, from Case 8, all axes lying in a plane perpendicular to Rx_1 are equivalent. Without loss in generality we may therefore identify Rx_1' with Rx_2, Rx_2' with Rx_3, and Rx_3' with Rx_1. Equation (5–5.26) then becomes

$$(\mathbf{C}_R) = \begin{pmatrix} C_{22}^{(R)} & \cdot & \cdot \\ \cdot & C_{33}^{(R)} & \cdot \\ \cdot & \cdot & C_{33}^{(R)} \end{pmatrix} \tag{5–5.27}$$

Comparison with Eq. (5–5.24) shows that

$$C_{11}^{(R)} = C_{22}^{(R)} = C_{33}^{(R)} = C, \text{ say,} \tag{5–5.28}$$

in which case the coupling tensor is isotropic at R. Similarly, by considering the orthogonality of the principal axes of the translation tensor, and of the rotation tensor at R, it is easily proved that $\mathbf{K}$ and $\mathbf{\Omega}_R$ are also isotropic. We may therefore write

$$\mathbf{K} = \mathbf{I}K, \qquad \mathbf{\Omega}_R = \mathbf{I}\Omega, \qquad \mathbf{C}_R = \mathbf{I}C \tag{5–5.29}$$

In the context of ideal fluid theory, Kelvin[31] calls such a body an *isotropic helicoid.* We shall retain this terminology, though its physical significance in

Stokes flow is quite different than in potential flow. It follows that any body possessing helicoidal symmetry about two different axes is helicoidally isotropic. Isotropy of this type should be clearly distinguished from spherical isotropy, since $\mathbf{C}_R = \mathbf{0}$ in the latter case. Helicoidally isotropic bodies require knowledge of the three scalars K, Ω, and C for a complete characterization of their hydrodynamic properties. These three constants must satisfy the inequality set forth in Eq. (5–4.25). For reasons made clear in the next section, bodies for which $C < 0$ are right-handed whereas bodies for which $C > 0$ are left-handed. The mirror-image of a helicoidally isotropic body in any plane is itself helicoidally isotropic, the two bodies having the same K and Ω values, differing only in the algebraic sign of the pseudoscalar C.

As specific geometric examples of isotropic helicoids we cite the following: According to Kelvin[31]: "An isotropic helicoid can be made by attaching projecting vanes to the surface of a globe in proper positions; for instance cutting at 45° each, at the middles of the twelve quadrants of any three great circles dividing the globe into eight quadrantal triangles" (see also Lamb[32, p.179] for a slightly more detailed description of this body). According to Larmor[35]: "If we take a regular tetrahedron (or other regular solid) and replace the edges by skew bevel faces placed in such wise that when looked at from any corner they all slope the same way, we have an example of an isotropic helicoid. This would also be the result if three plagiedral faces sloping the same way were imposed on each vertex of the tetrahedron. The first process probably gives the simplest form that a solid of this class can have. A form equivalent to the second is obtained by fixing four equal symmetrical screw-propellers on the surface of a sphere at the corners of an inscribed regular tetrahedron."

5-6 Nonskew bodies

When a body moves through a viscous fluid under the influence of an external force it will, in general, experience a hydrodynamic torque. Thus it will not generally be possible to select a point of application of the force in such a way as to eliminate this torque and thereby prevent the body from rotating as it translates*. For those bodies for which $\mathbf{C}_R = \mathbf{0}$, however, the center of reaction provides just such a point. For, as is clear from Eq. (5–4.17b), a translating body in any orientation will then experience no hydrodynamic torque about R. Hence, if the line of action of the body forces (for example, gravity) acting on the particle passes through R, the total external torque

*It may be possible, however, in certain cases, to eliminate the torque by suitably orienting the body relative to the direction of the applied force. For example, the two-bladed "screw-propeller" shown in Fig. 5–4.1 will not rotate if the external force lies along the Rx_3 axis.

about this point will be zero, and there will be no tendency for the particle to rotate about R. The possible modes of behavior of such particles are intrinsically simpler than that of any other class of particles.

As the general theory for completely arbitrary bodies tends to be somewhat abstract, we shall employ this special class of bodies to provide simple examples illustrating the physical significance of the more general theory. To simplify our terminology, it is convenient to refer to these bodies as *nonskew* particles, and to say that they possess a "center of hydrodynamic stress," C. Of course, this point is also the center of reaction, R, of the nonskew body. The entire hydrodynamic force on the body (in any orientation) behaves as if it acts through this point. From the discussion of Section 5–5, common examples of nonskew particles are orthotropic bodies (for example, ellipsoids), spherically isotropic bodies, and bodies of revolution.

In addition to the hydrodynamic force experienced by a nonskew particle, it also experiences a hydrostatic or buoyant force, yet to be taken into account. The resultant of all the external forces exerted on the particle by the fluid is due to the action of both types of forces. As is well known, the buoyant force acts through the center of buoyancy (B) of the body, located at the center of gravity of the displaced fluid. For an incompressible, homogeneous fluid in a uniform gravitational field, the location of B is thus an intrinsic property of the particle itself, dependent solely on the external geometry of the particle.

The center of hydrodynamic stress plays a fundamental role in the theory of nonskew bodies. Not only is the hydrodynamic torque about this point zero for a pure translational motion, but the converse is equally true; namely, a body rotating in an otherwise quiescent fluid about any axis which contains this point will experience no hydrodynamic force. Thus, the torque required to maintain rotation about such an axis is in the nature of a couple—see Eq. (5–4.17).

As a simple but instructive example of a nonskew body, consider the dumbbell-like body in Fig. 5–6.1, cf. Case 5, Section 5–5, formed by joining together two spheres along their line of centers by means of a thin, rigid rod. The spheres are assumed to be of unequal radii, a_1

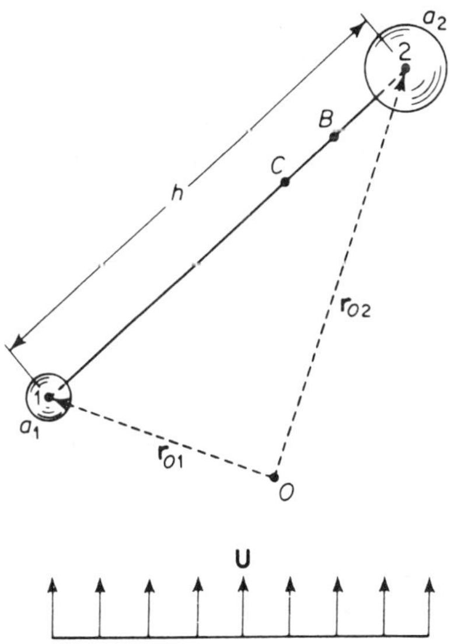

Figure 5-6.1. Streaming flow past an asymmetric dumbbell.

and a_2, respectively, the latter being the larger sphere. The center-to-center distance is h. It is assumed that the mass and hydrodynamic resistance of the connecting rod are nil compared to those of the spheres. Moreover, it is further assumed that the spheres are sufficiently far apart for hydrodynamic interaction between them to be negligible.

Consider a streaming flow with velocity $\mathbf{U}$ past this dumbbell and suppose that the body is not allowed to rotate. As the spheres are assumed to behave independently, the hydrodynamic force experienced by each is, by Stokes' law,

$$\mathbf{F}_1 = 6\pi\mu a_1\mathbf{U} \quad \text{and} \quad \mathbf{F}_2 = 6\pi\mu a_2\mathbf{U} \tag{5-6.1}$$

The total force experienced by the dumbbell is therefore

$$\mathbf{F} = 6\pi\mu(a_1 + a_2)\mathbf{U} \tag{5-6.2}$$

whence the translation tensor is

$$\mathsf{K} = \mathsf{I}\, 6\pi(a_1 + a_2) \tag{5-6.3}$$

This shows the dumbbell to be translationally isotropic. This is true, of course, only in the zeroth approximation, where h/a_1 and h/a_2 are large. More accurate approximations are given in Chapter 6 for the case of two unequal spheres moving either parallel or perpendicular to their line of centers. These show the body to be anisotropic when the spheres are relatively close.

It is clear from symmetry considerations that the forces $\mathbf{F}_1$ and $\mathbf{F}_2$ act through the centers of each of the respective spheres. Thus, the moment of these forces about an arbitrary origin O is

$$\mathbf{T}_O = \mathbf{r}_{O1} \times \mathbf{F}_1 + \mathbf{r}_{O2} \times \mathbf{F}_2 = 6\pi\mu(a_1\mathbf{r}_{O1} + a_2\mathbf{r}_{O2}) \times \mathbf{U} \tag{5-6.4}$$

The location of the center of hydrodynamic stress C, about which $\mathbf{T}_C$ vanishes is thereby governed by the equation

$$\mathbf{r}_{C1} = -\frac{a_2}{a_1}\mathbf{r}_{C2} \tag{5-6.5}$$

This relation requires that $\mathbf{r}_{C1}$ and $\mathbf{r}_{C2}$ be parallel. Hence, C lies along the line of centers joining the two spheres. If $\mathbf{e}$ is a unit vector along the line of centers of the spheres, directed from the larger sphere (2) to the smaller sphere (1), then $\mathbf{r}_{2C} + \mathbf{r}_{C1} = \mathbf{e}h$, where we have noted that $\mathbf{r}_{2C} = -\mathbf{r}_{C2}$. It follows that C is situated at the point

$$\mathbf{r}_{2C} = -\frac{h}{1 + (a_2/a_1)} \tag{5-6.6}$$

The center of buoyancy, B, of the dumbbell lies along the line of centers of the two spheres at the point

$$\mathbf{r}_{2B} = \mathbf{e}\frac{h}{1 + (a_2/a_1)^3} \tag{5-6.7}$$

Thus, B lies closer to the center of sphere 2 than does C.

For later reference, we note here that the center of mass, M, of the dumbbell may occupy any number of possible positions depending upon the distribution of matter within the spheres. If the spheres are homogeneous and of equal density, the centers of mass and buoyancy coincide—otherwise not. In the latter case, for example, if the two spheres are homogeneous, but of unequal densities, ρ_1 and ρ_2, respectively, M will be located at the point

$$\mathbf{r}_{2M} = \mathbf{e}\frac{h}{1 + (\rho_2/\rho_1)(a_2/a_1)^3} \tag{5-6.8}$$

We wish to demonstrate, using this dumbbell example, the validity of the general theorem that rotation of a nonskew body about any axis passing through C will result in no net hydrodynamic force. Suppose the dumbbell rotates with angular velocity $\boldsymbol{\omega}$ about an axis passing through C as in Fig. 5-6.2. At each instant the center of sphere 1 is translating through the surrounding fluid with velocity $\mathbf{U}_1$, variable in direction but constant in magnitude. Likewise, the center of sphere 2 is translating through the fluid with velocity $\mathbf{U}_2$. In consequence of these motions, the spheres experience forces

$$\mathbf{F}_1 = -6\pi\mu a_1\mathbf{U}_1 \quad \text{and} \quad \mathbf{F}_2 = -6\pi\mu a_2\mathbf{U}_2 \tag{5-6.9}$$

respectively, acting through their centers. But,

$$\mathbf{U}_1 = \boldsymbol{\omega} \times \mathbf{r}_{C1} \quad \text{and} \quad \mathbf{U}_2 = \boldsymbol{\omega} \times \mathbf{r}_{C2} \tag{5-6.10}$$

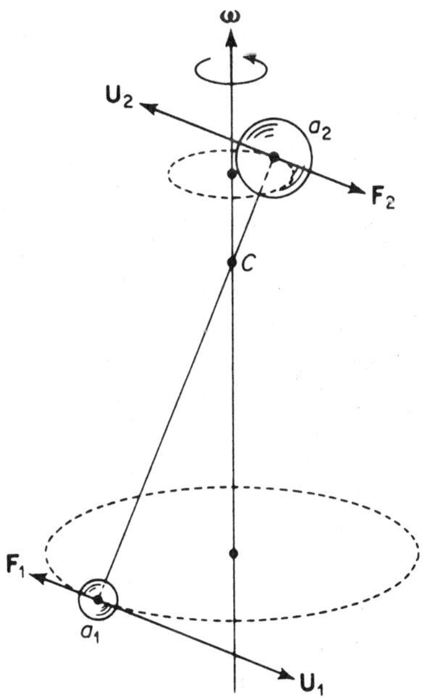

Figure 5-6.2. Rotation of a dumbbell about an axis through its center of hydrodynamic stress.

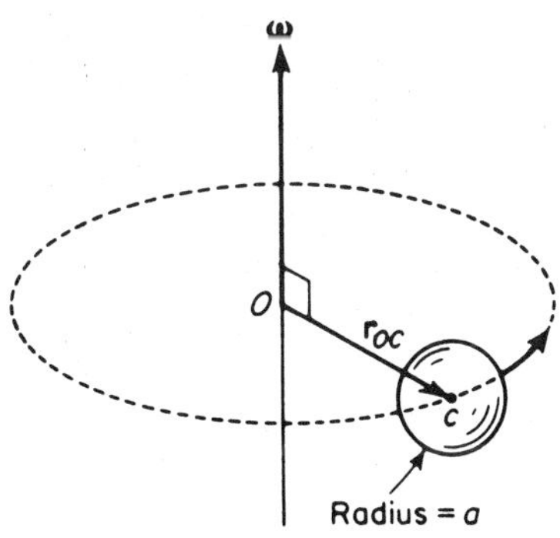

Figure 5-6.3. Rotation of a sphere about a noncentrally located axis.

Thus, the total force experienced by the dumbbell is

$$\mathbf{F} = -6\pi\mu\boldsymbol{\omega} \times (a_1\mathbf{r}_{C1} + a_2\mathbf{r}_{C2}) \tag{5–6.11}$$

Since the location of C is governed by Eq. (5–6.5), the term in parentheses vanishes. This makes $\mathbf{F} = \mathbf{0}$, in accordance with the general theorem.

As has already been noted in Eq. (5–4.18), the dependence of the rotation tensor on origin, for a body possessing a center of hydrodynamic stress, is governed by the equation

$$\boldsymbol{\Omega}_O = \boldsymbol{\Omega}_C - \mathbf{r}_{OC} \times \mathbf{K} \times \mathbf{r}_{OC} \tag{5–6.12}$$

In the special case of a body which is isotropic with respect to translation, that is, $\mathbf{K} = \mathbf{I}K$, the last term becomes

$$-\mathbf{r}_{OC} \times \mathbf{K} \times \mathbf{r}_{OC} = [\mathbf{I}(r_{OC})^2 - \mathbf{r}_{OC}\mathbf{r}_{OC}]K \tag{5–6.13}$$

in which $(r_{OC})^2 = \mathbf{r}_{OC}\cdot\mathbf{r}_{OC}$.

As a simple example of the physical significance and utility of Eq. (5–6.12), consider the problem of a sphere rotating about an axis which does not pass through its center, as in Fig. 5–6.3. The sphere is assumed to be joined to the stationary point O by a thin, rigid rod of negligible resistance. The point O lies at the intersection of the axis of rotation and the perpendicular dropped from C to this axis. Since the sphere is isotropic with respect to both translation and rotation about its center, we find with the aid of Eqs. (5–2.22) and (5–3.13) that

$$\boldsymbol{\Omega}_O = \mathbf{I}[8\pi a^3 + 6\pi a(r_{OC})^2] - \mathbf{r}_{OC}\mathbf{r}_{OC}\,6\pi a \tag{5–6.14}$$

Hence, from Eq. (5–4.2), since $\mathbf{U}_O = \mathbf{0}$ and $\mathbf{r}_{OC}\cdot\boldsymbol{\omega} = 0$, the torque about O is found to be

$$\mathbf{T}_O = -8\pi\mu a^2\boldsymbol{\omega} - 6\pi\mu a(r_{OC})^2\boldsymbol{\omega} \tag{5–6.15}$$

The first term is the torque due to the rotation of the sphere about its own axis. The second term arises from the translation of the sphere center through the fluid with velocity $\mathbf{U}_C = \boldsymbol{\omega} \times \mathbf{r}_{OC}$, relative to the fluid at infinity.

It is worthwhile noting that necessary criteria for the applicability of this relation are that the two dimensionless groups, $a^2|\boldsymbol{\omega}|\rho/\mu$ and $r_{OC}a|\boldsymbol{\omega}|\rho/\mu$, *both* be small, the last group arising from the translational motion of the sphere center.

There is at least one detailed solution available in the literature which provides confirmation of Eq. (5–6.12). Roscoe[45] theoretically studied the slow motion of an elliptic disk: (**I**) rotating about its minor axis, (**II**) rotating broadside-on about a tangent at one end of its minor axis, and (**III**) translating broadside-on. The center of hydrodynamic stress is obviously at the center of the disk. We choose O at the extreme end of the minor axis. When the couple associated with **I** and force arising from **III** are appropriately combined via Eq. (5–6.12), the resultant torque obtained from Eq. (5–4.2) (with $\mathbf{U}_O = \mathbf{0}$) agrees identically with Roscoe's calculation for case **II**.

5-7 Terminal Settling Velocity of an Arbitrary Particle

We shall confine our discussion to the "steady" motion of a particle in a gravitational field, in which neither its linear nor angular velocity changes with time. Such motion will be characterized by the existence of some point Q which translates through the fluid at a constant velocity $\mathbf{U}_Q$ and through which passes an axis about which the particle spins with a constant angular velocity $\boldsymbol{\omega}$. In such cases the net external forces and torques about Q must vanish.

The external forces for motion in a gravitational field are;

(i) the quasi-static hydrodynamic force:

$$\mathbf{F} = -\mu(\mathbf{K}\cdot\mathbf{U}_Q + \mathbf{C}_Q^{\dagger}\cdot\boldsymbol{\omega}) \tag{5–7.1}$$

(ii) the buoyant force:

$$\mathbf{F}^{(\text{buoy})} = -m_f\mathbf{g} \tag{5–7.2}$$

(iii) the gravitational force:

$$\mathbf{F}^{(\text{grav})} = m_p\mathbf{g} \tag{5–7.3}$$

in which m_f and m_p are, respectively, the masses of displaced fluid and particle, and $\mathbf{g}$ is the acceleration of gravity vector, having the magnitude of the local acceleration of gravity, g, and directed vertically downward. At equilibrium we have

$$\mathbf{F} + \mathbf{F}^{(\text{buoy})} + \mathbf{F}^{(\text{grav})} = \mathbf{0} \tag{5–7.4}$$

or

$$\mu(\mathbf{K}\cdot\mathbf{U}_Q + \mathbf{C}_Q^{\dagger}\cdot\boldsymbol{\omega}) = (m_p - m_f)\mathbf{g} \tag{5–7.5}$$

Similarly, the external torques about Q are;

(i) the quasi-static hydrodynamic torque:

$$\mathbf{T}_Q = -\mu(\mathbf{C}_Q\cdot\mathbf{U}_Q + \boldsymbol{\Omega}_Q\cdot\boldsymbol{\omega}) \tag{5–7.6}$$

(ii) the buoyant torque:

$$\mathbf{T}_Q^{(\text{buoy})} = -\mathbf{r}_{QB}\times(m_f\mathbf{g}) \tag{5–7.7}$$

(iii) the gravitational torque:

$$\mathbf{T}_Q^{(\text{grav})} = \mathbf{r}_{QM}\times(m_p\mathbf{g}) \tag{5–7.8}$$

Here, B is the center of buoyancy and M is the center of mass. Again, at equilibrium we have that

$$\mathbf{T}_Q + \mathbf{T}_Q^{(\text{buoy})} + \mathbf{T}_Q^{(\text{grav})} = \mathbf{0} \tag{5–7.9}$$

or

$$(\mathbf{C}_Q\cdot\mathbf{U}_Q + \boldsymbol{\Omega}_Q\cdot\boldsymbol{\omega}) = (m_p\mathbf{r}_{QM} - m_f\mathbf{r}_{QB})\times\mathbf{g} \tag{5–7.10}$$

Not all bodies are capable of attaining steady terminal motions of this type. (With unsymmetrical bodies, spiraling or wobbling movement may occur.) Questions of the stability of such motions would require the introduction of unsteady state terms. We shall discuss here only those terminal states which are dynamically possible in the sense of Eqs. (5–7.5) and (5–7.10). We note that in all cases the rotational Reynolds number

$$N_{\text{Re}}^{(r)} = \frac{c^2|\boldsymbol{\omega}|\rho}{\mu} \tag{5–7.11}$$

as well as the translational Reynolds number must be small to allow application of the quasi-steady creeping motion equations (see Section 2–10).

Consider first the class of *spin-free* terminal states characterized by $\boldsymbol{\omega}_\infty = \mathbf{0}$. The subscript infinity refers to the terminal state. All points of the particle then move with the same terminal velocity, $\mathbf{U}_\infty$. It is convenient in this case to refer all parameters to an origin at the center of reaction, in which case the pertinent equations become

$$\mu\mathbf{K}\cdot\mathbf{U}_\infty = (m_p - m_f)\mathbf{g} \tag{5–7.12a}$$

$$\mu\mathbf{C}_R\cdot\mathbf{U}_\infty = (m_p\mathbf{r}_{RM} - m_f\mathbf{r}_{RB}) \times \mathbf{g} \tag{5–7.12b}$$

Eliminating $\mathbf{U}_\infty$ between these two equations we obtain

$$(\mathbf{C}_R\cdot\mathbf{K}^{-1} - \mathbf{R}_R \times \mathbf{I})\cdot\mathbf{g} = \mathbf{0} \tag{5–7.13}$$

in which $\mathbf{K}^{-1}$ is the dyadic reciprocal to $\mathbf{K}$, and

$$\mathbf{R}_R = \frac{m_p\mathbf{r}_{RM} - m_f\mathbf{r}_{RB}}{m_p - m_f} \tag{5–7.14}$$

It should be noted that in Eq. (5–7.13) the dyadic in parentheses involves parameters which depend only on the external and internal geometry of the particle and on the distribution of its mass. These parameters are constant relative to body axes; $\mathbf{g}$ is constant only relative to space axes.

Equation (5–7.13) constitutes the necessary and sufficient condition that the terminal state of the particle be spin-free. We now consider the various ways in which this condition can be satisfied.

CASE I: $\mathbf{C}_R = \mathbf{0}$ and $\mathbf{R}_R = \mathbf{0}$

We inquire first as to the class of bodies for which spin-free, terminal states are possible for *all* orientations of the particle relative to the direction of the gravity field. This can occur if and only if the quantity in parentheses in Eq. (5–7.13) vanishes identically. If we put this dyadic equal to zero and postmultiply by $\mathbf{K}$ we obtain

$$\mathbf{C}_R - \mathbf{R}_R \times \mathbf{K} = \mathbf{0} \tag{5–7.15}$$

As $\mathbf{C}_R$ is symmetric, this relation can be satisfied only if $\mathbf{R}_R \times \mathbf{K}$ is symmetric. But since $\mathbf{K}$ is itself symmetric, $\mathbf{R}_R \times \mathbf{K}$ can never be symmetric, unless $\mathbf{R}_R = \mathbf{0}$. From Eq. (5–7.15) this, in turn, requires that $\mathbf{C}_R = \mathbf{0}$. Hence, only those bodies for which both $\mathbf{C}_R$ and $\mathbf{R}_R$ vanish can attain spin-free terminal states for all orientations.

An ellipsoid of uniform density is an example of such a body, for not only is $\mathbf{C}_R = \mathbf{O}$ (see Section 5–5, Case 3) but R, M, and B coincide, so that $\mathbf{R}_R = \mathbf{0}$.

Bodies of revolution which possess fore-aft symmetry and are of uniform density (see Section 5–5, Case 5) will also meet this requirement.

Spherically isotropic bodies (see Section 5–5, Case 4) of uniform density not only have the property of falling stably in any orientation (*neutrally stable* in the terminology of buoyancy theory[4]), but are also isotropic with respect to translation. Bodies of this type will therefore always move vertically with velocity

$$\mathbf{U}_\infty = \frac{(m_p - m_f)}{\mu K}\mathbf{g} \tag{5–7.16}$$

Neutrally stable *anisotropic* bodies, such as ellipsoids, behave more interestingly than their isotropic counterparts. Although the former fall stably in any orientation they do not generally fall vertically downward unless, perchance, they are dropped with a principal axis of translation parallel to the gravity field; otherwise, such bodies drift off to the side in the course of settling. A quantitative example of this type of behavior is furnished by the fall of a circular disk, discussed later in this section.

Case II: $\mathbf{C}_R = \mathbf{O}$ and $\mathbf{R}_R \neq \mathbf{0}$

Ellipsoids of nonuniform density, and homogeneous or inhomogeneous bodies of revolution lacking fore-aft symmetry are examples of bodies for which $\mathbf{C}_R = \mathbf{O}$ but for which $\mathbf{R}_R \neq \mathbf{0}$, since $\mathbf{r}_{RM} \neq \mathbf{0}$. Bodies of this class are capable of attaining spin-free terminal conditions for only certain orientations. From Eq. (5–7.13), these orientations are governed by the requirement that

$$\mathbf{R}_R \times \mathbf{g} = \mathbf{0} \tag{5–7.17}$$

Such bodies can therefore settle without rotation only when $\mathbf{R}_R$ is parallel to $\mathbf{g}$.

The condition that $\mathbf{R}_R$ be parallel to $\mathbf{g}$ corresponds to two possible orientations of the body: one in which $\mathbf{R}_R$ points vertically downward and the other in which it points vertically upward. Of the two possibilities, only the latter corresponds to a *stable* equilibrium. The former state is *unstable* in the sense that a small departure from this orientation produces a couple which acts in such a direction as further to increase the departure of the particle from its original orientation. Conversely, the latter state is stable because the direction of the couple produced by a slight change in orientation tends to restore the original state. The proof of these remarks follows closely that employed in questions of the static stability of submerged bodies to overturning[4]. In fact, the condition that $\mathbf{R}_R$ be directed vertically upward is completely analogous to the static stability requirement that the center of gravity lie vertically above the center of mass.

The ultimate orientation achieved by a nonskew settling particle is therefore that for which the following two conditions are met:

$$\mathbf{R}_R \text{ is parallel to } \mathbf{g} \tag{5–7.18}$$

and

$$\mathbf{R}_R \cdot \mathbf{g} \geq 0 \tag{5–7.19}$$

It is interesting to note from Eq. (5–7.14) that the ultimate orientation of a settling particle is not generally a property of the particle alone. Rather, this orientation will generally differ in fluids of different density, owing to the dependence of $\mathbf{R}_R$ on m_f. If, however, the density of the body is uniform, so that $\mathbf{r}_{MB} = \mathbf{0}$, this ceases to be the case; for then $\mathbf{r}_{RM} = \mathbf{r}_{RB}$.

A simple example of the application of these criteria is furnished by considering the fall of the asymmetric dumbbell previously discussed in Section 5–6, see Fig. 5–6.1. It is assumed that the two unequal spheres are separately homogeneous, but that their respective densities are not necessarily equal. After considerable reduction, Eq. (5–7.14) here takes the form

$$\mathbf{R}_R = \mathbf{e}h\frac{4}{3}\pi\frac{a_1 a_2}{a_1 + a_2}[(\rho_2 - \rho)a_2^3 - (\rho_1 - \rho)a_1^3] \tag{5–7.20}$$

The condition Eq. (5–7.18) requires that the body fall in such a way that $\mathbf{e}$, the unit vector drawn from the larger to the smaller sphere, be parallel to $\mathbf{g}$; that is, the dumbbell moves parallel to the line of centers, an obviously correct inference. The stability condition, Eq. (5–7.19), requires that

$$[(\rho_2 - \rho)a_2^3 - (\rho_1 - \rho)a_1^3](\mathbf{e}\cdot\mathbf{g}) \geq 0 \tag{5–7.21}$$

Thus, if the term in brackets is positive (as, for example, if $\rho_2 \geq \rho_1 > \rho$), the dumbbell will move in such a way that $(\mathbf{e}\cdot\mathbf{g}) < 0$. This implies that the larger sphere will be situated *below* the smaller one when steady motion is ultimately achieved. The converse is equally true.

By an appropriate choice of densities and sizes, it is always possible to arrange matters such that a larger, heavier sphere falls above a smaller, lighter one.

If the term in brackets vanishes, that is to say, if $\mathbf{R}_R = \mathbf{0}$ as in Case I, the dumbbell ceases to have any unique terminal orientation.

Bodies of uniform density (that is, homogeneous bodies) are the type most often encountered in practice. Here, the centers of mass and buoyancy coincide, so that $\mathbf{r}_{BM} = \mathbf{0}$. This simplifies the preceding orientation questions considerably. In particular, the terminal attitude of the body is now simply determined by the condition that $\mathbf{r}_{RM}$ be parallel to $\mathbf{g}$. A nonskew homogeneous body will therefore fall with an orientation such that the line joining its centers of hydrodynamic stress (that is, center of reaction) and mass lies parallel to the direction of the gravity field. Of the two possible directions along this line, that direction for which

$$(m_p - m_f)\mathbf{r}_{RM}\cdot\mathbf{g} \geq 0 \tag{5–7.22}$$

is the direction of stable fall. If the particle is denser than the fluid, M will lie directly below R, and conversely.

CASE III: $\mathbf{C}_R \neq \mathbf{0}$ and $\mathbf{R}_R = \mathbf{0}$

Spin-free terminal settling for this case requires that

$$\mathbf{C}_R \cdot \mathbf{K}^{-1} \cdot \mathbf{g} = \mathbf{0}$$

As neither $\mathbf{C}_R$ nor $\mathbf{K}$ is zero, this equation will be satisfied if and only if $\mathbf{C}_R \cdot \mathbf{K}^{-1}$ is an *incomplete* dyadic[22]*. Since $\mathbf{K}$ and, hence, $\mathbf{K}^{-1}$ are complete, the satisfaction of this relation requires that $\mathbf{C}_R$ be incomplete[+].

An example of this behavior is furnished by the screw-propeller of Fig. 5–4.1. If the disks are homogeneous and of equal density then $\mathbf{R}_R = \mathbf{0}$. Also, from Eqs. (5–4.43) and (5–4.33)

$$\mathbf{C}_R = \tfrac{32}{3} ch(\mathbf{i}_1 \mathbf{i}_1 - \mathbf{i}_2 \mathbf{i}_2) \sin\theta \cos\theta \tag{5–7.23}$$

and
$$\mathbf{K}^{-1} = \frac{3}{32c}\left(\frac{\mathbf{i}_1 \mathbf{i}_1}{2 + \cos^2\theta} + \frac{\mathbf{i}_2 \mathbf{i}_2}{2 + \sin^2\theta} + \frac{\mathbf{i}_3 \mathbf{i}_3}{2}\right) \tag{5–7.24}$$

Thus, $\mathbf{C}_R \cdot \mathbf{K}^{-1} \cdot \mathbf{g}$ will be zero if the orientation of the propeller is such that the x_3 axis is parallel to the gravity field. In this case the spin-free terminal motion is

$$\mathbf{U}_\infty = \mathbf{g}\,\frac{3(m_p - m_f)}{64\mu c} \tag{5–7.25}$$

CASE IV: $\mathbf{C}_R \neq \mathbf{0}$ and $\mathbf{R}_R \neq \mathbf{0}$

Rotation-free terminal motions are possible if and only if

$$\det(\mathbf{C}_R \cdot \mathbf{K}^{-1} - \mathbf{R}_R \times \mathbf{I}) = 0$$

This criterion applies, of course, to all cases.

A particular example leading to spin-free motion in the present case occurs when the following two conditions are *simultaneously* satisfied:

$$\mathbf{C}_R \cdot \mathbf{K}^{-1} \cdot \mathbf{g} = \mathbf{0} \quad \text{and} \quad \mathbf{R}_R \times \mathbf{g} = \mathbf{0} \tag{5–7.26}$$

*A dyadic $\mathbf{Q}$ is complete if its determinant

$$\det \mathbf{Q} = \tfrac{1}{6}\mathbf{Q} \,{}^{\times}_{\times}\, \mathbf{Q} : \mathbf{Q}$$

does not vanish. Conversely, it is incomplete if its determinant does vanish. Our multiple dot and cross notation is that of Gibbs[22]; namely,

$$\mathbf{ab} \,{}^{\times}_{\times}\, \mathbf{cd} = (\mathbf{a} \times \mathbf{c})(\mathbf{b} \times \mathbf{d})$$

$$\mathbf{ef} : \mathbf{gh} = (\mathbf{e} \cdot \mathbf{g})(\mathbf{f} \cdot \mathbf{h})$$

where $\mathbf{a}, \mathbf{b}, \ldots$ are any vectors.

[+]This follows from the general relation

$$\det(\mathbf{A} \cdot \mathbf{B}) = (\det \mathbf{A})(\det \mathbf{B})$$

valid for any dyadics $\mathbf{A}$ and $\mathbf{B}$.

Since the latter condition requires that $\mathbf{R}_R$ be parallel to $\mathbf{g}$, the former condition may be written as

$$\mathbf{C}_R \cdot \mathbf{K}^{-1} \cdot \mathbf{R}_R = \mathbf{0} \tag{5–7.27}$$

As $\mathbf{R}_R$ is non-zero and $\mathbf{K}^{-1}$ is necessarily complete, this condition requires, among other things, that $\mathbf{C}_R$ be an incomplete dyadic.

The "impeller" shown in Fig. 5–4.4 is an example of a body of this class. If the disks are of uniform and equal densities then, from Eq. (5–4.45), since

$$\mathbf{r}_{RM} = \mathbf{r}_{RB} = \mathbf{i}_1 \frac{h \sin\theta \cos\theta}{4 + \sin^2\theta}$$

Eq. (5–7.14) yields

$$\mathbf{R}_R = \mathbf{i}_1 \frac{h \sin\theta \cos\theta}{4 + \sin^2\theta} \tag{5–7.28}$$

Also, $\mathbf{C}_R$ is given by Eq. (5–4.46) whereas $\mathbf{K}^{-1}$ has (coincidentally) the same value as in Eq. (5–7.24). Equation (5–7.27) is therefore satisfied. We conclude, then, that a dynamically possible, spin-free, terminal motion of the impeller occurs when it settles in a manner such that the x_1 axis is parallel to the gravity field. The terminal motion under these circumstances is

$$\begin{aligned} \mathbf{U}_\infty &= \mathbf{g}\frac{3(m_p - m_f)}{32(2 + \cos^2\theta)\mu c} \\ \boldsymbol{\omega}_\infty &= \mathbf{0} \end{aligned} \tag{5–7.29}$$

Only by applying an appropriate stability criterion can we determine whether the actual terminal motion is such that the positive or negative x_1 axis is directed vertically downward, assuming, of course, that at least one of these two modes is stable.

Next, we consider some simple examples of steady motion for which $\boldsymbol{\omega}_\infty \neq \mathbf{0}$. Rather than attempt a general theory of such motions, we content ourselves with examining a few special cases which are dynamically possible in the sense of Eqs. (5–7.5) and (5–7.10).

For example, in the case of the screw-propeller of Fig. 5–4.1, a possible mode of steady spinning motion occurs when the x_1 axis is parallel to the gravity field, this motion being of the form

$$\begin{aligned} (\mathbf{U}_M)_\infty &= \mathbf{g}\frac{(m_p - m_f)(2 + \sin^2\theta)}{64\mu c} \\ \boldsymbol{\omega}_\infty &= -\mathbf{g}\frac{(m_p - m_f)\sin\theta\cos\theta}{64\mu c h} \end{aligned} \tag{5–7.30}$$

For $-\pi/2 < \theta < 0$ the propeller is right-handed with respect to the x_1 axis in the sense that the spin vector points in the same direction as the particle is moving. Conversely, for $\pi/2 > \theta > 0$ the propeller is left-handed with respect to this axis.

The case in which the x_2 axis of the screw-propeller in Fig. 5–4.1 lies parallel to the gravity field affords another example of a steady, spinning, terminal motion, the details of which are left to the interested reader.

It is interesting to note that the impeller in Fig. 5–4.4 (assumed homogeneous) cannot attain a terminal spinning motion with the x_2 axis parallel to $\mathbf{g}$. For $\mathbf{R}_R$ would not be parallel to $\mathbf{g}$, and the resulting buoyant and gravitational torques about R would cause the body to turn about the x_3 axis, eventually attaining a steady, spin-free terminal motion with the x_1 axis parallel to $\mathbf{g}$.

Of the large class of bodies which lack a center of hydrodynamic stress, that is, those for which $\mathbf{C}_R \neq \mathbf{O}$, the most fascinating are those which are helicoidally isotropic. Their terminal motion is independent of orientation, even if they are inhomogeneous. This ultimate motion consists of a vertical translation of the centroid with velocity

$$(\mathbf{U}_R)_\infty = \mathbf{g}\frac{(m_p - m_f)\Omega}{\mu(K\Omega - C^2)} \tag{5–7.31a}$$

and a spin about a vertical axis through the centroid at an angular velocity

$$\boldsymbol{\omega}_\infty = -\mathbf{g}\frac{(m_p - m_f)C}{\mu(K\Omega - C^2)} \tag{5–7.31b}$$

This motion is such that the trajectory of every point in the body, except for those along the spin axis, is a helix whose pitch is $|C|/\Omega$.

We note from Eq. (5–4.25) that the preceding denominators are both positive. Moreover, Ω is positive. Hence, mirror-image isotropic helicoids, which differ only in the algebraic sign of C, settle at the same velocity but with opposite spins. For $C < 0$ the body is right-handed whereas for $C > 0$ the body is left-handed.

The idea of two particles differing only in sense is not restricted to isotropic bodies. Thus, the screw-propeller in Fig. 5–4.1 will, for the two different choices $\theta = +\alpha$ and $\theta = -\alpha$, give rise to mirror-image bodies, identical in all properties except sense. Another example is furnished by two threaded screws differing only in their sense of threading.

Ultimate trajectory of a nonskew particle

Consider a nonskew ($\mathbf{C}_R = \mathbf{O}$) particle settling in its terminal state of motion through an unbounded fluid. The ultimate velocity of this particle may be determined from Eq. (5–7.12a) if both its translation tensor and the equilibrium orientation of its principal axes of translation relative to the gravity field are known. Thus

$$\mathbf{U}_\infty = \frac{(m_p - m_f)}{\mu}\mathbf{K}^{-1}\cdot\mathbf{g} \tag{5–7.32}$$

Let (x', y', z') be a set of cartesian axes, fixed in the particle, lying parallel to the principal axes of translation of the latter. Moreover, let (x, y, z) be a second set, fixed in space, chosen in such a way that the positive z axis points vertically downward. This makes $\mathbf{g} = \mathbf{k}g$. The translation dyadic referred to its principal axes is then given by Eq. (5–2.21). A simple calculation yields the reciprocal dyadic,

$$\mathbf{K}^{-1} = \mathbf{i}'\mathbf{i}'K_1^{-1} + \mathbf{j}'\mathbf{j}'K_2^{-1} + \mathbf{k}'\mathbf{k}'K_3^{-1} \tag{5–7.33}$$

where $(\mathbf{i}', \mathbf{j}', \mathbf{k}')$ are unit vectors parallel to the principal axes. In Eq. (5–7.32) we note that $m_p - m_f = V\,\Delta\rho$ where V is the particle volume and $\Delta\rho = \rho_p - \rho$ is the algebraic difference between particle and fluid densities. If, now, one sets

$$\mathbf{U}_\infty = \mathbf{i}(U_x)_\infty + \mathbf{j}(U_y)_\infty + \mathbf{k}(U_z)_\infty \tag{5–7.34}$$

the component velocities eventually obtained from Eq. (5–7.32) are found to be

$$\begin{aligned}(U_x)_\infty &= (l_{11}l_{31}K_1^{-1} + l_{12}l_{32}K_2^{-1} + l_{13}l_{33}K_3^{-1})gV\,\Delta\rho/\mu \\ (U_y)_\infty &= (l_{21}l_{31}K_1^{-1} + l_{22}l_{32}K_2^{-1} + l_{23}l_{33}K_3^{-1})gV\,\Delta\rho/\mu \\ (U_z)_\infty &= (l_{31}^2K_1^{-1} + l_{32}^2K_2^{-1} + l_{33}^2K_3^{-1})gV\,\Delta\rho/\mu\end{aligned} \tag{5–7.35}$$

where the l-values are the appropriate direction cosines,

$$l_{31} = (\mathbf{k}\cdot\mathbf{i}') = \cos(z, x'), \quad l_{13} = (\mathbf{i}\cdot\mathbf{k}') = \cos(x, z'), \quad \text{etc.} \tag{5–7.36}$$

As a definite example, consider a thin circular disk of radius c and thickness b ($c \gg b$) falling obliquely, as in Fig. 5–7.1. The disk, assumed homogeneous, is neutrally stable (that is, $\mathbf{R}_R = \mathbf{0}$) and can fall stably in *any* orientation. The principal axes of translation obviously lie in the plane of the disk and along its normal. The normal to the plane of the disk corresponds to the z' axis, and the angle between this and $\mathbf{g}$ is denoted by Φ ($0 \le \Phi \le \pi/2$). The y and y' axes coincide and lie in the plane of the disk. They are directed out of the plane of the paper. By these means the x' and z' axes may be regarded as deriving from the x and z axes, respectively, by a rotation about the y, y' axis through the angle Φ. The x and z axes then lie in the same plane as the corresponding primed axes. To be unambiguous we also specify that the angle between the x' and z axes lie between 0 and $\pi/2$.

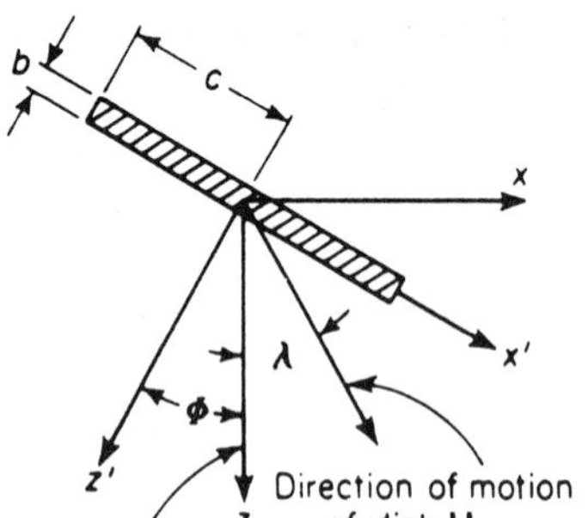

Figure 5-7.1. Oblique fall of a circular disk.

The direction cosines appropriate to this rotation are

$$l_{11} = l_{33} = \cos\Phi; \qquad -l_{13} = l_{31} = \sin\Phi; \qquad l_{22} = 1 \tag{5–7.37}$$

all others being zero. The principal translational resistances of the disk are

$$K_1 = K_2 = \tfrac{32}{3}c, \qquad K_3 = 16c \tag{5–7.38}$$

As the disk volume is $V = \pi c^2 b$ the components of the settling velocity are, from Eq. (5–7.35),

$$(U_x)_\infty = \frac{\pi cbg\,\Delta\rho}{64\mu}\sin 2\Phi, \qquad (U_y)_\infty = 0$$

$$(U_z)_\infty = \frac{\pi cbg\,\Delta\rho}{64\mu}(5 - \cos 2\Phi) \tag{5–7.39}$$

which indicate that the disk depicted in Fig. 5–7.1 moves down and to the right in the course of settling.

If λ denotes the angle between the downward vertical and the direction of motion of the disk, then

$$\lambda = \tan^{-1}\left(\frac{dx}{dz}\right) = \tan^{-1}\left(\frac{\sin 2\Phi}{5 - \cos 2\Phi}\right) \tag{5–7.40}$$

This angle is a maximum when the disk orientation is

$$\Phi = \tfrac{1}{2}\cos^{-1}\tfrac{1}{5} = 39.2^\circ \tag{5–7.41}$$

and is

$$\lambda = \tan^{-1}\frac{\sqrt{6}}{12} = 11.5^\circ \tag{5–7.42}$$

corresponding to a maximum ratio of horizontal to vertical velocities of

$$\left[\frac{(U_x)_\infty}{(U_z)_\infty}\right]_{\max} = \frac{\sqrt{6}}{12} = 0.204 \tag{5–7.43}$$

5-8 Average Resistance to Translation

Repeated experiments performed on a nonskew, neutrally stable, anisotropic particle will in general show a spread in terminal settling velocities, owing to the dependence of settling velocity on orientation. It is of some interest to calculate the average value, $\overline{\mathbf{U}}_\infty$, which would be observed in a long sequence of experiments in which the particle was dropped with random orientation into the fluid. The result is also of interest in connection with certain phenomena relating to Brownian motion.

In order to specify the orientation of the body we introduce the three *Eulerian* angles (ϕ, ψ, θ)[23] relating body coordinates, (x', y', z'), rigidly affixed to the particle, to space

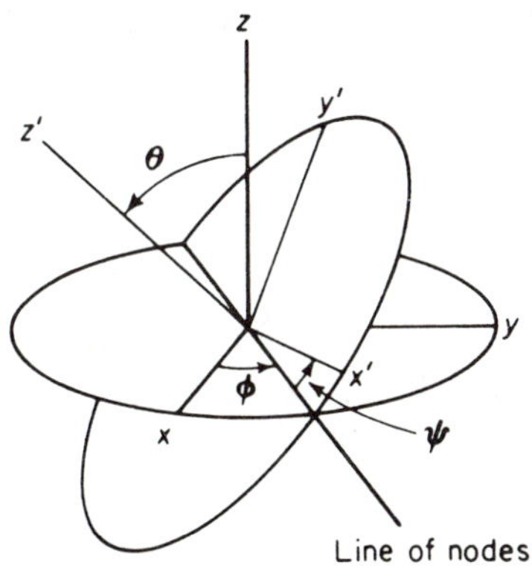

Figure 5-8.1. Eulerian angles.

coordinates, (x, y, z), fixed in the fluid. The latter system may be regarded as deriving from the former by successive rotations through the angles ϕ, θ, and ψ, as in Fig. 5–8.1. These two systems are connected through the matrix transformation[23]

$$\begin{pmatrix} x \\ y \\ z \end{pmatrix} = \begin{pmatrix} \cos\psi\cos\phi - \cos\theta\sin\phi\sin\psi & -\sin\psi\cos\phi - \cos\theta\sin\phi\cos\psi & \sin\theta\sin\phi \\ \cos\psi\sin\phi + \cos\theta\cos\phi\sin\psi & -\sin\psi\sin\phi + \cos\theta\cos\phi\cos\psi & -\sin\theta\cos\phi \\ \sin\theta\sin\psi & \sin\theta\cos\psi & \cos\theta \end{pmatrix} \begin{pmatrix} x' \\ y' \\ z' \end{pmatrix} \tag{5–8.1}$$

But, if l_{ij} $(i, j = 1, 2, 3)$ are the direction cosines referred to in Eq. (5–7.36), the two coordinate systems are also connected by the matrix relation

$$\begin{pmatrix} x \\ y \\ z \end{pmatrix} = \begin{pmatrix} l_{11} & l_{12} & l_{13} \\ l_{21} & l_{22} & l_{23} \\ l_{31} & l_{32} & l_{33} \end{pmatrix} \begin{pmatrix} x' \\ y' \\ z' \end{pmatrix} \tag{5–8.2}$$

The direction cosines required in Eq. (5–7.35) may therefore be expressed in terms of the three Eulerian angles by equating corresponding elements in Eqs. (5–8.1) and (5–8.2).

The orientation of the body is described by specifiying the Eulerian angles relative to the space coordinates fixed in the fluid. As the Eulerian angles are independent of each other, the probability $P(\phi, \psi, \theta)\, d\theta\, d\psi\, d\phi$ that the orientation of the body lies between θ and $\theta + d\theta$, ψ and $\psi + d\psi$, and ϕ and $\phi + d\phi$ is

$$P(\phi, \psi, \theta)\, d\theta\, d\psi\, d\phi = \frac{1}{8\pi^2} \sin\theta\, d\theta\, d\psi\, d\phi \tag{5–8.3}$$

where all orientations have been assumed equally probable. The extraneous factor of $8\pi^2$ results from normalizing the probability distribution function, that is,

$$\int_{\phi=0}^{2\pi} \int_{\psi=0}^{2\pi} \int_{\theta=0}^{\pi} P(\phi, \psi, \theta)\, d\theta\, d\psi\, d\phi = 1 \tag{5–8.4}$$

The average settling velocity may now be obtained from the relation

$$\overline{\mathbf{U}}_\infty = \frac{1}{8\pi^2} \int_0^{2\pi} \int_0^{2\pi} \int_0^{\pi} \mathbf{U}_\infty(\phi, \psi, \theta) P(\phi, \psi, \theta)\, d\theta\, d\psi\, d\phi \tag{5–8.5}$$

where $\mathbf{U}_\infty(\phi, \psi, \theta)$ is obtained from Eqs. (5–7.34), (5–7.35), (5–8.1), and (5–8.2). A straightforward integration then gives

$$\overline{\mathbf{U}}_\infty = \mathbf{k} \frac{gV\,\Delta\rho}{3\mu} \left(\frac{1}{K_1} + \frac{1}{K_2} + \frac{1}{K_3} \right) \tag{5–8.6}$$

This shows that, on the average, the direction of settling is parallel to the gravitational field. Thus, if we write

$$\mathbf{F} = -\mu \bar{K} \mathbf{U}_\infty \tag{5–8.7}$$

the average resistance, $\bar{K}$, is the scalar given by

$$\frac{1}{\bar{K}} = \frac{1}{3}\left(\frac{1}{K_1} + \frac{1}{K_2} + \frac{1}{K_3}\right) \tag{5–8.8}$$

In the case of a circular disk, for example, we find from Eq. (5–7.38) that

$$\bar{K} = 12\, c \tag{5–8.9}$$

Equation (5–8.8) also gives the proper average resistance to be used in formulas of the Stokes-Einstein type[19], relating to the translational Brownian motion of colloidal particles of arbitrary shape, such particles moving under the influence of *random* impacts by solvent molecules.

5-9 The Resistance of a Slightly Deformed Sphere

One problem encountered in practical applications of the preceding theory arises because many particles encountered in industrial practice and in nature are irregular, and thus do not possess a shape which is amenable to expression in a simple coordinate system. As a simple model of such bodies*, Brenner[13] has developed the case for a slightly deformed sphere, which we reproduce here. In principle, the method should be applicable to asymmetric bodies, but as yet it has not been carried forward to considering deformed bodies such as would spin during settling under the influence of gravity.

It is assumed that the surface of the deformed sphere may be described by an equation of the form

$$r = a\,[1 + \epsilon f(\theta, \phi)] \tag{5–9.1}$$

in which (r, θ, ϕ) are spherical coordinates having their origin at the center of the undeformed sphere of radius $r = a$; $|\epsilon| \ll 1$ is a small, dimensionless parameter and $f(\theta, \phi)$ is an arbitrary function of angular position which is of $O(1)$ with respect to the parameter ϵ. The "axis" $|\cos\theta| = 1$ may be chosen arbitrarily. The parameter ϵ is, of course, an artificial parameter, being inseparable from $f(\theta, \phi)$. It is introduced merely as an aid in directing attention to the order of the approximation of the results ultimately obtained. In the final results ϵ and $f(\theta, \phi)$ appear in the combination $\epsilon f(\theta, \phi)$, so it is immaterial how ϵ is chosen. Any single-valued function of θ and ϕ may be expanded in a series of surface spherical harmonics, $f_k(\theta, \phi)$; hence, no loss in generality results from taking the surface of the body to be of the form

$$r = a[1 + \epsilon \sum_{k=0}^{\infty} f_k(\theta, \phi)] \tag{5–9.2}$$

*An independent, though less complete, treatment of the same problem is given by A. Acrivos and T. D. Taylor, Chem. Eng. Sci. **19** (1964), 445.

Nonconvex bodies, for which r is not a single-valued function of θ and ϕ, cannot, of course, be described by such an equation.

At small translational and/or angular Reynolds numbers, $a|\mathbf{U}|\rho/\mu$ and $a^2|\boldsymbol{\omega}|\rho/\mu$, respectively, the fluid motion is governed by Stokes' equations,

$$\nabla^2\mathbf{v} = \frac{1}{\mu}\nabla p, \qquad \nabla\cdot\mathbf{v} = 0 \tag{5–9.3a,b}$$

For streaming flow past the deformed particle the appropriate boundary conditions are

$$\mathbf{v} = \mathbf{0} \quad \text{on } S_d, \qquad \mathbf{v} = \mathbf{U} \quad \text{at } r = \infty \tag{5–9.4a,b}$$

where S_d denotes the surface of the deformed sphere, Eq. (5–9.1) or Eq. (5–9.2). When the particle rotates about an axis passing through the origin ($r = 0$) in a fluid at rest at infinity the boundary conditions are

$$\mathbf{v} = \boldsymbol{\omega}\times\mathbf{r} \quad \text{on } S_d, \qquad \mathbf{v} = \mathbf{0} \quad \text{at } r = \infty \tag{5–9.5a,b}$$

where $\mathbf{r}$ is the position vector of a point relative to an origin at the center of the undeformed sphere, and $r = |\mathbf{r}|$. Because of the linearity of the Stokes equations and the boundary conditions, the case in which translational and rotational motions occur simultaneously may be synthesized by superposition of the results for the two separate cases.

By utilizing Stokes stream function, Sampson (see Section 4–25) succeeded in obtaining a solution of Eq. (5–9.3) to the first order in ϵ, for the special case of axisymmetric streaming flow parallel to the axis of the spheroid of revolution

$$r = a\left[1 + \epsilon \sum_{k=0}^{\infty} \alpha_k P_k(\cos\theta)\right] \tag{5–9.6}$$

where α_k is an arbitrary parameter of $O(1)$ with respect to ϵ; P_k is the Legendre polynomial of order k, and $\theta = 0$ (and its continuation, $\theta = \pi$) is the axis of the spheroid. Though Sampson did not utilize his expression for the stream function to calculate the force on the spheroid, it is easily obtained by application of a formula due to Payne and Pell[41] relating the force on an axisymmetric body to the asymptotic expansion of the stream function at great distances from the body, as in Eq. (4–14.20). In this way the force is found to be

$$\mathbf{F} = 6\pi\mu a\mathbf{U}\,[1 + \epsilon(\alpha_0 - \tfrac{1}{5}\alpha_2) + O(\epsilon^2)] \tag{5–9.7}$$

It is interesting to observe that only the terms corresponding to $k = 0$ and $k = 2$ contribute to the force.

Sampon's techniques admit of no direct generalization to asymmetric flows. As will appear in the following, the problem of solving Stokes' equations so as to satisfy the boundary conditions on the deformed sphere, to any order in ϵ, may be reduced to a sequence of related problems requiring the satisfaction of more complicated boundary conditions on the *undeformed* sphere. In this context the general solution of the creeping motion equations in terms of spherical harmonics, given in Section 3–2, is ideally suited to our purposes. For exterior problems, in which the fluid motion takes place in the infinite space external to the sphere $r = a$, the general solution is given by Eq. (3–2.31).

Streaming flow past a deformed sphere

In the present instance we wish to obtain a solution of the Stokes equations, Eq. (5–9.3), satisfying the boundary conditions Eq. (5–9.4) on a *deformed sphere.* It is assumed that the velocity and pressure fields can be expanded in powers of ϵ in the form

$$\mathbf{v} = \sum_{i=0}^{\infty} \epsilon^i \mathbf{v}^{(i)}, \qquad p = \sum_{i=0}^{\infty} \epsilon^i p^{(i)} \tag{5–9.8a,b}$$

Upon substituting these into Eq. (5–9.3) and equating terms involving like powers of ϵ, one finds that each of the individual perturbation fields, $(\mathbf{v}^{(i)}, p^{(i)})$, satisfies Stokes' equations,

$$\nabla^2 \mathbf{v}^{(i)} = \frac{1}{\mu} \nabla p^{(i)}, \qquad \nabla \cdot \mathbf{v}^{(i)} = 0 \tag{5–9.9a,b}$$

Likewise, substituting Eq. (5–9.8a) into the boundary condition at infinity, Eq. (5–9.4b), and equating terms involving like orders of ϵ it is found that

$$\mathbf{v}^{(0)} = \mathbf{U} \quad \text{at } r = \infty \tag{5–9.10}$$

and

$$\mathbf{v}^{(i)} = \mathbf{0} \quad \text{at } r = \infty \quad (i = 1,2,3, \ldots) \tag{5–9.11}$$

The remaining boundary condition, Eq. (5–9.4a), may be written in the form

$$\sum_{i=0}^{\infty} \epsilon^i \mathbf{v}^{(i)} = \mathbf{0} \quad \text{on } S_d \tag{5–9.12}$$

Now, by a Taylor series expansion about $r = a$ we obtain

$$\mathbf{v}^{(i)} = (\mathbf{v}^{(i)})_{r=a} + \sum_{j=1}^{\infty} \frac{1}{j!} (r - a)^j \left(\frac{\partial^{(j)} \mathbf{v}^{(i)}}{\partial r^{(j)}} \right)_{r=a}$$

Since $r - a = \epsilon a f(\theta, \phi)$ on S_d, the boundary condition Eq. (5–9.12) takes the form

$$\sum_{i=0}^{\infty} \epsilon^i \left[\mathbf{v}^{(i)} + \sum_{j=1}^{\infty} \frac{1}{j!} \epsilon^j a^j f^j(\theta, \phi) \left(\frac{\partial^{(j)} \mathbf{v}^{(i)}}{\partial r^{(j)}} \right) \right] = \mathbf{0} \quad \text{at } r = a$$

or, grouping together terms involving like powers of ϵ,

$$\mathbf{v}^{(0)} + \sum_{i=1}^{\infty} \epsilon^i \left[\mathbf{v}^{(i)} + \sum_{j=1}^{i} \frac{1}{j!} a^j f^j(\theta,\phi) \left(\frac{\partial^{(j)} \mathbf{v}^{(i-j)}}{\partial r^{(j)}} \right) \right] = \mathbf{0} \quad \text{at } r = a$$

Thus, the boundary condition on the deformed sphere may be satisfied to any order in ϵ by requiring that the various perturbation fields satisfy the following boundary conditions:

$$\mathbf{v}^{(0)} = \mathbf{0} \quad \text{at } r = a \tag{5–9.13}$$

and for $i = 1, 2, 3, \ldots,$

$$\mathbf{v}^{(i)} = -\sum_{j=1}^{i} \frac{1}{j!} a^j f^j(\theta, \phi) \left(\frac{\partial^{(j)} \mathbf{v}^{(i-j)}}{\partial r^{(j)}} \right) \quad \text{at } r = a \tag{5–9.14}$$

In explicit form, the first few terms in the expansion are

$$\mathbf{v}^{(0)} = \mathbf{0} \quad \text{at } r = a \tag{5–9.15}$$

$$\mathbf{v}^{(1)} = -af(\theta, \phi) \frac{\partial \mathbf{v}^{(0)}}{\partial r} \quad \text{at } r = a \tag{5–9.16}$$

$$\mathbf{v}^{(2)} = -\left[af(\theta, \phi) \frac{\partial \mathbf{v}^{(1)}}{\partial r} + \frac{1}{2} a^2 f^2(\theta, \phi) \frac{\partial^2 \mathbf{v}^{(0)}}{\partial r^2} \right] \quad \text{at } r = a \tag{5–9.17}$$

etc., etc.

Beginning then with the zeroth-order field, each higher-order field may be successively determined by satisfying appropriate boundary conditions at the surface of the *un*deformed sphere. As general methods have been developed (Section 3–2) for carrying out this program, the problem can in principle be solved to any order of approximation in ϵ. In practice, of course, the algebraic manipulations quickly become unwieldy. We limit ourselves therefore to calculating only the first-order correction to Stokes' law. Furthermore, we make no attempt to justify the proposed perturbation scheme. Questions of convergence are far too complex to be investigated here.

The leading term in the expansion, $(\mathbf{v}^{(0)}, p^{(0)})$, satisfying boundary conditions Eqs. (5–9.10) and (5–9.15) is, of course, Stokes' solution for *streaming flow* past the undeformed sphere. This solution is[32]

$$\mathbf{v}^{(0)} = \mathbf{U} + \nabla \phi_{-2}^{(0)} + \frac{1}{2\mu} r^2 \nabla p_{-2}^{(0)} + \frac{2}{\mu} \mathbf{r} p_{-2}^{(0)} \tag{5–9.18a}$$

$$p^{(0)} = p_{-2}^{(0)} \tag{5–9.18b}$$

where

$$p_{-2}^{(0)} = -\frac{3\, a\mu \mathbf{r} \cdot \mathbf{U}}{2r^3}, \qquad \phi_{-2}^{(0)} = -\frac{a^3 \mathbf{r} \cdot \mathbf{U}}{4r^3} \tag{5–9.19a,b}$$

Alternatively, this velocity field may be written in the invariant form

$$\mathbf{v}^{(0)} = \mathbf{U} \cdot \left[\mathbf{I} - \frac{3}{4} a \left(\frac{\mathbf{I}}{r} + \frac{\mathbf{rr}}{r^3} \right) - \frac{1}{4} a^3 \left(\frac{\mathbf{I}}{r^3} - 3 \frac{\mathbf{rr}}{r^5} \right) \right] \tag{5–9.20}$$

The boundary conditions to be satisfied by the next perturbation, $\mathbf{v}^{(1)}$, are given by Eqs. (5–9.11) and (5–9.16). Since $f(\theta, \phi) = \sum_{k=0}^{\infty} f_k(\theta, \phi)$, the linearity of the Stokes equations and boundary conditions enables us to write

$$\mathbf{v}^{(1)} = \sum_{k=0}^{\infty} \mathbf{v}_k^{(1)}, \qquad p^{(1)} = \sum_{k=0}^{\infty} p_k^{(1)} \tag{5–9.21a,b}$$

where each $\mathbf{v}_k^{(1)}$ $(k = 0, 1, 2, \ldots)$ satisfies Stokes' equations and the boundary conditions

$$\mathbf{v}_k^{(1)} = \mathbf{0} \quad \text{at } r = \infty \tag{5–9.22}$$

and

$$\mathbf{v}_k^{(1)}(a, \theta, \phi) = -\frac{3}{2} \mathbf{U} \cdot \left(\mathbf{I} - \frac{\mathbf{rr}}{r^2} \right) f_k(\theta, \phi) \tag{5–9.23}$$

We note that $\mathbf{r}/r = \mathbf{i}_r$ is a unit vector drawn outward from the origin. The latter is independent of the distance r, but is a function of angular position (θ, ϕ). Physically, the significance of the symmetric dyadic operator $\mathbf{I} - (\mathbf{rr}/r^2)$ stems from the decomposition theorem

$$\mathbf{v} = \frac{\mathbf{rr}}{r^2} \cdot \mathbf{v} + \left(\mathbf{I} - \frac{\mathbf{rr}}{r^2}\right) \cdot \mathbf{v}$$

in which the lead term is the normal (vector) component of $\mathbf{v}$ on the surface of a sphere and the last term is the tangential (vector) component.

The general methods of Section 3–2 may be employed here to find the fields $(\mathbf{v}_k^{(1)}, p_k^{(1)})$ satisfying the foregoing boundary conditions. Upon calculating from Eq. (5–9.23) the quantity $(\mathbf{r}/r)\cdot\mathbf{v}_k^{(1)}(a, \theta, \phi)$ and the two other requisite quantities required in Eqs. (3–2.22)–(3–2.24), we find with the aid of some elementary vector identities that

$$\sum_{n=1}^{\infty} {}_kX_n^{(1)} = 0 \tag{5–9.24}$$

$$\sum_{n=1}^{\infty} {}_kY_n^{(1)} = \tfrac{3}{2}\left(r\mathbf{U} \cdot \nabla f_k - 2f_k\mathbf{U} \cdot \frac{\mathbf{r}}{r}\right) \tag{5–9.25}$$

$$\sum_{n=1}^{\infty} {}_kZ_n^{(1)} = \tfrac{3}{2}\mathbf{U}\cdot\nabla \times (\mathbf{r}f_k) \tag{5–9.26}$$

The right-hand sides of each of these expressions must now be expanded in a series of surface spherical harmonics.

The result is trivial in the case of Eq. (5–9.24) and we obtain

$${}_kX_n^{(1)} = 0 \quad \text{for } n \geq 1 \tag{5–9.27}$$

Next consider Eq. (5–9.26). As we now prove, the right side of this expression is already a surface harmonic of order k. To prove this, define

$$T_k(\theta, \phi) = \tfrac{3}{2}\mathbf{U}\cdot\nabla \times (\mathbf{r}f_k)$$

To demonstrate the contention it suffices to prove that $r^k T_k$ is a solid spherical harmonic of degree k. Since $r^k T_k$ is homogeneous in r^k, it remains then only to show that it satisfies Laplace's equation. Now,

$$\nabla \times (\mathbf{r}f_k) = \nabla \times \left(\frac{\mathbf{r}\, r^k f_k}{r^k}\right) = \frac{1}{r^k}\nabla \times (\mathbf{r}r^k f_k)$$

whence $$r^k T_k = \tfrac{3}{2}\mathbf{U}\cdot\nabla \times (\mathbf{r}r^k f_k)$$

Consequently, $$\nabla^2(r^k T_k) = \tfrac{3}{2}\mathbf{U}\cdot\nabla \times \{\nabla^2(\mathbf{r}r^k f_k)\}$$

However, $$\nabla^2(\mathbf{r}r^k f_k) = \mathbf{r}\,\nabla^2(r^k f_k) + 2\nabla(r^k f_k) + r^k f_k\,\nabla^2\,\mathbf{r}$$

But since f_k is a surface spherical harmonic of order k then, by definition, $r^k f_k$ is a solid spherical harmonic of degree k, which satisfies Laplace's equation, $\nabla^2(r^k f_k) = 0$. Also, $\nabla^2\,\mathbf{r} = \mathbf{0}$.

Hence, $$\nabla^2(r^k T_k) = 3\mathbf{U}\cdot\nabla \times \nabla(r^k f_k)$$

But the curl operator annihilates the gradient of any function and so

$\nabla^2(r^k T_k) = 0$. Thus the contention is demonstrated. In view of this, it follows from Eq. (5–9.26) that

$$ {}_kZ_n^{(1)} = \begin{cases} \frac{3}{2}\mathbf{U}\cdot\nabla\times(\mathbf{r}f_k) & \text{for } n = k \\ 0 & \text{for } n \neq k \end{cases} \tag{5–9.28} $$

Lastly consider Eq. (5–9.25). By some simple vector identities it may be written in the form

$$ \sum_{n=1}^{\infty} {}_kY_n^{(1)} = \frac{3}{2(2k+1)}[(k-1)U_{k-1} + (k+2)V_{k+1}] \tag{5–9.29} $$

where

$$ \begin{aligned} U_{k-1}(\theta, \phi) &= r^{-(k-1)}(\mathbf{U}\cdot\nabla)r^k f_k \\ V_{k+1}(\theta, \phi) &= r^{k+2}(\mathbf{U}\cdot\nabla)r^{-(k+1)} f_k \end{aligned} \tag{5–9.30} $$

We shall now show that these two functions are surface harmonics of degrees $k-1$ and $k+1$, respectively. In the former case it suffices to show that

$$ r^{k-1}U_{k-1} = (\mathbf{U}\cdot\nabla)r^k f_k $$

is a solid spherical harmonic of order $k-1$. But the function is homogeneous in r^{k-1}. Also

$$ \nabla^2(r^{k-1}U_{k-1}) = (\mathbf{U}\cdot\nabla)\nabla^2(r^k f_k) = 0 $$

Thus the contention is demonstrated. Similarly, the proof that V_{k+1} is a surface harmonic of degree $k+1$ follows by showing that the function $r^{-(k+2)}V_{k+1}$ is a solid spherical harmonic of order $-(k+2)$. In consequence of these observations it is clear form Eq. (5–9.29) that

$$ {}_kY_n^{(1)} = \begin{cases} \dfrac{3(k-1)}{2(2k+1)} r^{-(k-1)}(\mathbf{U}\cdot\nabla)r^k f_k & \text{for } n = k-1 \\ \dfrac{3(k+2)}{2(2k+1)} r^{k+2}(\mathbf{U}\cdot\nabla)r^{-(k+1)} f_k & \text{for } n = k+1 \\ 0 & \text{for all other } n \end{cases} \tag{5–9.31} $$

For example when $k = 2$, only Y_1 and Y_3 are non-zero.

Upon substituting Eqs. (5–9.27), (5–9.28), and (5–9.31) into Eqs. (3–2.33)–(3–2.35), we obtain

$$ \frac{{}_kp_{-(n+1)}^{(1)}}{\mu} = \begin{cases} \dfrac{3(k-1)(2k-3)a^{k-1}}{2k(2k+1)} r^{-(2k-1)}(\mathbf{U}\cdot\nabla)r^k f_k & \text{for } n = k-1 \\ \frac{3}{2} a^{k+1}(\mathbf{U}\cdot\nabla)r^{-(k+1)} f_k & \text{for } n = k+1 \\ 0 & \text{for all other } n \end{cases} \tag{5–9.32} $$

$$ {}_k\Phi_{-(n+1)}^{(1)} = \begin{cases} \dfrac{3(k-1)a^{k+1}}{4k(2k+1)} r^{-(2k-1)}(\mathbf{U}\cdot\nabla)r^k f_k & \text{for } n = k-1 \\ \dfrac{3a^{k+3}}{4(2k+1)}(\mathbf{U}\cdot\nabla)r^{-(k+1)} f_k & \text{for } n = k+1 \\ 0 & \text{for all other } n \end{cases} \tag{5–9.33} $$

$$
{}_k\chi^{(1)}_{-(n+1)} = \begin{cases} \dfrac{3a^{k+1}}{2k(k+1)} r^{-(k+1)} \mathbf{U}\cdot\nabla \times (\mathbf{r} f_k) & \text{for } n = k \\ 0 \quad \text{for all other } n \end{cases} \tag{5–9.34}
$$

In conjunction with Eq. (3–2.31) (with $\mathbf{v}_\infty = \mathbf{0}$) these expressions furnish the velocity and pressure fields $(\mathbf{v}_k^{(1)}, p_k^{(1)})$. When the latter are summed over k in accordance with Eq. (5–9.21), appropriate expressions are obtained for $(\mathbf{v}^{(1)}, p^{(1)})$.

The force on the deformed sphere may now be obtained from Eq. (3–2.42). Write

$$
\mathbf{F} = \mathbf{F}^{(0)} + \epsilon \mathbf{F}^{(1)} + O(\epsilon^2) \tag{5–9.35}
$$

where $\mathbf{F}^{(0)} = 6\pi\mu a\mathbf{U}$ is the Stokes force on the undeformed sphere and

$$
\mathbf{F}^{(1)} = -4\pi\nabla\left(r^3 \sum_{k=0}^{\infty} {}_k p^{(1)}_{-2}\right) \tag{5–9.36}
$$

But from Eq. (5–9.32), since $n = 1$, only the terms $k = 0$ and $k = 2$ contribute to this infinite sum. Hence, since $f_0 =$ constant, we obtain

$$
\sum_{k=0}^{\infty} {}_k p^{(1)}_{-2} = -\tfrac{3}{2}\mu a r^{-3}[f_0 \mathbf{U}\cdot\mathbf{r} - \tfrac{1}{10}(\mathbf{U}\cdot\nabla)(r^2 f_2)]
$$

whence

$$
\mathbf{F} = 6\pi\mu a\mathbf{U} + 6\pi\mu a\epsilon[\mathbf{U} f_0 - \tfrac{1}{10}(\mathbf{U}\cdot\nabla)\nabla(r^2 f_2)] + O(\epsilon^2) \tag{5–9.37}
$$

The Stokes translation dyadic, $\mathbf{K}$, for the deformed sphere is defined by the relation (5–2.7) (where $\mathbf{U}$ is now the *streaming velocity*, whereas $\mathbf{U}_O$ was designated in Eq. (5–2.7) as particle velocity),

$$
\mathbf{F} = \mu \mathbf{K}\cdot\mathbf{U} \tag{5–9.38}
$$

Comparing this with Eq. (5–9.37), we find that

$$
\mathbf{K} = 6\pi a\{\mathbf{I} + \epsilon[\mathbf{I} f_0 - \tfrac{1}{10}\nabla\nabla(r^2 f_2)] + O(\epsilon^2)\} \tag{5–9.39}
$$

The dyadic $\nabla\nabla(r^2 f_2)$ is, of course, a *constant* dyadic.

In addition to the foregoing force, the body also experiences a hydrodynamic torque about the origin. This torque may be obtained from Eq. (3–2.45). Write

$$
\mathbf{T}_O = \mathbf{T}_O^{(0)} + \epsilon \mathbf{T}_O^{(1)} + O(\epsilon^2) \tag{5–9.40}
$$

where $\mathbf{T}_O^{(0)} = \mathbf{0}$ is the torque on the undeformed sphere and

$$
\mathbf{T}_O^{(1)} = -8\pi\mu\nabla\left(r^3 \sum_{k=0}^{\infty} {}_k\chi^{(1)}_{-2}\right) \tag{5–9.41}
$$

From Eq. (5–9.34) only the term $k = 1$ need be considered in this infinite sum. Thus,

$$
\sum_{k=0}^{\infty} {}_k\chi^{(1)}_{-2} = \tfrac{3}{4} a^2 r^{-2} \mathbf{U}\cdot\nabla \times (\mathbf{r} f_1)
$$

which leads eventually to the following result:

$$\mathbf{T}_O = -6\pi\mu a^2 \epsilon \mathbf{U} \times \nabla(rf_1) + O(\epsilon^2) \tag{5-9.42}$$

in which $\nabla(rf_1)$ is a *constant* vector.

Rotation of a deformed sphere

The boundary conditions appropriate to the *rotation* of a deformed sphere are given by Eq. (5–9.5). An expansion in powers of ϵ, as in Eq. (5–9.8), is again assumed. The same general procedure is followed as previously to determine the force on the rotating body and the torque (about the origin), except that $\mathbf{F}^{(0)} = \mathbf{0}$ and $\mathbf{T}_O^{(0)} = -8\pi\mu a^3 \boldsymbol{\omega}$ in the present instance[32]. It is found that[13]

$$\mathbf{F} = -6\pi\mu a^2 \epsilon \boldsymbol{\omega} \times \nabla(rf_1) + O(\epsilon^2) \tag{5-9.43}$$

and

$$\mathbf{T}_O = -8\pi\mu a^3 \boldsymbol{\omega} - 24\pi\mu a^3 \epsilon [\boldsymbol{\omega} f_0 - \tfrac{1}{10}(\boldsymbol{\omega}\cdot\nabla)\nabla(r^2 f_2)] + O(\epsilon^2) \tag{5-9.44}$$

If we define the rotation dyadic at the origin by Eq. (5–3.4),

$$\mathbf{T}_O = -\mu \boldsymbol{\Omega}_O \cdot \boldsymbol{\omega}$$

then Eq. (5–9.44) may be expressed in the alternate form,

$$\boldsymbol{\Omega}_O = 8\pi a^3 \{\mathbf{I} + 3\epsilon[\mathbf{I} f_0 - \tfrac{1}{10}\nabla\nabla(r^2 f_2)] + O(\epsilon^2)\} \tag{5-9.45}$$

Example: Streaming flow past a spheroid

In the axisymmetric case for streaming flow, Eq. (5–9.37) reduces to Sampson's result, Eq. (5–9.7). To illustrate the usefulness of the result, consider the problem of streaming flow past the spheroid

$$\frac{x^2 + y^2}{a^2} + \frac{z^2}{a^2(1-\epsilon)^2} = 1 \tag{5-9.46}$$

For $\epsilon > 0$ the spheroid is oblate; for $\epsilon < 0$ it is prolate. To $O(\epsilon)$, Eq. (5–9.46) can be written in polar form[36]

$$r = a[1 - \epsilon\{\tfrac{1}{3}P_0(\cos\theta) + \tfrac{2}{3}P_2(\cos\theta)\}] + O(\epsilon^2) \tag{5-9.47}$$

where $\cos\theta = z/r$ and $P_0(\cos\theta) = 1$, $P_2(\cos\theta) = \frac{1}{2}(3\cos^2\theta - 1)$. Comparison with Eq. (5–9.2) shows that

$$f_0 = -\tfrac{1}{3}, \qquad f_2 = -\tfrac{2}{3}P_2(\cos\theta) \tag{5-9.48; 49}$$

The latter yields $r^2 f_2 = \frac{1}{3}(x^2 + y^2 - 2z^2) = \mathbf{r}\cdot(\frac{1}{3}\mathbf{I} - \mathbf{kk})\cdot\mathbf{r}$, where $\mathbf{k}$ is a unit vector in the z direction. Thus,

$$\nabla\nabla(r^2 f_2) = 2(\tfrac{1}{3}\mathbf{I} - \mathbf{kk}) \tag{5-9.50}$$

Substitution of Eqs. (5–9.48) and (5–9.50) into Eq. (5–9.39) yields

$$\mathbf{K} = 6\pi a\{\mathbf{I}(1 - \tfrac{2}{5}\epsilon) + \tfrac{1}{5}\epsilon\mathbf{kk} + O(\epsilon^2)\} \tag{5-9.51}$$

From Eq. (5–9.38) the force on the spheroid when fluid streams past it with velocity $\mathbf{U}$ is therefore

$$\mathbf{F} = 6\pi\mu a[\mathbf{U}(1 - \tfrac{2}{5}\epsilon) + \tfrac{1}{5}\epsilon(\mathbf{U}\cdot\mathbf{k})\mathbf{k}] + O(\epsilon^2) \tag{5–9.52}$$

For streaming flow *parallel* to the spheroid axis, we have successively, $\mathbf{U} = \mathbf{k}U$ and $(\mathbf{U}\cdot\mathbf{k})\,\mathbf{k} = U\mathbf{k} = \mathbf{U}$, whence (see also Eq. (4–25.23))

$$\mathbf{F} = 6\pi\mu a[1 - \tfrac{1}{5}\epsilon + O(\epsilon^2)]\mathbf{U} \tag{5–9.53}$$

To the order indicated, this agrees with the exact solutions of Payne and Pell[41] for axisymmetric flow past oblate and prolate spheroids. The comparable result for flow *perpendicular* to the spheroid axis ($\mathbf{U}\cdot\mathbf{k} = 0$) is

$$\mathbf{F} = 6\pi\mu a[1 - \tfrac{2}{5}\epsilon + O(\epsilon^2)]\mathbf{U} \tag{5–9.54}$$

At any other angles of incidence the force on the spheroid is no longer parallel to the stream velocity vector.

Though Eqs. (5–9.53) and (5–9.54) are presumably valid only for small values of ϵ, they are, in fact, surprisingly accurate for even large departures from the spherical shape. For example, the eccentricity e of an oblate spheroid is $e = [1 - (b/a)^2]^{1/2}$ where b and a are, respectively, its polar and equatorial radii. Comparison with Eq. (5–9.46) shows that

$$\epsilon = 1 - (1 - e^2)^{1/2}$$

The eccentricity of a flat circular disk of radius a is unity, whence $\epsilon = 1$. For this case, Eqs. (5–9.53) and (5–9.54) yield, respectively,

$$\frac{F}{\mu a U} = \begin{cases} 15.1 & \text{(broadside-on)} \\ 11.3 & \text{(edge-on)} \end{cases}$$

The *exact* solutions for this case are[32]

$$\frac{F}{\mu a U} = \begin{cases} 16 & \text{(broadside-on)} \\ 10\tfrac{2}{3} & \text{(edge-on)} \end{cases}$$

so the maximum errors for any oblate spheroid are less than 6 per cent. These errors decrease rapidly with decreasing eccentricity. For $e = 0.8$ (that is, $b/a = 0.6$) the discrepancy is less than 0.5 per cent.

The rotation dyadic for the spheroid Eq. (5–9.46) is

$$\mathbf{\Omega}_O = 8\pi a^3\{\mathbf{I}(1 - \tfrac{6}{5}\epsilon) + \tfrac{3}{5}\epsilon\mathbf{kk} + O(\epsilon^2)\} \tag{5–9.55}$$

so that the torque on the rotating spheroid is in general

$$\mathbf{T}_O = -8\pi\mu a^3[\boldsymbol{\omega}(1 - \tfrac{6}{5}\epsilon) + \tfrac{3}{5}\epsilon(\boldsymbol{\omega}\cdot\mathbf{k})\mathbf{k}] + O(\epsilon^2) \tag{5–9.56}$$

For rotation about the symmetry axis this yields

$$\mathbf{T}_O = -8\pi\mu a^3[1 - \tfrac{3}{5}\epsilon + O(\epsilon^2)]\boldsymbol{\omega} \tag{5–9.57}$$

The analogous result for rotation about an equatorial diameter is

$$\mathbf{T}_O = -8\pi\mu a^3[1 - \tfrac{6}{5}\epsilon + O(\epsilon^2)]\boldsymbol{\omega} \tag{5–9.58}$$

Equation (5–9.57) agrees to $O(\epsilon)$ with Jeffery's[27] exact solutions for the rotation of oblate and prolate spheroids about their axes of revolution.[18]

Average resistance to translation

The preceding development may be applied to the determination of the average resistance to translation of a deformed sphere. According to Eq. (5–8.8),

$$\frac{1}{\bar{K}} = \frac{1}{3}\left(\frac{1}{K_1} + \frac{1}{K_2} + \frac{1}{K_3}\right) \tag{5–9.59}$$

in which K_1, K_2, and K_3 are the eigenvalues of $\mathbf{K}$. If $\mathbf{i}_1$, $\mathbf{i}_2$, $\mathbf{i}_3$ denote the normalized eigenvectors of $\mathbf{K}$ we then have

$$K_j = \mathbf{i}_j \cdot \mathbf{K} \cdot \mathbf{i}_j \quad (j = 1, 2, 3) \tag{5–9.60}$$

Hence from Eq. (5–9.39)

$$\frac{K_j}{6\pi a} = 1 + \epsilon\left[f_0 - \frac{1}{10}\frac{\partial^2}{\partial x_j^2}(r^2 f_2)\right] + O(\epsilon^2)$$

whence

$$\frac{6\pi a}{K_j} = 1 - \epsilon\left[f_0 - \frac{1}{10}\frac{\partial^2}{\partial x_j^2}(r^2 f_2)\right] + O(\epsilon^2)$$

and

$$\frac{6\pi a}{\bar{K}} = 1 - \epsilon f_0 + \frac{\epsilon}{30}\left(\frac{\partial^2}{\partial x_1^2} + \frac{\partial^2}{\partial x_2^2} + \frac{\partial^2}{\partial x_3^2}\right) r^2 f_2 + O(\epsilon^2)$$

or, since $\nabla^2(r^2 f_2) = 0$,

$$\frac{6\pi a}{\bar{K}} = 1 - \epsilon f_0 + O(\epsilon^2)$$

so finally

$$\bar{K} = 6\pi a[1 + \epsilon f_0 + O(\epsilon^2)] \tag{5–9.61}$$

Now the surface area and volume[36] of the deformed sphere are, respectively,

$$S = 4\pi a^2(1 + 2\epsilon f_0) + O(\epsilon^2) \tag{5–9.62}$$

and

$$V = \frac{4\pi a^3}{3}(1 + 3\epsilon f_0) + O(\epsilon^2) \tag{5–9.63}$$

Thus, an *undeformed* sphere of the same *volume* as the particle would have a radius α determined by

$$\frac{4\pi\alpha^3}{3} = \frac{4\pi a^3}{3}(1 + 3\epsilon f_0)$$

This makes

$$\alpha = a(1 + \epsilon f_0)$$

so that

$$\bar{K} = 6\pi\alpha \tag{5–9.64}$$

The sphericity, that is, the ratio of the area of a sphere of the same volume as the particle divided by the area of the particle, becomes equal to unity in this first-order approximation. Thus the calculation furnishes partial theoretical justification for the sphericity concept, in that it demonstrates that, for slight deformations, use of a radius based on that of a sphere of the same *volume* will yield a correct result for the *average* resistance (see Section 5–10).

The centroid of a deformed sphere

The centroid (V) of the deformed sphere is located at the point whose position vector $\mathbf{r}_{OV}$ relative to the origin O is

$$\mathbf{r}_{OV} = \frac{1}{V}\int_V \mathbf{r}\, dV \tag{5-9.65}$$

This integral may be evaluated by writing it in the form

$$\mathbf{r}_{OV} = \frac{1}{(4\pi a^3/3)(1 + 3\epsilon f_0) + O(\epsilon^2)}\int_{\phi=0}^{2\pi}\int_{\theta=0}^{\pi}\int_{r=0}^{r=a(1+\epsilon\Sigma f_k)} \frac{\mathbf{r}}{r}\, r^3\, dr \sin\theta\, d\theta\, d\phi$$

Note that $\mathbf{r}/r$ is independent of r. On performing the r integration, we obtain

$$\int_{r=0}^{r=a(1+\epsilon\Sigma f_k)} r^3\, dr = \tfrac{1}{4}a^4\left(1 + 4\epsilon \sum_{k=0}^{\infty} f_k\right) + O(\epsilon^2)$$

whence

$$\mathbf{r}_{OV} = \frac{3a}{16\pi}(1 - 3\epsilon f_0)\int_{S_1} \frac{\mathbf{r}}{r}\left(1 + 4\epsilon\sum_{k=0}^{\infty} f_k\right) d\Omega + O(\epsilon^2) \tag{5-9.66}$$

where $d\Omega = \sin\theta\, d\theta\, d\phi$ is an element of surface area on a unit sphere, S_1. The vector $\mathbf{r}$ is homogeneous in r. Moreover, it satisfies Laplace's equation. Hence $\mathbf{r}/r$ is a vector surface spherical harmonic of order 1. Because of the orthogonality properties of surface spherical harmonics of different orders with respect to integration over the unit sphere, only the f_1 term in the preceding integrand will contribute to the final result. Equation (5–9.66) therefore reduces to

$$\mathbf{r}_{OV} = \frac{3a\epsilon}{4\pi}\int_{S_1} \frac{\mathbf{r}}{r} f_1\, d\Omega + O(\epsilon^2) \tag{5-9.67}$$

By Euler's theorem, $rf_1 = \mathbf{r}\cdot\nabla(rf_1)$. Since $\nabla(rf_1)$ is a constant vector, the integral may be written in the form

$$\mathbf{r}_{OV} = \frac{3a\epsilon}{4\pi}\left\{\int_{S_1} \frac{\mathbf{r}\mathbf{r}}{r^2}\, d\Omega\right\}\cdot\nabla(rf_1) + O(\epsilon^2) \tag{5-9.68}$$

But, it can be shown that

$$\frac{1}{r^2}\int_{S_1} \mathbf{r}\mathbf{r}\, d\Omega = \frac{4\pi}{3}\mathbf{I} \tag{5-9.69}$$

from which it follows that

$$\mathbf{r}_{OV} = \epsilon a\nabla(rf_1) + O(\epsilon^2) \tag{5-9.70}$$

With the aid of the preceding relation, Eq. (5–9.42) for the torque on a translating body* and Eq. (5–9.43) for the force on a rotating body may be written, respectively, as

$$\mathbf{T}_O = 6\pi\mu a\mathbf{U}_O \times \mathbf{r}_{OV} + O(\epsilon^2)$$

$$\mathbf{F} = -6\pi\mu a\boldsymbol{\omega} \times \mathbf{r}_{OV} + O(\epsilon^2)$$

*The transition from the formula for the torque on a body immersed in a streaming flow $\mathbf{U}$ to the torque on a body translating with velocity $\mathbf{U}_O$ has been made by replacing $\mathbf{U}$ by $-\mathbf{U}_O$ in Eq. (5–9.42).

Since the translation dyadic is

$$\mathbf{K} = 6\pi a\mathbf{I} + O(\epsilon^2)$$

these may be written alternatively in the forms

$$\mathbf{T}_O = -\mu(\mathbf{r}_{OV} \times \mathbf{K})\cdot\mathbf{U}_O + O(\epsilon^2)$$

and

$$\mathbf{F} = \mu(\mathbf{r}_{OV} \times \mathbf{K})\cdot\boldsymbol{\omega} + O(\epsilon^2)$$

Now, in the present instance, Eqs. (5–4.1) and (5–4.2) reduce to

$$\mathbf{T}_O = -\mu\mathbf{C}_O\cdot\mathbf{U}_O, \qquad \mathbf{F} = -\mu\mathbf{C}_O^\dagger\cdot\boldsymbol{\omega}$$

Comparison with the preceding formulas shows that

$$\mathbf{C}_O = \mathbf{r}_{OV} \times \mathbf{K} + O(\epsilon^2)$$

From Eq. (5–2.11), however, we note that the coupling dyadic at the centroid V is related to the coupling dyadic at the center O of the undeformed sphere by the general expression

$$\mathbf{C}_V = \mathbf{C}_O - \mathbf{r}_{OV} \times \mathbf{K}$$

Thus, in the present circumstances $\mathbf{C}_V = O(\epsilon^2)$. This result permits us to draw the following conclusions: (1) the center of reaction of the slightly deformed sphere coincides with its centroid to at least the first order in the deformation parameter ϵ; (2) to at least $O(\epsilon)$, the slightly deformed sphere is a nonskew body, possessing a center of hydrodynamic stress.

Given a body of irregular shape, there will be in general no obvious choice for a set of axes from which the angles θ and ϕ in Eq. (5–9.2) are to be measured. Accordingly, it should be possible to recast Eq. (5–9.39) for the translation tensor into an intrinsic form in which these angles do not appear explicitly. Similar considerations apply to the rotation tensor. Brenner[13] has shown that this may be done as follows: If S and V are the surface area and volume of the irregular particle, and if $\mathbf{r}_V$ is the position vector of a point measured relative to an origin at the centroid (V), then

$$\mathbf{K} = 6\pi a\left[\mathbf{I}\frac{11}{10} - \left(\frac{3S}{V}\right)^5 \frac{3}{8\pi}\int_V \mathbf{r}_V\mathbf{r}_V\, dV + O(\epsilon^2)\right] \tag{5–9.71}$$

and

$$\boldsymbol{\Omega}_V = 8\pi a^3\left[\mathbf{I}\frac{13}{10} - \left(\frac{3S}{V}\right)^5 \frac{9}{8\pi}\int_V \mathbf{r}_V\mathbf{r}_V\, dV + O(\epsilon^2)\right] \tag{5–9.72}$$

It is interesting to note that for a body of uniform density ρ_p, the moment of inertia dyadic about the centroid is (cf. Burgers[16])

$$\mathbf{m}_V = \rho_p \iiint_V (\mathbf{I}r_V^2 - \mathbf{r}_V\mathbf{r}_V)\,dV \tag{5–9.73}$$

the integral being closely related to that appearing in Eqs. (5–9.71) and (5–9.72).

Since $\mathbf{rr} = \frac{1}{3}\nabla\cdot(\mathbf{rrr})$, Eqs. (5–9.71) and (5–9.72) could, by Gauss' divergence theorem, also be expressed in the form of surface integrals.

Brenner[9-11] has also shown with the aid of the reciprocal theorem (see Section 3–5) that the macroscopic properties of the deformed sphere may be obtained directly to the first order in ϵ from a knowledge of the Stokes velocity field for the *undeformed* body, without requiring a solution of the equations of motion. In principle this method is quite general and should be applicable to bodies of other than spherical shape (for example, a slightly deformed ellipsoid). Like other integral procedures, however, the method furnishes no detailed description of the flow field such as would be required for extending the results to higher orders in the deformation parameter ϵ (which would probably be necessary to develop results for the case where $\mathbf{C}_R \neq \mathbf{0}$), or for theoretical calculations of heat and mass transfer coefficients.

5-10 The Settling of Spherically Isotropic Bodies

As has been noted in Section 5–5, Case 4, any body which possesses spherical isotropy and is of uniform density will have the same translational resistance in any orientation. Such a body will also be isotropic as regards the couple arising from rotation about any axis passing through its center. If such a body is initially placed in any orientation in a fluid and allowed to fall without initial spin, it will not rotate to a different position but fall *vertically* in its original orientation.

Pettyjohn and Christiansen[42] in an extensive experimental study of the effect of particle shape on free settling rates of a number of isometric particles concluded that the sphericity ψ constituted an appropriate parameter for predicting the average resistance of such particles. The *sphericity* for a nonspherical particle is defined as the ratio of the area of a sphere of the same *volume* as the particle divided by the area of the particle itself:

$$\psi = \frac{A_s}{A_{ns}} \tag{5–10.1}$$

At the same time, d_s is taken as the diameter of the sphere of the same volume as the particle. Under creeping flow conditions ($N_{\mathrm{Re}} < 0.05$) Stokes' law as written below can be applied to within an accuracy of ± 2 per cent:

$$\mathbf{U} = \frac{K(\rho_p - \rho)\, d_s^2 \mathbf{g}}{18\mu} \tag{5–10.2}$$

where

$$K = 0.843 \log_{10} \frac{\psi}{0.065} \tag{5–10.3}$$

Table 5–10.1 gives data on the sphericity of particles investigated by Pettyjohn and Christiansen.

TABLE 5–10.1
SPHERICITY OF PARTICLE SHAPES INVESTIGATED

Particle Shape	*Sphericity* (ψ)
Sphere	1.000
Cube-octahedron	0.906
Octahedron	0.846
Cube	0.806
Tetrahedron	0.670

These particles are all isometric, that is, all edges are of equal length. The last three are also regular polyhedra, with all faces equal. In addition to these three, only two additional regular polyhedra are possible, namely, the dodecahedron (with twelve regular pentagonal faces) and the icosahedron (with twenty equilateral triangular faces). As noted previously, other isotropic bodies are possible in addition to the isometric ones. Though these have not been investigated experimentally there is no reason to believe that their behavior would be greatly different.

5-11 The Settling of Orthotropic Bodies

This class includes bodies which possess three mutually perpendicular symmetry planes. It includes ellipsoids, right elliptic cylinders, and rectangular parallelepipeds. Exact solution is available only for the case of the translation of an ellipsoid, and this will therefore be considered first.

Ellipsoids

The problem of steady translation of an ellipsoid in a viscous liquid was originally solved by Oberbeck.[39] We proceed to outline the method, following that given in Lamb's[32] treatise. The equation of the surface of an ellipsoid is

$$\frac{x^2}{a^2} + \frac{y^2}{b^2} + \frac{z^2}{c^2} = 1 \qquad (5\text{–}11.1)$$

in which a, b, and c are the semiaxes in the direction of the x, y, and z coordinates respectively. As a first step it is necessary to find a solution of Laplace's equation which reduces to a constant on the ellipsoidal surface. The gravitational potential for a solid homogeneous ellipsoid of unit density is given by Dirichlet's formula[30]

$$\Omega = \pi abc \int_\lambda^\infty \left(\frac{x^2}{a^2 + \lambda} + \frac{y^2}{b^2 + \lambda} + \frac{z^2}{c^2 + \lambda} - 1\right) \frac{d\lambda}{\Delta} \qquad (5\text{–}11.2)$$

where
$$\Delta = \sqrt{(a^2 + \lambda)(b^2 + \lambda)(c^2 + \lambda)} \qquad (5\text{–}11.3)$$

The lower limit of integration is the positive root of the cubic equation

$$\frac{x^2}{a^2+\lambda}+\frac{y^2}{b^2+\lambda}+\frac{z^2}{c^2+\lambda}=1 \tag{5–11.4}$$

This relationship makes

$$\frac{\partial\Omega}{\partial x}=2\pi\alpha x: \qquad \frac{\partial\Omega}{\partial y}=2\pi\beta y; \qquad \frac{\partial\Omega}{\partial z}=2\pi\gamma z \tag{5–11.5}$$

where

$$\alpha=abc\int_\lambda^\infty\frac{d\lambda}{(a^2+\lambda)\,\Delta}; \qquad \beta=abc\int_\lambda^\infty\frac{d\lambda}{(b^2+\lambda)\,\Delta}; \qquad \gamma=abc\int_\lambda^\infty\frac{d\lambda}{(c^2+\lambda)\,\Delta} \tag{5–11.6}$$

A second solution of Laplace's equation, also given by Lamb, corresponds to the velocity potential for an ellipsoid translating in a frictionless fluid. Thus the equation $\nabla^2\chi=0$ is satisfied by

$$\chi=abc\int_\lambda^\infty\frac{d\lambda}{\Delta} \tag{5–11.7}$$

Without loss in generality, we need only treat the special case where the fluid streams with speed U parallel to the x axis. For once we have the resistance along a principal axis, we may, by superposition, treat the general case where the ellipsoid has any orientation relative to the stream. For the special case taken, we assume, following Oberbeck's[39] original treatment,

$$\begin{aligned} u&=A\frac{\partial^2\Omega}{\partial x^2}+B\left(x\frac{\partial\chi}{\partial x}-\chi\right)+U\\ v&=A\frac{\partial^2\Omega}{\partial x\,\partial y}+Bx\frac{\partial\chi}{\partial y}\\ w&=A\frac{\partial^2\Omega}{\partial x\,\partial z}+Bx\frac{\partial\chi}{\partial z}\end{aligned} \tag{5–11.8}$$

These satisfy the equation of continuity because $\nabla^2\Omega=0$ and $\nabla^2\chi=0$. They also make $u=U$, $v=0$, and $w=0$ at infinity. Furthermore

$$\nabla^2u=2B\frac{\partial^2\chi}{\partial x^2}; \qquad \nabla^2v=2B\frac{\partial^2\chi}{\partial x\,\partial y}; \qquad \nabla^2w=2B\frac{\partial^2\chi}{\partial x\,\partial z} \tag{5–11.9}$$

so that the creeping motion equations are satisfied by taking

$$p=2B\mu\frac{\partial\chi}{\partial x}+\text{constant} \tag{5–11.10}$$

It remains to establish the values of A and B which will make $u, v, w=0$ at the surface of the ellipsoid. The conditions $v=0$, $w=0$ require

$$\left[2\pi A\frac{d\alpha}{d\lambda}+B\frac{d\chi}{d\lambda}\right]_{\lambda=0}=0 \quad\text{or}\quad 2\pi\frac{A}{a^2}+B=0 \tag{5–11.11}$$

With the help of this relation, the condition for $u=0$ reduces to

$$2\pi A\alpha_0 - B\chi_0 + U = 0 \tag{5–11.12}$$

where the suffix denotes that the lower limit of integrals Eq. (5–11.6) and Eq. (5–11.7) is to be replaced by zero. Thus

$$\pi A = -\frac{1}{2}Ba^2, \qquad B = \frac{U}{\chi_0 + \alpha_0 a^2} \tag{5–11.13}$$

At a great distance r from the origin we have

$$\Omega = -\frac{4}{3}\pi\frac{abc}{r}, \qquad \chi = 2\frac{abc}{r}$$

and it appears, upon comparison with Stokes' law for the velocity field produced by the movement of fluid past a sphere, that the disturbance is the same as would be produced by a sphere of radius R, determined by

$$\frac{3}{4}UR = 2abcB \quad \text{or} \quad R = \frac{8}{3}\frac{abc}{\chi_0 + \alpha_0 a^2} \tag{5–11.14}$$

The resistance experienced by the ellipsoid is therefore

$$6\pi\mu RU \tag{5–11.15}$$

Thus, we have only to evaluate the integrals,

$$\chi_0 = abc\int_0^\infty \frac{d\lambda}{\Delta} \tag{5–11.16}$$

and

$$\alpha_0 = abc\int_0^\infty \frac{d\lambda}{(a^2+\lambda)\,\Delta} \tag{5–11.17}$$

For our purposes, the most interesting cases are those involving ellipsoids of revolution. Thus, platelike objects can be approximated by oblate spheroids, the circular disk being the limiting case. Needle- or fiber-shaped objects can be represented by prolate spheroids.

Where fluid motion is parallel to the axis of symmetry, we may take the semiaxes $b = c$, and define a length to diameter ratio $\phi = a/c$. When the length-to-diameter ratio ϕ is greater than unity, the radius of the "equivalent sphere" becomes

$$R = \frac{8c}{3}\frac{1}{\left[-\dfrac{2\phi}{\phi^2-1} + \dfrac{2\phi^2-1}{(\phi^2-1)^{3/2}}\ln\left(\dfrac{\phi+\sqrt{\phi^2-1}}{\phi-\sqrt{\phi^2-1}}\right)\right]} \tag{5–11.18}$$

When ϕ becomes very large with respect to unity, this becomes approximately

$$R \approx \frac{2c\phi}{3[\ln(2\phi) - 1/2]} \tag{5–11.19}$$

Note in this case if $c \to 0$, no matter how long the object, its resistance approaches zero. Values corresponding to formula (5–11.18), which checks

TABLE 5–11.1

VALUES OF EQUIVALENT RADIUS FOR AN ELLIPSOID OF REVOLUTION

[($a = b$) Flow perpendicular to axis, R/a]

$c/a < 1$		$c/a > 1$	
Discoid Objects (Oblate spheroids)		*Needlelike Objects* (Prolate spheroids)	
Length/dia. c/a	Equiv. radius/actual radius R/a	Length/dia. c/a	Equiv. radius/actual radius R/a
10^{-6}	0.5659	1.01	1.004
10^{-5}	0.5659	1.05	1.020
10^{-4}	0.5659	1.1	1.040
10^{-3}	0.5664	1.5	1.194
10^{-2}	0.5707	2	1.379
10^{-1}	0.6133	5	2.371
1.5×10^{-1}	0.6366	10	3.812
2×10^{-1}	0.6596	20	6.365
3×10^{-1}	0.7049	50	13.06
4×10^{-1}	0.7492	10^2	23.00
5×10^{-1}	0.7927	2×10^2	41.08
6×10^{-1}	0.8355	5×10^2	90.00
7×10^{-1}	0.8775	10^3	164.6
8×10^{-1}	0.9189	10^4	1281
9×10^{-1}	0.9597	10^5	10,493
9.9×10^{-1}	0.9960	10^6	88,838

[($b = c$) Flow parallel to axis, R/c]

$a/c < 1$		$a/c > 1$	
Discoid Objects (Oblate Spheroids)		*Needlelike Objects* (Prolate Spheroids)	
Length/dia. a/c	Equiv. radius/actual radius R/c	Length/dia. a/c	Equiv. radius/actual radius R/c
10^{-6}	0.8488	1.01	1.002
10^{-5}	0.8488	1.05	1.010
10^{-4}	0.8488	1.10	1.020
10^{-3}	0.8488	1.50	1.102
10^{-2}	0.8489	2.0	1.204
10^{-1}	0.8525	5.0	1.785
1.5×10^{-1}	0.8564	10	2.647
2×10^{-1}	0.8615	20	4.172
3×10^{-1}	0.8739	50	8.117
4×10^{-1}	0.8888	10^2	13.895
5×10^{-1}	0.9053	2×10^2	24.280
6×10^{-1}	0.9230	5×10^2	52.022
7×10^{-1}	0.9415	10^3	93.881
8×10^{-1}	0.9606	10^4	708.92
9×10^{-1}	0.9801	10^5	5,695.2
9.9×10^{-1}	0.9980	10^6	47,590

Eq. (4–30.12), Table 4–26.1, are found in Table 5–11.1; and formula (5–11.19) is identical with Eq. (4–31.4). When the length-to-diameter ratio ϕ is less than unity, we obtain

$$R = \frac{8c}{3}\frac{1}{\left[\dfrac{2\phi}{1-\phi^2} + \dfrac{2(1-2\phi^2)}{(1-\phi^2)^{3/2}}\tan^{-1}\left(\dfrac{\sqrt{1-\phi^2}}{\phi}\right)\right]} \tag{5–11.20}$$

In this case as $\phi \to 0$,

$$R_{a=0} = \frac{8c}{3\pi} \tag{5–11.21}$$

corresponding to the motion of a circular disk broadside-on. Values corresponding to formula (5–11.20), which checks Eq. (4–26.38), are to be found in Table 5–11.1; and formula (5–11.21) is identical with Eq. (4–27.2).

Fluid motion perpendicular to the axis of symmetry cannot, of course, be treated by the methods for axisymmetric motion and so is of special interest as an application of this method of treatment. In this case we let the semiaxes $a = b$ and define $\phi = c/a$. When the length-to-diameter ratio is greater than unity, the radius of the "equivalent sphere" is found to be

$$R = \frac{8a}{3}\frac{1}{\left[\dfrac{\phi}{\phi^2-1} + \dfrac{2\phi^2-3}{(\phi^2-1)^{3/2}}\ln\left(\phi+\sqrt{\phi^2-1}\right)\right]} \tag{5–11.22}$$

When ϕ is large compared to unity, this becomes approximately

$$R \approx \frac{4a\phi}{3[\ln(2\phi)+1/2]} \tag{5–11.23}$$

For a length-to-diameter ratio ϕ which is less than unity, it is found that

$$R = \frac{8a}{3}\frac{1}{\left[-\dfrac{\phi}{1-\phi^2} - \dfrac{2\phi^2-3}{(1-\phi^2)^{3/2}}\sin^{-1}\sqrt{1-\phi^2}\right]} \tag{5–11.24}$$

The limiting value as $\phi \to 0$ is

$$R_{c=0} = \frac{16a}{9\pi} \tag{5–11.25}$$

corresponding to the case for a disk moving edgewise.

Numerical values for motion perpendicular to the axis of symmetry are given[38] in Table 5–11.1.

Oseen[40], using his linearized equations (2–6.4) to determine the effect of inertia on the resistance to the uniform translation of an ellipsoid in a viscous fluid, obtained results similar to those obtained in the case of a sphere, Eq. (2–6.5). Thus, for the case of motion of a disk broadside-on, corresponding to Eq. (5–11.21), he obtained for a disk of radius c

$$F = 16\mu c U\left(1 + \frac{N_{Re}}{\pi}\right) \tag{5–11.26}$$

where $N_{Re} = cU\rho/\mu$.

Breach[5] extended the approximation to include inertial effects,[47] employing the method of Proudman and Pearson[43]. His treatment is confined to the axial motion of ellipsoids of revolution, both prolate and oblate. For the case of a disk moving broadside-on to a stream, he obtains

$$F = 16\mu c U\left[1 + \frac{N_{\text{Re}}}{\pi} + \frac{8N_{\text{Re}}^2}{5\pi^2}\ln N_{\text{Re}} + O(N_{\text{Re}}^2)\right] \tag{5-11.27}$$

Venkates'[50] investigation of motion of a viscous liquid past an ellipsoid involves a unitary treatment which includes potential flow, creeping motion, and a boundary layer of small thickness as limiting cases.

Squires and Squires[48] have experimentally studied the limiting case of circular disks moving both edgewise and broadside-on for very thin aluminum disks. Agreement with theory is reasonably good.

An example of the method for calculation of the motion of an ellipsoid in an infinite medium has already been given (Section 5–7) for the special case of a circular disk. Its translation tensor is

$$\mathbf{K} = \left((\mathbf{i}'\mathbf{i}'\frac{32}{3} + \mathbf{j}'\mathbf{j}'\frac{32}{3} + \mathbf{k}'\mathbf{k}'16\right)c \tag{5-11.28}$$

For the case of a needle-shaped ellipsoid of length l with its axis lying in the z' direction, and of radius c, the corresponding translation tensor is

$$\mathbf{K} = \left\{\mathbf{i}'\mathbf{i}'\left[\frac{8\pi\phi}{\ln(2\phi) + 1/2}\right] + \mathbf{j}'\mathbf{j}'\left[\frac{8\pi\phi}{\ln(2\phi) + 1/2}\right] + \mathbf{k}'\mathbf{k}'\left[\frac{4\pi\phi}{\ln(2\phi) - 1/2}\right]\right\}c \tag{5-11.29}$$

Here $\phi = l/2c$, the ratio of the length to diameter of the needle.

We will consider here the case of oblique fall of a needle as a further illustration, using the terminology of Section 5–7, as shown in Fig. 5–11.1. For this case, the principal translational resistances of the needle are—analogous to Eq. (5–7.38)—

$$K_1 = K_2 = \left[\frac{8\pi\phi}{\ln(2\phi) + 1/2}\right]c,$$

$$K_3 = \left[\frac{4\pi\phi}{\ln(2\phi) - 1/2}\right]c \tag{5-11.30}$$

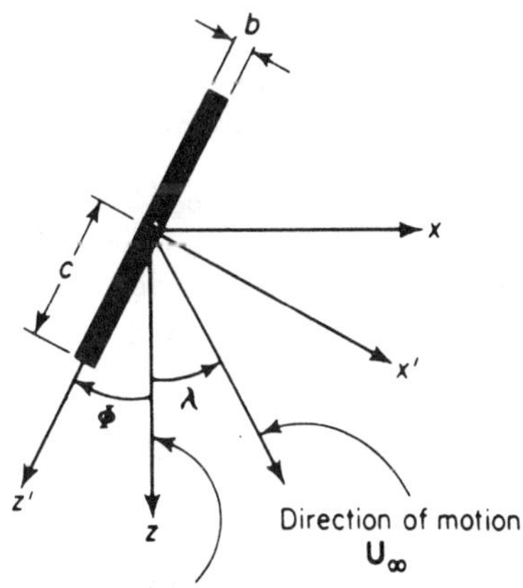

Figure 5-11.1. Oblique fall of a needle-shaped object.

If the needle radius is c and its length l, its volume will be $V = \pi c^2 l$ and the components of settling velocity are therefore, from Eq. (5–7.35),

$$(U_x)_\infty = \left(\frac{3}{2} - \ln\frac{l}{c}\right)\frac{gc^2\,\Delta\rho}{8\mu}\sin 2\Phi, \qquad (U_y)_\infty = 0 \tag{5-11.31}$$

$$(U_z)_\infty = -\left(\frac{3}{2} - \ln\frac{l}{c}\right)\frac{gc^2\,\Delta\rho}{8\mu}\left[\frac{(1/2) - 3\ln(l/c)}{(3/2) - \ln(l/c)} + \cos 2\Phi\right] \quad (5\text{–}11.32)$$

These relationships show that the needle moves down and to the left, as depicted in Fig. 5–11.1. If l/c is a very large number, so that the constants in the preceding formulas become small with respect to the logarithmic terms, we can simplify them. It is found that λ, the angle between the downward vertical and the direction in which the needle moves, is given by

$$\lambda = \tan^{-1}\left(\frac{-\sin 2\Phi}{3 + \cos 2\Phi}\right) \quad (5\text{–}11.33)$$

This angle will be a maximum when the needle orientation with the vertical is

$$\Phi = \tfrac{1}{2}\cos^{-1}(-\tfrac{1}{3}) = 54.8^\circ \quad (5\text{–}11.34)$$

and the direction of movement will then be

$$\lambda = \tan^{-1}(-0.354) = -19.5^\circ \quad (5\text{–}11.35)$$

so $\lambda = 90^\circ - 2\Phi$. This compares with $\lambda = 11.5^\circ$ for the case of sedimenting disks. Therefore, it would be anticipated that a dilute sedimenting group of needle-shaped objects would show a greater degree of dispersion than would occur in the case of sedimentation of a cluster of plate-shaped objects. Note that in all cases the needles or plates fall stably in their initial orientation at sufficiently small Reynolds numbers.

For the fall of intermediate-shaped ellipsoids of revolution, appropriate values of the constants to employ in the translation tensor may be computed readily from the listing in Table 5–11.1.

Brenner[10,11] has shown that a generalization of Faxen's law can be obtained to calculate the resistance of an ellipsoid suspended in an *arbitrary flow* $\mathbf{u} = \mathbf{u}(\mathbf{r})$ which satisfies the creeping motion equations. The translation dyadic $\mathbf{K}$ for an ellipsoid is, from preceding results, expressed as

$$\mathbf{K} = 16\pi abc\left(\frac{\mathbf{ii}}{\chi_0 + a^2\alpha_0} + \frac{\mathbf{jj}}{\chi_0 + b^2\beta_0} + \frac{\mathbf{kk}}{\chi_0 + c^2\gamma_0}\right) \quad (5\text{–}11.36)$$

with α_0, β_0, γ_0, and χ_0 defined in Eqs. (5–11.6) and (5–11.7). The force on a *stationary* ellipsoid immersed in the undisturbed Stokes flow $\mathbf{u}$ at infinity is then

$$\mathbf{F} = \mu\mathbf{K}\cdot\left[\mathbf{u}_O + \frac{1}{3!}(D^2\mathbf{u})_O + \frac{1}{5!}(D^4\mathbf{u})_O + \frac{1}{7!}(D^6\mathbf{u})_O + \cdots\right] \quad (5\text{–}11.37)$$

where the subscript O refers to evaluation of the unperturbed field at the center of the ellipsoid and D^2 is the operator

$$D^2 = a^2\frac{\partial^2}{\partial x^2} + b^2\frac{\partial^2}{\partial y^2} + c^2\frac{\partial^2}{\partial z^2} \quad (5\text{–}11.38)$$

Similarly, the torque $\mathbf{T}_O$ about the center of the stationary ellipsoid is

$$\mathbf{T}_o = \mu \mathbf{Q} \cdot \left[(\Box \times \mathbf{u})_o + \frac{3!2}{5!}(D^2 \Box \times \mathbf{u})_o + \frac{3!3}{7!}(D^4 \Box \times \mathbf{u})_o + \frac{3!4}{9!}(D^6 \Box \times \mathbf{u})_o + \cdots \right] \tag{5-11.39}$$

in which $\mathbf{Q}$ is the constant dyadic

$$\mathbf{Q} = \frac{16\pi abc}{3}\left(\frac{\mathbf{ii}}{b^2\beta_0 + c^2\gamma_0} + \frac{\mathbf{jj}}{c^2\gamma_0 + a^2\alpha_0} + \frac{\mathbf{kk}}{a^2\alpha_0 + b^2\beta_0}\right) \tag{5-11.40}$$

and $\Box$ is the vector operator

$$\Box = \mathbf{i}a^2 \frac{\partial}{\partial x} + \mathbf{j}b^2 \frac{\partial}{\partial y} + \mathbf{k}c^2 \frac{\partial}{\partial z} \tag{5-11.41}$$

($\mathbf{i}, \mathbf{j}, \mathbf{k}$) are unit vectors parallel to the corresponding principal axes (x, y, z) of the ellipsoid.

In the event that the ellipsoid is not stationary one has only to add to these results the formulas for the force and torque on an ellipsoid moving through a fluid at rest at infinity.

These results may be checked against Faxen's law as follows: For the case of a sphere,

$$a = b = c \quad (=a, \text{ say})$$

and thus $D^2 = a^2 \nabla^2$ and $\Box = a^2 \nabla$. Also $\mathbf{K} = \mathbf{I}6\pi a$ and $\mathbf{Q} = \mathbf{I}4\pi a$ where $\mathbf{I}$ is the idemfactor. Hence, in this case

$$\mathbf{F} = 6\pi\mu a\left[\mathbf{u}_o + \frac{a^2}{3!}(\nabla^2 \mathbf{u})_o + \frac{a^4}{5!}(\nabla^4 \mathbf{u})_o + \cdots\right]$$

$$\mathbf{T}_o = 4\pi\mu a\left[a^2(\nabla \times \mathbf{u})_o + a^4 \frac{3!2}{5!}(\nabla^2 \nabla \times \mathbf{u})_o + \cdots\right]$$

But, in creeping flow, $\nabla^4\mathbf{u} = 0$, $\nabla^6\mathbf{u} = 0$, etc., and $\nabla^2(\nabla \times \mathbf{u}) = \nabla^4(\nabla \times \mathbf{u}) = 0$, etc. Hence,

$$\mathbf{F} = 6\pi\mu a\left[\mathbf{u}_o + \frac{a^2}{6}(\nabla^2 \mathbf{u})_o\right]$$

$$\mathbf{T}_o = 4\pi\mu a^3(\nabla \times \mathbf{u})_o$$

which are equivalent to Eqs. (3–2.46) and (3–2.47), Faxen's laws for the sphere.

Long finite cylinders

As discussed earlier, the resistance of an infinitely long cylinder moving in an otherwise unbounded medium cannot be treated using the creeping motion equations. Exact solutions have not yet been obtained for the case of finite cylinders, but since these resemble ellipsoids in shape, approximate methods have been employed. In particular, the method developed by Burgers[15] and discussed in Section 3–4 may be employed to calculate the resistance of long cylindrical bodies. For that purpose we assume that the body may be replaced by a system of forces distributed in an appropriate way along the points which were originally situated upon the axis of the body. One can write down expressions for the components of velocity which will be produced by these point forces, and then attempt to determine the intensity of the forces in such a way that the mean value of the resultant velocity vanishes approximately on the surface originally occupied by the surface of the body. This method was previously illustrated to derive Stokes' law.

Consider a thin spindle-shaped body of revolution of length $L = 2a$ moving in the direction of its axis. The axis of rotation will be taken as the x axis, the origin being taken at the center of the body. When at an element $d\xi$ of the axis a force $f(\xi)\, d\xi$ is introduced, the u component of the velocity produced by this force at a point on the surface of the body, situated on the section made by the plane $x =$ constant, is given by

$$du = \frac{1}{8\pi\mu} f(\xi)\, d\xi \left[\frac{1}{\{(x-\xi)^2 + b^2\}^{1/2}} + \frac{(x-\xi)^2}{\{(x-\xi)^2 + b^2\}^{3/2}} \right] \tag{5–11.42}$$

where b is the radius of the section considered. When the body is a true cylinder, b will be a constant; in the more general case, b will be different for the various sections and must be considered as a function of x. When similar forces act at every element $d\xi$ of the axis, the total value of u will be given by the integral of expression (5–11.42) from $\xi = -a$ to $\xi = +a$, assuming, of course, that the proper function of ξ has been inserted. The approximation thus assumes that the point force approximation in Eq. (5–11.42) will be developed for a line. Adding to the result the original velocity U, we obtain the following expression for the resultant velocity at a point on the surface of the body:

$$U + u = U + \frac{1}{8\pi\mu} \int_{-a}^{+a} f(\xi)\, d\xi \left[\frac{1}{\{(x-\xi)^2 + b^2\}^{1/2}} + \frac{(x-\xi)^2}{\{(x-\xi)^2 + b^2\}^{3/2}} \right] \tag{5–11.43}$$

This expression must vanish for every value of x. There is no need to consider the v and w components, as the mean values of these would vanish in consequence of the approximation made.

If the expression (5–11.43) is set equal to zero, we have an integral equation for the unknown function $f(\xi)$, which cannot be solved explicity. Burgers employed an approximate method of solution which involves assuming a simple polynomial form for $f(\xi)$,

$$f(\xi) = -8\pi\mu U \left[A_0 + A_1 \left(\frac{\xi}{a}\right)^2 + A_2 \left(\frac{\xi}{a}\right)^4 \right] \tag{5–11.44}$$

If Eq. (5–11.44) is inserted in Eq. (5–11.43), the integration can be effected. When terms of order b^2/a^2 are neglected in comparison with unity and the result is divided by U, we obtain

$$
\begin{gathered}
2\left\{A_0 + A_1\left(\frac{x}{a}\right)^2 + A_2\left(\frac{x}{a}\right)^4\right\}\ln\frac{4(a^2 - x^2)}{b^2} \\
-(2A_0 - 2A_1 - A_2) - (8A_1 - 2A_2)\left(\frac{x}{a}\right)^2 - \frac{31}{3}A_2\left(\frac{x}{a}\right)^4 - 1 = 0
\end{gathered}
\tag{5–11.45}
$$

As a simple example, this may be applied to the case of a long ellipsoid of revolution moving parallel to its axis. For this case, we have

$$\frac{x^2}{a^2} + \frac{b^2}{b_0^2} = 1 \tag{5–11.46}$$

where b_o is the equatorial radius of the ellipsoid; thus

$$\frac{4(a^2 - x^2)}{b^2} = \frac{4a^2}{b_0^2} \tag{5–11.47}$$

It is found that Eq. (5–11.45) is satisfied by taking

$$A_0 = \frac{1}{2\ln(4a^2/b_0^2) - 2}, \quad A_1 = 0, \quad A_2 = 0 \tag{5–11.48}$$

The resultant force is obtained from the integral

$$F = -\int_{-a}^{+a} f(\xi)\,d\xi = 16\pi\mu Ua(A_0 + \tfrac{1}{3}A_1 + \tfrac{1}{5}A_2) \tag{5–11.49}$$

Substituting the appropriate values of A_0, A_1, and A_2, we find

$$F = \frac{4\pi\mu Ua}{\ln(2a/b_o) - 0.5} \tag{5–11.50}$$

which corresponds to Oberbeck's result, Eq. (5–11.19).

For the case of a cylindrical body with b = constant, Burgers replaces $\ln 4(a^2 - x^2)/b^2$ by the approximation $\ln(4a^2/b^2) - x^2/a^2 - \frac{1}{2}x^4/a^4$, which is inserted into Eq. (5–11.45). Then, neglecting powers of x/a higher than the fourth, we equate to zero the coefficients of $(x/a)^0$, $(x/a)^2$, and $(x/a)^4$ and thus obtain three linear equations for the determination of the three unknown constants. In this way expressions are obtained which can be used for all values of the ratio b/a. The following approximate result is thus obtained

$$A_0 + \frac{1}{3}A_1 + \frac{1}{5}A_2 \approx \frac{1}{4[\ln(2a/b) - 0.72]} \tag{5–11.51}$$

The resistance of the cylinder is therefore

$$F = \frac{4\pi\mu Ua}{\ln(2a/b) - 0.72} \tag{5–11.52}$$

Thus, the resistance experienced by the cylinder is slightly higher than that of an ellipsoid having its equatorial radius b_o equal to the radius b of the cylinder. The result given in Eq. (5–11.52) may be approximated from the ellipsoid formula by taking the equatorial radius

$$b_o \approx 1.25\,b \tag{5–11.53}$$

If a long straight cylinder moves perpendicularly to its axis, a similar approach can be employed. Burgers gives only the roughest approximation for this case, requiring the vanishing of the resultant velocity $U + u$ only at the central section of the cylinder. This leads to the result

$$F \approx \frac{8\pi\mu U a}{\ln(2a/b) + 0.5} \tag{5-11.54}$$

This is equivalent to Oberbeck's formula (5–11.23) for the resistance of an elongated ellipsoid of revolution moving in a direction perpendicular to the axis of symmetry, where b_o is taken as the equatorial radius. It is likely that a more exact calculation would again give a resistance for the cylinder which is slightly higher than the preceding value.

Broersma[14] extended Burgers' method in an attempt to obtain a more accurate approximation for the resistance of a finite cylinder, but experimental data on the rate of fall of finite cylinders are not sufficiently accurate to establish whether an improved result is obtained. Data of White[51] and of Jones and Knudsen[26] indicate the importance of the dimensions of the containing vessel on the drag coefficient of elongated cylinders in the creeping motion range. Unfortunately, the wall correction factor has not been very well established as yet for elongated cylinders. Precise evaluation of the resistance in an infinite medium is thus difficult, though agreement with the theoretical formulas is reasonable.

Tchen[49] made an interesting study of the resistance experienced by a curved and elongated small particle based on Burgers' method of velocity perturbations. As his model he chose shapes varying from a straight ellipsoid to those obtained by bending the ellipsoid in the form of the arc of a circle, to include such shapes as a half circle and, at the extreme case, a circular ring. Table 5–11.2 gives values of the resistance obtained for flow in the x, y, and z directions, referring to Fig. 5–11.2. The article also gives results for tangential and radial fluid motion relative to the particles. Here l is taken as the particle length and b_o its maximum radius. Note that the formulas for the resistance of a straight ellipsoid check the Oberbeck formulas previously discussed. Though the half circle is not an orthotropic body, its resistances to flow in the xy plane do not depend on orientation.

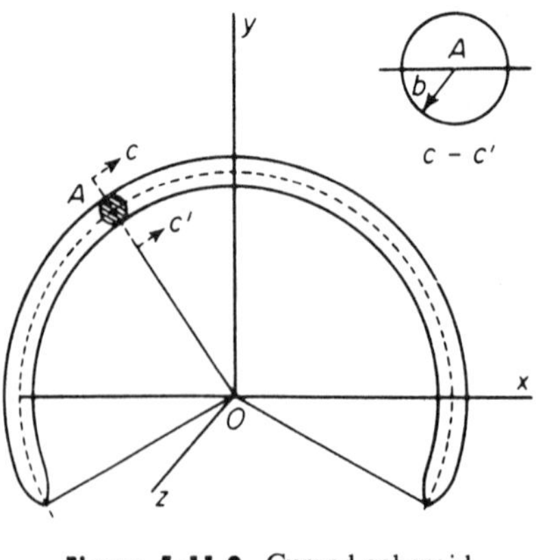

Figure 5-11.2. Curved spheroid.

TABLE 5–11.2

RESISTANCE COEFFICIENTS FOR CURVED ELLIPSOIDS OF REVOLUTION

	F_x	F_y	F_z
Straight ellipsoid	$\dfrac{2\pi\mu U}{\ln(l/b_o) - 0.50}$	$\dfrac{4\pi\mu U}{\ln(l/b_o) + 0.50}$	$\dfrac{4\pi\mu U}{\ln(l/b_o) + 0.50}$
Half circle	$\dfrac{3\pi\mu U}{\ln(l/b_o) - 0.68}$	$\dfrac{3\pi\mu U}{\ln(l/b_o) - 0.68}$	$\dfrac{4\pi\mu U}{\ln(l/b_o) + 0.56}$
Circular ring	$\dfrac{3\pi\mu U}{\ln(l/b_o) - 2.09}$	$\dfrac{3\pi\mu U}{\ln(l/b_o) - 2.09}$	$\dfrac{4\pi\mu U}{\ln(l/b_o) + 0.75}$

Short finite cylinders and parallelepipeds

No theoretical treatment is available for creeping motion relative to solid objects possessing plane boundaries, or for short cylinders, but a considerable amount of experimental data has been obtained for the settling rates of such particles.

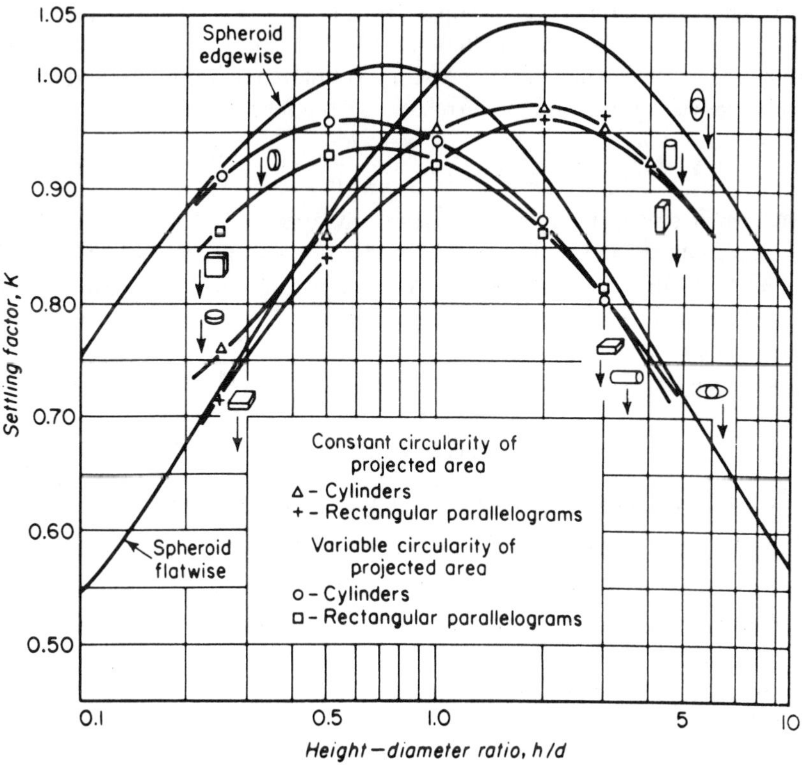

Figure 5-11.3. Settling factor K versus height-diameter ratio for cylinders, rectanguler parallelepipeds, and spheroids.

Heiss and Coull's[24] comprehensive investigation includes a review of settling rates of rectangular parallelepipeds and circular cylinders and a comparison with Gans' theoretical settling equations for spheroids[21], which agree substantially with those presented in this and the previous chapter. On this basis, and using new experimental data, the authors developed empirical relationships for these objects for both edgewise and flatwise fall, Fig. 5–11.3. For the case of flatwise flow, where the cross-sectional area does not change for a range of height-to-diameter ratios, the following equation applies

$$\log_{10} K = \log_{10}\left(\frac{d_s}{d_n}\sqrt{\psi}\right) - 0.25\sqrt{\psi}\,\frac{d_s}{d_n}\left(\frac{d_s}{d_n} - 1\right) \qquad (5\text{–}11.55)$$

Here K and ψ are as defined in Eqs. (5–10.3) and (5–10.1) respectively, and d_s is again taken as the dimaeter of a sphere with the same volume as the object. The variable d_n is a new one, taken as the diameter of a *circle* of an area equal to the projected area of the particle perpendicular to the direction in which it falls (always a principal axis in these experiments). The ratio d_s/d_n is a measure of the "circularity" of the particle. For orientations in which the particle falls edgewise, with variable circularity as the ratio of h/d increases, the results were correlated on the following basis:

$$\log_{10} K = \log_{10}\left(\frac{d_s}{d_n}\sqrt{\psi}\right) - \frac{0.270}{\sqrt{\psi}\,(d_s/d_n)^{0.345}}\left(\frac{d_s}{d_n} - 1\right) \qquad (5\text{–}11.56)$$

Heiss and Coull in their original article give graphs for evaluation of the two preceding equations. Jones and Knudsen[26] present some additional data on short cylinders (see also Bart[2]). A more recent paper by Becker[3] gives experimental data for freely oriented bodies of various shapes at higher Reynolds numbers.

BIBLIOGRAPHY

1. Acrivos, A., and N. R. Amundson, Ind. Eng. Chem. **47** (1955), 1533.
2. Bart, E., M. S. Thesis, New York Univ. New York, N. Y., 1959.
3. Becker, H. A., Can. J. Chem. Eng. **37** (1959), 85.
4. Binder, R. C., *Fluid Mechanics*, 3rd ed. Englewood Cliffs, N. J.: Prentice-Hall, Inc., 1955.
5. Breach, D. R., J. Fluid Mech. **10** (1961), 306.
6. Brenner, H., Chem. Eng. Sci. **18** (1963), 1.
7. ———, Chem. Eng. Sci. **18** (1963), 557.
8. ———, Chem. Eng. Sci. **19** (1964), 599.
9. ———, Chem. Eng. Sci. **19** (1964), 631.

10. ———, Chem. Eng. Sci. **19** (1964), 703.

11. ———, Chem. Eng. Sci. (in press, 1965).

12. ———, in *Recent Advances in Chemical Engineering*, Vol. 6. T. B. Drew, J. W. Hoopes, Jr., and T. Vermeulen, eds. New York: Academic Press, 1966.

13. ———, Chem. Eng. Sci. **19** (1964), 519.

14. Broersma, S., J. Chem. Phys. **32** (1960), 1632.

15. Burgers, J. M., in *Second Report on Viscosity and Plasticity*. Amsterdam: North Holland Publ. Co., 1938.

16. ———, Proc. Koningl. Akad. Wetenschap (Amsterdam) **44** (1941), 1045.

17. Drew, T. B., *Handbook of Vector and Polyadic Analysis*. New York: Reinhold, 1961.

18. Edwardes, D., Quart. J. Math. **26** (1892), 70.

19. Einstein, A., *Investigations on the Theory of the Brownian Movement*. New York: Dover, 1956.

20. Frazer, R. A., W. J. Duncan, and A. R. Collar, *Elementary Matrices*. Cambridge: Cambridge Univ. Press, 1938.

21. Gans, R., Sitzber. math-physik Klasse bayer. Akad. Wissen. (Munich) **41** (1911), 191.

22. Gibbs, J. W., and E. B. Wilson, *Vector Analysis*. New York: Dover, 1960.

23. Goldstein, H., *Classical Mechanics*. Reading, Mass.: Addison-Wesley, 1950.

24. Heiss, J. F., and J. Coull, Chem. Eng. Prog. **48** (1952), 133.

25. Hildebrand, F. B., *Methods of Applied Mathematics*. Englewood Cliffs, N. J.: Prentice-Hall, Inc., 1952.

26. Jones, A. M., and J. G. Knudsen, A. I. Ch. E. Jour. **7** (1961), 20.

27. Jeffery, G. B., Proc. London Math. Soc. **14** (1915), 327.

28. Kanwal, R. P., J. Appl. Mech. **26** (1959), 485.

29. ———, J. Fluid Mech. **10** (1961), 17.

30. Kellogg, O. D., *Foundations of Potential Theory*. New York: Dover, 1953.

31. Kelvin, Lord, Phil. Mag. **42** (1871), 362.

32. Lamb, H., *Hydrodynamics*, 6th ed. Cambridge: Cambridge Univ. Press, 1932.

33. Landau, L. D., and E. M. Lifshitz, *Fluid Mechanics*. Reading, Mass.: Addison-Wesley, 1959.

34. ——— and ———, *Statistical Physics*. Reading, Mass.: Addison-Wesley, 1958

35. Larmor, J., Quart. J. Math. **20** (1884), 261.

36. MacRobert, T. M., *Spherical Harmonics*, 2nd ed. New York: Dover, 1947, pp. 174, 355.

37. Milne-Thomson, L. M., *Theoretical Hydrodynamics*, 4th ed. New York: Macmillan, 1960.

38. Nelson, R. W., Physics Group, Institute of Paper Chemistry, Appleton, Wis., (private communication), 1962.

39. Oberbeck, H. A., Crelles J. **81** (1876), 79.

40. Oseen, C. W., *Hydrodynamik*. Leipzig: Akademische Verlagsgesellshaft, 1927. See also Arkiv. Mat. Astr. Fys. **24** (1915), 108.

41. Payne, L. E., and W. H. Pell, J. Fluid Mech. **7** (1960), 529.

42. Pettyjohn, E. A., and E. B. Christiansen, Chem. Eng. Prog. **44** (1948), 157.

43. Proudman, I., and J. R. A. Pearson, J. Fluid Mech. **2** (1957), 237.

44. Rayleigh, Lord, *The Theory of Sound*, 2nd ed. Vol. I. New York: Dover, 1945, p. 102.

45. Roscoe, R., Phil. Mag. **40** (1949), 338.

46. Sachs, M., *Solid State Theory*. New York: McGraw-Hill, 1963.

47. Saffman, P. G., J. Fluid Mech. **1** (1956), 540.

48. Squires, L., and W. Squires, Jr., Trans. Am. Inst. Chem. Engrs. **33** (1937), 1.

49. Tchen, C., J. Appl. Phys. **25** (1954), 463.

50. Venkates, H. G., Phys. Fluids **4** (1961), 33.

51. White, C. M., Proc. Roy. Soc. (London) **186** (1946), 472.

Interaction between Two or More Particles 6

6-1 Introduction

We shall consider in this chapter the behavior of a small number of rigid particles moving slowly through a viscous fluid under the influence of external forces, typically gravity. The particles are supposed sufficiently close to interact hydrodynamically. It is assumed that the particles are sufficiently distant from boundary walls for the surrounding fluid to be regarded as unbounded. Attention will be predominantly directed to situations where the fluid at infinity is at rest. The magnitude of the interaction among the particles is, in general, governed by the following variables: (a) their shapes and sizes; (b) the distances between them; (c) their orientations with respect to each other; (d) their individual orientations relative to the direction of the gravitational field; (e) their velocities and spins relative to the fluid at infinity.

For prescribed translational and angular particle velocities, the macroscopic parameters of primary physical interest are the hydrodynamic forces and torques exerted by the fluid on the particles. Once these parameters are known for a given particle array, one may immediately solve the inverse problem of determining the state of motion of the particles from the known gravitational body forces and torques acting on them.

The local fluid motion is assumed to satisfy the quasi-static Stokes equations,

$$\nabla^2 \mathbf{v} = \frac{1}{\mu} \nabla p \tag{6–1.1}$$

$$\nabla \cdot \mathbf{v} = 0 \tag{6–1.2}$$

where $\mathbf{v}$ is the fluid velocity, p the dynamic pressure, and μ the viscosity. The most general motions which rigid particles may undergo are translation and rotation. Because of the linearity of the governing equations of motion and boundary conditions, these two modes of motion may be separately investigated, and the results superposed. Initially, therefore, we shall restrict our attention to cases where the particles translate, without rotation, as they move through the fluid. This is, by far, the simpler of the two possible modes of rigid body motion.

Consider a rigid particle of arbitrary shape translating through an unbounded fluid which is at rest at infinity. If we identify this particle by the label a, the appropriate boundary conditions are

$$\mathbf{v} = \mathbf{U}_a \quad \text{on } a \tag{6–1.3}$$

and

$$\mathbf{v} \to \mathbf{0} \quad \text{as } r \to \infty \tag{6–1.4}$$

If other particles (b, c, etc.) are present, one must satisfy the additional conditions:

$$\mathbf{v} = \mathbf{U}_b \quad \text{on } b \tag{6–1.5}$$

$$\mathbf{v} = \mathbf{U}_c \quad \text{on } c, \text{ etc.} \tag{6–1.6}$$

The only exact solution of a multiparticle problem of this class is that of Stimson and Jeffery[30], for the slow motion of two spheres parallel to their line of centers (an axisymmetric flow). The system of bipolar coordinates (see Section A–19) employed by them is unique in that it permits one simultaneously to satisfy boundary conditions on two external spheres. For larger collections of particles, or for pairs of nonspherical bodies, it is not generally possible to find coordinate systems which permit simultaneous satisfaction of all boundary conditions. Accordingly, we seek a systematic scheme of successive iterations, whereby the boundary-value problem may be solved to any degree of approximation by considering boundary conditions associated with one particle at a time. Such a scheme is provided by the "method of reflections." Application of this technique to a system of n spheres was inaugurated by Smoluchowski[29]. The method employed here is similar, through it must be pointed out that no rigorous proof exists that the iteration scheme converges to the desired solution. Rather, we must content ourselves at present with the limited empirical proof afforded, for example, by the agreement of the technique with the exact results of Stimson and Jeffery for the axisymmetric two-sphere case, and by the agreement of results with actual experimental data in a limited number of other situations.

Consider a system of n translating particles. In accordance with previous notation, let $\mathbf{U}_k$ ($k = a, b, c, \ldots, n$) denote the velocity of the kth particle. Of particular interest to us are the forces, $\mathbf{F}_k$, necessary to maintain each particle in its state of uniform motion, and the restraining torques, $\mathbf{T}_k$, required to keep the particles from rotating under the influence of the hydrodynamic stresses developed at their surfaces.

To solve the boundary-value problem posed by Eqs. (6–1.1)–(6–1.6), we proceed as follows: Since the equations of motion and boundary conditions are linear, the local velocity and pressure fields may be decomposed into a sum of fields; thus,

$$\mathbf{v} = \mathbf{v}^{(1)} + \mathbf{v}^{(2)} + \mathbf{v}^{(3)} + \mathbf{v}^{(4)} + \cdots \tag{6–1.7}$$

$$p = p^{(1)} + p^{(2)} + p^{(3)} + p^{(4)} + \cdots \tag{6–1.8}$$

each term of which, $(\mathbf{v}^{(j)}, p^{(j)})$, separately satisfies the equations of motion and vanishes at infinity. Again, because of linearity, we may further subdivide each of these into a finite sum of terms, $(\mathbf{v}_k^{(j)}, p_k^{(j)})$, also satisfying the governing differential equations and vanishing at infinity. New, focus attention on any particle in the system, say, a, and define $(\mathbf{v}^{(1)}, p^{(1)})$ by the boundary condition

$$\mathbf{v}^{(1)} = \mathbf{U}_a \quad \text{on } a \tag{6–1.9}$$

The "reflection" of this field from particle b is then defined by the boundary condition

$$\mathbf{v}_b^{(2)} = \mathbf{U}_b - \mathbf{v}^{(1)} \quad \text{on } b \tag{6–1.10}$$

In general, the reflection of $\mathbf{v}^{(1)}$ from any of the $n - 1$ particles is defined by

$$\mathbf{v}_k^{(2)} = \mathbf{U}_k - \mathbf{v}^{(1)} \quad \text{on } k \qquad (k = b, c, \ldots, n) \tag{6–1.11}$$

Thus, the reflection of $\mathbf{v}^{(1)}$ from *all* the remaining $n - 1$ particles is given approximately by

$$\mathbf{v}^{(2)} = \sum_{k=b}^{n} \mathbf{v}_k^{(2)} \tag{6–1.12}$$

The first-order effect arising from the interaction of particle a with the remaining particles may now be obtained by finding the reflection of $\mathbf{v}^{(2)}$ from a; namely, the field $\mathbf{v}^{(3)}$ vanishing at infinity and satisfying the boundary condition

$$\mathbf{v}^{(3)} = -\mathbf{v}^{(2)} \quad \text{on } a \tag{6–1.13}$$

That $\mathbf{v}^{(3)}$ does indeed provide the first-order interaction correction is evident from the following argument: The field $\mathbf{v}^{(1)}$ introduces the characteristic particle dimension c through terms of the form c/r raised to some positive power. Furthermore, since $r = O(l)$, where l is the characteristic distance between particles, the even-numbered field $\mathbf{v}^{(2)}$, determined by the boundary conditions on $b, c, \ldots, n$, introduces the characteristic distance dimension, l. The field $\mathbf{v}^{(3)}$, therefore, will possess terms in c/l. These provide a first approximation to the interaction effect arising from the presence of the other $n - 1$ particles.

However, the field $\mathbf{v}^{(3)}$ has been computed only at the location of particle a. In order to complete the approximation for the entire fluid field it is necessary for us to make a calculation for particle b to establish the magnitude of $\mathbf{v}^{(3)}$ in the vicinity of b, as Eq. (6–1.13) establishes $\mathbf{v}^{(3)}$ in the vicinity of a. This calculation is repeated for all n particles and thus an approximation of the field $\mathbf{v} = \mathbf{v}^{(1)} + \mathbf{v}^{(3)}$ is available for all n particles.

To obtain the next order of interaction, we proceed again as in Eq. (6–1.9) to satisfy the boundary-value problem on b,

$$\mathbf{v}_b^{(4)} = -\mathbf{v}^{(3)} \quad \text{on } b \tag{6–1.14}$$

The field $\mathbf{v}^{(4)}$ is developed by establishing the value at the surface of all $n - 1$ spheres of the $-\mathbf{v}^{(3)}$ field arising from the presence of particle a, as was done in the case of developing Eq. (6–1.12). The field $\mathbf{v}^{(5)}$ at the position of particle a is obtained as before by the condition

$$\mathbf{v}^{(5)} = -\mathbf{v}^{(4)} \quad \text{on } a \tag{6–1.15}$$

This velocity field at the location of particle a represents terms correct to $O(c^2/l^2)$. The reflection process may be continued as far as necessary to obtain satisfaction of all boundary conditions to the desired accuracy. Except for simple arrangements, numerical evaluation will be much easier than general analytical representation of the results. It is necessary that the field be capable, in the vicinity of each particle, of being represented in terms of polynomial series in increasing powers of c/l. This appears to be possible because the interaction of two spheres touching each other ultimately can be represented by such a series, and empirical data on concentrated assemblages can be represented by a few terms in power series of this type.

For a completely rigorous treatment it is necessary to have available a solution of the creeping motion equations for the case of a single particle with an arbitrary velocity field prescribed on its surface. Good approximations are possible, however, by assuming that when the particles are sufficiently separated: (a) the field produced by a given particle will be the same as that produced by a point force acting at the center of the particle; (b) the drag resulting from the field reflected at a given particle can be approximated by considering the field to be equivalent to a uniform velocity field whose magnitude and direction are the same as what actually would exist at the location particle center if it were not present.

Before proceeding to a development of some typical cases it is of interest to note that since the resulting fields involve only positive contributions (assuming all velocities are in the same direction), the resultant interaction will result in increasing velocities. Thus, as Smoluchowski pointed out, the larger the swarm of particles the faster it will move. The method is thus not applicable for an assemblage of particles which increases without limit, because it presumes that the velocity field produced will vanish at infinity. Modification of the procedure to take into consideration boundaries, or cases in which the particle assemblage itself extends indefinitely, will be considered in later chapters.

Once the velocity field is computed to any degree of approximation desired, the force $\mathbf{F}_a$ exerted on the particle by the fluid is obtained by summing the drag contributions of the individual fields. A velocity field which is free from singularities in the interior of the volume occupied by a particle can produce neither a resultant force nor couple on the particle (that is, it satisfies boundary conditions which imply absence of the particle). Thus no contribution to the drag on a particle is made by fields which do not involve satisfying the boundary conditions on its surface, and there remain only the contributions from the odd-numbered fields; hence

$$\mathbf{F}_a = \mathbf{F}_a^{(1)} + \mathbf{F}_a^{(3)} + \mathbf{F}_a^{(5)} + \cdots \tag{6–1.16}$$

where $\mathbf{F}_a^{(j)}$ is the force on particle a associated with the jth reflection.

At the same time that a drag is experienced by each of the particles in an assemblage, rotational effects may also occur. If the particles are free to rotate as they move, they will do so, and no net couple will be exerted on any particle when steady state motion is attained. Under such conditions the particles will not exert any rotational forces on the fluid so that the mutual disturbances can still be approximated by point forces, neglecting dilational effects. A point force is, of course, not capable of representing fluid rotation at the location of its origin. If, however, the particles are not free to rotate by being held rigid, or do rotate due to external couples exerted upon them, rotational moments will be developed.

Such possibilities are implied in the general development of the velocity fields as outlined in the previous equations. The torque exerted on a particle by a given fluid motion is again that derived from the odd-numbered fields,

$$\mathbf{T}_a = \mathbf{T}_a^{(1)} + \mathbf{T}_a^{(3)} + \mathbf{T}_a^{(5)} + \cdots \tag{6–1.17}$$

Approximations suitable for rotational effects are obtained by the employment of a point couple at the origin of a given field, analogous to the employment of a point force.

In the following sections, we will develop cases of increasing complexity, starting first with methods applicable to two particles at a substantial distance from each other. More complicated systems, with particles closer together, will be considered next. Finally, some of the limitations of particle shape and neglect of inertial terms in the equations of motion will be discussed.

6-2 Two Widely Spaced Spherically Isotropic Particles

We consider two particles of characteristic dimensions a and b, which are isotropic both with respect to translation and rotation, moving with instantaneous velocities $\mathbf{U}_a$ and $\mathbf{U}_b$ in an otherwise unbounded medium which is at rest at infinity. By a *spherically isotropic body*, it will be recalled, we mean one which exhibits the same resistance to translation no matter what orientation it has relative to a uniformly moving fluid, and which will not rotate if suspended freely in any orientation in a uniformly moving fluid. Particles which are spherical in shape meet this requirement. As discussed in Section 5–5, all regular polyhedra and bodies derived from them by symmetrically cutting or rounding the corners and/or edges and/or faces are spherically isotropic. A particle with equal resistance along each of its three principal axes will also be isotropic.

It is convenient to choose one axis of the reference system of coordinates along a line connecting the centers of the two particles. Also, we will assume that it is necessary to specify only one additional coordinate, implying that the particles move in a plane, under the influence of a body force which is

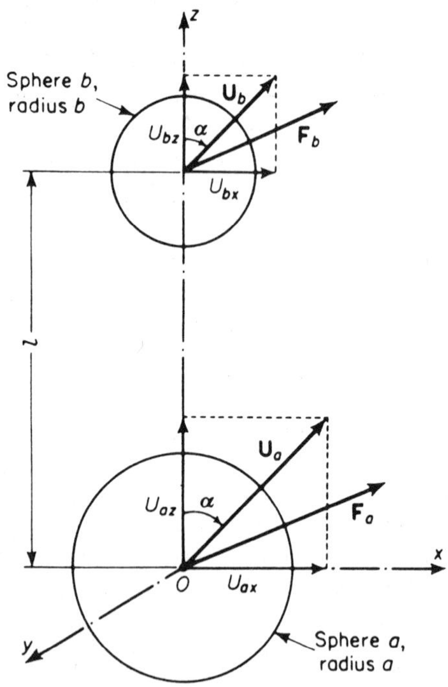

Figure 6-2.1. Coordinate system for two-particle interactions.

independent of particle location. We will choose the xz plane, as shown in Fig. 6–2.1. In this case, in addition to equations of motion (6–1.1) and (6–1.2), we require the boundary condition at infinity, Eq. (6–1.4), and the following conditions to be satisfied:

$$\mathbf{v}^{(1)} = \mathbf{U}_a \quad \text{on } a \tag{6–2.1}$$

$$\mathbf{v}^{(2)} = -\mathbf{v}^{(1)} + \mathbf{U}_b \quad \text{on } b \tag{6–2.2}$$

$$\mathbf{v}^{(3)} = -\mathbf{v}^{(2)} \quad \text{on } a \tag{6–2.3}$$

$$\mathbf{v}^{(4)} = -\mathbf{v}^{(3)} \quad \text{on } b, \text{ etc.} \tag{6–2.4}$$

The initial field $\mathbf{v}^{(1)}$ obviously will correspond to the setting of particle a in an unbounded fluid. Associated with this motion is the force (see Section 5–5, Case 4 for a discussion of the resistance coefficient)

$$\mathbf{F}_a^{(1)} = -\mu K_a \mathbf{U}_a = -\mu K_a(\mathbf{i}U_{ax} + \mathbf{k}U_{az}) \tag{6–2.5}$$

which is exerted by the fluid on the particle. Since particle a is assumed to be located at a relatively large distance (several diameters at least) from particle b, we may compute the translational effect of particle a by assuming that it generates the same field as would be produced by a *point force* situated at the center of the particle[24]:

$$\mathbf{v}^{(1)} = -\frac{\mathbf{F}_a^{(1)}}{6\pi\mu r} - \frac{r^2}{24\pi\mu}\nabla(\mathbf{F}_a^{(1)}\cdot\nabla)\frac{1}{r} \tag{6–2.6}$$

and

$$p = \frac{1}{4\pi}(\mathbf{F}_a^{(1)}\cdot\nabla)\frac{1}{r} \tag{6–2.7}$$

where r is measured from the center of particle a. To the degree of approximation to which Eq. (6–2.6) is valid, the initial field is independent of the shape of the particle a, being determined entirely by its resistance. If we express Eq. (6–2.6) in cartesian coordinates appropriate to Fig. 6–2.1, we obtain

$$\mathbf{v}^{(1)} = \frac{K_a U_{ax}}{8\pi r}\left(\mathbf{i} + \mathbf{r}\frac{x}{r^2}\right) + \frac{K_a U_{az}}{8\pi r}\left(\mathbf{k} + \mathbf{r}\frac{z}{r^2}\right) \tag{6–2.8}$$

Since the center of particle b has the coordinates $x = 0$, $y = 0$, $z = l$, the value of $\mathbf{v}^{(1)}$ at this point is

$$[\mathbf{v}^{(1)}]_b = \frac{K_a}{8\pi l}(\mathbf{i}U_{ax} + 2\mathbf{k}U_{az}) \tag{6–2.9}$$

From this we compute the force exerted on particle b

$$\mathbf{F}_b^{(2)} = -\mu K_b(\mathbf{U}_b - [\mathbf{v}^{(1)}]_b) = -\mathbf{i}\mu K_b\left(U_{bx} - \frac{K_a U_{ax}}{8\pi l}\right) - \mathbf{k}\mu K_b\left(U_{bz} - \frac{K_a U_{az}}{4\pi l}\right) \tag{6–2.10}$$

Using the same technique as previously, we can calculate the velocity field generated by the force $\mathbf{F}_b^{(2)}$ acting at the location of particle b. In this case, the origin of the coordinate system will now be at the center of particle b, and $\mathbf{v}^{(2)}$ at particle a will be

$$[\mathbf{v}^{(2)}]_a = \mathbf{i}\,\frac{K_b}{8\pi l}\left(U_{bx} - \frac{K_a U_{ax}}{8\pi l}\right) + \mathbf{k}\,\frac{K_b}{4\pi l}\left(U_{bz} - \frac{K_a U_{az}}{4\pi l}\right) \tag{6–2.11}$$

In turn, the force contribution $\mathbf{F}_a^{(3)} = \mu K_a[\mathbf{v}^{(2)}]_a$ will be

$$\mathbf{F}_a^{(3)} = \mathbf{i}\,\frac{\mu K_a K_b}{8\pi l}\left(U_{bx} - \frac{K_a U_{ax}}{8\pi l}\right) + \mathbf{k}\,\frac{\mu K_a K_b}{4\pi l}\left(U_{bz} - \frac{K_a U_{az}}{4\pi l}\right) \tag{6–2.12}$$

Similarly,

$$\begin{aligned}\mathbf{F}_a^{(5)} = \mathbf{i}\mu K_a\left(\frac{K_a}{8\pi l}\right)\left(\frac{K_b}{8\pi l}\right)^2\left(U_{bx} - \frac{K_a U_{ax}}{8\pi l}\right)\\ + \mathbf{k}\,\mu K_a\left(\frac{K_a}{4\pi l}\right)\left(\frac{K_b}{4\pi l}\right)^2\left(U_{bz} - \frac{K_a U_{az}}{4\pi l}\right)\end{aligned} \tag{6–2.13}$$

Thus

$$\begin{aligned}\mathbf{F}_a &= \mathbf{F}_a^{(1)} + \mathbf{F}_a^{(3)} + \mathbf{F}_a^{(5)} + \cdots\\ &= -\mathbf{i}\,\mu K_a\left\{U_{ax} - \frac{K_b}{8\pi l}\left(U_{bx} - \frac{K_a U_{ax}}{8\pi l}\right)\times\right.\\ &\qquad \left.\times\left[1 + \left(\frac{K_a}{8\pi l}\right)\left(\frac{K_b}{8\pi l}\right) + \left(\frac{K_a}{8\pi l}\right)^2\left(\frac{K_b}{8\pi l}\right)^2 + \cdots\right]\right\}\\ &\quad -\mathbf{k}\,\mu K_a\left\{U_{az} - \frac{K_b}{4\pi l}\left(U_{bz} - \frac{K_a U_{az}}{4\pi l}\right)\times\right.\\ &\qquad \left.\times\left[1 + \left(\frac{K_a}{4\pi l}\right)\left(\frac{K_b}{4\pi l}\right) + \left(\frac{K_a}{4\pi l}\right)^2\left(\frac{K_b}{4\pi l}\right)^2 + \cdots\right]\right\}\end{aligned} \tag{6–2.14}$$

Thus, noting that a geometric series can be expressed as a fraction, and combining terms, we find

$$\frac{\mathbf{F}_a}{\mu K_a} = -\mathbf{i}\,\frac{U_{ax} - (K_b U_{bx}/8\pi l)}{1 - (K_a K_b)/(8\pi l)^2} - \mathbf{k}\,\frac{U_{az} - (K_b U_{bz}/4\pi l)}{1 - (K_a K_b)/(4\pi l)^2} \tag{6–2.15}$$

The force exerted on particle b, $\mathbf{F}_b$, is obtainable by modifying Eq. (6–2.15) so that the subscripts b and a are interchanged. These relationships constitute four scalar equations relating force and velocity components. It is convenient to take the scalars positive when both particles are settling in the same direction. This is generally the case, and thus the resistance of either particle is reduced by the presence of the other. If, for example, the particles are falling under the influence of gravity, $\mathbf{F}_a$ and $\mathbf{F}_b$ will be known quantities, depending on particle shape, size, and specific gravity as well as specific gravity of the suspending medium. We will then have four equations with the four unknowns corresponding to the unknown velocity components to be determined.

As an illustration, consider the case of two equal-sized spheres of radii a. Equation (6–2.15) then becomes

$$-\frac{\mathbf{F}}{6\pi\mu a} = \mathbf{i}\,\frac{U_x}{1 + (3/4)(a/l)} + \mathbf{k}\,\frac{U_z}{1 + (3/2)(a/l)} \tag{6–2.16}$$

Note that the force exerted by the fluid on each particle is the same and that their motion is parallel and with the same velocity. They can move sidewise, but will maintain the same distance between each other.

If α is the angle which the velocity **U** makes with the line of centers of the two particles, and the velocity **U** is determined, we may express the force exerted on each particle by the fluid in terms of drag components in the direction of the velocity, $\mathbf{F}_{V}$, and perpendicular to the direction of velocity, $\mathbf{F}_{D}$. To do this, note that

$$F_{V} = F_z \cos\alpha + F_x \sin\alpha \tag{6-2.17}$$

where

$$F_x = -\frac{6\pi\mu a U \sin\alpha}{1 + (3/4)(a/l)}, \qquad F_z = -\frac{6\pi\mu a U \cos\alpha}{1 + (3/2)(a/l)} \tag{6-2.18}$$

Here U is the scalar velocity corresponding to the vector **U**. By combining these relationships we obtain

$$F_{V} = -6\pi\mu a U\left[\frac{1}{1 + (3/4)(a/l)} - \left\{\frac{1}{1 + (3/4)(a/l)} - \frac{1}{1 + (3/2)(a/l)}\right\}\cos^2\alpha\right] \tag{6-2.19}$$

Similarly,

$$F_{D} = -6\pi\mu a U \sin\alpha\cos\alpha\left[\frac{1}{1 + (3/2)(a/l)} - \frac{1}{1 + (3/4)(a/l)}\right] \tag{6-2.20}$$

On the other hand, if we know the angle β which gravity makes with the line of centers, and require a description of the velocity of fall, it is convenient to express the latter in terms of U_F, the velocity in the direction of gravity, and U_{H}, the velocity of drift in the horizontal direction. The same type of resolution of velocity components gives

$$U_F = -\frac{F}{6\pi\mu a}\left[1 + \frac{3}{4}\frac{a}{l}(1 - \cos^2\beta)\right] \tag{6-2.21}$$

$$U_{H} = -\frac{F}{6\pi\mu a}\frac{3a}{4}\sin\beta\cos\beta \tag{6-2.22}$$

where F is the absolute value of the gravitational force **F**. Thus, a drift occurs only when the angle $\beta \neq 0$ or 90°, that is, when the spheres fall one behind the other along their line of centers, or side by side perpendicular to their line of centers. Smoluchowski[29] gives similar relationships, but some confusion may arise in using them because he refers to force components along the line of centers and the direction of motion, which directions are not perpendicular to each other.

These relationships may be used to describe the motion of two spheres of different sizes with respect to each other, on the assumption that the steady state form will describe the velocities at any position of particles relative to each other—see also Eqs. (6–2.34) and (6–2.35). For example, we may wish to inquire whether a larger sphere following a smaller one vertically can touch the smaller sphere. Equation (6–2.15) is applicable, and we need consider only the vertical component corresponding to the z axis and unit vector $\mathbf{k}$. If we apply this relationship to particles b and a, noting that $F_a \propto -(4/3)\pi a^3$ and $F_b \propto -(4/3)\pi b^3$, we obtain after some algebraic manipulation

$$\frac{U_a}{U_b} = \frac{(a^3/b^3) + (3/2)(a/l)}{1 + (3/2)(a^2/b^2)(a/l)} \tag{6–2.23}$$

For the case where the spheres touch $l = a + b$, whereupon

$$\frac{U_a}{U_b} = \frac{(a^3/b^3) + (a^2/b^2) + (3/2)}{1 + (b/a) + (3a^2/2b^2)} \tag{6–2.24}$$

Where $a = b$, this of course makes $U_a/U_b = 1$, as it should. Where $b \gg a$, $U_a/U_b \rightarrow (3a)/(2b)$. Thus, the velocity of the smaller sphere, a, will always be much less than that of the larger. Anderson[1] treats this case theoretically using a point force to represent the disturbance, but concludes that with a larger sphere overtaking a smaller one there will be a minimum distance beyond which the particles cannot approach each other, so that they will not touch. The calculation of relative velocities when spheres are close to each other is a problem which, in principle, involves the unsteady form of the equations of motion. Care must therefore be exercised in drawing conclusions from approximation procedures which do not include this refinement.

An improvement is possible in this approximation procedure if the solution to the equations of motion for a uniform field moving with respect to the body in question is available. The disturbance can be represented more accurately by such a solution than by using the point force approximation. The reflected fields will constitute a geometric series and the entire effect may be represented by superposition of components along and perpendicular to the line of centers, as previously. Thus, corresponding to spherical particles the usual Stokes field[24] (see Section 4–17) for motion in the z direction is applicable:

$$u_z = U\left(1 - \frac{a}{r}\right) + \frac{1}{4}\,Ua(r^2 - a^2)\left(\frac{1}{r^3} - \frac{3z^2}{r^5}\right) \tag{6–2.25}$$

The same equation with x substituted for z will apply to motion entirely in the x direction. Thus for two equal-sized spheres, in place of Eq. (6–2.16), we will have

$$-\frac{\mathbf{F}}{6\pi\mu a} = \frac{\mathbf{i}U_x}{1 + (3/4)(a/l) + (1/4)(a^3/l^3)} + \frac{\mathbf{k}U_z}{1 + (3/2)(a/l) - (1/2)(a^3/l^3)} \tag{6–2.26}$$

Further improvement in the approximation is possible in the case of spherical particles, since we can determine the resistance of a sphere exactly in terms of the undisturbed field surrounding it by Faxen's law, Eqs. (3–2.46) and (3–2.47). Thus we may calculate exactly the resistance contribution to sphere b resulting from an original Stokes field generated at the location of sphere a. As Burgers[8] has noted, this amounts to using the velocity u_m at the location of particle b, where

$$u_m = u_c + \tfrac{1}{6}a^2\nabla^2 u \tag{6–2.27}$$

u_c is the Stokes field evaluated at the center of particle b (which we have employed in the previous formulas). If this procedure is followed using Eq. (6–2.25) for the Stokes velocity, we obtain, for two equal-sized spheres, analogous to Eqs. (6–2.16) and (6–2.26),

$$-\frac{\mathbf{F}}{6\pi\mu a} = \frac{\mathbf{i}U_x}{1 + (3/4)(a/l) + (1/2)(a^3/l^3)} + \frac{\mathbf{k}U_z}{1 + (3/2)(a/l) - (a^3/l^3)} \tag{6–2.28}$$

Burgers obtained a resistance formula essentially the same as Eq. (6–2.28) He later[9] obtained an improved approximation to the velocity field by an additional reflection, but this is not directly useful in deriving a better approximation for the resistance because this form of equation already amounts to taking the sum of an infinite number of reflections. A different mathematical form is thus necessary to take advantage of calculations involving additional reflections. Such calculations are carried out in the next section. Numerical comparison of these approximation methods is deferred until after development of more exact methods to serve as the basis for judgment.

In order to apply the last two approximation procedures to other than spherical particles, it is necessary not only to locate centers of the particles involved, but also to ascribe to each a characteristic "radius," which may be taken the same as that of a sphere exhibiting the same Stokes resistance as the particle.

If two *orthotropic* particles (possessing three mutually perpendicular planes of symmetry) are allowed to fall freely in the vicinity of each other, the original fields they generate can be represented by point forces, provided that the particles are sufficiently far apart. By resolving the resistances in terms of appropriate coordinate systems, it is possible to calculate their mutual effect on each other. A numerical procedure is most convenient in this case because nomenclature is complicated by the fact that *two* independent orientations are involved—that of each particle with reference to the direction of gravity. As the particles are brought closer together, interactions between platelike or rodlike bodies can no longer be satisfactorily approximated by point forces, but require detailed solutions of the boundary value problems involved.

In general, as the particles become separated by a few diameters, all the approximation procedures give comparable results and particle interactions can be established on the basis of sedimentation data or calculated drag coefficient values for single particles, using the point force approximation procedure.

Generalization for translation

In the case of particles which are not spherically isotropic, their trajectories will in general not be parallel, as assumed in the previous development Brenner[6] has shown that the reflection technique may be generalized for two particles moving in arbitrary directions, provided that the point force approximation is applicable. Free rotation of such anisotropic particles as they settle in a fluid in the presence of each other will give rise to velocity components which cannot be represented simply by point forces and point couples, so that extension to higher reflections than $O(c/l)$ is, in general, not allowable. Also, a rotating anisotropic particle falling under the influence of gravity will experience changes in magnitude and direction of its instantaneous velocity **U**. Therefore, we will not develop the treatment in general here but simply give the result obtained. Simpler results are obtained if we are concerned only with the motion of a single particle in the presence of a stationary wall, as discussed in Section 7–2.

For the case of two particles with characteristic dimensions c_a and c_b, we may write Eq. (6–2.5) in terms of the Stokes translation tensor in the case of an arbitrary particle which is not spherically isotropic:

$$(\mathbf{F}_\infty)_a = -\mu(\mathbf{K}_\infty)_a \cdot \mathbf{U}_a = -6\pi\mu c_a (\boldsymbol{\phi}_\infty)_a \cdot \mathbf{U}_a \tag{6–2.29}$$

where $(\mathbf{F}_\infty)_a$ is the force which body a would experience if it moved through an unbounded fluid with a velocity $\mathbf{U}_a$. In particular for a sphere of radius $c_a = a$, $(\boldsymbol{\phi}_\infty)_a = \mathbf{I}$. Now, if we let $\mathbf{e}$ be a unit vector along the line of centers of the two particles, the relationship obtained by Brenner, analogous to Eq. (6–2.15), is

$$\frac{\mathbf{F}_a}{6\pi\mu c_a} = -\left[(\boldsymbol{\phi}_\infty)_a^{-1} - \frac{9}{16}\frac{c_a}{l}\frac{c_b}{l}(\mathbf{I}+\mathbf{ee})\cdot(\boldsymbol{\phi}_\infty)_b\cdot(\mathbf{I}+\mathbf{ee})\right]^{-1} \left[\mathbf{U}_a - \frac{3}{4}\frac{c_b}{l}(\mathbf{I}+\mathbf{ee})\cdot(\boldsymbol{\phi}_\infty)_b\cdot\mathbf{U}_b\right] \tag{6–2.30}$$

The corresponding force on particle b is obtained by interchanging the subscripts b and a; l is the center-to-center distance.

In most applications of interest, $\mathbf{F}_a$ and $\mathbf{F}_b$ are given and one desires to calculate the settling velocities $\mathbf{U}_a$ and $\mathbf{U}_b$. This is readily done by simultaneously solving Eq. (6–2.30) and its counterpart for these velocities. As a simple example, let the particles be spheres of radii a and b, respectively. The dimensionless translation tensors are now isotropic and have the values $(\boldsymbol{\phi}_\infty)_a = (\boldsymbol{\phi}_\infty)_b = \mathbf{I}$. Substitution in Eq. (6–2.30) yields

$$-\frac{\mathbf{F}_a}{6\pi\mu a} = \left[\mathbf{I} - \frac{9}{16}\frac{a}{l}\frac{b}{l}(\mathbf{I}+3\mathbf{ee})\right]^{-1} \cdot \left[\mathbf{U}_a - \frac{3}{4}\frac{b}{l}(\mathbf{I}+\mathbf{ee})\cdot\mathbf{U}_b\right] \tag{6–2.31}$$

Now let $\mathbf{V}_a$ be the Stokes' law velocity with which sphere a would settle under the influence of gravity in an infinite medium from which sphere b was absent. Then $\mathbf{F}_a = -6\pi\mu a\mathbf{V}_a$, a relation that we employ to eliminate

$\mathbf{F}_a$ from the foregoing. A second relation of similar form is then obtained by interchanging the indices. Upon solving the two resulting equations simultaneously for $\mathbf{U}_a$ and $\mathbf{U}_b$ with the aid of the identity

$$\left[\mathbf{I} - \frac{9}{16}\frac{a}{l}\frac{b}{l}(\mathbf{I} + 3\mathbf{ee})\right]^{-1} = (\mathbf{I} - \mathbf{ee})\left(1 - \frac{9}{16}\frac{a}{l}\frac{b}{l}\right)^{-1} + \mathbf{ee}\left(1 - \frac{9}{4}\frac{a}{l}\frac{b}{l}\right)^{-1} \tag{6–2.32}$$

we obtain, after considerable reduction, the expression

$$\mathbf{U}_a = \mathbf{V}_a + \frac{3}{4}\frac{b}{l}(\mathbf{V}_b + \mathbf{ee}\cdot\mathbf{V}_b) \tag{6–2.33}$$

with a comparable formula for $\mathbf{U}_b$. To express this result in component form, consider the situation depicted in Fig. 6–2.2 If $\mathbf{i}$ and $\mathbf{k}$ denote unit vectors in the x and z directions, respectively, then $\mathbf{V}_a = \mathbf{i}V_a$, $\mathbf{V}_b = \mathbf{i}V_b$ and $\mathbf{e} = \mathbf{i}\sin\theta + \mathbf{k}\cos\theta$. Thus we obtain

$$(U_a)_x = U_F = V_a + \frac{3}{4}\frac{b}{l}V_b(1 + \sin^2\theta) \tag{6–2.34}$$

$$(U_a)_z = U_H = \frac{3}{4}\frac{b}{l}V_b\sin\theta\cos\theta \tag{6–2.35}$$

When $a = b$ we obtain Eqs. (6–2.21) and (6–2.22).

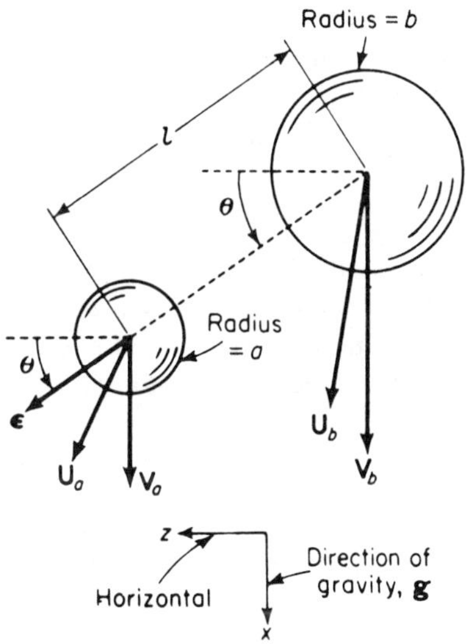

Figure 6-2.2. Two unequal spheres settling in a viscous fluid.

Rotational effects

Thus far we have not considered rotational effects in detail. As mentioned earlier, the point force approximation will not be strictly applicable in the form previously outlined unless the particles involved are free to rotate and are spherically isotropic. If this is the case we can develop the rotation of the field $\mathbf{v}^{(1)}$ which, expressed by Eq. (6–2.8), is as follows:

$$\nabla \times \mathbf{v}^{(1)} = \frac{K_a}{4\pi r^3}[U_{ax}(\mathbf{k}y - \mathbf{j}z) + U_{az}(\mathbf{j}x - \mathbf{i}y)] \tag{6–2.36}$$

Since, in general, the fluid rotation at the location of particle b is

$$\boldsymbol{\omega}_b^{(1)} = \tfrac{1}{2}(\nabla \times \mathbf{v}^{(1)})_b \tag{6–2.37}$$

and the location of particle b is taken at $x = 0$, $y = 0$, $z = l$, we have

$$\boldsymbol{\omega}_b^{(1)} = -\mathbf{j}\frac{K_a U_{ax}}{8\pi l^2} \tag{6–2.38}$$

Now, if the particle b is not fixed it will simply rotate at the velocity $\boldsymbol{\omega}_b^{(1)}$ and the only rotational effect experienced by particle a will be that due to the point force located at particle b, the curl of which can be obtained in a similar fashion to Eq. (6–2.36). Thus the rotation developed at particle a by the field $\mathbf{v}^{(2)}$ will be

$$\boldsymbol{\omega}_a^{(2)} = \mathbf{j}\frac{K_b}{8\pi l^2}\left(U_{bx} - \frac{K_a U_{ax}}{8\pi l}\right) \tag{6–2.39}$$

This is opposite in direction to the rotation of particle b, but the lead terms are the same in magnitude for equal-sized particles. These terms will increase in the same way as the additional point force contributions, in a geometric series, so that ultimately

$$\boldsymbol{\omega}_a = \boldsymbol{\omega}_a^{(2)} + \boldsymbol{\omega}_a^{(4)} + \cdots = \mathbf{j}\left(\frac{K_b}{8\pi l^2}\right)\frac{U_{bx} - (K_a U_{ax}/8\pi l)}{1 - (K_a K_b)/(8\pi l)^2} \tag{6–2.40}$$

The rotation of particle b, $\boldsymbol{\omega}_b$, may be obtained from the foregoing by interchanging the subscripts a and b and by changing the sign of the rotation. The rotations occur about axes perpendicular to the xz plane. If the particles are the same size and spherical ($K = 6\pi a$), we have for a and b

$$\boldsymbol{\omega} = \mp\mathbf{j}\frac{3aU_x}{4l^2}\left[\frac{1}{1 + (3a/4l)}\right] \tag{6–2.41}$$

Note that $U_x = U\sin\alpha$, the velocity component perpendicular to the line connecting the particle centers, is the only one contributing to rotation. If two particles fall along their line of centers no rotation will result.

Now, if the particles are held rigidly a more complicated situation results because a resisting torque will be required at the location of particle b. The effect of this torque can be computed by employing a point couple at particle b. It will give rise upon reflection to terms involving l^6 in the denominator of the expression for $\mathbf{T}$ and l^5 in the denominator of the expression for $\mathbf{F}$ and, therefore, will be important only if the particles are close together. If the particles are close, a more exact approximation procedure than that involving point forces and couples should be applied; hence, this procedure is not elaborated here.

As a first approximation to the torque exerted on the body, we can assume that in the case of a regular particle the rotational moment will be independent of the axis of rotation and will be characterized by a scalar; that is,

$$\mathbf{T}_a = d_a\boldsymbol{\omega}_a \tag{6–2.42}$$

where d_a is a torque factor characteristic of the body in question and $\mathbf{T}_a$ is the torque required to maintain an arbitrary body rotating with the velocity $\boldsymbol{\omega}_a$. If the particle is a sphere of radius a, $d_a = 8\pi\mu a^3$.

We shall assume that if a torque is opposed to the fluid rotation $\boldsymbol{\omega}_a$ exerted on a particle a, so that the particle a is prevented from rotating, the magnitude of this torque can be given to a first approximation by taking the lead term of Eq. (6–2.40); namely,

$$\mathbf{T}_a = \left(\mathbf{j}\,\frac{K_b U_{bz}}{8\pi l^2}\right) d_a \tag{6–2.43}$$

In the case of spherical particles, this becomes

$$\mathbf{T}_a = \mathbf{j}\,6\pi\mu a^2 \left(\frac{a}{l}\right)\left(\frac{b}{l}\right) U_{bz} \tag{6–2.44}$$

This is the torque exerted by the fluid on the particle in order to prevent rotation. Smoluchowski[29] arrives at essentially the same type of formulation, but owing to a computational error, gives the numerical constant in Eq. (6–2.44) as 4 instead of 6.

6-3 Two Spheres by the Method of Reflections and Similar Techniques

Where more accurate results are required than are obtainable by the approximations noted in Section 6–2, it is possible in principle to solve the boundary value problem of two objects to any desired degree of accuracy by the method of reflections. In this case it is necessary to employ a general solution for each body subject to arbitrarily specified boundary conditions at its surface. Such solutions become complicated; consequently, the method has been applied only to the case of two spheres in an arbitrary position relative to each other.

We shall define the coordinates of a point in space, with respect to an origin situated at the center of sphere a, as (x, y, z) in cartesian coordinates and as (r, θ, ϕ) in spherical coordinates. Similarly (X, Y, Z) and (R, Θ, Φ) are the corresponding coordinates in terms of an origin at the center of sphere b. The systems of cartesian and spherical coordinates are related by (see Figs. 6–3.1 and 6–3.2)

$$\begin{aligned} x &= r\sin\theta\cos\phi \\ y &= r\sin\theta\sin\phi \\ z &= r\cos\theta \\ r^2 &= x^2 + y^2 + z^2 \end{aligned}$$

with similar relationships applying to the terms with capital symbols. The center of sphere b has the coordinates (x_o, y_o, z_o) and (r_o, θ_o, ϕ_o) with respect to an origin at the center of sphere a; the sphere a has the location (X_o, Y_o, Z_o) and (R_o, Θ_o, Φ_o) with respect to an origin at the center of sphere b. Without loss of generality the centers of the spheres can be taken along the z axis so that these sets of points become:

$$
\begin{aligned}
X_o &= -x_o = 0, & R_o &= r_o = l, \quad \text{say} \\
Y_o &= -y_o = 0, & \Theta_o &= \pi - \theta_o = \pi \\
Z_o &= -z_o, & \Phi_o &= \pi + \phi_o = \pi
\end{aligned}
\tag{6–3.1}
$$

We will designate unit vectors in the (r, θ, ϕ) system by $(\mathbf{i}_r, \mathbf{i}_\theta, \mathbf{i}_\phi)$ and in the (R, Θ, Φ) system by $(\mathbf{i}_R, \mathbf{i}_\Theta, \mathbf{i}_\Phi)$.

Because of the linearity of the original differential equations (6–1.1) and (6–1.2), solutions to the complete problem may be obtained by superposition of partial solutions. Thus we may write

$$
\begin{aligned}
\mathbf{v} &= \mathbf{v}^* + \mathbf{v}^{**} \\
p &= p^* + p^{**}
\end{aligned}
\tag{6–3.2}
$$

where $\mathbf{v}^*$ and p^* represent the solutions which satisfy boundary conditions on sphere b and are regular outside sphere b, whereas $\mathbf{v}^{**}$ and p^{**} represent solutions which satisfy the boundary conditions on a and are regular outside of sphere a. Thus

$$
\mathbf{v}^* = \begin{cases} \mathbf{U}_b & \text{on sphere } b \\ \mathbf{0} & \text{on sphere } a \\ \mathbf{0} & \text{at } \infty \end{cases}
\tag{6–3.3}
$$

and

$$
\mathbf{v}^{**} = \begin{cases} \mathbf{0} & \text{on sphere } b \\ \mathbf{U}_a & \text{on sphere } a \\ \mathbf{0} & \text{at } \infty \end{cases}
\tag{6–3.4}
$$

The fields constituting the desired solution will be obtained by adding an appropriate series of partial solutions:

$$
\begin{aligned}
\mathbf{v}^* &= \sum_{i=1}^{\infty} \mathbf{v}_i^*, & p^* &= \sum_{i=1}^{\infty} p_i^* \\
\mathbf{v}^{**} &= \sum_{i=1}^{\infty} \mathbf{v}_i^{**}, & p^{**} &= \sum_{i=1}^{\infty} p_i^{**}
\end{aligned}
\tag{6–3.5}
$$

Each of the fields $\mathbf{v}_i^*, p_i^*$ and $\mathbf{v}_i^{**}, p_i^{**}$ is a solution of the original differential equations, and all solutions, of course, meet the requirement

$$
\mathbf{v}_i^* = \mathbf{v}_i^{**} = \mathbf{0} \quad \text{at } \infty
$$

Confining our attention for the moment to $\mathbf{v}^*$, we note that on the surface of sphere b,

$$
\mathbf{U}_b = \sum_{i=1}^{\infty} \mathbf{v}_i^*|_b
\tag{6–3.6}
$$

whereas on the surface of sphere a,

$$
\mathbf{0} = \sum_{i=1}^{\infty} \mathbf{v}_i^*|_a
\tag{6–3.7}
$$

These two relationships are satisfied by

$$\begin{aligned} \mathbf{v}_1^*|_b &= \mathbf{U}_b \\ \mathbf{v}_2^*|_a &= -\mathbf{v}_1^*|_a \\ \mathbf{v}_3^*|_b &= -\mathbf{v}_2^*|_b \\ \mathbf{v}_4^*|_a &= -\mathbf{v}_3^*|_a, \text{ etc.} \end{aligned} \tag{6–3.8}$$

or, in general, for $j = 1, 2, 3, \ldots$

$$\begin{aligned} \mathbf{v}_{2j}^* &= -\mathbf{v}_{2j-1}^* \quad \text{on sphere } a \\ \mathbf{v}_{2j+1}^* &= -\mathbf{v}_{2j}^* \quad \text{on sphere } b \end{aligned} \tag{6–3.9}$$

When the original velocity $\mathbf{U}_b$ is prescribed as a result of the motion of sphere b, the conditions stated serve to define uniquely each of the fields $\mathbf{v}_i^*$.

In an entirely analogous manner we find that

$$\mathbf{v}_1^{**} = \mathbf{0} \quad \text{everywhere,} \qquad \mathbf{v}_2^{**} = \mathbf{U}_a \quad \text{on sphere } a, \text{ etc.}$$

so that in general for $j = 1, 2, 3, \ldots$

$$\begin{aligned} \mathbf{v}_{2j+1}^* &= -\mathbf{v}_{2j}^* \quad \text{on sphere } b \\ \mathbf{v}_{2j+2}^* &= -\mathbf{v}_{2j+1}^* \quad \text{on sphere } a \end{aligned} \tag{6–3.10}$$

Although the general technique to be discussed is suitable for solving the problem of two spheres in any orientation with respect to the coordinate system chosen, it is simpler to construct the solution to this problem by a combination of the particular solutions corresponding to the situations in which the spheres move (a) parallel to their line of centers and (b) perpendicular to their line of centers, as was done in the previous section.

First, we shall consider the problem of two spheres moving along their line of centers, that is along the z axis, as shown in Fig. 6–3.1.

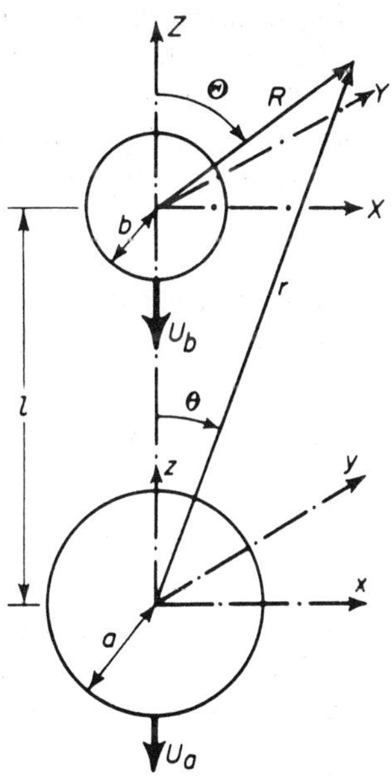

Figure 6-3.1. Two spheres falling along their line of centers.

Motion along line of centers

For this situation, if we assume that the direction of fall of the spheres constitutes a negative direction with respect to the coordinate system (instead of taking the sphere motion positive as in the previous section), we have the boundary conditions

$$\mathbf{v} = \begin{cases} -\mathbf{i}_z U_b & \text{on sphere } b \\ -\mathbf{i}_z U_a & \text{on sphere } a \\ \mathbf{0} & \text{at } \infty \end{cases} \tag{6–3.11}$$

with U_a and U_b corresponding to the scalar numbers representing the velocity of each sphere (taken positive). Thus,

$$\mathbf{v}^* = \begin{cases} -\mathbf{i}_z U_b & \text{on sphere } b \\ \mathbf{0} & \text{on sphere } a \\ \mathbf{0} & \text{at } \infty \end{cases} \tag{6–3.12}$$

and

$$\mathbf{v}^{**} = \begin{cases} \mathbf{0} & \text{on sphere } b \\ -\mathbf{i}_z U_a & \text{on sphere } a \\ \mathbf{0} & \text{at } \infty \end{cases} \tag{6–3.13}$$

The general solution for component velocity and pressure fields is given by Lamb[24] (also Section 3–2). When the spheres move along their line of centers the solution will be independent of the angle $\phi = \Phi$, and the field components $v_\phi = v_\Phi$ will be identically zero, due to axial symmetry.

For a field with symmetry about the z axis, we have, in general, in terms of (R, Θ, Φ) coordinates

$$\begin{aligned} \mathbf{v}_i &= \sum_{N=0}^{\infty} \nabla \Psi^i_{-N-1} + \frac{2-N}{2N(2N-1)} \nabla (R^2 \Pi^i_{-N-1}) + \frac{1}{N} \mathbf{R}\, \Pi^i_{-N-1} \\ p_i &= \mu \sum_{N=0}^{\infty} \Pi^i_{-N-1} \end{aligned} \tag{6–3.14}$$

The harmonic functions Ψ^i_{-N-1} and Π^i_{-N-1} are each solid spherical harmonics of order $-(N+1)$ in the (R, Θ, Φ) coordinate system.

We begin the development of the solution by considering first the fields $\mathbf{v}_i^*$. From the boundary condition given by Eq. (6–3.8), we note that $\mathbf{v}_1^*$ is required to satisfy the boundary condition

$$\mathbf{v}_1^* = -\mathbf{i}_z U_b \quad \text{at } R = b \tag{6–3.15}$$

Thus, the field $\mathbf{v}_1^*$ is simply the Stokes' law velocity field in the (R, Θ, Φ) system. For this field the harmonic functions in Eq. (6–3.14) are as follows:

$$\Psi^{*1}_{-N-1} = \begin{cases} -\frac{1}{4} U_b b^3 R^{-2} P_1(\cos\Theta) & \text{for } N = 1 \\ 0 & \text{for all other } N \end{cases} \tag{6–3.16}$$

$$\Pi^{*1}_{-N-1} = \begin{cases} -\frac{3}{2} U_b b R^{-2} P_1(\cos\Theta) & \text{for } N = 1 \\ 0 & \text{for all other } N \end{cases} \tag{6–3.17}$$

where $P_1(\cos\Theta) = \cos\Theta$.

For convenience in deriving an iterative solution, we designate the harmonic functions in Eq. (6–3.14) in the following form:

$$\begin{aligned} \Psi^i_{-N-1} &= (-1)^N b^{N+2} B^i_N R^{-(N+1)} P_N(\cos\Theta) \\ \Pi^i_{-N-1} &= (-1)^N b^N C^i_N R^{-(N+1)} P_N(\cos\Theta) \end{aligned} \tag{6–3.18}$$

Thus from Eqs. (6–3.16) and (6–3.17) we have

$$
\begin{aligned}
B_N^{*1} &= \begin{cases} \frac{1}{4}U_b & \text{for } N = 1 \\ 0 & \text{for all other } N \end{cases} \\
C_N^{*1} &= \begin{cases} \frac{3}{2}U_b & \text{for } N = 1 \\ 0 & \text{for all other } N \end{cases}
\end{aligned}
\tag{6–3.19}
$$

In order to proceed from the field $\mathbf{v}_1^*$ to obtain the field $\mathbf{v}_2^*$, we must convert the velocity field $\mathbf{v}_1^*$ from the (R, Θ, Φ) system to the (r, θ, ϕ) system. It can be shown that in the (r, θ, ϕ) system the harmonic functions in Eq. (6–3.18) may be expressed as follows:

$$
\begin{aligned}
\Psi_{-N-1}^i &= \left(\frac{b}{l}\right)^{N+2} B_N^i \sum_{n=0}^{\infty} \frac{f_n^N r^n P_n(\cos\theta)}{l^{n-1}} \\
\Pi_{-N-1}^i &= \left(\frac{b}{l}\right)^{N} C_N^i \sum_{n=0}^{\infty} \frac{f_n^N r^n P_n(\cos\theta)}{l^{n+1}}
\end{aligned}
\tag{6–3.20}
$$

where
$$f_n^N = \frac{(n+N)!}{n!\,N!}$$

In addition,

$$R^2 = r^2 + l^2 - 2rl\cos\theta \tag{6–3.21}$$

and
$$\mathbf{R} = \mathbf{i}_r(r - l\cos\theta) + \mathbf{i}_\theta l\sin\theta \tag{6–3.22}$$

If these expressions are substituted into Eq. (6–3.14), we then obtain $\mathbf{v}_1^*$ in the (r, θ, ϕ) system. For convenience, these transformations are carried out in a general fashion to obtain the field $\mathbf{v}_{i+1}$ from that for $\mathbf{v}_i$ in Eq. (6–3.14). The field $\mathbf{v}_{i+1}$ will have the same general form as the field $\mathbf{v}_i$, but it is expressed in the (r, θ, ϕ) system, appropriate for sphere a.

Thus, in general, for axial symmetry

$$
\begin{aligned}
\mathbf{v}_{i+1} &= \sum_{n=0}^{\infty} \nabla\psi_{-n-1}^{i+1} + \frac{2-n}{2n(2n-1)}\nabla(r^2\pi_{-n-1}^{i+1}) + \frac{1}{n}\mathbf{r}\,\pi_{-n-1}^{i+1} \\
p_{i+1} &= \mu\sum_{n=0}^{\infty}\pi_{-n-1}^{i+1}
\end{aligned}
\tag{6–3.23}
$$

where the harmonic functions assume the form

$$
\begin{aligned}
\psi_{-n-1}^{i+1} &= -a^{n+2}b_n^{i+1}r^{-(n+1)}P_n(\cos\theta) \\
\pi_{-n-1}^{i+1} &= -a^{n}c_n^{i+1}r^{-(n+1)}P_n(\cos\theta)
\end{aligned}
\tag{6–3.24}
$$

By means of the transformations in Eqs. (6–3.20)–(6–3.22) we are able to determine the constants b_n^{i+1}, c_n^{i+1}, which determine the field $\mathbf{v}_{i+1}$, in terms of the constants B_N^i and C_N^i, which determine the field $\mathbf{v}_i$. The general boundary condition in Eqs. (6–3.9) and (6–3.10), $\mathbf{v}_{i+1} = -\mathbf{v}_i$ on sphere a, is also employed. The desired relationships are as follows:

$$b_n^{i+1} = \frac{n}{2(n+1)}\left(\frac{a}{l}\right)^{n-1} \sum_{N=0}^{\infty} \frac{(n+N)!}{n!\,N!}\left(\frac{b}{l}\right)^N$$
$$\left[(2n-1)\left(\frac{b}{l}\right)^2 B_N^i - (2n-1)\frac{2nN+2-n-N}{2(2N-1)(2n-1)(n+N)} C_N^i \right. \tag{6–3.25}$$
$$\left. + \frac{2n+1}{2(2n+3)}\left(\frac{a}{l}\right)^2 C_N^i\right]$$

$$c_n^{i+1} = \frac{n(2n-1)}{n+1}\left(\frac{a}{l}\right)^{n-1} \sum_{N=0}^{\infty} \frac{(n+N)!}{n!\,N!}\left(\frac{b}{l}\right)^N$$
$$\left[(2n+1)\left(\frac{b}{l}\right)^2 B_N^i - (2n+1)\frac{2nN+2-n-N}{2(2N-1)(2n-1)(n+N)} C_N^i \right. \tag{6–3.26}$$
$$\left. + \frac{1}{2}\left(\frac{a}{l}\right)^2 C_N^i\right]$$

We can again reflect the field $\mathbf{v}_{i+1}$ given in Eq. (6–3.23) to find the field $\mathbf{v}_{i+2}$ which satisfies the boundary condition $\mathbf{v}_{i+2} = -\mathbf{v}_{i+1}$ on sphere b. In order to accomplish this we require a set of inverse transformations, similar to those given by Eqs. (6–3.20)–(6–3.22), which will enable us to transform the field $\mathbf{v}_{i+1}$ from the (r, θ, ϕ) system to the (R, Θ, Φ) system of coordinates appropriate for sphere b. If B_N^{i+2} and C_N^{i+2} are now taken in equations similar to (6–3.14) and (6–3.18) to represent the constants which determine the field $\mathbf{v}_{i+2}$, then in terms of the constants b_n^{i+1} and c_n^{i+1} we will have the following recurrence formulas similar to Eqs. (6–3.25) and (6–3.26) to accomplish the necessary transformation to the field $\mathbf{v}_{i+2}$:

$$B_N^{i+2} = \frac{N}{2(N+1)}\left(\frac{b}{l}\right)^{N-1} \sum_{n=0}^{\infty} \frac{(n+N)!}{n!\,N!}\left(\frac{a}{l}\right)^n$$
$$\left[(2N-1)\left(\frac{a}{l}\right)^2 b_n^{i+1} - (2N-1)\frac{2nN+2-n-N}{2(2N-1)(2n-1)(n+N)} c_n^{i+1} \right. \tag{6–3.27}$$
$$\left. + \frac{(2N+1)}{2(2N+3)}\left(\frac{b}{l}\right)^2 c_n^{i+1}\right]$$

$$C_N^{i+2} = \frac{N(2N-1)}{N+1}\left(\frac{b}{l}\right)^{N-1} \sum_{n=0}^{\infty} \frac{(n+N)!}{n!\,N!}\left(\frac{a}{l}\right)^n$$
$$\left[(2N+1)\left(\frac{a}{l}\right)^2 b_n^{i+1} - (2N+1)\frac{2nN+2-n-N}{2(2N-1)(2n-1)(n+N)} c_n^{i+1} \right. \tag{6–3.28}$$
$$\left. + \frac{1}{2}\left(\frac{b}{l}\right)^2 c_n^{i+1}\right]$$

The relationships (6–3.25) and (6–3.26), would thus be used in conjunction with Eq. (6–3.23) to obtain $\mathbf{v}_2^*$ from the field $\mathbf{v}_1^*$. In turn, Eqs. (6–3.27) and (6–3.28) would be employed in conjunction with Eq. (6–3.14) to obtain the field $\mathbf{v}_3^*$ from the field $\mathbf{v}_2^*$. This iterative procedure will enable all the $\mathbf{v}^*$ fields to be derived.

Similarly, the general relationships apply equally well to the fields $\mathbf{v}_k^*$ and $\mathbf{v}_k^{**}$. To distinguish the constants which apply separately to these fields we shall employ the symbols b_n^{*k}, c_n^{*k}, or B_N^{*k} and C_N^{*k} for the former, and b_n^{**k}, c_n^{**k}, or B_N^{**k}, C_N^{**k} for the latter.

We now proceed with the derivation that started with the development of a Stokes field $\mathbf{v}_1^*$ as noted in Eq. (6–3.19). In these developments it is necessary at the outset to establish the degree of approximation which will ultimately be required. Thus some arbitrary power of $(a/l)^\alpha\,(b/l)^\beta$ will be selected. For illustration we shall, in what follows, limit ourselves to $\alpha + \beta \leq 5$.

Thus, if we now wish to establish the constants for the field $\mathbf{v}_2^*$, we commence by determining the constants b_n^{*2} from Eq. (6–3.25):

$$b_0^{*2} = 0$$

$$\frac{b_1^{*2}}{U_b} = -\frac{3}{8}\frac{b}{l} + \frac{1}{8}\left(\frac{b}{l}\right)^3 + \frac{9}{40}\left(\frac{a}{l}\right)^2\left(\frac{b}{l}\right)$$

$$\frac{b_2^{*2}}{U_b} = -\frac{3}{4}\left(\frac{a}{l}\right)\left(\frac{b}{l}\right) + \frac{3}{4}\left(\frac{a}{l}\right)\left(\frac{b}{l}\right)^3 + \frac{15}{28}\left(\frac{a}{l}\right)^3\left(\frac{b}{l}\right)$$

$$\frac{b_3^{*2}}{U_b} = -\frac{9}{8}\left(\frac{a}{l}\right)^2\left(\frac{b}{l}\right) + \frac{15}{8}\left(\frac{a}{l}\right)^2\left(\frac{b}{l}\right)^3 + \frac{7}{8}\left(\frac{a}{l}\right)^4\left(\frac{b}{l}\right) \tag{6–3.29}$$

$$\frac{b_4^{*2}}{U_b} = -\frac{3}{2}\left(\frac{a}{l}\right)^3\left(\frac{b}{l}\right)$$

$$\frac{b_5^{*2}}{U_b} = -\frac{15}{8}\left(\frac{a}{l}\right)^4\left(\frac{b}{l}\right)$$

$$b_6^{*2} = 0$$

The c_n^{*2} coefficients are determined in a similar fashion from Eq. (6–3.26), but are not repeated here. Next we proceed to calculate the constants B_N^{*3} and C_N^{*3} by using Eqs. (6–3.27) and (6–3.28) in conjunction with the set of values of b_n^{*2} and c_n^{*2} which have been computed as previously indicated. This process is continued to determine the b_n^{*4} and c_n^{*4} constants and similarly the B_n^{*5} and C_n^{*5} constants. The evaluation is completed by determination of the b_n^{*6} and c_n^{*6} constants, since all higher reflections, that is, $i > 6$, are zero to the order of approximation which we have selected.

To determine the second series of constants we note that the field $\mathbf{v}_1^{**} = \mathbf{0}$, which makes

$$B_N^{**1} = C_N^{**1} = 0 \quad \text{for all } N$$

Meanwhile the field $\mathbf{v}_2^{**}$ is required to satisfy the boundary condition $\mathbf{v}_2^{**} = -\,\mathbf{i}_z U_a$ at $r = a$. The field which satisfies this condition is given by Eq. (6–3.23) with the following values for the harmonic functions, corresponding again to the Stokes' field:

$$\psi_{-n-1}^{**2} = \begin{cases} -\frac{1}{4}U_a a^3 r^{-2} P_1(\cos\theta) & \text{for } n = 1 \\ 0 \quad \text{for all other } n \end{cases} \tag{6–3.30}$$

$$\pi^{**2}_{-n-1} = \begin{cases} -\frac{3}{2} U_a a r^{-2} P_1(\cos\theta) & \text{for } n = 1 \\ 0 & \text{for all other } n \end{cases} \tag{6–3.31}$$

If these relationships are compared with Eq. (6–3.24) we find that

$$b_n^{**2} = \begin{cases} \frac{1}{4} U_a & \text{for } n = 1 \\ 0 & \text{for all other } n \end{cases} \tag{6–3.32}$$

$$c_n^{**2} = \begin{cases} \frac{3}{2} U_a & \text{for } n = 1 \\ 0 & \text{for all other } n \end{cases} \tag{6–3.33}$$

Utilizing these constants to begin the reflection procedure, we first compute the B_n^{**3} and C_n^{**3} constants from Eqs. (6–3.27) and (6–3.28), respectively. The procedure is carried forward as before to the evaluation of the B_n^{**7} and C_n^{**7} constants, which are the highest order necessary when retaining a sum of the powers of (a/l) and (b/l) equal to 5.

Thus the constants which determine the entire velocity field, $\mathbf{v} = \mathbf{v}^* + \mathbf{v}^{**}$, satisfying the desired boundary conditions can be obtained from the relationships

$$b_n = \sum_{i=1}^{\infty} b_n^i = \sum_{i=1}^{\infty} (b_n^{*i} + b_n^{**i}) \tag{6–3.34}$$

with similar relationship applying for the constants c_N, B_N, and C_N. Thus, to our present approximation,

$$\begin{aligned} b_1 = U_b\Bigg[&-\frac{3}{8}\left(\frac{b}{l}\right) + \frac{1}{8}\left(\frac{b}{l}\right)^3 + \frac{9}{40}\left(\frac{a}{l}\right)^2\left(\frac{b}{l}\right) - \frac{27}{32}\left(\frac{a}{l}\right)\left(\frac{b}{l}\right)^2 \\ &- \frac{9}{16}\left(\frac{a}{l}\right)\left(\frac{b}{l}\right)^4 - \frac{27}{80}\left(\frac{a}{l}\right)^3\left(\frac{b}{l}\right)^2 - \frac{243}{128}\left(\frac{a}{l}\right)^2\left(\frac{b}{l}\right)^3\Bigg] \\ + U_a\Bigg[&\frac{1}{4} + \frac{9}{16}\left(\frac{a}{l}\right)\left(\frac{b}{l}\right) - \frac{21}{40}\left(\frac{a}{l}\right)^3\left(\frac{b}{l}\right) \\ &+ \frac{3}{2}\left(\frac{a}{l}\right)\left(\frac{b}{l}\right)^3 + \frac{81}{64}\left(\frac{a}{l}\right)^2\left(\frac{b}{l}\right)^2\Bigg] \end{aligned} \tag{6–3.35}$$

$$\begin{aligned} b_2 = U_b\Bigg[&-\frac{3}{4}\left(\frac{a}{l}\right)\left(\frac{b}{l}\right) + \frac{3}{4}\left(\frac{a}{l}\right)\left(\frac{b}{l}\right)^3 \\ &+ \frac{15}{28}\left(\frac{a}{l}\right)^3\left(\frac{b}{l}\right) - \frac{27}{16}\left(\frac{a}{l}\right)^2\left(\frac{b}{l}\right)^2\Bigg] \\ + U_a\Bigg[&\frac{9}{8}\left(\frac{a}{l}\right)^2\left(\frac{b}{l}\right) - \frac{33}{28}\left(\frac{a}{l}\right)^4\left(\frac{b}{l}\right) \\ &+ 6\left(\frac{a}{l}\right)^2\left(\frac{b}{l}\right)^3 + \frac{81}{32}\left(\frac{a}{l}\right)^3\left(\frac{b}{l}\right)^2\Bigg] \end{aligned} \tag{6–3.36}$$

$$\begin{aligned} b_3 = U_b\Bigg[&-\frac{9}{8}\left(\frac{a}{l}\right)^2\left(\frac{b}{l}\right) + \frac{15}{8}\left(\frac{a}{l}\right)^2\left(\frac{b}{l}\right)^3 \\ &+ \frac{7}{8}\left(\frac{a}{l}\right)^4\left(\frac{b}{l}\right) - \frac{81}{32}\left(\frac{a}{l}\right)^3\left(\frac{b}{l}\right)^2\Bigg] + U_a\left[\frac{27}{16}\left(\frac{a}{l}\right)^3\left(\frac{b}{l}\right)\right] \end{aligned} \tag{6–3.37}$$

$$b_4 = U_b\left[-\frac{3}{2}\left(\frac{a}{l}\right)^3\left(\frac{b}{l}\right)\right] + U_a\left[\frac{9}{4}\left(\frac{a}{l}\right)^4\left(\frac{b}{l}\right)\right] \tag{6-3.38}$$

$$b_5 = U_b\left[-\frac{15}{8}\left(\frac{a}{l}\right)^4\left(\frac{b}{l}\right)\right] \tag{6-3.39}$$

$$\begin{aligned} c_1 = U_b\Bigg[&-\frac{9}{4}\left(\frac{b}{l}\right) + \frac{3}{4}\left(\frac{b}{l}\right)^3 + \frac{3}{4}\left(\frac{a}{l}\right)^2\left(\frac{b}{l}\right) - \frac{81}{16}\left(\frac{a}{l}\right)\left(\frac{b}{l}\right)^2 \\ &-\frac{27}{8}\left(\frac{a}{l}\right)\left(\frac{b}{l}\right)^4 - \frac{27}{8}\left(\frac{a}{l}\right)^3\left(\frac{b}{l}\right)^2 - \frac{729}{64}\left(\frac{a}{l}\right)^2\left(\frac{b}{l}\right)^3\Bigg] \\ + U_a\Bigg[&\frac{3}{2} + \frac{27}{8}\left(\frac{a}{l}\right)\left(\frac{b}{l}\right) - \frac{9}{4}\left(\frac{a}{l}\right)^3\left(\frac{b}{l}\right) \\ &+ \frac{27}{8}\left(\frac{a}{l}\right)\left(\frac{b}{l}\right)^3 + \frac{243}{32}\left(\frac{a}{l}\right)^2\left(\frac{b}{l}\right)^2\Bigg] \end{aligned} \tag{6-3.40}$$

$$\begin{aligned} c_2 = U_b\Bigg[&-\frac{15}{2}\left(\frac{a}{l}\right)\left(\frac{b}{l}\right) + \frac{15}{2}\left(\frac{a}{l}\right)\left(\frac{b}{l}\right)^3 \\ &+ \frac{9}{2}\left(\frac{a}{l}\right)^3\left(\frac{b}{l}\right) - \frac{135}{8}\left(\frac{a}{l}\right)^2\left(\frac{b}{l}\right)^2\Bigg] \\ + U_a\Bigg[&\frac{45}{4}\left(\frac{a}{l}\right)^2\left(\frac{b}{l}\right) - \frac{21}{2}\left(\frac{a}{l}\right)^4\left(\frac{b}{l}\right) \\ &+ \frac{45}{2}\left(\frac{a}{l}\right)^2\left(\frac{b}{l}\right)^3 + \frac{405}{16}\left(\frac{a}{l}\right)^3\left(\frac{b}{l}\right)^2\Bigg] \end{aligned} \tag{6-3.41}$$

$$\begin{aligned} c_3 = U_b\Bigg[&\frac{63}{4}\left(\frac{a}{l}\right)^2\left(\frac{b}{l}\right) + \frac{105}{4}\left(\frac{a}{l}\right)^2\left(\frac{b}{l}\right)^3 \\ &+ \frac{45}{4}\left(\frac{a}{l}\right)^4\left(\frac{b}{l}\right) - \frac{567}{16}\left(\frac{a}{l}\right)^3\left(\frac{b}{l}\right)^2\Bigg] + U_a\left[\frac{189}{8}\left(\frac{a}{l}\right)^3\left(\frac{b}{l}\right)\right] \end{aligned} \tag{6-3.42}$$

$$c_4 = U_b\left[-27\left(\frac{a}{l}\right)^3\left(\frac{b}{l}\right)\right] + U_a\left[\frac{81}{2}\left(\frac{a}{l}\right)^4\left(\frac{b}{l}\right)\right] \tag{6-3.43}$$

$$c_5 = U_b\left[-\frac{165}{4}\left(\frac{a}{l}\right)^4\left(\frac{b}{l}\right)\right] \tag{6-3.44}$$

The constants B_N and C_N can be obtained from the preceding constants by exchanging a for b in both subscripts and radii of spheres involved.

Thus, the velocity field approximately satisfying the given boundary conditions is obtained by summing the two velocity fields (each with a *different origin*):

$$\begin{aligned} \mathbf{v} = &\sum_{n=1}^{\infty} \nabla\psi_{-n-1} + \frac{2-n}{2n(2n-1)}\nabla(r^2\pi_{-n-1}) + \frac{1}{n}\mathbf{r}\pi_{-n-1} \\ &+ \sum_{N=1}^{\infty} \nabla\Psi_{-N-1} + \frac{2-N}{2N(2N-1)}\nabla(R^2\Pi_{-N-1}) + \frac{1}{N}\mathbf{R}\Pi_{-N-1} \end{aligned} \tag{6-3.45}$$

where, in terms of the known constants b_n, c_n, B_N, and C_N,

$$\begin{aligned} \psi_{-n-1} &= -b_n a^{n+2} r^{-(n+1)} P_n(\cos\theta) \\ \pi_{-n-1} &= -c_n a^n r^{-(n+1)} P_n(\cos\theta) \\ \Psi_{-N-1} &= (-1)^N B_N b^{N+2} R^{-(N+1)} P_N(\cos\Theta) \\ \Pi_{-N-1} &= (-1)^N C_N b^N R^{-(N+1)} P_N(\cos\Theta) \end{aligned} \tag{6–3.46}$$

This velocity field could be employed in estimating the effect of the motion of two spheres in the vicinity of other bodies to a first approximation. For accurate results, when a third body is close to the two spheres, it would be necessary to employ the iteration formulas to develop the disturbance of each additional body or sphere separately.

These relationships may be employed to determine the drag on either of the two spheres. Thus, in Section 3–2, it is shown that the drag exerted on sphere a, $\mathbf{F}_a$, is given by

$$\mathbf{F}_a = -4\pi\mu\,\nabla\left(r^3 \sum_{i=1}^{\infty} \pi_{-2}^i\right) \tag{6–3.47}$$

where, here,

$$\pi_{-2} = \sum_{i=1}^{\infty} \pi_{-2}^i = -c_1 a r^{-2} P_1(\cos\theta) \tag{6–3.48}$$

This makes

$$\mathbf{F}_a = 4\pi\mu a c_1 \nabla(r\cos\theta) \tag{6–3.49}$$

Noting that $\nabla(r\cos\theta) = \nabla z = \mathbf{i}_z$, we find

$$\mathbf{F}_a = \mathbf{i}_z 4\pi\mu a c_1 \tag{6–3.50}$$

We may insert the appropriate value of c_1 from Eq. (6–3.40) and obtain the following, where the subscript $a1$ indicates the drag on sphere a when the spheres are moving along their line of centers.

$$\begin{aligned} \mathbf{F}_{a1} = \mathbf{i}_z 6\pi\mu a \Bigg\{ & U_a\left[1 + \frac{9}{4}\frac{ab}{l^2} + \frac{3}{4}\left(-2\frac{a^3 b}{l^4} + \frac{27}{4}\frac{a^2 b^2}{l^4} + \frac{3ab^3}{l^4}\right)\right] \\ & + U_b\left[-\frac{3}{2}\frac{b}{l} + \frac{1}{2}\left(\frac{a^2 b}{l^3} - \frac{27}{4}\frac{ab^2}{l^3} + \frac{b^3}{l^3}\right)\right. \\ & \left.\left. - \frac{9}{4}\left(\frac{a^3 b^2}{l^5} + \frac{27}{8}\frac{a^2 b^3}{l^5} + \frac{ab^4}{l^5}\right)\right]\right\} \\ = \mathbf{i}_z 6\pi\mu a T_1 & \end{aligned} \tag{6–3.51}$$

Here, U_a and U_b are positive numbers representing the absolute magnitude of the velocities of the spheres a and b. If we had taken the direction of fall of the spheres to be *positive*, F_{a1} would be negative in sign, so that the right-hand side of Eq. (6–3.51) would then be negative. The drag exerted on sphere b is obtained from Eq. (6–3.51) by exchanging the letters b and a where they appear both in dimensions and as subscripts.

Faxen[14], in an extension of Smoluchowski's[29] treatment referred to earlier, employed a reflection procedure somewhat similar to the foregoing for the case of two spheres following each other along their line of centers.

His result is the same as given by Eq. (6–3.51). Faxen's technique differs somewhat from ours and that of Smoluchowski. In Faxen's method the transformation of the reflected fields from coordinates at the center of one sphere to those at the center of the other is accomplished by means of a conformal transformation involving the well-known "inversion in a sphere", whereas in Smoluchowski's treatment, this is accomplished in an approximate way by means of Taylor series expansions about the points corresponding to the two coordinate system origins.

For the case of equal-sized spheres, $a = b$, Dahl[14, appendix] carried Faxen's computations to the ninth power and obtained the following expression for the drag exerted on a sphere a:

$$\begin{aligned}\mathbf{F}_{a1} = \mathbf{i}_z 6\pi\mu a\{U_a[1 + 9x^2 + 93x^4 + 1197x^6 + 19{,}821\,x^8] \\ - U_b[3x + 19x^3 + 387x^5 + 5{,}331x^7 + 76{,}115x^9]\}\end{aligned} \tag{6–3.52}$$

where $x = a/(2l)$. In order to obtain the drag exerted on sphere b, $\mathbf{F}_{b1}$, one must interchange the velocities U_a and U_b in Eq. (6–3.52). This alternating series converges very rapidly for small values of the parameter x. As the spheres approach each other until they touch, that is, $x \to \frac{1}{4}$, the convergence is poor, and Dahl concludes that the method is not applicable.

One can, however, employ an empirical procedure based on the assumption that the last terms represent a slowly converging geometric series. Thus, if we rewrite Dahl's equation (6–3.52) letting $s = 2a/l$, for two equal-sized spheres falling at the same velocity U we obtain

$$\mathbf{F}_1 = \mathbf{i}_z 6\pi\mu a U\lambda \tag{6–3.53}$$

where

$$\lambda = \sum_{n=0}^{9} (-1)^n \alpha_n s^n$$

in which the values of the coefficients α_n are as follows:

$\alpha_0 = 1$	$\alpha_5 = 0.37792968$
$\alpha_1 = 0.75$	$\alpha_6 = 0.29223632$
$\alpha_2 = 0.5625$	$\alpha_7 = 0.32537841$
$\alpha_3 = 0.296875$	$\alpha_8 = 0.30244445$
$\alpha_4 = 0.36328125$	$\alpha_9 = 0.29035568$

If we assume that for terms in λ corresponding to $n > 9$, $\alpha_n = \text{constant} \approx \frac{1}{3}$, we will have

$$\lambda = \sum_{n=0}^{9} (-1)^n \alpha_n s^n + \tfrac{1}{3} \sum_{n=10}^{\infty} (-1)^n s^n \tag{6–3.54}$$

The last term here is a geometric series, whose sum is

$$\frac{1}{3} \sum_{n=10}^{\infty} (-1)^n s^n = \frac{1}{3}\left(\frac{s^{10}}{1 + s}\right) \tag{6–3.55}$$

which for the case of spheres touching, $s = 1$, results in the addition of $\frac{1}{6}$ to Dahl's value, giving

$$T_1 = \lambda = 0.647 \tag{6–3.56}$$

This is in good agreement with more exact values, as we shall note later in this chapter [$\lambda = 0.645$ from Eq. (6–4.15)].

Motion perpendicular to line of centers

We shall retain substantially the same nomenclature as previously, referring to Fig. 6–3.2, except that we will assume that though the particle velocities will still be along the z axis, the spheres will be located along the x axis. Thus the boundary conditions Eqs. (6–3.11)–(6–3.13) are still the same as before for the case where the spheres are acted upon by a *restraining torque which prevents them from rotating.*

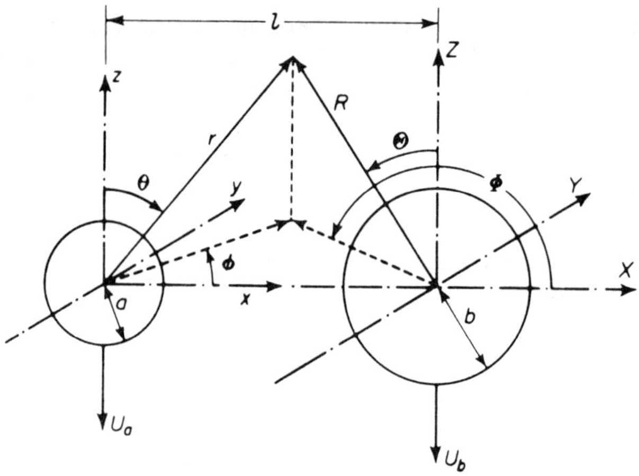

Figure 6-3.2. Motion of two spheres perpendicular to their line-of-centers.

A more general form than Eq. (6–3.14) is required for the $\mathbf{v}_i$ fields; namely,

$$\mathbf{v}_i = \sum_{N=0}^{\infty} \nabla \Psi^i_{-N-1} + \nabla \times (\mathbf{R}\,\Omega^i_{-N-1}) + \frac{2-N}{2N(2N-1)} \nabla (R^2\, \Pi^i_{-N-1}) + \frac{1}{N}\,\mathbf{R}\,\Pi^i_{-N-1} \tag{6–3.57}$$

with the solid spherical harmonic functions having the following forms:

$$\begin{aligned}
\Psi^i_{-N-1} &= (-1)^N b^{N+2} R^{-(N+1)} \sum_{M=-N}^{N} B^i_{N,M} \cos M\Phi\, P^M_N(\cos\Theta) \\
\Pi^i_{-N-1} &= (-1)^N b^{N} R^{-(N+1)} \sum_{M=-N}^{N} C^i_{N,M} \cos M\Phi\, P^M_N(\cos\Theta) \\
\Omega^i_{-N-1} &= (-1)^N b^{N+1} R^{-(N+1)} \sum_{M=-N}^{N} D^i_{N,M} \sin M\Phi\, P^M_N(\cos\Theta)
\end{aligned} \tag{6–3.58}$$

For the fields $\mathbf{v}_{i+1}$ we have, corresponding to Eq. (6–3.23),

$$\mathbf{v}_{i+1} = \sum_{n=0}^{\infty} \nabla \psi_{-n-1}^{i+1} + \nabla \times (\mathbf{r}\omega_{-n-1}^{i+1}) + \frac{2-n}{2n(2n-1)} \nabla (r^2 \pi_{-n-1}^{i+1}) + \frac{1}{n} \mathbf{r} \pi_{-n-1}^{i+1} \tag{6–3.59}$$

with solid spherical harmonic functions of the following forms:

$$\begin{aligned} \psi_{-n-1}^{i+1} &= -a^{n+2} r^{-(n+1)} \sum_{m=-n}^{n} (-1)^m b_{n,m}^{i+1} \cos m\phi \, P_n^m(\cos\theta) \\ \pi_{-n-1}^{i+1} &= -a^{n} r^{-(n+1)} \sum_{m=-n}^{n} (-1)^m c_{n,m}^{i+1} \cos m\phi \, P_n^m(\cos\theta) \\ \omega_{-n-1}^{i+1} &= -a^{n+1} r^{-(n+1)} \sum_{m=-n}^{n} (-1)^m d_{n,m}^{i+1} \sin m\phi \, P_n^m(\cos\theta) \end{aligned} \tag{6–3.60}$$

As before, we seek to determine the constants $b_{n,m}^{i+1}$, $c_{n,m}^{i+1}$, $d_{n,m}^{i+1}$ which determine the field $\mathbf{v}_{i+1}$, in terms of the constants $B_{N,M}^{i}$, $C_{N,M}^{i}$, and $D_{N,M}^{i}$, which determine the previous field, $\mathbf{v}_i$. Note that pairs of indices, (n, m) or (N, M), now appear.

The transformations required are more involved than in the case of spheres falling along their line of centers, but similar recurrence formulas can be obtained as follows:

$$\begin{aligned} b_{n,m}^{i+1} = \frac{2n-1}{2(n+1)} \left(\frac{a}{l}\right)^{n-1} \sum_{N=0}^{\infty} \sum_{M=-N}^{N} \left(\frac{b}{l}\right)^{N} F_{n,m}^{N,M} \Bigg[\left(\frac{b}{l}\right)^2 nB_{N,M}^{i} \\ + \frac{n(2-N)(2n-1) + q(Nn+1-2n-2N)}{2N(2n-1)(2N-1)} C_{N,M}^{i} \\ + \frac{n(2n+1)}{2(2n-1)(2n+3)} \left(\frac{a}{l}\right)^2 C_{N,M}^{i} \Bigg] \\ + \frac{2n-1}{2(n+1)} \left(\frac{a}{l}\right)^{n-1} \sum_{N=0}^{\infty} \sum_{M=-N}^{N} \left(\frac{b}{l}\right)^{N+1} F_{n,m+1}^{N,M} Q D_{N,M}^{i} \end{aligned} \tag{6–3.61}$$

$$\begin{aligned} c_{n,m}^{i+1} = \frac{(2n-1)(2n+1)}{(n+1)} \left(\frac{a}{l}\right)^{n-1} \sum_{N=0}^{\infty} \sum_{M=-N}^{N} \left(\frac{b}{l}\right)^{N} F_{n,m}^{N,M} \Bigg[\left(\frac{b}{l}\right)^2 nB_{N,M}^{i} \\ + \frac{n(2-N)(2n-1) + q(Nn+1-2n-2N)}{2N(2n-1)(2N-1)} C_{N,M}^{i} \\ + \frac{n}{2(2n+1)} \left(\frac{a}{l}\right)^2 C_{N,M}^{i} \Bigg] \\ + \frac{(2n-1)(2n+1)}{(n-1)} \left(\frac{a}{l}\right)^{n-1} \sum_{N=0}^{\infty} \sum_{M=-N}^{N} \left(\frac{b}{l}\right)^{N+1} F_{n,m+1}^{N,M} Q D_{N,M}^{i} \end{aligned} \tag{6–3.62}$$

$$\begin{aligned} d_{n,m}^{i+1} = -\frac{1}{(n-1)} \left(\frac{a}{l}\right)^{n} \sum_{N=0}^{\infty} \sum_{M=-N}^{N} \left(\frac{b}{l}\right)^{N+1} F_{n,m}^{N,M} N D_{N,M}^{i} \\ - \frac{1}{n(n+1)} \left(\frac{a}{l}\right)^{n} \sum_{N=0}^{\infty} \sum_{M=-N}^{N} \left(\frac{b}{l}\right)^{N} \frac{F_{n,m+1}^{N,M} Q}{N} C_{N,M}^{i} \end{aligned} \tag{6–3.63}$$

where

$$F_{n,m}^{N,M} = \begin{cases} (-1)^{(1/2)(n+m+N-M)} \left\{\dfrac{(n+m+N-M)!}{(n+m)!\,(N-M)!}\right\} \times \\ \times \left\{\dfrac{(n-m+N+M)!}{2^{n+N}[(n-m+N+M)/2]!\,[(n+m+N-M)/2]!}\right\} \\ \quad \text{for } (n+m+N-M) \text{ an even integer} \\ 0 \quad \text{for } (n+m+N-M) \text{ an odd integer} \\ 0 \quad \text{for } |M-m| > n+N \\ 0 \quad \text{for } M > N \end{cases} \tag{6–3.64}$$

and

$$q = \frac{(n+m)(n+m-1)}{(n+m+N-M-1)} + \frac{(n-m)(n-m-1)}{(n-m+N+M-1)} \tag{6–3.65}$$

$$Q = \frac{(Mn+Nm)(n+m+1)}{(n+m+N-M)} \tag{6–3.66}$$

By an entirely analogous procedure, it is possible to find the set of inverse relationships which enable us to find the field $\mathbf{v}_{i+2}$ from the known field $\mathbf{v}_{i+1}$ and the boundary condition $\mathbf{v}_{i+2} = -\mathbf{v}_{i+1}$ on sphere b. If $B_{N,M}^{i+2}$, $C_{N,M}^{i+2}$, and $D_{N,M}^{i+2}$ are the constants which determine the field $\mathbf{v}_{i+2}$, then in terms of the constants $b_{n,m}^{i+1}$, $c_{n,m}^{i+1}$, and $d_{n,m}^{i+1}$ we find that

$$\begin{aligned} B_{N,M}^{i+2} = \frac{2N-1}{2(N+1)}\left(\frac{b}{l}\right)^{N-1} \sum_{n=0}^{\infty}\sum_{m=-n}^{n}\left(\frac{a}{l}\right)^{n} f_{N,M}^{n,m}\Bigg[\left(\frac{a}{l}\right)^{2} N b_{n,m}^{i+1} \\ + \frac{N(2-n)(2N-1)+\Delta(nN+1-2n-2N)}{2n(2n-1)(2N-1)}\, c_{n,m}^{i+1} \\ + \frac{N(2N+1)}{2(2N-1)(2N+3)}\left(\frac{b}{l}\right)^{2} c_{n,m}^{i+1}\Bigg] \\ + \frac{2N-1}{2(N+1)}\left(\frac{b}{l}\right)^{N-1} \sum_{n=0}^{\infty}\sum_{m=-n}^{n}\left(\frac{a}{l}\right)^{n+1} f_{N,M+1}^{n,m}\,\delta\, d_{n,m}^{i+1} \end{aligned} \tag{6–3.67}$$

$$\begin{aligned} C_{N,M}^{i+2} = \frac{(2N-1)(2N+1)}{(N+1)}\left(\frac{b}{l}\right)^{N-1} \sum_{n=0}^{\infty}\sum_{m=-n}^{n}\left(\frac{a}{l}\right)^{n} f_{N,M}^{n,m}\Bigg[\left(\frac{a}{l}\right)^{2} N b_{n,m}^{i+1} \\ + \frac{N(2-n)(2N-1)+\Delta(nN+1-2n-2N)}{2n(2n-1)(2N-1)}\, c_{n,m}^{i+1} \\ + \frac{N}{2(2N+1)}\left(\frac{b}{l}\right)^{2} c_{n,m}^{i+1}\Bigg] \\ + \frac{(2N-1)(2N+1)}{(N+1)}\left(\frac{b}{l}\right)^{N-1} \sum_{n=0}^{\infty}\sum_{m=-n}^{n}\left(\frac{a}{l}\right)^{n+1} f_{N,M+1}^{n,m}\,\delta\, d_{n,m}^{i+1} \end{aligned} \tag{6–3.68}$$

$$\begin{aligned} D_{N,M}^{i+2} = -\frac{1}{(N+1)}\left(\frac{b}{l}\right)^{N} \sum_{n=0}^{\infty}\sum_{m=-n}^{n}\left(\frac{a}{l}\right)^{n+1} n f_{N,M}^{n,m}\, d_{n,m}^{i+1} \\ - \frac{1}{N(N+1)}\left(\frac{b}{l}\right)^{N} \sum_{n=0}^{\infty}\sum_{m=-n}^{n}\left(\frac{a}{l}\right)^{n} \frac{1}{n}\,\delta f_{N,M+1}^{n,m} c_{n,m}^{i+1} \end{aligned} \tag{6–3.69}$$

where

$$\Delta = \frac{(N+M)(N+M-1)}{(n-m+N+M-1)} + \frac{(N-M)(N-M-1)}{(n+m+N-M-1)} \tag{6–3.70}$$

$$\delta = \frac{(Mn+Nm)(N+M+1)}{(n-m+N+M)} \tag{6–3.71}$$

and

$$f^{n,m}_{N,M} = \begin{cases} (-1)^{(1/2)(n+m+N-M)} \left\{ \dfrac{(n+m+N-M)!}{(n-m)!\,(N+M)!} \right\} \times \\ \times \left\{ \dfrac{(n-m+N+M)!}{2^{n+N}[(n-m+N+M)/2]!\,[(n+m+N-M)/2]!} \right\} \\ \quad \text{for } (n+m+N-M) \text{ an even integer} \\ 0 \quad \text{for } (n+m+N-M) \text{ and odd integer} \end{cases} \tag{6–3.72}$$

These relationships apply equally to the fields $\mathbf{v}_k^*$ and $\mathbf{v}_k^{**}$. To distinguish the constants which contribute to each of these fields we use the symbols b_n^{*k}, c_n^{*k}, d_n^{*k} and B_N^{*k}, C_N^{*k}, D_N^{*k} for the respective fields.

Consider for the moment the fields $\mathbf{v}_k^*$. From Eqs. (6–3.8) and (6–3.11), we note that $\mathbf{v}_1^*$ is required to satisfy the boundary condition

$$\mathbf{v}_1^* = -\mathbf{i}_z U_b \quad \text{at } R = b \tag{6–3.73}$$

This field is again simply the Stokes' law velocity field in the (R, Θ, Φ) system. In our present nomenclature this is given by Eq. (6–3.57), with the following values of the harmonic functions:

$$\Psi^{*1}_{-N-1} = \begin{cases} -\frac{1}{4} U_b b^3 R^{-2} P_1(\cos\Theta) & \text{for } N = 1 \\ 0 \quad \text{for all other } N \end{cases} \tag{6–3.74}$$

$$\Pi^{*1}_{-N-1} = \begin{cases} -\frac{3}{2} U_b b R^{-2} P_1(\cos\Theta) & \text{for } N = 1 \\ 0 \quad \text{for all other } N \end{cases} \tag{6–3.75}$$

Note that Eqs. (6–3.74) and (6–3.75) are the same as Eqs. (6–3.16) and (6–3.17). In addition,

$$\Omega^{*1}_{-N-1} = 0 \quad \text{for all } N \tag{6–3.76}$$

If these relationships are compared with Eqs. (6–3.58), we find

$$\begin{aligned} B^{*1}_{N,M} &= \begin{cases} \frac{1}{4} U_b & \text{for } N = 1 \text{ and } M = 0 \\ 0 & \text{for all other } N \text{ and } M \end{cases} \\ C^{*1}_{N,M} &= \begin{cases} \frac{3}{2} U_b & \text{for } N = 1 \text{ and } M = 0 \\ 0 & \text{for all other } N \text{ and } M \end{cases} \\ D^{*1}_{N,M} &= 0 \quad \text{for all } N \text{ and } M \end{aligned} \tag{6–3.77}$$

These are the constants which determine the field $\mathbf{v}_1^*$. Those for the field $\mathbf{v}_2^*$ are obtained from Eqs. (6–3.61), (6–3.62), and (6–3.63) using the foregoing values of $B^{*1}_{N,M}$, $C^{*1}_{N,M}$, and $D^{*1}_{N,M}$. In carrying out this calculation, it is necessary to limit the approximation to some arbitrary powers of $(b/l)^{\alpha}(a/l)^{\beta}$. We shall again limit ourselves to $\alpha + \beta \leq 5$.

As in the previous case, we first compute the constants $b^{*2}_{n,m}$, $c^{*2}_{n,m}$, and $d^{*2}_{n,m}$. Then using Eqs. (6–3.67), (6–3.68), and (6–3.69), we obtain the next set of constants $B^{*3}_{N,M}$, $C^{*3}_{N,M}$, $D^{*3}_{N,M}$. By an iterative procedure we obtain the necessary additional fields.

To obtain the constants $b^{**k}_{n,m}$, $c^{**k}_{n,m}$, etc., we note that, from Eq. (6–3.13),

$$\mathbf{v}_1^{**} = \mathbf{0} \tag{6–3.78}$$

whereas $\mathbf{v}_2^{**}$ is required to satisfy the boundary condition

$$\mathbf{v}_2^{**} = -\mathbf{i}_z U_a \quad \text{at } r = a \tag{6–3.79}$$

This is satisfied by the values

$$\begin{aligned}
\psi^{**2}_{-n-1} &= \begin{cases} -\frac{1}{4} U_a a^3 r^{-2} P_1(\cos\theta) & \text{for } n = 1 \\ 0 \quad \text{for all other } n \end{cases} \\
\pi^{**2}_{-n-1} &= \begin{cases} -\frac{3}{2} U_a a r^{-2} P_1(\cos\theta) & \text{for } n = 1 \\ 0 \quad \text{for all other } n \end{cases} \\
\omega^{**2}_{-n-1} &= 0 \quad \text{for all } n
\end{aligned} \tag{6–3.80}$$

If these constants are compared with Eq. (6–3.60) we find that

$$\begin{aligned}
b^{**2}_{n,m} &= \begin{cases} \frac{1}{4} U_a & \text{for } n = 1 \text{ and } m = 0 \\ 0 \quad \text{for all other } n \text{ and } m \end{cases} \\
c^{**2}_{n,m} &= \begin{cases} \frac{3}{2} U_a & \text{for } n = 1 \text{ and } m = 0 \\ 0 \quad \text{for all other } n \text{ and } m \end{cases} \\
d^{**2}_{n,m} &= 0 \quad \text{for all } n \text{ and } m
\end{aligned} \tag{6–3.81}$$

Starting with these values we can compute the required constants.

In general, the velocity field satisfying the boundary conditions of *no rotation* is given by

$$\begin{aligned}
\mathbf{v} = \sum_{n=1}^{\infty} \nabla\psi_{-n-1} + \nabla \times (\mathbf{r}\omega_{-n-1}) + \frac{2-n}{2n(2n-1)} \nabla(r^2 \pi_{-n-1}) \\
+ \frac{1}{n} \mathbf{r}\pi_{-n-1} \\
+ \sum_{N=1}^{\infty} \nabla\Psi_{-N-1} + \nabla \times (\mathbf{R}\Omega_{-N-1}) + \frac{2-N}{2N(2N-1)} \nabla(R^2 \Pi_{-N-1}) \\
+ \frac{1}{N} \mathbf{R}\Pi_{-N-1}
\end{aligned} \tag{6–3.82}$$

and the pressure field by

$$p = \mu\left(\sum_{n=1}^{\infty} \pi_{-n-1} + \sum_{N=1}^{\infty} \Pi_{-N-1}\right) \tag{6–3.83}$$

The harmonic functions are given by the calculated values of the constants,

$$B_{N,M} = \sum_{i=0}^{\infty} (B_{N,M}^{*i} + B_{N,M}^{**i}) \tag{6–3.84}$$

$$b_{n,m} = \sum_{i=0}^{\infty} (b_{n,m}^{*i} + b_{n,m}^{**i}) \tag{6–3.85}$$

with appropriate similar values for the $C_{N,M}$, $c_{n,m}$, $D_{N,M}$, $d_{n,m}$ constants.

In order to compute the drag, we require the π_{-2} harmonic. From its definition,

$$\pi_{-2} = -ar^{-2} \sum_{i=0}^{\infty} \sum_{m=-1}^{1} (-1)^m (c_{1,m}^{*i} + c_{1,m}^{**i}) \cos m\phi \, P_1^m(\cos\theta) \tag{6–3.86}$$

Noting that in all the reflected fields

$$c_{1,1}^{*i} = c_{1,1}^{**i} = 0 \tag{6–3.87}$$

and

$$c_{1,-1}^{*i} = c_{1,-1}^{**i} = 0 \tag{6–3.88}$$

we find that

$$\pi_{-2} = -ar^{-2} \sum_{i=0}^{\infty} (c_{1,0}^{*i} + c_{1,0}^{**i}) P_1(\cos\theta) \tag{6–3.89}$$

Since $P_1(\cos\theta) = \cos\theta$ and $z = r\cos\theta$, we find from Eq. (6–3.47) that

$$\mathbf{F}_{a2} = \mathbf{i}_z \pi\mu a \sum_{i=0}^{\infty} (c_{1,0}^{*i} + c_{1,0}^{**i}) \tag{6–3.90}$$

Using values of the constants $c_{1,0}^{*i}$ and $c_{1,0}^{**i}$ computed as just discussed, we obtain, denoting by the subscript $a2$ the situation that the spheres move perpendicular to their line of centers and do *not* rotate,

$$\begin{aligned}\mathbf{F}_{a2} = \mathbf{i}_z 6\pi\mu a \Big\{ & U_a\left[1 + \frac{9}{16}\frac{ab}{l^2} + \frac{3}{8}\left(\frac{a^3 b}{l^4} + \frac{27}{32}\frac{a^2 b^2}{l^4} + \frac{3ab^3}{l^4}\right)\right] \\ & - U_b\left[\frac{3}{4}\frac{b}{l} + \frac{1}{4}\left(\frac{a^2 b}{l^3} + \frac{27}{16}\frac{ab^2}{l^3} + \frac{b^3}{l^3}\right) \right. \\ & \left. + \frac{63}{64}\left(\frac{a^3 b^2}{l^5} + \frac{27}{112}\frac{a^2 b^3}{l^5} + \frac{ab^4}{l^5}\right)\right]\Big\} \\ = {} & \mathbf{i}_z 6\pi\mu a T_2\end{aligned} \tag{6–3.91}$$

Note, as before, had we taken the fall velocity as positive in sign, the right hand of Eq. (6–3.91) would be negative in sign. The drag exerted on sphere b can be obtained from the foregoing by exchanging the letters a and b where they appear as dimensions and subscripts.

In the case we have considered, rotation is prevented and thus a torque is developed on the spheres. This torque is obtained as follows—see Eq. (3–2.45):

$$\mathbf{T}_a = -8\pi\mu\nabla\left(r^3 \sum_{i=1}^{\infty} \omega_{-2}^i\right) \tag{6–3.92}$$

The harmonic function ω_{-2}^i is given by

$$\omega_{-2}^i = -a^2 r^2 \sum_{i=0}^{\infty} \sum_{m=-1}^{1} (-1)^m (d_{1,m}^{*i} + d_{1,m}^{**i}) \sin m\phi \, P_1^m(\cos\theta) \tag{6–3.93}$$

Noting that

$$P_1^1(\cos\theta) = \sin\theta$$

$$P_1^{-1}(\cos\theta) = -\tfrac{1}{2}\sin\theta$$

$$\sin(-\phi) = -\sin\phi$$

$$y = r\sin\theta\sin\phi$$

we find

$$\mathbf{T}_a = \mathbf{i}_y 8\pi\mu a^2 \left\{-\sum_{i=0}^{\infty} [(d_{1,1}^{*i} + d_{1,1}^{**i}) + \tfrac{1}{2}(d_{1,-1}^{*i} + d_{1,-1}^{**i})]\right\} \tag{6–3.94}$$

Employing the appropriate constants evaluated as previously discussed, we obtain

$$\begin{aligned}\mathbf{T}_a = \mathbf{i}_y 8\pi\mu a^3 \Bigg\{ & U_b\left[\frac{3}{4}\frac{b}{l^2} + \frac{27}{64}\frac{ab^2}{l^4} + \frac{9}{32}\left(\frac{3a^3b^2}{l^6} + \frac{27}{32}\frac{a^2b^3}{l^6} + \frac{2ab^4}{l^6}\right)\right] \\ & - U_a\left[\frac{9}{16}\frac{ab}{l^3} + \frac{3}{16}\left(\frac{a^3b}{l^5} + \frac{27}{16}\frac{a^2b^2}{l^5} + \frac{3ab^3}{l^5}\right)\right]\Bigg\}\end{aligned} \tag{6–3.95}$$

The torque required to prevent rotation will be of equal magnitude to $\mathbf{T}_a$ but of opposite direction and, therefore, opposite sign. The torque on b is obtained by exchanging subscripts b and a, and by changing the *sign* of the moment; that is, $\mathbf{T}_b$ acts in the opposite direction from $\mathbf{T}_a$.

Wakiya[33] has developed a very similar solution to the problem of two spheres moving close to each other, which involves the selection of a coordinate system the same as that shown in Fig. 6–2.1, so that it results in a form similar to the two solutions we have obtained for spheres falling along and perpendicular to their line of centers. His technique is again somewhat different, though involving spherical harmonics. The harmonics for a second sphere are developed directly with respect to the origin of the first sphere to obtain one set of relationships among the characteristic constants by using the boundary conditions on the first sphere, a. In the same way, another set of relationships is obtained from the boundary conditions on sphere b. By eliminating one set of these constants from the two sets of relations, it is possible to obtain a simultaneous system of equations of infinite order with respect to the other set of constants characterizing the appropriate harmonic functions. Wakiya solves this infinite series of equations by a method of successive approximation and so the numerical form of his result is the same as obtained by us. His results are in agreement with those of Faxen for two spheres following each other and with ours both for spheres following each other and moving perpendicular to their line of centers.

Wakiya also considers the case of two spheres which are free to rotate as they move. No restraining moment is developed but instead the spheres rotate at a velocity which is determined by the rotation of the fluid at the particle locations. If two particles follow each other, there will be no rotational effect and the drag is given by Eq. (6–3.51). If, however, they fall perpendicular to their line of centers the drag will be less than that given by Eq. (6–3.91), as follows:

$$\begin{aligned}\mathbf{F}_{a3} = \mathbf{i}_z 6\pi\mu a \Big\{ U_a \Big[1 + \frac{9}{16}\frac{ab}{l^2} + \frac{3}{8}\Big(\frac{a^3 b}{l^2} + \frac{27}{32}\frac{a^2 b^2}{l^4} + \frac{ab^3}{l^4}\Big)\Big] \\ - U_b \Big[\frac{3}{4}\frac{b}{l} + \frac{1}{4}\Big(\frac{a^2 b}{l^3} + \frac{27}{16}\frac{ab^2}{l^3} + \frac{b^3}{l^3}\Big) \\ + \frac{27}{64}\Big(\frac{a^3 b^2}{l^3} + \frac{9}{16}\frac{a^2 b^3}{l^5} + \frac{ab^4}{l^5}\Big)\Big]\Big\} \\ = \mathbf{i}_z 6\pi\mu a T_3 \end{aligned} \qquad (6\text{–}3.96)$$

where the subscript $a3$ refers to the situation where the spheres move perpendicular to their line of centers but are *free to rotate.*

At the same time, the speed of rotation of sphere a will be

$$\begin{aligned}\boldsymbol{\omega}_a = \mathbf{i}_y \Big\{ U_b \Big[\frac{3}{4}\frac{b}{l^2} + \frac{27}{64}\frac{ab^2}{l^4} + \frac{9}{32}\Big(\frac{a^3 b^2}{l^6} + \frac{27}{32}\frac{a^2 b^3}{l^6} + \frac{ab^4}{l^6}\Big)\Big] \\ - U_a \Big[\frac{9}{16}\frac{ab}{l^3} + \frac{3}{16}\Big(\frac{a^3 b}{l^5} + \frac{27}{16}\frac{a^2 b^2}{l^5} + \frac{ab^3}{l^5}\Big)\Big]\Big\}\end{aligned} \qquad (6\text{–}3.97)$$

The rotation of b is obtained from Eq. (6–3.97) by exchanging the dimensions and subscripts, b and a, and by changing the sign of the rotation $\boldsymbol{\omega}_b$ (Fig. 6–3.3).

For equal-sized spheres falling with the same velocity but *not* free to rotate we obtain, from Eq. (6–3.91),

$$F_2 = 6\pi\mu a U\Big(1 - \frac{3}{4}\frac{a}{l} + \frac{9}{16}\frac{a^2}{l^2} - \frac{59}{64}\frac{a^3}{l^3} + \frac{465}{256}\frac{a^4}{l^4} - \frac{15813}{7168}\frac{a^5}{l^5}\Big) \qquad (6\text{–}3.98)$$

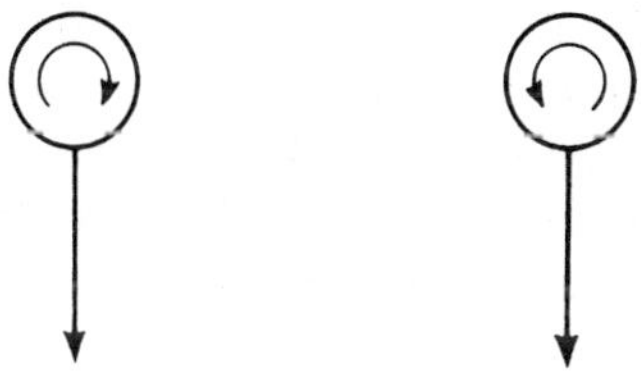

Figure 6-3.3. Direction of rotation of two spheres settling beside each other.

If the coefficients of the last terms in Eq. (6–3.98) are taken ≈ 2.0, we can complete the series as in the case of Eq. (6–3.54). For the situation where the two spheres touch, $a/l = \frac{1}{2}$, we find from Eq. (6–3.98) that

$$F_2 = 6\pi\mu a U(0.716) \qquad (6\text{–}3.99)$$

For the case where the spheres *are* free to rotate, we obtain similarly from Eq. (6–3.96),

$$F_3 = 6\pi\mu a U\Big(1 - \frac{3}{4}\frac{a}{l} + \frac{9}{16}\frac{a^2}{l^2} - \frac{59}{64}\frac{a^3}{l^3} + \frac{273}{256}\frac{a^4}{l^4} - \frac{1107}{1024}\frac{a^5}{l^5}\Big) \qquad (6\text{–}3.100)$$

In this case, if we assume that the last terms will have a constant coefficient ≈ 1.0, we can again complete the series. For the situation where the two spheres touch, $a/l = \frac{1}{2}$, we find, corresponding to Eq. (6–3.100),

$$F_3 = 6\pi\mu a U(0.694) \qquad (6\text{–}3.101)$$

These approximations should be better than in the case of two spheres following each other because convergence is much more rapid in this case. Freedom to rotate results in only a very small reduction of the drag of spheres in the vicinity of each other.

Kynch[22] also derived expressions for the velocity of two spheres moving slowly under external forces through a viscous fluid. His method is similar to Wakiya's in some respects in that he develops solutions for a second sphere with respect to the origin of the first directly. Instead of using spherical harmonic functions, however, the solution is developed in terms of derivatives of the basic solution. A series of equations with unknown constants is obtained, and these equations are solved by successive approximations. Numerical values are given for solutions to the problems where both particles are acted upon by forces along the line of centers, and perpendicular to the line of centers, respectively. Kynch notes that a general solution can be obtained by combining these two solutions. Kynch's numerical results involve approximations in order to obtain a series which can be summed analytically. His final results check quite well those which we have obtained by a similar procedure. Thus for two equal-sized spheres touching and falling along their line of centers, Kynch's equations yield $T_1 = \lambda = 0.642$— compared with a value of 0.647 from Eq. (6–3.56) and 0.645 from the exact solution, Eq. (6–4.15). For the case of two spheres falling perpendicular to their line of centers, Kynch's results give $T_3 = 0.710$ (compared with 0.694 from Eq. (6–3.101)). We have chosen to elaborate the treatment given previously because the recurrence formulas developed furnish a relatively simple procedure whereby we could develop the solutions further to any desired degree of accuracy, preferably on a digital computer.

General motion of two spheres

It is simple to apply the relationships developed for flow along and perpendicular to the line of centers of two spheres to obtain general relationships. Following the nomenclature of the last section, we obtain, instead of Eq. (6–2.15) (note the change in algebraic sign), for the case of two spheres which are prevented from rotating,

$$\frac{\mathbf{F}_a}{6\pi\mu a} = \mathbf{i}T_1 + \mathbf{k}T_2 \tag{6–3.102}$$

where T_1 and T_2 are obtained from Eqs. (6–3.51) and (6–3.91) respectively.

Similarly, for the case where the spheres are free to rotate the force exerted by particle a will be given by

$$\frac{\mathbf{F}_a}{6\pi\mu a} = \mathbf{i}T_1 + \mathbf{k}T_3 \tag{6–3.103}$$

where T_3 is obtained from Eq. (6–3.96). Using these relationships, it is possible to derive expressions similar to those in the previous section for the force exerted in terms of approach velocity and magnitude, or for the motion the particles will undergo when subjected to a gravitational field.

For two spheres of the same size, Table 6–3.1 gives values for the resistance coefficients. Geometric series were employed.

TABLE 6–3.1
RESISTANCE COEFFICIENTS FOR EQUAL-SIZED SPHERES

Length/Dia. $l/2a$	T_1, Eq. (6–3.54)	T_2, Eq. (6–3.98)	T_3, Eq. (6–3.100)
1.0	0.64658991	0.71581015	0.69367473
2.0	0.74226592	0.83858356	0.83655721
3.0	0.80472200	0.88751164	0.88705925
4.0	0.84412282	0.91362144	0.91346928
5.0	0.87060051	0.92986452	0.92979986
6.0	0.88948753	0.94095210	0.94092014
7.0	0.90360223	0.94900595	0.94898839
8.0	0.91453736	0.95512292	0.95511249
9.0	0.92325335	0.95992763	0.95992104
10.0	0.93036094	0.96380171	0.96379736

Meteorologists have shown considerable interest in applying relationships of this type in determining the initial rate at which small droplets can grow by coalescence in nonfreezing clouds. Thus, Hocking[19] has given a comprehensive treatment of the theoretical collision efficiency of small spheres of different radii falling under the influence of gravity. Hocking notes that the flow problem can be solved by superimposing the flows for two spheres moving along, and perpendicular to, their line of centers, and employs a method similar to Kynch's for finding the two basic solutions. He considers it sufficient to take terms up to and including the seventh power of a/l and b/l, the ratios of radii to distance between particles, and states that the next three powers make little difference even when the spheres are close together. Numerical evaluation of Hocking's resulting formulas for the resistance coefficient $T_3 = \lambda$, the case of two equal-sized spheres moving perpendicular to their line of centers and in contact, is in good agreement with other results. For the case of $T_1 = \lambda$, however, with two equal-sized spheres falling along their line of centers and touching, the resistance coefficient is only 0.256 compared with the exact Stimson and Jeffery[30] value of 0.645. Kynch[23] and Hocking[20] both indicate that the accuracy could be improved by taking additional reflections. As we have already seen, a very large number of terms would be necessary to secure convergence for two spheres following each other and touching—see Eqs. (6–3.52) and (6–3.54).

It is important to note[20] that, when two spheres are approaching each other, the steady state form of the creeping motion equations will not give correct results. A form involving at least the local acceleration terms might be employed for more accurate results but solutions are not available for two or more spheres. Hocking states that good agreement on collision efficiencies is obtained with his results and experimental data, so it is apparent that under some conditions the approximate treatment is satisfactory.

6-4 Exact Solution for Two Spheres Falling along Their Line of Centers

The problem of the motion of two solid spheres, equal or unequal, moving with equal small constant velocities along their line of centers has been solved by Stimson and Jeffery[30] and furnishes a convenient reference as to the accuracy of the other more approximate treatments discussed previously in this chapter. The solution is based on determining Stokes stream function for the motion of the fluid, and from this the forces necessary to maintain the motion of the spheres. This simplification is possible because the motion is axisymmetric.

The axisymmetric motion of a symmetric solid in a viscous fluid is discussed in Chapter 4, where the expression for the force on it is given as

$$F_z = \mu\pi \int \rho^3 \frac{\partial}{\partial n}\left(\frac{E^2\psi}{\rho^2}\right) ds \tag{6–4.1}$$

the integral being taken around the meridian section of the solid in a direction making a positive right angle with the direction n, where n is the normal drawn outwards from the solid; ρ is the distance from the axis.

For the problem of two spheres we take ξ, η as curvilinear coordinates in a meridian plane defined by the conformal transformation[25, p. 168] (see also Section A–19*)

$$z + i\rho = ic \cot \tfrac{1}{2}(\eta + i\xi) \tag{6–4.2}$$

or, equivalently, by Jeffery

$$\xi + i\eta = \ln \frac{\rho + i(z + c)}{\rho + i(z - c)} \tag{6–4.3}$$

where c is a positive constant; thus

$$\rho = c\frac{\sin \eta}{\cosh \xi - \cos \eta}, \qquad z = \frac{\sinh \xi}{\cosh \xi - \cos \eta} \tag{6–4.4}$$

and the surfaces obtained by rotating the curves ξ = constant about the axis of z are a family of spheres having $z = 0$ (or $\xi = 0$) for a common radical plane. Two spheres external to each other will be defined by $\xi = \alpha$,

*With regard to the definitions of ξ and η, we follow here the original notation of Stimson and Jeffery. This differs from that outlined in Section A–19 by the fact that the symbols ξ and η are interchanged.

$\xi = \beta \, (\alpha > 0, \beta < 0)$; α, β and the constant c may be chosen so that these spheres have any radii and any center distance greater than the sum of their radii. If the centers of the spheres are of radii r_1, r_2 and have their centers at distances d_1, d_2 from, and on opposite sides of the origin, then

$$\begin{aligned} r_1 &= c \operatorname{cosech} \alpha, & r_2 &= -c \operatorname{cosech} \beta \\ d_1 &= c \coth \alpha, & d_2 &= -c \coth \beta \end{aligned} \tag{6–4.5}$$

Neglecting the effects of fluid inertia and the time derivatives of velocity, the creeping motion equations become in terms of the stream function—see also Section 4–7—

$$E^4 \psi = 0 \tag{6–4.6}$$

A solution of Eq. (6–4.6) using bipolar coordinates is obtained by Stimson and Jeffery. With slight modifications their solution is

$$(\cosh \xi - \mu)^{3/2} \psi = \sum_{n=0}^{\infty} U_n(\xi) C_{n+1}^{-1/2}(\mu) \tag{6–4.7}$$

where for brevity we have put

$$\mu = \cos \eta \tag{6–4.8}$$

Here,

$$\begin{aligned} U_n(\xi) = a_n \cosh (n - \tfrac{1}{2})\xi + b_n \sinh (n - \tfrac{1}{2})\xi \\ + c_n \cosh (n + \tfrac{3}{2})\xi + d_n \sinh (n + \tfrac{3}{2})\xi \end{aligned} \tag{6–4.9}$$

and $C_{n+1}^{-1/2}(\mu)$ is the Gegenbauer polynomial of order $n + 1$ and degree $-\frac{1}{2}$. These latter functions are related to Legendre polynomials via the relation

$$C_n^{-1/2}(\mu) = \frac{P_{n-2}(\mu) - P_n(\mu)}{2n - 1} \tag{6–4.10}$$

The constants a_n b_n, . . . are to be determined from the boundary conditions.

They show that the force necessary to maintain the motion of the sphere $\xi = \alpha$, in the positive z direction is

$$F_1 = \frac{2\pi\mu\sqrt{2}}{c} \sum_{n=1}^{\infty} (a_n + b_n + c_n + d_n) \tag{6–4.11}$$

and for the sphere $\xi = \beta$,

$$F_2 = \frac{2\pi\mu\sqrt{2}}{c} \sum_{n=1}^{\infty} (a_n - b_n + c_n - d_n) \tag{6–4.12}$$

where, in these expressions, μ is the viscosity.

For equal spheres the coefficients b_n and d_n vanish and these forces are equal. The force may be conveniently written in the form

$$F = 6\pi\mu a U \lambda \tag{6–4.13}$$

where a is the radius of either sphere and λ is a coefficient which may be written in the form

$$\lambda = \frac{4}{3} \sinh \alpha \sum_{n=1}^{\infty} \frac{n(n+1)}{(2n-1)(2n+3)} \times \left\{ 1 - \frac{4 \sinh^2 (n+\frac{1}{2})\alpha - (2n+1)^2 \sinh^2 \alpha}{2 \sinh (2n+1)\alpha + (2n+1) \sinh 2\alpha} \right\} \tag{6–4.14}$$

Note that a typographical error in Eq. (6–4.14) occurs in the original Stimson and Jeffery article where the constant coefficient on the right-hand side is given as $\frac{2}{3}$ instead of $\frac{4}{3}$. Faxen[13], who called attention to this error, also calculated λ for the limiting case where the spheres touch, $\alpha = 0$. Faxen notes that λ is a continuous function of α so that even though the forces are not expressible in bipolar coordinates, Eq. (6–4.14) should still be applicable. Thus, he writes

$$\lim_{\alpha \to 0} \lambda = \frac{4}{3} \int_0^\infty \frac{1}{4} \left\{ 1 - \frac{4 \sinh^2 x - 4x^2}{2 \sinh 2x + 4x} \right\} dx \tag{6–4.15}$$

and by numerical evaluation obtains the value of $\lambda = 0.645$ for this case.

Comparison of values of λ obtained by the exact method using bipolar coordinates with the modification of Dahl's[14] solution given by Eq. (6–3.54) is of interest. When the spheres are far apart, both methods agree to extreme accuracy; even when the spheres touch, excellent agreement is possible using the modified Eq. (6–3.54). Typical results are shown in Table 6–4.1, applying to two spheres following each other along their line of centers. The values of λ are calculated to a greater degree of accuracy then originally given by Stimson and Jeffery.

TABLE 6–4.1
STOKES LAW CORRECTION FOR TWO EQUAL SPHERES MOVING PARALLEL TO THEIR LINE OF CENTERS

α	Ratio of Center-to-center Distance to Diameter $1/s = l/2a$	Method of Reflections, Eq. (6–3.54) λ	Exact Solution, Eqs. (6–4.14), (6–4.15) λ
0	1	0.64658991	0.645
0.5	1.1276260	0.65974465	0.65963461
1.0	1.5430806	0.70249745	0.70245211
1.5	2.3524096	0.76778115	0.76777914
2.0	3.7621957	0.83620498	0.83619387
2.5	6.1322895	0.89158529	0.89158513
3.0	10.067662	0.93079478	0.9307947
∞	∞	1.0000000	1.0000000

This close agreement is considered good theoretical justification for the method of reflections, which is extensively used in subsequent developments. Note, however, that for two spheres moving at different velocities, the Stimson-Jeffery treatment will give a drag force which tends to infinity as they approach each other, due to omission of unsteady motion terms in the basic equations.

6-5 Comparison of Theories with Experimental Data for Two Spheres

In addition to knowing that the methods we use lead to suitable approximate solutions of the creeping motion equations, it is desirable to know how well the predicted effects will be realized physically. Considerable data are now available which indicate good agreement for the case of two mutually interacting spheres. This is important in furnishing a sound basis for the treatment of more complicated assemblages. Since, as we have previously noted, the motion of two spheres can be resolved into the two problems of spheres falling along and perpendicular to their line of centers, it is convenient to consider these cases separately. Data for both cases are available from the work of Eveson, *et al.*[12] and Bart[2], and additional data for two spheres following each other from that of Happel and Pfeffer[18].

In all cases it is necessary to consider that, experimentally, spheres are dropped in a vessel, usually cylindrical, and not in an infinite medium. This surrounding boundary must be considered in order to obtain accurate results. This is especially important when the particles are not close together because such wall effects may be of substantial magnitude compared with particle interaction. If the particles are close together (less than 2–3 diameters between their centers) they may be considered as a single particle[5] and a correction factor applied the same as that for a sphere falling in a cylindrical vessel (Chapter 4). This presupposes that their center-to-center distance is small compared with the cylinder diameter. As they move further apart, however, their separate effects must be considered[18]. Thus if two spheres a and b are falling at different positions in a cylinder, the drag on sphere a will be the result of adding four velocities. These include the initial Stokes velocity of sphere a and the first reflection of this field from the cylinder wall. In addition, sphere b will disturb the motion of sphere a in two ways, first by a direct reflection of its own Stokes field and secondly by the reflection of this field from the cylinder wall and then to sphere a[18]. These mutual effects are discussed in some detail in Chapters 7 and 8. In general, if the drag force exerted on each sphere when two spheres fall in an infinite viscous fluid is compared with that exerted on a single sphere, we have

$$\lambda = \frac{U_{\mathrm{I}\infty}}{U_{\mathrm{II}\infty}} \tag{6–5.1}$$

To correct the experimental velocities for the effect of the cylindrical boundary, the factors K_{I} and K_{II} are used. If U_{II} is the experimentally measured velocity of the two spheres at a known fluid temperature and U_{I} is the velocity of a single sphere at the same temperature, then

$$\lambda = \frac{K_{\mathrm{I}}\, U_{\mathrm{I}}}{K_{\mathrm{II}}\, U_{\mathrm{II}}} \tag{6–5.2}$$

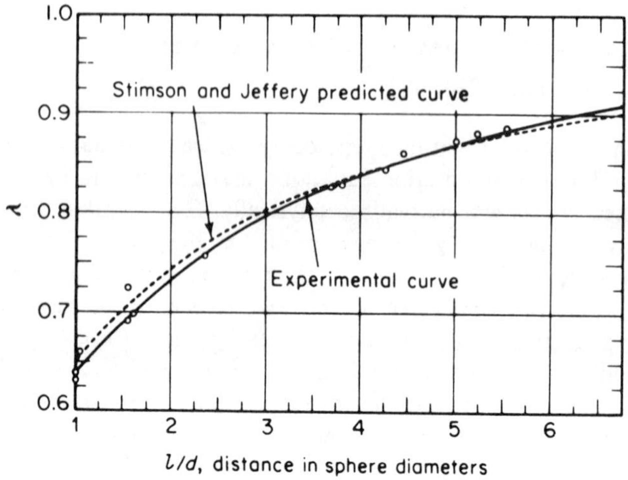

Figure 6-5.1. λ vs l/d for two spheres falling parallel to their line of centers.

The data of Bart taken at Reynolds numbers of less than 0.05 is perhaps the most precise since it involves a (sphere diameter)/(cylinder diameter) ratio of only 0.0163, so that, with particles close together, the correction factor $K_I/K_{II} = 0.99$. Figure 6–5.1 gives a plot of λ for a range of values of l/d, from 1 (spheres touching) to 6, the ratio of distance between centers to diameter, for two equal-sized spheres falling along their line of centers. Agreement with the theoretical value derived by Stimson and Jeffery[30] is excellent. When the spheres touch, an experimental value of $\lambda = 0.647$ is obtained as compared with the theoretical value of $\lambda = 0.645$. The Happel and Pfeffer data for the same case show reasonable agreement with the theory but are not as precise since the smallest (sphere diameter)/(cylinder diameter) employed was 0.0437, so that a larger correction factor is necessary. In the case of the Eveson data, the average (sphere diameter)/(cylinder diameter) was 0.0339 approximately, which corresponds, when the spheres touch, to $K_I/K_{II} = 0.98$. For this case an uncorrected value of λ is given as 0.660 so that the corrected value from Eq. (6–5.2) will be $\lambda = 0.660 \times 0.98 = 0.647$. This is in good agreement with Stimson and Jeffery. Kynch[22] has noted that at higher values of l/d, the apparent value of λ obtained by Eveson is somewhat higher than predicted. Thus for two spheres following each other at $l/d = 5$, Eveson obtains $\lambda = 0.894$. In this case $K_I/K_{II} \approx 0.96$ so that the corrected value of $\lambda \approx 0.86$, which is close to the value of $\lambda = 0.87$ predicted by Stimson and Jeffery. Thus the theoretical solution for this case appears to be well substantiated for the case of spheres close together. Additional data of Bart for spheres moving along their line of centers show good agreement with theory up to $l/d = 20$, at which point the precision of the data is not sufficient for accurate interpretation.

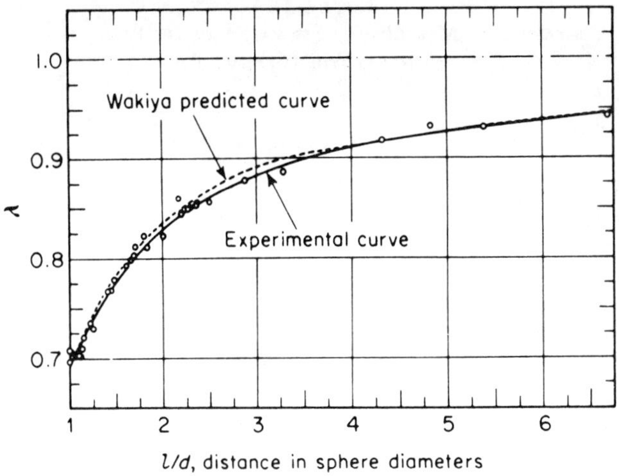

Figure 6-5.2. λ vs l/d for two spheres falling perpendicular to their line of centers.

For the case of two spheres falling perpendicular to their line of centers, Fig. 6–5.2 gives a similar correlation of Bart's data compared with the theoretical relationship of Wakiya, covering cases where the spheres are held rigid and are free to rotate, respectively (see also Fig. 6–3.3). Good agreement is obtained for the freely rotating case, except where the spheres touch. In this case the experimental value is $\lambda = 0.707$ compared with 0.694 for the freely rotating spheres and 0.716 for spheres prevented from rotating. The agreement obtained is thus again well within experimental error, but indicates that actual contact acts to restrict free rotation of two touching spheres. The data of Eveson for this case corresponds to $\lambda = 0.73 \times 0.98 = 0.716$, which is again in good agreement. Additional data obtained by Bart for spheres more widely separated shows good agreement with theory up to l/d ratios of 25.

Eveson, *et al.* also give data for spheres falling with various angles between the line of centers and the horizontal. These data are in good agreement with the theoretical premise that such motion can be determined from the two extreme cases, though the data are not as extensive as for fall along, and perpendicular to, the line of centers. Thus the general theoretical relationships derived from the creeping motion equations have been shown to apply to systems where the spheres may be up to 20–25 diameters apart, corresponding to very dilute assemblages.

A comparison of the various theoretical relationships and approximations is given in Table 6–5.1 for the case where the spheres touch. Agreement, using the exact reflection technique, is very good. When the spheres are separated slightly, any of the approximation procedures will give satisfactory results also.

TABLE 6–5.1
COMPARISON OF METHODS FOR ESTIMATION OF THE RESISTANCE COEFFICIENT (λ) WHEN SPHERES TOUCH

Method	Coefficient (λ)	
	One Sphere above the Other	Spheres Side by Side
Eq. (6–2.16)	0.572	0.728
Eq. (6–2.26)	0.592	0.714
Eq. (6–2.28)	0.616	0.695
Eq. (6–3.56)	0.647	. . .
Eq. (6–3.98) (no rotation)	. . .	0.716
Eq. (6–3.100) (free rotation)	. . .	0.694
Eq. (6–4.15) (exact)	0.645	. . .
Data (Bart[2])	0.647	0.707

6-6 More Than Two Spheres

A general method of solving the boundary value problem for a group of a number of spheres was developed in Section 6–1, following the procedure originally outlined by Smoluchowski[29]. Kynch[22] has presented general formulas indicating how analytical solutions can be obtained for the case of three or more particles, but the expressions are so complicated that generalizations are possible only for special arrangements of the particles. Brenner[7] has also developed general relationships for the treatment of multisphere problems, similar to those presented by Kynch. Since the relationships are sufficiently general to include wall effects, we shall defer development of the treatment to Chapter 8 (Section 8–5).

Kynch discusses one special arrangement of three spheres where one particle A is in a vertical plane midway between the others. If the three particles are the same size, the two outside ones separate to allow A to pass between them and then close up behind it. Another three-sphere problem, discussed by Happel and Pfeffer[18] and studied experimentally by them, is depicted in Fig. 6–6.1. If spheres C, B, and A are falling along the same vertical line, so that the distance between spheres A and B is less than the distance between spheres B and C, the two spheres A and B constitute a doublet and move faster than C. At some time

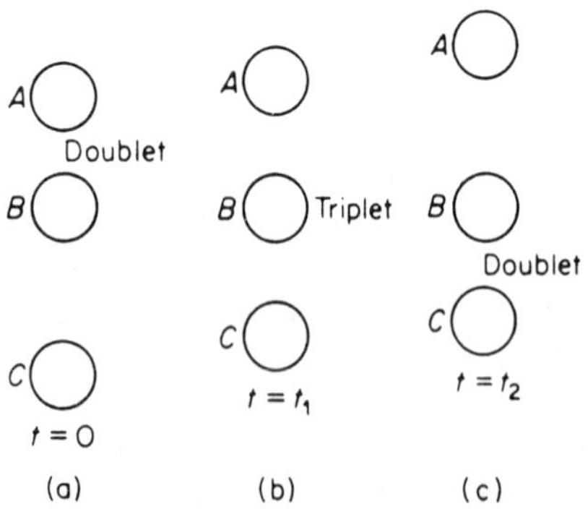

Figure 6-6.1. The motion of three spheres.

$t = t_1$, spheres A and B will catch up to sphere C and the distance between all three will be equal. The triplet will not, however, be stable because the center sphere is influenced by interaction effects from both spheres A and C. These interaction effects cause sphere B to move closer to sphere C, which produces a doublet between the latter. At time $t = t_2$, the doublet of spheres B and C is moving away from sphere A. These phenomena will occur even if inertial effects are negligible.

Slack and Matthews[28] have studied experimentally the sedimentation of freely falling clusters of uniform spheres for a range of terminal Reynolds numbers of individual spheres ranging from 10^{-1} to 10. If there are no more than six spheres in the cluster, they arrange themselves in the same horizontal plane at the vertices of a regular polyhedron and then diverge slowly as they fall, owing to inertial effects similar to those observed in the case of two falling spheres. Thus three spheres in a horizontal equilateral triangle will represent a stable arrangement in creeping motion.

Burgers[10] made calculations based on a first reflection technique for assemblages consisting of small numbers of spheres rigidly held in position with respect to each other (the link is assumed not to influence the motion of the liquid), in connection with the determination of the shape of protein molecules from viscosity data. In this case it is assumed that a/l, the ratio of particle radius to distance between particles, will be very small.

The drag on such an assemblage is given by the relationship:

$$F = 6\pi\mu a U \eta \tag{6–6.1}$$

where a is the radius of each particle, U is the velocity of sedimentation of the group, and η is a resistance coefficient. Burgers reports η for the case where the sedimenting assemblage can take all possible directions in space. He has computed the following cases involving several spheres in a line;

System of 2 spheres in a line:

$$\eta = \frac{2}{1 + (a/l)} \tag{6–6.2}$$

This result is similar to that expressed in Eq. (6–2.28) when allowance for the various orientations is made.

System of 3 spheres in a line:

$$\eta = \frac{3}{1 + (10/3)(a/l) - (1/4)(a/l)^2} \tag{6–6.3}$$

Here l is the distance between the centers of the outer spheres and a is the radius of each sphere.

System of 4 spheres in a line:

$$\eta = \frac{4}{1 + (13/2)(a/l) - (9/8)(a/l)^2} \tag{6–6.4}$$

where again l denotes the distance between centers of the outer spheres.

Note that the resistance per sphere decreases as the number increases, as would be expected.

For the case of four spheres lying at the corners of a square, it is also necessary to consider various possible orientations of the assemblage. The final result is

$$\eta = \frac{4}{1 + 2.7(a/l) - 0.04(a/l)^2} \tag{6–6.5}$$

Burgers also considered the case for eight spheres located at the corners of a cube. In the case of a cubic system the total resistance does not change with the direction of orientation (that is, it is in effect a regular body). In this case we have

$$\eta = \frac{8}{1 + 5.70(a/l) - 0.34(a/l)^2} \tag{6–6.6}$$

It is not likely that the results for larger numbers of spheres will give results as accurate as Eq. (6–2.28) does for the two-sphere problem. This is because the factor in the denominator of the expression for η does not take into account multiple interparticle reactions, as it does in the case of Eq. (6–6.2).

6-7 Two Spheroids in a Viscous Liquid

An investigation by S. Wakiya, as yet unpublished, considers the problem of two spheroids moving along their line of centers in the direction of the negative x axis, as shown in Fig. 6–7.1.

Cartesian coordinates (x_1, y_1, z_1) are so chosen that one of the spheroids with its origin at O_1 is represented by

$$\frac{x_1^2 + z_1^2}{a^2} + \frac{y_1^2}{b^2} = 1 \tag{6–7.1}$$

Another set of cartesian coordinates (x_2, y_2, z_2) is further chosen so that a second spheriod, which has the same shape as that whose center is located at O_1, has its center at the origin O_2 and is represented by

$$\frac{x_2^2 + z_2^2}{a^2} + \frac{y_2^2}{b^2} = 1 \tag{6–7.2}$$

Here a is the radius and b the half-length or half-thickness of the spheroids. The relationship between (x_1, y_1, z_1) and (x_2, y_2, x_2) is

$$\begin{aligned} x_2 &= x_1 - x_0 \\ y_2 &= y_1 \cos\theta + z_1 \sin\theta \\ z_2 &= -y_1 \sin\theta + z_1 \cos\theta \end{aligned} \tag{6–7.3}$$

where x_0 is the x coordinate of O_2 based upon the (x_1, y_1, z_1) system. Each spheroid moves with a constant velocity $-U$ along the x axis. The boundary conditions to be satisfied in either coordinate system are

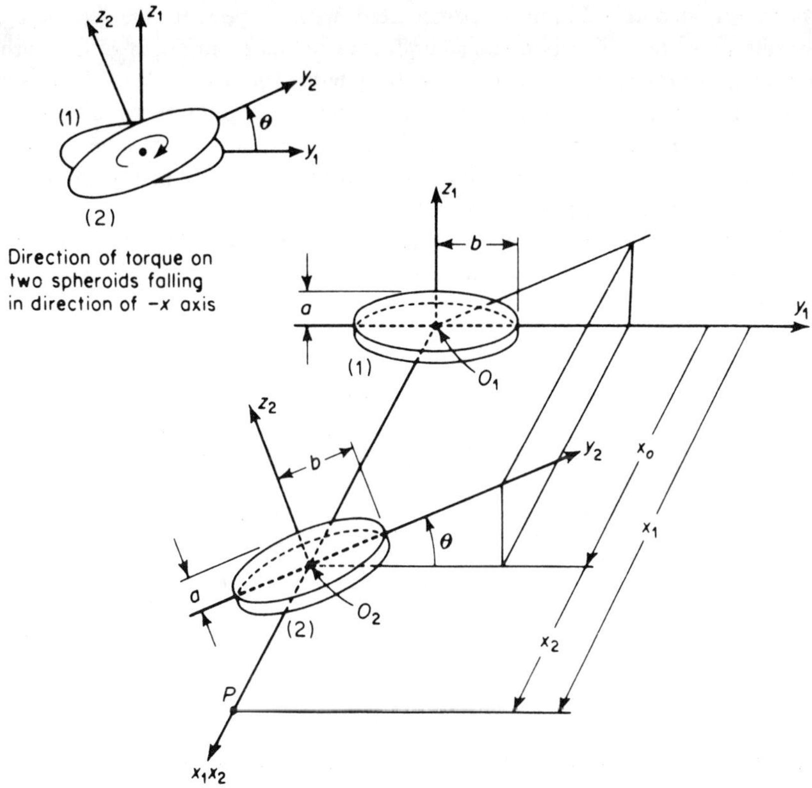

Figure 6-7.1. Interaction between two spheroids.

$$\begin{aligned} &u = -U, \quad v = w = 0 \quad \text{on the spheroids } O_1 \text{ and } O_2 \\ &u = v = w = 0 \quad \text{at infinity} \end{aligned} \tag{6–7.4}$$

The creeping motion equations are assumed to describe the fluid motion. Thus, referring to the coordinate system (x_1, y_1, z_1),

$$\begin{gathered} \frac{\partial p}{\partial x_1} = \mu \nabla^2 u_1, \qquad \frac{\partial p}{\partial y_1} = \mu \nabla^2 v_1, \qquad \frac{\partial p}{\partial z_1} = \mu \nabla^2 w_1 \\ \frac{\partial u_1}{\partial x_1} + \frac{\partial v_1}{\partial y_1} + \frac{\partial w_1}{\partial z_1} = 0 \end{gathered} \tag{6–7.5}$$

Solutions are constructed, in a manner similar to that discussed in Section 7–5, by means of harmonic functions defined by

$$\chi_n = \int_\lambda^\infty \frac{1}{2^n} \left(\frac{x_1^2 + z_1^2}{a^2 + s} + \frac{y_1^2}{b^2 + s} - 1 \right)^n \frac{ds}{(a^2 + s)\sqrt{b^2 + s}} \tag{6–7.6}$$

Here $\lambda(x_1, y_1, z_1)$ is the positive root of the cubic equation

$$\frac{x_1^2 + z_1^2}{a^2 + \lambda} + \frac{y_1^2}{b^2 + \lambda} = 1 \tag{6–7.7}$$

One solution is constructed with respect to the (x_1, y_1, z_1) coordinate system; a second solution is constructed with respect to the (x_2, y_2, z_2) coordinate system. The boundary conditions on spheroid O_1, together with the relationships between the coordinate systems, Eq. (6–7.3), lead to one set of infinite simultaneous equations. A second set of infinite simultaneous equations is obtained by utilizing the boundary conditions on spheriod O_2. From these two sets of equations, the arbitrary constants in the solutions are established for the case where $a, b \ll l = |x_o|$.

The drag and torque on each spheroid are the same. They may be computed by the following relationships, written for the spheroid O_1, when terms of the order of $(a/b)^2$ and higher are neglected:

$$D_{x_1} = 16\pi\mu A \tag{6–7.8}$$

$$M_{x_1} = \tfrac{16}{3}\pi\mu B \tag{6–7.9}$$

where

$$A = \frac{bU}{4\Omega}\left[1 - \frac{1}{\Omega}\frac{b}{l} + \frac{1}{\Omega^2}\frac{b^2}{l^2} - \frac{1}{\Omega}\left(\frac{1}{\Omega^2} - \frac{2}{3}\right)\frac{b^3}{l^3} + \frac{1}{\Omega}\left\{\frac{1}{\Omega^3} - \frac{4}{3\Omega} + \frac{1}{24(\Omega-1)}\right\}\frac{b^4}{l^4}\right] \tag{6–7.10}$$

$$B = -\frac{b^2 U}{8\Omega(2\Omega - 1)}\frac{b^4}{l^4}\left\{1 - \frac{1}{\Omega}\frac{b}{l} + O\left(\frac{b}{l}\right)^2\right\}\frac{x_o}{|x_o|}\sin\theta\cos\theta\,(6\cos^2\theta - 1) \tag{6–7.11}$$

in which

$$\Omega = \frac{1}{4}\left[2\ln\left(\frac{2b}{a}\right) + 1\right] \tag{6–7.12}$$

Equations (6–7.10)–(6–7.12) apply to the case of two prolate spheroids or needlelike objects.

For the case of two disks or circular plates, where $b = 0$, Wakiya's treatment reduces to

$$A = -\frac{2aU}{3\pi}\left[1 - \frac{8}{3\pi}\frac{a}{l} - \left(\frac{8}{3\pi}\right)^2\frac{a^2}{l^2} - \left(\frac{8}{3\pi}\right)^3\frac{a^3}{l^3} + \left(\frac{8}{3\pi}\right)^2\left\{\left(\frac{8}{3\pi}\right)^2 + \frac{7}{8}\right\}\frac{a^4}{l^4}\right] \tag{6–7.13}$$

$$B = -\frac{4}{3\pi^2}a^2 U\frac{a^4}{l^4}\left[1 - \frac{8}{3\pi}\frac{a}{l} + \left\{\left(\frac{8}{3\pi}\right)^2 + \frac{1}{2}\right\}\frac{a^2}{l^2}\right] \times \times \frac{x_o}{|x_o|}\sin\theta\cos\theta\,(6\cos^2\theta - 1) \tag{6–7.14}$$

Note that the torque on these spheroids would result in screwlike motion, shown in Fig. 6–7.1 for oblate spheroids. The expression $\sin\theta\cos\theta(6\cos^2\theta - 1)$ is positive from $\theta = 0$ to $\theta = 65.8°$, corresponding to the direction of torque indicated in Fig. 6–7.1. It is negative in the range from $\theta = 65.8°$ to $\theta = 90°$, which results in a reversal of torque direction. In all cases the effect first enters as the fourth power of a/l or b/l. It is very small and in fact would not appear in lower-order approximations, as in the case of the similar problem for two disks discussed following Eq. (5–4.26). For the case where the axes of the spheroids are at right angles to each other the torque vanishes.

These effects all vanish as the spheroids approach the spherical shape $a = b$, for which case

$$A = \frac{3}{8}aU\left[1 - \frac{3}{2}\frac{a}{l} + \left(\frac{3}{2}\right)^2\frac{a^2}{l^2} - \frac{19}{8}\frac{a^3}{l^3} + \frac{93}{16}\frac{a^4}{l^4}\right] \qquad (6\text{–}7.15)$$

which is in agreement with the treatment for two spheres moving along their line of centers, Eq. (6–3.51). Wakiya has also treated the more general case in which the axes of the spheroids still lie in yz planes, but they are no longer oriented so that their line of centers is parallel to the x axis.

6-8 Limitations of Creeping Motion Equations

When two spheres fall at Reynolds numbers over 0.25, inertial effects are experimentally noted, in that the spheres no longer maintain a fixed position relative to each other as they do in the creeping motion regime.

Oseen's[26] equations present a possible basis for considering the mutual interaction of particles where small inertial forces are present. For this purpose it would be necessary to take into account that the Oseen equations would not correctly approximate boundary conditions in the vicinity of the spheres themselves, as pointed out by Proudman and Pearson[27]. For this reason, Oseen's theoretical results for the two-sphere problem, discussed in subsequent paragraphs, must be regarded with suspicion. Brenner[4]* has shown by their procedure how the Oseen resistance of a single particle of arbitrary shape may be obtained, once one knows its Stokes resistance. In the method used, however, the classical Oseen equations are not used to describe the velocity field in the vicinity of the particle.

Oseen[26] has presented a treatment for the case of two spheres at low Reynolds numbers in which he simply uses Smoluchowski's development for two spheres in creeping motion, but substitutes Oseen velocity fields for Stokes fields to approximate particle interaction. For two equal-sized spheres following each other along the z axis, the force exerted by the *leading* sphere on the fluid is given by Oseen as

$$F_{z1} = 6\pi\mu aU\left(1 - \frac{3}{4}\frac{a}{l}e^{-2\sigma l}\right) - \frac{9}{2}\pi\mu a^2U\left[\frac{1-(1+\sigma l)e^{-2\sigma l}}{\sigma l^2}\right] \qquad (6\text{–}8.1)$$

where

$$\sigma = \frac{\rho U}{2\mu} \qquad (6\text{–}8.2)$$

*See also the following additional papers: W. Chester, J. Fluid Mech. **13** (1962), 557; H. Brenner and R. G. Cox, J. Fluid Mech. **17** (1963), 561.

If σl is small, Eq. (6–8.1) becomes

$$F_{z1} = 6\pi\mu aU\left(1 - \frac{3}{2}\frac{a}{l} + \frac{3}{8}N_{\text{Re}}\right) \tag{6–8.3}$$

The Reynolds number N_{Re} is based on sphere radius and l is the distance between the centers of the spheres. The force exerted by the *following* sphere will be the same as in the case of the creeping motion equations, namely,

$$F_{z2} = 6\pi\mu aU\left(1 - \frac{3}{2}\frac{a}{l}\right) \tag{6–8.4}$$

Thus, the leading sphere is slowed up to a greater extent than the following sphere.

For two spheres falling side by side along the x axis, the force on each sphere will be the same, but it will not be directed entirely in the direction of fall as in the case of the creeping motion equations. Thus, for this case, the force exerted by each sphere along the x axis will be

$$F'_{x1} = F'_{x2} = 6\pi\mu aU - \frac{9}{2}\pi\mu a^2 U \frac{e^{-\sigma l}}{l} \tag{6–8.5}$$

whence, when σl is small,

$$F'_{x1} = F'_{x2} = 6\pi\mu aU\left(1 - \frac{3}{4}\frac{a}{l} + \frac{3}{16}N_{\text{Re}}\right) \tag{6–8.6}$$

The force exerted in the z direction, that is, along the line connecting the centers of two particles, will be one corresponding to *repulsion*. Its magnitude is

$$F'_{z1} = -F'_{z2} = \frac{9}{2}\pi\mu a^2 U \frac{1 - (1 + \sigma l)e^{-\sigma l}}{\sigma l^2} \tag{6–8.7}$$

In case σl is small this becomes

$$F'_{z1} = -F'_{z2} = 6\pi\mu aU\left(\frac{3N_{\text{Re}}}{32}\right) \tag{6–8.8}$$

Though Oseen's equations are linear, one cannot treat the case of two spheres in *arbitrary orientation* by simple vector addition of the separate results for motion parallel and perpendicular to the line of centers; for in the Oseen equations the velocity $\mathbf{U}$, appearing in the inertial approximation $\rho\mathbf{U}\cdot\nabla\mathbf{v}$, is different for the two cases.

Oseen's original derivation deals with the case of two spheres of arbitrary size, but the inertial correction term in the relationship presented is proportional to the product $a_1 a_2$ of the sphere radii; hence no inertial correction at all is obtained if either a_1 or a_2 is equal to zero. This is, of course, an untenable conclusion and probably arises because, when two spheres are relatively far apart, one must match the boundary conditions separately on each of the two spheres as well as at infinity, as is done in the Proudman and Pearson matching procedure discussed in Section 2–6.

Happel and Pfeffer[18] report experiments with two equal-sized spheres following each other in the Reynolds number range of 0.27 to 0.73. Qualitatively, the effect previously predicted by the Oseen equations was observed, but better correlation was found using the constant preceding N_{Re} in Eq. (6–8.3) equal to 0.11 instead of the theoretical value of $\frac{3}{8}$. Experiments with single spheres indicate the same proportionate discrepancy. It should be noted that the simplification presented in Eqs. (6–8.3), (6–8.6), and (6–8.8) will not be applicable when the spheres are very large distances apart, because it involves the approximation $e^{\chi} = 1 + \chi$. In view of the questionable applicability of the Oseen equations themselves, however, this further refinement is probably not justified.

As was noted in Section 2–7, no creeping motion solution exists to the problem of streaming flow of a viscous fluid past a single cylinder. Flow past a cylinder has, however, been successfully treated using the Oseen[26] linearized equations, with qualitative agreement with experimental data. Since cylinders constitute an extreme of particle shape as compared with spheres it is also of interest that there are a number of studies on the forces acting on two cylinders placed near each other in a slow viscous flow. Theoretical studies by Fujikawa[15,16,17] include cases for fluid flowing perpendicular to the axes of cylinders of unequal radii, when placed with their axes parallel to each other. Results are in some ways similar to what has been found in the case of spheres. Thus with two cylinders of equal radii with one behind the other, the front cylinder experiences a greater drag than the rear cylinder. When the cylinders are placed side by side they tend to repel each other. In both cases the drag acting on each of the two cylinders is less than when they are present alone. Taneda[32] conducted experimental studies and was able to report agreement with the theoretical conclusions deduced from the Oseen equations. Further calculations along these lines are also reported by Kuwabara[21]. The forces on two parallel elliptic cylinders in a uniform flow at small Reynolds numbers have been treated by Takaisi[31], giving numerous calculated results.

One other point deserves consideration: namely, in applying the creeping motion equations, we have assumed that the drag is the same as that given by the steady motion of the particles. In cases where particles are nearly the same size, it will always be possible to attain this condition. When particles are unequal in size it would be necessary, for exact results, to take into account particle and fluid acceleration. This can be done for the case of single particles as discussed by Landau and Lifshitz (Ref. 35 in Chapter 2).

BIBLIOGRAPHY

1. Anderson, O., Svensk Papperstidning **59** (1956), 540.
2. Bart, E., M. Ch. E. Thesis, New York University, 1959.
3. Brenner, H., Chem. Eng. Sci. **18** (1963), 1.
4. ———, J. Fluid Mech. **11** (1961), 604.
5. ———, J. Fluid Mech. **12** (1962), 35.
6. ———, J. Fluid Mech. **18** (1964), 144.
7. ———, Chem. Eng. Sci. **19** (1964), 599.
8. Burgers, J. M., Proc. Koningl. Akad. Wetenschap. (Amsterdam) **44** (1941), 1045.
9. ———, Proc. Koningl. Akad. Wetenschap. (Amsterdam) **44** (1941), 1177.
10. ———, Proc. Koningl. Akad. Wetenschap. (Amsterdam) **43** (1940), 425, 646.
11. ———, Chap. III, *Second Report on Viscosity and Plasticity*. Amsterdam: North Holland Publishing Co., 1938.
12. Eveson, G. F., E. W. Hall, and S. G. Ward, Brit. J. Appl. Phys. **10** (1959), 43.
13. Faxen, H., Z. Angew. Math. Mech. **7** (1927), 79.
14. ———, Arkiv. Mat. Astron. Fys. **19A,** no 13 (1925).
15. Fujikawa, H., J. Phys. Soc. Japan **11** (1956), 558.
16. ———, J. Phys. Soc. Japan **11** (1956), 690.
17. ———, J. Phys. Soc. Japan **12** (1957), 423.
18. Happel, J., and R. Pfeffer, A. I. Ch. E. Jour. **6** (1960), 129.
19. Hocking, L. M., Quart. J. Roy. Meteorol. Soc. **85** (1959), 44; Ph. D. Thesis, Univ. London, 1958.
20. ———, Private communication.
21. Kuwabara, S., J. Phys. Soc. Japan **12** (1957), 291.
22. Kynch, G. J., J. Fluid Mech. **5** (1959), 193.
23. ———, in *Aerodynamic Capture of Particles*, ed. E. G. Richardson. New York: Pergamon, 1960, p. 194.
24. Lamb, H., *Hydrodynamics*, 6th ed. Cambridge: Cambridge Univ. Press, 1932, pp. 596, 598, 606.
25. Milne-Thomson, L. M., *Theoretical Hydrodynamics*, 3rd ed. New York: Macmillan, 1950.
26. Oseen, C. W., *Hydrodynamik*. Leipzig: Akademische Verlag., 1927.
27. Proudman, I., and J. R. A. Pearson, J. Fluid Mech. **2** (1957), 237.

28. Slack, G. W., and H. W. Mathews, Porton Technical Paper No. 797, Chemical Defence Exptl. Establishment, Porton, Wilts., England, 1961.

29. Smoluchowski, M., Bull. Inter. acad. Polonaise sci. lett. **1A** (1911), 28.

30. Stimson, M., and G. B. Jeffery, Proc. Roy. Soc. (London) **A111** (1926), 110.

31. Takaisi, Y., J. Phys. Soc. Japan **13** (1959), 506.

32. Taneda, S., J. Phys. Soc. Japan **12** (1957), 291.

33. Wakiya, S., Niigata Univ. (Nagaoka, Japan) Coll. Eng. *Res. Report* No. 6, Mar. 30, 1957.

Wall Effects on the Motion of a Single Particle

7

7-1 Introduction

The effect of containing walls on the rate of settling of a particle is, in many ways, similar to the effect of a second particle, discussed in Chapter 6. In order to treat the behavior of a group of particles settling in a container, it is necessary first to establish the effect of walls on the particles separately. These effects may then be combined with those due to particle interaction by an extension of the method of reflections discussed in Section 8–3. Rapid convergence appears to be obtained when the dimensions of the particle are small compared with those of the containing wall, and when the wall effect can be developed for each reflection.

The interaction of a particle with walls will depend on the particle shape, orientation, and position, as well as the geometry of the containing walls. As we shall establish in the following discussion, it is possible in the case of dilute systems to separate these effects, so that wall effects developed for spherical particles may be applied to other shapes placed in positions of corresponding exterior geometry. The quasi-static form of the creeping motion equations again furnishes the main basis for the development.

As in the previous chapter, we shall first develop the simplest approximations, using the point force treatment; these are generally applicable to situations where the particle dimensions are small with respect to particle distance from container walls. Both in the case of these very "dilute" situations and where higher approximation is necessary, the method of reflections will be employed essentially as outlined in Section 6–1.

As discussed in Chapter 3, it is possible to obtain a first approximation to the increased pressure drop arising from the presence of a particle in a bounded flow fluid, solely on the basis of a solution for the case of fluid flow through the container in the absence of the particle. Momentum and energy theorems (Section 3–6) applied to the relative motion of a particle confined within a cylindrical boundary of arbitrary cross-section lead to the relationship[4]

$$\mathbf{F}_w^+ = \left(\frac{v_o^{(0)}}{V_m} - 1\right)\mathbf{F} \tag{7–1.1}$$

where $\mathbf{F}_w^+$ is the shearing force on the cylinder walls above that which would be observed in the absence of the particle from the cylinder, V_m = mean velocity of flow with respect to the walls, $v_o^{(0)}$ = approach velocity to particle, $\mathbf{F}$ = drag on particle. The result is correct to zeroth and first powers of the ratio of characteristic particle/cylinder dimensions. Since the ratio $v_o^{(0)}/V_m$ is independent of V_m, this result also applies to pure settling, $V_m = 0$, in which case $\mathbf{F}_w^+$ is the actual force on the wall. This relationship is equivalent to

$$\Delta\mathbf{P}^+ A = \frac{v_o^{(0)}}{V_m}\mathbf{F} \tag{7–1.2}$$

where $\Delta\mathbf{P}^+$ refers to the additional pressure drop caused by the presence of the particle, provided that $\mathbf{F}$ is understood to refer to only that component of the force on the particle which is parallel to the walls of the cylinder. A is the cross-sectional area of the duct.

Equation (7–1.1) shows that both the magnitude and direction of the force exerted on the wall depend on the relative location of the particle with respect to the cylinder axis.

For a circular cylinder of radius R_o,

$$\frac{v_o^{(0)}}{V_m} = 2\left[1 - \left(\frac{b}{R_o}\right)^2\right]$$

where b is the distance from the axis to the center of the particle. Thus, from Eq. (7–1.1), as we move from the cylinder axis ($b/R_o = 0$) toward the wall ($b/R_o = 1$), the direction of the extra force changes from being parallel to the drag to being oppositely directed from it. For the special case

$$\frac{b}{R_o} = \frac{1}{\sqrt{2}} = 0.707 \tag{7–1.3}$$

the force vanishes.

Section 2–5 gives solutions of the rectilinear flow equations for several ducts of noncircular cross-section. Thus for a particle situated at the axis of a duct having the cross-section of an equilateral triangle

$$\frac{v_o^{(0)}}{V_m} = \frac{v_{\max}}{V_m} = \frac{20}{9} \tag{7–1.4}$$

It is clear that the additional force or pressure drop is thus a function of the shape of the container, even though its boundaries are infinitely far removed from the particle.

7-2 The Translation of a Particle in Proximity to Container Walls

In this section we consider the first-order effect of boundaries on the hydrodynamic force experienced by a rigid, translating particle of arbitrary shape. The analysis follows that of Brenner[5,9]. To make the arguments clear, the analysis is given in two parts. In the first part the details are carried through first, only for the case of motion parallel to a principal axis of translation of the particle, and then only for the special case where the motion is parallel to a "principal axis of the boundary." In this case, quantities which are intrinsically vectors and dyadics may be treated as scalars. Thus, the basic simplicity of the ideas and results is not lost by the mathematical abstractness required to treat the more general case. In the second, more general, part we remove all symmetry restrictions and give the result in its complete generality.

In either case, the advantage of the first-order analysis lies in the ability to treat the wall effect quantitatively in such a way that it is the same for all particles, irrespective of their shape. Thus, known theoretical or experimental results pertaining to wall effects for spherical particles may immediately be applied to nonspherical particles.

A word of caution! The analysis applies generally only to the case where the particle translates *without rotation*. As we shall discuss, however, the analysis of the hydrodynamic force may also be applied to nonskew rotating bodies, provided that we properly interpret the significance of certain terms appearing in the final formula.

The boundary conditions appropriate to the translation of a solid in a bounded fluid require that

$$\mathbf{v} = \mathbf{U} \quad \text{on } P, \qquad \mathbf{v} = \mathbf{0} \quad \text{on } S, \qquad \mathbf{v} \to \mathbf{0} \quad \text{at infinity}$$

where P refers to the surface of the particle, and S refers to the boundary surface. If S completely bounds P, then $\mathbf{v}$ need not vanish at infinity.

As in Section 6–1, we solve the problem by the method of reflections. Thus,

$$\mathbf{v} = \mathbf{v}^{(1)} + \mathbf{v}^{(2)} + \mathbf{v}^{(3)} + \cdots$$

$$p = p^{(1)} + p^{(2)} + p^{(3)} + \cdots$$

where $\mathbf{v}^{(1)} = \mathbf{U}$ on P, $\mathbf{v}^{(1)} \to \mathbf{0}$ at infinity

and $\mathbf{v}^{(2)} = -\mathbf{v}^{(1)}$ on S, $\mathbf{v}^{(2)} \to \mathbf{0}$ at infinity

etc.

As shown in Eq. (6–1.16), the even-numbered fields do not contribute to the hydrodynamic force on the particle. Hence,

$$\mathbf{F} = \mathbf{F}^{(1)} + \mathbf{F}^{(3)} + \mathbf{F}^{(5)} + \cdots \tag{7–2.1}$$

The initial field $\mathbf{v}^{(1)}$ corresponds to the translational motion of the particle in an unbounded fluid. Associated with this motion is the force

$$\mathbf{F}^{(1)} = \mathbf{F}_\infty \tag{7–2.2}$$

A detailed knowledge of $\mathbf{v}^{(1)}$ is required only insofar as it enters into the computation of the next reflection, $\mathbf{v}^{(2)}$, through the boundary condition $\mathbf{v}^{(2)} = -\mathbf{v}^{(1)}$ on the boundary S. Now, at large distances from any translating particle, the velocity field must be asymptotically equal to that which would be generated by the action of a *point force* of strength $\mathbf{F}_\infty$ situated near the center of the particle. Thus, from Eqs. (6–2.6) and (6–2.7), we have, at large distances from any finite particle,

$$\mathbf{v}^{(1)} = -\frac{\mathbf{F}_\infty}{6\pi\mu r} - \frac{r^2}{24\pi\mu}(\mathbf{F}_\infty \cdot \nabla)\nabla\frac{1}{r} + o(r^{-1}) \tag{7–2.3}$$

and

$$p^{(1)} = \frac{1}{4\pi}(\mathbf{F}_\infty \cdot \nabla)\frac{1}{r} + o(r^{-2}) \tag{7–2.4}$$

Once the field $\mathbf{v}^{(1)}$ has been established, it is necessary to determine the field $\mathbf{v}^{(2)}$ canceling this field on the wall, S, of the containing vessel. Having computed $\mathbf{v}^{(2)}$ in detail for a given particle and wall geometry, we are then in a position to establish the field $\mathbf{v}^{(3)}$ obtained by canceling $\mathbf{v}^{(2)}$ at the particle surface P; that is, $\mathbf{v}^{(3)}$ satisfies the boundary conditions

$$\mathbf{v}^{(3)} = -\mathbf{v}^{(2)} \quad \text{on } P$$

$$\mathbf{v}^{(3)} \to \mathbf{0} \quad \text{at infinity}$$

Once $\mathbf{v}^{(3)}$ is known, one may determine the force $\mathbf{F}^{(3)}$ experienced by the particle in consequence of the motion. This may be done in a general way, at least to the order of the approximation of interest to us here.

For a particle translating without rotation, we have from Chapter 5 that

$$\mathbf{F} = -\mu\mathbf{K}\cdot\mathbf{U} \tag{7–2.5}$$

where $\mathbf{K}$ is the translation tensor. Since the field $\mathbf{v}^{(2)}$ is regular at all points of the fluid, it can be expanded in a Taylor series about the "center" O of the particle. By an application of the reciprocal theorem it may then be shown[5] that, to terms of lowest order in the particle-to-wall dimension ratio,

$$\mathbf{F}^{(3)} = \mu\mathbf{K}\cdot\mathbf{v}_o^{(2)} \tag{7–2.6}$$

where the subscript "o" implies that the field is to be evaluated at the center of the particle.

The treatment thus far has been completely general. We now specialize the development to cases where the particle moves parallel to a principal axis of translation with velocity $\mathbf{U} = \mathbf{i}U$ ($\mathbf{i}$ = unit vector in the direction of motion). The force on the particle moving through the unbounded fluid is then antiparallel to its direction of motion; hence we may write

$$\mathbf{F}_\infty = -\mathbf{i}F_\infty \tag{7–2.7}$$

where the scalar F_∞ is essentially positive. Furthermore, from Eq. (7–2.3), we now have

$$\mathbf{v}^{(1)} = \frac{F_\infty}{8\pi\mu}\left[\mathbf{i}\left(\frac{1}{r} + \frac{x^2}{r^3}\right) + \mathbf{j}\frac{yz}{r^3} + \mathbf{k}\frac{zx}{r^3}\right] \tag{7–2.8}$$

for particle motion in the x direction.

It remains yet to establish the direction of the constant vector $\mathbf{v}_o^{(2)}$. For the moment we shall confine ourselves to situations where the bounding surfaces possess mutually perpendicular planes of reflection symmetry and/or symmetry axes (for example, a circular cylinder, a plane wall, etc.), though we shall remove this restriction in our subsequent generalization. Suppose the principal axis along which the particle is moving lies parallel to a symmetry axis of the boundary, or lies perpendicular to a plane of symmetry of the boundary. In this case, $\mathbf{v}_o^{(2)}$ will have but a single component, and this will lie parallel to the direction of motion of the particle. This follows by observing that this is the only velocity component of the field $\mathbf{v}^{(2)}$ which will not be an *odd function* of at least one of the coordinates x, y, or z. And any components which are odd functions vanish at the center of the particle, ($x = 0$, $y = 0$, $z = 0$). Some simple cases in which the interested reader can quickly establish this fact occur for a particle falling parallel or perpendicular to a plane wall. Thus, we may write $\mathbf{v}_o^{(2)} = -\mathbf{i}v_o^{(2)}$ where the scalar $v_o^{(2)}$ may be either positive or negative.

Under the assumed conditions of symmetry, Eq. (7–2.6) shows that $\mathbf{F}^{(3)}$ is parallel to the direction of motion of the particle. Thus, we may write $\mathbf{F}^{(3)} = -\mathbf{i}F^{(3)}$. Upon combining Eqs. (7–2.5) and (7–2.6), we now obtain

$$F^{(3)} = \frac{F_\infty}{U}v_o^{(2)} \tag{7–2.9}$$

in which F_∞ and U are essentially positive. The algebraic sign of $F^{(3)}$ is then positive or negative according as $v_o^{(2)}$ is positive or negative. For later reference we also write $\mathbf{F} = -\mathbf{i}F$, as it is now obvious that the *total* force on the particle is parallel to its direction of motion and oppositely directed. The scalar F is essentially positive.

As $\mathbf{F}^{(3)}$ is now known, the velocity field $\mathbf{v}^{(3)}$ may be computed at large distances from the particle via an equation of the form (7–2.3) in which $\mathbf{v}^{(3)}$ replaces $\mathbf{v}^{(1)}$ and $\mathbf{F}^{(3)}$ replaces $\mathbf{F}_\infty$. This then furnishes the boundary conditions to determine $\mathbf{v}^{(4)}$ in accord with the condition $\mathbf{v}^{(4)} = -\mathbf{v}^{(3)}$ on S. By analogy to our previous calculations it is at once apparent that

$$F^{(5)} = F_\infty\left[\frac{v_o^{(2)}}{U}\right]^2, \qquad F^{(7)} = F_\infty\left[\frac{v_o^{(2)}}{U}\right]^3, \text{ etc.} \tag{7-2.10}$$

Upon summing the individual drags in accordance with Eq. (7–2.1), the following expression is obtained for the total drag on the particle:

$$F = F_\infty + F_\infty\left[\frac{v_o^{(2)}}{U}\right] + F_\infty\left[\frac{v_o^{(2)}}{U}\right]^2 + F_\infty\left[\frac{v_o^{(2)}}{U}\right]^3 + \cdots \tag{7-2.11}$$

This geometric series may be summed, whereupon

$$\frac{F}{F_\infty} = \frac{1}{1 - (v_o^{(2)}/U)} \tag{7-2.12}$$

From Eq. (7–2.3), the initial field $\mathbf{v}^{(1)}$ is directly proportional to F_∞. Since $\mathbf{v}^{(2)}$ is linearly connected to $\mathbf{v}^{(1)}$ through the fact that $\mathbf{v}^{(2)} = -\mathbf{v}^{(1)}$ on S, it follows that the same must be true of $\mathbf{v}^{(2)}$; hence, $\mathbf{v}^{(2)}$ is directly proportional to F_∞. In addition, it is clear that $\mathbf{v}^{(2)}$ must be independent of the viscosity of the fluid. Since F_∞ is directly proportional to the viscosity, this requires that $\mathbf{v}^{(2)}$ be proportional to F_∞/μ. Finally, $\mathbf{v}^{(2)}$ must obviously go to zero as S recedes infinitely far from P, that is, as $l \to \infty$; l is taken as a characteristic dimension proportional to the distance of the particle from the container wall. For the sake of definiteness we shall often take l as the distance from the center of the particle to the bounding wall; l may vary according to particle location and wall geometry in specific cases. By simple dimensional arguments it follows that we must have

$$v_o^{(2)} = k\frac{F_\infty}{6\pi\mu l} \tag{7-2.13}$$

where k is a dimensionless constant of $O(1)$, dependent solely on the nature of the boundary S. The boundary correction is then of the form

$$\frac{F}{F_\infty} = \frac{1}{1 - k(F_\infty/6\pi\mu Ul)} \tag{7-2.14}$$

It remains yet to estimate the degree of approximation inherent in the preceding relation. Since F_∞ is proportional to μUc, the result is certainly correct to first powers of c/l, c being a characteristic particle dimension. The error will not therefore exceed terms of $O(c/l)^2$ in the denominator of the foregoing. However, there always exists an origin such that the error term will be of $O(c/l)^3$. In this event we are led to the following form for the drag correction:

$$\frac{F}{F_\infty} = \frac{1}{1 - k(F_\infty/6\pi\mu Ul) + O(c/l)^3} \tag{7-2.15}$$

It should be clearly understood that c refers to the *maximum* particle dimension. For example, in the case of a needle-like body falling in a circular cylinder it is not sufficient that the diameter of the needle be small compared with the cylinder diameter. Rather, the length of the needle must be small compared with the latter. A specific example of this is considered in Eq. (7–5.21).

It is of some interest to examine, initially, the case in which the particle is a *sphere* of radius c. For here, the *exact* forms of some of our approximate relationships are known. In the first place, the Stokes velocity field[39] shows that the error in Eq. (7–2.3) is of $O(c/r)^3$. Furthermore, for a spherical particle, the analog of Eq. (7–2.6) is given exactly by Faxen's law[46].

$$\mathbf{F}^{(3)} = \frac{F_\infty}{U}\mathbf{v}_o^{(2)} + \mu\pi c^3(\nabla^2 \mathbf{v}^{(2)})_o \tag{7–2.16}$$

With the aid of these observations, the validity of the error estimate in Eq. (7–2.15) follows at once for spherical particles. The general form of Eq. (7–2.15) for spheres is confirmed in all particulars by the many detailed solutions which are available for a wide variety of different boundaries. Typical k values are cited in Table 7–6.1 from detailed solutions which follow.

It is worthwhile to note that Eq. (7–2.15) is correct whether or not S is a solid surface to which fluid adheres. Careful reexamination of the previous development shows that the final result holds for any homogeneous *linear* boundary condition on S—for example, a free surface, that is one on which the normal velocity and tangential stresses vanish. It goes without saying that the value of k depends on the boundary conditions imposed on S.

Generalization for translational motions

In a later paper, Brenner[9] generalized the preceding treatment to cases where the principal axes of translation of the particle may have any orientation relative to the principal axes of the bounding walls. As we shall show, to the first order in the ratio of particle to boundary dimensions, the increased resistance of the particle to translational motions may be represented by a symmetric, second-rank tensor (dyadic) whose value is independent of particle shape and orientation.

At large distances from any particle, the asymptotic form of the initial field, corresponding to Eqs. (7–2.3) and (7–2.4), is

$$\mathbf{v}^{(1)} = -\left(\mathbf{I} + \frac{\mathbf{rr}}{r^2}\right)\cdot\frac{\mathbf{F}_\infty}{8\pi\mu r} + o\left(\frac{c}{r}\right) \tag{7–2.17}$$

$$p^{(1)} = -\frac{\mathbf{r}\cdot\mathbf{F}_\infty}{4\pi r^3} + o\left(\frac{c}{r}\right)^2 \tag{7–2.18}$$

where $\mathbf{I}$ is the idemfactor and $\mathbf{r}$ is the position vector of a point in the fluid relative to an origin at the "center" of the particle. The terms displayed explicitly in the foregoing are the same as would arise in the unbounded fluid from the action of a *point force* of strength $\mathbf{F}_\infty$ situated at the origin.

Since $r = O(l)$ on S it follows from Eq. (7–2.17) that the boundary condition to be satisfied by $\mathbf{v}^{(2)}$ on S is of the form

$$\mathbf{v}^{(2)} = \boldsymbol{\lambda}_s\cdot\frac{\mathbf{F}_\infty}{6\pi\mu l} + o\left(\frac{c}{l}\right) \quad \text{on } S \tag{7–2.19}$$

where

$$\boldsymbol{\lambda}_S\left(\frac{x_S}{l}, \frac{y_S}{l}, \frac{z_S}{l}\right) = \frac{3l}{4r_S}\left[\mathbf{I} + \frac{(\mathbf{r}_S/l)(\mathbf{r}_S/l)}{(r_S/l)^2}\right] \tag{7–2.20}$$

in which $\mathbf{r}_S \equiv (x_S, y_S, z_S)$ refers to a point on S. The dimensionless, variable dyadic $\boldsymbol{\lambda}_S$ is of $O(1)$ with respect to the parameter c/l. It is clear that, at any point on S, $\boldsymbol{\lambda}_S$ depends only upon the location of the center of P relative to S and upon the geometrical shape of the latter. It is independent of the size of S.

In the Stokes regime, $\mathbf{F} = O(6\pi\mu c\mathbf{U})$. It follows that $\mathbf{v}^{(2)}$ must be of the form

$$\mathbf{v}^{(2)} = \boldsymbol{\lambda}\cdot\frac{\mathbf{F}_\infty}{6\pi\mu l} + o\left(\frac{c}{l}\right) \tag{7–2.21}$$

where $\boldsymbol{\lambda} = \boldsymbol{\lambda}(x/l, y/l, z/l)$ is a dimensionless, variable dyadic of $O(1)$ which reduces to $\boldsymbol{\lambda}_S$ on S. Thus, at any point in the fluid, it too depends only upon the location of the center of P relative to S and upon the shape of the latter.

Now, from Eq. (7–2.6), the force $\mathbf{F}^{(3)}$ arising from $(\mathbf{v}^{(3)}, p^{(3)})$ is, in our present notation,

$$\mathbf{F}^{(3)} = 6\pi\mu c\boldsymbol{\phi}_\infty\cdot\mathbf{v}_0^{(2)} + o\left(\frac{c}{l}\right) \tag{7–2.22}$$

where $\boldsymbol{\phi}_\infty = \mathbf{K}/6\pi c$ is the dimensionless Stokes translation tensor for the particle, and $\mathbf{v}_0^{(2)}$ refers to the value of $\mathbf{v}^{(2)}$ at the center of the space presently occupied by the particle. But, from Eq. (7–2.21), we have

$$\mathbf{v}_0^{(2)} = \mathbf{k}\cdot\frac{\mathbf{F}_\infty}{6\pi\mu l} + o\left(\frac{c}{l}\right) \tag{7–2.23}$$

where $\mathbf{k}$ denotes the value of $\boldsymbol{\lambda}$ at the center of the particle, $x = 0$, $y = 0$, $z = 0$. This relation constitutes the generalization of Eq. (7–2.13). $\mathbf{k}$ is therefore a dimensionless dyadic of $O(1)$, dependent only upon the relative location of the center of P with respect to S and upon the shape of the latter. It is shown in the original paper[9] that $\mathbf{k}$ is symmetric.

Upon combining Eqs. (7–2.22) and (7–2.23), we obtain

$$\mathbf{F}^{(3)} = \boldsymbol{\phi}_\infty\cdot\mathbf{k}\cdot\mathbf{F}_\infty\frac{c}{l} + o\left(\frac{c}{l}\right) \tag{7–2.24}$$

Though $\boldsymbol{\phi}_\infty$ and $\mathbf{k}$ are symmetric, their product is not generally symmetric and one must be careful to preserve the proper order of these dyadics in the multiplication.

Higher-order reflections may be obtained as outlined previously. One thereby obtains for $m = 0, 1, 2, \ldots$

$$\mathbf{F}^{(2m+1)} = \left(\boldsymbol{\phi}_\infty\cdot\mathbf{k}\frac{c}{l}\right)^m\cdot\mathbf{F}_\infty + o\left(\frac{c}{l}\right)^m \tag{7–2.25}$$

If we now substitute Eq. (7–2.25) into Eq. (7–2.1) and sum the resulting geometric series, we obtain, corresponding to the scalar equation (7–2.15),

$$\mathbf{F} = \left[\mathbf{I} - \boldsymbol{\phi}_\infty \cdot \mathbf{k}\frac{c}{l} + o\left(\frac{c}{l}\right)\right]^{-1} \cdot \mathbf{F}_\infty \tag{7–2.26}$$

If $\mathbf{F}_\infty$ is eliminated from this expression via the definition of the dimensionless translation dyadic,

$$\mathbf{F}_\infty = -6\pi\mu c\boldsymbol{\phi}_\infty \cdot \mathbf{U}$$

we obtain our principal result,

$$\mathbf{F} = -6\pi\mu c\left[\boldsymbol{\phi}_\infty^{-1} - \mathbf{k}\frac{c}{l} + o\left(\frac{c}{l}\right)\right]^{-1} \cdot \mathbf{U} \tag{7–2.27}$$

This constitutes the generalization of Eq. (7–2.15).

It is immaterial how the characteristic dimensions c and l in Eq. (7–2.27) are defined. Different choices for these quantities merely produce differing definitions of the dyadics $\boldsymbol{\phi}_\infty$ and $\mathbf{k}$ in such a way that the over-all result is unaffected.

Equation (7–2.27) was derived only for situations in which the particle translates without rotation. In this case the result is applicable to particles of arbitrary shape. Most particles settling in a container under the influence of gravity rotate as they fall. It is natural therefore to inquire as to the applicability of Eq. (7–2.27) under more general circumstances. This question is partially discussed elsewhere[9]. We complete the discussion by observing that only in utilizing the relation $\mathbf{F}_\infty = -\mu\mathbf{K}\cdot\mathbf{U}$ have we explicitly assumed the particle not to be rotating. In more general cases we have

$$\mathbf{F}_\infty = -\mu(\mathbf{K}\cdot\mathbf{U}_R + \mathbf{C}_R\cdot\boldsymbol{\omega})$$

where $\mathbf{C}_R$ is the coupling dyadic at the center of reaction R. Since $\mathbf{C}_R = \mathbf{O}$ for nonskew bodies, we conclude that Eq. (7–2.27) may be applied to any nonskew particle, provided that $\mathbf{U}$ in that relation refers to the velocity, $\mathbf{U}_R$, of the center of reaction (that is, center of hydrodynamic stress) of the body.

Application to a settling particle with an arbitrary orientation

For applications involving settling particles, the hydrodynamic force $\mathbf{F}$ is known *a priori* (provided that we neglect the inertial force on the particle resulting from its acceleration, if any) and its instantaneous velocity $\mathbf{U}$ is sought. Solving Eq. (7–2.27) explicitly for this velocity we obtain

$$\mathbf{U} = \mathbf{U}_\infty + \frac{\mathbf{k}\cdot\mathbf{F}}{6\pi\mu l} + o\left(\frac{c}{l}\right) \tag{7–2.28}$$

where

$$\mathbf{U}_\infty = \frac{\boldsymbol{\phi}_\infty^{-1}\cdot\mathbf{F}}{6\pi\mu c} \tag{7–2.29}$$

denotes the velocity with which the particle would settle (for the same orientation) in an unbounded fluid. To compute $\mathbf{F}$, let $\mathbf{g}$ be the local acceleration of gravity vector, directed vertically downward, and let m_p and m_f, respectively, be the mass of the particle and displaced fluid. Thus, upon neglecting the inertial force on the particle mass, the gravitational, hydrostatic, and hydrodynamic forces on the particle are in equilibrium. The constant hydrodynamic force on the particle is, therefore,

$$\mathbf{F} = -(m_p - m_f)\mathbf{g}$$

These equations show that even an isotropic particle will not generally settle vertically in a bounded fluid unless one of the principal axes of $\mathbf{k}$ lies parallel to Earth's gravity field.

As a simple example of the application of Eq. (7–2.28), consider the motion of a thin, homogeneous, circular disk of thickness b and radius c ($c \gg b$) falling under the influence of gravity in a semi-infinite viscous fluid bounded below by a infinitely extended, rigid plane wall, as in Fig. 7–2.1. The instantaneous position of the center of the disk from the wall is l. We let x_j, $\bar{x}_j$, and $\bar{\bar{x}}_j$ ($j = 1, 2, 3$) be coordinates fixed in space, fixed in the disk, and fixed in the wall, respectively. The x_1 coordinate is directed toward the center of Earth and the x_2 coordinate is parallel to Earth's surface. The "3" coordinates are all directed out of the plane of the paper. Unit vectors in the three different coordinate systems are denoted by $\mathbf{i}_j$, $\bar{\mathbf{i}}_j$ and $\bar{\bar{\mathbf{i}}}_j$. The angles made by the plane of the disk and wall, respectively, with the horizontal are denoted by ξ and η. We propose to calculate the components of the instantaneous settling velocity of the disk in the x_j system.

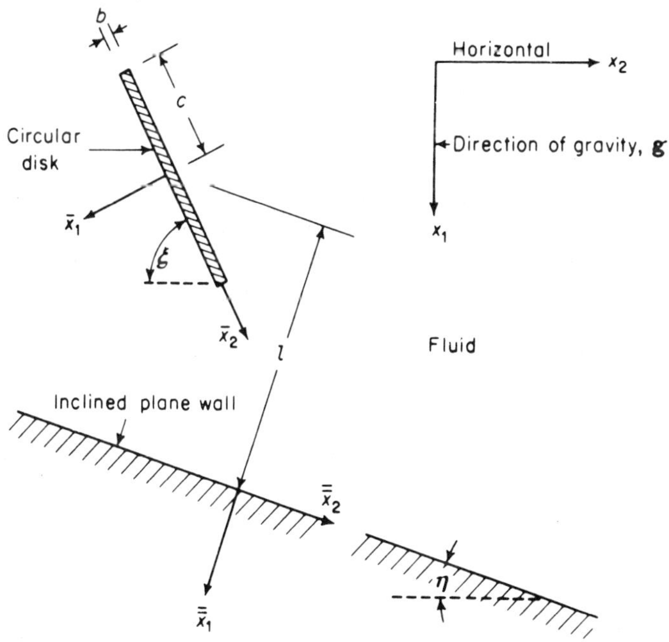

Figure 7-2.1. Circular disk settling asymmetrically near an inclined plane wall.

It is clear from symmetry that the principal axes of $\boldsymbol{\phi}_\infty$ lie parallel to the $\bar{x}_j$ axes. The Stokes force experienced by a circular disk moving broadside-on with velocity $\mathbf{U}$ in an unbounded medium is $\mathbf{F}_\infty = -16\mu c\mathbf{U}$. The corresponding expression for edge-on motion is $\mathbf{F}_\infty = -\frac{32}{3}\mu c\mathbf{U}$. Hence, we obtain—cf. Eq. (5–4.29),

$$\boldsymbol{\phi}_\infty = \frac{8}{9\pi}(\bar{\mathbf{i}}_1\bar{\mathbf{i}}_1 3 + \bar{\mathbf{i}}_2\bar{\mathbf{i}}_2 2 + \bar{\mathbf{i}}_3\bar{\mathbf{i}}_3 2) \tag{7–2.30}$$

It also follows from symmetry that the principal axes of $\mathbf{k}$ are everywhere parallel to the $\bar{\bar{x}}_j$ axes. Hence, from Table 7–6.1, we find that*

$$\mathbf{k} = \tfrac{9}{16}(\bar{\bar{\mathbf{i}}}_1\bar{\bar{\mathbf{i}}}_1 2 + \bar{\bar{\mathbf{i}}}_2\bar{\bar{\mathbf{i}}}_2 + \bar{\bar{\mathbf{i}}}_3\bar{\bar{\mathbf{i}}}_3) \tag{7–2.31}$$

We also note that $\mathbf{g} = \mathbf{i}_1 g$. Since $m_p - m_f = \pi c^2 b\Delta\rho$ ($\Delta\rho$ = difference in density of disk and fluid), a straightforward calculation yields $U_3 = 0$ and

$$\begin{aligned} U_1 &= \frac{\pi cbg\Delta\rho}{32\mu}\left[(2 + \sin^2\xi) - \frac{3}{\pi}(2 - \sin^2\eta)\frac{c}{l} + o\left(\frac{c}{l}\right)\right] \\ U_2 &= \frac{\pi cbg\Delta\rho}{32\mu}\left[\sin\xi\cos\xi + \frac{3}{\pi}\sin\eta\cos\eta\left(\frac{c}{l}\right) + o\left(\frac{c}{l}\right)\right] \end{aligned} \tag{7–2.32}$$

The only other boundary and position for which the wall-effect tensor is *completely* known is at the center of a hollow sphere filled with viscous liquid. It is self-evident that $\mathbf{k}$ must be isotropic for this situation. Hence, if l is the radius of the hollow sphere, we find from Table 7–6.1 that

$$\mathbf{k} = \mathbf{I}\tfrac{9}{4} \tag{7–2.33}$$

Moving boundary or net flow at infinity

Equations (7–2.15) and (7–2.27) are applicable only to the case where the boundary S is at rest and the fluid at infinity is at rest. It is of some interest to modify them so as to remove these restrictions. The case in which S is in motion arises, for example, when S is itself a particle falling in proximity to the original particle P. The case in which a net flow occurs at infinity arises, for example, during Poiseuille flow through a circular tube containing a particle. Problems of the latter type are of interest in connection with the observed radial migration of particles in Poiseuille flow (Segré and Silberberg[50]; Goldsmith and Mason[24]).

In the event that either S moves or the fluid at infinity is in net flow, let $(\mathbf{v}^{(0)}, p^{(0)})$ denote the local velocity and pressure fields which would arise from either of these motions if P were absent from the fluid. It is assumed that these fields satisfy the creeping motion equations. To the previous

*The comparable expression for a *free* plane surface is

$$\mathbf{k} = \tfrac{3}{8}(\bar{\bar{\mathbf{i}}}_1\bar{\bar{\mathbf{i}}}_1 2 - \bar{\bar{\mathbf{i}}}_2\bar{\bar{\mathbf{i}}}_2 - \bar{\bar{\mathbf{i}}}_3\bar{\bar{\mathbf{i}}}_3)$$

The plane surface, either free or rigid, appears to be the only boundary for which $\mathbf{k}$ has the same value at *all* points of the fluid.

problem, we now add initial fields $(\mathbf{v}^{(0)}, p^{(0)})$. Among the original boundary conditions only the definition of the field $\mathbf{v}^{(1)}$ requires modification. In its stead it is now required that

$$\mathbf{v}^{(1)} = \mathbf{U} - \mathbf{v}^{(0)} \quad \text{on } P \tag{7–2.34}$$

The initial field $\mathbf{v}^{(0)}$ can be expanded in a Taylor series about the center of P:

$$\mathbf{v}^{(0)} = \mathbf{v}_o^{(0)} + o\left(\frac{c}{l}\right) \tag{7–2.35}$$

where $\mathbf{v}_o^{(0)}$ denotes the value of $\mathbf{v}^{(0)}$ at the center of the space presently occupied by the particle. Thus, in place of Eq. (7–2.34), $\mathbf{v}^{(1)}$ is now uniquely defined to the first order in c/l by the relation

$$\mathbf{v}^{(1)} = \mathbf{U} - \mathbf{v}_o^{(0)} + o\left(\frac{c}{l}\right) \quad \text{on } P \tag{7–2.36}$$

Upon repeating the analysis which led to Eq. (7–2.27), we now find that the force on the particle is given correctly to the first order by the expression

$$\mathbf{F} = -6\pi\mu c\left[\boldsymbol{\phi}_\infty^{-1} - \mathbf{k}\frac{c}{l} + o\left(\frac{c}{l}\right)\right]^{-1} \cdot \left[\mathbf{U} - \mathbf{v}_o^{(0)} + o\left(\frac{c}{l}\right)\right] \tag{7–2.37}$$

where $\mathbf{k}$ has the same value as previously, that is, in the case where $\mathbf{v}^{(0)}$ was identically zero.

Equation (7–2.37) in conjunction with the relation $\mathbf{F} = -(m_p - m_f)\,\mathbf{g}$ yields the instantaneous particle velocity

$$\mathbf{U} = \mathbf{v}_o^{(0)} + \frac{m_p - m_f}{6\pi\mu c}\left(\boldsymbol{\phi}_\infty^{-1} - \mathbf{k}\frac{c}{l}\right) \cdot \mathbf{g} + o\left(\frac{c}{l}\right) \tag{7–2.38}$$

This relation shows that a particle moving within a Poiseuille field* of flow in a vertical circular tube will, if its density differs from that of the surrounding fluid, undergo radial movement whenever no one of its three principal axes of translation lies parallel to the gravity field.

For the simple case of motion parallel to a principal axis, Eq. (7–2.37) is equivalent to Eq. (7–2.15), where the only modification required in the latter is the replacement of the denominator of the left-hand side by F'_∞, the infinite medium drag based on the *approach velocity* to the particle, that is,

$$F'_\infty = F_\infty\left[1 - \left(\frac{v_o^{(0)}}{U}\right)\left\{1 + O\left(\frac{c}{l}\right)^2\right\}\right] \tag{7–2.39}$$

The methods of this section may be applied to a variety of combinations of particle and wall geometry, based on knowledge of the separate solutions for particles in infinite media and for spheres in the proximity of container walls of various shape.

*Here, $\mathbf{v}_o^{(0)} = 2\mathbf{V}_m[1 - (b/l)^2]$, where $\mathbf{V}_m$ is the mean velocity of flow through the tube, b = distance of center of particle from cylinder axis, l = tube radius.

7-3 Sphere Moving in Axial Direction in a Circular Cylindrical Tube

In this section we consider the slow translation of a single spherical particle moving parallel to the longitudinal axis of an infinitely long circular cylinder through which a viscous fluid may be flowing. The sphere may occupy any preassigned position. The general method using a reflection treatment has been developed to a first approximation[6]. Haberman[27] and others have treated more exactly the axisymmetrical case, in which the sphere is restricted to the cylinder axis. These solutions will be briefly considered at the end of this section. Note that we consider the case here where the sphere is kept from rotating as it falls. Since we develop only first-order corrections, the effect of rotation on drag will not be significant.

Nomenclature and boundary conditions

In the treatment to follow, it is necessary to resort to a variety of different coordinate systems. These are cartesian coordinates (x, y, z); spherical coordinates, (r, θ, ϕ); and cylindrical coordinates, (ρ, ϕ, z), each having a common origin at the sphere center. It is also necessary to utilize cartesian coordinates (X, Y, Z), and cylindrical coordinates, (R, Φ, Z), both originating along the cylinder axis and chosen so as to make $z = Z$. Relations between these various coordinate systems are depicted in Fig. 7–3.1.*

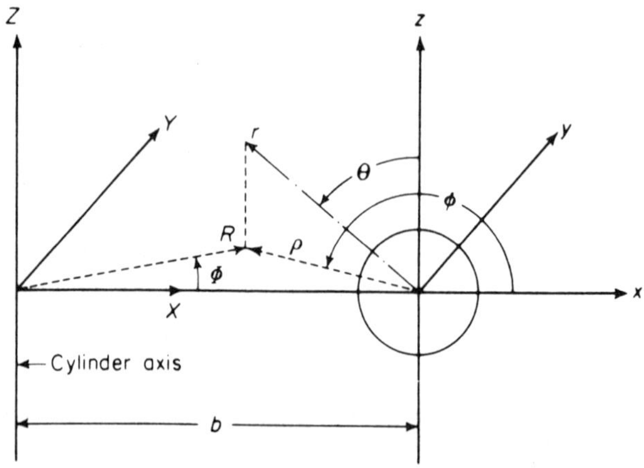

Figure 7-3.1. Coordinate systems for a sphere in a circular cylinder.

*In the original article[6], we inadvertantly employed a left-handed coordinate system. This oversight is corrected here.

The sphere moves with an arbitrary constant velocity, U, relative to the cylinder wall in the direction of z positive, parallel to the cylinder axis. At the same time, the fluid flows in laminar flow with a superficial velocity of $\frac{1}{2}U_o$, in the same direction. The sphere radius is a, the cylinder radius is R_o, and the center of the sphere is situated at a distance b from the cylinder axis, as shown in Fig. 7–3.2.

In terms of a coordinate system which moves with the sphere, the usual hypothesis of no relative motion at fluid-solid interfaces results in the following boundary conditions, which define the fluid velocity field, $\mathbf{v}$:

$$\mathbf{v} = \mathbf{0} \quad \text{at } r = a \tag{7–3.1}$$

and

$$\mathbf{v} = -\mathbf{i}_z U \quad \text{at } R = R_o \tag{7–3.2}$$

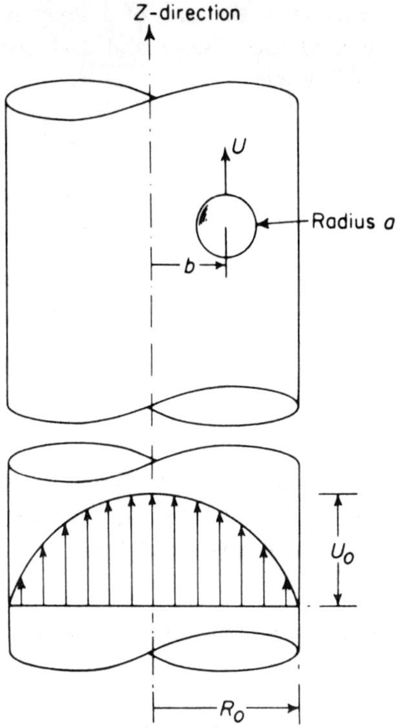

Figure 7-3.2. Particle and fluid motion in a circular cylinder.

where $\mathbf{i}_z$ is a unit vector in the z direction. At large distances from the sphere, $z = \pm\infty$, the disturbance created by the sphere vanishes and the fluid velocity distribution becomes Poiseuillian. This gives rise to the additional boundary condition,

$$\mathbf{v} = \mathbf{i}_z\left[U_o\left(1 - \frac{R^2}{R_o^2}\right) - U\right] \quad \text{at } z = \pm\infty \tag{7–3.3}$$

The equations of motion to be satisfied are

$$\nabla^2\mathbf{v} = \frac{1}{\mu}\nabla p \tag{7–3.4}$$

together with the continuity equation for incompressible fluids,

$$\nabla\cdot\mathbf{v} = 0 \tag{7–3.5}$$

Use of the creeping motion equations restricts the validity of the final results to situations in which the relative particle Reynolds number, $2a|U_o(1 - b^2/R_o^2) - U|/\nu$ is small; ν is the kinematic viscosity.

As in a previous study by Happel and Byrne[30], the foregoing boundary value problem is solved by the method of "reflections." Thus, the solution consists of the sum of a series of velocity fields, all of which satisfy Eqs. (7–3.4) and (7–3.5), and each partially satisfying the boundary conditions as follows:

$$\mathbf{v}^{(0)} = \mathbf{i}_z\left[U_o\left(1 - \frac{R^2}{R_o^2}\right) - U\right] \text{ everywhere} \tag{7–3.6}$$

$$\mathbf{v}^{(1)} = \begin{cases} -\mathbf{v}^{(0)} & \text{at } r = a \\ \mathbf{0} & \text{at } z = \pm\infty \text{ (that is, } r = \infty) \end{cases} \tag{7–3.7}$$

$$\mathbf{v}^{(2)} = \begin{cases} -\mathbf{v}^{(1)} & \text{at } R = R_o \\ \mathbf{0} & \text{at } Z = \pm\infty \end{cases} \tag{7–3.8}$$

$$\mathbf{v}^{(3)} = \begin{cases} -\mathbf{v}^{(2)} & \text{at } r = a \\ \mathbf{0} & \text{at } z = \pm\infty \text{ (that is, } r = \infty) \end{cases} \tag{7–3.9}$$

etc., etc.

with as many fields taken as needed for an appropriate degree of approximation. The field $\mathbf{v}$ satisfying the boundary conditions, Eqs. (7–3.1)–(7–3.3), is then obtained in the form

$$\mathbf{v} = \mathbf{v}^{(0)} + \mathbf{v}^{(1)} + \mathbf{v}^{(2)} + \mathbf{v}^{(3)} + \cdots \tag{7–3.10}$$

and the corresponding pressure field is given by

$$p = p^{(0)} + p^{(1)} + p^{(2)} + p^{(3)} + \cdots \tag{7–3.11}$$

The success of this method of solution depends upon the linearity of the equations of motion. It has the advantage that it is necessary to consider boundary conditions associated with only one surface at a time.

If we let $\mathbf{F}$ and $\mathbf{T}$, respectively, represent the frictional force and torque (about the sphere center) experienced by the sphere, there are analogous relations of the form

$$\mathbf{F} = \mathbf{F}^{(0)} + \mathbf{F}^{(1)} + \mathbf{F}^{(2)} + \mathbf{F}^{(3)} + \cdots \tag{7–3.12}$$

and

$$\mathbf{T} = \mathbf{T}^{(0)} + \mathbf{T}^{(1)} + \mathbf{T}^{(2)} + \mathbf{T}^{(3)} + \cdots \tag{7–3.13}$$

Furthermore, if ΔP_S represents the additional pressure drop (above that due to the original Poiseuillian field, ΔP_0) experienced by the fluid as a result of the presence of the sphere, then

$$\Delta P_S = \Delta P_1 + \Delta P_2 + \Delta P_3 + \Delta P_4 + \cdots \tag{7–3.13a}$$

where the pressure drop, ΔP_i, associated with field i is simply the difference between the viscous pressures, $p^{(i)}$, at $Z = -\infty$ and $Z = +\infty$. There is no ambiguity in this definition of pressure drop, since as $Z \to \pm\infty$ the pressure becomes constant across the tube and, hence, independent Φ and R. It will prove convenient, therefore, in calculating pressure drop to evaluate the pressures, $p^{(i)}$, at $R = R_o$. Except for an arbitrary additive constant, which can be taken to be zero without loss of generality, the conditions of symmetry about the plane $Z = 0$ demand that pressure be an odd function of Z. These observations lead to the formula

$$\Delta P_i = -2 \lim_{Z \to +\infty} [p^{(i)}]_{R=R_o} \tag{7–3.14}$$

where the subscript denotes evaluation of the function at the cylinder wall.

The first reflected field

The unperturbed field, $\mathbf{v}^{(0)}$, is defined in Eq. (7–3.6). It is easy to demonstrate that

$$\mathbf{F}^{(0)} = \mathbf{0} \tag{7–3.15}$$

and

$$\mathbf{T}^{(0)} = \mathbf{0} \tag{7–3.16}$$

The first reflected field, $\mathbf{v}^{(1)}$, defined in Eq. (7–3.7), has already been obtained by Simha[51] in connection with suspension viscosity. Using Lamb's[39] general solution in spherical coordinates, Simha's result can be expressed in the form

$$\mathbf{v}^{(1)} = \nabla \times (\mathbf{r}\chi_{-2}^{(1)}) + \nabla\left(\Phi_{-2}^{(1)} + \Phi_{-3}^{(1)} + \Phi_{-4}^{(1)} + \frac{1}{2\mu} r^2 p_{-2}^{(1)} - \frac{1}{30\mu} r^2 p_{-4}^{(1)}\right) + \frac{\mathbf{r}}{\mu}\left(p_{-2}^{(1)} + \frac{1}{2} p_{-3}^{(1)} + \frac{1}{3} p_{-4}^{(1)}\right) \tag{7–3.17}$$

and

$$p^{(1)} = p_{-2}^{(1)} + p_{-3}^{(1)} + p_{-4}^{(1)} \tag{7–3.18}$$

where $\mathbf{r}$ is the radius vector drawn from the sphere origin, and $p_{-(n+1)}$, $\Phi_{-(n+1)}$ and $\chi_{-(n+1)}$ are solid spherical harmonics of degree $-(n + 1)$ defined as follows (cf. the derivation on pp. 68–71):

$$\begin{aligned}
p_{-2}^{(1)}/\mu &= A r^{-2} P_1(\cos\theta) \\
p_{-3}^{(1)}/\mu &= B R_o r^{-3} \cos\phi\, P_2^1(\cos\theta) \\
p_{-4}^{(1)}/\mu &= C R_o^2 r^{-4} P_3(\cos\theta) \\
\Phi_{-2}^{(1)} &= D R_o^2 r^{-2} P_1(\cos\theta) \\
\Phi_{-3}^{(1)} &= E R_o^3 r^{-3} \cos\phi\, P_2^1(\cos\theta) \\
\Phi_{-4}^{(1)} &= F R_o^4 r^{-4} P_3(\cos\theta) \\
\chi_{-2}^{(1)} &= G R_o r^{-2} \sin\phi\, P_1^1(\cos\theta),
\end{aligned} \tag{7–3.19}$$

with the constants given by

$$\begin{aligned}
A &= \frac{3}{2} a\left[U - U_o\left(1 - \frac{b^2}{R_o^2} - \frac{2}{3}\frac{a^2}{R_o^2}\right)\right] \\
B &= \frac{10}{3} a U_o \left(\frac{b}{R_o}\right)\left(\frac{a}{R_o}\right)^2 \\
C &= -\frac{7}{2} a U_o \left(\frac{a}{R_o}\right)^4 \\
D &= \frac{1}{4} a \left(\frac{a}{R_o}\right)^2 \left[U - U_o\left(1 - \frac{b^2}{R_o^2} - \frac{6}{5}\frac{a^2}{R_o^2}\right)\right]
\end{aligned} \tag{7–3.20a}$$

$$E = \frac{1}{3} aU_o \left(\frac{b}{R_o}\right)\left(\frac{a}{R_o}\right)^4$$
$$F = -\frac{1}{4} aU_o \left(\frac{a}{R_o}\right)^6 \qquad (7\text{-}3.20\text{b})$$
$$G = -aU_o \left(\frac{b}{R_o}\right)\left(\frac{a}{R_o}\right)^2$$

The quantities $P_n(\cos\theta)$ and $P_n^m(\cos\theta)$ are Legendre polynomials of order n, and associated Legendre polynomials (of the first kind) of order n and rank m, respectively. For reference, the following values appear in Eq. (7–3.19):

$$\begin{aligned} P_1(\cos\theta) &= \cos\theta \\ P_1^1(\cos\theta) &= \sin\theta \\ P_2^1(\cos\theta) &= 3\sin\theta\cos\theta \\ P_3(\cos\theta) &= \tfrac{1}{2}(5\cos^3\theta - 3\cos\theta) \end{aligned} \qquad (7\text{-}3.21)$$

The force and torque experienced by the sphere can be obtained from the relations (3–2.42) and (3–2.45),

$$\mathbf{F}^{(1)} = -4\pi\nabla(r^3 p_{-2}^{(1)}) \qquad (7\text{-}3.22)$$

and

$$\mathbf{T}^{(1)} = -8\pi\mu\nabla(r^3 \chi_{-2}^{(1)}) \qquad (7\text{-}3.23)$$

These relations apply quite generally to Lamb's solution. Noting from Eq. (7–3.19) that $r^3 p_{-2}^{(1)} = \mu A r\cos\theta = \mu A z$, and introducing the value of A from Eq. (7–3.20), results in

$$\mathbf{F}^{(1)} = -\mathbf{i}_z 6\pi\mu a\left[U - U_o\left(1 - \frac{b^2}{R_o^2} - \frac{2}{3}\frac{a^2}{R_o^2}\right)\right] \qquad (7\text{-}3.24)$$

In a like manner, application of the relation $r\sin\theta\sin\phi = y$ yields

$$\mathbf{T}^{(1)} = \mathbf{i}_y 8\pi\mu a^2 U_o \left(\frac{b}{R_o}\right)\left(\frac{a}{R_o}\right) \qquad (7\text{-}3.25)$$

If Fig. 7–3.2 represents a meridian plane passing through both the sphere origin and cylinder axis, the tendency of the couple is to rotate the sphere in a clockwise direction.

We shall limit ourselves in the subsequent development to an approximate solution in which the ratio of sphere-to-cylinder radii, a/R_o, is small. The complete expression for the field $\mathbf{v}^{(2)}$ without neglecting any terms is given in the supplement to a paper by Happel and Brenner[1]. The first approximation covers most situations of practical interest and illustrates the procedure employed. The terms in Eq. (7–3.19) are such that a final expression for the drag, correct to zeroth and first powers of a/R_o, can ultimately be obtained by retaining only the $p_{-2}^{(1)}$ harmonic in the present solution. This is equivalent to the use of a point-force approximation for $\mathbf{v}^{(1)}$. Thus, in place of the previous results, we now resort to the approximate solution,

$$\mathbf{v}^{(1)} \approx \frac{1}{2\mu}\nabla(r^2 p_{-2}^{(1)}) + \frac{\mathbf{r}}{\mu}p_{-2}^{(1)} \qquad (7\text{-}3.26)$$

and
$$p^{(1)} \approx p^{(1)}_{-2} \tag{7–3.27}$$
where
$$\frac{p^{(1)}_{-2}}{\mu} = Hr^{-2}\cos\theta = H\frac{z}{r^3} \tag{7–3.28}$$
with the constant H defined by
$$H = \frac{3}{2}a\left[U - U_o\left(1 - \frac{b^2}{R_o^2}\right)\right] \tag{7–3.29}$$
To the present degree of approximation, the drag is now given by
$$\mathbf{F}^{(1)} \approx -\mathbf{i}_z 6\pi\mu a\left[U - U_o\left(1 - \frac{b^2}{R_o^2}\right)\right] \tag{7–3.30}$$
whereas the torque remains unaltered.

Transformation to cylindrical coordinates

In order to compute the field $\mathbf{v}^{(2)}$, defined in Eq. (7–3.8), it is necessary to establish the form taken by $\mathbf{v}^{(1)}$ at the cylinder wall, $R = R_o$. This is best done by transforming the latter field to cylindrical coordinates (R, Φ, Z), originating along the cylinder axis. With the expression for $p^{(1)}_{-2}/\mu$, given in Eq. (7–3.28), and the aid of the relation
$$\nabla\left(\frac{1}{r}\right) = -\frac{\mathbf{r}}{r^3} \tag{7–3.31}$$
it is easy to show that an alternative form for the first reflected field is
$$\mathbf{v}^{(1)} = \frac{H}{2}\left[2\mathbf{i}_z\left(\frac{1}{r}\right) - \nabla\left(\frac{z}{r}\right)\right] \tag{7–3.32}$$
In terms of the desired coordinates we have
$$\mathbf{i}_z = \mathbf{i}_Z \tag{7–3.33}$$
and
$$\nabla = \mathbf{i}_R\frac{\partial}{\partial R} + \mathbf{i}_\Phi\frac{1}{R}\frac{\partial}{\partial\Phi} + \mathbf{i}_Z\frac{\partial}{\partial Z}, \tag{7–3.34}$$
so that our objective may be attained by transforming the scalar functions $1/r$ and z/r to these coordinates.

Watson[64; pp. 361, 389] gives the relation
$$\frac{1}{r} = (\rho^2 + z^2)^{-1/2} = \frac{2}{\pi}\int_0^\infty K_0(\lambda\rho)\cos\lambda z\, d\lambda \tag{7–3.35}$$
where K_0 is the modified Bessel function of the second kind of order zero. Although this transformation is valid only for $\rho > 0$, the restriction is inconsequential since it is sufficient for our immediate purposes to evaluate $\mathbf{v}^{(1)}$ only in the vicinity of the cylinder wall. Noting that
$$\rho = (R^2 + b^2 - 2bR\cos\Phi)^{1/2} \tag{7–3.36}$$
we can avail ourselves of the further transformation,
$$K_0(\lambda\rho) = \sum_{k=-\infty}^{\infty} K_k(\lambda R)I_k(\lambda b)\cos k\Phi \tag{7–3.37}$$

given by Watson. Here, I_k and K_k are modified Bessel functions of the first and second kinds, respectively, of order k. The relation is valid only for $R > b$. For the reason previously outlined, this restriction is of no consequence. These expressions combine to give

$$\frac{1}{r} = \frac{2}{\pi} \sum_{k=-\infty}^{\infty} \cos k\Phi \int_0^\infty K_k(\lambda R) I_k(\lambda b) \cos \lambda Z \, d\lambda \tag{7-3.38}$$

To obtain the second scalar transformation we observe, upon performing the indicated differentiations, that the following identity is valid:

$$ZK_k(\lambda R)I_k(\lambda b)\cos\lambda Z = \frac{\partial}{\partial\lambda}[K_k(\lambda R)I_k(\lambda b)\sin\lambda Z] - \sin\lambda Z \frac{\partial}{\partial\lambda}[K_k(\lambda R)I_k(\lambda b)] \tag{7-3.39}$$

Thus, if Eq. (7-3.38) is multiplied by z, we obtain with the aid of the foregoing,

$$\frac{z}{r} = \frac{2}{\pi} \sum_{k=-\infty}^{\infty} \cos k\Phi [K_k(\lambda R)I_k(\lambda b)\sin\lambda Z]_{\lambda=0}^{\lambda=\infty} - \frac{2}{\pi} \sum_{k=-\infty}^{\infty} \cos k\Phi \int_0^\infty \frac{\partial}{\partial\lambda}[K_k(\lambda R)I_k(\lambda b)] \sin\lambda Z \, d\lambda \tag{7-3.40}$$

The first term in brackets vanishes at the upper and lower limits of evaluation, $\lambda = 0$ and ∞. Upon performing the indicated differentiation in the second expression we arrive at the requisite transformation,

$$\frac{z}{r} = -\frac{2}{\pi} \sum_{k=-\infty}^{\infty} \cos k\Phi \int_0^\infty [RK_k'(\lambda R)I_k(\lambda b) + bK_k(\lambda R)I_k'(\lambda b)] \sin\lambda Z \, d\lambda \tag{7-3.41}$$

The differentiations denoted by primes are with respect to the entire argument.

It is now relatively simple to obtain the desired transformation,

$$[\mathbf{v}^{(1)}]_{R_o} = \mathbf{i}_R \frac{1}{\pi} \sum_{k=-\infty}^{\infty} \cos k\Phi \int_0^\infty \alpha_k(\lambda) \sin\lambda Z \, d\lambda + \mathbf{i}_\Phi \frac{1}{\pi} \sum_{k=-\infty}^{\infty} \sin k\Phi \int_0^\infty \beta_k(\lambda) \sin\lambda Z \, d\lambda + \mathbf{i}_Z \frac{1}{\pi} \sum_{k=-\infty}^{\infty} \cos k\Phi \int_0^\infty \gamma_k(\lambda) \cos\lambda Z \, d\lambda \tag{7-3.42}$$

where we have written

$$\alpha_k(\lambda) = H\left[\lambda R_o K_k(\lambda R_o) I_k(\lambda b) + \lambda b K_k'(\lambda R_o) I_k'(\lambda b) + \frac{k^2}{\lambda R_o} K_k(\lambda R_o) I_k(\lambda b)\right] \tag{7-3.43}$$

$$\beta_k(\lambda) = -Hk\left[\frac{b}{R_o} K_k(\lambda R_o) I_k'(\lambda b) + K_k'(\lambda R_o) I_k(\lambda b)\right] \tag{7-3.44}$$

$$\gamma_k(\lambda) = H[\lambda R_o K_k'(\lambda R_o) I_k(\lambda b) + \lambda b K_k(\lambda R_o) I_k'(\lambda b) + 2K_k(\lambda R_o) I_k(\lambda b)] \tag{7-3.45}$$

These relations have been simplified to some extent by the use of Bessel's modified equation,

$$K_k''(\lambda R) = -\frac{K_k'(\lambda R)}{\lambda R} + \left(1 + \frac{k^2}{\lambda^2 R^2}\right) K_k(\lambda R) \tag{7-3.46}$$

An analogous expression for the initial pressure field at the cylinder wall can be obtained from Eqs. (7–3.27), (7–3.28), and (7–3.38) with the assistance of the relation $z/r^3 = -\partial(1/r)/\partial z$, yielding

$$[p^{(1)}]_{R_o} = \frac{2\mu H}{\pi} \sum_{k=-\infty}^{\infty} \cos k\Phi \int_0^{\infty} \lambda K_k(\lambda R_o) I_k(\lambda b) \sin \lambda Z \, d\lambda \tag{7-3.47}$$

The second reflection, $\mathbf{v}^{(2)}$

A general solution of Eqs. (7–3.4) and (7–3.5) in cylindrical coordinates, suitable for the field $\mathbf{v}^{(2)}$, as given in Section 3–3, is

$$\mathbf{v}^{(2)} = \sum_{k=-\infty}^{\infty} \left[\nabla \times (\mathbf{i}_Z \Omega_k^{(2)}) + \nabla \Psi_k^{(2)} + R \frac{\partial}{\partial R} \nabla \Pi_k^{(2)} + \mathbf{i}_Z \frac{\partial \Pi_k^{(2)}}{\partial Z} \right] \tag{7-3.48}$$

and

$$p^{(2)} = -2\mu \sum_{k=-\infty}^{\infty} \frac{\partial^2 \Pi_k^{(2)}}{\partial Z^2} \tag{7-3.49}$$

where Ω_k, Ψ_k, and Π_k are arbitrary cylindrical harmonic functions of order k in (R, Φ, Z) coordinates. These equations are perfectly general and would be applicable for the fields $\mathbf{v}^{(4)}$, $\mathbf{v}^{(6)}$, etc. In the present instance, the harmonic functions are assumed to have the forms

$$\begin{aligned}
\Pi_k^{(2)} &= -\frac{1}{\pi} \cos k\Phi \int_0^{\infty} \frac{\pi_k(\lambda)}{\lambda} I_k(\lambda R) \sin \lambda Z \, d\lambda \\
\Psi_k^{(2)} &= -\frac{1}{\pi} \cos k\Phi \int_0^{\infty} \frac{\psi_k(\lambda)}{\lambda} I_k(\lambda R) \sin \lambda Z \, d\lambda \\
\Omega_k^{(2)} &= -\frac{1}{\pi} \sin k\Phi \int_0^{\infty} \frac{\omega_k(\lambda)}{\lambda} I_k(\lambda R) \sin \lambda Z \, d\lambda
\end{aligned} \tag{7-3.50}$$

The functions $\pi_k(\lambda)$, $\psi_k(\lambda)$ and $\omega_k(\lambda)$ are to be determined from the boundary conditions expressed by Eq. (7–3.8). Upon substituting values from Eq. (7–3.50) into Eq. (7–3.48), we eventually find

$$\begin{aligned}
\mathbf{v}^{(2)} = &-\mathbf{i}_R \frac{1}{\pi} \sum_{k=-\infty}^{\infty} \cos k\Phi \int_0^{\infty} \left[\omega_k(\lambda) \frac{k I_k(\lambda R)}{\lambda R} + \psi_k(\lambda) I_k'(\lambda R) \right. \\
&\qquad \left. + \pi_k(\lambda) \lambda R I_k''(\lambda R) \right] \sin \lambda Z \, d\lambda \\
&- \mathbf{i}_\Phi \frac{1}{\pi} \sum_{k=-\infty}^{\infty} \sin k\Phi \int_0^{\infty} \left[\pi_k(\lambda) \frac{k I_k(\lambda R)}{\lambda R} - \pi_k(\lambda) k I_k'(\lambda R) \right. \\
&\qquad \left. - \omega_k(\lambda) I_k'(\lambda R) - \psi_k(\lambda) \frac{k I_k(\lambda R)}{\lambda R} \right] \sin \lambda Z \, d\lambda \\
&- \mathbf{i}_Z \frac{1}{\pi} \sum_{k=-\infty}^{\infty} \cos k\Phi \int_0^{\infty} \left[\psi_k(\lambda) I_k(\lambda R) + \pi_k(\lambda) R I_k'(\lambda R) \right. \\
&\qquad \left. + \pi_k(\lambda) I_k(\lambda R) \right] \cos \lambda Z \, d\lambda
\end{aligned} \tag{7-3.51}$$

Evaluation of the preceding equation at $R = R_o$, and comparison to Eq. (7–3.42) by means of the boundary condition,

$$[\mathbf{v}^{(2)}]_{R_o} = -[\mathbf{v}^{(1)}]_{R_o} \tag{7–3.52}$$

leads to three simultaneous equations involving the functions $\pi_k(\lambda)$, $\psi_k(\lambda)$, and $\omega_k(\lambda)$ in terms of $\alpha_k(\lambda)$, $\beta_k(\lambda)$, and $\gamma_k(\lambda)$. Solving these equations and introducing values for the latter functions from Eqs. (7–3.43)–(7–3.45) gives

$$\pi_k(\lambda) = H\frac{K_k(\lambda R_o)I_k(\lambda b)}{I_k(\lambda R_o)} + \xi_k(\lambda) \tag{7–3.53}$$

$$\begin{aligned}\psi_k(\lambda) = &-\left[1 + \frac{\lambda R_o I_k'(\lambda R_o)}{I_k(\lambda R_o)}\right]\pi_k(\lambda) \\ &+ \frac{H}{I_k(\lambda R_o)}[\lambda R_o K_k'(\lambda R_o)I_k(\lambda b) + \lambda b K_k(\lambda R_o)I_k'(\lambda b) \\ &+ 2K_k(\lambda R_o)I_k(\lambda b)]\end{aligned} \tag{7–3.54}$$

and

$$\omega_k(\lambda) = \frac{2k}{\lambda R_o I_k'(\lambda R_o)}[I_k(\lambda R_o)\pi_k(\lambda) - HK_k(\lambda R_o)I_k(\lambda b)] \tag{7–3.55}$$

where $\xi_k(\lambda)$ in Eq. (7–3.53) is given by

$$\xi_k(\lambda) = \frac{H\{[I_k(\lambda b)I_k'(\lambda R_o)/I_k(\lambda R_o)] - (b/R_o)I_k'(\lambda b)\}}{\dfrac{2k^2 I_k^3(\lambda R_o)}{(\lambda R_o)^2 I_k'(\lambda R_o)} + \left(\lambda R_o + \dfrac{k^2}{\lambda R_o}\right)I_k^2(\lambda R_o) - 2I_k(\lambda R_o)I_k'(\lambda R_o) - \lambda R_o[I_k'(\lambda R_o)]^2} \tag{7–3.56}$$

These equations have been simplified to some extent by means of the relations

$$I_k''(\lambda R_o) = -\frac{I_k'(\lambda R_o)}{\lambda R_o} + \left(1 + \frac{k^2}{\lambda^2 R_o^2}\right)I_k(\lambda R_o)$$

and

$$K_k'(\lambda R_o)I_k(\lambda R_o) - K_k(\lambda R_o)I_k'(\lambda R_o) = -\frac{1}{\lambda R_o} \tag{7–3.57}$$

The additional boundary condition, $\mathbf{v}^{(2)} \to \mathbf{0}$ as $Z \to \pm\infty$, is met because $\mathbf{v}^{(1)} \to \mathbf{0}$ as $Z \to \pm\infty$ and $\mathbf{v}^{(1)}$ and $\mathbf{v}^{(2)}$ are related by Eq. (7–3.52).

The pressure drop due to the first two reflected fields can now be computed. Substitution of the value of $\Pi_k^{(2)}$, given in Eq. (7–3.50), into the expression for $p^{(2)}$, given by Eq. (7–3.49), and evaluation at $R = R_o$ using the value of $\pi_k(\lambda)$ from Eq. (7–3.53), yields

$$\begin{aligned}[p^{(2)}]_{R_o} = &-\frac{2\mu H}{\pi}\sum_{k=-\infty}^{\infty}\cos k\Phi\int_0^\infty \lambda K_k(\lambda R_o)I_k(\lambda b)\sin\lambda Z\,d\lambda \\ &-\frac{2\mu}{\pi}\sum_{k=-\infty}^{\infty}\cos k\Phi\int_0^\infty \lambda\xi_k(\lambda)I_k(\lambda R_o)\sin\lambda Z\,d\lambda\end{aligned} \tag{7–3.58}$$

As evidenced by Eq. (7–3.47), the leading term in the preceding expression is simply $-[p^{(1)}]_{R_o}$, so that

$$[p^{(1)} + p^{(2)}]_{R_o} = -\frac{2\mu}{\pi}\sum_{k=-\infty}^{\infty}\cos k\Phi\int_0^\infty \xi_k(\lambda)I_k(\lambda R_o)\sin\lambda Z\,d\lambda \tag{7–3.59}$$

From Eq. (7–3.14), the pressure drop due to these first two reflections is, therefore,

$$(\Delta P_1 + \Delta P_2) = \frac{4\mu}{\pi R_o^2} \lim_{Z\to\infty} \sum_{k=-\infty}^{\infty} \cos k\Phi \int_0^\infty \eta_k(\lambda) \frac{\sin \lambda Z}{\lambda}\, d\lambda \tag{7–3.60}$$

where we have written

$$\eta_k(\lambda) = (\lambda R_o)^2 \xi_k(\lambda) I_k(\lambda R_o) \tag{7–3.61}$$

By Dirichlet's theorem,

$$\lim_{Z\to\infty} \int_0^\infty \eta_k(\lambda) \frac{\sin \lambda Z}{\lambda}\, d\lambda = \frac{\pi}{2}\eta_k(0+) \tag{7–3.62}$$

This limit, as $\lambda \to 0+$, can easily be obtained by expanding the Bessel functions in series for small values of their arguments; hence,

$$\eta_k(0+) = \eta_k(0) = \begin{cases} -4H\left(1 - \dfrac{b^2}{R_o^2}\right) & \text{for } k = 0 \\ 0 \quad \text{for } k \neq 0 \end{cases} \tag{7–3.63}$$

from which, with the definition of H in Eq. (7–3.29), we finally obtain

$$(\Delta P_1 + \Delta P_2) = \frac{12\mu a}{R_o^2}\left(1 - \frac{b^2}{R_o^2}\right)\left[U_o\left(1 - \frac{b^2}{R_o^2}\right) - U\right] \tag{7–3.64}$$

A more physically meaningful representation is

$$[(\Delta P_1 + \Delta P_2)\pi R_o^2][\tfrac{1}{2}U_o] = \left[6\pi\mu a\left\{U_o\left(1 - \frac{b^2}{R_o^2}\right) - U\right\}\right]\left[U_o\left(1 - \frac{b^2}{R_o^2}\right)\right] \tag{7–3.65}$$

For the small values of a/R_o for which the present approximation is valid, the first term in brackets represents the additional force (above Poiseuille's law) required to force fluid through the tube. The second term is the mean velocity with which fluid traverses the cylinder. Their product gives the additional energy dissipation incurred by the presence of the sphere in the original field of flow. On the right-hand side of this expression, the first term in brackets is the frictional force experienced by the sphere, Eq. (7–3.30), whereas the second bracketed term corresponds to the *local* velocity of the unperturbed parabolic field in the vicinity of the sphere. It appears then that, for a sufficiently small sphere, the product of the drag and *local* velocity also gives the additional energy dissipation caused by the presence of an obstacle in the field of flow. This conclusion is confirmed by a more direct and general development by Brenner[4] based on the reciprocal theorem derived in Section 3–5 and applied in Section 3–6.

Within the volume of space presently occupied by the particle, the field $\mathbf{v}^{(2)}$ has no singularities. As such, the sphere can experience no frictional forces or torques in virtue of this field; hence

$$\mathbf{F}^{(2)} = \mathbf{0} \tag{7–3.66}$$

$$\mathbf{T}^{(2)} = \mathbf{0} \tag{7–3.67}$$

The third reflection

The results obtained thus far for drag and pressure drop may be regarded as the *zeroth* approximation in powers of a/R_o; that for torque is correct to the first power. To evaluate each of these quantities correctly to the next highest powers in a/R_o, it is necessary to consider the third and fourth reflections. Fortunately, exact solutions for $\mathbf{v}^{(3)}$ and $\mathbf{v}^{(4)}$ are not required to arrive at exact values for these initial corrections.

The frictional force and torque associated with $\mathbf{v}^{(3)}$ can be calculated exactly by means of Faxen's laws. In the present application, these laws take the form

$$\mathbf{F}^{(3)} = 6\pi\mu a[\mathbf{v}^{(2)}]_o + \pi a^3[\nabla p^{(2)}]_o$$

and

$$\mathbf{T}^{(3)} = 4\pi\mu a^3[\nabla \times \mathbf{v}^{(2)}]_o \tag{7-3.68}$$

where the subscript "o" indicates that the function in brackets is to be evaluated at the sphere center. The relations are valid for arbitrary fields $\mathbf{v}^{(2)}$ and $\mathbf{v}^{(3)}$ provided that each is in accord with the equations of slow motion and that they are related by Eq. (7–3.9). To preserve the consistency of the present results regarding powers of a/R_o, we omit the last term of $\mathbf{F}^{(3)}$ appearing in Eq. (7–3.68).

In terms of the (R, Φ, Z) system of coordinates, the center of the sphere is situated at the point $(R = b, \Phi = 0, Z = 0)$. Thus, from Eqs. (7–3.51) and (7–3.68),

$$\mathbf{F}^{(3)} = -\mathbf{i}_Z 6\mu\left(\frac{a}{R_o}\right)\sum_{k=-\infty}^{\infty}\int_0^{\infty}[\psi_k(\lambda)I_k(\lambda b) + \pi_k(\lambda)\lambda b I_k'(\lambda b) + \pi_k(\lambda)I_k(\lambda b)]\,d(\lambda R_o) \tag{7-3.69}$$

Employing the values of $\psi_k(\lambda)$ and $\pi_k(\lambda)$ given by Eqs. (7–3.53) and (7–3.54), and putting $\alpha = \lambda R_o$, this becomes

$$\mathbf{F}^{(3)} = -\mathbf{i}_Z 6\pi\mu a\left[U - U_o\left(1 - \frac{b^2}{R_o^2}\right)\right]f\left(\frac{b}{R_o}\right)\left(\frac{a}{R_o}\right) \tag{7-3.70}$$

where

$$f\left(\frac{b}{R_o}\right) = f(\beta) = -\frac{3}{2\pi}\sum_{k=-\infty}^{\infty}\int_0^{\infty}\left\{\delta_k(\alpha)\left[\frac{I_k'(\alpha)I_k(\alpha\beta)}{I_k(\alpha)} - \beta I_k'(\alpha\beta)\right]^2 + \left[\frac{I_k(\alpha\beta)}{I_k(\alpha)}\right]^2 - \frac{2K_k(\alpha)I_k(\alpha\beta)}{I_k(\alpha)}[I_k(\alpha\beta) + \alpha\beta I_k'(\alpha\beta)]\right\}d\alpha \tag{7-3.71}$$

in which

$$\delta_k(\alpha) = \frac{1}{\dfrac{2k^2 I_k^3(\alpha)}{\alpha^3 I_k'(\alpha)} + \left(1 + \dfrac{k^2}{\alpha^2}\right)I_k^2(\alpha) - \dfrac{2I_k(\alpha)I_k'(\alpha)}{\alpha} - [I_k'(\alpha)]^2} \tag{7-3.72}$$

For brevity we have put $\beta = b/R_o$. Equation (7–3.57) has been employed in simplifying the foregoing.

The function $f(b/R_o)$ has been evaluated[6] as a power series in even powers of b/R_o. The first two terms, obtained by numerical integration, are

$$f\left(\frac{b}{R_o}\right) = 2.10444 - 0.6977\left(\frac{b}{R_o}\right)^2 + O\left(\frac{b}{R_o}\right)^4 \tag{7-3.73}$$

This development is valid for values of $b/R_o \to 0$, near the cylinder axis. In the vicinity of the cylinder wall, where $b/R_o \to 1$, it is possible to make an exact calculation, provided that the fractional distance of the particle from the cylinder wall, $a/(R_o - b)$, still remains small:

$$\lim_{b/R_o \to 1} \left(1 - \frac{b}{R_o}\right) f\left(\frac{b}{R_o}\right) = \frac{9}{16} \tag{7-3.74}$$

This limit is approached asymptotically and corresponds to a sphere near a *plane* wall. In the original paper[6], details of the computation are incorrect because only the lead term in the Stokes field was employed. Famularo[13] has evaluated the integrals appearing in Eq. (7-3.71) and obtained the results indicated in Table 7-3.1. A plot of the function $[1 - (b/R_o)]f(b/R_o)$ versus b/R_o, is presented in Fig. 7-3.3.

TABLE 7-3.1
VALUES OF ECCENTRICITY FUNCTION $f(b/R_o)$

b/R_o	$f(b/R_o)$
0.0	2.10444 ± 0.00002
0.01	2.10436 ± 0.00002
0.03	2.10381 ± 0.00002
0.05	2.10270 ± 0.00002
0.10	2.09763 ± 0.00002
0.20	2.07942 ± 0.00002
0.30	2.05691 ± 0.00002
0.35	2.04805 ± 0.00002
0.37	2.04567 ± 0.00002
0.39	2.04426 ± 0.00002
0.40	2.04401 ± 0.00002
0.41	2.04407 ± 0.00002
0.43	2.04530 ± 0.00002
0.45	2.04825 ± 0.00002
0.50	2.06566 ± 0.00002
0.60	2.16965 ± 0.00002
0.70	2.45963 ± 0.00002
0.80	3.2316 ± 0.0003
0.90	5.905 ± 0.003

Equation (7-3.51) results in the expression

$$[\nabla \times \mathbf{v}^{(2)}]_o = \mathbf{i}_\nu \frac{1}{\pi} \sum_{k=-\infty}^{\infty} \int_0^\infty \left[2\lambda\pi_k(\lambda)I_k'(\lambda b) - \frac{k\omega_k(\lambda)I_k(\lambda b)}{b}\right] d\lambda \tag{7-3.75}$$

where we have noted that $\Phi = 0$ at the sphere center, and therefore

$$[\mathbf{i}_\phi]_o = \mathbf{i}_\nu \tag{7-3.76}$$

Substituting in Eq. (7-3.68) and employing both Eq. (7-3.53) and Eq. (7-3.55) eventually gives

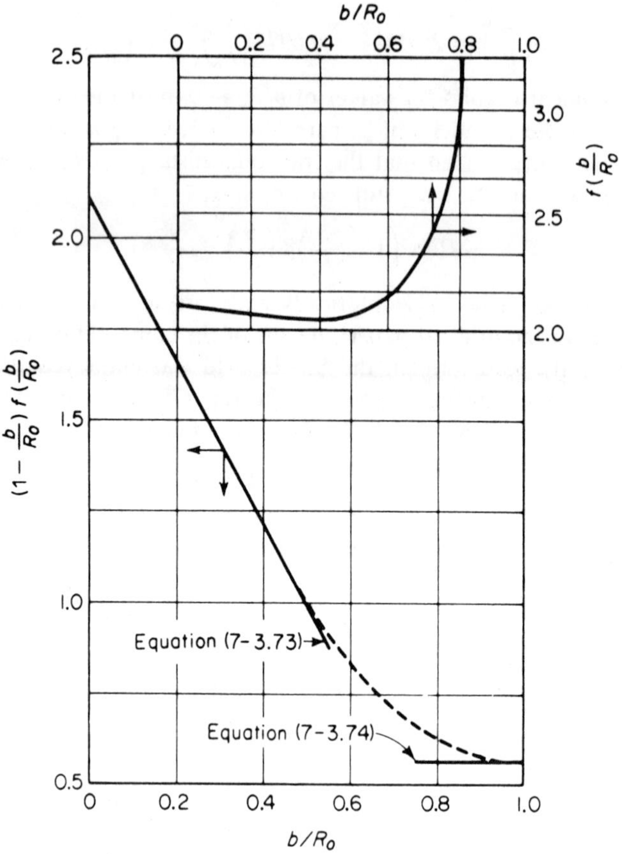

Figure 7-3.3. Eccentricity function for translation of a sphere in a circular cylinder.

$$\mathbf{T}^{(3)} = -\mathbf{i}_y 8\pi\mu a^2 \left[U - U_o\left(1 - \frac{b^2}{R_o^2}\right)\right] g\left(\frac{b}{R_o}\right)\left(\frac{a}{R_o}\right)^2 \tag{7–3.77}$$

where

$$\begin{aligned} g\left(\frac{b}{R_o}\right) = g(\beta) = -\frac{3}{2\pi}\sum_{k=-\infty}^{\infty}\int_0^\infty \Bigg\{ & \frac{\alpha K_k(\alpha) I_k(\alpha\beta) I_k'(\alpha\beta)}{I_k(\alpha)} \\ & + \delta_k(\alpha)\left[I_k'(\alpha\beta) - \frac{k^2 I_k(\alpha) I_k(\alpha\beta)}{\alpha^2 \beta I_k'(\beta)}\right]\left[\frac{I_k(\alpha\beta) I_k'(\alpha)}{I_k(\alpha)} - \beta I_k'(\alpha\beta)\right]\Bigg\} d\alpha \end{aligned} \tag{7–3.78}$$

The function $\delta_k(\alpha)$ is defined in Eq. (7–3.72).

For small values of the eccentricity b/R_o, numerical integration of the foregoing yields

$$g\left(\frac{b}{R_o}\right) = 1.296\left(\frac{b}{R_o}\right) + O\left(\frac{b}{R_o}\right)^3 \tag{7–3.79}$$

and the expansion proceeds in odd powers of b/R_o. As the wall is approached, it appears likely from Eqs. (7–3.100) and (7–3.106) that

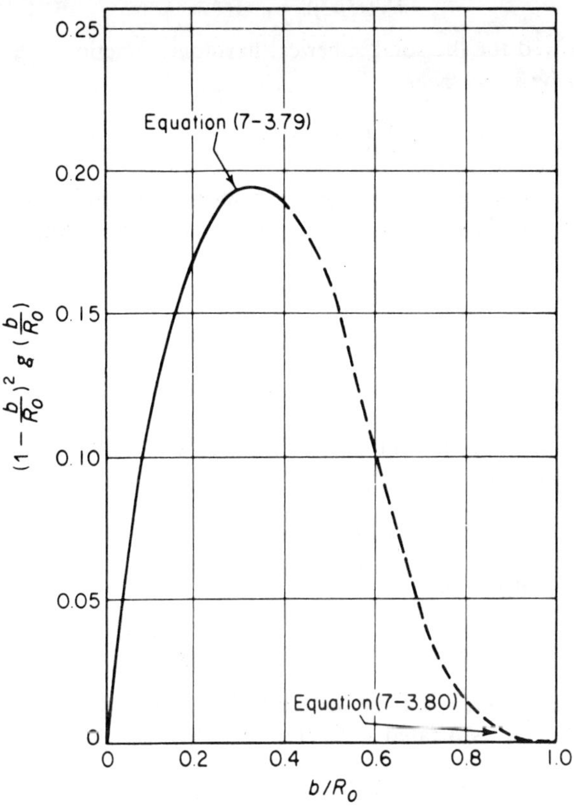

Figure 7-3.4. Eccentricity function for the torque on a sphere in a circular cylinder.

$$\lim_{b/R_0\to 1}\left(1-\frac{b}{R_0}\right)^2 g\left(\frac{b}{R_0}\right)=0 \qquad (7\text{–}3.80)$$

A preliminary graph of $[1-(b/R_0)]^2\, g(b/R_0)$ versus b/R_0, prepared with the aid of these limiting forms, is shown in Fig. 7–3.4. Here, again, the original paper does not give the correct limiting constant because of failure to use the complete Stokes field.

The fourth reflection, $\mathbf{v}^{(4)}$

In order to obtain the first correction to the *zeroth* approximation for pressure drop it is necessary to consider the field $\mathbf{v}^{(4)}$. Here, again, an exact calculation to the first power in a/R_0 can be made without a corresponding knowledge of $\mathbf{v}^{(4)}$. This can be seen from the following argument: If we resort to Lamb's general solution for the field $\mathbf{v}^{(3)}$ reflected from the sphere, then by analogy to Eq. (7–3.22) we have

$$4\pi\nabla(r^3 p_{-2}^{(3)}) = -\mathbf{F}^{(3)} \tag{7–3.81}$$

This can be solved for the solid spherical harmonic function, $p_{-2}^{(3)}$, by multiplying scalarly by $\mathbf{r}$; whence,

$$4\pi r\frac{\partial}{\partial r}(r^3 p_{-2}^{(3)}) = -\mathbf{r}\cdot\mathbf{F}^{(3)} \tag{7–3.82}$$

However, $p_{-2}^{(3)}$ is, among other things, a homogeneous polynomial in r of order -2. Thus, by Euler's theorem on homogeneous polynomials,

$$r\frac{\partial}{\partial r}(p_{-2}^{(3)}) = -2p_{-2}^{(3)} \tag{7–3.83}$$

This makes

$$p_{-2}^{(3)} = -\frac{1}{4\pi r^3}\mathbf{r}\cdot\mathbf{F}^{(3)} \tag{7–3.84}$$

Using the expression for $\mathbf{F}^{(3)}$ in Eq. (7–3.70) and noting that $\mathbf{r}\cdot\mathbf{i}_z = z$ yields the harmonic function,

$$\frac{1}{\mu}p_{-2}^{(3)} = Jr^{-2}\cos\theta \tag{7–3.85}$$

where we have written

$$J = \frac{3}{2}a\left[U - U_o\left(1 - \frac{b^2}{R_o^2}\right)\right]f\left(\frac{b}{R_o}\right)\left(\frac{a}{R_o}\right) \tag{7–3.86}$$

As in Eqs. (7–3.26) and (7–3.27), those parts of the entire velocity and pressure fields, $\mathbf{v}^{(3)}$ and $p^{(3)}$, associated with this particular harmonic function are

$$\mathbf{v}^{(3)} \approx \frac{1}{2\mu}\nabla(r^2 p_{-2}^{(3)}) + \frac{\mathbf{r}}{\mu}p_{-2}^{(3)} \tag{7–3.87}$$

and

$$p^{(3)} \approx p_{-2}^{(3)} \tag{7–3.88}$$

Of all the terms in Lamb's general solution, the only harmonic function which introduces the sphere radius a to the first power in the velocity and pressure fields is the p_{-2} harmonic. This is the dominant term as $r \to \infty$. Thus, as regards powers of a and, ultimately, in the next reflection, powers of a/R_o, this is the only term which need be retained to obtain results which are correct to the first power in a/R_o. To this degree of approximation, Eqs. (7–3.87) and (7–3.88) may be regarded as exact representations of the velocity and pressure fields. Comparing these expressions to those given in Eqs. (7–3.26)–(7–3.29), we find

$$\mathbf{v}^{(3)} = \frac{J}{H}\mathbf{v}^{(1)} \tag{7–3.89}$$

and

$$p^{(3)} = \frac{J}{H}p^{(1)} \tag{7–3.90}$$

To the same degree of approximation, it is easy to see that

$$\mathbf{v}^{(4)} = \frac{J}{H}\mathbf{v}^{(2)} \tag{7–3.91}$$

and

$$p^{(4)} = \frac{J}{H}p^{(2)} \tag{7–3.92}$$

This immediately leads to the result

$$(\Delta P_3 + \Delta P_4) = \frac{J}{H}(\Delta P_1 + \Delta P_2) \tag{7–3.93}$$

With the values for J and H, and the expression for the pressure drop due to the first two reflected fields quoted in Eq. (7–3.64), this becomes

$$(\Delta P_3 + \Delta P_4) = \frac{12\mu a}{R_0^2}\left(1 - \frac{b^2}{R_0^2}\right)\left[U_0\left(1 - \frac{b^2}{R_0^2}\right) - U\right]f\left(\frac{b}{R_0}\right)\left(\frac{a}{R_0}\right) \tag{7–3.94}$$

The fourth reflected field can have no singularities within the volume occupied by the sphere, so that

$$\mathbf{F}^{(4)} = \mathbf{0}, \quad \text{and} \quad \mathbf{T}^{(4)} = \mathbf{0} \tag{7–3.95}$$

Final results for off-center sphere

Upon summing the individual results for drag, torque, and pressure drop we find

$$\mathbf{F} = \mathbf{i}_z 6\pi\mu a\left[U_0\left(1 - \frac{b^2}{R_0^2}\right) - U\right]\left[1 + f\left(\frac{b}{R_0}\right)\left(\frac{a}{R_0}\right) + \cdots\right] \tag{7–3.96}$$

$$\mathbf{T} = \mathbf{i}_y 8\pi\mu a^2\left[U_0\left(\frac{b}{R_0}\right)\left(\frac{a}{R_0}\right) + \left\{U_0\left(1 - \frac{b^2}{R_0^2}\right) - U\right\}g\left(\frac{b}{R_0}\right)\left(\frac{a}{R_0}\right)^2 + \cdots\right] \tag{7–3.97}$$

and

$$\Delta P_S = \frac{12\mu a}{R_0^2}\left(1 - \frac{b^2}{R_0^2}\right)\left[U_0\left(1 - \frac{b^2}{R_0^2}\right) - U\right]\left[1 + f\left(\frac{b}{R_0}\right)\left(\frac{a}{R_0}\right) + \cdots\right] \tag{7–3.98}$$

These expressions are correct to the highest powers of a/R_0 quoted. The case of a sedimenting sphere in a quiescent fluid is obtained by putting $U_0 = 0$ in the foregoing. In the absence of detailed information on higher-order reflections, a more accurate form of Eqs. (7–3.96) and (7–3.98) would be obtained by placing the eccentricity term in the denominator with a change in sign of $f(b/R_0)\, a/R_0$, equivalent to a geometric series summation for higher reflections; that is,

$$1 + f\left(\frac{b}{R_0}\right)\frac{a}{R_0} + \cdots \approx \frac{1}{1 - f(b/R_0)a/R_0}$$

A form of the preceding equations suitable for examining situations in which the sphere is near the container walls can be obtained by expressing the previous results in terms of the ratio of sphere radius to minimum distance (of the sphere center) from the wall, $a/(R_o - b)$. This results in

$$\mathbf{F} = \mathbf{i}_z 6\pi\mu a\left[U_o\left(1 - \frac{b^2}{R_o^2}\right) - U\right]\left[1 + \left(1 - \frac{b}{R_o}\right)f\left(\frac{b}{R_o}\right)\left(\frac{a}{R_o - b}\right) + \cdots\right] \tag{7-3.99}$$

$$\begin{aligned}\mathbf{T} = \mathbf{i}_y 8\pi\mu a^2\Big[&U_o\frac{b}{R_o}\left(1 - \frac{b}{R_o}\right)\left(\frac{a}{R_o - b}\right)\\ &+ \left\{U_o\left(1 - \frac{b^2}{R_o^2}\right) - U\right\}\left(1 - \frac{b}{R_o}\right)^2 g\left(\frac{b}{R_o}\right)\left(\frac{a}{R_o - b}\right)^2 + \cdots\Big]\end{aligned} \tag{7-3.100}$$

$$\begin{aligned}\Delta P_S = \frac{12\mu a}{R_o^2}\left(1 - \frac{b^2}{R_o^2}\right)&\left[U_o\left(1 - \frac{b^2}{R_o^2}\right) - U\right] \times\\ &\times \left[1 + \left(1 - \frac{b}{R_o}\right)f\left(\frac{b}{R_o}\right)\left(\frac{a}{R_o - b}\right) + \cdots\right]\end{aligned} \tag{7-3.101}$$

As previously observed, the quantities $[1 - (b/R_o)]f(b/R_o)$ and $[1 - (b/R_o)]^2 g(b/R_o)$ attain limiting values $\frac{9}{16}$ and 0, respectively, as $b/R_o \to 1$. It should be noted that in these cases we assume

$$\frac{a}{R_o - b} \ll 1 \tag{7-3.102}$$

It appears from the results of this investigation that both the drag and pressure drop are even functions of the eccentricity, b/R_o. This is to be expected, since symmetry requires that if the sphere is moved from its present location, $R = b$, to the opposite side of the cylinder axis, $R = -b$, neither the direction nor the magnitude of the drag should be altered. Likewise, the pressure drop should be unaffected by this transition, in accord with present calculations. On the other hand, Eq. (7-3.97) shows that the torque is an odd function of b/R_o. This conclusion is correct, since the sphere will tend to rotate in a direction opposite to its original direction, without alteration in magnitude, when it is placed in mirror-image position on the opposite side of the cylinder axis. Of course, at the cylinder axis we have

$$\mathbf{T} = \mathbf{0} \tag{7-3.103}$$

Further support for the validity of the present results is provided by observing that, in the absence of external forces acting on the sphere, the viscous pressure drop due to the presence of a particle in the original field of flow is an essentially positive quantity. This stems from a one-to-one correspondence between pressure drop in rectilinear flow and energy dissipation. Thus, Eq. (7-3.98) shows that, for net flow in the $+z$ direction, a positive pressure drop implies the inequality

$$U_o\left(1 - \frac{b^2}{R_o^2}\right) \geqq U$$

Inasmuch as $U_o(1 - b^2/R_o^2)$ is the *local* fluid velocity, this shows that the sphere necessarily lags the fluid, an obviously correct inference. Moreover, Eq. (7–3.96) correctly indicates that in these circumstances the frictional force on the sphere is always in the direction of net flow, and the sphere is thus dragged along by the fluid.

An interesting conclusion to be drawn from the present calculations is that the drag experienced by a small sphere sedimenting in a quiescent fluid does not increase monotonically as we proceed outward from the cylinder axis towards the wall. Rather, it attains a minimum value at some intermediate point. This can be seen quite clearly from Eq. (7–3.96) and Fig. 7–3.3. In present circumstances, the former has the form

$$\mathbf{F} = -\mathbf{i}_z 6\pi\mu a U\left[1 + f\left(\frac{b}{R_o}\right)\left(\frac{a}{R_o}\right) + \cdots\right] \tag{7–3.104}$$

As we proceed away from the axis the function $f(b/R_o)$, and hence the drag, decreases initially, as required by Eq. (7–3.73); however, as the wall is approached, Eq. (7–3.74) shows that $f(b/R_o)$ ultimately increases like $1/(1 - b/R_o)$. There is thus an intermediate value of eccentricity at $b/R_o \approx 0.4$ for which the drag is least. Experimental confirmation of this location of minimum resistance is difficult because it is small. Data of Craig[12] and Bart[1] confirm that very little increase in resistance to sedimentation of a single particle occurs until an eccentricity of $b/R_o \approx 0.6$ is exceeded.

The results for the sedimentation of a small particle through a quiescent fluid near the cylinder wall are exactly the same as for a sphere moving in the positive z direction in the vicinity of a plane wall. Lorentz'[42] results for the latter problem are

$$\mathbf{F} = -\mathbf{i}_z 6\pi\mu a U\left[1 + \frac{9}{16}\left(\frac{a}{h}\right) + \cdots\right] \tag{7–3.105}$$

whereas Faxen gives, in his thesis[15],

$$\mathbf{T} = \mathbf{i}_y 8\pi\mu a^2 U\left[\frac{3}{32}\left(\frac{a}{h}\right)^4 - \frac{9}{256}\left(\frac{a}{h}\right)^5 + \cdots\right] \tag{7–3.106}$$

where h is the perpendicular distance from the sphere center to the wall, and the positive x direction is from the sphere center normal to the wall. It is emphasized that for these relationships to apply, both a/h and a/R_o must be small. These should be compared with Eqs. (7–3.99) and (7–3.100), respectively, which, when we set $U_o = 0$ and utilize Eqs. (7–3.74) and (7–3.80), take the following forms near the cylinder walls:

$$\mathbf{F} = -\mathbf{i}_z 6\pi\mu a U\left[1 + \frac{9}{16}\left(\frac{a}{R_o - b}\right) + \cdots\right] \tag{7–3.107}$$

and

$$\mathbf{T} = o\left(\frac{a}{R_o - b}\right)^2 \tag{7–3.108}$$

The distances from the wall, h and $R_o - b$, are comparable.

Treatment of the present problem by means of the creeping motion equations fails to reveal the presence of any sidewise forces tending to move the sphere perpendicular to the tube axis. In any real situation, "Bernoulli" forces tending to produce this result must exist. This lack of sidewise forces is a characteristic failing of the creeping motion equations and stems from a neglect of fluid inertia in the original equations of motion.

Recent calculations by Rubinow and Keller[49] based on a matching procedure using the Stokes and Oseen expansions indicate that a spinning sphere moving in an otherwise stationary fluid experiences a lift force $\mathbf{F}_L$ orthogonal to its direction of motion. This force is given by

$$\mathbf{F}_L = \pi a^3 \rho \mathbf{\Omega} \times \mathbf{U}[1 + O(N_{\text{Re}})] \tag{7–3.109}$$

Here a is the radius of the sphere, $\mathbf{\Omega}$ is its angular velocity, $\mathbf{U}$ is its velocity, ρ is the fluid density, and N_{Re} is the Reynolds number, $N_{\text{Re}} = \rho U a/\mu$. For small values of N_{Re} the transverse force is independent of the viscosity.

Rubinow and Keller then attempted to apply their results to the problem of the lateral motion of a neutrally buoyant sphere freely suspended in a Poiseuillian flow. They replaced $\mathbf{\Omega}$ in Eq. (7–3.109) by one-half the curl of the unperturbed parabolic flow field, that is, the local fluid spin, which is presumably the angular velocity with which a small solid sphere would spin. They also replaced $\mathbf{U}$ in Eq. (7–3.109) by the slip velocity. The latter is obtained by setting $\mathbf{F} = \mathbf{0}$ in Eq. (7–3.24), the magnitude of the slip velocity thus obtained being* $\frac{2}{3}U_o(a/R_o)^2$. According to this interpretation of Eq. (7–3.109), the direction of the lateral force is always such as to drive the sphere to the center of the tube. But this conclusion is at odds with the experiments of Segré and Silberberg[50], according to which the equilibrium position of a spinning, neutrally buoyant sphere is about 0.6 tube radii from the axis.

It is of interest to note that the Oseen equations have been employed by Faxen[15] for the case of a sphere moving along the axis of a cylinder through a quiescent fluid. As we have already noted, however, there is a fundamental objection to using the Oseen equations to estimate inertial effects; and experimental data do not confirm the Faxen solution at the higher Reynolds numbers for which it is supposed to apply.

Thus, Fayon and Happel[23] experimentally investigated the effect of a cylindrical boundary on a spherical particle suspended in a moving viscous liquid. Pressure drop due to motion of fluid past the sphere and drag on the sphere itself were both measured in the particle Reynolds number range from

*Though this formula for the slip velocity is indeed correct, it cannot properly be obtained from Eq. (7–3.24). For the wall-effect term in Eq. (7–3.24) is of $O(a/R_0)^2$, whereas, according to Eq. (7–3.96), there is a more dominant wall-effect term of $O(a/R_0)$. Thus, there is an inconsistency. Further unpublished work by the authors, in which Eq. (7–3.96) was extended to $O(a/R_0)^2$, shows, rather surprisingly, that the preceding formula for the slip velocity is correct.

0.1 to 40.0 based on sphere diameter. Their experiments included data at small eccentricities for which it was not possible to arrive at a value of the eccentricity function $f(b/R_o)$ differing significantly from 2.104. The experimental data were correlated in such a way as to separate wall effect from inertial effect. Good correlation was obtained within the limits of accuracy of the data. McNown[44] also obtained data for the sedimentation of a sphere along the axis of a cylinder, which is well correlated by the relationship developed by Fayon and Happel. For sedimentation this takes the form

$$\frac{F}{6\pi\mu Ua} = \frac{1}{1 - 2.104(a/R_o) + 2.087(a/R_o)^3} + \left(\frac{C_A}{C_S} - 1\right) \qquad (7\text{–}3.110)$$

Here C_A = actual drag coefficient of a sphere in an unbounded medium, C_S = drag coefficient according to Stokes' law = $24/N_{\text{Re}}$ (with N_{Re} based on sphere diameter). Thus $(C_A/C_S) - 1$ is a measure of fractional deviation of the actual drag in an infinite medium from that calculated by Stokes' law. Figure 7–3.5 gives a comparison of McNown's data with Eq. (7–3.110) and Faxen's equation based on a theoretical solution using the Oseen approximation:

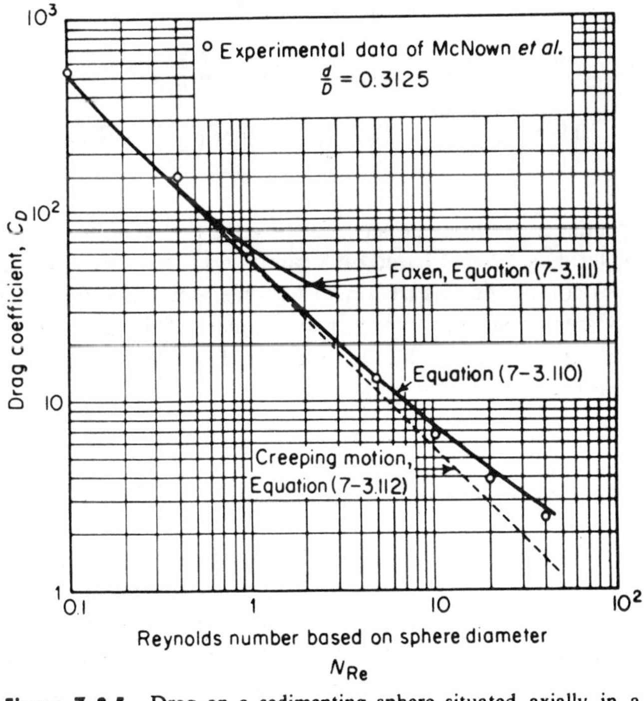

Figure 7-3.5. Drag on a sedimenting sphere situated axially in a circular cylinder.

$$\frac{F}{6\pi\mu Ua} = \frac{1}{1 - \frac{3}{16}N_{Re} - (d/D)f(N_{Re}, D/d) + 2.09(d/D)^3} \quad (7\text{-}3.111)$$

where N_{Re} = Reynolds number based on the diameter of the sphere, and d and D refer, respectively, to sphere and cylinder diameters. Faxen theoretically calculated $f(N_{Re}, D/d)$ as equal to 2.104 for N_{Re} equal to zero. He obtained values for $f(N_{Re}, D/d)$ which decreased rapidly as N_{Re} increased. Examination of Fig. 7–3.5 indicates that too much weight is given to the effect of increasing Reynolds number on the drag coefficient. This conclusion is also supported by the data of Fidleris and Whitmore[25] at $N_{Re} = 100$. These authors investigated wall effects from $N_{Re} = 0.05$ to 20,000 (based on sphere diameter).

Sphere in axial position

When a sphere is at the cylinder axis, $b/R_o = 0$, Eqs. (7–3.96)–(7–3.98) reduce to the first approximation of more accurate results previously given by Faxen[15], Happel and Byrne[30], and Wakiya[57].

Bohlin[3], using an extension of the method of reflections as originally developed by Faxen, carried the approximation further for a sphere in the axial position. He gives the formula

$$\mathbf{F} = -\mathbf{i}_z 6\pi\mu a U K_1 \quad (7\text{-}3.112)$$

where
$$K_1 = 1 \Big/ \Big[1 - 2.10443\left(\frac{a}{R_o}\right) + 2.08877\left(\frac{a}{R_o}\right)^3 - 0.94813\left(\frac{a}{R_o}\right)^5 - 1.372\left(\frac{a}{R_o}\right)^6 + 3.87\left(\frac{a}{R_o}\right)^8 - 4.19\left(\frac{a}{R_o}\right)^{10} + \cdots \Big] \quad (7\text{-}3.113)$$

In his appendix to Bohlin's paper[3], Faxen gives the value for the coefficient of the a/R_o term as 2.10444 based on a careful recalculation, and suggests the possibility of combining this treatment with his previous work, based on the use of the Oseen equations, to obtain an estimate of the effect of inertia.

For the case where the liquid moves in Poiseuille flow with an axial approach velocity U_o, Bohlin obtains

$$\mathbf{F} = -\mathbf{i}_z 6\pi\mu a(UK_1 - U_o K_2) \quad (7\text{-}3.114)$$

where
$$K_2 = K_1 \Big[1 - \frac{2}{3}\left(\frac{a}{R_o}\right) - 0.1628\left(\frac{a}{R_o}\right)^3 - 0.4059\left(\frac{a}{R_o}\right)^7 + 0.5236\left(\frac{a}{R_o}\right)^9 + 1.51\left(\frac{a}{R_o}\right)^{10} + \cdots \Big] \quad (7\text{-}3.115)$$

Haberman and Sayre[27] also considered the axisymmetric case for "large" a/R_o. They employed general solutions of the creeping motion equations in terms of the stream function for both cylindrical and spherical coordinate systems. For the satisfaction of the boundary conditions on the cylinder walls, the cylindrical coordinate solution for the stream function is used. The expression thus obtained represents the flow inside a circular cylinder, not as yet fully specified but satisfying the boundary conditions on the cylinder. This expression is then transformed into spherical coordinates. By comparing termwise the constants in the foregoing expression with the constants in the stream function expansion obtained directly in spherical coordinates, a relationship between the constants is obtained. The boundary conditions on the sphere yield a relationship between the constants in the spherical coordinate solution. By substituting the previous relationships into those obtained from the boundary conditions on the sphere, an infinite set of linear algebraic equations is obtained for evaluating the coefficients appearing in the expansion of the stream function.

Wall correction factors for rigid spheres moving within a still liquid inside an infinitely long cylinder were determined by numerically solving the algebraic system. The number of equations employed was systematically increased (at most up to eight) until only minor changes occurred in the values of the constants. A very good approximation for the drag was obtained by retaining only the first two equations of the infinite set, corresponding to Eq. (7–3.114) with the following constants:

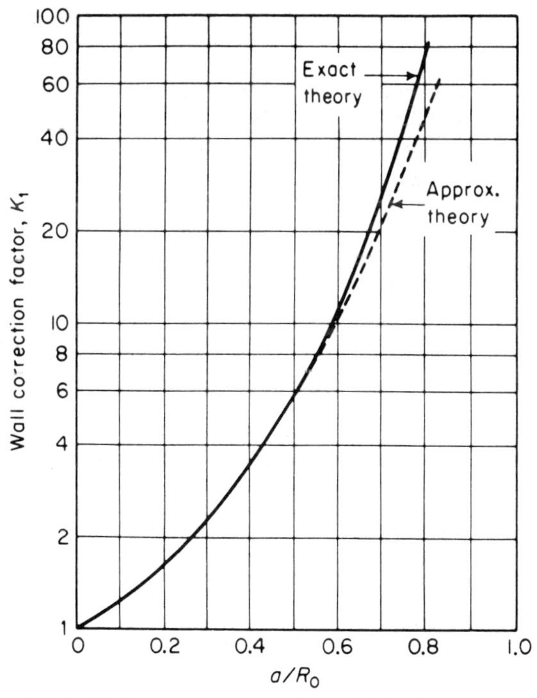

Figure 7-3.6. Wall correction factor for a rigid sphere moving axially in a circular cylindrical container.

$$K_1 = \frac{1 - 0.75857(a/R_o)^5}{1 - 2.1050(a/R_o) + 2.0865(a/R_o)^3 + 0.72603(a/R_o)^6} \tag{7-3.116}$$

and

$$K_2 = \frac{1 - (2/3)(a/R_o)^2 - 0.20217(a/R_o)^5}{1 - 2.1050(a/R_o) + 2.0865(a/R_o)^3 - 1.7068(a/R_o)^5 + 0.72603(a/R_o)^6} \tag{7-3.117}$$

Actual numerical values based on the complete set of equations give more accurate results for values of $a/R_o > 0.6$. Figure 7–3.6 gives a comparison of values obtained by Haberman's approximate and exact solutions. Fidleris and Whitmore[25] obtained data up to $a/R_o = 0.6$ with experimental values of K_1 several per cent lower than those indicated in Table 7–3.2 at the higher values of a/R_o. These differences are perhaps due to end effects. Tables 7–3.2 and 7–3.3 provide a comparison of K_1 and K_2 as determined by Haberman and Bohlin.

Bohlin's values for K_1 lie between Haberman's formula and his "exact values" up to $a/R_o = 0.6$. Thus the agreement with Haberman's method is very good for the interval $0 < a/R_o < 0.6$ where Bohlin's formula should be

TABLE 7–3.2
WALL CORRECTION FACTOR (K_1) FOR RIGID SPHERES MOVING IN A STILL LIQUID IN A CYLINDRICAL TUBE

a/R_o	Haberman Exact Theory	Haberman Eq. (7–3.116)	Bohlin Eq. (7–3.113)
0.0	1.000	1.000	1.000
0.1	1.263	1.263	1.263
0.2	1.680	1.680	1.680
0.3	2.371	2.370	2.370
0.4	3.596	3.582	3.588
0.5	5.970	5.871	5.923
0.6	11.135	10.591	11.057
0.7	24.955	21.406	(36.598)
0.8	73.555	48.985	(−11.757)

TABLE 7–3.3
WALL CORRECTION FACTOR (K_2) FOR FIXED RIGID SPHERES IN POISEUILLE FLOW

a/R_o	Haberman Exact Theory	Haberman Eq. (7–3.117)	Bohlin Eq. (7–3.115)
0.0	1.000	1.000	1.000
0.1	1.255	1.255	1.255
0.2	1.635	1.635	1.635
0.3	2.231	2.231	2.227
0.4	3.218	3.218	3.198
0.5	5.004	4.973	4.902
0.6	8.651	8.377	8.298
0.7	17.671	15.686	(24.765)
0.8	43.301	33.056	...

valid. For higher values of a/R_0 the term $(a/R_0)^{11}$ is no longer negligible and probably accounts for the impossible value of K_1 at $a/R_0 = 0.8$.

Agreement between the K_2 values tabulated in Table 7–3.3 shows similarly good agreement for $a/R_0 < 0.6$.

Haberman and Sayre also considered the case of *fluid* particles moving within a Poiseuille flow, neglecting the effect of surface tension in the stress equations. They showed that the assumption of a spherical shape for a fluid drop moving inside a cylinder cannot lead to an exact solution, although in many cases it appears to be a good approximation on the basis of experimental data they obtained. These authors also studied the motion of a sphere at the instant it passes the center of a spherical container, discussed in Section 4–22. This case is of interest as an upper bound for the resistance to motion within a cylindrical container, since the wall effects for spherical boundaries exceed these for infinite cylinders of equal radii. This problem is not strictly a steady state one, in contrast to the axial fall of a sphere in an infinitely long cylinder.

Famularo[14] also considered the case of a spherical particle settling inside a spherical container, using the method of reflections. A first reflection was obtained for the case where the particle may occupy any position inside the container. Lamb's[39] solution of the creeping motion equations in terms of spherical harmonics (see Section 3–2) were employed for this purpose, utilizing coordinate transformations similar to those discussed earlier in this section. To facilitate computations the instantaneous motion of the particle at an arbitrary point was resolved into (a) a motion through the center of the spherical container, and (b) a motion perpendicular to the direction of the latter. In the special case where the motion is axially symmetric, it is possible to obtain an *exact* solution in bipolar coordinates for any ratio of inner to outer sphere radii.

Smythe[53] studied the potential flow of an *ideal* fluid through a tube containing a concentric spherical obstacle. The vector potential between the sphere and cylinder was found for ratios of sphere to tube radii, a/R_0, from 0 to 0.95. Haberman[28,29] also considered this and similar potential problems related to a sphere in a circular cylindrical container.

Happel and Ast[32] studied the axial motion of a rigid sphere in a "frictionless" cylinder on the basis of the creeping motion equations. The theoretical development follows that employed by Haberman. Sphere-to-cylinder radii ratios from 0 to 0.7 were investigated. Solution of the problem is carried out under the assumption that the fluid shear on the cylinder walls is everywhere equal to zero. This model is discussed further in Section 8–4 as a basis for theoretical studies of assemblages of particles.

To date, no studies are available on motion of a sphere perpendicular to the axial direction in a cylinder. Such motion can occur with single spheres if the cylinder axis is not parallel to the gravity field. The forces developed when more than one sphere is present in a tube can also lead to sidewise components.

7-4 Sphere Moving Relative to Plane Walls

The motion of a sphere parallel to a single plane wall is of interest as the limiting case of motion of a small sphere in a cylindrical container when the sphere approaches the cylinder wall. This problem and the more general one of the motion of a sphere parallel to two external plane walls was treated some time ago by Faxen[15]. Extensions of the theory to nonspherical bodies, and to shear and parabolic flows, have been developed by generalization of Faxen's original technique.

In order to illustrate the basic technique involved, we will first consider the general problem of a sphere moving between, and parallel to, two parallel walls extending indefinitely in a viscous fluid. Following this, results for various special cases will be given.

The basis of Faxen's[15] method is to express the fundamental solution of Laplace's equation

$$\nabla^2 p = 0 \tag{7–4.1}$$

which is

$$p = -\frac{1}{4\pi r} \tag{7–4.2}$$

in integral form. This is done by expressing $1/r$ in the following form:

$$\frac{1}{r} = \frac{1}{2\pi}\int_{-\infty}^{+\infty}\int_{-\infty}^{+\infty} e^{i(\alpha x+\beta y)-k|z|}\frac{d\alpha\, d\beta}{k} \tag{7–4.3}$$

where $i = \sqrt{-1}$. Here, α and β are dummy variables of integration and $k = \sqrt{\alpha^2 + \beta^2}$. This function and its cartesian partial derivatives may be employed to express Stokes' original solution for a sphere in an infinite medium entirely in terms of a cartesian coordinate system.

Sphere moving parallel to one or two stationary parallel walls

The creeping motion equations in cartesian coordinates assume the form

$$\mu\nabla^2 u = \frac{\partial p}{\partial x}$$
$$\mu\nabla^2 v = \frac{\partial p}{\partial y} \tag{7–4.4}$$
$$\mu\nabla^2 w = \frac{\partial p}{\partial z}$$

$$\frac{\partial u}{\partial x} + \frac{\partial v}{\partial y} + \frac{\partial w}{\partial z} = 0 \tag{7–4.5}$$

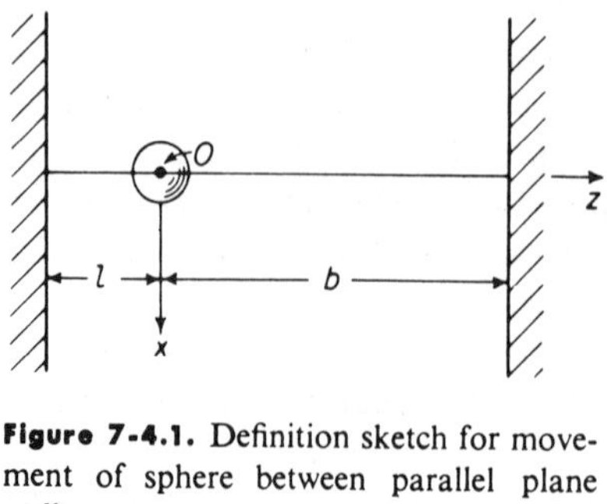

Figure 7-4.1. Definition sketch for movement of sphere between parallel plane walls.

The velocity field is to satisfy the conditions $\mathbf{v} = \mathbf{U}$ at the sphere surface and $\mathbf{v} = \mathbf{0}$ on the plane wall(s). See the definition sketch, Fig. 7–4.1.

Based on the expression for $1/r$ in Eq. (7–4.3), Faxen[15,16] established that a general solution of Eqs. (7–4.4)–(7–4.5) is

$$u = \frac{1}{2\pi}\int_{-\infty}^{+\infty}\int_{-\infty}^{+\infty} e^{i(\alpha x+\beta y)-k|z|} \left\{\frac{i\alpha}{k}g_2 + \frac{2g_1}{k} - \frac{g_1\alpha^2}{k^3}(k|z|+1) + \frac{\alpha^2 z}{k}g_3\right\}d\alpha\, d\beta \tag{7–4.6}$$

$$v = \frac{1}{2\pi}\int_{-\infty}^{+\infty}\int_{-\infty}^{+\infty} e^{i(\alpha x+\beta y)-k|z|}\, i\beta \left\{\frac{g_2}{k} + \frac{i}{k^3}g_1(k|z|+1) - \frac{i\alpha x}{k}g_3\right\}d\alpha\, d\beta \tag{7–4.7}$$

$$w = \frac{1}{2\pi}\int_{-\infty}^{+\infty}\int_{-\infty}^{+\infty} e^{i(\alpha x+\beta y)-k|z|} \left\{-\frac{z}{|z|}g_2 - \frac{zi\alpha}{k}g_1 + \frac{i\alpha}{k}g_3(k|z|+1)\right\}d\alpha\, d\beta \tag{7–4.8}$$

and

$$p = \frac{\mu}{\pi}\int_{-\infty}^{+\infty}\int_{-\infty}^{+\infty} e^{i(\alpha x+\beta y)-k|z|}\frac{i\alpha}{k}\left\{-g_1 + \frac{z}{|z|}kg_3\right\}d\alpha\, d\beta \tag{7–4.9}$$

In general, the quantities g_1, g_2, and g_3 may be any functions of α and β. This representation is valuable because it is naturally adapted to satisfy boundary conditions on *both* spherical and plane boundaries. The problem is to establish the values of g_1, g_2, and g_3 satisfying the appropriate boundary conditions.

The method of reflections is utilized. First, the motion of a sphere in an unbounded medium is expressed in the preceding form. This may be accomplished by setting

$$\begin{aligned} g_1 &= \tfrac{3}{4}aU = c, \text{ say} \\ g_2 &= -\tfrac{1}{3}ci\alpha a^2 \\ g_3 &= 0 \end{aligned} \tag{7–4.10}$$

where $\mathbf{U} = \mathbf{i}_x U$.

Next, we wish simultaneously to cancel, on the plane walls, the velocity due to the sphere falling with velocity U. That is, if the basic solution with the functions given by Eq. (7–4.10) is reflected at the walls, we must have at the right wall, $z = b$, and the left wall, $z = -l$, independent solutions of the creeping motion equations which will cancel the original Stokes field. This is accomplished by selecting two additional solutions, each with three arbitrary functions analogous to Eqs. (7–4.6)–(7–4.9). Thus, we define

$$u^* = \frac{1}{2\pi}\int_{-\infty}^{+\infty}\int_{-\infty}^{+\infty} e^{i(\alpha x+\beta y)-kz} \left\{\frac{i\alpha}{k}g_5 + \frac{2}{k}g_4 - \frac{\alpha^2}{k^3}g_4(zk+1) + \frac{z\alpha^2}{k}g_6\right\}d\alpha\, d\beta \tag{7–4.11}$$

$$v^* = \frac{1}{2\pi}\int_{-\infty}^{+\infty}\int_{-\infty}^{+\infty} e^{i(\alpha x+\beta y)-kz} i\beta \left\{\frac{g_5}{k} + \frac{i\alpha}{k^3}g_4(zk+1) - \frac{zi\alpha}{k}g_6\right\} d\alpha\, d\beta \tag{7–4.12}$$

$$w^* = \frac{1}{2\pi}\int_{-\infty}^{+\infty}\int_{-\infty}^{+\infty} e^{i(\alpha x+\beta y)-kz} \left\{-g_5 - \frac{i\alpha z}{k}g_4 + \frac{i\alpha}{k}g_6(zk+1)\right\} d\alpha\, d\beta \tag{7–4.13}$$

$$p^* = \frac{\mu}{\pi}\int_{-\infty}^{+\infty}\int_{-\infty}^{+\infty} e^{i(\alpha x+\beta y)-kz}\frac{i\alpha}{k}\{-g_4 + kg_6\} d\alpha\, d\beta \tag{7–4.14}$$

and
$$u^{**} = \frac{1}{2\pi}\int_{-\infty}^{+\infty}\int_{-\infty}^{+\infty} e^{i(\alpha x+\beta y)+kz} \left\{\frac{i\alpha}{k}g_8 + \frac{2}{k}g_7 + \frac{\alpha^2}{k}g_7(zk-1) + \frac{\alpha^2 z}{k}\right\} d\alpha\, d\beta \tag{7–4.15}$$

$$v^{**} = \frac{1}{2\pi}\int_{-\infty}^{+\infty}\int_{-\infty}^{+\infty} e^{i(\alpha x+\beta y)+kz} i\beta \left\{\frac{1}{k}g_8 + \frac{i\alpha}{k^3}g_7(1-zk) - \frac{zi\alpha}{k}g_9\right\} d\alpha\, d\beta \tag{7–4.16}$$

$$w^{**} = \frac{1}{2\pi}\int_{-\infty}^{+\infty}\int_{-\infty}^{+\infty} e^{i(\alpha x+\beta y)+kz} \left\{g_8 - \frac{zi\alpha}{k}g_7 - \frac{i\alpha}{k}g_9(zk-1)\right\} d\alpha\, d\beta \tag{7–4.17}$$

$$p^{**} = \frac{\mu}{\pi}\int_{-\infty}^{+\infty}\int_{-\infty}^{+\infty} e^{i(\alpha x+\beta y)+kz}\frac{i\alpha}{k}\{-g_7 - kg_9\} d\alpha\, d\beta \tag{7–4.18}$$

We must now determine the g_j functions such that when $z = b$ and $z = -l$,

$$u + u^* + u^{**} = v + v^* + v^{**} = w + w^* + w^{**} = 0 \tag{7–4.19}$$

The vanishing of $u + u^* + u^{**}$ at $z = -l$ requires that

$$\begin{aligned} 0 &= [u + u^* + u^{**}]_{z=-l} \\ &= \frac{1}{2\pi}\int_{-\infty}^{+\infty}\int_{-\infty}^{+\infty} e^{i(\alpha x+\beta y)} \left\{\frac{i\alpha}{k}(g_2 e^{-kl} + g_5 e^{kl} + g_8 e^{-kl})\right. \\ &\quad + \left(\frac{2}{k} - \frac{\alpha^2}{k^3}\right)(g_1 e^{-kl} + g_4 e^{kl} + g_7 e^{-kl}) \\ &\quad - \frac{\alpha^2 l}{k^2}(g_1 e^{-kl} - g_4 e^{kl} + g_7 e^{-kl}) \\ &\quad \left. - \frac{\alpha^2 l}{k}(g_3 e^{-kl} + g_6 e^{kl} + g_9 e^{-kl})\right\} d\alpha\, d\beta \end{aligned} \tag{7–4.20}$$

Since this equation applies for all values of x and y, it follows that the expression in the vinculum must vanish. Two similar equations are obtained from the condition that $v + v^* + v^{**} = 0$ and $w + w^* + w^{**} = 0$ at $z = -l$. Three analogous equations arise from the conditions on the other wall, $z = b$. These six simultaneous equations determine the values of the six functions g_4 to g_9 in terms of the known functions g_1 to g_3. The velocity components arising from the fields $u^* + u^{**}$, $v^* + v^{**}$, $w^* + w^{**}$ may now be evaluated at the center of the spherical particle, and Faxen's law used to evaluate the increased drag due to the presence of the walls.

The force on the particle lies along the x axis and is, in general, given by the expression

$$\mathbf{F} = \frac{-\mathbf{i}_x 6\pi\mu a U}{1 + (3/4)(a/c)(u^* + u^{**})_o + (a^3/8\mu c)[(\partial/\partial x)(p^* + p^{**})]_o} \tag{7-4.21}$$

The subscript "o" indicates that the values of u^*, etc. are to be evaluated at the location of the sphere center. In the present case, we obtain

$$\mathbf{F} = \frac{-\mathbf{i}_x 6\pi\mu a U}{1 - A(a/l) + B(a/l)^3 - C(a/l)^5 + \cdots} \tag{7-4.22}$$

where the constants A and B are as follows[19]:

$$\begin{aligned} A = \frac{3(1-h)}{4}\int_0^\infty \frac{x}{N(st-1)}\{&s^2t^3(1-h)[(1-h)x-1] \\ &+ s^3t^2(1+h)[(1+h)x-1] + 4s^2t^2[2(1-h^2)x^2 - 2x + 1] \\ &+ 2st^2[(3-h^2)x - h] + 2s^2t[(3-h^2)x + h] - 4st[2(1-h^2)x^2 \\ &+ 2x + 1] + s(1-h)[(1-h)x + 1] + t(1+h)[(1+h)x + 1]\}\,dx \\ &+ \frac{9(1-h)}{8}\int_0^\infty \frac{s+t-2}{st-1}\,dx \end{aligned} \tag{7-4.23}$$

$$\begin{aligned} B = \frac{(1-h)^2}{8}\int_0^\infty \frac{x^2}{N}\{&st^2[2(1-h)x-1] + s^2t[2(1+h)x-1] \\ &+ st[(4x-1)^2+1] + t[2(1+h)x+1] + s[2(1-h)x+1] - 2\}\,dx \\ &+ \frac{(1-h)^3}{4}\int_0^\infty \frac{x^3}{(st-1)N}\{s^2t^3(1-h) + s^3t^2(1+h) + s^2t^2(8x-4) \\ &- st^2(8x-2h) - s^2t(8x+2h) + st(8x+4) - s(1-h) - t(1+h)\}\,dx \\ &- \frac{(1-h)^3}{8}\int_0^\infty x^2\frac{s+t-2}{st-1}\,dx \end{aligned} \tag{7-4.24}$$

Here, $N = (st-1)^2 - 16x^2st$, $s = e^{2x(1-h)}$, $t = e^{2x(1+h)}$, and $k^2 = \alpha^2 + \beta^2$. x is a new variable of integration, not to be confused with the corresponding cartesian coordinate. Finally, $l = (1-h)L$, and $b = (1+h)L$, where $2L$ is the distance between the two parallel walls.

Faxen did not develop a general expression for the constant C in terms of a single variable of integration. He applied these formulas for the case

$h = \frac{1}{2}$, corresponding to $l = \frac{1}{2}L$ and $b = \frac{3}{2}L$, so that the distance to one wall is three times as great as to the other wall; that is, $b = 3l$. He also evaluated the constant C numerically for this case. In addition, he carried the approximation further by an additional reflection of the field developed at the sphere to the two parallel walls, thus obtaining two additional correction terms for the resistance formula.

Thus, for the case of a sphere falling at the position $b = 3l$, Faxen obtained the following resistance formula:

$$\mathbf{F} = \frac{-\mathbf{i}_x 6\pi\mu a U}{1 - 0.6526(a/l) + 0.1475(a/l)^3 - 0.131(a/l)^4 - 0.0644(a/l)^5 + O(a/l)^6} \tag{7–4.25}$$

Here, the constants corresponding to Eq. (7–4.22) are $A = 0.6526$ as calculated from Eq. (7–4.23); $B = 0.1475$ as calculated from Eq. (7–4.24); $C = 0.0644$ calculated directly from the relationships derived from the first reflected field. The coefficient 0.131 of the $(a/l)^4$ term is derived from the second reflected field. The latter field would, of course, also give rise to a correction term in $(a/l)^6$, but this term is neglected in Faxen's treatment.

A sphere located such that $b = 3l$ will rotate about an axis which is perpendicular to the direction of fall and parallel to the walls. The direction of rotation is opposite to that which would occur if the sphere touched the nearer wall (see Fig. 7–4.2). If $(a/l)^2$ is small compared with unity the rate of rotation is given by

$$\omega = \frac{0.025aU}{l^2[1 - 0.6526(a/l)]} \tag{7–4.26}$$

If the sphere is prevented from rotating it will experience a couple about the y axis of magnitude $|T_y| = 8\pi\mu a^3\omega$. Naturally, if the sphere were moved sufficiently close to the nearer wall, its direction of rotation would change.

For the special case where the sphere lies midway between the two plane

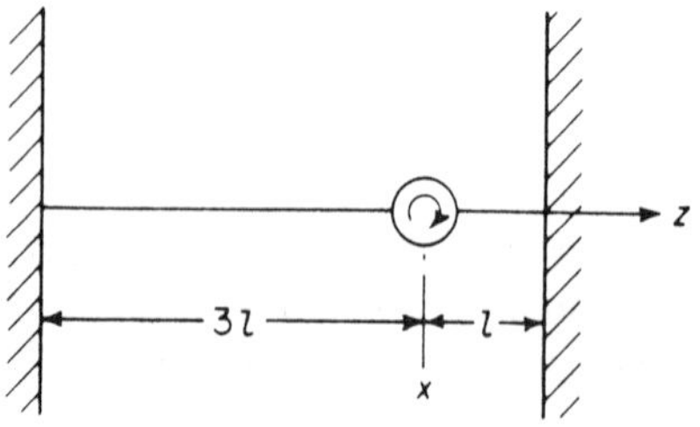

Figure 7-4.2. Direction of rotation of a sphere settling in eccentric position between parallel walls.

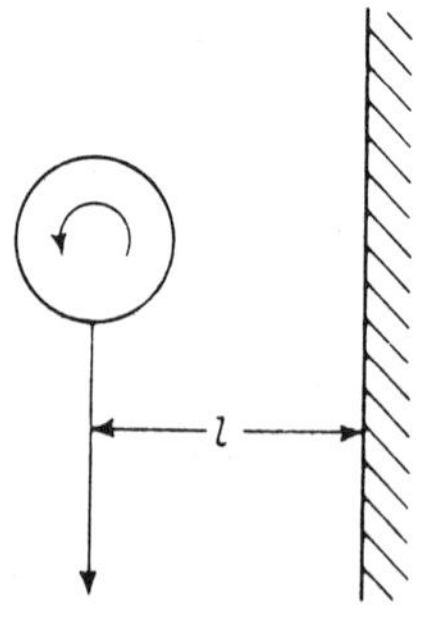

Figure 7-4.3. Direction of rotation of a sphere settling in the presence a single plane wall.

walls Eq. (7–4.22) becomes indeterminate. The general method simplifies considerably for this case, however, and Faxen[17] obtains

$$\mathbf{F} = \frac{-\mathbf{i}_x 6\pi\mu a U}{1 - 1.004(a/l) + 0.418(a/l)^3 + 0.21(a/l)^4 - 0.169(a/l)^5} \tag{7–4.27}$$

where $2l$ is the distance between walls. Of course, no torque is experienced by a sphere sedimenting in the midposition.

The motion of a sphere parallel to a *single* plane wall may be treated more simply than when two plane walls are involved (Fig. 7–4.3). Only one additional arbitrary solution, similar to Eqs. (7–4.11)–(7–4.14), corresponding to three additional functions g_4, g_5, and g_6, is now required instead of six additional functions as is the case with two plane walls. Faxen[15] worked this case out in his dissertation and obtained, for a sphere and a single plane wall,

$$\mathbf{F} = \frac{-\mathbf{i}_x 6\pi\mu a U}{1 - (9/16)(a/l) + (1/8)(a/l)^3 - (45/256)(a/l)^4 - (1/16)(a/l)^5} \tag{7–4.28}$$

This expression includes the effects of a second reflection. If the sphere is free to rotate it will do so at an angular velocity

$$\omega = \frac{3U}{32a}\left(\frac{a}{l}\right)^4\left(1 - \frac{3}{8}\frac{a}{l}\right) \tag{7–4.29}$$

in the same direction as if it were rolling along the wall. In these relationships, Faxen also obtained corrections for the inertial effects on the basis of the Oseen equations. This inertial correction is subject to the same basic criticism as presented previously in the case of a sphere falling in a cylindrical container (Section 7–3).

The Lorentz resistance formula, discussed in Section 3–5, is a special case of Eq. (7–4.28) in which only the coefficient of the term of $O(a/l)$ is given explicitly.

Wakiya[63] considered the influence of a single plane wall on the *unsteady* motion of a sphere parallel to it whose initial velocity is zero. He concludes that the steady state is attained more rapidly in the presence of a wall than without it (see Section 7–9). This solution is of interest in that no other treatments of unsteady flow in the presence of boundaries appear to be available.

One might be tempted to assume that, in the limit, cases for two walls could be built up in a simple fashion from the solution for a single plane wall and a sphere, as suggested by Oseen[41, p. 144], by separately adding the effects. This leads to the formula

$$\mathbf{F} = \frac{-\mathbf{i}_x 6\pi\mu a U}{1 - (9a/16)[(1/l_1) + (1/l_2)]} \tag{7–4.30}$$

where l_1 and l_2 are the distances of the middle of the sphere from each wall. Thus, this procedure results in $\frac{3}{4}$ (a/l) for the lead term of the wall-correction in Eq. (7–4.25), rather than the correct value of 0.6526 (a/l); similarly we would obtain $\frac{9}{8}\,(a/l)$ for the lead correction term in Eq. (7–4.27) instead of 1.004 (a/l). This approximation procedure obviously gives too large a correction.

The remarks of the preceding paragraph emphasize that the method of reflections cannot be employed in multiple boundary problems, using only one reflection, when some of the boundaries are close together or, what amounts to the same thing, if they are infinite in extent. Thus, we can correctly obtain the influence of two plane walls on the resistance of a sphere only by considering the effect of the two walls *simultaneously*. Similarly, in the problem of *two* spheres close together in the presence of, say, a distant plane wall, it would be necessary to *simultaneously* reflect velocity fields from the surface of both spheres. This could be done, for example, in bipolar coordinates.

In his final paper on the plane wall problem, Faxen[19] again considered the case of a sphere falling parallel to, and between, two parallel walls. He assumed that the sphere was free to move back and forth between the walls, and that all distances from sphere to wall were equally probable. Equations (7–4.23) and (7–4.24) were employed to obtain expressions for average values of the coefficients, and an average sedimentation velocity thereby calculated for a particle which Brownian motion causes to move sidewise.

A sphere in a shearing flow between two parallel walls

Wakiya[58,59], using the general method developed by Faxen, considered the case of a sphere in a shear flow between two parallel planes. In one case the plates were assumed stationary and a Poiseuille flow occurred between them. In the other case, one plane was held stationary and the other was moved parallel to itself to produce a uniform shear field between the two plates. In all cases the motion of the sphere was assumed to be parallel to the walls.

For the two-dimensional Poiseuille flow in which the sphere is placed eccentrically at a location $l = \frac{1}{2}L$—one-fourth the distance between the plates, as in Faxen's numerical example illustrated by Eqs. (7–4.25) and (7–4.26)—we have, for the undisturbed Poiseuille flow, taking the origin $z = 0$ at the sphere center,

$$u_\infty = U + \frac{2U}{3l}z - \frac{U}{3l^2}z^2 \qquad v_\infty = w_\infty = 0 \tag{7–4.31}$$

This corresponds to a maximum velocity of $4U/3$ at the midpoint $z = +l$, and an approach velocity to the sphere of U at the location $z = 0$. For this flow the resistance is

$$F_x = \frac{6\pi\mu a U[1 - (1/9)(a/l)^2]}{1 - 0.6526(a/l) + 0.3160(a/l)^3 - 0.242(a/l)^4} \tag{7–4.32}$$

If the sphere is prevented from rotating, it will experience a torque about the y axis given by

$$T_y = \frac{8}{3}\pi\mu a^2 U \frac{a}{l}\left[1 + 0.0758\left(\frac{a}{l}\right) + 0.049\left(\frac{a}{l}\right)^2\right] \qquad (7\text{–}4.33)$$

For Couette type flow, Fig. 7–4.4 depicts the pertinent geometry involved. A sphere is placed between two plane walls which lie parallel to the xy plane, located at $z = -l$ and $z = b = 3l$. At a sufficient distance in the x direction away from the sphere, the flow is represented by

$$u_\infty = \frac{U}{4} + \frac{U}{4l}z \qquad (7\text{–}4.34)$$

$$v_\infty = w_\infty = 0$$

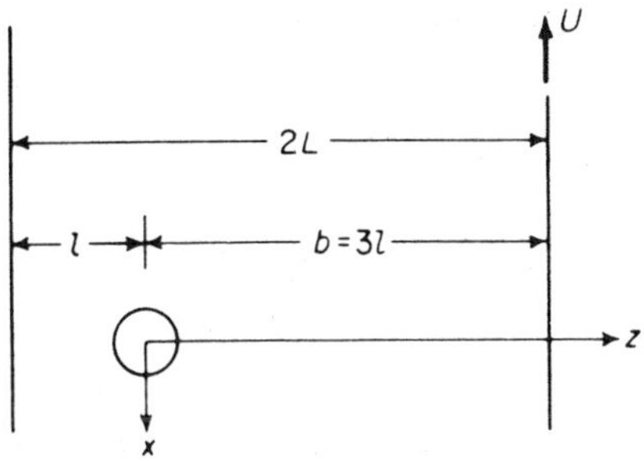

Figure 7-4.4. Definition sketch for a sphere in a shearing flow.

Thus at $z = -l$, the motion of the fluid and the wall is zero, whereas at $z = 3l$ the motion of the fluid and wall is at a velocity U in its own plane.

For this flow, the resistance of a sphere which is held fixed is

$$F_x = \frac{(3/2)\pi\mu a U}{1 - 0.6526(a/l) + 0.4003(a/l)^3 - 0.297(a/l)^4} \qquad (7\text{–}4.35)$$

Note that the approach velocity to the sphere is $U/4$. If the sphere is prevented from rotating it will experience a couple about the y axis given by

$$T_y = 4\pi\mu a^2 U \frac{a}{l}\left[1 + 0.0506\left(\frac{a}{l}\right) + 0.033\left(\frac{a}{l}\right)^2\right] \qquad (7\text{–}4.36)$$

Wakiya made further investigations for situations such that the sphere could rotate freely around its center and that the sphere could move freely in the flow. Equations for the angular velocity were derived. A relationship was also developed for the case where a sphere moves in a stationary fluid. It was found to agree with Faxen's result. The treatment for a freely moving sphere should be applicable to a treatment of suspension viscosity taking wall effects into consideration.

A sphere moving perpendicular to a plane wall

In the present section we consider the motion of a spherical particle toward or away from a single plane surface in an otherwise unlimited fluid. Two distinct cases are of interest: (a) the plane surface is rigid as, for example, when it constitutes the bottom of the container in which the particle falls; (b) the plane is a free surface, as for example, when it corresponds to the interface between a liquid and the atmosphere.

The rigid wall case has been treated by Lorentz, as discussed in Section

3–5, for the case where the sphere radius is small compared with the instantaneous distance of its midpoint from the plane. Solutions* which are not restricted by this limitation have been developed by Brenner[7] using the general bipolar coordinate solution of the creeping motion equations employed by Stimson and Jeffery[54] in their solution of the problem of two spheres falling along their line of centers (Section 6–4).

If we denote the sphere radius by a and the distance of its midpoint from the plane by h, motion toward the wall results in a resistance

$$F = 6\pi\mu U a\lambda \tag{7–4.37}$$

where $\lambda = \lambda(a/h)$ is a correction to Stokes' law given by the expression

$$\lambda = \frac{4}{3}\sinh\alpha \sum_{n=1}^{\infty} \frac{n(n+1)}{(2n-1)(2n+3)} \left[\frac{2\sinh(2n+1)\alpha + (2n+1)\sinh 2\alpha}{4\sinh^2(n+\frac{1}{2})\alpha - (2n+1)^2\sinh^2\alpha} - 1\right] \tag{7–4.38}$$

The parameter α is given in terms of the ratio of the sphere radius a to the distance h by the relationship $\alpha = \cosh^{-1}(h/a)$. Values of λ are given in Table 7–4.1.

Experimental confirmation of this Stokes' law correction for a sphere approaching a plane wall has been reported by MacKay, Suzuki, and Mason[43].

A treatment by Wakiya[60], using the method of reflections as applied to two spheres in the presence of a plane wall, gives, as a specialized case, the resistance due to motion of a single sphere toward a plane wall to $O(a/h)^3$:

$$F = \frac{6\pi\mu a U}{1 - (9/8)(a/h) + (1/2)(a/h)^3} \tag{7–4.39}$$

At dimensionless distances of $h/a > 10$ this formula agrees very closely with the values computed by Brenner in Table 7–4.1.

When the plane toward which the sphere falls is a planar free surface, so that the tangential stresses and normal velocity vanish, the frictional force experienced by the sphere is

$$F = 6\pi\mu a U\beta \tag{7–4.40}$$

where the Stokes' correction β is given by[7]

$$\beta = \frac{4}{3}\sinh\alpha \sum_{n=1}^{\infty} \frac{n(n+1)}{(2n-1)(2n+3)} \left[\frac{4\cosh^2(n+\frac{1}{2})\alpha + (2n+1)^2\sinh^2\alpha}{2\sinh(2n+1)\alpha - (2n+1)\sinh 2\alpha} - 1\right] \tag{7–4.41}$$

Values of β calculated from this expression are given in Table 7–4.2. Independent confirmation of these results is provided by the work of Faxen and Dahl[20]—see Eq. (6–3.52)—which, for the case of two equal spheres moving

*An independent solution agreeing with that outlined here is also given by A. D. Maude, Brit. J. Appl. Phys. **12** (1961), 293.

TABLE 7–4.1
STOKES' LAW CORRECTION FOR A SOLID PLANE

α	h/a	λ
0	1	∞
0.5	1.1276260	9.2517663
1.0	1.5430806	3.0360641
1.5	2.3524096	1.8374749
2.0	3.7621957	1.4128629
2.5	6.1322895	1.2219882
3.0	10.067662	1.1252465
∞	∞	1

TABLE 7–4.2
STOKES' LAW CORRECTION FOR A FREE SURFACE

α	h/a	β
0	1	∞
0.5	1.1276260	3.98670
1.0	1.5430806	1.97369
1.5	2.3524096	1.4636
2.0	3.7621957	1.247131
2.5	6.1322895	1.1388563
3.0	10.067662	1.0803758
∞	∞	1

toward each other, is equivalent to the motion of a single sphere approaching a free surface.[7] Their results for the first few terms [they give a relationship correct to $(a/h)^9$] are equivalent to

$$F = 6\pi\mu aU\left[1 + \frac{3}{4}\left(\frac{a}{h}\right) + \frac{9}{16}\left(\frac{a}{h}\right)^2 + \frac{19}{64}\left(\frac{a}{h}\right)^3 + \cdots\right] \qquad (7\text{–}4.42)$$

As discussed by Brenner[7], it is not correct to apply these results to the case of a sphere simultaneously bounded by a plane wall and a cylindrical surface—for example, a sphere falling near the bottom of a falling ball viscometer. Use of cylindrical and planar correction factors as suggested by Ladenburg[38] and Brenner[7] will yield, at best, only an estimate of the additional resistance experienced by a spherical particle.*

7-5 Spheroid Moving Relative to Cylindrical and Plane Walls

Wakiya[59], following Faxen's methods for treating the interaction of spherical particles with plane walls, has developed more general solutions in which spheroidal particles may be involved.

*See also the pertinent discussion and numerical results of R. I. Tanner, J. Fluid Mech. **17** (1963), 161.

A solution of the creeping motion equation in cartesian coordinates is first developed by extending Oberbeck's[45] solution. Solutions which are regular outside the ellipsoid

$$\frac{x^2}{a^2} + \frac{y^2}{b^2} + \frac{z^2}{c^2} = 1 \tag{7–5.1}$$

and which vanish at infinity may be constructed in terms of cartesian derivatives of the harmonic functions defined by

$$\chi_n = \int_\lambda^\infty \frac{1}{2^n}\left(\frac{x^2}{a^2+s} + \frac{y^2}{b^2+s} + \frac{z^2}{c^2+s} - 1\right)^n \frac{ds}{\sqrt{(a^2+s)(b^2+s)(c^2+s)}}$$
$$(n = 0, 1, 2 \cdots) \tag{7–5.2}$$

where $\lambda(x, y, z)$ is the positive root of the cubic equation

$$\frac{x^2}{a^2+\lambda} + \frac{y^2}{b^2+\lambda} + \frac{z^2}{c^2+\lambda} = 1 \tag{7–5.3}$$

Instead of a uniform external flow, as was utilized by Oberbeck for streaming flow past an ellipsoid, we assume a more general flow at infinity of the form:

$$\begin{aligned} u_o &= a_0 + a_1 z + a_2 z^2 + a_3 x^2 \\ v_o &= b_2 xy \\ w_o &= c_1 x + c_2 xz \end{aligned} \tag{7–5.4}$$

The general problem of an ellipsoid in an unbounded medium which satisfies the boundary condition Eq. (7–5.4) is solved in terms of appropriate values of the constants in the basic solution. In order to determine the effect of boundaries, this solution for an ellipsoid in an infinite medium, for the case where the ellipsoid has axial symmetry ($b = c$), is expressed in the form of infinite integrals (similar to Faxen's treatment for plane walls). Thus, for example, for $a < c$

$$\chi_0 = \frac{1}{\pi^2} \iiint_{-\infty}^{+\infty} \frac{\sin(\sigma\sqrt{c^2 - a^2})}{\sigma\sqrt{c^2 - a^2}} e^{i(\xi x + \eta y + \zeta z)} \frac{d\xi\, d\eta\, d\zeta}{\rho^2} \tag{7–5.5}$$

where $i = \sqrt{-1}$, $\sigma = \sqrt{\eta^2 + \zeta^2}$, $\rho = \sqrt{\xi^2 + \eta^2 + \zeta^2}$

Similar forms are available for $x\chi_0$, χ_1 and $x\chi_2$, and another set of integrals is developed for the case where $a > c$.

Spheroid between two parallel walls

This method is applied to the case of flow past a spheroid ($b = c$) (axis of symmetry parallel to the direction of flow) placed between two plane walls which are parallel to the xy plane (see Fig. 7–5.1). The surfaces of the walls correspond to the equations

$$z = -d_1, \qquad z = d_2$$

where d_1 and d_2 are positive constants, and $z = 0$ lies at the center of the spheroid. At large distances from the spheroid, the flow is represented by

$$u_\infty = \alpha + \beta z + \gamma z^2$$
$$v_\infty = w_\infty = 0 \tag{7-5.6}$$

where α, β, and γ are arbitrary constants. Thus, the boundary conditions to be satisfied are

$$\begin{aligned} &u = u_\infty, \quad v = w = 0 \\ &\qquad \text{at infinity} \\ &u = u_\infty, \quad v = w = 0 \\ &\qquad \text{for } z = -d_1 \text{ and } z = d_2 \\ &u = v = w = 0 \\ &\qquad \text{on the spheroid} \end{aligned} \tag{7-5.7}$$

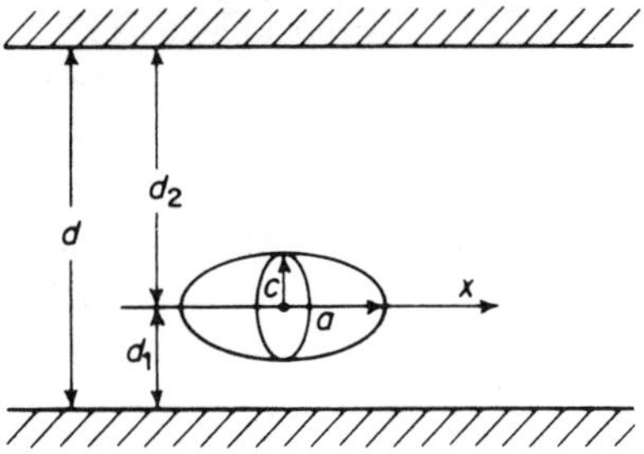

Figure 7-5.1. A spheroid between two plane walls.

Numerical calculations were carried forward for the case where $d_1 = d/4$ and $d_2 = 3d/4$ (similar to Faxen's case discussed in Section 7–4), where d is the distance between the walls. Solutions were obtained for the following special cases:

CASE A: A spheroid moving with a constant velocity $(-U)$ along the x axis: $\alpha = U$, $\beta = 0$, $\gamma = 0$.

CASE B: A spheroid is in the Couette flow obtained by the steady translation of one of the two walls in its own plane: $\alpha = U$, $\beta = U/d_1$, $\gamma = 0$. Thus, the spheroid is closest to the stationary wall.

CASE C: A fixed spheroid is immersed in a two-dimensional Poiseuille flow: $\alpha = U$, $\beta = 2U/d_2$, $\gamma = -U/d_1d_2$. The approach velocity to the spheroid is U.

Some values of the drag parameter, giving the ratio of spheroid resistance to that in an unbounded medium,

$$\frac{F_x(K + a^2L_1)}{16\pi\mu U} \tag{7-5.8}$$

and torque parameter

$$T_y/\mu c^2 U \tag{7-5.9}$$

are tabulated in Tables 7–5.1 and 7–5.2. Note that

$$K = \int_0^\infty \frac{ds}{\sqrt{(a^2 + s)(b^2 + s)(c^2 + s)}} \tag{7-5.10}$$

$$L_1 = \int_0^\infty \frac{ds}{(a^2 + s)\sqrt{(a^2 + s)(b^2 + s)(c^2 + s)}} \tag{7-5.11}$$

TABLE 7–5.1

RESISTANCE OF A SPHEROID AT AN ECCENTRIC POSITION BETWEEN PLANE WALLS

Values of $F_x \Big/ \dfrac{16\pi\mu U}{K + a^2 L_1}$

$\left(b/l = c/l = \dfrac{\text{radius}}{\text{distance to near wall}};\quad a/c = a/b = \dfrac{\text{half length}}{\text{radius}};\quad l = d_1 = d/4\right)$

CASE A: $u_\infty = U$

c/l \ a/c	0*	0.5	1†	2	4
0.05	1.03	1.03	1.03	1.04	1.05
0.01	1.06	1.06	1.07	1.08	1.11
0.2	1.12	1.13	1.15	1.18	1.23
0.3	1.20	1.21	1.24	1.29	1.34
0.4	1.28	1.31	1.34	1.40	. . .
0.5	1.38	1.41	1.46	1.53	. . .
$\dfrac{16\pi}{c(K + a^2 L_1)}$	16	17.06	18.85	22.69	30.12

CASE B: $u_\infty = U(1 + z/d_1)$

c/l \ a/c	0	0.5	1	2	3
0.05	1.03	1.03	1.03	1.04	1.05
0.1	1.06	1.06	1.07	1.08	1.11
0.2	1.12	1.13	1.15	1.18	1.23
0.3	1.19	1.21	1.23	1.27	1.31
0.4	1.27	1.29	1.32	1.37	. . .
0.5	1.36	1.38	1.42	1.45	. . .

CASE C: $u_\infty = U(1 + z/d_1)(1 - z/d_2)$

c/l \ a/c	0	0.5	1	2	4
0.05	1.03	1.03	1.03	1.04	1.05
0.1	1.06	1.06	1.07	1.08	1.11
0.2	1.12	1.13	1.14	1.17	1.22
0.3	1.18	1.20	1.22	1.27	1.30
0.4	1.25	1.27	1.30	1.36	. . .
0.5	1.32	1.35	1.39	1.44	. . .

*Circular disk perpendicular to main stream and to walls.

†Sphere.

correspond to constants already determined in connection with the Oberbeck treatment for the motion of an ellipsoid in an unbounded medium (Chapter 5, Section 11). The notation in these tables is as follows: $l = d/4$ denotes the distance from the center of the spheroid to the nearest wall; $c =$ the radius of the cross section (which lies perpendicular to the x axis); and $a =$ the semiaxis along the x axis.

TABLE 7–5.2
TORQUE EXERTED ON A SPHEROID AT AN ECCENTRIC POSITION BETWEEN PLANE WALLS
Values of $T_y/\mu c^2 U$

$\left(b/l = c/l = \dfrac{\text{radius}}{\text{distance to near wall}}; \quad a/c = a/b = \dfrac{\text{half length}}{\text{radius}}; \quad l = d_1 = d/4\right)$

CASE A: $u_\infty = U$

c/l \ a/c	0	0.5	1	2	4
0.01	−0.00	−0.00	0.00	0.00	0.06
0.05	−0.02	−0.01	0.00	0.12	1.03
0.1	−0.06	−0.05	0.02	0.54	4.33
0.2	−0.26	−0.21	0.09	2.16	18.9
0.3	−0.62	−0.49	0.21	5.2	...

CASE B: $u_\infty = U(1 + z/d_1)$

c/l \ a/c	0	0.5	1	2	4
0.01	0.32	0.34	0.38	0.46	0.64
0.05	1.58	1.70	1.89	2.39	4.24
0.1	3.14	3.37	3.79	5.04	10.35
0.2	6.14	6.62	7.63	11.24	31.0
0.3	8.98	9.75	11.51	18.8	...

CASE C: $u_\infty = U(1 + z/d_1)(1 - z/d_2)$

c/l \ a/c	0	0.5	1	2	4
0.01	0.21	0.23	0.25	0.31	0.44
0.05	1.05	1.13	1.26	1.63	3.04
0.1	2.07	2.23	2.53	3.53	8.34
0.2	4.00	4.43	5.11	8.22	27.0
0.3	5.78	6.33	7.74	14.3	...

Note from Table 7–5.2, Case A, that the torque on a sedimenting spheroid changes its sign when the ratio of a/c exceeds a certain value. Thus the edge of a disk falling broadside will tend to dip on the side nearest the near wall. When the ratio exceeds the value

$$(a/c)_{\text{crit}} \approx 0.901 \tag{7–5.12}$$

[assuming we can neglect terms of higher order than $(a/l)^3$ or $(c/l)^3$] the direction of the torque becomes such that it tends to rotate the object as if it were rolling on the near wall, as is also the case with a sphere sedimenting in an eccentric position. As long as the spheroid moves in the direction of its symmetry axis, it will experience no sidewise thrust. But when an eccentrically located particle rotates due to the torque it experiences, a sidewise thrust will be experienced. This will tend to move it towards the nearer wall, for both disk- and needle-shaped objects (see Chapter 5).

Wakiya[61] also studied, via the same general analysis, the behavior of a spheroid located midway between two parallel walls. In this case he carried forward a numerical development for the case $a = c$ (that is, the circular cross section is perpendicular to the walls but parallel to the direction of flow; the axis of symmetry is perpendicular to the flow direction). Two cases were considered:

CASE A: A spheroid moving with a constant velocity $(-U)$ along the x axis. (Similar to Case A for the off-center spheroid).

CASE B: A fixed spheroid immersed in a two-dimensional Poiseuille flow with approach velocity $(+U)$ along the x axis.

No torque arises in these circumstances. Some values of the drag parameter

$$\frac{F_x(K + a^2L_1)}{16\pi\mu U}$$

are given in Table 7–5.3.

TABLE 7–5.3
RESISTANCE OF A SPHEROID AT A CENTRAL POSITION BETWEEN PLANE WALLS

Values of $F_x \Big/ \dfrac{16\pi\mu U}{K + a^2L_1}$

$\left(a/l = c/l = \dfrac{\text{radius}}{\text{distance to wall}};\quad b/c = b/a = \dfrac{\text{half length}}{\text{radius}}\right)$

CASE A: $u_\infty = U$

c/l \ b/c	0	0.5	1	2	3
0.01	1.01	1.01	1.01	1.02	1.02
0.05	1.03	1.04	1.05	1.06	1.08
0.1	1.06	1.08	1.10	1.14	1.19
0.2	1.13	1.17	1.24	1.37	1.47
0.3	1.20	1.30	1.41	1.61	. . .
0.4	1.28	1.44	1.60	. . .	. . .

CASE B: $u_\infty = U[1 - (z/l)^2]$

c/l \ b/c	0	0.5	1	2	3
0.01	1.01	1.01	1.01	1.02	1.02
0.05	1.03	1.04	1.05	1.06	1.08
0.1	1.06	1.08	1.10	1.14	1.19
0.2	1.11	1.15	1.22	1.35	1.45
0.3	1.15	1.25	1.36	1.56	. . .
0.4	1.21	1.36	1.52	. . .	. . .

The values tabulated in Table 7–5.3, Case B, for a Poiseuille flow, correspond to the following formula:

$$F_x = \frac{16\pi\mu U[1 - (1/3)(a/l)^2]/(K + a^2L_1)}{1 - [1/(K + a^2L_1)l][2.6776 - 0.4244(b/l)^2 - 0.6902(a/l)^2]} \qquad (7\text{–}5.13)$$

The foregoing is correct to $O(a \text{ or } b/l)^5$. Values of the resistance of the spheroid in an infinite medium, that is, $16\pi\mu U/(K + a^2L_1)$, are available either from Table 7–5.1 or the tabulation in Table 5–11.1.

For the case of a two-dimensional Poiseuille flow approaching a disk held edgewise to the flow, with its axis of symmetry parallel to the plane walls and midway between them, we set $b = 0$ in Eq. (7–5.13). Since, for this case, $K = \pi/a$, $L_1 = \pi/(2a^3)$, we obtain

$$F_x = \frac{(32/3)\mu a U[1 - (1/3)(a/l)^2]}{1 - 0.568(a/l) + 0.146(a/l)^3} \qquad (7\text{–}5.14)$$

For the case where $a = b$ we have a sphere in a two-dimensional Poiseuille flow, midway between the walls. When $a = b$, $K = 2/a$ and $L_1 = 2/(3a^3)$, whence

$$F_x = \frac{6\pi\mu a U[1 - (1/3)(a/l)^2]}{1 - 1.004(a/l) + 0.418(a/l)^3} \qquad (7\text{–}5.15)$$

This may be compared with Faxen's formula Eq. (7–4.27) for a sphere sedimenting between two plane walls. The denominators of both expressions are identical to $O(a/l)^3$.

Spheroid moving parallel to a plane wall with its symmetry axis at an arbitrary angle of attack

In another study, Wakiya[62] considered the motion of a spheroid parallel to a single plane wall, but with the axis of symmetry making an arbitrary angle to the wall instead of being parallel to it, as in the two cases previously discussed for motion between parallel planes. This case is more complicated than the previous ones. It requires a transformation of coordinates from the x, y, z system appropriate to the ellipsoid to the X, Y, Z, system appropriate to the direction of motion of the spheroid (see Fig. 7–5.2).

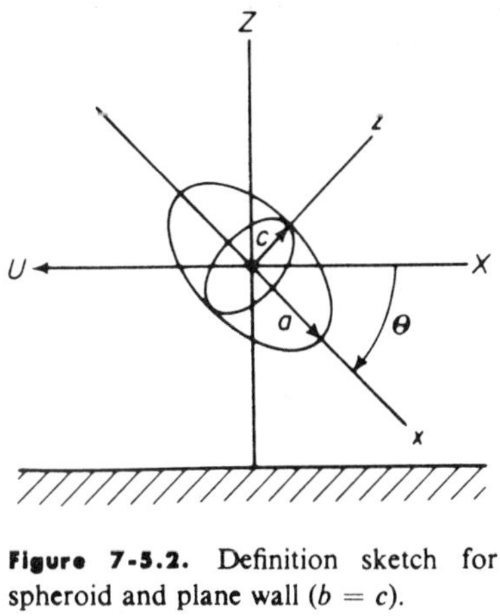

Figure 7-5.2. Definition sketch for spheroid and plane wall ($b = c$).

In the previous case, where an ellipsoid moves in an off-center position between two walls, we noted that the direction of the torque may change, depending on the shape of the ellipsoid. This fact suggests the existence of a stable orientation for a spheroid; and this problem motivated Wakiya's study. Motion along a single wall was chosen to reduce the computational work. It is found that in addition to the drag on the ellipsoid, it also experiences a lift force, at right angles to the plane wall. Within the accuracy of the approximation, $[(a \text{ or } c)/l]^3$, the direction of the torque changes in algebraic sign, depending on both shape and inclination. This torque vanishes for $a/c = 1$ (a sphere). Note that this is slightly different from the critical ratio $a/c = 0.901$ obtained for a spheroid eccentrically located between two parallel planes. Thus it is concluded that if a spheroid is allowed to rotate freely around its center, it will finally adopt a definite inclination and will not continue to rotate like a sphere. Moreover, it will then tend to move laterally, owing to the sidewise thrust.

Spheroid at the center of a circular cylinder

Utilizing the same general solution for flow relative to an ellipsoid in an infinite medium as previously, Wakiya[59] considered axisymmetric flow past a spheroid in an infinitely long pipe of circular cross section, its axis coinciding with the axis of symmetry of the spheroid (the x axis). Thus, for the spheroid, $b = c$. Numerical evaluation was carried out for two special cases:

CASE A: The spheroid moves with constant velocity $(-U)$ along the pipe axis.
CASE B: The spheroid is immersed in a Poiseuille flow through the pipe.

In these cases (in contrast to flow between plates) an additional pressure difference between the ends of the pipe is necessary to maintain the same total flux as when the spheroid is absent.

Insofar as terms of the order of $[(a \text{ or } c)/R_0]^3$ are negligible, the additional pressure drop is related to the drag on the spheroid by the same relationship as applies for a sphere; namely,

$$\Delta P = \frac{2F}{\pi R_0^2}\left[1 - \frac{2}{3}\left(\frac{c}{R_0}\right)^2\right]$$

R_0 is the cylinder radius and c the spheroid radius. If U_0 denotes the centerline velocity of the Poiseuille flow, the general relationship for the drag is

$$F = \frac{16\pi\mu[U_0 + U - (2/3)U_0(c/R_0)^2]/(K + a^2L_1)}{1 - [1/(K + a^2L_1)R_0][5.6112 - 2.0211(a/R_0)^2 - 3.5431(c/R_0)^2]} \qquad (7\text{–}5.16)$$

Values of the drag and pressure drop parameters are given in Tables 7–5.4 to 7–5.6.

In particular, for the case of a circular disk ($a \doteq 0$) we have

$$F = \frac{16\mu c[U + U_0 - (2/3)U_0(c/R_0)^2]}{1 - 1.786(c/R_0) + 1.128(c/R_0)^3} \qquad (7\text{–}5.17)$$

TABLE 7–5.4
RESISTANCE OF A SPHEROID SEDIMENTING IN A CYLINDRICAL TUBE

$$\left(\text{Values of } F\Big/\frac{16\pi\mu U}{K + a^2L_1} \text{ for } u_\infty = U\right)$$

c/R_o \ a/c	0	0.5	1	2	4
0.01	1.02	1.02	1.02	1.03	1.03
0.05	1.10	1.11	1.12	1.14	1.20
0.1	1.22	1.23	1.26	1.33	1.46
0.2	1.53	1.59	1.68	1.87	2.00
0.3	2.02	2.15	2.35	2.62	...

TABLE 7–5.5
RESISTANCE OF A SPHEROID IN POISEUILLE FLOW, OR PRESSURE DROP DUE TO A SEDIMENTING SPHEROID IN A CYLINDRICAL TUBE

$$\left(\begin{array}{c}\text{Values of } \dfrac{F}{16\pi\mu U_o/(K + a^2L_1)} \text{ for } u_\infty = U_o\,[1 - (R/R_o)^2] \\ \text{or} \\ \text{Values of } \dfrac{\Delta P\pi R_o^2}{32\pi\mu U/(K + a^2L_1)} \text{ for } u_\infty = U\end{array}\right)$$

c/R_o \ a/c	0	0.5	1	2	4
0.01	1.02	1.02	1.02	1.03	1.03
0.05	1.10	1.10	1.12	1.14	1.20
0.1	1.21	1.22	1.25	1.32	1.45
0.2	1.49	1.55	1.63	1.82	1.95
0.3	1.90	2.02	2.21	2.46	...

TABLE 7–5.6
PRESSURE DROP DUE TO A SPHEROID IN POISEUILLE FLOW IN A CYLINDER

$$\left(\text{Values of } \frac{\Delta P\pi R_o^2}{32\pi\mu U_o/(K + a^2L_1)} \text{ for } u_\infty = U_o\,[1 - (R/R_o)^2]\right)$$

c/R_o \ a/c	0	0.5	1	2	4
0.01	1.02	1.02	1.02	1.03	1.03
0.05	1.09	1.10	1.11	1.14	1.19
0.1	1.20	1.22	1.25	1.31	1.44
0.2	1.45	1.50	1.59	1.77	...
0.3	1.79	1.90	2.08	...	...

and
$$\Delta P = \frac{(32\mu c/\pi R_o^2)[U + U_o - (2/3)(2U_o + U)(c/R_o)^2]}{1 - 1.786(c/R_o) + 1.128(c/R_o)^3} \tag{7–5.18}$$

As an illustration, consider the sedimentation of a spheroid in a circular cylinder. For the present case we have $b = c$; a is the half-length of the symmetry axis. Referring to Section 5–11 (note nomenclature change, since there the same case is taken as $a = c$), in an unbounded fluid

$$F_\infty = \frac{16\pi\mu U}{(K + a^2 L_1)} = 6\pi\mu R U \tag{7–5.19}$$

where R is the "equivalent sphere radius." For a ratio of $a/c = 10$, we have

$$\frac{R}{c} = 3.812 \tag{7–5.20}$$

Thus $1/(K + a^2 L_1) = 1.43c$. Applying Eq. (7–5.16), we obtain

$$\frac{F}{F_\infty} = \frac{1}{1 - 1.43(c/R_0)[5.612 - 2.0211(a/R_0)^2 - 3.5431(c/R_0)^2]} \tag{7–5.21}$$

or, since $a = 10c$,

$$\frac{F}{F_\infty} = \frac{1}{1 - 1.43(c/R_0)[5.612 - 205.65(c/R_0)^2]} \tag{7–5.22}$$

Because of the large value of the coefficient of the $(c/R_0)^2$ term in square brackets, the effect of length of the object will be appreciable except at very small ratios of c/R_0. Thus if $c/R_0 = 0.1$, $F/F_\infty = 2.04$. Neglect of the $(c/R_0)^2$ term would have yielded $F/F_\infty = 5.10$. The lead correction term could, of course, be obtained directly from Eq. (7–2.15) by using the point force approximation with $k = 2.104$. Thus,

$$\frac{F}{F_\infty} = \frac{1}{1 - k(F_\infty/6\pi\mu U R_0)} = \frac{1}{1 - 2.104(3.812 \times 6\pi\mu U c/6\pi\mu U R_0)}$$

$$= \frac{1}{1 - 8.02(c/R_0)} \tag{7–5.23}$$

in agreement with Eq. (7–5.21).

7-6 k-Coefficients for Typical Boundaries

The previous treatments for spheres in the presence of various boundaries enable us to establish the values of k in Eq. (7–2.15), the point force approximation for drag as influenced by the presence of walls. Higher-order approximations require an explicit solution of the boundary value problem, and this requires that the geometry of the particle as well as the walls be considered. Table 7–6.1 gives a collection of k values from previous sections of this chapter as a function of particle location, direction of fall, and shape of boundary. Other first-order approximations involving free surfaces, Poiseuille, and shearing flows may also be handled without an explicit consideration of the particle geometry.[5]

TABLE 7–6.1
k-COEFFICIENTS FOR THE DRAG ON A TRANSLATING PARTICLE IN THE PRESENCE OF RIGID BOUNDARIES

Shape of External Boundary	Location of Particle Center	Direction of Motion	Value of k
Circular cylinder	Cylinder axis; l = radius of cylinder	Axial	2.1044
	Eccentric	Axial	See Table 7–3.1; $k = f(b/R_o)$
Parallel plane walls	Midway between walls; l = distance to either wall	Parallel to walls	1.004
	1/4 distance between walls; l = distance to nearer wall	Parallel to walls	0.6526
Single plane wall	l = distance to wall	Perpendicular to wall	9/8
	l = distance to wall	Parallel to wall	9/16
Spherical	Center of sphere; l = radius of sphere	Radial	9/4

7-7 One- and Two-Dimensional Problems

Though not directly related in most instances to particulate media, several problems involving one- and two-dimensional situations in which an outside wall and an inside boundary occur are of interest in providing analogies to the more complicated three-dimensional problems encountered with particles.

A cylindrical rod moving axially inside a stationary circular cylinder

We assume in this case that the motion of the inside cylinder is axial, as shown in Fig. 7–7.1. In this case, the governing equation is

$$\frac{1}{R}\frac{d}{dR}\left(R\frac{dw}{dR}\right) = \frac{1}{\mu}\frac{dp}{dz} \tag{7–7.1}$$

where $w = w(R)$ is the z component of velocity. The general solution of this equation is

$$w = \frac{1}{4\mu}\frac{dp}{dz}R^2 + C_1 \ln R + C_2 \tag{7–7.2}$$

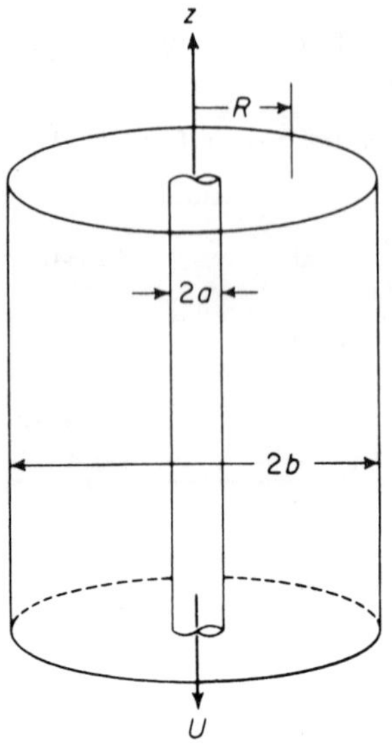

Figure 7-7.1. Definition sketch for axial movement of a cylinder.

The boundary conditions are

$$w = -U \quad \text{at } R = a \tag{7–7.3}$$

$$w = 0 \quad \text{at } R = b \tag{7–7.4}$$

For the case of a sedimenting cylinder, the net flow Q through the annulus is zero; thus

$$Q = 2\pi \int_a^b wR\,dR = 0 \tag{7–7.5}$$

If we set

$$C_3 = \frac{1}{4\mu}\frac{dp}{dz} \tag{7–7.6}$$

the three constants C_1, C_2, and C_3 appearing in Eq. (7–7.2) may be calculated from the three boundary conditions, Eqs. (7–7.3)–(7–7.5). Knowledge of these may then be utilized to calculate the drag F per unit length on the inner cylinder via the formula

$$F = -2\pi\mu a\left[\frac{dw}{dR}\right]_{R=a} \tag{7–7.7}$$

The result is

$$\frac{F}{2\pi\mu U} = \frac{1 - [2a^2/(b^2 - a^2)][1 - \{2a^2/(b^2 - a^2)\}\{\ln (b/a)\}]}{(b^2 + a^2)\{\ln (b/a) - 1\}/(b^2 - a^2)} \tag{7–7.8}$$

If $b \gg a$, we may neglect terms of order $(a/b)^2$ and obtain

$$F = \frac{2\pi\mu U}{\ln (b/a) + 1} \tag{7–7.9}$$

This result may be compared with the limiting case of flow through an annulus, given by Rothfus, *et al*[48]. This is based on Lamb's[39] derivation. Here neither the inside nor outside tube moves, but in the limit the resistance should be equivalent to the case just considered with $2V$ substituted for $-U$, where V is the *average* velocity of flow in the annulus. It is found that with $b \gg a$, Rothfus' result becomes

$$F = \frac{4\pi\mu V}{\ln (b/a) + 1} \tag{7–7.10}$$

Another interesting case occurs when the outer cylindrical shell is frictionless. Here the pressure gradient times the cross-sectional area must equal the drag per unit length on the inside cylinder since there is no shear on the outside cylinder. With the inner cylinder moving with velocity U and the outer one stationary, the limiting case of $b \gg a$ yields

$$F = \frac{2\pi\mu U}{\ln (b/a) + 3/4} \tag{7–7.11}$$

The behavior of a rod in an eccentric annular position has been considered in Section 2–5. Brenner[4] has also discussed the drag on the rod for this case.

Cylindrical rod located along the axis of another cylinder and moving perpendicular to the axis

This case is considered in greater detail in the following chapter, since it may be applied to the case of flow through an assemblage using a cell model technique. The basic treatment is developed by Slezkin[52].

When a cylinder moves transversely in a tube whose outer wall is solid, the limiting case of $b \gg a$ yields

$$F = \frac{4\pi\mu U}{\ln (b/a) - 1} \tag{7–7.12}$$

Alternatively, if the outside wall is frictionless

$$F = \frac{4\pi\mu U}{\ln (b/a) + 1/2} \tag{7–7.13}$$

Experimental data for rods in circular cylinders

It is of interest that experimental data for circular cylinders falling sidewise but *perpendicular* to the axis of the containing cylinder appear to be in reasonable agreement with Eq. (7–7.12), though the derivation of the latter refers to sidewise fall *parallel* to the axis of the containing cylinder. Thus, White[66] reported data at very low Reynolds numbers for wires falling sidewise as they settled in a vertical cylindrical container. White correlated his data by the formula

$$\frac{F}{\mu U} = \frac{14.5}{\ln (0.8b/a)} \tag{7–7.14}$$

In this same form, for $b \gg a$, Slezkin's formula (7–7.12) becomes

$$\frac{F}{\mu U} = \frac{4\pi}{\ln (b/a) - 1} \tag{7–7.15}$$

For example, if $b/a = 100$, Eq. (7–7.14) gives $F/\mu U = 3.3$, whereas Eq. (7–7.15) gives $F/\mu U = 3.5$. At higher values of b/a, agreement is still closer. White's work was conducted on wires of length-to-diameter ratios of 50 and less, and a correction factor for the finite length employed, which is substantial in the cases of short wires.

A limited amount of additional data[17] indicates that for long wires falling axially in a cylindrical container with $b \gg a$, Eq. (7–7.9) is applicable. Thus with values of $b/a = 267$, and rods 200 diameters in length, the predicted $F/\mu U$ values were only 10 per cent less than observed.

It thus appears that the drag per unit length on a long cylinder falling in a container can often be approximated by the one- and two-dimensional solutions with boundaries. It will be recalled that Lamb's[39] (Section 2–7) formula for a cylinder, which involves taking inertial effects into consideration, gives an analytical expression for the fall of a cylinder in an infinite medium. No finite solution is possible for this case using the creeping motion equations. At slow speeds, however, Lamb's formula is found to apply only when the boundaries are very far away indeed. For example[66] when $N_{Re} = 0.001$, boundaries 500 diameters away completely dominate the drag (as computed by the creeping motion approximation), and it is not until they are some 10,000 diameters away that their influence disappears.

Cylinder moving between two parallel plane walls

A circular cylinder moving perpendicular to its axis, with its axis midway between and parallel to the walls, gives rise to a two-dimensional problem. Two cases are of interest: (1) the cylinder moves parallel to the walls; (2) the cylinder moves perpendicular to the walls.

The definition sketch, Fig. 7–7.2, depicts the geometry of the first case. Faxen[21,22] worked this problem out in great detail. Since the problem is two-dimensional, it can be simplified by use of the stream function ψ, where the x and y velocity components are, respectively,

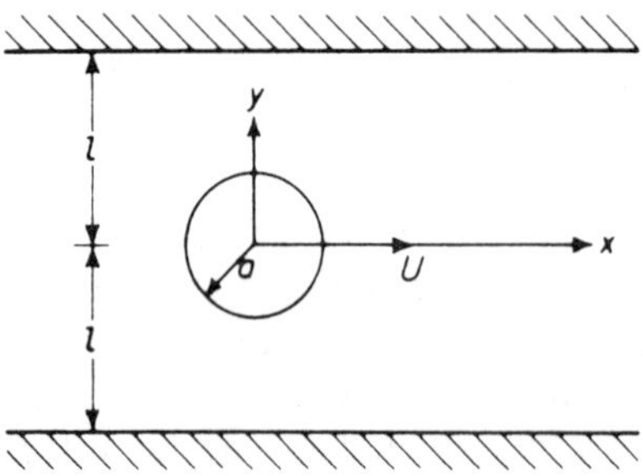

Figure 7-7.2. Definition sketch for cylinder moving parallel to walls.

$$u = \frac{\partial \psi}{\partial y}, \quad v = -\frac{\partial \psi}{\partial x} \tag{7–7.16}$$

In terms of the stream function, the creeping motion equations are

$$\nabla^4 \psi = \left(\frac{\partial^2}{\partial x^2} + \frac{\partial^2}{\partial y^2}\right)\left(\frac{\partial^2}{\partial x^2} + \frac{\partial^2}{\partial y^2}\right)\psi = 0 \tag{7–7.17}$$

as discussed in Section 3–1.

In terms of the component velocities, the boundary conditions are

$$\begin{aligned} &u = v = 0 \quad \text{on the planes } y = \pm l \\ &u = U = \text{constant}, \; v = 0 \quad \text{on the cylinder } x^2 + y^2 = a^2 \\ &u = v \to 0 \quad \text{as } |x| \to \infty \end{aligned} \tag{7–7.18}$$

Equivalently, in terms of the stream function these boundary conditions require

$$\psi \to 0 \quad \text{as } |x| \to \infty \tag{7–7.19}$$

On the cylinder $x^2 + y^2 = a^2$,

$$\psi = Uy, \qquad \frac{\partial \psi}{\partial x} = 0 \tag{7–7.20}$$

On the planes $y = \pm l$,

$$\psi = 0, \qquad \frac{\partial \psi}{\partial y} = 0 \tag{7-7.21}$$

Equations (7–7.17), (7–7.19) and (7–7.21) are satisfied by the following choice of stream function:

$$\begin{aligned}\psi(x, y) = &\int_0^\infty (A_0 + A_1\alpha^2 + A_2\alpha^4 + \cdots)\Big[ye^{-\alpha|y|} \\ &+ \frac{(2\alpha l - 1 + e^{-2\alpha l})y \cosh \alpha y - 2\alpha l^2 \sinh \alpha y}{\sinh 2\alpha l - 2\alpha l}\Big] \frac{\cos \alpha x}{\alpha}\, d\alpha \\ &+ \int_0^\infty (B_0 + B_1\alpha^2 + B_2\alpha^4 + \cdots)\Big[\frac{y}{|y|} e^{-\alpha|y|} \\ &+ \frac{2\alpha y \cosh \alpha y - (1 + 2\alpha l + e^{-2\alpha l}) \sinh \alpha y}{\sinh 2\alpha l - 2\alpha l}\Big] \cos \alpha x\, d\alpha\end{aligned} \tag{7-7.22}$$

The foregoing integrals may be divided into a series of simpler convergent integrals, and the constants $A_0, A_1, A_2 \ldots, B_0, B_1, B_2 \ldots$ determined so as to satisfy condition (7–7.20). For this purpose it is convenient to introduce series expansions for A_n and B_n. An infinite number of simultaneous equations is obtained and these are solved for the constants. For a unit length of cylinder the resistance is ultimately found to be

$$F = \frac{4\pi\mu U}{\ln(l/a) - 0.9157 + 1.7244(a/l)^2 - 1.7302(a/l)^4} \tag{7-7.23}$$

In a later treatment, Faxen[22] extended the analysis to powers of $(a/l)^8$, and provided an expression for the additional pressure drop due to movement of the cylinder in the space between the plane walls. Comparable results were also presented for the case of Poiseuille flow past the stationary cylinder. It is interesting to note that in the case where $l \gg a$, Eq. (7–7.23) gives a result very similar to that observed in the case of cylindrical boundary, Eq. (7–7.12).

Westberg[65], in an extended treatment of similar problems, including that for potential flow, considers the second case, in which the cylinder moves perpendicular to the planes in a viscous incompressible fluid. The magnitude of the force acting per unit length is computed to powers of $(a/l)^{12}$. The following formula gives the first few terms:

$$F = \frac{4\pi\mu U}{\ln(l/a) - 0.62026 + 1.04207(a/l)^2} \tag{7-7.24}$$

Takaisi[34] considered the problem of motion of a circular cylinder parallel to a single plane wall, when moving perpendicular to its own axis. In the limiting case of large b/a (b = distance from wall to center of cylinder; a = radius of cylinder), the drag per unit length is

$$F = \frac{4\pi\mu U}{\ln(2b/a)} \tag{7–7.25}$$

Takaisi's treatment is based on the Oseen equations, and the result cited is obtained by putting the Reynolds number equal to zero in his more general formula. When inertial effects are retained, one finds a lift force tending to move the cylinder away from the wall.

In a later paper, Takaisi[56] also considered the motion of a cylinder parallel to, and at the midpoint between, two planes. The results are in general agreement with Eq. (7–7.23).

7-8 Solid of Revolution Rotating Symmetrically in a Bounded Fluid

In this section, we consider the first-order effects of wall proximity on the hydrodynamic torque experienced by a rotating body.

Equation (7–2.15) has an analog for the torque experienced by a particle rotating near a boundary. The methods for obtaining it are, on the whole, quite similar to those already utilized in Section 7–2 for a translating particle. For reasons which will be made clear, however, we must here limit the scope of our investigation to bodies of revolution rotating about their symmetry axes. Furthermore, the shape and orientation of the container boundaries must be such that the streamlines arising from the slow rotation of the body in their presence lie in concentric circles. This arises, for example, from the rotation of an axially symmetric body about the axis of a circular cylindrical container filled with viscous fluid. The treatment follows Brenner's original investigation[5].

Let $\mathbf{T}$ denote the torque about any point on the axis of revolution of the body when the latter rotates about this axis with angular velocity $\boldsymbol{\omega}$. Because of the symmetrical disposition of the body relative to the boundaries, both $\mathbf{T}$ and $\boldsymbol{\omega}$ will lie parallel to this symmetry axis.

The boundary conditions to be satisfied are as follows:

$$\mathbf{v} = \boldsymbol{\omega} \times \mathbf{r} \quad \text{on } P \tag{7–8.1}$$

$$\mathbf{v} = \mathbf{0} \quad \text{on } S \tag{7–8.2}$$

and

$$\mathbf{v} \to \mathbf{0} \quad \text{as } r \to \infty \tag{7–8.3}$$

where P and S represent the particle and bounding surface, respectively. In the case of a particle rotating inside a closed surface, it is not necessary to satisfy Eq. (7–8.3). The velocity and pressure fields can be decomposed and successive reflections determined as in Section 7–2, except that the initial field in the unbounded fluid is now defined by

$$\mathbf{v}^{(1)} = \boldsymbol{\omega} \times \mathbf{r} \quad \text{on } P, \qquad \mathbf{v}^{(1)} \to \mathbf{0} \qquad \text{as} \quad r \to \infty \tag{7–8.4}$$

Let $\mathbf{T}$ denote the couple exerted on the particle by the fluid, and let $\mathbf{T}^{(n)}$ denote the couple associated with the field $\mathbf{v}^{(n)}$. The couples deriving from the even-numbered fields, reflected from S, vanish identically; therefore, analogously to Eq. (7–2.1), one may write

$$\mathbf{T} = \mathbf{T}^{(1)} + \mathbf{T}^{(3)} + \mathbf{T}^{(5)} + \cdots \tag{7–8.5}$$

The initial field obviously corresponds to rotation of the particle in the unbounded fluid with angular velocity $\boldsymbol{\omega}$. Associated with this motion is the couple

$$\mathbf{T}^{(1)} = \mathbf{T}_\infty \tag{7–8.6}$$

where $\mathbf{T}_\infty$ denotes the couple in the absence of S.

Now, when any nonskew body (that is, one for which the coupling dyadic vanishes at its center of reaction) rotates about any axis through its center of reaction, it experiences no net hydrodynamic force, at least in an unbounded fluid. Under these circumstances it can be shown that the velocity field at great distances is of the general form

$$\mathbf{v}^{(1)} = -\frac{\mathbf{T}_\infty \times \mathbf{r}}{8\pi\mu r^3} + \mathbf{r}\frac{p^{(1)}}{2\mu} + o(r^{-2}) \tag{7–8.7}$$

where $p^{(1)} = p_\infty$ is the pressure field generated by the rotation of the particle in the unbounded fluid. [More explicitly, $p^{(1)}$ is that part of the pressure field which is of $O(r^{-2})$.] When the rotating body is a solid of revolution rotating about its symmetry axis, the pressure is everywhere constant; hence

$$p^{(1)} = 0 \tag{7–8.8}$$

which leads to the following asymptotic expression for the velocity field:

$$\mathbf{v}^{(1)} = -\frac{\mathbf{T}_\infty \times \mathbf{r}}{8\pi\mu r^3} + o(r^{-2}) \tag{7–8.9}$$

The terms displayed explicitly correspond to the motion created by the action of a symmetrical *point couple* concentrated at the origin.

In general, if $\boldsymbol{\Omega}$ is the rotation dyadic at the point on the axis about which we are taking moments, then

$$\mathbf{T}_\infty = -\mu\boldsymbol{\Omega}\cdot\boldsymbol{\omega} \tag{7–8.10}$$

We next focus attention on the calculation of $\mathbf{T}^{(3)}$ in Eq. (7–8.5). It can be shown, by an application of the reciprocal theorem[5], that for symmetrical flows of the type under consideration,

$$\boldsymbol{\omega}\cdot\mathbf{T}^{(3)} = -\boldsymbol{\omega}_o^{(2)}\cdot\mathbf{T}_\infty \tag{7–8.11}$$

where $\boldsymbol{\omega}^{(2)} = \frac{1}{2}\nabla \times \mathbf{v}^{(2)}$, and the subscript o refers to evaluation at the location of the point about which moments are being taken.

Because of the prevailing symmetry, vector torques and spins may equally well be represented by comparable scalar quantities. Let $\mathbf{k}$ be a unit vector parallel to the axis of rotation in a sense such that $\boldsymbol{\omega} = \mathbf{k}\omega$ where

$\omega = |\boldsymbol{\omega}| > 0$. Then we may write $\mathbf{T} = -\mathbf{k}T$, $\mathbf{T}_\infty = -\mathbf{k}T_\infty$, $\mathbf{T}^{(n)} = -\mathbf{k}T^{(n)}$ and $\boldsymbol{\omega}_o^{(2)} = -\mathbf{k}\omega_o^{(2)}$. The scalars appearing here are all positive with the possible exceptions of $T^{(n)}$ and $\omega_o^{(2)}$. From Eq. (7–8.11), we now obtain

$$T^{(3)} = \frac{T_\infty}{\omega}\omega_o^{(2)} \tag{7–8.12}$$

$T^{(3)}$ is thus positive or negative according as $\omega_o^{(2)}$ is positive or negative. This relation is the analog of Eq. (7–2.9).

From here on, the analysis is sufficiently similar to that given in Section 7–2 for details to be omitted. This eventually leads to the relation

$$\frac{T}{T_\infty} = \frac{1}{1 - (\omega_o^{(2)}/\omega)} \tag{7–8.13}$$

analogous to Eq. (7–2.12). By appropriate dimensional arguments one may conclude that

$$\omega_o^{(2)} = K\frac{T_\infty}{8\pi\mu l^3} \tag{7–8.14}$$

where K is a dimensionless constant of $O(1)$ which is independent of the size and shape of the particle, depending only on S. Note that l is, as before, a characteristic distance from particle to boundary. Different definitions of l give rise to different numerical values of K in such a way that the over-all result is unaffected. By analysis of the errors incurred in our various approximations, and by invoking arguments similar to those in Section 7–2, we are led to the final result

$$\frac{T}{T_\infty} = \frac{1}{1 - K(T_\infty/8\pi\mu l^3\omega) + O(c/l)^5} \tag{7–8.15}$$

where c is a characteristic particle dimension. This result is the counterpart of Eq. (7–2.15), which gives the influence of walls on a translating particle.

In general, $T_\infty = O(8\pi\mu c^3\omega)$. Thus, it may be seen from the preceding relation that the magnitude of the wall effect for a rotating particle depends on terms of $O(c/l)^3$. This is in contrast to the wall effect for a translating particle, which depends on terms of $O(c/l)$. The wall effect is therefore very much smaller in the former case. The smallness of the effect has already been commented on by Jeffery[36] in reference to the problem of a sphere rotating near a plane wall. In consequence of this, Eq. (7–8.15) applies to very much larger c/l ratios than does its counterpart for translation. For example, when a spherical particle of radius c rotates about an axis perpendicular to a solid, infinite plane wall situated at a distance l from its center, Eq. (7–8.15) gives $T/T_\infty = 1.055$ and 1.110 for $c/l = 0.7477$ and 0.925, respectively. The *exact* values tabulated by Jeffery[36] for these two cases are 1.057 and 1.126, respectively.

The general *form* of Eq. (7–8.15), including the error estimate, is confirmed by Jeffery's[36] solutions for spheroidal particles rotating at the center of confocal spheroidal shells. Analogous agreement exists for spherical particles in diverse situations, except that the actual error is very much smaller for such particles than that indicated by Eq. (7–8.15). For example, Lamb's[39] solution for a spherical particle rotating at the center of a concentric spherical shell yields Eq. (7–8.15) exactly, with no error whatsoever, regardless of the ratio of inner to outer sphere radii. On the other hand, Jeffery's[36] solution for a sphere near a plane boundary leads to an error of $O(c/l)^8$, rather than the one shown in Eq. (7–8.15). It appears that spherical particles constitute a degenerate case with respect to the magnitude of the error implicit in the approximate solution.

In the case of a spherical particle of radius a, $T_\infty = 8\pi\mu a^3\omega$; Equation (7–8.15) therefore becomes

$$\frac{T}{T_\infty} = \frac{1}{1 - K(a/l)^3 + \cdots} \tag{7–8.16}$$

The K value for any particular surface S can be obtained by comparing the foregoing with the known solution for a spherical particle. Results are tabulated in Table 7–8.1. Only a very few solutions of the equations of motion are known for spheres rotating near boundaries. These are considered subsequently. The first two cases in Table 7–8.1 are drawn from Jeffery's[36] article on the rotation of bodies of revolution; the third case, from recent independent investigations by Haberman[29] and Brenner and Sonshine[10].

Other K values can be deduced from Jeffery's solutions for a particle at the center of a spheroidal boundary and for a non-centrally situated particle in a spherical boundary.

No experimental data appear to exist for nonspherical particles against which Eq. (7–8.15) may be checked. This seems surprising as the subject appears to be of some interest in the general theory of rotational viscometers.

As an example of other possible applications to which the theory may be put, consider the problem of two equal circular disks (radii $= c$), rotating parallel to each other about their line of centers in an infinite fluid. Let $2l$ denote the distance between them and suppose that they rotate with the same angular speed ω, either in the same or opposite directions. The plane midway between them then behaves as either a free surface or solid plane, according as they rotate in the same or opposite directions. This problem falls within the scope of Case II (Table 7–8.1). Inasmuch as $T_\infty = (32/3)\mu c^3\omega$ (Jeffery[36]) we obtain, for the couple on either disk required to maintain the uniform rotation,

$$T = \frac{(32/3)\mu c^3\omega}{1 \pm [4/(3\pi)](c/l)^3 + O(c/l)^5} \tag{7–8.17}$$

the upper or lower sign being taken according as the disks rotate in the same or opposite directions. In the first case the couple is decreased by the presence of the other disk, whereas it is increased in the second.

TABLE 7–8.1
VALUES OF K FOR USE IN EQ. (7–8.15)

CASE I: Particle rotating at the center of an outer sphere with solid walls; l = radius of outer sphere:

$$K = 1$$

CASE II: Particle rotating about an axis which is perpendicular to a single, infinite plane surface; l = distance from center of particle to plane:

(a) *Solid* plane:

$$K = 1/8$$

(b) *Frictionless* plane (free surface):*

$$K = -1/8$$

CASE III: Particle rotating about the longitudinal axis of a circular cylinder; l = radius of cylinder:

$$K = 0.79682417$$

*This is obtained from Jeffery's solution for the rotation, about their line of centers, of two equal spheres, external to each other in an infinite fluid, when they rotate at the same angular velocity. The plane of symmetry midway between them is then a "frictionless" plane.

Rotation of a sphere inside a second sphere

Landau and Lifshitz[40] discuss the slow motion of fluid contained in the space between two concentric spheres of radii a_1 and a_2, respectively ($a_2 > a_1$). Both spheres rotate uniformly about generally different diameters with angular velocities $\boldsymbol{\omega}_1$ and $\boldsymbol{\omega}_2$. The angular Reynolds number, $\rho a^2\omega/\mu$, is assumed small compared with unity. Because of the linearity of the pertinent equations, it is possible to solve the problem by superposition of the two motions obtained when one sphere is at rest and the other rotates. The pressure field is zero; the velocity field is found to be

$$\mathbf{v} = \frac{1}{(1/a_1)^3 - (1/a_2)^3}\left[\left(\frac{1}{r^3} - \frac{1}{a_2^3}\right)\boldsymbol{\omega}_1 + \left(\frac{1}{a_1^3} - \frac{1}{r^3}\right)\boldsymbol{\omega}_2\right] \times \mathbf{r} \tag{7–8.18}$$

where $\mathbf{r}$ is measured from the center.

A simple calculation shows that the torque $\mathbf{T}_1$ on the inner sphere is

$$\mathbf{T}_1 = -8\pi\mu\frac{\boldsymbol{\omega}_1 - \boldsymbol{\omega}_2}{(1/a_1)^3 - (1/a_2)^3} \tag{7–8.19}$$

or, equivalently,

$$\mathbf{T}_1 = -\frac{8\pi\mu a_1^3(\boldsymbol{\omega}_1 - \boldsymbol{\omega}_2)}{1 - (a_1/a_2)^3} \tag{7–8.20}$$

The torque $\mathbf{T}_2$ on the outer sphere is equal and opposite.

Ghildyal[26] discusses the *unsteady* slow motion of a viscous liquid contained between two concentric spheres. One case considered, for example, is that in which the outer sphere is given an impulsive twist so as to rotate with a uniform velocity while the inner sphere is kept fixed. A general unsteady state solution of equations of creeping motion for an incompressible fluid is obtained by integral transform methods. After a sufficiently long time the steady part dominates and the solution reduces to that obtainable from Eq. (7–8.18).

Sphere rotating in a viscous liquid inside a coaxial circular cylinder

Haberman[29] and, independently, Brenner and Sonshine[10] studied the slow symmetric rotation of a sphere (radius $= a$) in a viscous fluid bounded externally by an infinitely long circular cylinder (radius $= R_o$) having the sphere at its center. The latter solution[10] is exact and completely covers the range from $a/R_o = 0$ to 1. For $a/R_o < 0.9$, the torque on the sphere rotating with angular velocity ω is accurately expressed by the formula[10]

$$T = \frac{8\pi\mu a^3\omega}{1 - 0.79682417(a/R_0)^3 - 0.060047040(a/R_0)^{10} + O(a/R_o)^{14}} \tag{7–8.21}$$

If the cylinder itself is also rotating, one need only interpret ω in the preceding equation to be the algebraic difference between the angular velocities of sphere and cylinder. In either case, the torque on the cylinder is equal and opposite to that on the sphere.

Slow viscous rotation of an axisymmetric body in a circular cylinder of finite length

Brenner[8] has studied the effect of boundary proximity on the rotation of any axially symmetric body whose size is small compared with the exterior boundary dimensions, using the point couple approximation procedure previously outlined.

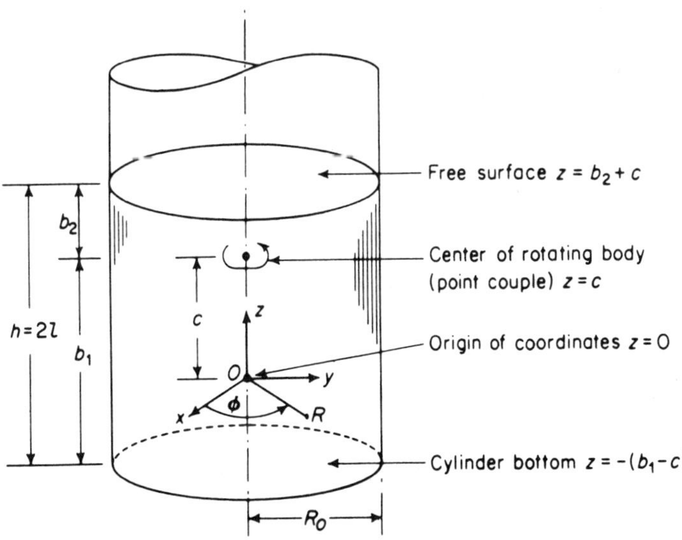

Figure 7-8.1. Definition sketch for a body rotating in a finite cylinder.

The creeping motion equations are employed to solve the case in which the particle rotates symmetrically about the axis of a circular cylinder filled to a *finite* depth with viscous liquid. The body may occupy any position along the cylinder axis, as shown in the definition sketch, Fig. 7–8.1. The body rotates about the longitudinal axis (z axis) of a vertical circular cylindrical vessel of radius R_o filled to a finite depth $h = 2l$ with liquid. Circular cylindrical coordinates (R, ϕ, z) are employed. The center of the body is situated at a distance $z = c$ from the origin of coordinates, O. The free surface of the liquid is located at a vertical distance b_2 above the center of the body. Likewise, the bottom of the cylinder is situated at a distance b_1 below the center of the body.

Suppose the body rotates with the same angular velocity ω in both the unbounded and bounded fluids. The body then experiences couples T and T_∞, respectively. The problem consists of finding the torque ratio T/T_∞ in terms of the three relative particle-to-boundary dimensions, a/R_o, a/b_1, and a/b_2. Here, a represents a characteristic particle dimension. These three particle-to-boundary dimensions are each assumed small.

The calculation is accomplished by replacing the actual body by a *point* couple concentrated at its center of hydrodynamic stress. Equation (7–8.13) is then employed to find the increased couple. The principal problem is that of calculating the field $\mathbf{v}^{(2)}$, the *simultaneous* reflection of the point couple field $\mathbf{v}^{(1)}$ from *all* the bounding surfaces. With $\mathbf{v} = \mathbf{v}^{(1)} + \mathbf{v}^{(2)}$, the equations of motion reduce to the single scalar equation,

$$\nabla^2 v - \frac{v}{R^2} = \frac{\partial^2 v}{\partial R^2} + \frac{1}{R}\frac{\partial v}{\partial R} - \frac{v}{R^2} + \frac{\partial^2 v}{\partial z^2} = 0 \qquad (7\text{–}8.22)$$

The vector velocity $\mathbf{v} = \mathbf{i}_\phi v$, in which $\mathbf{i}_\phi$ is a unit vector in the ϕ direction and $v = v(R, z)$ is independent of the azimuthal angle ϕ. The boundary conditions to be satisfied are:

(i) at the cylinder wall, $R = R_o$,

$$v = 0 \qquad (7\text{–}8.23)$$

(ii) at the bottom of the container, $z = -(b_1 - c)$,

$$v = 0 \qquad (7\text{–}8.24)$$

(iii) at the free surface at the top of the container, $z = b_2 + c$,

$$\frac{\partial v}{\partial z} = 0 \qquad (7\text{–}8.25)$$

(iv) in addition we require that as R_o, b_1, and $b_2 \to \infty$,

$$v \to v^{(1)} \qquad (7\text{–}8.26)$$

The original point couple field, $v^{(1)}$, given by Eq. (7–8.9), is reflected from all three boundaries simultaneously. The final result obtained is expressed in the form of Eq. (7–8.15) in which $l = R_o$ and

$$K = V\left(\frac{b_1}{h}, \frac{h}{R_o}\right) \tag{7–8.27}$$

where $h = b_1 + b_2$ is the total depth of liquid, and

$$V\left(\frac{b_1}{h}, \frac{h}{R_o}\right) = C_1 + \sum_{n=1}^{\infty} \frac{k_n}{J_0^2(k_n)} \times \\ \times \frac{2\exp(-2k_n h/R_o) + \exp(-2k_n b_1/R_o) - \exp(-2k_n b_2/R_o)}{1 + \exp(-2k_n h/R_o)} \tag{7–8.28}$$

where $C_1 = 0.79682417$ and the k_n are the roots of the transcendental equation $J_1(k_n) = 0$. The J's denote Bessel functions of the first kind. This series converges rapidly for large values of h/R_o. For small values of h/R_o an alternative form is given for numerical calculation. As we let b_1 and/or b_2 and/or R_o become infinite, the formula (7–8.27) reduces to various limiting cases, many of which may be obtained directly by using the K values tabulated in Table 7–8.1.

Brenner gives a large number of numerical results, a few of which are reproduced as appropriate values of $V(b_1/h, h/R_o)$ in Table 7–8.2.

An interesting feature of the treatment is that even though a/b_1, a/b_2, and a/R_o are each small, the calculation shows explicitly that one cannot obtain the combined wall-effect by simply superposing the wall-effects arising from the presence of each boundary separately. Rather, there is an interaction among the boundaries which depends on the ratios b_1/h, b_2/h, and h/R_o. And these ratios may be large or small without violating the condition that a/b_1, a/b_2, and a/R_o each be small compared with unity.

TABLE 7–8.2

VALUES OF $V(b_1/h, h/R_o)$ FOR USE IN EQ. (7–8.27)

b_1/h \ h/R_o	1	3	6	10
0.1	125.01654	4.8109122	1.0527617	0.8079821
0.5	0.8191296	0.7968242	0.7968242	0.7968242
0.9	−123.37828	−3.2172638	0.5408867	0.7856663

Rotation of circular cylinders

The motion of a fluid between two infinite coaxial cylinders, rotating about their common axis with constant angular velocities is discussed by Landau and Lifshitz[40, p. 60]. The stability of such flows has been the subject of considerable investigation.[41] The solution of the more complex problem of the motion of a viscous fluid in a narrow space between cylinders whose axes are parallel but not coincident may be found in Kochin, *et al.*[37, p. 419] and Sommerfeld[55].

Howland and Knight[33] have dealt with the case where a cylinder rotates about its own axis using a method developed by Howland for the state of stress in an infinite strip with a single circular hole.

Citron[11] has treated the case of slightly deformed but axisymmetric cylinders rotating. There is also a similar paper dealing with *axial* (axisymmetric) flow through a slightly deformed pipe with sinusoidal variations in cross section, but with no inner pipe.[2] The latter investigation is not restricted to the low Reynolds numbers characterizing creeping flows.

7-9 Unsteady Motion of a Sphere in the Presence of a Plane Wall

Wakiya[63] considered the case of unsteady motion of a sphere parallel to a plane wall, where the flow can be described by the unsteady form of the creeping motion equations,

$$\rho \frac{\partial \mathbf{v}}{\partial t} = -\nabla p + \mu \nabla^2 \mathbf{v} \tag{7-9.1}$$

$$\nabla \cdot \mathbf{v} = 0 \tag{7-9.2}$$

The equations are solved for the situation where the fluid is initially at rest, and the sphere velocity is some arbitrarily prescribed function of elapsed time. The method of solution involves the reflection technique and the use of the Laplace transform with respect to time. The transformed problem is similar in mathematical form to the Oseen equations of motion for the steady state.

For the problem in which the motion is impulsive, so that the sphere is given a motion of constant velocity U, the force on the sphere at time t is expressed by the relation

$$\frac{F}{6\pi\mu a U} = 1 + \frac{a}{\sqrt{\pi \nu t}} + \frac{9}{16}\frac{a}{l} K\left(\frac{l}{\sqrt{\nu t}}\right) \tag{7-9.3}$$

The third term on the right-hand side of Eq. (7–9.3) is given in terms of the dimensionless group $\xi = l/\sqrt{\nu t}$, where l is the distance of the center of the sphere from the wall and ν the kinematic viscosity:

$$K(\xi) = 1 - \frac{16}{9\sqrt{\pi}}\xi + \frac{8}{9\sqrt{\pi}}\xi^3 - \frac{1}{6}\xi^4 + O(\xi^5) \quad \text{for } \xi < 1 \tag{7-9.4}$$

$$K(\xi) = \frac{1}{3}\xi^{-2} + \frac{4}{3\sqrt{\pi}}\xi^{-3} + O(\xi^{-4}) \quad \text{for } \xi > 1 \tag{7-9.5}$$

Equation (7–9.4) will apply to the situation in which, for a given geometry, a very long time is allowed to elapse. The second term on the right-hand side of Eq. (7–9.4) will cancel the second term on the right side of Eq. (7–9.3), so that to a first approximation

$$\frac{F}{6\pi\mu a U} = 1 + \frac{9}{16}\frac{a}{l} + \frac{1}{2}\frac{a l^2}{\sqrt{\pi}\,(\sqrt{\nu t})^3} \tag{7-9.6}$$

For sufficiently large t this reduces to the steady state case. Likewise, if we consider the case where a relatively short time is involved and the wall is at a great distance from the particle, substitution of Eq. (7–9.5) into Eq. (7–9.3) gives

$$\frac{F}{6\pi\mu aU} = 1 + \frac{a}{\sqrt{\pi\nu t}} + \frac{9}{16}\frac{a}{l}\left(\frac{\nu t}{3l^2}\right) \tag{7–9.7}$$

which is in agreement with known results[40] for the case where no wall is present.

Equations (7–9.6) and (7–9.7) may be employed to investigate the rate at which steady state is approached for a given geometry. In the presence of a wall as $t \to \infty$ the correction for unsteady state will vary as $1/t^{3/2}$. In the absence of a wall the correction for unsteady state will vary as $1/t^{1/2}$. Thus, the presence of a wall results in a more rapid approach to steady state at very long times. For the intermediate case where $\xi = 1$, a more general development is necessary. Such a development is also given in Wakiya's paper in terms of expressions involving the incomplete gamma function.

BIBLIOGRAPHY

1. Bart, E., M. S. Thesis, New York University, 1959.
2. Belinfante, D. C., Proc. Camb. Phil. Soc. **58** (1962), 405.
3. Bohlin, T., Trans. Roy. Inst. Technol. (Stockholm), No. 155 (1960); see also comments by H. Faxen, Kolloid Z. **167** (1959), 146.
4. Brenner, H., Chem. Eng. Sci. **17** (1962), 435.
5. ———, J. Fluid Mech. **12** (1962), 35.
6. ———, and J. Happel, J. Fluid Mech. **4** (1958), 195.
7. ———, Chem. Eng. Sci. **16** (1961), 242.
8. ———, Appl. Sci. Res. (ser. A) **13** (1964), 81.
9. ———, J. Fluid Mech. **18** (1964), 144.
10. ———, and R. M. Sonshine, Quart. J. Mech. Appl. Math. **17** (1964), 55.
11. Citron, S. J., J. Appl. Mech. **29** (1962), 188.
12. Craig, F. F., Ph. D. Thesis, Univ. Pittsburgh, 1951; see Univ. Pittsburgh *Bull.* **48**, No. 10 (1952).
13. Famularo, J., D. Eng. Sci. Thesis, New York University, 1962.
14. ———, Private communications (1963).
15. Faxen, H., Arkiv. Mat. Astron. Fys. **17**, No. 27 (1923); dissertation, Uppsala Univ., 1921; see also Oseen's[46] treatise.
16. ———, Arkiv. Mat. Astron. Fys. **20**, No. 8 (1927).
17. ———, Ann. Phys. **68** (1922), 89; the term in $(a/l)^4$ has been added only recently (Faxen, H., private communication, 1964).
18. ———, Arkiv. Mat. Astron. Fys. **18**, No. 29 (1924).

19. ———, Arkiv. Mat. Astron. Fys. **19**, No. 22 (1925).

20. ———, and H. Dahl, Arkiv. Mat. Astron. Fys. **19A**, No. 13 (1925).

21. ———, Neuvième Congrès des Mathemeticiens Scandinaves, Helsingfors (1938).

22. ———, Proc. Roy. Swedish Inst. Eng. Res. (Stockholm), No. 187 (1946).

23. Fayon, A. M., and J. Happel, A. I. Ch. E. Jour. **6** (1960), 55.

24. Goldsmith, H. L., and S. G. Mason, J. Coll. Sci. **17** (1962), 448.

25. Fidleris, V., and R. L. Whitmore, Brit. J. Appl. Phys. **12** (1961), 490.

26. Ghildyal, C. D., Z. Angew. Math. Mech. **42** (1962), 508.

27. Haberman, W. L. and R. M. Sayre, David Taylor Model Basin *Report* No. 1143. Washington, D.C.: U. S. Navy Dept., 1958.

28. ———, David Taylor Model Basin *Report* No. 1563. Washington, D.C.: U.S. Navy Dept., 1961.

29. ———, David Taylor Model Basin *Report* No. 1578. Washington, D.C.: U.S. Navy Dept., 1961.

30. Happel, J., and B. J. Byrne, Ind. Eng. Chem. **46** (1954), 1181; corrections-Ind. Eng. Chem. **49** (1957), 1029.

31. ———, and H. Brenner, A.I.Ch.E. Jour. **3** (1957), 506; also supplement deposited as document 5441 with the American Documentation Institute, Library of Congress, Washington, D.C.

32. ———, and P. A. Ast, Chem. Eng. Sci. **11** (1960), 286.

33. Howland, R. C. J., and R. C. Knight, Proc. Cambr. Phil. Soc. **29** (1933), 277.

34. Takaisi, Y., J. Phys. Soc. Japan **10** (1955), 407; **11** (1956), 1004.

35. Jeffery, G. B., Proc. Roy. Soc. **A102** (1922), 161.

36. ———, Proc. Lond. Math. Soc. **14** (1915), 327.

37. Kochin, N. E., I. A. Kibel, and N. U. Rose, *Theoretical Hydrodynamics*, Vol. 2, 3rd ed., (in Russian). Moscow, 1948.

38. Ladenburg, R., Ann. Phys. **23** (1907), 447.

39. Lamb, H., *Hydrodynamics*, 6th ed. Cambridge: Cambridge Univ. Press, 1932.

40. Landau, L. P., and E. M. Lifshitz, *Fluid Mechanics*. Reading, Mass.: Addison-Wesley, 1959.

41. Lin, C. C., *The Theory of Hydrodynamic Stability*. Cambridge: Cambridge Univ. Press, 1955.

42. Lorentz, H. A., Abhand. theor. Phys., Leipzig **1** (1907), 23.

43. MacKay, G. D. M., M. Suzuki, and S. G. Mason, J. Coll. Sci. **18** (1963), 103.

44. McNown, J. S., H. M. Lee, M. B. McPherson, and S. M. Engez., Proc. Seventh Intern. Cong. Appl. Mech., London, 1948.

45. Oberbeck, A., J. reine angew. Math. **81** (1876), 62.

46. Oseen, C., *Neuere Methoden und Ergebnisse in der Hydrodynamik*. Leipzig: Akademische Verlagsgesellschaft, 1927.

47. Pfeffer, R., and J. Happel, Unpublished experiments (1958).

48. Rothfus, R. R., C. C. Monrad, and V. E. Senecal, Ind. Eng. Chem. **42** (1950), 2511.

49. Rubinow, S. I., and J. B. Keller, J. Fluid Mech. **11** (1961), 447.

50. Segré, G., and A. Silberberg, J. Fluid Mech. **14** (1962), 115, 136.

51. Simha, R., Kolloid Z. **76** (1936), 16.

52. Slezkin, N. A., *Dynamics of Viscous Incompressible Fluids* (in Russian). Moscow: Gos. Izdat. Tekh.-Teor. Lit., 1955.

53. Smythe, W. R., Phys. Fluids **4** (1961), 756.

54. Stimson, M., and G. B., Jeffery, Proc. Roy. Soc. **A111** (1926), 110.

55. Sommerfeld, A., *Mechanics of Deformable Bodies*. New York: Academic Press, 1950.

56. Takaisi, Y., J. Phys. Soc. Japan **10** (1955), 685; **11** (1956), 1092.

57. Wakiya, S., J. Phys. Soc. Japan **8** (1953), 254.

58. ———, Res. Rep. Fac. Eng. Niigata Univ. (Japan) **5** (1956), 1.

59. ———, J. Phys. Soc. Japan **12** (1957), 1130.

60. ———, Res. Rep. Fac. Eng. Niigata Univ. (Japan) **9** (1960), 31.

61. ———, Res. Rep. Fac. Eng. Niigata Univ. (Japan) **7** (1958), 1.

62. ———, Res. Rep. Fac. Eng. Niigata Univ. (Japan) **8** (1959), 17.

63. ———, Res. Rep. Fac. Eng. Niigata Univ. (Japan) **10** (1961), 15.

64. Watson, G. N., *A Treatise on the Theory of Bessel Functions*. Cambridge: Cambridge Univ. Press, 1922.

65. Westberg, R., Proc. Roy. Swedish Academy Eng. Sci. (Stockholm) No. 197, (1948).

66. White, C. M., Proc. Roy. Soc. **A186** (1946), 472.

Flow Relative to Assemblages of Particles

8

8-1 Introduction

Fluid flow relative to assemblages of particles represents an area of interest in many fields of science and technology[16] as discussed in Chapter 1. An extremely wide range of problems is involved and, as would be expected, the number of possible variables is also large. A hydrodynamic treatment therefore constitutes a considerable idealization.

In order to characterize a fluid-particle system the following factors must usually be considered:

(a) The nature of the fluid: This may often be specified by information on the viscosity and density at the temperatures and pressures involved. In some cases, non-newtonian behavior may obtain. In more complicated instances, more than one fluid may be involved, in which case such properties as surface or interfacial tension may be important.

(b) The nature of the solid medium: The density will be a significant parameter, unless the solids are supported by the walls of the containing vessel rather than by hydrodynamic forces. The size and shape of the particles involved, and the variation of these items, is of great importance. Differences in shape may be evaluated in many cases by appropriate shape factors, which empirically relate irregular particles to equivalent spheres[37]. The concept of a shape factor is normally valid only for smooth, regularly shaped particles. Surface roughness also may have an important effect on flow characteristics. This effect is difficult to evaluate independently of shape, especially in the case of small particles.

Nonuniformity of particle size is usually taken into account by use of an average diameter. Many methods have been proposed for properly establishing this average[37]. It is difficult to obtain definitive data to determine which method is best, because changes in mixture components usually result in simultaneous changes in other properties of the system, such as fractional void volume.

(c) The proportion of solid to fluid: This is usually specified in terms of the fractional void volume. The term *fractional void volume* here includes only the proportion of voids which exists between granules through which the fluid moves. It should be distinguished from *porosity*, which is often defined as the fractional pore space within granules of a porous material. Investigations of flow through assemblages of particles have covered a wide range of fractional void volumes, limited at one extreme by fluid passing around a single particle, and at the other by flow through irregular channels in a solid body.

(d) Motion of solids and fluid: Motion of the solid and fluid phases relative to the walls of the containing vessel serves further to classify the type of system involved. Motion of fluid and solids relative to each other, and independent motion of parts of the fluid or solid phases, is also of importance. The velocity of the fluid flow relative to various boundaries is of special importance. At low velocities, the creeping motion equations will apply. With the onset of inertial effects, it is necessary to employ the complete Navier-Stokes equations or modifications thereof which include the convective terms. Mathematical analysis under conditions of turbulent flow is difficult even in the case of a single spherical particle.

Other variables, such as the effect of containing vessel walls, Brownian movement, solids friction, fluid slip, and electrostatic effects, are important in special circumstances. Thus, it is evident that no complete theoretical treatment is possible covering all situations.

The creeping motion equations do, however, furnish a practically useful basis for achieving a uniform treatment and better understanding of the principles underlying these phenomena. This basic treatment is best presented in terms of the classification scheme of proportion of solid to fluid—see item (c). Mathematical techniques applicable to *low solids content systems* (*dilute*) are first considered. In this case the effect of the boundary walls of the containing vessel cannot always be neglected. This is followed by consideration of techniques applicable in more *concentrated systems*.

In order to construct tractable mathematical models of the flow systems involved, it is necessary to resort to a number of simplifications. In addition to omission of inertia terms from the Navier-Stokes equations, other simplifications include the assumption of equal-sized spherical particles, fluid adherence without slippage on particle surfaces, neglect of Brownian motion effects, and the absence of extraneous forces other than gravity. In solving the boundary value problems involved, we are interested in predicting the motion of particles and fluid or, in some cases, resistance to flow due to the presence of particles. As will be seen, a concise mathematical treatment of many fluid-solids flow relationships is possible within the scope of these restrictions.

The results are applicable to the considerable volume of existing experimental data. It is expected that these theoretical results will either prove directly useful, or will serve as a guide in the development of semiempirical relationships, often employed in practical situations involving particulate systems. Practical applications are discussed essentially in terms of the classification scheme based on relative motion of solids and fluid—see item (d). Where the particles settle under the influence of gravity, but the fluid experiences no net motion, the particle movement is designated as *sedimentation.* Conversely, when the particles remain stationary, we are interested in the pressure drop experienced by passage of fluid through the particle assemblage. In such cases the particles are usually held immobile by contact with each other, forming *packed beds.* In such beds the solids concentration is normally higher than in most sedimenting systems. At intermediate concentrations, particles may move relative to each other, as well as with respect to the fluid, since they are not held immobile by interparticle contacts. The phenomena of *fluidization* and *hindered settling* fall into this category.

Naturally, it is not possible to treat these various technological applications comprehensively within the scope of this book. Emphasis here is on the basic principles underlying the systems treated, from the standpoint of the creeping motion equations.

8-2 Dilute Systems—No Interaction Effects

In order to study the behavior of a multiple assemblage of spheres, it is necessary to consider the effects of the wall of the containing vessel. For this work a cylindrical boundary is chosen because it is the simplest one which can completely surround the fluid stream parallel to the direction of flow, and because it is the shape usually encountered in practice. As will be seen, this enables one to treat both pressure drop and sedimentation.

We propose first to develop the "zeroth" order approximation, wherein it is assumed that the spheres are very far apart, that is, $a/l \ll 1$ (where a is sphere radius and l is the distance between any two spheres), and that the radius of each sphere is small compared with that of the containing vessel, $a/R_o \ll 1$ (R_o is the radius of the cylindrical container). Two limiting conditions can be realized for very dilute systems, depending on the ratio of surface area of solid particles to the walls of the containing vessel. This ratio will be proportional to $(a/l)^3(R_o/a)$. In order for the surface of the particles to be small with respect to the area of container walls, we thus require that $(a/l)^3(R_o/a) \ll 1$. In any case, in order for a suspension to exist, there should be at least a single column of particles falling at a distance apart no greater than the distance of each from the wall. Thus, we have the additional restriction that $a/l < R_o/a$. If, say, $a/R_o = 0.01$ and we wish $(a/l)^3(R_o/a) = 0.01$, we must have $(a/l)^3 = 0.0001$ and $a/l = 0.0464$. Hence, should it prove necessary to restrict ourselves to particle/wall area ratios below 0.01, the "zeroth" approximation for this limiting case would be restricted to very dilute suspensions. The other limiting condition arises when $(a/l)^3(R_o/a) \gg 1$. It is readily seen that this condition is realized when the radius of the containing vessel is large with respect to particle radius.

In the following treatment we assume that the velocity field in the cylinder will consist of the original undisturbed Poiseuille flow (empty tube), except possibly for such distribution effects as may be postulated at the ends. For Poiseuille's law to apply in a dilute system, it is necessary that the particle-wall areal ratio be small. Our treatment is thus restricted to cases where $(a/l)^3(R_o/a) \ll 1$. But without a more involved development taking a/l and a/R_o into account, it is impossible to specify exactly how small this surface ratio must be. A nontrivial case which arises at the other limiting condition, where $(a/l)^3(R_o/a) \gg 1$, occurs when $R_o/a \to \infty$, corresponding to flow through an unbounded assemblage of particles. For this situation we may use a uniform approach velocity to approximate the dynamics of the system corresponding to the absence of walls. This second case may be approximated even in very dilute suspensions where small particle sizes are involved. It is apparent that if the slip velocity is sufficiently small, so particles are carried along with the fluid, the disturbance to the parabolic pattern induced by the walls will be minimized. For conditions other than the limiting ones just cited, where $(a/l)^3(R_o/a) = O(1)$, it would be necessary to take into account both a/l and a/R_o. The "zeroth" approximation discussed below would not then be applicable.

In all cases, the drag exerted on each sphere will be given by Stokes' law with the approach velocity evaluated at the location of the center of the particle. The force on each particle will thus be given as follows:

$$F = -6\pi\mu a\left[U - U_{OF}\left(1 - \frac{b^2}{R_o^2}\right)\right] = 6\pi\mu a\left[(U_{OF} - U) - U_{OF}\left(\frac{b}{R_o}\right)^2\right] \tag{8-2.1}$$

The notation is explained in the next paragraph.

The rate of energy dissipation stems from three sources[6]: (a) the translation of the particles relative to the surrounding fluid, (b) the rotation of the particles relative to the fluid, (c) the inability of a solid particle to undergo deformation and thereby accommodate itself to the dilational components of the original fluid motion. In the case of small spherical particles, the rotational contribution to energy dissipation will normally vanish. In many macroscopic systems, where the densities of particles and fluid are not matched, the dilational contribution, which constitutes an effective suspension "viscosity," will also be small with respect to that due to frictional drag. In what follows here, the former contribution will therefore be neglected. The matter of the dilational contribution is developed in detail in Chapter 9. Energy dissipation gives rise to pressure drop due to fluid flow relative to a suspension.[40] For the case of a single spherical particle immersed in a Poiseuille flow, this pressure drop is as follows:

$$\Delta P_S = \frac{12\mu a}{R_o^2}\left(1 - \frac{b^2}{R_o^2}\right)\left[(U_{OF} - U) - U_{OF}\left(\frac{b}{R_o}\right)^2\right] \tag{8–2.2}$$

This expression is taken from the development in Section 7–3. Here, b refers to the radial distance at which the center of a sphere (radius $= a$) is located from the cylinder axis. U is the velocity with which the sphere moves in a direction parallel to the axis of a cylinder. The fluid is in laminar flow with a center-line velocity U_{OF}. (The velocity U_{OF} is taken at a sufficiently great distance from the sphere so that the parabolic velocity pattern is not disturbed. At this point the average or superficial velocity is $U_{MF} = \frac{1}{2}U_{OF}$). All velocities are measured relative to the cylinder walls. As just noted, the assumptions inherent in Eqs. (8–2.1) and (8–2.2) correspond to the limiting case of a low particle-wall areal ratio. It is important to observe that the pressure drop will be different from that required to support the particle itself, essentially equivalent to multiplication of Stokes' law drag by a factor $2[1 - (b^2/R_o^2)]$. Thus the additional pressure drop due to a small particle situated at the cylinder axis will be just double that required hydrodynamically to support the weight of the particle.

For the limiting case $R_o/a \to \infty$, where a parabolic flow pattern is not maintained because of the occurrence of a high particle-wall areal ratio, the assumption of a cell model[35] with perfect slip at the walls of each cell, corresponding to the complete absence of container walls, yields an average pressure drop force equal to the sum of the Stokes' law drags on the particles.

Several additional items are worth noting in connection with the application of this simplified model. First, since there is no interaction between particles, one can make no prediction of the effect of changes in the void volume between particles. It is assumed, in effect, that the particles are sufficiently far apart that such changes will occur only over a very long period of time, and hence may be neglected. Actual bed depth, corresponding to a given number of particles and a prescribed mean fluid velocity, must be specified independently. Similarly, on purely hydrodynamic grounds, each particle will move only in an axial direction. No collisions are assumed, and the pattern of motion continues unchanged along the entire tube length.

It is assumed that the particles suspended in the cylinder extend for an effectively infinite distance axially. Boundary conditions at the inlet and outlet control the radial particle distribution prevailing in the tube. This particle distribution, together with the particle terminal settling velocity and fluid velocity, constitute the three basic variables which influence the behavior of an assemblage. In the following discussion, three basic cases are selected for specific development: (a) the particles do not move relative to each other and are uniformly distributed; (b) the particles are free to move relative to each other but are still uniformly distributed; (c) the particles move relative to each other but are not uniformly distributed.

First the simplest type of assemblage behavior will be considered[40]; namely, the case in which the particles suspended in a cylinder do not move relative to each other, and in which they are randomly distributed throughout the cylinder cross section for an infinite distance axially. In practice, this case is realized in sedimentation of a mass of particles in a quiescent fluid. The particles all fall at the same velocity and so maintain a fixed position relative to each other. Where fluid motion is involved, bridging or solid-solid contacts between particles must be maintained or the particles will tend to move relative to each other. Happel and Epstein[43] studied the case of bridging in dilute systems in experiments where pressure drop through rigid assemblages of spheres mounted on wires was measured. In such assemblages the pressure drop is obtainable from Eq. (8–2.2) as follows:

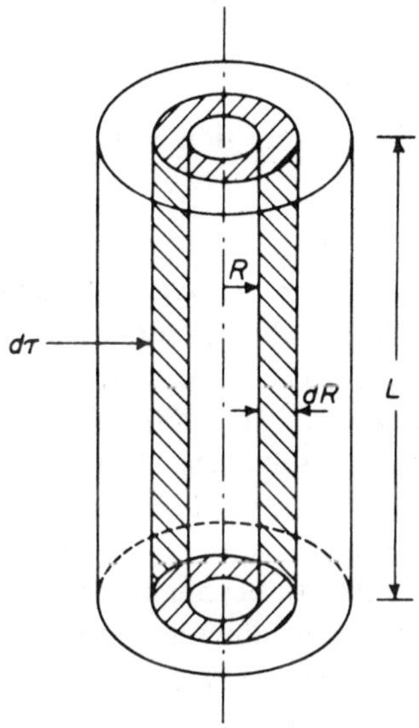

Figure 8-2.1. Annular element of volume.

Referring to Fig. 8-2.1, consider the contribution to pressure drop by the particles in the annular volume $d\tau$ bounded by radii R and $R + dR$. The volume is $d\tau = L(2\pi R\, dR)$ where L is the bed length. The number of spheres contained in this volume is $NL(2\pi R\, dR)$, where N is the local number of particles per unit volume. Thus, the pressure drop resulting from the particles in $d\tau$ is

$$d(\Delta P_S) = NL(2\pi R\, dR)\left(\frac{12\mu a}{R_o^2}\right)\left(1 - \frac{R^2}{R_o^2}\right)\left[(U_{OF} - U) - U_{OF}\frac{R^2}{R_o^2}\right] \quad (8\text{–}2.3)$$

Integration between the limits $R = 0$ and R_o gives the total pressure drop,

$$\Delta P_S = N_M L 8\pi\mu a\left(U_{MF} - \frac{3U}{4}\right) \quad (8\text{–}2.4)$$

for the case where $N = N_M =$ constant is independent of R. Here, as in the remainder of this discussion, pressure drops arising from fluid flow alone (Poiseuille's law) and from the effect of static head of the fluid are not included in the term ΔP_S, which refers only to the disturbance due to the presence of particles. For the case of sedimentation, where $U_{MF} = 0$, the pressure drop $\Delta P_S = -N_M L\, 6\pi\mu a U$, corresponds to a summation of the drag on the individual particles. For the case where the particle velocity $U = 0$, we have $\Delta P_S = N_M L\, 8\pi\mu a U_{MF}$.

These equations may be expressed in terms of the prevailing fractional void volume as follows: If there are q particles contained in a volume of cylinder of length L and radius R_o, the fractional void volume ϵ will be

$$\epsilon = 1 - \frac{q(4/3)\pi a^3}{\pi R_o^2 L} \tag{8–2.5}$$

Thus the number N_M of particles per unit volume is

$$N_M = \frac{q}{\pi R_o^2 L} = \frac{3(1-\epsilon)}{4\pi a^3} \tag{8–2.6}$$

If this expression is substituted for N in Eq. (8–2.4), we obtain

$$\Delta P_S = 6(1-\epsilon)\frac{\mu L}{a^2}\left(U_{MF} - \frac{3U}{4}\right) \tag{8–2.7}$$

Thus, for the case of sedimentation (Stokes' law), $\Delta P_S = -4.5(1-\epsilon)\mu UL/a^2$ whereas for the case of pressure drop through a rigid assemblage, $\Delta P_S = 6(1-\epsilon)\mu U_{MF} L/a^2$. Actual data by Happel and Epstein[43], when extrapolated to infinite dilution, appear to agree with this latter value for a rigid assemblage. Note, however, that for a dilute assemblage in which $(a/l)^3/(R_o/a) \gg 1$ a uniform, rather than parabolic, flow pattern will be approached, resulting in $\Delta P_S = 4.5(1-\epsilon)\mu U_{MF} L/a^2$.

A variational approach[76,107] to creeping viscous flow through an isotropic porous medium develops lower bounds for the pressure drop. Particle interaction effects vanish in a bed of widely spaced spheres, and a constant of 3.51 is computed as a lower bound, in contrast with the coefficient 4.5, corresponding to Stokes' law, in the preceding equations.

Next the case will be considered in which the spheres are free to move relative to each other in the axial direction. In situations of practical interest, radial distribution of particles may be nonuniform depending on conditions at the ends, where particles and fluid are assumed to enter or leave the system. The general case for an arbitrary radial distribution will first be developed. The frictional force in the direction of flow experienced by a sphere translating in the same direction with a constant velocity U, when the fluid flows with a mean velocity $U_{MF} = \frac{1}{2}U_{OF}$, is given by Eq. (8–2.1), Stokes' law. The gravitational force (corrected for the buoyancy of the fluid) experienced by each particle is

$$F_g = \frac{4\pi a^3(\rho_S - \rho_F)g}{3} \tag{8–2.8}$$

where ρ_S and ρ_F are, respectively, the solids and fluid density, and g is the local acceleration of gravity. This gravitational force is equivalent to $F_g = 6\pi\mu a U_{TS}$, where U_{TS} is the terminal settling velocity of the sphere in an unbounded, quiescent fluid. When no net force acts on the particle it will move with a velocity U, obtained by equating the drag and gravitational forces:

$$U = U_{OF}\left(1 - \frac{R^2}{R_o^2}\right) - U_{TS} \tag{8–2.9}$$

Where all particles are free to move, a parabolic particle-velocity pattern in the radial direction thus results. There is a critical radius for which the particle velocity $U = 0$, as shown in Fig. 8-2.2.

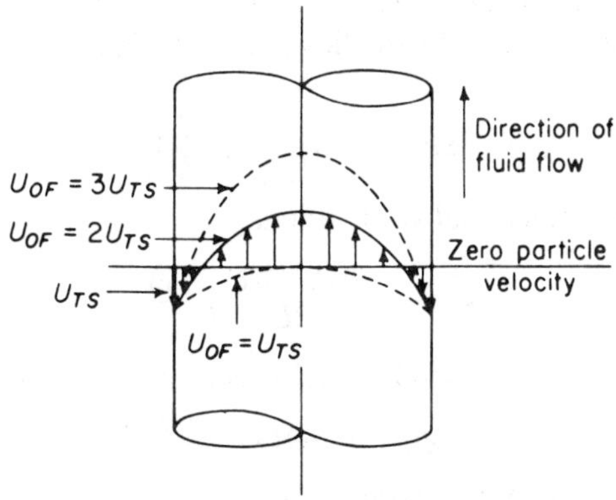

Figure 8-2.2. Particle-velocity pattern.

The pressure drop through such an assemblage is given for each sphere as

$$\Delta P_S = \frac{12\mu a}{R_o^2}\left(1 - \frac{R^2}{R_o^2}\right)U_{TS} \tag{8–2.10}$$

Hence the pressure drop due to particles present in the differential volume element $d\tau$ is

$$d(\Delta P_S) = \frac{24\pi\mu a L}{R_o^2} U_{TS}\left(1 - \frac{R^2}{R_o^2}\right) N R\, dR \tag{8–2.11}$$

Integration between the limits of zero and R_o, will, as before, give the overall pressure drop. The number N of particles per unit volume at any radial location R is determined from the entering conditions. The integration may, however, be effected without specific information regarding the dependence of N upon R. If ψ is the net number of particles transported per unit time through the pipe, then

$$\int_0^{R_o} N U 2\pi R\, dR = \psi \tag{8–2.12}$$

whence,

$$2\pi \int_0^{R_o} \left[U_{OF}\left(1 - \frac{R^2}{R_o^2}\right) - U_{TS}\right] N R\, dR = \psi \tag{8–2.13}$$

This makes

$$\int_0^{R_o}\left(1-\frac{R^2}{R_o^2}\right)NR\,dR=\frac{\psi}{2\pi U_{OF}}+\frac{U_{TS}}{2\pi U_{OF}}\int_0^{R_o}N2\pi R\,dR \tag{8–2.14}$$

If, however, N_M is taken as the mean number of particles per unit volume averaged over the entire assemblage, we have

$$N_M\pi R_o^2=\int_0^{R_o}N2\pi R\,dR \tag{8–2.15}$$

so that the pressure drop is generally expressible as follows:

$$\Delta P_S=\frac{24\pi\mu aL}{R_o^2}U_{TS}\left[\frac{\psi}{2\pi U_{OF}}+\frac{U_{TS}N_M R_o^2}{2U_{OF}}\right] \tag{8–2.16}$$

For the case of no net particle transport, where total upward flow of particles balances the downward flow, $\psi=0$. Equation (8–2.16) therefore reduces to

$$\Delta P_S=12\pi\mu aN_M L\frac{U_{TS}^2}{U_{OF}} \tag{8–2.17}$$

or, in terms of the fractional void volume,

$$\Delta P_S=\frac{9\mu LU_{TS}^2(1-\epsilon)}{2U_{MF}a^2} \tag{8–2.18}$$

The weight w of the bed of solid particles, corrected for buoyancy, is given by

$$w=g(\rho_S-\rho_F)L(1-\epsilon)\pi R_o^2 \tag{8–2.19}$$

Hence Eq. (8–2.18) may be simplified to

$$\Delta P_S=\frac{U_{TS}}{U_{MF}}\frac{w}{\pi R_o^2} \tag{8–2.20}$$

$\Delta P_S\pi R_o^2$ is the force F_S required to maintain flow through the bed of spheres (above the Poiseuille law pressure drop force). Hence,

$$F_S=w\frac{U_{TS}}{U_{MF}} \tag{8–2.21}$$

The simplest particle distribution, with particles moving relative to each other, is that in which the radial distribution is uniform. This condition may be realized by mixing particles and fluid uniformly at both ends of the tube and assuming that a steady state flow pattern is immediately established. The fluid entering a given zone in the tube will then possess a uniform particle concentration. This situation may be approximated in practice when both fluid and particles are introduced into the system simultaneously, as in pneumatic conveying, pumping of slurries, and some sedimenting systems. It is also likely that in fluidized systems, where the velocity of the fluid is close to the particle velocity, motion of particles at ends and other locations will cause mixing, resulting in a uniform particle distribution. In such cases the variable concentration $N = N(R)$ is replaced by the constant concentration N_M. Equation (8–2.13) becomes

$$2\pi \int_0^{R_0} \left[U_{OF}\left(1 - \frac{R^2}{R_0^2}\right) - U_{TS}\right] N_M R\, dR = \psi \qquad (8\text{–}2.22)$$

which is readily integrated. Upon setting $\frac{1}{2}U_{OF} = U_{MF}$, the simple result

$$U_{MF} - U_{TS} = \frac{\psi}{N_M \pi R_0^2} \qquad (8\text{–}2.23)$$

is obtained. Since for the case of no net particle transport (that is, the teeter condition), ψ is zero, we obtain the important result

$$U_{MF} = U_{TS} \qquad (8\text{–}2.24)$$

For no net particle transport, Eq. (8–2.17) therefore yields

$$\Delta P_S = 6\pi\mu a N_M L U_{TS} = 6\pi\mu a N_M L U_{MF} \qquad (8\text{–}2.25)$$

Thus, $F_S = w$; that is, the pressure drop force is simply equal to the Stokes' law resistance corresponding to the terminal velocity of the solid particles, or, in turn, is equal to the apparent bed weight corrected for buoyancy.

As noted previously, any radial distribution of particles is possible, depending on conditions at the bed entrance and exit. If we assume random mixing of particles alone, rather than both particles and fluid, and a condition of no net transport, a particle leaving the inner cylindrical space (that is, the region where all particles are moving upward) at the top of the bed has an equal probability of entering the outer cylindrical space (where particles are moving downward) at any point. It can be shown[40] that the particle distribution in such a system depends on the relative velocities U_{TS} and U_{OF}. The case where $U_{MF} = U_{TS}$ is of interest as it corresponds to the teeter condition for a bed of uniformly distributed particles. Using Eq. (8–2.16) with $\psi = 0$ and $U_{TS} = U_{MF}$, we see from Eq. (8–2.25) that the pressure drop is the same, though the particle distribution, as shown by Fig. 8-2.3, is decidedly different. There is no theoretical upper limit to the value of

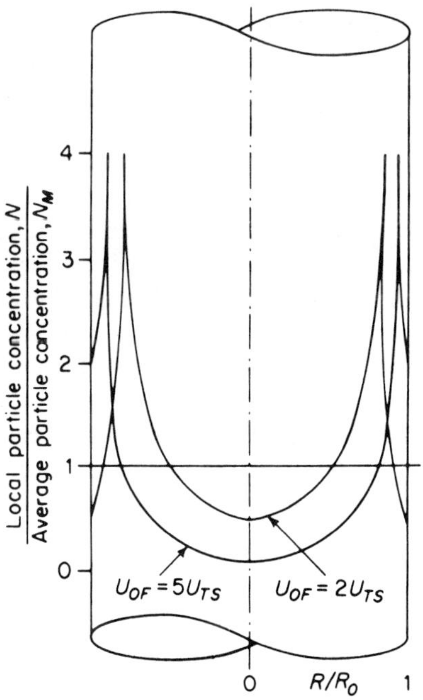

Figure 8-2.3. Particle distribution: uniform particle flux.

U_{MF}, except that caused by difficulty in maintaining appropriate end conditions. Particle build-up near the walls becomes pronounced as also shown in Fig. 8-2.3. It is interesting to note that, with this mechanism for redistribution, an increase in fluid velocity will result in a *reduction* in pressure drop, for a given total bed weight supported, once U_{MF} exceeds U_{TS}. Such reduction in pressure drop would, in each case, result in a corresponding increase in particle segregation. In practical situations, readjustment will occur, owing to the movement of particles back toward the middle of the tube because of solid-solid contacts, collisions between particles, mixing effects, or hydrodynamic forces tending to move particles toward the center of the tube. The latter may arise due to inertial forces, neglected in our treatment. All these possibilities have been neglected in the simple hydrodynamic analysis just presented.

The preceding analysis shows how recirculation effects and distribution of particles can be examined in a semiquantitative fashion. Pressure drops are shown to be greatest relative to bed weight supported in the case of fixed assemblages. Recirculation of a uniformly dispersed suspension results in a lowering of pressure drop. If redistribution as well as recirculation occurs, still lower pressure drops are attainable, but an unstable condition may result.

Some interesting theoretical and experimental work has recently appeared regarding the development of hydrodynamic forces tending to move particles sidewise, across the streamlines. The effects previously discussed are all due to the presence of a boundary surrounding the fluid, so that the laminar flow which passes the particles is essentially a shear flow. Such a flow may give rise to lateral forces, perpendicular to the direction of flow[69], similar to the Magnus effect caused by rotating-translating bodies. The phenomenon is due to inertial effects, neglected in the creeping motion equations. Slezkin[88] has developed equations of motion for suspensions of particles which *hypothetically* superpose this effect on the forces already calculated by the creeping motion equations. Slezkin and Shustov[90] subsequently applied these equations to the determination of forces exerted on particles suspended in laminar flows. In all cases, interaction effects between particles are not considered, so that this treatment should apply especially in dilute systems.

In terms of the motion of a unit volume of particles of true density ρ_S immersed in a fluid stream, the momentum balance is*

$$\rho_S \frac{d\mathbf{V}}{dt} = \mathbf{P} + \mathbf{P}_1 + \mathbf{P}_2 \tag{8–2.26}$$

where $\mathbf{P}$ is the gravitational force corrected for buoyancy; if gravity is exerted in the x direction,

$$\mathbf{P} = -\mathbf{i}g(\rho_S - \rho_F) \tag{8–2.27}$$

*As the original work is in Russian, we have thought it worthwhile to retain the original notation.

$\mathbf{P}_1$ is the drag per unit volume as determined by Stokes' law;

$$\mathbf{P}_1 = k_1(\mathbf{U} - \mathbf{V}) = \frac{9\mu}{2a^2}(\mathbf{U} - \mathbf{V}) \tag{8–2.28}$$

$\mathbf{U}$ is the fluid velocity, a is the particle radius, and $\mathbf{V}$ is the particle velocity; $\mathbf{P}_2$ is the sidewise force per unit volume,

$$\mathbf{P}_2 = k_2\rho_F(\mathbf{U} - \mathbf{V}) \times (\nabla \times \mathbf{U}) \tag{8–2.29}$$

Equation (8–2.29) is based on an approximate solution of the Euler equations by Zierep[110]. It predicts a sidewise thrust on a sphere in a shear field in the direction of increasing velocity. Since the Euler equations predict no tangential stresses and hence no torque on the sphere, it is not possible to assess the relative effects of rotation induced by the shear on a particle, and external forces (not associated with fluid motion) tending to cause particle rotation. Theodore[102] studied the effect of rotation on the sidewise force exerted on a stationary particle immersed in a fluid flowing in a cylindrical tube and found it to be quite small. Unfortunately, his experiments are not sufficiently conclusive either to confirm or deny Zierep's theoretical expression for sidewise force.

For a uniform shear flow in the xy plane we have

$$\begin{aligned} \mathbf{U} &= \mathbf{i}\frac{Uy}{h} \\ \nabla \times \mathbf{U} &= -\mathbf{k}\frac{U}{h} \end{aligned} \tag{8–2.30}$$

Thus, the equations of motion for the particles in cartesian coordinates will be

$$\begin{aligned} \rho_S\frac{dV_x}{dt} &= -g(\rho_S - \rho_F) + k_1\left(\frac{Uy}{h} - V_x\right) - k_2\rho_F\frac{U}{h}V_y \\ \rho_S\frac{dV_y}{dt} &= -k_1V_y - k_2\rho_F\frac{U}{h}\left(\frac{Uy}{h} - V_x\right) \\ \rho_S\frac{dV_z}{dt} &= -k_1V_z \end{aligned} \tag{8–2.31}$$

The following example will serve to illustrate the relatively small magnitude of the sidewise effect. We employ a consistent set of units involving lb-force, feet, seconds, so that mass must be specified in slugs. Consider a small spherical particle of catalyst 2×10^{-5} ft in diameter. It is fluidized in a vertical air stream, $\rho_F = 0.067$ lb/cu ft $= 0.0021$ slug/cu ft. The air viscosity is $\mu = 3.7 \times 10^{-7}$ (lb-force) (sec)/sq ft. Particle density is $\rho_S = 100$ lb/cu ft $= 3.12$ slugs/cu ft. It will be assumed that the parameter h is 10 ft. Roughly speaking, the velocity U is then attained 10 ft from the wall of the vessel in which fluidization takes place. Assuming steady state conditions, we wish to compare the sidewise velocity with the velocity U required to suspend the particles at $y = h$. The vertical particle velocity, V_x, is zero at the teeter condition. From Eq. (8–2.31), in the vertical direction (x axis), we obtain

$$g(\rho_s - \rho_F) = 100 = \frac{9\mu}{2a^2} U$$

$$U = \frac{2 \times 10^{-10} \times 100}{9 \times 3.7 \times 10^{-7}} = 6 \times 10^{-3} \text{ ft/sec}$$

Also, in the horizontal direction (y axis),

$$k_1 V_y = -k_2 \rho_F \frac{U^2}{h}$$

Taking $k_2 \approx 1$, this gives

$$V_y = -\frac{2 \times 10^{-10}}{9 \times 3.7 \times 10^{-7}} \times \frac{0.067 \times 36 \times 10^{-6}}{10} = -1.44 \times 10^{-11} \text{ ft/sec}$$

Thus, the ratio of fluidizing velocity to sidewise velocity is

$$\frac{U}{V_y} = -\frac{6 \times 10^{-3}}{1.44 \times 10^{-11}} = -4.15 \times 10^8$$

The negative sign indicates sidewise motion away from the wall.

Higher velocities than that corresponding to Stokes' law are often encountered in fluidized beds, so that larger particles are often present. These factors would tend to decrease the relative ratio of U/V_y, but in all cases the sidewise velocity induced by shear appears to be very small.

Zierep[110] studied the problem of the behavior of a spherical droplet of water in a horizontal air stream, the velocity of which changes with altitude. He concluded that vertical motion of the droplet due to shear and slip combined was normally negligible compared with vertical motion due to gravity.

Slezkin and Shustov[90] utilized Eq. (8–2.31) in its original unsteady state form, to develop a theory for the stability of motion of particles suspended in laminar flow. They thus developed a criterion for determining when flow will become unstable. Because of the uncertainty of the value of the constant k_2, it is not possible to draw quantititative conclusions, since the factor $1/\sqrt{k_2(1 - k_2)}$ is involved.

It must be emphasized that the preceding treatment of sidewise forces on spherical particles in shear flows is both approximate and empirical. No rigorous method is presently available for taking into account sidewise forces on particles in shear fields. Indeed, recent experimental data indicate that exact theoretical treatment may be very complicated. Thus, Segré and Silberberg[87] recently observed that neutrally buoyant spheres in dilute suspensions transported in laminar flow through circular tubes are subject to radial forces which tend to move the central ones outward and those near the wall inward. Regardless of their initial radial positions, particles tend to concentrate into an annular region about halfway between the tube center and the wall. These findings have been confirmed by Oliver[73] in an independent experimental investigation conducted with single spheres. Forces tending to move spheres away from the axis are in apparent contradiction to theoretical studies which indicate the opposite effect. A theoretical study by Rubinow and Keller[79] [see Eq. (7–3.109)] points out that Poiseuille flow in a tube is not the same as uniform shear, but rather involves a variable shear field. They suggest that this nonuniformity may explain the results of Segré and Silberberg. Another possibility lies in the presence of a wall effect which, as noted in Section 7–3, causes a solid sphere to lag behind the fluid stream in which it is suspended. And any spherical particle which moves relative to the surrounding fluid and rotates will experience a lateral, Magnus force. Finally, a critical discussion of these effects by Saffman (1956)[81] suggests that non-newtonian as well as inertial effects may be of importance in some circumstances.

The discussion of the preceding paragraph is limited to rigid spherical particles. Solid, neutrally buoyant, nonspherical particles may migrate across streamlines, even in creeping flow.[9] The same is true of deformable liquid droplets in creeping flow.*

8-3 Dilute Systems—First-Order Interaction Effects

In order to consider the effect of particle interactions, we must be concerned with two magnitudes—a/l, the ratio of a sphere radius to distance between spheres [which ratio is also proportional to $(1 - \epsilon)^{1/3}$], and a/R_o, the ratio of the sphere radius to that of the containing cylinder. It is apparent that, since the particle size can be made indefinitely small, any given concentration of particles can be made to satisfy the condition $a/l \ll a/R_o$. Thus it might be

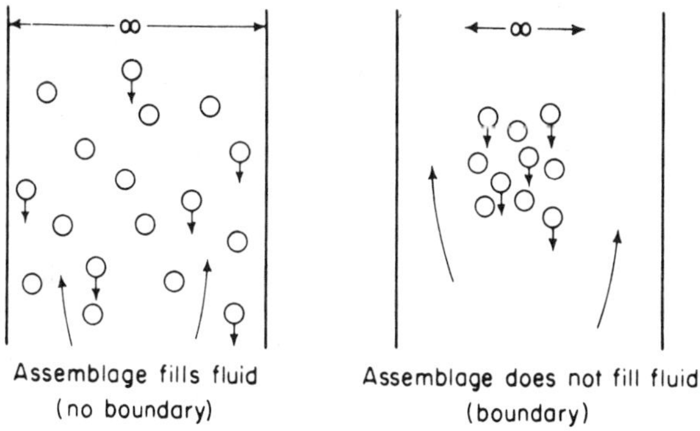

Figure 8-3.1. Infinite number of particles.

*Goldsmith, H. L., and S. G. Mason, J. Coll. Sci. **17** (1962), 447.

expected under these conditions that sedimentation velocity, even in dilute suspensions, will be uniquely related to particle concentration. This does not mean, however, that wall effects need not be considered, even though the containing wall be indefinitely distant from any particle specified at random. Figure 8-3.1 illustrates the situation involved. It is important to distinguish between two cases: (a) an assemblage which completely fills the fluid, so that both assemblage and fluid extend indefinitely; and (b) one in which the assemblage does not fill the fluid container completely, even though the former is indefinitely large in extent. The descent of a given particle creates a velocity field throughout the fluid which tends to increase the velocity of all particles near it by dragging them along. In opposition to this, the downward motion of each particle plus its adjacent fluid must be compensated by an equal upward flow. The latter tends to retard the settling velocity of more distant particles. In other words, an over-all continuity condition must be satisfied for the domain chosen. If the particles are not distributed uniformly throughout the fluid, the velocity tending to make the assemblage move faster can greatly exceed the retarding effect arising from the compensating upward flow. For the latter will occur predominantly in the region where there are few or no particles. On the other hand, if the particles are more or less uniformly distributed through the fluid, each will be retarded to much the same extent since no bypassing of return flow will now be possible.

The method of reflections, discussed in Section 6–1, is especially applicable to multisphere problems of this type. It involves a piecewise matching of boundary conditions using a number of partial solutions. The method may be visualized physically by supposing an initial disturbance to be reflected from the boundaries involved and to produce succeedingly smaller effects with each successive reflection. Thus if we consider a suspension consisting of n spherical particles each moving with velocity U in an otherwise unlimited fluid, the first reflection will be the velocity field created by each sphere as if it were alone in the fluid, and moving with velocity U. A disturbance will be set up, which in turn will produce effects on each of the other particles. A first-order approximation of this effect on the ith particle will be obtained by "reflecting" from it the sum of the "first reflections" for all the other $n - 1$ particles. The effect will be to make the ith particle move more rapidly than it would alone. In order to formalize this approach, we shall employ a nomenclature which identifies by three subscripts: (a) the particle at which a disturbance originates; (b) the number of times that reflections have occurred to produce this disturbance; (c) the particle at which the disturbance produced is evaluated. Thus U_{ijk} will mean a velocity field* produced by reflection for the jth time from particle i and evaluated at particle k.

*Strictly speaking, the fields U_{ijk} are only the scalar components of the velocity fields parallel to the direction of motion, that is, parallel to the x-axis.

In these terms, the first reflection of an assemblage on particle No. 1 will be

$$\sum_{i=2}^{i=n} U_{i01} \tag{8–3.1}$$

In order to satisfy the condition of a constant velocity U at the location of particle No. 1, the field of flow represented by Eq. (8–3.1) must be balanced by an equal and opposite "reflection" from this particle,

$$U_{111} = -\sum_{i=2}^{i=n} U_{i01} \tag{8–3.2}$$

The fields $U_{101} = U_{202} = U$ so that the first reflection from the ith particle in the assemblage will be

$$U_{i1i} = U - \sum_{k=1}^{n} U_{k0i} \tag{8–3.3}$$

The fields U_{i1i} will be fields which cancel the disturbance of the first reflection due to each of the other particles. In general they will involve terms to the first power of a/l when the system is sufficiently dilute. These fields will each cause a force to be exerted on each particle which is smaller than the force associated with the original velocity U in an undisturbed medium, but in the opposite direction, tending to make the resistance less. Further reflections are treated in analogous fashion since the fields U_{i1i} may be considered in the same way as U_{i0i}. Thus, if a second reflection is desired,

$$U_{i2i} = U_{i1i} - \sum_{k=1}^{k=n} U_{i1k} \tag{8–3.4}$$

The process may be continued as far as required, the fields U_{i2i} giving rise to terms in the drag force involving the second power of a/l. The net equivalent velocity field at any particle i will be the vector sum of the original particle velocity and the velocities of the induced fields. It will be given by $\sum_{j=0}^{j=m} U_{iji}$, where a total of m reflections is employed.

As discussed in connection with two-body problems, if a suitable coordinate system can be found, each of these reflections may be accurately determined and expressed in analytical form. If the particles are sufficiently far apart, it is necessary to consider only the lead term in the development of the Stokes field at a distance. Also, the effect at any particle may be computed by simply evaluating the magnitude of the field at the particle center. The re-reflections are then computed in the same manner as the first reflection. Such a procedure will give accurate results for the first two reflections. (See the references relating to the two-sphere problems of Chapter 6.) The general principle applies to any degree of accuracy, provided that the two-body problem can be solved.

This principle was first employed by Smoluchowski[91] who considered an assemblage of n spheres falling in an otherwise unlimited medium. If we confine our attention to the vertical forces exerted on such spheres in the direction of net motion, the equation corresponding to Eq. (8–3.2) is

$$U_{111} = -\sum_{i=2}^{i=n} \frac{3aU}{4r_i}\left(1 + \frac{x_i^2}{r_i^2}\right) \tag{8–3.5}$$

Thus the force acting on particle No. 1 due to the first reflection will be

$$F_1 = 6\pi\mu aU\left[1 - \frac{3}{4}a\sum_{i=2}^{n}\frac{1}{r_i}\left(1 + \frac{x_i^2}{r_i^2}\right)\right] \tag{8–3.6}$$

where r_i is the distance between particle i and particle No. 1, and x_i is the vertical displacement of particle i from particle No. 1 (see Fig. 8-3.2). It is apparent that the greater the number of spheres, the smaller will be the resistance to movement of each sphere. Or, stating it in other terms, the larger the number of spheres, the greater will be the velocity of sedimentation.

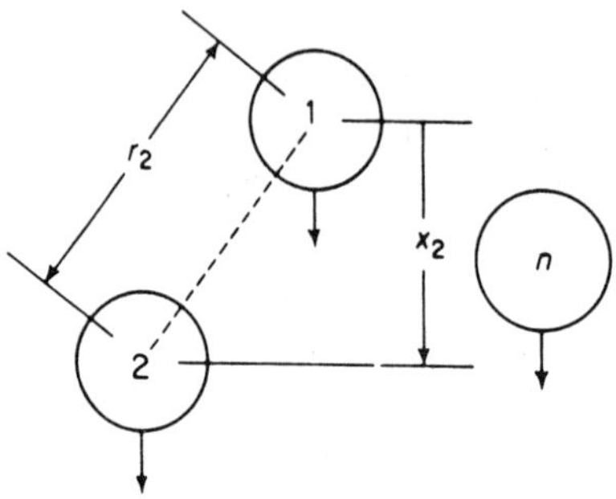

Figure 8-3.2. Definition sketch. A number of spheres in an unlimited field.

Smoluchowski[92] also considered the case of an assemblage which falls in a closed container. For convenience he chose a cubic lattice and assumed that all the particles approach a plane wall perpendicularly, which reflects the induced fluid velocity caused by sedimentation. Burgers[11] attacked the same problem and used a "diffuse" field to approximate the return flow resulting from the presence of a boundary. McNown and Lin[67] also employed a technique similar to Burgers', but in addition used the Oseen equations of motion to evaluate the first reflection in estimating particle interaction.

If it is assumed that the return flow due to a single particle will finally be balanced at a sufficient distance by the induced effect of other particles, it is then possible to obtain the total effect on a single particle by considering a finite volume V. The field U'_{111} (the prime is used to distinguish the symbol from the case where the assemblage falls on an infinite medium) may now be written, by analogy to Eq. (8–3.5),

$$U'_{111} = -\frac{3aU}{4}\left[\sum_{V}\frac{1}{r_i}\left(1 + \frac{x_i^2}{r_i^2}\right) - \frac{1}{l^3}\iiint_V \frac{1}{r}\left(1 + \frac{x^2}{r^2}\right) dx\, dy\, dz\right] \tag{8–3.7}$$

The numerical value of the preceding summation depends on the particle arrangement chosen, but in any case it is clear that downward flow close to the particles being considered will be much larger than that at any location at a distance l, representing the average distance of other particles. For convenience in making this summation most studies use a cubic lattice. If l

designates the side length of the assumed cubical spacing (or the distance between particles), r the distance from the reference particle No. 1 to any other, and x the distance measured in the vertical direction, the summation is begun with the single particles cited, ($r = l$, $x = \pm l$), and the four other nearest particles, ($r = l$, $x = 0$), in the horizontal plane through the reference particle. The summation may then be continued outward as indicated in the following tabular arrangement. Though both the summation and the integral in Eq. (8–3.7) increase indefinitely as the volume under consideration is increased, the difference between them (that is, the expression in brackets) remains finite. This difference already closely approaches its limiting value when a cube of only $3l$ units per side is selected.[64] This involves 27 cubic cells with a sphere at the center of each. The chosen particle is thus surrounded by 26 others, as indicated in Table 8–3.1.

TABLE 8–3.1
SCHEME FOR SUMMATION EVALUATION

Number of Particles	2	4	4	8	8
r/l	1	1	2	2	3
x/l	± 1	0	0	± 1	± 1

Physically, this amounts to the observation that if a finite number of particles fall in an unlimited medium, though the assemblage itself may fall faster as the number of particles is increased, the flow pattern in the vicinity of a particle in the interior will not change significantly with a further increase in the number of particles. Lin[64] found that using a larger assemblage (a cube $5l$ on a side, comprising 125 cells) did not appreciably alter the value of the bracketed expression in Eq. (8–3.7).

It must be admitted that these assumptions are to some extent based on physical intuition. In the investigations just cited[92, 11, 67], however, they lead to substantially the same result for a dilute assemblage in a simple cubic arrangement; namely,

$$F_1 = 6\pi\mu a U\left(1 + 2.6\frac{a}{l}\right) \tag{8–3.8}$$

To express this result in terms of the settling velocity U_0 of a noninteracting assemblage, where Stokes' law applies to the individual particles, we note that since $F_1 = 6\pi\mu a U_0$, then

$$\frac{U}{U_0} = \frac{1}{1 + 2.6(a/l)} \tag{8–3.9}$$

In a cubic assemblage, the concentration of solid particles by volume is $\phi = \frac{4}{3}\pi(a/l)^3$. Hence $a/l = 0.62\phi^{1/3}$, whereupon

$$\frac{U}{U_0} = \frac{1}{1 + k\phi^{1/3}} \tag{8–3.10}$$

where $k \approx 1.6$. Burgers[11] also considered the case for a random assemblage of particles and obtained $U/U_0 = 1/(1 + 6.88\phi)$. It is difficult to see why random arrangement should result in entirely different dependence of sedimentation rate on concentration. One possible explanation is that Burgers assumes that particles may occupy all positions around the reference particle with equal probability. This does not correspond to a random assemblage because, once the location of a given particle is fixed, others will be less likely to occupy sites in its immediate neighborhood if an over-all random distribution is assumed.

In addition to the preceding treatments based on the reflection technique, leading to numerical results for sedimentation velocity in suspensions of particles, there is an elegant formal analysis of the problem of a number of spheres in an unbounded medium presented by Kynch[57]. This study includes numerical evaluation and comparison with data for the two-sphere problem. General comments are given on three-particle interactions and sedimentation. Kynch's treatment is too involved to elaborate here. He concludes that Burgers[11] treatment overestimates the effect of concentration on sedimentation velocity, on the grounds that higher-order terms in concentration must be considered. This emphasizes the need for a treatment in the dilute region which specifically fixes the range of concentrations for which the first-order interaction correction is sufficient.

The generalized treatment by Brenner[8] of the Stokes resistance of a single arbitrary particle, discussed in Chapter 5, also includes an extension to systems consisting of any number of interacting particles. Numerical application to systems containing more than two particles will be complicated but the general treatment enables interesting conclusions to be drawn, even for concentrated systems. We shall therefore consider this treatment in more detail in Section 8-5.

Another technique which has been employed successfully to predict the effect of concentration on sedimentation rate, when wall effects are not important, is the use of a so-called cell model. This involves the concept that an assemblage can be divided into a number of identical cells, one sphere occupying each cell. The boundary value problem is thus reduced to a consideration of the behavior of a single sphere and its bounding envelope. This technique, which will be discussed further in the next section, applies best where symmetry of the particle assemblage is more or less complete. It is thus of great application in concentrated assemblages, where the effect of container walls will not be important. Uchida[104] investigated flow through a simple cubic lattice of spheres using this technique. He simplified the boundary conditions somewhat in order to obtain a closed solution. The significance of his result is therefore questionable, especially for concentrated suspensions. In the dilute region he obtained an expression of the same form as Eq. (8-3.10) but with $k = 2.1$. Kawaguchi[52] also employed a cell model involving a sphere in a frictionless cylindrical tube rather than a cubic cell. He employed the reflection technique to solve the boundary value problem involved, so that his results are applicable only in dilute suspensions. It was also necessary to assume a distance between the top and bottom of the "cell" on an empirical basis, since the basic solution obtained applied to an infinitely long tube. A form slightly different from Eq. (8–3.10) was obtained; namely,

$$\frac{U}{U_0} = 1 - c\phi^{1/3} \tag{8–3.11}$$

where $c = 1.6$. Note that as $\phi \to 0$, $1/(1 + c\phi^{1/3}) \to 1 - c\phi^{1/3}$, so that the two forms are very similar. Happel[35] employed a cell model, using an outer concentric spherical fluid cell, and obtained an expression for dilute suspensions of the same form as Eq. (8-3.11) with $c = 1.5$ [see Eq. (8–4.14) for the expression of this limiting result]. Thus the cell technique, which avoids the problem of evaluating return flow, gives good agreement with the reflection technique making use of the diffuse return field.

Hasimoto[47] handled the problem of flow through an array of particles using another interesting technique. He constructed a periodic array by repeating a basic cell defined by three linearly independent vectors, $\mathbf{a}^{(1)}$, $\mathbf{a}^{(2)}$, $\mathbf{a}^{(3)}$, as in Fig. 8-3.3. With the origin at a corner of a basic cell, the center of any particle is given by the vector

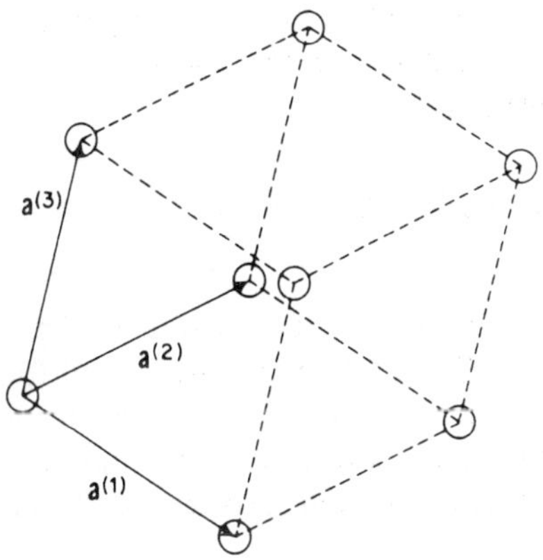

Figure 8-3.3. Basic cell for periodic array.

$$\mathbf{r}_n = n_1\mathbf{a}^{(1)} + n_2\mathbf{a}^{(2)} + n_3\mathbf{a}^{(3)}, \quad (n_1, n_2, n_3 = 0, \pm 1, \pm 2, \cdots) \tag{8–3.12}$$

Rather than solve the creeping motion equations subject to the boundary condition of zero slip velocity on the surface of each particle, Hasimoto restricted his treatment to dilute suspensions by replacing each particle by a point force retarding the motion of fluid. The creeping motion equations were then modified to include the discontinuous external force field consisting of point forces at every cell corner. Hasimoto chose to work with the reaction force which the fluid exerts on each particle and modified the creeping motion equations as follows:

$$\mu\nabla^2\mathbf{v} = \nabla p + \mathbf{F}\sum_{n_1}\sum_{n_2}\sum_{n_3}\delta(\mathbf{r} - \mathbf{r}_n) \tag{8–3.13}$$

$$\nabla\cdot\mathbf{v} = 0 \tag{8–3.14}$$

where $\mathbf{r}$ is the position vector of any point in the fluid relative to an arbitrary origin. The quantity $\delta(\mathbf{r} - \mathbf{r}_n)$ is the Dirac delta function, which satisfies the conditions,

$$\int_\tau \delta(\mathbf{r} - \mathbf{r}_n)\, d\tau = \begin{cases} 1 \text{ for } \mathbf{r}_n \text{ included in } \tau \\ 0 \text{ for } \mathbf{r}_n \text{ not included in } \tau \end{cases} \tag{8–3.15}$$

and

$$\delta(\mathbf{r} - \mathbf{r}_n) = 0 \text{ for } \mathbf{r} \neq \mathbf{r}_n$$

Here, τ is any volume.

Owing to the periodic arrangement of the particles, Hasimoto assumed that the velocity fields and pressure gradients would also be periodic. Accordingly, he expanded these fields in triple Fourier series of period $\mathbf{a}^{(j)}$ as follows:

$$\mathbf{v} = \sum_{n_1} \sum_{n_2} \sum_{n_3} \mathbf{V}_n e^{-2\pi i(\mathbf{k}\cdot\mathbf{r})} \tag{8–3.16}$$

$$-\nabla p = \sum_{n_1} \sum_{n_2} \sum_{n_3} \mathbf{P}_n e^{-2\pi i(\mathbf{k}\cdot\mathbf{r})} \tag{8–3.17}$$

The vector $\mathbf{k}$ in the preceding equations may be considered to be the position vector drawn to the nodes of a reciprocal array with base vectors $\mathbf{b}^{(1)}$, $\mathbf{b}^{(2)}$, $\mathbf{b}^{(3)}$. Hence,

$$\mathbf{k} = n_1\mathbf{b}^{(1)} + n_2\mathbf{b}^{(2)} + n_3\mathbf{b}^{(3)} \tag{8-3.18}$$

where $\mathbf{b}^{(1)}$, $\mathbf{b}^{(2)}$, and $\mathbf{b}^{(3)}$, are defined such that

$$\mathbf{k} \cdot \mathbf{a}^{(j)} = n_j \qquad \text{for } j = 1,2,3 \tag{8-3.19}$$

Hasimoto then substituted the series expressions for $\mathbf{v}$ and ∇p into Eqs. (8-3.13) and (8-3.14) and solved for the constant vector coefficients $\mathbf{V}_n$ and $\mathbf{P}_n$ as functions of the unknown force $\mathbf{F}$. The unknown force was then obtained by following a procedure used by Burgers[11], requiring that the mean velocity over the surface of each spherical particle vanish; that is,

$$\langle\mathbf{v}\rangle = \frac{1}{4\pi a^2} \iint_{r=a} \mathbf{v}\, dS = 0 \tag{8–3.20}$$

By this procedure Hasimoto obtained the following drag formula for flow through a simple cubic array of spheres of radii a:

$$F = \frac{6\pi\mu a U}{1 - 1.7601\,\phi^{1/3}} + O(a^3) \tag{8–3.21}$$

where ϕ is the volume fraction of solids, and U is the superficial velocity of flow. For dilute suspensions, $\phi \ll 1$, whence

$$\frac{1}{1 - 1.7601\,\phi^{1/3}} \approx 1 + 1.7601\,\phi^{1/3} + O(\phi^{2/3})$$

Equation (8-3.21) thus takes the form

$$F = 6\pi\mu a U(1 + 1.7601\,\phi^{1/3}) + O(\phi^{2/3}) \tag{8–3.22}$$

If Eq. (8-3.22) is applied to the computation of the settling velocity of a cubic suspension, the ratio of the suspension settling velocity, U, to the settling velocity of a single sphere, U_0, is given by

$$\frac{U}{U_0} = \frac{1}{1 + 1.7601\phi^{1/3}} \tag{8-3.23}$$

Hasimoto also obtained numerical values of the $\phi^{1/3}$ coefficient for body-centered and face-centered cubic arrays. His reported values are 1.791 for both these cases.

It is interesting that all these techniques give results in rather good agreement, indicating a rapid decrease in relative velocity when only a few tenths of 1 per cent of solids by volume are present in suspension. Without a treatment taking wall effects into consideration, however, it is not possible to state how large $(a/l)^3(R_0/a)$ must be to provide a sufficiently large particle-wall area ratio for them to be applicable. Note that with all these models, the pressure drop force experienced by a fluid traversing the array may be obtained by simply summing the drags on each particle.

Faxen[26] treated the special case of sedimentation of a single particle between parallel plates, for the case where Brownian motion would cause it to occupy any position between them. The average sedimentation rate thus obtained does not correspond to any situation involving the sedimentation of more than one particle.

The general problem of the sedimentation of a dilute system of spherical particles within a circular cylinder can be treated conveniently via the two-body solutions for two spheres and the solution for a single off-center sphere in a cylinder—Section7-3. The situation for n spheres settling in a cylinder is shown in Fig. 8-3.4. We see that not only must the direct interaction from all spheres to the reference sphere No. 1 be considered, but in addition, the interaction of all spheres, including the reference sphere, with the cylinder walls must be considered. This latter condition insures that there will be no net flow of fluid through the container walls. Even though the latter constitutes a second reflection, it does not increase the exponent of a/l, but simply introduces the additional multiplier a/R_0.

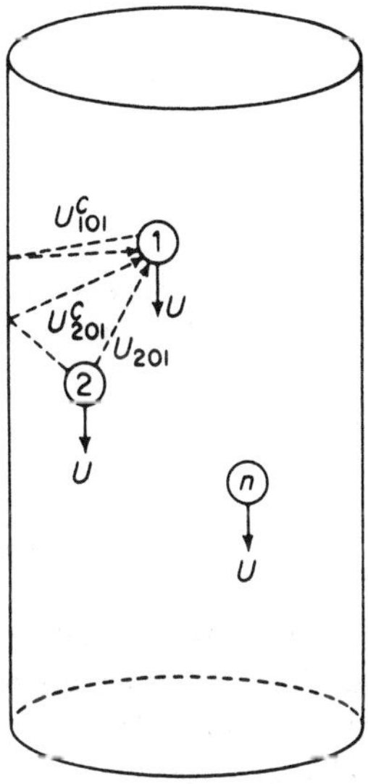

Figure 8-3.4. Suspension of spheres in a cylinder.

For the first reflection of an assemblage from particle No. 1, analogous to Eq. (8-3.2), the following expression results:

$$U^T_{111} = U - \sum_{i=1}^{i=n} (U_{i01} + U^C_{i01}) \tag{8-3.24}$$

This may, in turn, be generalized to any reflection from the ith particle, similar to Eq. (8-3.4). Thus, for $j \geq 1$

$$U^T_{ij1} = U_{1(j-1)1} - \sum_{k=1}^{k=n} (U + U^C)_{1(j-1)k} \tag{8-3.25}$$

Here U^T refers to the entire effect of cylinder plus direct particle interaction. U^C refers to the reflections which originate at a sphere and are canceled out at the cylinder walls by an equal and opposite field which, in turn, is reflected to a second sphere. Formally, we see that the simple addition of the cylinder reflection enables substantially the same technique to be employed for a bounded assemblage as for one which extends indefinitely.

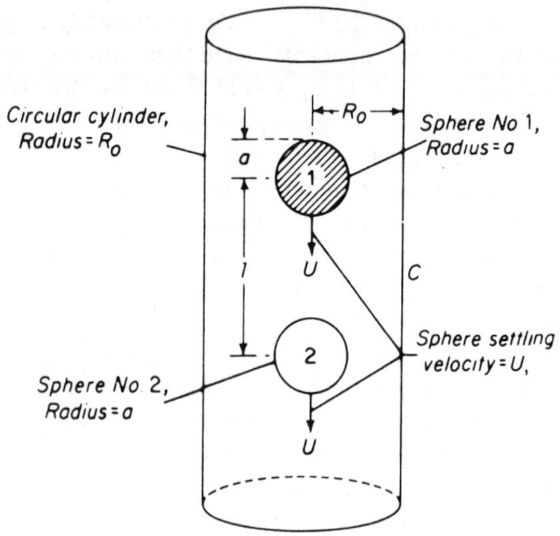

Figure 8-3.5. Effect of boundary on settling.

As a simple illustration, consider the effect of a cylindrical boundary on the forces experienced by two equal spheres separated by a distance l, falling axially with velocity U, as in Fig. 8-3.5. From the two-sphere and sphere-cylinder solutions, we have

$$\frac{U_{201}}{U} = 1.5\frac{a}{l}, \qquad \frac{U^C_{201}}{U} = -k\frac{a}{R_0}, \qquad \frac{U^C_{101}}{U} = -2.1\frac{a}{R_0}$$

Thus,

$$\frac{U^T_{111}}{U} = -1.5\frac{a}{l} + (2.1 + k)\frac{a}{R_0}$$

so that the force exerted on sphere No. 1 is

$$F_1 = 6\pi\mu a U\left[1 - 1.5\frac{a}{l} + (2.1 + k)\frac{a}{R_0}\right]$$

If, say, $l = R_0$ and $k = 1.3$, this makes

$$F_1 = 6\pi\mu a U\left(1 + 1.9\frac{a}{R_0}\right) \tag{8-3.26}$$

This shows the increase in resistance due to the presence of the wall. The force on sphere No. 2 will be the same as on sphere No. 1.

The "pressure drop" caused by the motion of the two spheres is computed by utilizing the same apparent velocity as produces the drag. To the present approximation, using the single sphere-cylinder solution to obtain the pressure drop, we find for the force necessary to move fluid past each sphere:

$$F_1 = 12\pi\mu a U\left(1 + 1.9\frac{a}{R_0}\right) \tag{8–3.27}$$

The total force $F_T = F_1 + F_2$ is equal to $\Delta P \pi R_0^2$. Similarly, if the spheres remain stationary at the axis of the cylinder, and fluid flows through the tube with an *axial* velocity equal to U, the pressure drop will again be given by the above. If, however, the spheres are not at the tube axis, Eq. (8-2.2) will be required to take into account the effect of eccentricity.

In cases where the spheres are not axially located, simple analytical representations for the fields U_{101}^C and U_{201}^C are not possible. By using these values the technique was extended to a number of spheres distributed in a cylinder, as discussed in the following development by Famularo[24].

Settling velocities of dilute suspensions of spheres

In the study of suspensions, where the spheres may assume arbitrary positions, it is not possible to obtain simple mathematical expressions for the field reflected from the container. Numerical values for these fields may, however, be computed from the solution to the single, eccentrically-positioned sphere problem discussed in Section 7-3. The field U_{i0k}^C is the axial component of the velocity field represented by Eq. (7-3.51). At the container wall it is equal to the negative of the fluid motion due to the settling of a sphere in an infinite medium.

With reference to Eq. (7–3.51), it is apparent that U_{i0k}^C may be represented as

$$U_{i0k}^C = \frac{a}{R_0} U_i \sum_{n=0}^{\infty} \int_0^{\infty} f(R_i, R_k, Z, \Phi, n, \lambda) d\lambda \tag{8–3.28}$$

where U_i = settling velocity of sphere i; R_i, R_k = radial distances to positions i and k; Z = axial distance from position i to k; Φ = azimuthal angle between i and k. Since the sum of integrals in Eq. (8–3.28) depends only on the positions i and k, it can be written compactly as

$$F_{ik} = F(\mathbf{r}_i, \mathbf{r}_k) = \sum_{n=0} \int_0^{\infty} f(R_i, R_k, Z, \Phi, n, \lambda) d\lambda \tag{8–3.29}$$

In terms of F_{ik},

$$U_{i0k}^C = \frac{a}{R_0} U_i F_{ik} \tag{8–3.30}$$

For the sake of uniformity, the direct (Stokes) field may be represented in a similar fashion. Neglecting terms of order greater than a/r, the Stokes' velocity field is given by

$$U_{i0k} = \frac{a}{R_o} U_i \frac{3}{4} \frac{R_o}{r} \left[1 + \left(\frac{Z}{r}\right)^2\right] \tag{8–3.31}$$

in which $Z = Z_i - Z_k$ and $r^2 = (X_i - X_k)^2 + (Y_i - Y_k)^2 + (Z_i - Z_k)^2$. Upon substituting,

$$G_{ik} = G(\mathbf{r}_i, \mathbf{r}_k) = \frac{3}{4} \frac{R_o}{r} \left[1 + \left(\frac{Z}{r}\right)^2\right] \tag{8–3.32}$$

the field U_{i0k} takes the form

$$U_{i0k} = \frac{a}{R_o} U_i G_{ik} \tag{8–3.33}$$

Examination of the functions $F(\mathbf{r}_i, \mathbf{r}_k)$ and $G(\mathbf{r}_i, \mathbf{r}_k)$ reveals symmetry with respect to interchange of $\mathbf{r}_i$ and $\mathbf{r}_k$; that is,

$$\begin{aligned} F_{ik} &= F_{ki} \\ G_{ik} &= G_{ki} \end{aligned} \tag{8–3.34}$$

Equations (8–3.30) and (8–3.33) may now be employed to compute the drag force acting on particles in a dilute suspension. A *dilute suspension* will be defined as one in which the shortest distance between two adjacent solid surfaces is large compared to sphere radii. In such a suspension, the fluid motion in the neighborhood of each sphere can be regarded as uniform motion. The drag on each sphere may now be computed as if it were settling in an infinite medium moving at the calculated uniform velocity.

The drag on sphere k in a dilute suspension of n spheres is therefore given by

$$F_k = 6\pi\mu a\left(U_k - \sum_{\substack{i=1 \\ i \neq k}}^{n} U_{i0k} - \sum_{i=1}^{n} U_{i0k}^{C}\right) \tag{8–3.35}$$

In terms of F_{ik} and G_{ik}, Eq. (8–3.35) takes the form

$$F_k = 6\pi\mu a \left\{U_k - \frac{a}{R_o}\left(\sum_{\substack{i=1 \\ i \neq k}}^{n} U_i G_{ik} + \sum_{i=1}^{n} U_i F_{ik}\right)\right\} \tag{8–3.36}$$

In principle, Eq. (8–3.36) could now be employed to obtain n simultaneous equations for the drag on each particle as a function of the settling velocities of each particle. The extra condition that the drag on every particle be equal to the gravitational force less buoyancy would then permit the computation of the n settling velocities $U_1, U_2, \ldots, U_n$.

The preceding procedure is impractical for studying a real suspension; therefore, an alternate procedure is adopted which allows computation of settling velocities to an accuracy of order a/R_o. The settling velocity of every particle is assumed to be equal and the drag is computed following this premise. The resulting values for the drag are then employed to obtain a first-order correction to the settling velocity.

The procedure will be illustrated for the simple case of two spheres, with an additional analysis of the accuracy of the method. Assuming $U_1 = U_2 = U$, the drag on each particle, obtained from Eq. (8–3.36) is

$$F_1 = 6\pi\mu a U\left\{1 - \frac{a}{R_o}(G_{12} + F_{11} + F_{12})\right\} \tag{8–3.37}$$

and

$$F_2 = 6\pi\mu a U\left\{1 - \frac{a}{R_o}(G_{12} + F_{22} + F_{12})\right\} \tag{8–3.38}$$

Since, however, the true terminal settling velocities must be inversely proportional to the drag acting on each particle, it follows that

$$\frac{U_2}{U_1} = \frac{F_1}{F_2} = \frac{1 - (a/R_o)(G_{12} + F_{11} + F_{12})}{1 - (a/R_o)(G_{12} + F_{22} + F_{12})} \tag{8–3.39}$$

In order to establish whether Eq. (8–3.39) represents U_2/U_1 to sufficient accuracy, it is necessary to substitute this value of U_2/U_1 into Eq. (8–3.36) and compare the resulting values of F_1 and F_2. To facilitate this process, first expand the right-hand side of Eq. (8–3.39) in a Taylor series in powers of a/R_o as follows:

$$\frac{U_2}{U_1} = 1 + \frac{a}{R_o}(F_{22} - F_{11}) + O\left(\frac{a}{R_o}\right)^2 \tag{8–3.40}$$

Neglecting the term of order $(a/R_o)^2$, elimination of U_2 via Eq. (8–3.40) results in

$$\begin{aligned} F_1 &= 6\pi\mu a U_1\left\{1 - \left(\frac{a}{R_o}\right) \times \right. \\ &\quad \left. \times \left[G_{12} + G_{12}\frac{a}{R_o}(F_{22} - F_{11}) + F_{11} + F_{12} + F_{12}\frac{a}{R_o}(F_{22} - F_{11})\right]\right\} \\ F_2 &= 6\pi\mu a U_1\left\{1 + \frac{a}{R_o}(F_{22} - F_{11}) - \left(\frac{a}{R_o}\right) \times \right. \\ &\quad \left. \times \left[G_{12} + F_{22} + F_{22}\frac{a}{R_o}(F_{22} - F_{11}) + F_{12}\right]\right\} \end{aligned} \tag{8–3.41}$$

By grouping together terms of order $(a/R_o)^2$ we obtain

$$\begin{aligned} F_1 &= 6\pi\mu a U_1\left\{1 - \frac{a}{R_o}[G_{12} + F_{11} + F_{12}]\right\} + O\left(\frac{a}{R_o}\right)^2 \\ F_2 &= 6\pi\mu a U_1\left\{1 - \frac{a}{R_o}[G_{12} + F_{11} + F_{12}]\right\} + O\left(\frac{a}{R_o}\right)^2 \end{aligned} \tag{8–3.42}$$

from which it follows that

$$F_1 = F_2 + O\left(\frac{a}{R_o}\right)^2 \tag{8–3.43}$$

Thus the technique of computing settling velocities starting from an initial assumption of equal settling velocities is accurate to the first power of a/R_o.

This technique has been applied by Famularo[24] to obtain formulas for the settling velocity of ordered and random suspensions as a function of volume concentration of solids. Integrals appearing in the function $F(\mathbf{r}_i, \mathbf{r}_k)$ were evaluated numerically on a digital computer. For suspensions containing a large number of particles per container volume (greater than 500 spheres per cylinder radius cubed), Famularo has shown that there is a tendency toward a flat settling profile in the central portion of the cylinder. Accordingly, he regards the settling velocity of a suspension as being equivalent to the settling velocity of a particle at the axis of the cylinder containing the suspension. In the case of randomly distributed spheres, the suspension settling velocity is equal to the mean velocity of a particle located at the cylinder axis.

The results of Famularo's investigation are summarized in the following equations:

Cubic suspension:

$$\frac{U}{U_0} = \frac{1}{1 + 1.91\phi^{1/3}} \tag{8–3.44}$$

Rhombohedral suspension:

$$\frac{U}{U_0} = \frac{1}{1 + 1.79\phi^{1/3}} \tag{8–3.45}$$

Random suspension:

$$\frac{U}{U_0} = \frac{1}{1 + 1.30\phi^{1/3}} \tag{8–3.46}$$

In Eq. (8–3.46) the coefficient 1.30 is a mean value of 231 suspensions investigated. A statistical analysis of the distribution of this coefficient places the 99.5 per cent confidence limits at 1.30 ± 0.24.

The small difference between Eqs. (8–3.44) and (8–3.45) indicates particle arrangement is not very important in uniform suspensions. To explore this premise further, Famularo obtained the settling rates of suspensions which were essentially rhombohedral, except that the vertical distance between adjacent horizontal planes was varied from the vertical spacing in a true rhombohedral suspension. If l is the side of the equilateral triangle in a horizontal plane of the rhombohedral suspension, and h the height of the tetrahedra (vertical spacing between adjacent horizontal layers), then $h/l = \sqrt{\frac{2}{3}}$ in the true rhombohedral suspension. The comparable h/l value for a cubic suspension is unity. For the distorted rhombohedral suspension, in which $h/l = 1$,

$$\frac{U}{U_0} = \frac{1}{1 + 1.88\phi^{1/3}} \tag{8–3.47}$$

which is quite close to Eq. (8–3.44). Smaller values of h/l were investigated, and it was found that the coefficient of $\phi^{1/3}$ lay between 1.72 and 1.88 for $0.2 \leq h/l \leq 1$. It dipped sharply for $h/l < 0.2$. In the range $h/l < 0.2$, however, the suspension begins to take on the appearance of vertical strings of particles and can hardly be regarded as uniform. Thus, it may be concluded that settling velocity will depend solely on particle concentration in the dilute range, *provided that the distance between adjacent particles is not greatly different from the mean interparticle spacing.*

It is interesting to compare Eq. (8–3.44) with the result obtained by Hasimoto, Eq. (8–3.23), for the settling rate of a cubic suspension. Hasimoto treats only the forces exerted by the particles on the fluid, without regard for a physical boundary. He therefore omits the tangential forces which a container would exert on the fluid continuum. It might be argued that, since the boundary conditions occur at infinity for an infinite array, they cannot influence the resistance of an "interior" sphere. In a practical sense, however, a so-called infinite suspension is constructed by increasing the number of particles in a fixed-sized container. A reduction in particle size must coincide with an increase in number for the suspension to remain dilute. To a particle-sized observer, this suspension would appear infinite in extent; however, now it is no longer clear that the boundary conditions on the container surface will not influence the settling velocity. In fact, a comparison of the work of Kawaguchi[52] and Faxen[25], as noted in Section 7–3, Eq. (7–3.113), suggests a definite effect of container boundary conditions on the first-order correction to settling velocity. Kawaguchi determined the settling velocity for a single sphere at the axis of a frictionless cylinder and obtained [see Eq. (8–3.11)].

$$\frac{U}{U_0} = \frac{1}{1 + 1.64931(a/R_o)} \tag{8–3.48}$$

Faxen solved the same problem subject to the boundary condition of zero velocity at the container wall and obtained the expression just given, with a coefficient 2.10444 in place of 1.64931. Hence, the settling velocity of an axially located sphere in a no-slip container is given by

$$\frac{U}{U_0} = \frac{1}{1 + 2.10444(a/R_o)} \tag{8–3.49}$$

Equations (8–3.48) and (8–3.49) demonstrate that the wall effect on a single particle is greater when return flow is impeded at the container wall. It is reasonable to assume that the return flow at the tube center, in a suspension of sedimenting particles, will also be greater when fluid cannot flow freely through the container, but rather is forced into the central region of the suspension. The direct interaction between particles due to the Stokes fields is not, however, dependent on the boundary conditions; therefore, the net effect of the no-slip boundary condition at the container wall is to reduce the settling velocity of the suspension. This accounts for the difference in coefficients of $\phi^{1/3}$, 1.91 versus 1.76, obtained by Famularo and Hasimoto. Based on the preceding argument it appears that Hasimoto's correction to the Stokes settling velocity might possibly be the correct formula for settling of a cubical suspension in a frictionless container.

Investigation of first-order interaction effects has been confined for the most part to spherical bodies, though Burgers[11] suggests a procedure for considering the effect of shape. Some treatments involving arrays of cylinders are available. These are of interest not only because of their direct value in studies involving flow past tube bundles in heat exchangers and through woven or fibrous materials, but also because they represent an extreme of particle shape gradation.

The method of reflections cannot be directly applied to problems involving arrays of cylinders because no solution of the creeping motion equations exists for a single cylinder in an unbounded medium. The cell technique has, however, been employed for such problems, in much the same way as for flow relative to assemblages of spherical particles, using the free surface model[36]. This technique is also applicable in relatively concentrated systems and is considered in greater detail in the following section. For a dilute assemblage it is not possible to express the result in terms of the velocity relative to that of a cylinder settling in an infinite medium, because the velocity in the latter case does not exist. It is convenient to use a form similar to Eq. (8–3.8), which simply expresses the drag in terms of the spacing between cylinders. Thus for flow with superficial velocity U perpendicular to a dilute square assemblage of circular cylinders of radii a, the free surface model gives, for the force per unit length on a single cylinder,

$$F_1 = \frac{4\pi\mu U}{\ln (l/a) - 1.07} \tag{8–3.50}$$

where l is the distance between cylinder centers. Hasimoto[47] also applied his periodic model, discussed earlier in this section in connection with arrays of spherical particles, to two-dimensional flow past a widely spaced square array of circular cylinders. The value he obtained for the constant in Eq. (8–3.50) was 1.3105 instead of 1.07.

For flow parallel to a square array of cylinders, the free surface model for dilute assemblages gives

$$F_1 = \frac{2\pi\mu U}{\ln (l/a) - 1.32} \tag{8–3.51}$$

This result is in agreement with that obtained by Emersleben[20] and Sparrow and Loeffler[95], discussed in the next section. Hasimoto has not treated this problem using his technique.

Other treatments involving two-dimensional arrays of objects include that of Tamada and Fujikawa[100] on perpendicular flow through a single column of parallel cylinders. This study, based on Oseen's equations, showed that the drag on one of the cylinders tended to that obtained on the basis of the creeping motion equations in the limit of small Reynolds numbers, $\rho a U/\mu$. Hasimoto[48] discussed flow through a thin screen, and obtained an exact solution of the creeping motion equations for a periodic series of flat plates set perpendicular to a uniform flow. Kuwabara[55] and Miyagi[69], respectively, also treated, on the basis of the creeping motion equations, flow past a row of parallel flat plates and flow past a row of parallel circular cylinders.

8-4 Concentrated Systems

As indicated previously, the cell model appears to be especially useful for obtaining significant numerical results in the case of concentrated suspensions. In the mathematical treatment which follows[35], it is assumed that a three-dimensional assemblage may be considered to consist of a number of identical unit cells, each of which contains a particle surrounded by a fluid envelope containing a volume of fluid sufficient to make the fractional void volume in the cell identical to that in the entire assemblage. For simplicity, we assume a typical cell envelope to be spherical. Furthermore, the outside surface of each cell is assumed to be frictionless. Thus, the entire disturbance due to each particle is confined to the cell of fluid with which it is associated. It is then possible to obtain a closed solution describing the concentration dependence of the sedimentation rate.

A similar model for study of the terminal velocity of spherical particles through a viscous fluid was employed by Cunningham[15], who assumed that each particle in the suspension would be effectively limited to motion within a concentric mass of fluid. He, however, assumed that the boundary of the outer fluid envelope was solid, corresponding in some way to the surfaces of the other spheres present in the cloud. This model presents the difficulty that, since the cell volumes are not mutually exclusive, the size of a representative spherical envelope must be fixed by some additional empirical consideration.

Brenner[7] tried to extend the treatment of a cubic lattice of spheres into the concentrated region. The boundary value problem, involving a sphere in a cubical cell of fluid, was solved in principle by the method of collocations. Equations were developed which satisfied the required boundary values on the sphere and at individual selected points on the cubical envelope. Numerical evaluation of the resulting sets of simultaneous equations proved intractable.

In the treatment presented here, two concentric spheres are considered: the inner one, of radius a, is solid; the outer sphere, of radius b, is frictionless. The boundary value problem to be solved involves satisfying the creeping motion equations with appropriate boundary conditions. The equations of motion are thus

$$\mu\nabla^2\mathbf{v} = \nabla p \tag{8–4.1}$$

$$\nabla\cdot\mathbf{v} = 0 \tag{8–4.2}$$

The internal sphere moves along the x axis with a velocity U inside a fluid sphere with a free surface, on which the normal velocity and tangential stresses vanish; hence, the boundary conditions are

$$u = U, \quad v = w = 0 \qquad \text{at } r = a \tag{8–4.3}$$

$$v_r = 0, \qquad \Pi_{r\theta} = \mu\left(\frac{1}{r}\frac{\partial v_r}{\partial \theta} + \frac{\partial v_\theta}{\partial r} - \frac{v_\theta}{r}\right) = 0 \qquad \text{at } r = b \tag{8-4.4}$$

where (r, θ, ϕ) are spherical coordinates. The motion is axially symmetric. Satisfaction of the boundary condition $v_r = 0$ at $r = b$ will automatically insure that $\partial v_r/\partial\theta = 0$ at $r = b$. Hence, the tangential stress condition requires only that $\partial v_\theta/\partial r - v_\theta/r$ vanish at $r = b$.

Lamb's[59] general solution of the creeping motion equations in spherical coordinates, given in Eq. (3–2.3), will be employed to obtain a solution. Thus,

$$\mathbf{v} = \sum_{n=-\infty}^{n=+\infty} \nabla \times (\mathbf{r}\chi_n) + \nabla\phi_n + \frac{n+3}{2(n+1)(2n+3)} r^2 \nabla p_n - \mathbf{r}\frac{np_n}{(n+1)(2n+3)} \tag{8-4.5}$$

$$p = \mu \sum_{n=-\infty}^{n=+\infty} p_n \tag{8-4.6}$$

The hydrodynamic force on a sphere, given by Eq. (3–2.42), is

$$\mathbf{F} = -4\pi\mu\nabla(r^3 p_{-2}) \tag{8-4.7}$$

Solution is also possible employing the Stokes stream function, but for consistency with the companion treatment for viscosity employing this model[38], discussed in Section 9–4, the vector development is used here.

An appropriate form for satisfying the present boundary value problem results from setting $\chi_n = 0$ and retaining solid spherical harmonics p_n and ϕ_n of orders -2 and $+1$. Thus

$$\mathbf{v} = \nabla\phi_1 + \nabla\phi_{-2} + \tfrac{1}{5}r^2\nabla p_1 - \tfrac{1}{10}\mathbf{r}p_1 + \tfrac{1}{2}r^2\nabla p_{-2} + 2\mathbf{r}p_{-2} \tag{8-4.8}$$

On symmetry grounds we are led to assume that

$$\begin{aligned} \phi_1 &= Ax \\ p_1 &= Bx \\ \phi_{-2} &= -C\frac{\partial}{\partial x}\left(\frac{1}{r}\right) = \frac{Cx}{r^3} \\ p_{-2} &= -D\frac{\partial}{\partial x}\left(\frac{1}{r}\right) = \frac{Dx}{r^3} \end{aligned} \tag{8-4.9}$$

With the boundary conditions expressed by Eqs. (8–4.3) and (8–4.4), it is found that four independent relationships are available for determining the constants, A, B, C, and D. For our present purposes it is necessary to determine only D, corresponding to the p_{-2} harmonic required to compute the drag from Eq. (8–4.7). It is as follows:

$$D = \frac{aU(3 + 2\gamma^5)}{2 - 3\gamma + 3\gamma^5 - 2\gamma^6} \tag{8-4.10}$$

where $\gamma = a/b$. From Eqs. (8–4.7) and (8–4.9), the force on the moving sphere is $F = -4\pi\mu D$. For the model under consideration, the drag divided by the cell volume $\frac{4}{3}\pi b^3$ will equal $-\Delta P/L$, the pressure drop per unit length of bed due to passage of fluid through it. Use of this relationship gives

$$U = \left[\left\{\frac{3 - (9/2)\gamma + (9/2)\gamma^5 - 3\gamma^6}{3 + 2\gamma^5}\right\}\frac{2a^2}{9\gamma^3}\right]\frac{\Delta P}{\mu L} \tag{8-4.11}$$

where the expression in brackets is the permeability coefficient in Darcy's law and U is the superficial fluid velocity through the assemblage. As discussed in Section 1–2 this law was originally obtained empirically by Darcy in experiments with water flow through sand filters.[17]

For comparison, the analogous expressions based on a dilute medium, where Stokes' law applies, may be employed. A sphere moving with velocity U_0 in an infinite medium will experience the same force F as previously if F and U_0 are related by the expression

$$F = -6\pi\mu a U_0 \tag{8-4.12}$$

Where wall effects may be neglected, the pressure drop ΔP_0, obtained by simply adding the resistances due to individual spheres on the assumption of no interaction, will be given by

$$U_0 = \frac{2}{9}\left(\frac{a^2}{\gamma^3\mu}\right)\frac{\Delta P_0}{L} \tag{8-4.13}$$

Equations (8–4.11) and (8–4.13) may be combined to give

$$\frac{U}{U_0} = \left[\frac{3 - (9/2)\gamma + (9/2)\gamma^5 - 3\gamma^6}{3 + 2\gamma^5}\right]\frac{\Delta P}{\Delta P_0} \tag{8-4.14}$$

In sedimentation problems, $\Delta P = \Delta P_0$. The velocity U is then the actual settling velocity of an assemblage of spheres in an otherwise stationary fluid. In the calculation of resistance to flow through porous media, $U = U_0$. In this case U is the superficial velocity of the fluid passing through an assemblage.

TABLE 8–4.1
RELATIVE VELOCITY FUNCTION, Eq. (8–4.14)

Sphere Radius Ratio, $a/b = \gamma$	Volume Fraction of Solids, $\gamma^3 = \phi = (1 - \epsilon)$	Relative Velocity, $(U/U_0)_{\Delta P = \Delta P_0}$
0.1	0.001	0.8500
0.2154	0.01	0.6773
0.3684	0.05	0.4526
0.4642	0.10	0.3215
0.5848	0.20	0.1773
0.6694	0.30	0.09867
0.7368	0.40	0.05287
0.7937	0.50	0.02638
0.8434	0.60	0.01175
0.8879	0.70	0.004340
0.9283	0.80	0.001132

In practical problems it is often convenient to express U/U_0 in terms of either the volume solids concentration, $\phi = \gamma^3$, or the void volume fraction, $\epsilon = 1 - \phi$. Some typical values of $(U/U_0)_{\Delta P = \Delta P_0}$ are given in Table 8–4.1.

It is, of course, evident that $(U/U_0)_{\Delta P = \Delta P_0}$ will be equal to $(\Delta P_0/\Delta P)_{U=U_0}$.

One should mention Kuwabara's[56] treatment, using a model similar to the free surface model, for establishing the forces experienced by randomly distributed parallel circular cylinders or spheres in a viscous flow at low Reynolds numbers. For the treatment of spheres, two concentric spheres are again considered. But, in place of the condition of no tangential stress on the outer spherical envelope, Eq. (8–4.4), Kuwabara uses the condition that the *vorticity* should vanish (see Milne-Thomson[68]). This requires that

$$\frac{\partial v_\theta}{\partial r} - \frac{1}{r}\frac{\partial v_r}{\partial \theta} + \frac{v_\theta}{r} = 0 \qquad \text{at } r = b$$

Kuwabara's solution results in a stronger concentration dependence of relative settling velocity or pressure drop than that indicated by the free surface model. This occurs because the vorticity model yields a larger energy dissipation in the envelope than that due to particle drag alone, owing to the additional work done by the stresses at the outer boundary. It is believed that a cell model should not require energy exchange with its surroundings, for the assemblage cannot then be considered as properly decomposed into *independent* cells. Thus, the assumption of zero vorticity does not appear to be a satisfactory boundary condition.

Equation (8–4.11) provides theoretical support for the validity of Darcy's law. It not only demonstrates the simple linear relationship between flow rate and pressure drop observed by Darcy, but for the idealized system chosen, gives numerical values for the proportionality coefficient. A convincing proof of this relationship has presented a challenge for theoretical investigators, and numerous theoretical and experimental studies have been concerned with expressing the coefficient of proportionality in terms of the gross properties of the porous medium. Convenient reviews are available[12,84] summarizing many of these treatments.

One of the more widely quoted theories aimed at obtaining a relationship between fractional void volume and relative settling velocity which is uniformly applicable in both dilute and concentrated systems is that of Brinkman[10]. His relationship for relative velocity corresponds to

$$\frac{U}{U_0} = 1 + \frac{3}{4}(1 - \epsilon)\left[1 - \sqrt{\frac{8}{(1 - \epsilon)} - 3}\,\right] \tag{8–4.15}$$

The model underlying his treatment is that of a spherical particle embedded in a porous medium. The flow through this porous mass is described by a modification of Darcy's equation. Since the latter is empirical, the result cannot be regarded as a rigorous solution to the problem. Furthermore, when

$\epsilon = \frac{1}{3}$ the Brinkman equation yields $U/U_0 = 0$, corresponding to zero permeability.

Richardson and Zaki[78] developed a cell-type model for sedimenting spherical particles in which the particles were assumed to be arrayed in a hexagonal-type pattern in horizontal planes. Particles were assumed to be settling in such a way that they were vertically aligned, one above the other, but the distance between adjacent horizontal layers was an additional variable parameter for which two cases were considered. The specification of this distance which provided the best agreement with Richardson and Zaki's own data in the intermediate concentration range (designated by them as Configuration No. 2), assumes that the spheres in adjacent horizontal layers actually touch. The other case studied (designated as Configuration No. 1) assumes that the spheres are the same distance apart vertically as horizontally. As would be expected, this uniform spatial distribution gives lower values for U/U_0, at corresponding values of fractional void volume, than in the case where the spheres are assumed to touch each other. This is indeed to be expected on theoretical grounds since a doublet consisting of two spheres in contact, following each other, will acquire a relative velocity $U/U_0 = 1.55$; that is, the doublet will fall 55 per cent faster than the spheres falling separately[27]. Richardson and Zaki's result for the uniform spatial arrangement is in good agreement with Brinkman's theory in the intermediate concentration range, $\epsilon = 0.6$ to 0.95. Equation (8–4.14), the free surface model, gives results in this range which are 25 per cent lower than comparable results for the uniform distribution, and 50 per cent lower than predicted by the Richardson and Zaki treatment for the case where the spheres touch. As will be seen subsequently, experimental data also show great variance in this region of concentration, indicating effects due to differences in particle distribution and segregation. Note that Richardson and Zaki do not provide the exact solution of the boundary value problem proposed by them. Their approximate solution leads to certain inconsistencies. Thus, at infinite dilution, $\epsilon = 1.0$, the relative velocity U/U_0 becomes infinite, rather than approaching the correct value of $U/U_0 = 1.0$, as does Eq. (8–4.14).

Another investigation, by Happel and Ast[39], was undertaken to develop an exact solution for a sphere-in-cylinder type of cell model, applicable to concentrated systems. In their model a sphere was assumed to be settling at the axis of an infinitely long "frictionless" cylinder, on whose surface the normal velocity and tangential stresses vanished. This differs from the Richardson and Zaki model in that spheres in an assemblage are not assumed to be lined up directly one above the other. It is again necessary to make an arbitrary assumption in order to establish the appropriate volume of a unit cell. The value $a/R_o = \lambda$ fixes the ratio of sphere-to-cylinder radii. It was assumed that $\phi = \lambda^3$, which is the same relationship as was previously employed in the case of a frictionless spherical boundary; this is equivalent to taking the length l of a unit cell to be $l = \frac{4}{3}a$. The solution of the sphere-cylinder boundary value problem was obtained in the form of an infinite set of linear simultaneous equations, utilizing a technique developed by Haberman[33]. Values of U/U_0 predicted on the basis of this model agree reasonably well with values predicted on the basis of the concentric sphere model, presented in Table 8–4.1, up to a solids concentration $\phi = 0.216$ (corresponding to $a/R_o = 0.6$). At this concentration, U/U_0 predicted on the basis of the sphere-cylinder model is 11 per cent less than that using the concentric sphere model. It thus appears that, with such cell models for predicting sedimentation dynamics, the particular shape assumed for the outer fluid boundary is not of great importance up to substantial particle concentrations. Numerical calculation of the velocity distribution for the case $a/R_o = 0.6$ shows that the fluid motion within the cylinder becomes negligible at a distance of about 1.3 sphere diameters.

Yet another cell model which may be applied in relatively concentrated porous media involves the assumption of arrays of cylinders instead of spheres. In this case the analysis[36] is based on the premise that two concentric circular cylinders can serve as a model for flow through an assemblage of cylinders. The inner cylinder consists of one of the rods in the assemblage and the outer cylinder is a fluid envelope with a free surface. The relative volume of fluid to solid in the cell model is taken to be the same as the relative volume of fluid to solid in the assemblage of cylinders. The conditions of no shear stress and no normal velocity along the walls of the fluid envelope is maintained.

Two basic cases are considered: in the first, flow is assumed to be parallel to the axes of the cylinders; in the second, flow is at right angles to the axes. In the case of random assemblages, where the cylinders are not parallel to each other, one must employ a weighted average of these separate results. For flow perpendicular to cylinders, the model does not distinguish between crossed or parallel arrangement of the cylinders; hence, in random assemblages it is necessary in averaging to give twice the weight to the correlation for flow perpendicular to cylinders as for flow parallel to cylinders.

For flow parallel to circular cylinders, the basic differential equation to be solved is

$$\frac{1}{r}\frac{d}{dr}\left(r\frac{du}{dr}\right) = \frac{1}{\mu}\frac{dp}{dx} \tag{8–4.16}$$

where (r, θ, x) are cylindrical coordinates and u is the x component of velocity. The general solution of this equation for constant dp/dx is

$$u = \frac{1}{4\mu}\frac{dp}{dx}r^2 + A\ln r + B \tag{8–4.17}$$

The fluid is assumed to move through the annular space bounded between the stationary solid cylinder of radius a and the fluid envelope of radius b ($b > a$).* These boundary conditions require that

$$\begin{aligned} u &= 0 \qquad \text{at } r = a \\ \frac{du}{dr} &= 0 \qquad \text{at } r = b \end{aligned} \tag{8–4.18}$$

From these we obtain

$$u = -\frac{1}{4\mu}\frac{dp}{dx}\left[(a^2 - r^2) + 2b^2 \ln \frac{r}{a}\right] \tag{8–4.19}$$

The flow rate through the annulus is then

$$Q = 2\pi \int_a^b ur\,dr = -\frac{\pi}{8\mu}\frac{dp}{dx}\left(4a^2b^2 - a^4 - 3b^4 + 4b^4 \ln \frac{b}{a}\right) \tag{8–4.20}$$

If $U = Q/\pi b^2$ denotes the *superficial* velocity, and if Darcy's equation for flow through a porous medium is written as $U = -(K/\mu)dp/dx$, then we find

$$K = \frac{1}{8b^2}\left(4a^2b^2 - a^4 - 3b^4 + 4b^4 \ln \frac{b}{a}\right) \tag{8–4.21}$$

The well-known Carman-Kozeny[12] equation, derived on the basis of semi-empirical reasoning, also gives an expression for the Darcy permeability constant,

$$K = \frac{\epsilon m^2}{k} \tag{8–4.22}$$

Here m is the hydraulic radius, defined for a porous medium as (free volume)/(wetted area), and k is the so-called Kozeny constant; k is a dimensionless number which, presumably, has the same numerical value for all random porous media, independently of the size of the particles and the void volume ϵ. This relationship will be discussed further in this chapter; suffice it to state here that the constancy of k is supported by a considerable volume of experimental data.

In the present case, for flow parallel to cylinders, $m = (b^2 - a^2)/2a$; thus, the Kozeny constant becomes

$$k = \frac{2\epsilon^3}{(1-\epsilon)[2 \ln \{1/(1-\epsilon)\} - 3 + 4(1-\epsilon) - (1-\epsilon)^2]} \tag{8–4.23}$$

It should be noted that Eq. (8–4.19) applies to the complete Navier-Stokes equations. The inertia terms vanish identically because of the unidirectionality of the flow.

When the flow is perpendicular to the cylinders we omit the inertia terms from the Navier-Stokes equations. Expressed in cylindrical coordinates, the resultant creeping motion equations for the two-dimensional flow are

*The same ultimate permeability expression results by allowing the inner cylinder to move with constant velocity U while requiring that the net flow rate in the annulus be zero.

$$\frac{\partial p}{\partial r} = \mu\left(\nabla^2 v_r - \frac{v_r}{r^2} - \frac{2}{r^2}\frac{\partial v_\theta}{\partial \theta}\right)$$
$$\frac{1}{r}\frac{\partial p}{\partial \theta} = \mu\left(\nabla^2 v_\theta - \frac{v_\theta}{r^2} + \frac{2}{r^2}\frac{\partial v_r}{\partial \theta}\right) \qquad (8\text{–}4.24)$$

whereas the continuity equation is

$$\frac{\partial v_r}{\partial r} + \frac{v_r}{r} + \frac{1}{r}\frac{\partial v_\theta}{\partial \theta} = 0$$

It is convenient to employ the stream function, defined by the relations

$$v_r = \frac{1}{r}\frac{\partial \psi}{\partial \theta} \quad \text{and} \quad v_\theta = -\frac{\partial \psi}{\partial r} \qquad (8\text{–}4.25)$$

Equations (8–4.24) then assume the form of the two-dimensional biharmonic equation,

$$\nabla^4 \psi = 0 \qquad (8\text{–}4.26)$$

For our present purposes, we observe that a solution of this equation is

$$\psi = \sin\theta\left[\frac{1}{8}Cr^3 + \frac{1}{2}Dr\left(\ln r - \frac{1}{2}\right) + Er + \frac{F}{r}\right] \qquad (8\text{–}4.27)$$

In this case, a solid cylinder of radius a is assumed to be moving with velocity U perpendicular to its axis in a fluid cell of radius b. (The same ultimate permeability result would be obtained if one assumed that fluid moved with mean velocity U perpendicular to a stationary cylinder of radius a.) As before, there is assumed to be no shearing stress or normal velocity on the outside cylindrical fluid surface. Under these assumptions the boundary conditions are:

At $r = a$,

$$v_r = U\cos\theta, \qquad v_\theta = -U\sin\theta \qquad (8\text{–}4.28a)$$

At $r = b$,

$$v_r = 0, \qquad \frac{\partial v_\theta}{\partial r} + \frac{1}{r}\frac{\partial v_r}{\partial \theta} - \frac{v_\theta}{r} = 0 \qquad (8\text{–}4.28b)$$

These conditions provide four simultaneous equations to establish the values of the four constants in Eq. (8–4.27). As is easily shown[89], the drag per unit length on the solid cylinder $r = a$ is

$$F = 2\pi\mu D \qquad (8\text{–}4.29)$$

The constant D is found to be

$$D = \frac{-2U}{\ln(b/a) + [a^4/(b^4 + a^4)] - 1/2} \qquad (8\text{–}4.30)$$

If the cylinder remains stationary while fluid passes by it, the force per unit length associated with a single cylindrical cell may be equated to the pressure gradient; hence

$$\frac{F}{\pi b^2} = \frac{dp}{dx} \qquad (8\text{–}4.31)$$

As previously stated, the Darcy equation may be written as $U = -(K/\mu)(dp/dx)$ where U is the *superficial* velocity. Thus, the Darcy constant is found to be

$$K = \frac{b^2}{4}\left[\ln\left(\frac{b}{a}\right) - \frac{1}{2}\left(\frac{b^4 - a^4}{b^4 + a^4}\right)\right] \tag{8–4.32}$$

Similarly, on the basis of the Carman-Kozeny theory, $K = \epsilon m^2/k$, whence the Kozeny constant becomes

$$k = \frac{2\epsilon^3}{(1-\epsilon)[\ln\{1/(1-\epsilon)\} - \{1-(1-\epsilon)^2\}/\{1+(1-\epsilon)^2\}]} \tag{8–4.33}$$

Table 8–4.2 provides a comparison of values of the Kozeny "constant" computed from the foregoing equations, and Eq. (8–4.11) for an assemblage of spherical particles. It is not possible to employ the ratio U/U_0, since there is no solution of the creeping motion equations for a single cylinder settling in an infinite medium, either parallel or perpendicular to its axis. Although the Kozeny constant for spheres lies between those for parallel and perpendicular cylinder flows in the range of ϵ between 0.4 and 0.8, at higher values of voidage the Kozeny constant for spheres is higher than for both types of cylinder cell models. The model does not appear to be applicable to fractional void volumes less than about 0.4 to 0.5, as noted below.[95]

The k values given in Table 8–4.2 for random orientation of cylinders are obtained by taking two-thirds of those for perpendicular flow plus one-third of those for parallel flow, at equal void volumes. It is interesting to note that the values obtained are close to those for spheres from $\epsilon = 0.40$–0.80, and not far from the experimentally determined constant of $k = 5.0$ in the range $\epsilon = 0.4$–0.70. Since cylinders represent an extreme departure in shape from spherical particles, this agreement lends support to the Carman-Kozeny theory of the permeability of porous media. Furthermore, the actual particle diameter does not enter into the determination of m, the hydraulic radius; thus, the constancy of the Kozeny factor k furnishes some justification for

TABLE 8–4.2

THEORETICAL VALUES OF THE KOZENY CONSTANT (k) FROM DIFFERENT CELL MODELS

Fractional Void Volume, ϵ	Flow Parallel to Cylinders	Flow Perpendicular to Cylinders	Flow through Random Orientation of Cylinders	Flow through Assemblages of Spheres
0.99	31.10	53.83	46.25	71.63
0.90	7.31	11.03	9.79	11.34
0.80	5.23	7.46	6.72	7.22
0.70	4.42	6.19	5.60	5.79
0.60	3.96	5.62	5.07	5.11
0.50	3.67	5.38	4.97	4.74
0.40	3.44	5.28	4.66	4.54

using a method of size-averaging nonuniformly sized particles in an assemblage, based on maintaining a constant hydraulic radius. This concept, of replacing an assemblage of nonuniformly sized particles by one of uniformly sized particles having the same (total surface)/(volume) as the original assemblage (but not the same number of particles), leads to the so-called reciprocal mean diameter, $\bar{D} = 1/\sum_i(w_i/D_i)$, where w_i is the weight fraction of material of diameter D_i. Of course, the use of an average diameter may be completely avoided by expressing the correlation directly in terms of hydraulic radius. It would be of interest to develop the free surface model for other shapes, for example, ellipsoids, to test the validity of the Kozeny-constant approach.

Of special interest is Emersleben's analytical solution of the Navier-Stokes equations for flow parallel to circular cylinders of equal radii in a square array. Emersleben represented the square array of circular sections by contours of constant value for a special periodic function, namely the second-order Epstein Zeta function[22]. This function, although it represents the contours well above $\epsilon = 0.8$, becomes a progressively poorer approximation at lower porosities. At $\epsilon = 0.9$, for example, the Emersleben equation corresponds to $k = 6.3$. This compares favorably with the value $k = 7.3$ from Table 8–4.2. At lower porosities agreement is poorer, but as porosity is increased, it becomes exceptionally good. Hasimoto[47], as noted earlier, has applied similar periodic solutions to dilute arrays of spheres and cylinders. In his treatment Hasimoto employs the Madelung constant, which derives from the Epstein Zeta function of the third order. No exact solution for concentrated assemblages of spheres is as yet available based on this general procedure.

An analytical solution in series has also been presented by Sparrow and Loeffler[95] for longitudinal laminar flow between cylinders arranged in equilateral triangular, or square, array, the latter being the same case studied by Emersleben. For this case the equation of motion may be expressed in terms of cylindrical coordinates (r, θ, x) as follows:

$$\nabla^2 u = \frac{\partial^2 u}{\partial r^2} + \frac{1}{r}\frac{\partial u}{\partial r} + \frac{1}{r^2}\frac{\partial^2 u}{\partial \theta^2} = \frac{1}{\mu}\frac{dp}{dx} \tag{8–4.34}$$

In order to solve this, it is convenient first to introduce a reduced velocity u^* defined by

$$u^* = u - \frac{r^2}{4}\left(\frac{1}{\mu}\frac{dp}{dx}\right) \tag{8–4.35}$$

By substitution into Eq. (8–4.34), we find that u^* must obey Laplace's equation,

$$\frac{\partial^2 u^*}{\partial r^2} + \frac{1}{r}\frac{\partial u^*}{\partial r} + \frac{1}{r^2}\frac{\partial^2 u^*}{\partial \theta^2} = 0 \tag{8–4.36}$$

The general solution may be written as

$$u^* = A + B\ln r + \sum_{k=1}^{\infty}(C_k r^k + D_k r^{-k})(E_k\cos k\theta + F_k\sin k\theta) \tag{8–4.37}$$

where k takes on integral values to insure that the velocity is single valued. Thus, in terms of u,

$$u = A + B \ln r - \frac{r^2}{4}\left(-\frac{1}{\mu}\frac{dp}{dx}\right) + \sum_{k=1}^{\infty} (C_k r^k + D_k r^{-k})(E_k \cos k\theta + F_k \sin k\theta) \tag{8–4.38}$$

The constants are to be determined from the boundary conditions.

For cylinders in square array, an end view of the configuration involved is shown in Fig. 8–4.1. Symmetry considerations permit us to confine our attention to the crosshatched element of the sketch. If n is designated as the direction of the normal to the element boundaries, we require that $\partial u/\partial n = \partial u/\partial \theta = 0$ at $\theta = 0$ and $\theta = 45°$; one finds, respectively, that

$$F_k = 0 \tag{8–4.39}$$

and that k is restricted to the values

$$k = 4, 8, 12, \ldots \tag{8–4.40}$$

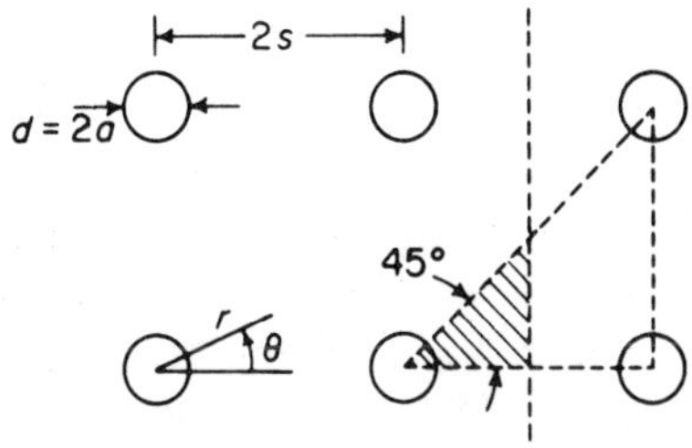

Figure 8-4.1. Definition sketch of a square array.

Imposing the requirement that $u = 0$ at the cylinder wall $r = a$ gives

$$D_k = -C_k a^{2k}, \qquad A = -B \ln a + \left(-\frac{1}{\mu}\frac{dp}{dx}\right)\frac{a^2}{4} \tag{8–4.41}$$

Further, it is required that the total drag force exerted on the fluid by the solid rod be balanced by the net pressure force acting over the entire cross section of the typical element. Thus,

$$\int_0^{\pi/4} \mu \left(\frac{\partial u}{\partial r}\right)_{r=a} a\, d\theta = \int_0^{\pi/4} \int_a^{s/\cos\theta} \left(\frac{dp}{dx}\right) r\, dr\, d\theta \tag{8–4.42}$$

Evaluation of this over-all force balance from Eq. (8–4.38) gives the value of B as follows:

$$B = \frac{2}{\pi} s^2 \left(-\frac{1}{\mu}\frac{dp}{dx}\right) \tag{8–4.43}$$

Introduction of these results into Eq. (8–4.38) results in the following equation for the velocity:

$$u = \frac{2}{\pi} s^2 \left(-\frac{1}{\mu}\frac{dp}{dx}\right) \ln \frac{r}{a} - \frac{1}{4}\left(-\frac{1}{\mu}\frac{dp}{dx}\right)(r^2 - a^2) + \sum_{j=1}^{\infty} G_j \left(r^{4j} - \frac{a^{8j}}{r^{4j}}\right) \cos 4j\theta \tag{8–4.44}$$

where the coefficients G_j $(= C_j E_j)$ still must be determined. At our disposal is the condition that $\partial u/\partial n = 0$ on the right-hand boundary of Fig. 8–4.1, on which $r = s/\cos\theta$. It is convenient to make use of the identity

$$\frac{\partial u}{\partial z} = \frac{\partial u}{\partial r}\cos\theta - \frac{\partial u}{\partial \theta}\frac{\sin\theta}{r} \tag{8–4.45}$$

where $z = r\cos\theta$. We may associate z with n on the boundary under consideration. Introducing Eq. (8–4.44) into the right-hand side of the foregoing, and setting $\partial u/\partial n = 0$ when $r = s/\cos\theta$, we find after rearrangement

$$\sum_{j=1}^{\infty} \delta_j(\cos\theta)^{1-4j}\left[\cos(4j-1)\theta + \left(\frac{a\cos\theta}{s}\right)^{8j}\cos(4j+1)\theta\right] + \frac{2\cos^2\theta}{\pi} - \frac{1}{2} = 0 \tag{8–4.46}$$

where

$$\delta_j = G_j\frac{4js^{4j}}{[-(1/\mu)(dp/dx)]s^2} \tag{8–4.47}$$

Equation (8–4.46) provides a means for determining the δ_j (that is, G_j). The first thought in applying this equation would be to apply the techniques of Fourier series. Such an approach would provide an infinite set of δ_j while satisfying Eq. (8–4.46) at all points along the boundary (that is, all $0° \le \theta \le 45°$). Unfortunately, the nature of Eq. (8–4.46) does not lend itself to Fourier analysis, and instead the equation was applied at a finite number of points along the boundary. Choosing l points results in l simultaneous equations for $\delta_1, \delta_2, \ldots, \delta_l$. Sufficient terms are retained in the series to assure good accuracy in shear stress and velocity distribution calculations. Numerical values of the δ_j obtained in this manner enable the velocity distribution to be computed, and ultimately, as well, the macroscopic parameters of engineering interest. The authors present similar derivations for the case of a configuration of cylinders arrayed at the vertices of equilateral triangles.

In order to obtain a pressure drop-flow relationship, the rate of flow is computed by integrating the velocity over the flow area, using, as before, only a typical element of the configuration as shown in Fig. 8–4.1. It is found that, for a given array, the pressure drop is directly proportional to the flow rate, as must be true. For convenience, the ratio s/a is replaced by its equivalent in terms of the fractional void volume ϵ. Sparrow and Loeffler's numerical results are presented in terms of the product of a friction factor f and Reynolds number N_{Re_d} based on tube diameter. This form is related to the Kozeny constant k via the relation

$$fN_{\mathrm{Re}_d} = \frac{8(1-\epsilon)}{\epsilon}k \tag{8–4.48}$$

In order to facilitate comparison, their results are presented in terms of the Kozeny constant in Fig. 8–4.2, along with values computed from the free surface model and Emersleben's theory. At a fractional void volume of $\epsilon = 0.5$, Sparrow and Loeffler find $k = 3.5$ for equilateral and $k = 2.9$ for square arrays, compared with $k = 3.7$ for the free surface model. Numerical values given by Sparrow and Loeffler at lower fractional void volumes show substantial deviations between equilateral and square arrangements and also

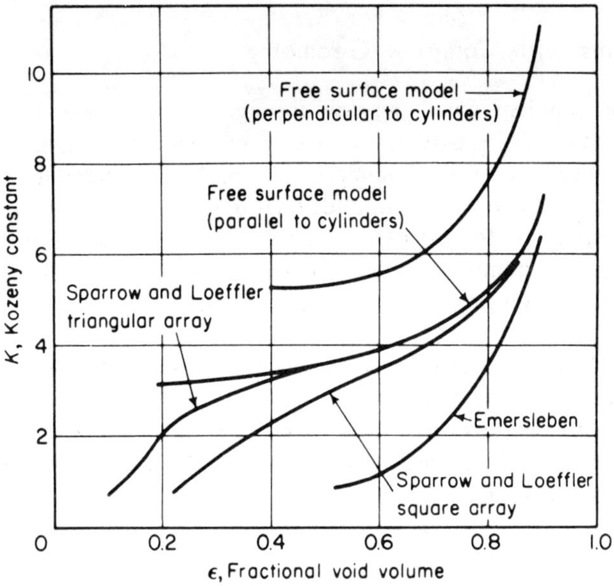

Figure 8-4.2. Comparison of theories for flow relative to circular cylinders.

from the free surface model. Sullivan's[98] experiments for rods in contact give a value of $k = 0.83$ for square array ($\epsilon = 0.215$) and $k = 0.81$ for equilateral triangular array ($\epsilon = 0.093$), in substantial agreement with Sparrow and Loeffler's theoretical treatment, providing good support for it. Sullivan's[98] experiments with bundles of fibers combed parallel to the direction of flow show a value of $k \approx 2.4$ for values of ϵ between 0.55 and 0.8, substantially lower than the theoretical relationships represented by Sparrow and Loeffler and the free surface model. Sparrow and Loeffler suggest that this may be due to a shortcoming in the experimental apparatus, the most likely factor being that the cylindrical filaments were not tightly packed against the wall of the bounding tube. Such an occurrence would lead to relatively large, open flow areas near the walls with a consequent increased permeability (lower k). The same type of discrepancy occurs in the behavior of fluidized beds of dilute suspensions of particles; namely, permeability is greater than indicated by theory. As noted earlier in this chapter, a possible explanation is particle segregation, resulting in the agglomeration of particles in small groups. Agglomeration might also explain the discrepancy noted in the case of cylindrical arrays.

It is of special interest to note that, at high fractional void volumes, particle arrangement does not affect permeability according to these theories. In this limit, Sparrow and Loeffler's result reduces exactly to Eq. (8–4.23). A similar result for square spacing was also obtained by Leibenson[62].

8-5 Systems with Complex Geometry

In most real problems involving flow through porous media, the arrangement and shape of the particles vary in a manner which makes an exact geometrical specification of a boundary value problem impossible. It is possible, however, on the basis of similitude relationships, to establish the plausibility of Darcy's law for the case of a viscous incompressible fluid at low particle Reynolds numbers, without having to obtain an explicit value of the permeability coefficient for a particular geometry.

Thus, consider two steady state systems, the first a reference state and the second a system which is geometrically similar, but with a different local fluid velocity at corresponding points. We wish to determine the relationship between pressure drop and fluid velocity for such systems. If we assume that inertial terms may be neglected, we obtain the creeping motion equations as applicable to each system:

$$\nabla p_0 = \mu_0 \nabla^2 \mathbf{v}_0 \tag{8–5.1}$$

$$\nabla p_1 = \mu_1 \nabla^2 \mathbf{v}_1 \tag{8–5.2}$$

Here p refers to the dynamic pressure at a given location; $\mathbf{v}$ is the local velocity at a location corresponding to p.

For one system to be dynamically similar to the other it is first necessary that the two systems be geometrically similar, so that each may be characterized by a characteristic length, l. Similarity is then maintained if corresponding terms in one differential equation can be obtained by multiplying the same terms in the other by a constant coefficient. If we restrict ourselves also to corresponding characteristic velocities having a common direction, we have

$$\frac{p_0}{l_0} = \alpha \frac{p_1}{l_1}; \qquad \frac{\mu_0 v_0}{l_0^2} = \alpha \frac{\mu_1 v_1}{l_1^2} \tag{8–5.3}$$

We can eliminate α and obtain a dimensionless parameter by dividing one of the equations in (8–5.3) by the other; hence,

$$\frac{p_0 l_0}{\mu_0 v_0} = \frac{p_1 l_1}{\mu_1 v_1} = c, \text{ say} \tag{8–5.4}$$

Thus, in this case, constancy of the dimensionless group $(pl)/(\mu v)$ characterizes all solutions of the creeping motion equations.

Now for flow through a bed of finite length confined by walls so that flow cross section is constant, we can take pressure at the outlet to be zero. Hence, the pressure gradient, or pressure drop per unit length will be $\Delta P/l$ in each case. Thus, in general, all solutions of the creeping motion equations will correspond to

$$v = \left(\frac{l^2}{c}\right)\left(\frac{\Delta P}{\mu l}\right) \tag{8–5.5}$$

Note that P is no longer the local pressure p. The velocity v is measured at a corresponding point in each system. The velocity through any geometry may, however, equally well be characterized by an average velocity taken as a linear function of the corresponding conditions in the flow space of each system. Such a velocity is the superficial velocity or average velocity referred to a unit cross section of the entire medium and in a reference direction of flow. The factor (l^2/c) is the well-known Darcy constant, and it is seen that it varies as the square of a characteristic dimension for geometrically similar media.

This relationship may be generalized in terms of flow in a bed where direction may vary, provided that it is still assumed that the resistance will be experienced in the same direction as the flow. Such flows often involve differences in liquid elevation in the system involved, which have been neglected in previous theoretical treatments in this book. Where the external force potential is given by gh, such effects are accounted for by adding ρgh to the pressure. In this form, Eq. (8–5.5) becomes

$$\mathbf{U} = -\frac{K}{\mu}\nabla(P + \rho gh) \tag{8–5.6}$$

where $\mathbf{U}$ is the superficial velocity, h is the vertical elevation above an arbitrary datum, P is a mean pressure, and g is the local acceleration of gravity.

It is not possible on the basis of similarity to draw conclusions regarding compressible flow of gases through porous media. Note also that, where the inertial terms cannot be neglected, similarity requires the same Reynolds number for the cases being compared; hence it is not possible to vary velocity without varying particle size if the same fluid is employed in the reference and test cases.

As shown in Table 8–4.2, for flow relative to cylindrical shapes in a random orientation as compared with spheres, the Kozeny constant appears to be a useful way of characterizing resistance to flow in beds of particles. It is also clear, however, that for flow perpendicular to a single row of cylinders all aligned and close to each other, the resistance to flow could become very large. It is evident, therefore, that the Carman-Kozeny concept, based on a mean hydraulic radius, will be applicable only where orientation effects are absent, that is, in isotropic porous media.

Hubbert[49] provides a clear discussion of the interrelation between Darcy's law and creeping flows. He points out a common misunderstanding which has arisen in connection with earlier derivations of Darcy's law, most of which have been based on various models of capillary tubes or pipes.[12] It has been known since the classical studies of Osborne Reynolds that Poiseuille's law fails when the flow transition from laminar to turbulent motion occurs. By analogy, the conclusion most often reached as to the cause of failure of Darcy's law has been that the motion became turbulent. This represents a serious misinterpretation of Darcy's law. In Darcy flow, each particle moves along a continuously curvilinear path at a continuously varying speed and with a varying acceleration. In true Poiseuille flow, each particle moves along a straight path at constant velocity and therefore suffers no acceleration. Consequently, in Poiseuille flow, inertial forces cannot become significant until turbulence occurs. On the other hand, failure of Darcy's law results when the distortion that occurs in the streamlines, owing to changes in direction of motion, is great enough that inertial forces become significant. The incidence of turbulence will occur at much higher Reynolds numbers, if indeed such a phenomenon occurs at all. Fundamental theoretical studies pertaining to the stability of single-phase flow through porous media are, at present, lacking.

Although Eq. (8–5.6) has been demonstrated to apply only for a given geometry, it may often be assumed that the effect of containing boundaries or walls is negligible as far as their effect on K is concerned. Then the equation of continuity may be written as follows in terms of superficial velocity for given external boundaries:

$$\nabla \cdot \mathbf{U} = 0 \tag{8–5.7}$$

where $\mathbf{U}$ is the superficial velocity. If this relation is applied to Eq. (8–5.6), we find[94]

$$\nabla^2(P + \rho g h) = 0 \tag{8–5.8}$$

This equation has been widely used to predict pressure relationships for flow through porous materials. It should be noted that Eq. (8–5.8), though formally similar in appearance to the relationship which applies to a viscous incompressible fluid in a continuum, refers in this case to a porous medium. The associated velocity field described by Eq. (8–5.6) will not be the same as that which applies to a continuum between solid boundaries as prescribed by the creeping motion equations. When the porous medium is not isotropic, K may assume directional properties and Eq. (8–5.8) will not be applicable. Neither can it be applied, of course, to predict effects of pressure transmitted by the bed of particles itself, or hydrodynamic forces on objects other than those forces normal to their surfaces.

The distinction between the relationships which apply to a porous medium and those which apply to the fluid medium itself, which flows in the pores, has not always been clearly distinguished in the literature, leading to confusion regarding the applicability of the former. Thus Streeter[97] states that, on the basis of the creeping motion equations, since $\nabla^2(P + \rho g h) = 0$, a velocity potential exists. He then concludes that Darcy's law gives the relationship between velocity and pressure gradient. This conclusion draws no distinction between the different boundary value problems involved. Again, Morse and Feshbach[70] state that for a liquid seeping through a porous solid, viscosity and vorticity can be neglected compared with the "friction of seepage"; they state, too, that Darcy's law follows. As we have seen earlier, the opposite is true, since Darcy's law applies only when inertial effects can be neglected.

Many authors have considered extensions of Darcy's law to compressible flows, unsteady states, and to situations where the inertial terms cannot be neglected. In addition, a variety of attempts have been made, on semiempirical grounds, to obtain numerical values for the Darcy constant. Scheidegger[84] reviewed a number of these treatments and also presented a new one. All these theories are based on analogies which are not quantitatively justifiable, and none of them lead to satisfactory values of the numerical constants involved without invoking *ad hoc* assumptions or resorting to experimental data.

Perhaps the most widely accepted of the theories aimed at finding the numerical coefficient in the Darcy equation are those based upon the concept that the pore space is equivalent to a bundle of parallel capillaries with a common hydraulic radius and with a cross sectional shape representative of the average shape of a pore cross section.[12] The development, usually attributed to Kozeny[54], was elaborated by Carman[13] and derived independently a few years later by Fair and Hatch[23]. In the derivation, it is assumed that the path of a streamline through the pore space will be tortuous, with an average length L_e greater than the length of the bed L. The length L_e is then regarded as the average length of the capillaries to be employed in Poiseuille type capillary formulas.

The average velocity for viscous flow through any noncircular capillary of hydraulic radius m and actual length L can be written

$$u_e = \frac{m^2 \,\Delta P}{k_o \mu L_e}$$

$k_o = 2$ for a circular capillary, and the equation becomes Poiseuille's law. If the hydraulic radius concept were completely applicable, k_o would be independent of cross-sectional shape for viscous flow (as, in fact, it is for turbulent flow). For reasonable rectangular, elliptical, and annular shapes, k_o lies in the range 2.0–2.5, and is therefore reasonably independent of shape. By putting

$$u_e = \frac{U}{\epsilon} \tag{8–5.9}$$

and

$$m = \frac{\epsilon}{S}$$

where S is the particle surface per unit volume of bed (that is, the specific surface) we obtain

$$U = \frac{L}{L_e}\frac{\epsilon^3}{k_o\mu S^2}\frac{\Delta P}{L} = \frac{\epsilon^3}{k\mu S^2}\frac{\Delta P}{L} \tag{8–5.10}$$

Here, $k = (L_e/L)k_o$ is designated as the Kozeny constant. The ratio L_e/L is called the *tortuosity*. Carman, in discussing this factor, contends that the time taken for an element of fluid to pass through a tortuous track of length L_e is greater than if it passed directly over length L, by an amount L_e/L. Accordingly, he proposes that Eq. (8–5.9) be replaced by

$$u_e = \frac{U}{\epsilon}\frac{L_e}{L} \tag{8–5.11}$$

whence we now obtain

$$k = \left(\frac{L_e}{L}\right)^2 k_o \tag{8–5.12}$$

As noted earlier in this chapter, $k \approx 5.0$. The value of L_e/L is difficult to estimate, but, if the value of $\sqrt{2}$ adopted by Carman is accepted, then $k_o \approx 2.5$. This gives a reasonable explanation for the factors entering into the Kozeny constant, k.

The Kozeny equation may be expressed in terms of particle size by relating S to the particle diameter. For spheres of diameter d, $S = 6(1 - \epsilon)/d$ and Eq. (8–5.10) becomes

$$U = \left[\frac{\epsilon^3 d^2}{36(1 - \epsilon)^2 k}\right]\frac{\Delta P}{\mu L} \tag{8–5.13}$$

Here, the expression in brackets is the Darcy constant. In the previous section (see Table 8–4.2) the Kozeny constant k was derived theoretically on the basis of models involving drag on spherical- or cylindrical-shaped objects, and it was shown that the empirical constant can be predicted theoretically to a considerable degree of accuracy.

More elaborate derivations have been made in attempts to generalize the Darcy equation, but they all seem to assume a specific geometrical model in the final analysis. Thus, Scheidegger[83] develops a theory involving statistical hydrodynamics of porous media which takes into account sidewise dispersion which a fluid undergoes in passing through a porous medium. In order to derive Darcy's law for pressure drop, however, he finds it necessary to assume that flow occurs in microscopic channels, which is, of course, tantamount to assuming that a variation of Poiseuille's law applies. Hall[34], in a generalized tensorial approach, assumes a relationship between stress and velocity gradient, referred to a one-dimensional single surface element model, which thus does not solve any specific boundary-value problem involving the complicated geometrical boundaries surrounding actual fluid elements in a porous medium. Irmay[51], in an ingenious derivation which takes many factors into consideration, assumes that flow in a porous medium can ultimately be idealized to motion of fluid in a two-dimensional pore channel with a parabolic fluid velocity profile, thereby again assuming a Poiseuille's law model. Thus, none of the more generalized studies permit the Darcy constant to be calculated for an actual particle assemblage.

Generalized treatment of multiparticle systems

The techniques utilized in Chapter 5 to describe the intrinsic resistance of a single particle in a quasi-static Stokes flow have been extended by Brenner[8] to systems consisting of any number of interacting particles moving through a fluid at rest at infinity. Furthermore, the extension applies equally well to the case in which the fluid motion is bounded by the walls of a container.

Consider a fluid-particle system consisting of any finite number of rigid particles of arbitrary shape. Let S_i denote the surface of the ith particle, and let S_b denote the surface(s) of the container boundaries, if any. If no containing walls bound the fluid externally, let S_∞ be a fluid surface of indefinitely large size which completely surrounds the particles.* The boundaries S_b and S_∞ are at rest. We suppose further that the nature of the boundaries S_b and/or S_∞ is such that the stresses exerted on these surfaces can do no work on the fluid contained in the interior. This occurs, for example, when the velocity field, $\mathbf{v}$, vanishes on S_b and/or S_∞. It also occurs when either the normal velocity, $\mathbf{n}\cdot\mathbf{v}$, or normal stresses, $\mathbf{n}\cdot\mathbf{\Pi}\cdot\mathbf{n}$, and the tangential stresses, $\mathbf{n}\cdot\mathbf{\Pi}\cdot\mathbf{t}$, *both* vanish on S_b ($\mathbf{n}=$ unit normal vector, $\mathbf{t}=$ unit tangent vector). It also occurs when the former condition, $\mathbf{v}=\mathbf{0}$, is satisfied on one part of S_b and the latter conditions are satisfied on the remainder of S_b — for example, when a suspension settles in a cylinder of finite depth, the upper surface being a free surface. (Interfacial forces at the free surface are assumed to be negligible.)

Let O_i be any point fixed in the ith particle and denote by $\mathbf{U}_i$ the instantaneous translational velocity of this point. By $\boldsymbol{\omega}_i$ we represent the instantaneous angular velocity of the particle. If $\mathbf{r}_i$ is the position vector of any point relative to an origin at O_i, the boundary conditions to be satisfied are then

$$\mathbf{v} = \mathbf{U}_i + \boldsymbol{\omega}_i \times \mathbf{r}_i \qquad \text{on each } S_i \tag{8–5.14}$$

and either

$$\mathbf{v} = \mathbf{0} \qquad \text{on } S_b \text{ and/or } S_\infty$$

or

$$\left\{\begin{matrix} \mathbf{n}\cdot\mathbf{v} = 0 \text{ or } \mathbf{n}\cdot\mathbf{\Pi}\cdot\mathbf{n} = 0 \\ \text{and } \mathbf{n}\cdot\mathbf{\Pi}\cdot\mathbf{t} = 0 \end{matrix}\right\} \quad \text{on } S_b \tag{8–5.15}$$

Because of the linearity of the equations of motion and boundary conditions, we can separate $(\mathbf{v}, p)$ into a translational contribution, $(\bar{\mathbf{v}}, \bar{p})$, satisfying only the $\mathbf{U}_i$ portion of Eq. (8–5.14), and a rotational contribution, $(\bar{\bar{\mathbf{v}}}, \bar{\bar{p}})$, satisfying the $\boldsymbol{\omega}_i \times \mathbf{r}_i$ portion of Eq. (8–5.14). Thus,

$$\mathbf{v} = \bar{\mathbf{v}} + \bar{\bar{\mathbf{v}}} \tag{8–5.16a}$$

$$p = \bar{p} + \bar{\bar{p}} \tag{8–5.16b}$$

Again, because of linearity, we can further subdivide each of these contributions by the following device: Let $\bar{\mathbf{v}}_i$ be the velocity field created when the ith particle translates with instantaneous velocity $\mathbf{U}_i$ *while all other particles are at rest*. Likewise, let $\bar{\bar{\mathbf{v}}}_i$ be the velocity field created when the ith particle rotates with angular velocity $\boldsymbol{\omega}_i$ about an axis through O_i *while all other particles are at rest*. Thus, we have

*If S_b only *partly* bounds the fluid externally as in the case of the walls of an infinitely long circular cylinder, then S_∞ denotes the two open "ends" of the cylinder.

$$\bar{\mathbf{v}} = \sum_i \bar{\mathbf{v}}_i, \qquad \bar{\bar{\mathbf{v}}} = \sum_i \bar{\bar{\mathbf{v}}}_i \tag{8–5.17}$$

with similar expressions for the pressures. Each $\bar{\mathbf{v}}_i$ satisfies the boundary conditions:

$$\begin{cases} \bar{\mathbf{v}}_i = \mathbf{U}_i & \text{on } S_i \\ \bar{\mathbf{v}}_i = \mathbf{0} & \text{on } S_j \ (j \neq i) \\ \text{plus equations of the type of Eq. (8–5.15)} \end{cases} \tag{8–5.18}$$

whereas each $\bar{\bar{\mathbf{v}}}_i$ satisfies the boundary conditions

$$\begin{cases} \bar{\bar{\mathbf{v}}}_i = \boldsymbol{\omega}_i \times \mathbf{r}_i & \text{on } S_i \\ \bar{\bar{\mathbf{v}}}_i = \mathbf{0} & \text{on } S_j \ (j \neq i) \\ \text{plus equations of the type of Eq. (8–5.15)} \end{cases} \tag{8–5.19}$$

Let ${}_j\bar{\mathbf{F}}_k$ be the hydrodynamic force exerted by the fluid on the kth particle due to the translational motion $\bar{\mathbf{v}}_j$ of the jth particle (j and k may be equal). We then have

$${}_j\bar{\mathbf{F}}_k = \int_{S_k} \bar{\mathbf{\Pi}}_j \cdot d\mathbf{S} \tag{8–5.20}$$

Also, let ${}_j\bar{\mathbf{T}}_k$ be the hydrodynamic torque about O_k exerted by the fluid on the kth particle in consequence of the translational motion of the jth particle, $\bar{\mathbf{v}}_j$. We then have

$${}_j\bar{\mathbf{T}}_k = \int_{S_k} \mathbf{r}_k \times (\bar{\mathbf{\Pi}}_j \cdot d\mathbf{S}) \tag{8–5.21}$$

The total hydrodynamic force on the kth particle due to the translational motion of *all* particles, including itself (that is, due to the field $\bar{\mathbf{v}}$) is then

$$\bar{\mathbf{F}}_k = \sum_j {}_j\bar{\mathbf{F}}_k \tag{8–5.22}$$

whereas the corresponding hydrodynamic torque (about O_k) on the kth particle due to the translational motion of *all* particles is

$$\bar{\mathbf{T}}_k = \sum_j {}_j\bar{\mathbf{T}}_k \tag{8–5.23}$$

In similar fashion we obtain for the corresponding *rotational* motions,

$${}_j\bar{\bar{\mathbf{F}}}_k = \int_{S_k} \bar{\bar{\mathbf{\Pi}}}_j \cdot d\mathbf{S} \tag{8–5.24}$$

$${}_j\bar{\bar{\mathbf{T}}}_k = \int_{S_k} \mathbf{r}_k \times (\bar{\bar{\mathbf{\Pi}}}_j \cdot d\mathbf{S}) \tag{8–5.25}$$

$$\bar{\bar{\mathbf{F}}}_k = \sum_j {}_j\bar{\bar{\mathbf{F}}}_k \tag{8–5.26}$$

$$\bar{\bar{\mathbf{T}}}_k = \sum_j {}_j\bar{\bar{\mathbf{T}}}_k \tag{8–5.27}$$

Finally, the total force and torque about O_k on the kth particle due to the translational *and* rotational motions of all particles [that is, due to the original motion $(\mathbf{v}, p)$ satisfying Eqs. (8–5.14)–(8–5.15)] are, respectively,

$$\mathbf{F}_k = \bar{\mathbf{F}}_k + \bar{\bar{\mathbf{F}}}_k \tag{8–5.28}$$

$$\mathbf{T}_k = \bar{\mathbf{T}}_k + \bar{\bar{\mathbf{T}}}_k \tag{8–5.29}$$

Using arguments substantially identical to those in Chapter 5, it may be shown[8] that the following four relations are valid:

$$_j\bar{\mathbf{F}}_k = -\mu \mathbf{K}_{kj} \cdot \mathbf{U}_j \tag{8–5.30a}$$

$$_j\bar{\mathbf{T}}_k = -\mu \mathbf{C}_{kj} \cdot \mathbf{U}_j \tag{8–5.30b}$$

$$_j\bar{\bar{\mathbf{F}}}_k = -\mu \mathbf{D}_{kj} \cdot \boldsymbol{\omega}_j \tag{8–5.30c}$$

$$_j\bar{\bar{\mathbf{T}}}_k = -\mu \boldsymbol{\Omega}_{kj} \cdot \boldsymbol{\omega}_j \tag{8–5.30d}$$

where the dyadics $\mathbf{K}_{kj}$, $\mathbf{C}_{kj}$, $\mathbf{D}_{kj}$, and $\boldsymbol{\Omega}_{kj}$ are independent of the properties of the fluid and of the magnitudes and directions of all the $\mathbf{U}_j$ and $\boldsymbol{\omega}_j$. Rather, they are intrinsic properties of the system of particles (and boundary, if any) which depend only on the size and shape of the particles, the instantaneous configuration of the system, and the choice of origins, O_i.

By utilizing the reciprocal theorem, it may be shown that these dyadics have the following symmetry properties:[8]

$$\mathbf{K}_{kj} = \mathbf{K}_{jk}^{+} \tag{8–5.31a}$$

$$\boldsymbol{\Omega}_{kj} = \boldsymbol{\Omega}_{jk}^{+} \tag{8–5.31b}$$

$$\mathbf{D}_{kj} = \mathbf{C}_{jk}^{+} \tag{8–5.31c}$$

for *all* j and k, including $j = k$.

Subject to the symmetry restrictions set forth in Eq. (8–5.31), the results of the present subsection are, from Eqs. (8–5.28)–(8–5.30),

$$\mathbf{F}_k = \quad \mu \sum_{j=1}^{n} (\mathbf{K}_{kj} \cdot \mathbf{U}_j + \mathbf{D}_{kj} \cdot \boldsymbol{\omega}_j) \tag{8–5.32a}$$

$$\mathbf{T}_k = -\mu \sum_{j=1}^{n} (\mathbf{C}_{kj} \cdot \mathbf{U}_j + \boldsymbol{\Omega}_{kj} \cdot \boldsymbol{\omega}_j) \tag{8–5.32b}$$

for $k = 1, 2, 3, \ldots, n$, where n is the number of particles. Note that the summation is on adjacent indices. This relation can be concisely expressed and manipulated in matrix form. Define the partitioned column matrices $(\mathbf{F})$, $(\mathbf{T})$, $(\mathbf{U})$, $(\boldsymbol{\omega})$ by expressions of the general form

$$(\mathbf{F}) = \begin{pmatrix} (\mathbf{F}_1) \\ (\mathbf{F}_2) \\ \vdots \\ (\mathbf{F}_n) \end{pmatrix}, \text{ etc.} \tag{8–5.33}$$

in which $(\mathbf{F}_i)$ is itself a column matrix whose scalar elements are the three components of the vector force $\mathbf{F}_i$ on the ith particle. Also define the square, partitioned matrices $(\mathbf{K})$, $(\boldsymbol{\Omega})$, $(\mathbf{C})$ by expressions of the form

$$(\mathbf{K}) = \begin{pmatrix} (\mathbf{K}_{11}) & (\mathbf{K}_{12}) & \cdots & (\mathbf{K}_{1n}) \\ (\mathbf{K}_{21}) & (\mathbf{K}_{22}) & \cdots & (\mathbf{K}_{2n}) \\ \cdots & \cdots & \cdots & \cdots \\ (\mathbf{K}_{n1}) & (\mathbf{K}_{n2}) & \cdots & (\mathbf{K}_{nn}) \end{pmatrix}, \text{ etc.} \tag{8–5.34}$$

in which $(\mathbf{K}_{ij})$ is a 3×3 matrix whose nine scalar elements are the components of the dyadic $\mathbf{K}_{ij}$.

If we now further define the partitioned column matrices

$$(\mathscr{F}) = \begin{pmatrix} (\mathbf{F}) \\ (\mathbf{T}) \end{pmatrix}, \qquad (\mathscr{U}) = \begin{pmatrix} (\mathbf{U}) \\ (\boldsymbol{\omega}) \end{pmatrix} \tag{8–5.35a}$$

and the square, partitioned matrix

$$(\mathscr{K}) = \begin{pmatrix} (\mathbf{K}) & (\mathbf{C})^{+} \\ (\mathbf{C}) & (\boldsymbol{\Omega}) \end{pmatrix} \tag{8–5.35b}$$

then Eq. (8–5.32) may be written in the compact form

$$(\mathscr{F}) = -\mu(\mathscr{K})(\mathscr{U}) \tag{8–5.36}$$

It is reasonable to call $(\mathscr{F})$ the *wrench matrix*, $(\mathscr{U})$ the *screw-velocity matrix*, and $(\mathscr{K})$ the *grand resistance matrix*. Since $(\mathbf{K}) = (\mathbf{K})$ and $(\boldsymbol{\Omega}) = (\boldsymbol{\Omega})^{+}$, the matrix $(\mathscr{K})$ is symmetric. Note, however, that except for the diagonal submatrices $(\mathbf{K}_{11}), (\mathbf{K}_{22}), \ldots, (\boldsymbol{\Omega}_{11}), (\boldsymbol{\Omega}_{22}), \ldots$, the individual submatrices of $(\mathscr{K})$ are themselves not symmetric. As the rate of mechanical energy dissipation is essentially positive, the grand resistance matrix is positive-definite.

In passing from the vector-polyadic formulation to the matrix formulation remember that all the subscripted submatrices of $(\mathscr{F})$, $(\mathscr{K})$, and $(\mathscr{U})$ must be expressed in a common system of coordinates.

Equation (8–5.36) has the important advantage that, when the hydrodynamic forces and torques on the individual particles are known *a priori*, as is usually the case in particle systems settling under the influence of gravity, Eq. (8–5.36) may be explicitly solved for the particle velocities and spins by merely inverting the nonsingular matrix $(\mathscr{K})$ as follows:

$$(\mathscr{U}) = -\frac{1}{\mu}(\mathscr{K})^{-1}(\mathscr{F}) \tag{8–5.37}$$

In summary, the intrinsic resistance of a multiparticle system, consisting of any number of rigid particles of any shape, in any instantaneous configuration, at any concentration, and with bounding walls can be described in terms of a partitioned, positive-definite, symmetrical matrix termed the *grand resistance matrix*.

Brenner[8] illustrates the significance of the formalism by considering the hydrodynamic interaction of two spherical particles of different size. For this example, Eqs. (8–5.34) and (8–5.35b) show that 16 dyadics are required to establish the grand resistance matrix. Only ten of these are independent because of the six symmetry relations

$$\mathbf{K}_{21} = \mathbf{K}_{12}^{+}, \quad \boldsymbol{\Omega}_{21} = \boldsymbol{\Omega}_{12}^{+}, \quad \mathbf{D}_{11} = \mathbf{C}_{11}^{+}, \quad \mathbf{D}_{12} = \mathbf{C}_{21}^{+}, \quad \mathbf{D}_{21} = \mathbf{C}_{12}^{+}, \quad \mathbf{D}_{22} = \mathbf{C}_{22}^{+}$$

Except perhaps in the case of multiparticle systems which are dilute, determination of the grand resistance matrix will be a complicated matter. Recourse to methods such as discussed in Section 8–3 may be necessary to obtain numerical results. The general formulation, however, shows how particle velocities and spins may vary with particle geometry and arrangement, since the dyadics in $(\mathscr{K})$ are dependent only on these latter factors.

Thus, Darcy's law will apply to the flow of fluid through a packed bed when the flow is restricted to one direction by container walls, in the sense that pressure drop will be directly proportional to fluid throughput per unit time. It can be seen that any systematic rigid arrangement of particles may also result in a sidewise thrust on container walls (for example, an assemblage of spheroids placed obliquely to the direction of average fluid flow). Darcy's law, in the more general sense of the vectorial relationship given by Eq. (8–5.6), would not apply to flow through a nonisotropic medium in which the net direction of flow was not determined by external bounding walls.

A problem much less well understood than the prediction of pressure drop is that of longitudinal and radial dispersion of marked fluid particles in flow through porous media. In principle, some of the techniques discussed earlier in this chapter should prove useful. Thus the reflection technique provides the detailed velocity pattern associated with any type of particle assemblage. Such detailed knowledge is a necessary prerequisite to a rational, quantitative description of dispersion. Cell models like the "free surface" model, by establishing a microscopic picture of the flow pattern adjacent to particles, also permit an estimate of nonuniformities in axial and lateral displacement of fluid particles. Experimental studies[44] of streamline flow through cubic assemblages of particles indicate negligible dispersion effects due to mechanical or convective diffusion. Such effects would, of course, be expected in flow through an irregularly or randomly packed porous medium, owing to the irregular pattern of the streamlines through the pores and voids. A "random residence-time" model for longitudinal dispersion has been developed by several writers[66], based on flow through a sequence of cells in each of which there is complete mixing. This model does not lend itself to a study of lateral dispersion. Moreover, its validity at low Reynolds numbers appears open to question. Several writers[5,83] have investigated dispersion on the assumption that a fluid particle carries out a random walk consisting of a succession of statistically independent straight steps in equal small intervals of time. This treatment is not entirely satisfactory, partly because the time is broken up into equal small intervals whereas one would expect a particle to stay longer in a region where the velocity is small than where it is large.

Saffman (1959)[81] reviewed a number of theories and put forth a new development which overcomes some of their shortcomings. It is based on a model of flow through a porous medium which assumes that the latter may be represented by an assemblage of randomly orientated and randomly distributed straight pores, in each of which the flow is uniform. The pores are supposed to be connected with one another at the ends, and several pores may start or finish at these end points. The dimensions of the pores are taken as comparable with the size of particles composing the bed. The path of a fluid particle may then be regarded as a random walk in which the length, direction, and duration of each step are random variables. An involved mathematical analysis based on statistical considerations leads to approximate numerical results for both longitudinal and lateral dispersion effects. The author also includes a discussion of the possible effect of higher velocities, at which Darcy's law no longer applies.

Scheidegger[84] has reviewed contributions to the study of multiple-phase flow. Studies are fewer in number and, owing to the greater complexity of the phenomena involved, treatments are based for the most part on rather simple hydrodynamic models, which at best give only qualitative results. Whether the fluids are miscible or immiscible is an important parameter in multiple-phase flow, and it turns out that the study of immiscible fluids is much simpler than that of miscible fluids. Most of the literature is concerned with immiscible fluids and utilizes models based on capillaries or on the Kozeny theory. Miscible displacement in porous media is a type of two-phase flow in which the two phases are completely soluble in each other. Therefore, capillary forces between the two fluids do not come into effect as in the immiscible case. It would seem that miscible displacement could be described in a very simple fashion. For the mixture, under conditions of complete miscibility, might be thought to behave locally as a single phase which would obey Darcy's law. The displacement front has, however, been shown[82] to always become unstable when the penetrating fluid is less viscous than the displaced one. This results in the formation of "fingers" which penetrate the porous medium in an irregular fashion, a phenomenon which is highly undesirable in the application of processes of this type to oil recovery. Scheidegger[85], among others, studied the process of finger generation caused by the heterogeneities present in the porous medium, and presented some formulas which include a criterion for the condition for serious fingering to occur.

The three last sections of this chapter consider engineering applications broadly, without going into detail. Emphasis is placed on problems where hydrodynamics is applicable and on the limitations of the hydrodynamic approach. These sections also include a comparison of the theories developed earlier in this chapter with available experimental data.

8-6 Particulate Suspensions

The particles in a suspension can often be assumed to behave independently, without interactions among them and without significant wall effects. In such instances the methods and results of Chapters 4 and 5 will be applicable. In addition to considering whether interaction or wall effects may be of importance, several other factors must sometimes be taken into account.

A body falling under gravity will usually attain a constant terminal velocity when the net gravitational accelerating force, corrected for buoyancy, equals the resisting drag force. For flow of fluid around a sphere, Stokes' law applies. Comparable relationships exist for differently shaped bodies, as discussed in Chapters 4 and 5. At Reynolds numbers N_{Re_d} (based on diameter) less than 0.05, innumerable experiments performed with spheres in a variety of media show that the deviation from Stokes' law does not exceed 1 per cent[32]. A Reynolds number equal to 0.05 corresponds to a 77-micron sphere of unit density falling through air. At higher Reynolds numbers the deviation becomes greater and Stokes' law tends to overestimate the settling velocity. From considerations of dynamic similarity, numerous results for variations of shape and fluid can be coordinated[61,32] by plotting Reynolds number against another nondimensional parameter, the drag coefficient C_D, which is defined as the ratio of the particle resistance to the product of cross-section and kinetic pressure. In the simple case of Stokes' law,

$$C_D = \frac{24}{N_{\mathrm{Re}_d}} \tag{8-6.1}$$

For particles of a size comparable with the mean free path of the gas molecules, the medium can no longer be regarded as continuous, and the particles fall between the molecules faster than predicted by continuum hydrodynamics. To allow for this slip, Cunningham[15] deduced a correction to Stokes' law, expressible as follows—see Eq. (2–8.1):

$$F = \frac{3\pi\mu d}{k_m} \tag{8-6.2}$$

where

$$k_m = 1 + a_m \frac{\lambda}{d} \tag{8-6.3}$$

d is the sphere diameter, λ is the mean free path of the gas molecules, and a_m is a numerical factor which lies between 1.3 and 2.3 for different gases, particle sizes, and materials. Lapple[61] gives the following typical values (Table 8–6.1):

TABLE 8–6.1
VALUES OF k_m FOR SPHERICAL PARTICLES IN AIR

Particle Diameter (microns)	*Value of k_m* (70°F)	(500°F)
0.1	2.88	5.14
0.25	1.682	2.528
0.5	1.325	1.711
1.0	1.160	1.338
10.0	1.016	1.033

These calculations are all based on the assumption that particle interactions or containing walls exert no retarding influence on the settling velocity.

Another correction to the rate of settling of a particle is due to Brownian movement. This results in a random motion of the particles in addition to any net motion in a given direction due to the action of external forces such as gravity. Quantitatively, this random motion may be expressed as follows[19]:

$$\sqrt{\overline{(\Delta s)^2}} = \sqrt{\frac{4RTk_m t}{3\pi^2 \mu N d}} \tag{8–6.4}$$

The value $\sqrt{\overline{(\Delta s)^2}}$ is a scalar quantity representing the statistical average linear displacement of a particle in a given direction in time t. R is the gas constant, T the absolute temperature, and N is Avogadro's number. Table 8–6.2 gives a comparison of the magnitude of Brownian movement displacement with that due to gravitational settling[61].

TABLE 8–6.2
COMPARISON OF BROWNIAN AND GRAVITATIONAL DISPLACEMENTS FOR PARTICLES SUSPENDED IN AIR AND WATER

	Displacement in 1.0 sec (microns)			
	In Air at 70°F (1 atm)		In Water at 70°F	
Particle Diameter (microns)	Due to Brownian Movement*	Due to Gravitational Settling†	Due to Brownian Movement*	Due to Gravitational Settling†
0.1	29.4	1.73	2.36	0.005
0.25	14.2	6.30	1.49	0.0346
0.5	8.92	19.9	1.052	0.1384
1.0	5.91	69.6	0.745	0.554
2.5	3.58	400.0	0.471	3.46
5.0	2.49	1550.0	0.334	13.84
10.0	1.75	6096.0	0.236	55.4

*This is the mean displacement for $t = 1$ sec given by Eq. (8–6.4).

†This is the distance settled in 1 sec by a particle of specific gravity = 2.0, including the Cunningham correction factor, k_m, for settling in air.

For situations involving particle interaction, where Stokes' law applies to individual spherical particles settling in an unbounded medium, the methods developed in the present chapter are applicable. There have also been numerous experimental investigations of this phenomenon of "hindered settling." Maude and Whitmore[65] present a review of pertinent experimental work, listing many of the important references. They propose the relationship

$$\frac{U}{U_0} = (1 - \phi)^{\beta} = \epsilon^{\beta} \tag{8–6.5}$$

where β is a function of particle shape, size distribution, and Reynolds number; ϕ is the volume of solids per unit volume of suspension. This relationship is shown to satisfy the experimental results of other workers, and some theoretical validity is claimed for it. They report values of $\beta \approx 5$ for equisized, equi-density spheres in creeping flow; $\beta \approx 2$ to 4 for spheres in turbulent flow; and $\beta \approx 7$ to 10 for rough particles in creeping flow. Relationships similar to Eq. (8–6.5) have been investigated further by Harris[46]. As discussed in the sequel to Eq. (8–4.15), Brinkman[10] and Richardson and Zaki[78] have also developed correlations which are intended to apply from the dilute range to intermediate concentrations, $\epsilon = 0.5$. The correlation proposed by Richardson and Zaki is of the same form as Eq. (8–6.5), with $\beta = 4.65$ for spheres in creeping flow.

In the dilute range $\epsilon = 0.999$ to $\epsilon = 0.95$, available data from investigations by Cheng and Schachman[14] and Noda[72] are in rough agreement with the relationship given by Eq. (8–6.5). The theoretical studies leading to Eqs. (8–3.44)–(8–3.46) indicate a much higher concentration dependence of the rate of settling.

The scatter of settling rate data reported by various investigators suggests, however, that no single equation relating settling rate to concentration is adequate for dilute suspensions. A plot of some representative data in Fig. 8–6.1 illustrates the range of scatter likely in sedimentation studies.

The highest settling rates shown in Fig. 8–6.1 were reported by Kaye and

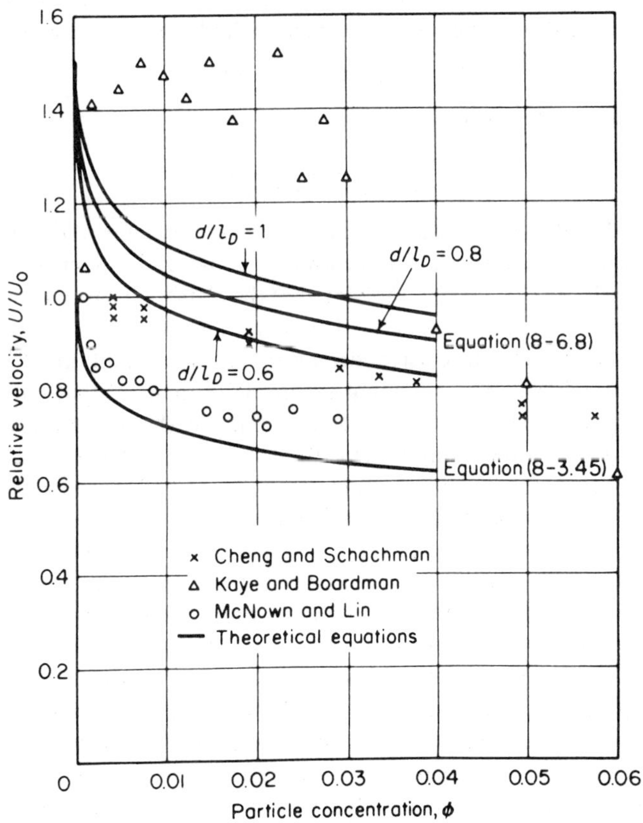

Figure 8-6.1. Comparison of theoretical equation with data in the dilùte concentration range.

Boardman[53] for experiments in which a considerable amount of cluster formation was observed. Since it is well established that clusters of particles will settle at a faster rate than the same number of particles uniformly distributed, theoretical equations based on uniform distribution of particles must be modified to predict settling rates for dilute suspensions.

In order to illustrate empirically how cluster formation can influence the theoretically derived equations, consider the simplest cluster, the doublet. The drag force on a single vertical doublet consisting of two equal spheres of radii a in contact in an infinite medium is given by the equation

$$F = 6\pi\mu a\lambda U \tag{8–6.6}$$

The resistance coefficient, λ, has been calculated by Stimson and Jeffery as a function of d/l_D (see Table 6–4.1), where $d = 2a$ is a sphere diameter and l_D is the center-to-center distance. λ is equal to 0.645 for $d/l_D = 1$ and approaches unity as $d/l_D \rightarrow 0$. For very dilute suspensions in which l_D is much smaller than the mean distance between doublets, each doublet may be considered as a point force with respect to the rest of the suspension; hence, the resistance of a dilute rhombohedral suspension of doublets can be represented by

$$F = 6\pi\mu a\lambda U(1 + 1.79\phi^{1/3}) \tag{8–6.7}$$

and its settling velocity by

$$\frac{U}{U_0} = \frac{1}{\lambda(1 + 1.79\phi^{1/3})} \tag{8–6.8}$$

Curves for $d/l_D = 0.6, 0.8, 1.0$, as well as a curve for a rhombohedral suspension of single particles, are shown in Fig. 8–6.1. These curves intersect the $\phi = 0$ axis at $U/U_0 = 1/\lambda$, implying that the elemental particle is a doublet composed of individual spheres. The data intersect the $\phi = 0$ axis at $U/U_0 = 1$, indicating that clusters are destroyed upon dilution and the elemental particle is a single, individual sphere. It is apparent that the presence of doublets can markedly influence settling rate; it should not, however, be inferred that real suspensions consist of doublets. The extent of cluster formation probably depends to a great extent on the technique of the experimenter and the nature of particle and fluid.

In the more concentrated range, $\epsilon = 0.95$ to $\epsilon = 0.60$, where more data are available, individual variations of up to 100 per cent exist from the lowest to highest values of U/U_0 obtained at a given concentration. Most of the data reported in this range show higher values of U/U_0 than predicted by the free surface model (see Table 8–4.1). It is believed that even with uniformly sized spherical particles, variables other than fractional void volume may be required to uniquely characterize the behavior of a sedimenting system. Circulation of particles, or effects tending to make them segregate into aggregates, will result in increased values of U/U_0, compared with those obtained for completely uniform distributions. It is possible that spheres suspended in random orientations may not maintain their positions relative to each other. Happel and Pfeffer[45], based on data obtained for the motion of two spheres in a viscous fluid, speculate that the formation of doublets, with their concomitantly larger settling velocitites, may be a contributing cause to the wide discrepancies observed in presently available data on fluidized and sedimenting systems.

When high concentrations are involved in sedimenting systems of particles, flocculation may often introduce another complication. Flocculation and dispersion have long been recognized as important factors in dealing with particles suspended in liquids. They are often important in aerosol systems too.

The kinetics of an assemblage of small, neutrally buoyant spherical particles, which coalesce upon collision to form progressively larger spheres, was originally worked out by Smoluchowski[93]. He assumed that the particles collide owing to Brownian movement and stick to each other in a certain fraction of these collisions, thereby causing a reduction in the total number of separate particles present. Green and Lane[32] present an outline of this theory together with other references. The Smoluchowski equation may be represented as follows:

$$\frac{1}{n_p} - \frac{1}{n_{p0}} = \frac{2}{3}\frac{RTs}{\mu N}t \tag{8-6.9}$$

where n_p is the number of particles per unit volume at time t; n_{p0}, the number of particles at zero time; R, the gas constant; T, the absolute temperature; s, the ratio of the sphere of influence to the radius of the particle (if $s = 2$, the spheres will coagulate only upon touching); μ, the viscosity of the medium; and N, Avogadro's number.

This equation assumes that initially all particles are the same size. Note that one consequence of Eq. (8–6.9) is that a linear relationship exists between particle volume and time during coagulation. This equation is independent of the size of the particle, so that sols of different particle size should coagulate at the same rate. Various modifications have been developed to allow for the mean free path of fluid molecules, electrical charge, heterogeneity of the dispersion, and failure of colliding particles to adhere. Flocculation may also arise by virtue of fluid stirring or turbulence.

In general, the simple form of Smoluchowski's equation has been found[61] to hold reasonably for dilute aerosols, indicating that adhesion generally occurs upon collision. Heterogeneity and mean free path effects for small particles both tend to give somewhat higher coagulation rates. In the case of liquid dispersions, the probability of adhesion upon collision may be very low because of the stabilizing effect of electric double layers or protective colloids or solvated adsorbed liquid layers. Thus it is often possible to stabilize, that is, prevent flocculation, of a liquid dispersion. Such stabilization has not been observed in the case of aerosols.

Another property of hydrodynamic interest is the formation of a sharp upper surface in batch sedimentation, especially for particles suspended in liquids. Kynch[58] has developed a mathematical theory of hindered settling based on the hypothesis that the speed of fall of particles in a dispersion is determined by *local* particle concentration only. The theory predicts that the existence of an upper surface, together with a knowledge of the initial distribution of particles, suffices to determine the variation of the velocity of fall with density for a particular dispersion. As he states, the validity of this hypothesis cannot be judged until the details of forces on the particles can be specified. In principle, the method of reflections discussed earlier in this chapter should provide the requisite information. Talmadge and Fitch[99] have applied Kynch's method to predict thickener capacity from batch settling tests.

The upper limit of concentration in sedimenting systems is reached when the settled bed is self-supporting due to forces exerted between particles by direct contact. At this limit, hydrodynamic forces no longer come into consideration; one is concerned rather with the geometric modes in which particles may be packed. Scheidegger[84] gives a review of various theories pertaining to sphere packing. He concludes that it is doubtful whether a unique stable packing with maximum porosity exists, since large arches of spheres which are stable can be constructed. The loosest stable packing described in the literature according to Scheidegger corresponds to a porosity $\epsilon = 0.875$. Wilson[105] prepared beds of spherical particles up to $\epsilon = 0.84$ by using dilute gelatin to promote aggregation. On the other hand, Steinour[96] was able to sediment nonflocculating suspensions of spherical particles down to voidages as low as $\epsilon = 0.3$. Thus, it would appear that such factors as the nature of the surface of the particles, their size, and the lubricating properties of the suspending fluid may be factors in determining the maximum solids concentration attainable by sedimentation.

More complicated problems occur when both hydrodynamic and interparticle contact forces must be considered. Exact theoretical analysis of such problems is difficult. An example of the problems encountered is given in the investigation by Eagleson, *et al.*[18] on the mechanics of the motion of discrete spherical bottom-settling particles due to shoaling waves on a beach. The theoretical analysis proceeds by considering the forces acting on a single spherical particle resting on a sloping beach, the top surface of which consists of spherical particles of uniform diameter. An equation obtained from this analysis is verified by laboratory measurements. One interesting hydrodynamic item is that the presence of a boundary (namely, the beach) seems to prolong the predominance of viscous forces, yielding so-called linear resistance laws, characteristic of the creeping motion equations, up to $N_{\mathrm{Re}_d} \approx 100$.

Various industrial and environmental aspects of particulate clouds of aerosols are reviewed in the book by Green and Lane[32], already referred to in connection with several of the topics treated earlier. DallaValle's[16] book on the technology of fine particles treats many aspects of the behavior of particulate matter in sedimenting systems from a practical viewpoint.

8-7 Packed Beds

A very extensive literature has developed on the subject of flow through packed beds. Excellent books by Scheidegger[84], on the physics of flow through porous media, and Carman[12], on flow of gases through porous media, deal exclusively with this subject. Zenz and Othmer[109] have also presented a thorough review from the applied viewpoint in a book covering many of the broader aspects of fluid particle technology, especially those related to fluidization.

The large number of variables involved in packed beds makes exact treatment difficult. In pressure drop correlations, the following factors must be considered: fractional void volume, type of fluid flow (creeping motion, laminar-inertial, or turbulent), particle shape, roughness, size distribution, and manner of packing. Theoretical treatments are usually confined to the simplest systems, involving uniformly sized spherical particles, as discussed earlier.

The densest possible packing of uniform spheres corresponds to the rhombohedral packing, where $\epsilon = 0.2595$. With uniform spheres, simple cubic packing yields $\epsilon = 0.476$. This number is of interest because it is close to the loose-packed density obtained in moving beds. For beds of randomly packed spheres, ϵ generally ranges from 0.38 to 0.47. The lower voidage is that exhibited by a random, gravity-packed bed of uniform spheres (prepared by dropping spheres into a chamber slowly, so that each sphere is at a position of repose before another sphere falls upon it). The higher voidage was obtained consistently by Oman and Watson[74] by inverting a bed in a small container and quickly righting it; they called the resulting bed a *random-loose arrangement*. Happel[37] obtained similar results by placing a weighed representative sample in a container and slowly inverting it several times until the volume remained constant. Packing density of beds of uniform spheres under a variety of other conditions is discussed by Rutgers[80]. When fines are present, it is possible to obtain considerably higher densities. As previously noted, much looser packings than the random-loose arrangement are also possible, especially with small particles and those of irregular shape.

For packed beds of uniform spheres the Carman-Kozeny[12] equation (8–4.22), within the range $\epsilon = 0.26$ to $\epsilon = 0.48$, gives excellent correlation with a Kozeny constant of $k = 4.8$. A recent study by Andersson[2], including additional sources, indicates that for uniform spheres $4.2 \leq k \leq 6.0$. Andersson proposes a refinement in which k is taken as a function of ϵ instead of being assumed constant. A large volume of data on beds consisting of a variety of nonspherical particles indicate that $k \approx 5.0$, independent of shape and porosity from $\epsilon = 0.26$ to $\epsilon = 0.8$. As shown in Table 8–4.2, agreement of the Carman-Kozeny relationship with hydrodynamic theory based on the free surface cell model is excellent.

Of particular interest in connection with the study of relationships between flow in packed and moving beds is the loose-packed condition, corresponding to $\epsilon \approx 0.47$. This point corresponds to both the void volume for incipient fluidization of large, smooth spheres and to the void volume for moving beds of particles. The value of U/U_0 from the Carman-Kozeny equation (with $k = 4.8$) at this voidage corresponds to 0.0216, as compared with 0.0221 based on the following formula derived from Eq. (8–4.11):

$$k = \frac{(1-\phi)^3(3+2\phi^{5/3})}{6-9\phi^{4/3}+9\phi^{8/3}-6\phi^3} \tag{8–7.1}$$

This is a difference of less than 3 per cent. From the high concentration range down to $\epsilon \approx 0.25$, other theoretical relationships discussed do not agree as well with data or with the Carman-Kozeny relationship.

Carman[12] also found that the Carman-Kozeny equation could be applied to mixtures of various particle sizes, using the hydraulic radius concept in place of particle diameter. As discussed following Table 8–4.2, this furnishes justification for applying the reciprocal mean diameter to treat nonuniformly sized, regularly shaped particles. Many other methods for obtaining average diameter have been proposed[109]. The Carman-Kozeny equation is not applicable to beds of highly irregular particles possessing stagnant pockets, or to beds comprised of shapes such as packed disks or plates. It is also inapplicable when changes in void volume are caused by extreme variations in particle size, as is true in moving beds[37].

It is believed[37] that a smaller change in permeability, resulting from changes in fractional void volume, occurs when the latter is caused by differences in shape and size gradation as compared with differences caused by changes in mode of arrangement or degree of compacting. Oman and Watson were the first to call attention to the disagreement caused by failure to prepare packed beds in a reproducible manner. They correlated a large amount of data on particles packed in a "random-dense" arrangement, and suggested a tentative factor for relating those results to the case of a "random-loose" arrangement. Zenz and Othmer suggest that one may, with equal accuracy, simply assume a 25 per cent decrease in a modified friction factor. Happel[37], on the other hand, obtained a correlation for moving beds in terms of a plot of a modified friction factor versus a modified Reynolds number. These beds occur in the loose-packed condition, so that the variation of pressure drop with fractional void volume reflects changes in size gradation. Happel and Epstein[42] suggested that, in order to calculate the compaction effect, in the direction of the denser arrangements that might be encountered in stationary packed beds, the Carman-Kozeny fractional void volume function might be employed. All correlations of this type are complicated because it is impossible on either theoretical or experimental grounds to independently determine the correct method of obtaining the average particle diameter and void volume function. In the case of the effect of variations in void volume fractions, Epstein[21] found that no less than twenty different relationships had been proposed by various investigators. A few of these are listed in Table 8–7.1[37], including the Carman-Kozeny relationship, Eq. (8–5.13). Here, n is the exponent of velocity in flow equations characterizing the type of flow as in Eq. (8–7.2); for viscous flow $n = 1$; for completely turbulent flow $n = 2$, as discussed below. The Oman and Watson void volume function in Table 8–7.1

TABLE 8–7.1
EXPRESSIONS FOR DEPENDENCE OF PRESSURE DROP ON FRACTIONAL VOID VOLUME

Function of Fractional Void Volume (proportional to ΔP)	Relative Pressure Drop (referred to $\epsilon = 0.40$)			
	for $n = 1$		for $n = 2$	
	($\epsilon = 0.3$)	($\epsilon = 0.5$)	($\epsilon = 0.3$)	($\epsilon = 0.5$)
Happel[37] $(1 - \epsilon)^{n+1}$	1.36	0.695	1.59	0.578
Bakhmeteff and Feodoroff[4] $\dfrac{1}{\epsilon^{(n+3)/3}}$	1.47	0.746	1.62	0.688
Oman and Watson[74] $\dfrac{(1 - \epsilon)^{3-n}}{\epsilon^{1.7}}$	2.20	0.475	1.90	0.574
Carman-Kozeny[12] $\dfrac{(1 - \epsilon)^{3-n}}{\epsilon^{3}}$	3.25	0.356	2.77	0.426

applies to changes in the random-dense arrangement. The Happel function applies only to differences in shape at the random-loose arrangement. The latter relationship, incidentally, must be regarded as purely empirical since it is based on a theoretical treatment of a random assemblage by Burgers[11] which, as pointed out following Eq. (8–3.10), is probably incorrect. The Bakhmeteff and Feodoroff function was also based on qualitative reasoning. It appears to agree closely with the Happel void volume function. The Carman-Kozeny relationship, which predicts comparatively large changes in pressure drop with variation in void volume function, appears to be associated with compacting of regularly shaped particles, as discussed previously.

Theoretical interpretation of the effect of higher fluid velocities, with its consequent increase in the importance of inertial effects, has also been difficult. As Zenz and Othmer point out, it is plausible to assume that at low Reynolds numbers the fluid trajectories hug the particle surfaces in a bed, filling the "tortuous" voids in a continuous fashion; but at high Reynolds numbers the fluid trajectories break away from the surfaces, so that there are dead regions and turbulent wakes. Based in part on dimensional reasoning, the effect of turbulence may be related to the exponent n as done in Table 8–7.1. The pressure drop due to a fluid traversing geometrically similar beds can be written as

$$\frac{\Delta P}{L} = \mu^{2-n}\tilde{v}^{n}\tilde{d}^{n-3}\rho^{n-1} \tag{8–7.2}$$

where $\tilde{v}$ = representative fluid velocity, $\tilde{d}$ = representative bed dimension, ρ = fluid density. The difficulty arises in properly defining $\tilde{v}$ and $\tilde{d}$, and combining them in an appropriate way with the fractional void volume. In the region $n > 1$ this difficulty has been resolved only on purely empirical grounds; for the theoretical basis of the Carman-Kozeny equation is not established in the region where inertial forces become important, and the capillary flow model on which it is based is open to some question in this region. It is usually assumed that the characteristic quantities $\tilde{v}$ and $\tilde{d}$ will be related to void volume in the same way for both viscous and turbulent flow. This leads to functions of the type shown in the first column of Table 8–7.1, where, for any given velocity, pressure drop variation is related to void volume simply by a change in the exponent n.

The effect of fine material in a packed bed cannot always be correctly evaluated by use of an average particle diameter. Gardner[29] has experimentally studied the effect of fine material in the structure and resultant pressure drop of a bed charged against an uprising gas stream. He found that very small amounts of material, fine enough to have a degree of mobility in and during the charging, will cause much smaller pressure drops than would normally be expected. The effect of fines introduced in this way is mostly due to an increase in the bed voidage.

As discussed in Section 2–8, the assumption of no slippage at particle surfaces is not always valid. The literature records considerable data which indicate that the permeability of a given porous media to gas flow through it is not strictly constant. This phenomenon has been explained in terms of molecular slippage of the gas at solid grain surfaces. The apparent permeability generally satisfies an equation of the form[3]

$$K_a = b + (m/\bar{P}) \tag{8–7.3}$$

where K_a = apparent permeability*, $\bar{P}$ = mean pressure in the bed; b and m are constants. The constant b, which is the limiting value of K_a at infinite pressure, is found to be equal, within experimental error, to the permeability as measured with liquids. The quantity b, which represents a property of the

*That is, it is the coefficient in the relation

$$\mathbf{U} = -\frac{K_a}{\mu}\nabla P$$

porous medium, should be independent of the particular fluid used to evaluate the permeability, and is therefore called the *true* or *slip-corrected permeability*. The quantity m is dependent on the properties of both the porous medium and the gas (mean free path). Wilson, *et al.*[106] present a brief theoretical review of flow of gases in porous media. These authors discuss essentially four types of gas flow in porous media: (a) normal gas dynamic flow, (b) viscous slip flow, (c) free molecular flow, (d) the transition zone from viscous to free molecular flow. The latter regime is, of course, not uniquely defined, and includes portions of (b) and (c). From experimental data they conclude that ordinary viscous flow will exist in a porous medium if the ratio of molecular mean free path to pore diameter is less than 0.025. Free molecular flow will exist if the ratio of molecular mean free path to pore diameter exceeds 1.6.

When an adsorbable gas flows through a porous medium under a constant pressure gradient, it is frequently found that the gas flows at an enhanced rate.[12] The explanation most widely accepted for this phenomenon is that, in addition to the normal flow rate in the gas phase, there is a steady parallel flow due to surface diffusion. Gilliland, *et al.*[30] have recently made a definitive study. They present an improved correlation of the main variables affecting transport in adsorbed layers.

Several books deal with detailed solutions of the Darcy differential equations of flow through porous media[54,71,75], for a variety of boundary conditions. This information is useful in applications, such as petroleum production and the flow of underground water in wells or dams. Two-dimensional problems are treated almost exclusively by complex variable theory. Three-dimensional problems are treated by general methods not too dissimilar from those found in the present text.

One important industrial application not extensively covered in books is the application of Carman-Kozeny and similar relationships to the resistance and compressibility of filter cakes. Grace[31] and Tiller[103] have each presented excellent reviews and developments which show the applications and limitations of basic hydrodynamic concepts to this problem. Grace showed that filtration resistance of cakes of compressible materials cannot be successfully predicted from data on the dry solids alone. Tiller succeeded in correlating data on the basis of the Carman-Kozeny equation by means of the following empirical equation for the pressure dependence of the cake porosity, ϵ:

$$\epsilon = \epsilon_0 p_s^{-c} \tag{8–7.4}$$

where ϵ_0 is the porosity at a unit reference pressure, p_s is the compressive pressure on the solids, and c is an empirical constant for a given system. In the pressure range 0.1 to 100 lb/sq in. this power function representation is useful with materials of moderate compressibility. Hutto[50] has observed experimentally that ϵ generally varies throughout the length of the filter cake. He concludes that porosity decreases relatively rapidly near the liquid face of the cake and then continues at a slower, almost linear rate of decrease down to minimum porosity at the septum.

8-8 Fluidization

In addition to the book on fluidization by Zenz and Othmer[109] mentioned earlier in the chapter, Leva[63] has written another book devoted exclusively to this subject. Several books in foreign languages have also appeared on the subject[77,86,101]. When a stream of gas or liquid is passed upward through a bed of solid particles, at a sufficiently high fluid velocity the bed will be physically lifted. At this point the particles may rearrange themselves in a random-loose condition. This porosity is characteristic of the flow of particles in a moving bed. Further increase in velocity will expand the bed further, so that it is no longer rigid and entirely supported by interparticle contacts. The bed is then said to be in the fluidized state. If the fluid passing through the particles is a liquid, the bed will usually expand uniformly, allowing greater ease of fluid passage through the mass. This is termed *particulate fluidization.* Under certain circumstances, especially at low velocities in the case of gas fluidization, a portion of the rising fluid may pass through the bed in the form of pockets or bubbles, giving rise to the so-called *aggregative* type of fluidization. Criteria for particulate versus aggregative fluidization are, at present, largely empirical, but there is some evidence that the ratio of particle to fluid density is an important variable[109].

As regards correlations, the most successful quantitative relationships have been made for particulate fluidization in the range of $\epsilon < 0.80$, using modifications of the Carman-Kozeny relationship. In this range, Leva suggests a Kozeny constant of $k = 5.55$, approximately 11 per cent higher than suggested by Carman—see Eq. (8–5.10) and Table 8–4.2. From his study of available data, Leva concludes that fixed and fluidized beds of equal porosity should, in this range, give the same pressure drop at the same flow rate. Zenz and Othmer, on the other hand, conclude that though relationships of the Carman-Kozeny type are applicable, pressure drop in a fluidized bed is of the order of 20 per cent less than that in a fixed bed of the same voidage, at all voidages where comparisons can be made. This is approximately in agreement with the work of Andersson[2].

Correlation at higher fractional void volumes, even at low Reynolds numbers, is difficult because there is considerable discrepancy between experimental results reported by different investigators.[35] These variations resemble those observed in sedimentation experiments, discussed following Eq. (8–6.8), for the porosity range $\epsilon = 0.95$ to $\epsilon = 0.60$. Although the fluidization and sedimentation data of a given study often agree with each other, there is considerable disagreement among different investigators. Values of relative velocity U/U_0 vary from 25 per cent to 100 per cent higher than predicted by the free surface model, with the exception of recent data specially obtained to avoid agglomeration effects.[1] The free surface model thus gives a lower bound for the relative velocity ratio U/U_0, and perhaps corresponds to a more completely uniform particle distribution than is normally obtained under conditions described as particulate fluidization.

No completely satisfactory correlation of the velocity-voidage relationship for aggregatively fluidized beds has been developed. In fact as we have already mentioned,[109] there are no satisfactory quantitative criteria for predicting whether a given fluid-solid system may be expected to fluidize aggregatively or particulately. Both Leva, and Zenz and Othmer suggest tentative procedures for correlation of aggregative fluidization data.

As in the case of flow through packed beds, theoretical correlation of fluidization data at high Reynolds numbers has not yet proved possible. In sedimentation problems, of course, high Reynolds numbers usually do not occur at high particle concentrations. In order to obtain insight into the fundamental hydrodynamics involved in fluidization problems of this type, Fayon and Happel[28] studied the flow of fluid past a single sphere located in a circular cylinder. They found that in the range of Reynolds numbers from 0.1 to 40.0, based on the approach velocity to the sphere and sphere diameter, the pressure drop due to the presence of the sphere, and the drag on it, may be represented by a semiempirical relationship containing two terms. The first, arising from the presence of the cylindrical boundary bounding the flow, is derived theoretically from the creeping motion equations, which neglect inertial effects. The second term, due to inertial effects, is established from data on the drag coefficient for uniform flow past a sphere in an unbounded medium [see Eq. (7–3. 110)].

In view of the successful application of the sphere-cylinder cell model for prediction of the effect of void volume in a bed of particles,[39] it seems worthwhile to attempt to exploit the findings of Fayon and Happel to derive a semi-theoretical relationship for flow through beds of particles taking inertial effects into consideration. Since the inertial effect is assumed in their correlation to be equivalent to that for each sphere in an unbounded medium, such a relationship might be expected to apply best in fluidized systems, where the spheres are sufficiently distant from each other that the wake is free to develop more or less fully. The drag due to creeping flow past a single sphere in an assemblage is given by the free surface model, Eq. (8–4.10), as

$$F_{\text{viscous}} = \frac{4\pi\mu a U(3 + 2\gamma^5)}{2 - 3\gamma + 3\gamma^5 - 2\gamma^6} \tag{8–8.1}$$

where a is the sphere radius, U is the superficial velocity, and $\gamma = (1 - \epsilon)^{1/3}$.

If we assume that the inertial drag is a function of bed porosity ϵ, the simplest postulate would be that the approach velocity increases inversely with the void volume. Thus,

$$F_{\text{inertial}} = \left(\frac{C - C_S}{C_S}\right)_{U/\epsilon} 6\pi\mu a U/\epsilon \tag{8–8.2}$$

where C is the drag coefficient for a single sphere at the velocity U/ϵ, and C_S is the drag coefficient computed as if Stokes' law applied at the same velocity U/ϵ. At low Reynolds numbers, $C = C_S$ whence the inertial contribution vanishes. The total drag is then

$$F = F_{\text{viscous}} + F_{\text{inertial}} \tag{8–8.3}$$

To obtain the pressure gradient, $\Delta P/L$, we note that from a force balance over a unit cell,

$$\frac{F}{\frac{4}{3}\pi b^3} = \frac{\Delta P}{L} \tag{8–8.4}$$

where $b = a(1 - \epsilon)^{1/3}$ is the outer radius of the unit cell. Substituting this relationship in Eq. (8–8.3), we obtain

$$\frac{U}{U_0} = \frac{1}{\left[\dfrac{3 + 2(1-\epsilon)^{5/3}}{3 - (9/2)(1-\epsilon)^{1/3} + (9/2)(1-\epsilon)^{5/3} - 3(1-\epsilon)^2}\right] + \dfrac{1}{\epsilon}\left[\left(\dfrac{C}{C_S}\right)_{U/\epsilon} - 1\right]} \tag{8–8.5}$$

where U_0 is the hypothetical Stokes velocity.

If $\epsilon \to 1$ and we designate the value U_1 at velocity U to distinguish it from Stokes velocity U_0, we note that

$$\frac{U_1}{U_0} = \left(\frac{C_S}{C}\right)_U \tag{8–8.6}$$

Using this relationship, we can obtain an expression similar to Eq. (8–8.5) in terms of U/U_1 as follows:

$$\frac{U}{U_1} = \frac{(C/C_S)_U}{\left[\dfrac{3 + 2(1-\epsilon)^{5/3}}{3 - (9/2)(1-\epsilon)^{1/3} + (9/2)(1-\epsilon)^{5/3} - 3(1-\epsilon)^2}\right] + \dfrac{1}{\epsilon}\left[\left(\dfrac{C}{C_S}\right)_{U/\epsilon} - 1\right]} \tag{8–8.7}$$

A few numerical values of the ratio U/U_1 obtained from the foregoing relation are compared in Table 8–8.1 with values obtained from the smoothed correlation of experimental data on particulate fluidization given by Zenz and Othmer[109, p. 236]. The agreement shown in the table is reasonable except when the Reynolds number is low and the porosity high, for example, $\epsilon > 0.7$. As already discussed, this discrepancy is probably due to agglomeration effects in this region, which are not considered in the theoretical derivation. Happel's[35] theoretical correlation for low Reynolds numbers is also plotted by Zenz and Othmer[109, p. 238] and compared with the experimental data used in Table 8–8.1, up to Reynolds numbers of 500. At these higher Reynolds numbers, inertial effects constitute a substantial proportion of the total drag. This effect is approximately shown in Table 8–8.1 where both the viscous and inertial effects are provisionally taken into account.

TABLE 8–8.1
COMPARISON OF FLUIDIZATION VELOCITY WITH VELOCITY TO LIFT A SINGLE SPHERE (N_{Re} based on sphere diameter)

Value of ϵ	U/U_1 from Eq. (8–8.7)	U/U_1 from Zenz-Othmer Correlation
$N_{Re} = 10$ for $\epsilon = 1$		
0.5	0.0426	0.0488
0.7	0.147	0.207
$N_{Re} = 50$ for $\epsilon = 1$		
0.5	0.0696	0.0828
0.7	0.221	0.272
$N_{Re} = 100$ for $\epsilon = 1$		
0.5	0.0915	0.1022
0.7	0.268	0.308
$N_{Re} = 200$ for $\epsilon = 1$		
0.5	0.115	0.1322
0.7	0.320	0.339
$N_{Re} = 500$ for $\epsilon = 1$		
0.5	0.1545	0.165
0.7	0.379	0.347

In problems involving mass transfer, heat transfer, and reaction kinetics, a knowledge of the exact fluid flow pattern through a fluidized medium would be desirable. Treatment from a theoretical viewpoint is very complicated. A qualitative idea of the radial distribution effects possible is shown in Fig. 8–2.3 and discussed on pp. 367–8. It is believed that such effects may be significant in laboratory reactors where the ratio of wall-to-particle surface area may be substantial.

A recent investigation by Lanneau[60] gives a practical approach to the problem of efficiency of gas-solids contacting under conditions where wall effects are of lesser importance. His work was conducted on a typical microsphere alumina catalyst. For such a material the observed initial fluidization point or first appearance of bubbles was observed at a gas velocity of about 0.02 ft/sec. At very low gas velocities, that is, below 0.1–0.2 ft/sec, the fluid bed is only "semifluid" and there is very little mobility of the solids. At fluidizing velocities from 0.1–1.0 ft/sec, the two-phase character of the fluid bed is prominent. Discrete, well-formed bubbles of very low density (containing little solids) are present. Under these conditions, the fluid bed consists of an "emulsion" phase through which the lower density phase (sometimes referred to as the "bubble" phase) flows. As velocity is increased to higher levels, the bubbles begin to break up, and smaller bubbles of higher relative density (containing more solids) are observed. At sufficiently high gas velocities, in the range of 3–5 ft/sec, a condition of more uniform or "particulate" fluidization is attained. This condition coincides with rapid entrainment of solids from the bed, and operation of a solids transfer line, as opposed to a captive fluid bed, is approached. The greater stability of the dilute phase is important in the design of equipment for conducting reactions in this region, and also for pneumatic conveying of free particles. Zenz[108] has presented a fluid-solids "phase diagram" which qualitatively describes some of the complicated phenomena associated with the distribution and stability of these systems.

BIBLIOGRAPHY

1. Adler, I. L. and J. Happel, Chem. Eng. Progr. Symposium Series **58** (1962), 98.
2. Andersson, K. E. B., Trans. Roy. Inst. Techn. (Stockholm) No. 131 (1959); No. 201 (1963); No. 202 (1963); Chem. Eng. Sci. **15** (1961), 276.
3. Arnofsky, J. S., J. Appl. Phys. **25** (1954), 48.
4. Bakhmeteff, B. A., and N. V. Feodoroff, J. Appl. Mech. **4** (1937), A 97; *Ibid.* **5** (1938), A 86; Proc. 5th Intern. Congr. Appl. Mech. (1938), p. 555; Trans Amer. Geophys. Union **24** (1943), 545.
5. Baron, T., Chem. Eng. Progr. **48** (1952), 118.
6. Brenner, H., Phys. Fluids **1** (1958), 338.
7. ———, Eng. Sc. D. Thesis, New York University, 1957.
8. ———, Chem. Eng. Sci. **19** (1964), 599.
9. Bretherton, F. P., J. Fluid Mech. **14** (1962), 284.
10. Brinkman, H. C., Appl. Sci. Res. **A1** (1947), 27; **A1** (1948), 81; **A2** (1949), 190.
11. Burgers, J. M., Proc. Koningl. Akad. Wetenschap. (Amsterdam) **44** (1941), 1045; **45** (1942), 9.
12. Carman, P. C., *Flow of Gases through Porous Media*. New York: Academic Press, 1956.
13. ———, Trans. Inst. Chem. Engrs. (London) **15** (1937), 150.
14. Cheng, P. Y., and H. K. Schachman, J. Polymer Sci. **16** (1955), 19.
15. Cunningham, E., Proc. Roy. Soc. (London) **A83** (1910), 357.
16. DallaValle, J. M., *Micromeritics—The Technology of Fine Particles*, 2nd ed. New York: Pitman, 1948.
17. Darcy, H. P. G., *Les Fontaines Publiques de la Ville de Dijon*. Paris: Victor Dalmont, 1856.
18. Eagleson, P. S., R. G. Dean, and L. A. Peralta, *Technical Memo* No. 104, Beach Erosion Board, Dept. of the Army, Corps of Engineers, 1958.

19. Einstein, A., Ann. Phys. **17** (1905), 549; **19** (1906), 371; **22** (1907), 569. See also *Investigations on the Theory of Brownian Movement*. New York: Dover, 1956.

20. Emersleben, O., Physik. Z. **26** (1925), 601.

21. Epstein, N., Eng. Sc. D. Thesis, New York University, 1954.

22. Epstein, P., Math. Ann. **60** (1903), 615; **63** (1906), 205.

23. Fair, G. M., and L. P. Hatch, J. Amer. Water Works Assoc. **25** (1933), 1551.

24. Famularo, J., Eng. Sc. D. Thesis, New York University, 1962.

25. Faxen, H., Ark. Mat. Astron. Fys. **17**, No. 27 (1923).

26. ———, Ark. Mat. Astron. Fys. **19A**, No. 22 (1925).

27. ———, Z. Angew. Math. Mech. **7** (1927), 79.

28. Fayon, A. M., and J. Happel, A.I.Ch.E. Jour. **6** (1960), 55.

29. Gardner, G. C., Chem. Eng. Sci. **7** (1957), 73.

30. Gilliland, E. R., R. F. Baddour, and J. L. Russell, A.I.Ch.E. Jour. **4** (1958), 90.

31. Grace, H. P., Chem. Eng. Progr. **49** (1953), 303, 367, 427; A.I.Ch.E. Jour. **2** (1956), 307.

32. Green, H. L., and W. R. Lane. *Particulate Clouds: Dusts, Smokes, and Mists*. London: Spon, Ltd., 1957.

33. Haberman, W., and R. M. Sayre, "Motion of Rigid and Fluid Spheres in Stationary and Moving Liquids Inside Cylindrical Tubes," *Report* 1143, David Taylor Model Basin, U.S. Navy Dept. Washington, D.C., October, 1958.

34. Hall, W. A., Trans. Amer. Geophys. Union **3** (1956), 185.

35. Happel, J., A.I.Ch.E. Jour. **4** (1958), 197.

36. ———, A.I.Ch.E. Jour. **5** (1959), 174.

37. ———, Ind. Eng. Chem. **41** (1949), 1161.

38. ———, J. Appl. Phys **28** (1957), 1288.

39. ———, and P. A. Ast, Chem. Eng. Sci. **11** (1960), 286.

40. ———, and H. Brenner, A.I.Ch.E. Jour. **3** (1957), 506.

41. ———, and B. J. Byrne, Ind. Eng. Chem. **46** (1954), 1181.

42. ———, and N. Epstein, Chem. Eng. **57** (1950), 137.

43. ———, and ———, Ind. Eng. Chem. **46** (1954), 1161.

44. ———, and ———, Ind. Eng. Chem. **46** (1954), 1187.

45. ———, and R. Pfeffer, A.I.Ch.E. Jour. **6** (1960), 129.

46. Harris, C. C., Nature **183** (1959), 530; **184** (1959), 716.

47. Hasimoto, H., J. Fluid Mech. **5** (1959), 317.

48. ———, J. Phys. Soc. Japan **13** (1958), 633.

49. Hubbert, M. K., Petrol. Trans. Amer. Inst. Min. Engrs. **207** (1956), 222.

50. Hutto, F. B., Chem. Eng. Progr. **53** (1957), 328.

51. Irmay, S., Trans. Amer. Geophys. Union **39** (1958), 702; see also discussion, Jour. Geophys. Res. **64** (1959), 485.

52. Kawaguchi, M., J. Phys. Soc. Japan **13** (1958), 209.

53. Kaye, B. H., and R. P. Boardman, *Proc. Symposium Interaction between Fluids and Particles*. London: Brit. Inst. Chem. Engrs., 1962.

54. Kozeny, J., *Hydraulik*. Wien: Springer Verlag, 1953, p. 390; Sitz.-Ber. Wiener Akad., Abt. IIa, **136** (1927), 271.

55. Kuwabara, S., J. Phys. Soc. Japan (to be published).

56. ———, J. Phys. Soc. Japan **14** (1959), 527.

57. Kynch, G. J., J. Fluid Mech. **5** (1959), 193.

58. ———, Trans. Farad. Soc. **48** (1952), 166.

59. Lamb, *Hydrodynamics*, 6th ed. Cambridge: Cambridge Univ. Press, 1932.

60. Lanneau, K. P., Paper presented at the meeting of the Institution of Chemical Engineers, Oct. 21, 1959.

61. Lapple, C. E., in *Chemical Engineers Handbook*, ed. J. H. Perry. New York: McGraw-Hill, 1950.

62. Leibenson, L. S., *Complete Works* (in Russian), Vol. 3. Moscow, 1955.

63. Leva, M., *Fluidization*. New York: McGraw-Hill, 1959.

64. Lin, P. N., Ph. D. Thesis, Iowa State University, 1951.

65. Maude, A. D., and R. L. Whitmore, Brit. J. Appl. Phys. **9** (1958), 477.

66. McHenry, K. W., and R. H. Wilhelm, A.I.Ch.E. Jour. **3** (1957), 83.

67. McNown, J. S., and P. N. Lin, "Proc. Second Midwestern Conf. Fluid Mechanics," *Reprint in Eng.* No. 109, Iowa State Univ., 1952.

68. Milne-Thomson, L. M., *Theoretical Hydrodynamics*, 3rd ed. New York: Macmillan, 1955, pp. 182, 540.

69. Miyagi, T., J. Phys. Soc. Japan **13** (1958), 209.

70. Morse, P. M., and H. Feshbach, *Methods of Theoretical Physics*. New York: McGraw-Hill, 1953.

71. Muskat, M., *The Flow of Homogeneous Fluids through Porous Media*. New York: McGraw-Hill, 1937; 2nd printing, Ann Arbor: Edwards, 1946.

72. Noda, H., Bull. Chem. Soc. Japan **30** (1957), 495.

73. Oliver, D. R., Nature **194** (1962), 1269.

74. Oman, A. O., and K. M. Watson, Natl. Petroleum News **36** (1944), R795.

75. Polubarinova-Kochina, P. Ya., *Theory of Ground-Water Movement*. Princeton: Princeton Univ. Press, 1962.

76. Prager, S., Phys. Fluids **4** (1961), 1477.

77. Reboux, P., *Phenomenes de Fluidisation*. Paris: Assoc. Francaise de Fluidization, 1954.

78. Richardson, J. F., and W. N. Zaki, Chem. Eng. Sci. **3** (1954), 65.

79. Rubinow, S. I., and J. B. Keller, J. Fluid Mech. **11** (1961), 447.

80. Rutgers, R., Nature **193** (1962), 465.

81. Saffman, P. G., J. Fluid Mech. **1** (1956), 540; *Ibid.* **6** (1959), 321.

82. ———, and G. I. Taylor, Proc. Roy. Soc. (London) **A 245** (1958), 312.

83. Scheidegger, A. E., J. Appl. Phys. **25** (1954), 994.

84. ———, *The Physics of Flow through Porous Media*, 2nd ed. New York: Macmillan, 1960.

85. ———, Phys. Fluids **3** (1960), 94.

86. Schytil, F., *Wirbelschttechnik*. Berlin: Springer Verlag, 1961.

87. Segré, G., and A. Silberberg, J. Fluid Mech. **14** (1962), 115, 136.

88. Slezkin, N. A., Dokladi Akad. Nauk. SSSR **86** (1952), 235.

89. ———, *Dynamics of Viscous Incompressible Fluids* (in Russian). Moscow: Gos. Izdat. Tekh.-Teor. Lit., 1955.

90. ———, and S. M. Shustov, Dokladi Akad. Nauk. SSSR **96** (1954), 933.

91. Smoluchowski, M., Bull. Acad. Sci. Cracow **1a** (1911), 28.

92. ———, Proc. 5th Intern. Congr. Math. **2** (1912), 192.

93. ———, Phys. Z. **17** (1916), 557, 585; Z. Phys. Chem. **92** (1917), 129.

94. Sokolnikoff, I. S., and R. M. Redheffer, *Mathematics of Physics and Modern Engineering*. New York: McGraw-Hill, 1958.

95. Sparrow, E. M., and A. L. Loeffler, Jr., A.I.Ch.E. Jour. **5** (1959), 325.

96. Steinour, H. H., Ind. Eng. Chem. **36** (1944), 618, 840, 901.

97. Streeter, V. L., *Fluid Dynamics*. New York: McGraw-Hill, 1948.

98. Sullivan, R. R., J. Appl. Phys. **13** (1942), 725.

99. Talmadge, W. P., and E. B. Fitch, Ind. Eng. Chem. **47** (1955), 38.

100. Tamada, K., and H. Fujikawa, Quart. J. Mech. Appl. Math. **10** (1957), 423.

101. Tesarik, J., *Proudeni tekuting porovityn prostredim* (Flow of Fluids in Porous Media) Prague: Artia, P.O.B. 790, 1961.

102. Theodore, L., Ph.D. Thesis, New York University, 1964.

103. Tiller, F. M., Chem. Eng. Progr. **49** (1953), 467; **51** (1955), 282.

104. Uchida, S., Dept. Inst. Sci. Technol. Univ. Tokyo (in Japanese), **3** (1949), 97; abstract, Ind. Eng. Chem. **46** (1954), 1194.

105. Wilson, B. W., Austr. J. Appl. Sci. **4** (1953), 300.

106. Wilson, L. H., W. L. Sabbitt, and M. Jakob, J. Appl. Phys. **22** (1951), 8.

107. Weissberg, H. L., and S. Prager, Phys. Fluids **5** (1962), 1390.

108. Zenz, F. A., Ind. Eng. Chem. **41** (1949), 2801.

109. Zenz, F. A., and D. Othmer, *Fluidization and Fluid-Particle Systems.* New York: Reinhold, 1960.

110. Zierep, J., Z. Flugwiss. **3** (1955), 22.

The Viscosity of Particulate Systems

9

9-1 Introduction

Our emphasis thus far has been primarily directed toward uniform, that is, translational, fluid-particle motions. In this chapter we shall examine phenomena arising from shearing motion of the fluid relative to the suspended solids. We adopt the point of view that, in a sense, a fluid-particle suspension may be regarded as a continuum. This attitude seems reasonable provided that the particle dimensions are very small compared with the dimensions of the apparatus containing the suspension. Thus, among other things, we shall seek to determine the apparent viscosity of such a suspension. Problems of suspension viscosity are important not only for the macroscopic particles involved in many industrial separation and reaction processes, but also in connection with the very small particles commonly described as colloidal, whose size approaches the molecular dimensions of the suspending fluid medium. The same basic variables characterize suspension viscosity as characterize sedimentation rates, namely: (a) the nature of the fluid; (b) the nature of the suspended particles; (c) the concentration of suspended particles; (d) the motion of particles and fluid–the shearing field of the latter being the prime distinguishing characteristic. Because of the small size of particles involved in viscosity problems, other properties, such as internal flexibility and ease of deformation, may also be important.

A number of excellent reviews of the general subject of suspension viscosity are available. Eirich's comprehensive three-volume treatise on rheology contains several monographs on disperse systems, as well as special chapters on industrial applications involving such materials. Of special interest from the viewpoint of the hydrodynamic treatment here is the contribution by H. L. Frisch and R. Simha, "The Viscosity of Colloidal Suspensions and Macromolecular Solutions."[13] Current quantitative theories, semiempirical approaches, and qualitative speculations are discussed, omitting most mathematical detail. Hermans' *Flow Properties of Disperse Systems*[20] contains an interesting chapter by Sadron, "Dilute Solutions of Impenetrable Rigid Particles," which considers the hydrodynamics of suspension viscosity in some detail.

In addition to the basic hydrodynamic problem, configurational aspects are very important in dealing with macromolecules. In many of these species, we not only encounter variations in shape but also variations in flexibility due to coiling effects. At the molecular scale, the simple hydrodynamic treatment must ultimately be supplanted by detailed molecular theories of flow in viscous fluids. So far these are not available—in part, because of the mathematical complexity involved. In this book we confine ourselves to the hydrodynamic aspects of the problem.

As was true for sedimentation problems discussed in Chapter 8, the creeping motion equations furnish an appropriate basis for a satisfactory treatment of suspension viscosity. In what follows, we shall first treat the simpler case of low solids concentration systems before discussing special techniques applicable to more concentrated systems. Special attention will be devoted to shapes other than spheres, because of the importance of shape in viscosity problems. A brief comparison with data, and consideration of the broader aspects of rheological properties of suspensions, will conclude our treatment. Detailed consideration of these two latter subjects is limited to topics having an immediate bearing on the basic hydrodynamic scope of this book.

An investigator engaged in the experimental determination of the viscosity of a suspension does so by means of certain macroscopic measurements made at the boundaries of his apparatus. Accordingly, the internal constitution of the material being tested does not immediately enter into the definition of viscosity. Based on the outcome of any single experiment performed on the suspension, a number called its *viscosity*, μ, may be assigned to it. In what follows, we shall attempt to provide a satisfactory operational definition of suspension viscosity. The arguments follow those of Brenner[3].

Fundamental experiments employed in determining the viscosities of homogeneous fluids are usually *linear* experiments, in that the inertial terms in the Navier-Stokes equations either: (a) vanish identically, as in shear flow between parallel planes, or in Pouiseuille flow in a capillary; (b) are negligibly small, as in the falling ball viscometer based on Stokes' law; (c) do not interact with the viscous terms, as for flow between rotating concentric cylinders, where the inertial forces are entirely balanced by pressure forces. These experiments are then linear in the sense that the viscosity coefficient appears as a proportionality factor in a linear relation existing between two macroscopic, experimentally measurable variables. Thus, in case (a), for a shear flow the force on a plane is directly proportional to its velocity; or the pressure drop in a capillary is directly proportional to the volumetric flow rate. In case (b), the terminal velocity of the ball is directly proportional to the force causing its motion. In case (c), the torque on either cylinder is directly proportional to the angular velocity. For a linear experiment of given geometry the viscosity appears as the proportionality coefficient between the dependent and independent macroscopic variables.

These linear experiments have the common feature that the work done in maintaining the motion of (or at) the external boundaries of the apparatus is in direct proportion to the viscosity of the fluid, at least when the latter is homogeneous. This then provides us with a satisfactory, self-consistent method for defining the viscosity of a suspension. In particular, suppose we conduct a linear experiment with a homogeneous Newtonian fluid of viscosity μ_0 and observe that the work done by the stresses acting over the apparatus boundaries is $W^{(0)}$. Now suspend particles in this fluid and repeat the preceding experiment by moving the boundaries S_0 of the apparatus in precisely the same way as previously. Thus, for the concentric cylinder case one would rotate the boundaries at the same rate of speed as previously. Similarly, in the capillary tube case one would pump the suspension through the tube at the same volumetric flow rate. Let W denote the rate at which the apparatus boundaries are doing work on the suspension. We then define the viscosity μ of the suspension by the relation

$$\frac{\mu}{\mu_0} = \frac{W}{W^{(0)}}$$

In linear experiments performed on homogeneous fluids the rate at which mechanical energy is dissipated in the apparatus is identical to the rate at which stresses acting over its surface are doing work. Accordingly, in such cases, we may speak synonomously of *energy dissipation* and *work*.

If $(\mathbf{v}, p)$ denote the velocity and pressure fields for fluid motion in the presence of the suspended matter, and $(\mathbf{v}^{(0)}, p^{(0)})$ denote the corresponding entities for the homogeneous fluid under identical boundary conditions, then the rate at which work is done on the fluid by the apparatus boundaries in the two cases is

$$W = \int_{S_0} d\mathbf{S}\cdot\mathbf{\Pi}\cdot\mathbf{v}, \qquad W^{(0)} = \int_{S_0} d\mathbf{S}\cdot\mathbf{\Pi}^{(0)}\cdot\mathbf{v}^{(0)}$$

where $\mathbf{\Pi}$ and $\mathbf{\Pi}^{(0)}$ are the respective pressure tensors and $d\mathbf{S}$ is a directed element of surface area parallel to the outer normal of the fluid domain; that is, directed into the surface. Hence, our definition of suspension viscosity is equivalent to

$$\frac{\mu}{\mu_0} = \frac{\int_{S_0} d\mathbf{S}\cdot\mathbf{\Pi}\cdot\mathbf{v}}{\int_{S_0} d\mathbf{S}\cdot\mathbf{\Pi}^{(0)}\cdot\mathbf{v}^{(0)}} \tag{9–1.1}$$

The equality of velocity on the apparatus boundaries,

$$\mathbf{v} = \mathbf{v}^{(0)} \quad \text{on } S_0 \tag{9–1.2}$$

constitutes the common basis upon which the two regimes may be compared.

The assumption of no relative motion at fluid-solid interfaces imposes boundary conditions at the particle surfaces. The most general motion which a rigid body can execute is that of translation and rotation about an axis. If $\mathbf{U}_i$ represents the translational velocity of a point on the instantaneous axis of rotation of the ith particle and $\mathbf{\Omega}_i$ refers to its angular velocity at any instant, then the supposition that fluid adheres to the particle surfaces requires that we set

$$\mathbf{v} = \mathbf{U}_i + \mathbf{\Omega}_i \times \mathbf{r}_i \quad \text{on each } S_i \tag{9–1.3}$$

In this equation, S_i denotes the surface of the ith particle and $\mathbf{r}_i$ denotes the position vector drawn from a point on the axis of rotation of the particle. If k particles are suspended in the volume Q_0 bounded by the apparatus boundaries S_0, and we denote by Q_i the volume occupied by the ith particle, the volume occupied by fluid will be

$$Q_0 - \sum_{i=1}^{k} Q_i$$

whereas the total surface area bounding this volume will be

$$S_0 + \sum_{i=1}^{k} S_i$$

For quasi-static creeping flows in the absence of external body forces, the kinetic energy of the fluid-particle system is negligible and the potential energy of the system remains constant as its configuration changes. Accordingly, the rate E at which energy is being dissipated within the confines of the apparatus is equal to the rate W' at which the stresses are doing work over *all* the surfaces bounding the fluid. In general, this includes both the apparatus boundaries and the particle surfaces. Thus we have

$$E = W' \tag{9–1.4}$$

and, hence,

$$E = \int_{S_0+\Sigma S_i} d\mathbf{S}\cdot\mathbf{\Pi}\cdot\mathbf{v} \tag{9–1.5}$$

or, upon employing Eq. (9–1.2),

$$E = \int_{S_0+\Sigma S_i} d\mathbf{S}\cdot\mathbf{\Pi}\cdot\mathbf{v}^{(0)} + \int_{\Sigma S_i} d\mathbf{S}\cdot\mathbf{\Pi}\cdot(\mathbf{v} - \mathbf{v}^{(0)}) \tag{9–1.6}$$

The original and perturbed fields satisfy the creeping motion and continuity equations in the fluid volume $Q_0 - \sum Q_i$. As such we may apply the reciprocal theorem of Section 3–5 in the form,

$$\int_{S_0+\sum S_i} d\mathbf{S}\cdot\mathbf{\Pi}\cdot\mathbf{v}^{(0)} = \int_{S_0+\sum S_i} d\mathbf{S}\cdot\mathbf{\Pi}^{(0)}\cdot\mathbf{v} \tag{9–1.7}$$

If this relationship is substituted for the first integral in Eq. (9–1.6), and Eq. (9–1.2) is applied once again, we obtain

$$E^* = \sum_i \left[\int_{S_i} d\mathbf{S}\cdot\mathbf{\Pi}^{(0)}\cdot\mathbf{v} + \int_{S_i} d\mathbf{S}\cdot\mathbf{\Pi}\cdot(\mathbf{v} - \mathbf{v}^{(0)})\right] \tag{9–1.8}$$

where we have defined

$$E^* = E - E^{(0)} \tag{9–1.9}$$

in which

$$E^{(0)} = \int_{S_0} d\mathbf{S}\cdot\mathbf{\Pi}^{(0)}\cdot\mathbf{v}^{(0)} \tag{9–1.10}$$

is the rate at which energy is being dissipated within the apparatus as a result of the homogeneous fluid motion, whereas E is the dissipation rate when suspended particles are present under comparable conditions prevailing at the apparatus boundaries. Thus, E^* is the *additional* rate at which dissipation occurs within the apparatus as a result of the presence of suspended solids.

The first integral on the right-hand side of Eq. (9–1.8) vanishes when inertial effects and body forces are absent. To demonstrate this, note that from Eq. (9–1.3),

$$\int_{S_i} d\mathbf{S}\cdot\mathbf{\Pi}^{(0)}\cdot\mathbf{v} = \left(\int_{S_i} d\mathbf{S}\cdot\mathbf{\Pi}^{(0)}\right)\cdot\mathbf{U}_i - \left(\int_{S_i} d\mathbf{S}\cdot\mathbf{\Pi}^{(0)}\times\mathbf{r}_i\right)\cdot\mathbf{\Omega}_i \tag{9–1.11}$$

But since the tensor $\mathbf{\Pi}^{(0)}$ has no singularities in the volume of space Q_i presently occupied by the ith particle, we may, upon application of Gauss' divergence theorem, convert these surface integrals into volume integrals over Q_i. Hence,

$$\int_{S_i} d\mathbf{S}\cdot\mathbf{\Pi}^{(0)} = -\int_{Q_i} \nabla\cdot\mathbf{\Pi}^{(0)}\, dQ = \mathbf{0} \tag{9–1.12}$$

and

$$\int_{S_i} d\mathbf{S}\cdot\mathbf{\Pi}^{(0)}\times\mathbf{r}_i = -\int_{Q_i} \nabla\cdot(\mathbf{\Pi}^{(0)}\times\mathbf{r}_i) = -\int_{Q_i} (\nabla\cdot\mathbf{\Pi}^{(0)})\times\mathbf{r}_i\, dQ = \mathbf{0} \tag{9–1.13}$$

The vanishing of these integrals depends upon the fact that

$$\nabla\cdot\mathbf{\Pi}^{(0)} = \mathbf{0} \tag{9–1.14}$$

for the inertialess homogeneous flows under consideration.

The final expression for the additional dissipation rate is thus of the form

$$E^* = \sum_i \int_{S_i} d\mathbf{S}\cdot\mathbf{\Pi}\cdot(\mathbf{v} - \mathbf{v}^{(0)}) \tag{9–1.15}$$

It is interesting to observe that the surface integrals in the preceding expression are over the surfaces of the *particles* only. Thus, their evaluation requires detailed knowledge of $\mathbf{v}$ and $\mathbf{v}^{(0)}$ only in the immediate vicinity of the particle surfaces.

In general, the original, unperturbed velocity field is regular at all points of the fluid. It may therefore be represented by a Taylor series expansion in the vicinity of the ith particle as follows:

$$\mathbf{v}^{(0)} = \mathbf{v}^{(0)}_{[i]} + \mathbf{r}_i\cdot(\nabla\mathbf{v}^{(0)})_{[i]} + \cdots \tag{9–1.16}$$

where $\nabla\mathbf{v}^{(0)}$ is the velocity gradient dyadic. The bracketed subscript implies that the function to which it is affixed is to be evaluated at the "center" of ith particle. It is convenient to choose this origin to lie along the axis of rotation, so that $\mathbf{r}_i$ has its previous significance. The undisturbed field $\mathbf{v}^{(0)}$ explicitly contains the characteristic apparatus dimension R_0. Furthermore, if a is a characteristic particle dimension, then $\mathbf{r}_i = O(a)$ in the vicinity of a particle. Hence, for sufficiently small a/R_0, one may neglect higher-order terms in the Taylor expansion, as we have done.

By writing,

$$\boldsymbol{\omega}^{(0)}_{[i]} = \tfrac{1}{2}(\nabla\times\mathbf{v}^{(0)})_{[i]} \tag{9–1.17}$$

for the local angular velocity of the fluid, and

$$\boldsymbol{\Delta}^{(0)}_{[i]} = \tfrac{1}{2}[\nabla\mathbf{v}^{(0)} + (\nabla\mathbf{v}^{(0)})^{\dagger}]_{[i]} \tag{9–1.18}$$

for the local rate of shear tensor, Eq. (9–1.16) may be written in the alternative form

$$\mathbf{v}^{(0)} = \mathbf{v}^{(0)}_{[i]} + \boldsymbol{\omega}^{(0)}_{[i]}\times\mathbf{r}_i + \boldsymbol{\Delta}^{(0)}_{[i]}\cdot\mathbf{r}_i \tag{9–1.19}$$

These three contributions to the original fluid motion correspond, respectively, to translation, rotation, and deformation of a fluid particle.

With the aid of the foregoing expansion, Eq. (9–1.3) now requires that the perturbed field $\mathbf{v}$ satisfy the boundary condition

$$\mathbf{v} - \mathbf{v}^{(0)} = -[(\mathbf{v}^{(0)}_{[i]} - \mathbf{U}_i) + (\boldsymbol{\omega}^{(0)}_{[i]} - \boldsymbol{\Omega}_i)\times\mathbf{r}_i + \boldsymbol{\Delta}^{(0)}_{[i]}\cdot\mathbf{r}_i] \quad \text{on each } S_i \tag{9–1.20}$$

This relation may be introduced into Eq. (9–1.15), yielding

$$E^* = \sum_i \left[(\mathbf{v}^{(0)}_{[i]} - \mathbf{U}_i)\cdot\mathbf{F}_i + (\boldsymbol{\omega}^{(0)}_{[i]} - \boldsymbol{\Omega}_i)\cdot\mathbf{T}_i - \left(\int_{S_i} d\mathbf{S}\cdot\boldsymbol{\Pi}\,\mathbf{r}_i\right):\boldsymbol{\Delta}^{(0)}_{[i]}\right] \tag{9–1.21}$$

Here,

$$\mathbf{F}_i = -\int_{S_i} d\mathbf{S}\cdot\boldsymbol{\Pi} \tag{9–1.22}$$

is the hydrodynamic force exerted by the fluid on the ith particle, and

$$\mathbf{T}_i = -\int_{S_i} \mathbf{r}_i\times(d\mathbf{S}\cdot\boldsymbol{\Pi}) \tag{9–1.23}$$

is the hydrodynamic torque exerted on the ith particle about the origin. The negative signs arise, of course, because $d\mathbf{S}$ is directed into the particle.

Evaluation of the various terms in Eq. (9–1.21) for multiparticle systems is a formidable problem in the general case. The translational terms are normally dominant in suspensions containing macroscopic particles whose densities differ from that of the surrounding fluid. If, however, either the densities are matched or the particles are sufficiently small, one may neglect the translational contribution, since the hydrodynamic force on the particle will then be either identically or approximately zero:

$$\mathbf{F}_i = \mathbf{0} \tag{9–1.24}$$

Furthermore, in the absence of external body torques on the particle, and for sufficiently small particles (that is, possessing essentially zero moment of inertia) the hydrodynamic torque must vanish;

$$\mathbf{T}_i = \mathbf{0} \tag{9–1.25}$$

We assume these conditions are met, in which case the additional energy dissipation rate becomes

$$E^* = \sum_i \left(\int_{S_i} d\mathbf{s} \cdot \mathbf{\Pi}\, \mathbf{r}_i \right) : \mathbf{\Delta}_{[i]}^{(0)} \tag{9–1.26}$$

where $d\mathbf{s} = -d\mathbf{S}$ is directed into the *fluid.* This additional energy dissipation ultimately determines the viscosity of the suspension via Eq. (9–1.1). Thus, the treatment of suspension viscosity is confined essentially to an evaluation of the *additional energy dissipation due to the dilational components of the original fluid motion.*

Before attempting to evaluate Eq. (9–1.26) for various idealized situations, it is convenient to establish an explicit relationship between the viscosity of the suspension and the additional and unperturbed dissipation rates. To do this we note, upon substituting Eq. (9–1.3) into Eq. (9–1.5) and utilizing Eqs. (9–1.22)–(9–1.23), that

$$E = \int_{S_0} d\mathbf{S} \cdot \mathbf{\Pi} \cdot \mathbf{v} - \sum_i (\mathbf{F}_i \cdot \mathbf{U}_i + \mathbf{T}_i \cdot \mathbf{\Omega}_i)$$

Since we have assumed the forces and torques on each particle to vanish, this becomes

$$E = \int_{S_0} d\mathbf{S} \cdot \mathbf{\Pi} \cdot \mathbf{v} \tag{9–1.27}$$

Introducing this relation and Eq. (9–1.10) into Eq. (9–1.1), we obtain

$$\frac{\mu}{\mu_0} = \frac{E}{E^{(0)}} \tag{9–1.28}$$

Alternatively, in consequence of the definition of E^* given in Eq. (9–1.9), we may write

$$\frac{\mu}{\mu_0} = 1 + \frac{E^*}{E^{(0)}} \tag{9–1.29}$$

This relation holds under fairly general circumstances.

The preceding formula for suspension viscosity ultimately leads to identical values of μ for each of the different types of "linear" viscometers–for example, Couette, capillary, falling ball, etc.

9-2 Dilute Systems of Spheres—No Interaction Effects

We shall restrict the treatment in this section to dilute systems of smooth spheres of equal size, reserving for a later section effects arising for non-spherical particles. Before embarking upon this program, it is important to note that, as in the sedimentation cases studied in the preceding chapter, it is necessary to examine possible boundary effects in dilute systems.

Let l denote the characteristic interparticle distance. Without repeating all the details of the previous discussion, it is clear that there exist two limiting possibilities in a dilute suspension: (a) the ratio of particle surface area to wall surface area is small, that is, $(a/l)^3(R_0/a) \ll 1$, thereby permitting us to assume that the particles do not interact with each other; (b) the particle-wall area ratio is large, that is, $(a/l)^3(R_0/a) \gg 1$. In this case particle interaction is more important than the effect of apparatus boundaries on individual particles, so that the sole function of the wall is to generate the original flow field. Theoretical treatments of suspension viscosity have, almost universally, implicitly restricted themselves to situations in which the first set of conditions prevail. Oddly enough, the reverse is true in theoretical studies of sedimentation, where it is commonly assumed that the pressure drop force will invariably reduce to the sum of Stokes' law drags on the individual particles. We shall first consider situations for which $(a/l)^3(R_0/a) \ll 1$.

We recapitulate the basic assumptions in our proposed investigation of dilute suspensions: (a) the suspended particles are spheres; (b) the dimensionless ratios a/l, a/R_0, and $(a/l)^3(R_0/a)$ are each very small; (c) inertial effects may be neglected; (d) the particles are rigid, and the fluid adheres perfectly to the suspended spheres.

The hydrodynamic force and torque on a small spherical particle of radius a_i suspended in an arbitrary field of flow are, by Faxen's laws,

$$\mathbf{F}_i = 6\pi\mu a_i(\mathbf{v}_{[i]}^{(0)} - \mathbf{U}_i) \qquad \mathbf{T}_i = 8\pi\mu a_i^3(\boldsymbol{\omega}_{[i]}^{(0)} - \boldsymbol{\Omega}_i)$$

Since $\mathbf{F}_i$ and $\mathbf{T}_i$ are assumed to be zero, this requires that

$$\mathbf{U}_i = \mathbf{v}_{[i]}^{(0)} \qquad \boldsymbol{\Omega}_i = \boldsymbol{\omega}_{[i]}^{(0)}$$

that is, the translational and angular slip velocities of the sphere relative to the surrounding fluid are zero. Consequently, Eq. (9–1.20) yields the boundary condition

$$\mathbf{v} - \mathbf{v}^{(0)} = -\boldsymbol{\Delta}_{[i]}^{(0)} \cdot \mathbf{r}_i \quad \text{at } r_i = a_i \tag{9–2.1}$$

whereas Eq. (9–1.2) requires that

$$\mathbf{v} - \mathbf{v}^{(0)} = \mathbf{0} \quad \text{on } S_0 \tag{9–2.2}$$

In order to evaluate the dilational contribution to the additional energy dissipation E^*, it is clear from Eq. (9–1.26) that, since we have only to perform an integration over the *particle* surface, we need accurately know the field $\mathbf{v}$ only in the immediate vicinity of the particle. Accordingly, since a/R_0 is small, it seems reasonable in the limit to replace Eq. (9–2.2) by the condition

$$\mathbf{v} - \mathbf{v}^{(0)} \to \mathbf{0} \quad \text{as } r_i \to \infty \tag{9–2.3}$$

This constitutes an important simplification of the problem.

The boundary-value problem posed by Eqs. (9–2.1) and (9–2.3) is readily solved by means of Lamb's[32] general solution of the creeping motion equations, Eq. (3–2.3):

$$\mathbf{v} - \mathbf{v}^{(0)} = \sum_{n=-\infty}^{\infty} \nabla \times (\mathbf{r}_i \chi_n) + \nabla \phi_n + \frac{(n+3)}{2(n+1)(2n+3)} r_i^2 \nabla\left(\frac{p_n}{\mu_0}\right) - \frac{n}{(n+1)(2n+3)} \mathbf{r}_i \frac{p_n}{\mu_0} \tag{9–2.4}$$

and

$$p - p^{(0)} = \sum_{n=-\infty}^{\infty} p_n \tag{9–2.5}$$

The solution is obtained by retaining only the p_{-3} and ϕ_{-3} harmonic functions, which are found to have the following values:

$$\frac{p_{-3}}{\mu_0} = -5a_i^3 \boldsymbol{\Delta}_{[i]}^{(0)} : \frac{\mathbf{r}_i \mathbf{r}_i}{r_i^5}$$
$$\phi_{-3} = -\frac{1}{2} a_i^5 \boldsymbol{\Delta}_{[i]}^{(0)} : \frac{\mathbf{r}_i \mathbf{r}_i}{r_i^5} \tag{9–2.6}$$

To evaluate Eq. (9–1.26) we note that

$$d\mathbf{s} = \mathbf{i}_r \, ds$$

where $\mathbf{i}_r$ is a unit normal vector drawn outward from the sphere, and $ds = r^2 d\Omega$ is a scalar element of surface area, in which $d\Omega = \sin\theta \, d\theta \, d\phi$ is an element of surface area on a sphere of unit radius. Thus

$$d\mathbf{s} \cdot \mathbf{\Pi} = \mathbf{\Pi}_r \, ds$$

where $\mathbf{\Pi}_r$ is the stress vector acting across the sphere surface. This vector may be obtained from the general formula in Eq. (3–2.37). In particular, we obtain

$$[\mathbf{\Pi}_{r_i} - \mathbf{\Pi}_{r_i}^{(0)}]_{r_i = a_i} = 3\mu_0 \boldsymbol{\Delta}_{[i]}^{(0)} \cdot \frac{\mathbf{r}_i}{r_i} \tag{9–2.7}$$

In order to evaluate $\mathbf{\Pi}_{r_i}^{(0)}$ in the proximity of the sphere i it suffices to use only the leading terms in the Taylor expansion of the velocity field shown in Eq. (9–1.19). Since $\nabla^2 \mathbf{v}^{(0)} = \mu^{-1} \nabla p^{(0)}$, we find that the Taylor expansion of the pressure field, comparable to Eq. (9–1.19), is $p^{(0)} =$ constant. Without loss in generality we may take this constant to be zero and write

$$p^{(0)} = 0 \tag{9–2.8}$$

Now, for an incompressible Newtonian fluid, the pressure tensor is

$$\mathbf{\Pi}^{(0)} = -\mathbf{I}p^{(0)} + 2\mu_0 \mathbf{\Delta}^{(0)}$$

where
$$\mathbf{\Delta}^{(0)} = \tfrac{1}{2}[\nabla\mathbf{v}^{(0)} + (\nabla\mathbf{v}^{(0)})^{+}]$$

Hence, from Eq. (9–1.19) and Eq. (9–2.8), to the same order of approximation, the Taylor expansion of $\mathbf{\Pi}^{(0)}$ about the sphere center gives $\mathbf{\Pi}^{(0)}$ the constant value

$$\mathbf{\Pi}^{(0)} = 2\mu_0 \mathbf{\Delta}^{(0)}_{[i]}$$

Since $\mathbf{i}_{r_i} = \mathbf{r}_i/r_i$, we obtain for the corresponding stress vector at the sphere surface,

$$[\mathbf{\Pi}^{(0)}_{r_i}]_{r_i=a_i} = 2\mu_0 \mathbf{\Delta}^{(0)}_{[i]}\cdot\frac{\mathbf{r}_i}{r_i}$$

Upon adding this to Eq. (9–2.7), we thus obtain

$$[\mathbf{\Pi}_{r_i}]_{r_i=a_i} = 5\mu_0 \mathbf{\Delta}^{(0)}_{[i]}\cdot\frac{\mathbf{r}_i}{r_i}$$

which makes

$$\int_{S_i} d\mathbf{s}\cdot\mathbf{\Pi}\mathbf{r}_i = 5\mu_0 a_i^3 \mathbf{\Delta}^{(0)}_{[i]}\cdot\int_{S_1}\frac{\mathbf{r}_i\mathbf{r}_i}{r_i^2}\,d\Omega$$

where S_1 is a sphere of unit radius. However, as is readily shown

$$\int_{S_1}\frac{\mathbf{r}_i\mathbf{r}_i}{r_i^2}\,d\Omega = \frac{4\pi}{3}\mathbf{I}$$

where $\mathbf{I}$ is the idemfactor. Thus,

$$\int_{S_i} d\mathbf{s}\cdot\mathbf{\Pi}\mathbf{r}_i = 5\mu_0\left(\frac{4\pi a_i^3}{3}\right)\mathbf{\Delta}^{(0)}_{[i]}$$

Substitution into Eq. (9–1.26) now gives for the additional energy dissipation

$$E^* = 5\mu_0 \sum_i \frac{4\pi a_i^3}{3}\mathbf{\Delta}^{(0)}_{[i]}:\mathbf{\Delta}^{(0)}_{[i]} \tag{9–2.9}$$

The preceding expression may be put into an alternative form by observing that the *local* rate of energy dissipation per unit time per unit volume is

$$\Phi^{(0)}_{[i]} = 2\mu_0 \mathbf{\Delta}^{(0)}_{[i]}:\mathbf{\Delta}^{(0)}_{[i]}$$

Thus,
$$E^* = \frac{5}{2}\sum_i Q_i \Phi^{(0)}_{[i]} \tag{9–2.10}$$

where $Q_i = 4\pi a_i^3/3$ is the volume of the ith sphere. This now provides us with the information needed to evaluate Eq. (9–1.29).

If we regard the particles as being continuously (but not necessarily uniformly) distributed throughout the fluid, the foregoing summation may be replaced by an appropriate integration. The volume dQ_i occupied by solid particles present in an element of volume dQ of space is

$$dQ_i = \phi\, dQ$$

where ϕ is the *local* volume fraction of suspended solids. Thus, for all the particles present in the volume of space Q_0 contained within the apparatus, we have, in place of Eq. (9–2.10),

$$E^* = \frac{5}{2}\int_{Q_0} \phi\,\Phi^{(0)}\,dQ \tag{9–2.11}$$

where $\Phi^{(0)}$ is the local dissipation rate. By definition, the total rate at which energy is being dissipated by the original, homogeneous fluid motion in the apparatus is

$$E^{(0)} = \int_{Q_0} \Phi^{(0)}\,dQ \tag{9–2.12}$$

Substitution into Eq. (9–1.29) then gives

$$\frac{\mu}{\mu_0} = 1 + \frac{5}{2}\,\frac{\int_{Q_0} \phi\,\Phi^{(0)}\,dQ}{\int_{Q_0} \Phi^{(0)}\,dQ} \tag{9–2.13}$$

The average volume fraction, $\bar{\phi}$, of solids in the apparatus is

$$\bar{\phi} = \frac{1}{Q_0}\int_{Q_0} \phi\,dQ \tag{9–2.14}$$

If the particles are uniformly distributed throughout the fluid, $\phi = \bar{\phi}$ has a constant value in Q_0, and we obtain

$$\frac{\mu}{\mu_0} = 1 + \frac{5}{2}\bar{\phi} \tag{9–2.15}$$

which is Einstein's formula[9] for the viscosity of a dilute suspension of spherical particles.

Alternatively, in the special case of a parallel plate viscometer, $\Phi^{(0)}$ is constant throughout the fluid, in which case we again obtain Eq. (9–2.15)—but this time without the added assumption of a random particle distribution. It is important to note that Eq. (9–2.15) is generally invalid for a nonuniform particle distribution in any other type of viscometer.

The foregoing derivation, given originally by Brenner[3], is believed to be an improvement of the original derivation of Eintein[9], and of subsequent derivations by Jeffery[21] and Burgers[5], who derived Eq. (9–2.15) by rather different arguments. Among other things it appears to show unequivocally that Einstein's formula holds for any type of linear viscometer. The derivation also emphasizes the necessity for a random particle distribution. In essence, the treatments by Einstein and Jeffery compute the additional energy dissipation by an unusual integration over the surface of a large, vaguely defined spherical surface concentric with the particle. The function of this surface is by no means clear. We have been able to avoid this integration by utilizing the reciprocal theorem to replace this by an alternative integration extended over the particle surfaces, as in Eq. (9–1.26).

Burgers'[5] treatment is novel because his development utilizes the dynamical definition of viscosity in terms of stresses. Thus he avoids what is essentially a definition of viscosity based on energy dissipation. Many investigators do not seem to be aware of these two alternative methods for defining (or actually measuring experimentally) suspension viscosity.

Despite the apparent agreement of these different schemes, there still exist important questions regarding the validity of Einstein's formula. A critical point in our analysis arises when the true boundary condition (9–2.2), $\mathbf{v} = \mathbf{v}^{(0)}$ on S_0, is replaced by the approximate boundary condition (9–2.3), $\mathbf{v} = \mathbf{v}^{(0)}$ as $r_i \to \infty$. An approximation of this type is introduced in all the theories of suspension viscosity mentioned. Its validity is argued on the grounds that the finiteness of the dimensions of S_0 merely introduces a conventional wall effect which will disappear in the limit as $a/R_0 \to 0$. It is possible, however, that this "wall effect" is not of the ordinary type, but rather may be of a paradoxical type, such as was discussed in Section 7–1 in connection with the pressure drop experienced by a sphere settling in a circular cylinder (where it was shown that the drag on the sphere was not balanced by the pressure drop forces even when the walls of the cylinder were at infinity).

In connection with the use of the approximate boundary condition, $\mathbf{v} = \mathbf{v}^{(0)}$ as $r_i \to \infty$, variations in energy dissipation due to the differences in position of each particle and the shape of the containing vessel are assumed to cancel in the process of integrating to determine the effect of a number of particles uniformly distributed through a suspension. In the case of a spherical particle moving in a circular cylindrical vessel, the energy dissipation has been calculated.[18] It depends upon the position of the particle in the vessel. Using this information, Happel and Lee[18] studied the motion of a sphere suspended in a parabolic field in a cylindrical tube, and summed the individual energy dissipations for a uniform suspension of particles to obtain an expression for suspension viscosity. The result was the same as given by Einstein, Eq. (9–2.15), thus indicating that this assumption is valid for a suspension of uniformly distributed particles.

Aside from this question regarding Einstein-type treatments of viscosity in dilute suspensions, there is the question of the validity of the assumption of an undisturbed original field when the ratio of particle-to-wall surface area is sufficiently large that $(a/l)^3(R_0/a) \gg 1$. In the limiting case of no wall it may well be that a cell model treatment similar to that employed for suspensions is applicable. A cell model, employing the boundary condition expressed by Eq. (9–2.3), that is, vanishing of the velocity components at $r_i = \infty$, was developed by Simha[48] in connection with a treatment for concentrated suspensions. His treatment in the case of dilute systems parallels Brenner's somewhat, except that energy dissipation E^* in the infinite domain selected is evaluated by integrating over the outer rather than inner spherical surface. The result is the same, namely, Eq. (9–2.15). Happel[16], in a treatment paralleling Simha's very closely, and also aimed primarily at concentrated assemblages, does not use the boundary condition of Eq. (9–2.3) but instead employs the conditions

$$\left.\begin{aligned} \Pi_{r\theta} - \Pi_{r\theta}^{(0)} = \Pi_{r\phi} - \Pi_{r\phi}^{(0)} = 0 \\ \text{and} \qquad v_r - v_r^{(0)} = 0 \end{aligned}\right\} \text{ at } r_i = \infty \tag{9–2.16}$$

corresponding to the vanishing of the "additional" tangential stresses and normal velocity at the boundary of the infinite domain. These theories are developed in greater detail in the Section 9–4. The counterpart of Eq. (9–2.15) obtained in this case is [see Eq. (9–4.22)]

$$\frac{\mu}{\mu_0} = 1 + 5.5\bar{\phi} \tag{9–2.17}$$

for very great dilution. This differs, of course, from Einstein's result. It is of some interest that the velocity fields $\mathbf{v}$ developed by both Happel and Simha are identical in the limiting case, $r_i \to \infty$, despite the different choice of boundary conditions. Moreover, their field $\mathbf{v}$ agrees with that of Einstein. The energy dissipation in the domain selected, however, is different and raises the question as to whether a unique viscosity versus solids concentration fraction formula is applicable in extremely dilute systems. It is possible that varying results may be obtained, depending on particle size and viscometer characteristics.

9-3 Dilute Systems—First-Order Interaction Effects

The "zeroth" order approximation developed in the preceding section may be regarded as an analog of Stokes' law as regards the extent of particle-particle interactions. In the case of pressure drop or energy dissipation arising solely from drag forces, a uniform suspension undergoing sedimentation yields the same result whether $(a/l)^3(R_0/a)$ is much smaller or greater than unity. It appears that, in the case of a shearing field, the same result may not be obtained for these two extremes of particle-wall area ratios. This uncertainty regarding the behavior of a sphere in a shearing field with arbitrary boundaries raises doubts regarding further extensions of the Einstein treatment to more concentrated systems.

In principle, the method of reflections, exploited in the previous chapter, could be rigorously applied to explore the simultaneous effects of variations in a/l and a/R_0, and the range, if any, in which one or the other of these parameters could be omitted from an appropriate viscosity correlation which takes particle concentration into account.

In a treatment extending the Einstein method to first-order interactions, Guth and Simha[15], using the method of reflections, were the first to consider the effect of the walls of a containing vessel on the viscosity of a suspension. They examined in detail the case of Couette flow. A dilational motion was assumed in an unbounded fluid parallel to the yz plane, as follows:

$$u^{(0)} = 0, \qquad v^{(0)} = By, \qquad w^{(0)} = -Bz, \qquad p^{(0)} = \text{constant} \tag{9-3.1}$$

This field was reflected from a single sphere of radius a, using the boundary condition of zero velocity at infinity for the reflected field, as in Einstein's derivation. Thus, the following reflected field was obtained:

$$\begin{aligned} u^{(1)} &= -\frac{5}{2} a^3 \frac{Bx}{r^5}(y^2 - z^2) + \frac{5}{2} a^5 \frac{Bx}{r^7}(y^2 - z^2) \\ v^{(1)} &= -\frac{5}{2} a^3 \frac{Bx}{r^5}(y^2 - z^2) + \frac{5}{2} a^5 \frac{Bx}{r^7}(y^2 - z^2) - \frac{a^5}{r^5} By \\ w^{(1)} &= -\frac{5}{2} a^3 \frac{Bx}{r^5}(y^2 - z^2) + \frac{5}{2} a^5 \frac{Bx}{r^7}(y^2 - z^2) + \frac{a^5}{r^5} Bz \\ p^{(1)} &= -5\mu_0 a^3 B \frac{(y^2 - z^2)}{r^5} + \text{constant} \end{aligned} \tag{9-3.2}$$

This field was then reflected from a wall placed at a distance d to the left of the origin of the sphere. The mirror image of Eq. (9–3.2) with respect to this was found, and the reflected field $\mathbf{v}^{(2)}$ expanded in the vicinity of the origin of the original sphere. Neglecting terms of fourth order in a/d one obtains

$$\begin{aligned} u^{(2)} &= 0 \\ v^{(2)} &= \frac{5}{16} B \frac{a^3}{d^3} y \\ w^{(2)} &= -\frac{5}{16} B \frac{a^3}{d^3} y \end{aligned} \tag{9-3.3}$$

In order to determine the behavior of the sphere in this additional field, we simply substitute for B in Eq. (9–3.1) the constant $5B(a/d)^3/16$. Now, suppose that a second wall is present in Couette flow. Denote by d_1 the distance of the given particle from one wall and by d_2 its distance from the other. Then, using the Einstein constant for the evaluation of the original field given by Eq. (9–3.1), we ultimately find, upon including the field taking the wall-effect into account, that

$$\frac{\mu}{\mu_0} = 1 + 2.5\phi\left[1 + \frac{5}{16} a^3 \left(\frac{1}{d_1^3} + \frac{1}{d_2^3}\right)\right] \tag{9-3.4}$$

This, of course, assumes that the two wall-effects are independently additive. As we have seen this is not strictly correct, at least for the case of sedimentation—see Eq. (7–4.30).

Since particles may occupy all positions between the walls, we can establish an average value of $(1/d_1^3) + (1/d_2^3)$ in the general case by integration. Thus, this average value will be

$$\frac{2}{2(d-a)}\int_{a}^{2d-a}\frac{dx}{x^3}=\frac{2d}{a^2(2d-a)^2} \tag{9–3.5}$$

where $2d$ is the distance between plates of the Couette apparatus. Substitution of this average value into Eq. (9–3.4) yields

$$\frac{\mu}{\mu_0}=1+2.5\phi\left[1+\frac{5}{8}\frac{ad}{(2d-a)^2}\right] \tag{9–3.6}$$

When $2d \gg a$ this reduces to

$$\frac{\mu}{\mu_0}=1+2.5\phi\left(1+\frac{5}{32}\frac{a}{d}\right) \tag{9–3.7}$$

The validity of this formula, of course, rests on the correctness of the Einstein constant 2.5 for evaluation of the zeroth approximation. There is also the question of whether the effect of the reflected field $\mathbf{v}^{(2)}$ from a given sphere to the wall should not be developed in the vicinity of *each* of the other particles present in the suspension, instead of only at the location of the particle originating the disturbance.

In the same paper, Guth and Simha[15] also considered the first-order effect of the spheres interacting with one another (that is, terms of order a/l), in a similar extension of the Einstein method using the method of reflections. A sphere of radius a is placed at the origin ($x'=0$, $y'=0$, $z'=0$) of a coordinate system, and an undisturbed dilational flow is assumed as follows:

$$\begin{aligned} u^{(0)'} &= A^{(0)}x' \\ v^{(0)'} &= B^{(0)}y' \\ w^{(0)'} &= C^{(0)}z' \end{aligned} \tag{9–3.8}$$

where, to satisfy the continuity equation,

$$A^{(0)}+B^{(0)}+C^{(0)}=0$$

This field reflected from the sphere, following the Einstein treatment, is $\mathbf{v}^{(1)'}$. Its x' component is

$$\begin{aligned} u^{(1)'} = &-\frac{5}{2}\frac{a^3}{r'^5}x'(A^{(0)}x'^2+B^{(0)}y'^2+C^{(0)}z'^2) \\ &+\frac{5}{2}\frac{a^5}{r'^7}x'(A^{(0)}x'^2+B^{(0)}y'^2+C^{(0)}z'^2)-\frac{a^5}{r'^5}A^{(0)}x' \end{aligned} \tag{9–3.9}$$

with similar expressions for the other components.

Now, at the point $P(x, y, z)$ we place a second sphere of the same radius. A Taylor series expansion gives, for the value of Eq. (9–3.9) on this sphere:

$$u^{(1)}=u_P^{(1)'}+\xi\left(\frac{\partial u^{(1)'}}{\partial x'}\right)_P+\eta\left(\frac{\partial u^{(1)'}}{\partial y'}\right)_P+\zeta\left(\frac{\partial u^{(1)'}}{\partial z'}\right)_P \tag{9–3.10}$$

where $\xi=x'-x$, $\eta=y'-y$, $\zeta=z'-z$. The field in the vicinity of the second sphere will be given by $\mathbf{v}^{(0)}+\mathbf{v}^{(2)}$ where $\mathbf{v}^{(2)}$ is the field equal to

$-\mathbf{v}^{(1)}$ on the surface of the second sphere and which vanishes at infinity. This field can be found by general solutions of the type given by Eq. (9–2.4). Energy dissipation due to the additional disturbance, when summed to take into account the number of spheres present in a unit volume of suspension, enables the effective suspension viscosity to be estimated. In essence, this is done by employing the basic definition of suspension viscosity given in Eq. (9–1.1). The calculation implicitly assumes that the unit volume contains a sufficiently large number of spheres to include all possible neighboring interactions. In evaluating the relationships involved, Guth and Simha inadvertantly omitted certain additional terms which, when properly taken into account[48], give

$$\frac{\mu}{\mu_0} = 1 + 2.5\phi + 14.1\phi^2 + \cdots \tag{9–3.11}$$

In deriving this formula, the problem of satisfying the over-all condition of conservation of fluid-particle volume was not explicitly considered. Rather, the Smoluchowski-type equation was applied in much the same manner as discussed in the previous chapter, without considering the problem of "return flow." Simha[48] states that a consideration taking into account the volume occupied by the spheres reduces the value of the last term in Eq. (9–3.11) to 12.6 ϕ^2. Saito[43], in further attempts to establish rigorously the coefficient of ϕ^2 in Eq. (9–3.11), concludes that, owing to the appearance of an indeterminate integral appearing in the Einstein treatment, the creeping motion equations are not applicable to this problem. He suggests that the neglect of inertia terms in the creeping motion equations is responsible for this difficulty and sets up a treatment based on the Oseen equations, which, unfortunately, have not been solved for this situation. Mooney[36], in further discussion, concludes that inertial terms are not important and that the difficulty is caused by an ambiguous setting of the appropriate boundary-value problem, a conclusion to which we subscribe.

The concentration-dependence of suspension viscosity not only depends on hydrodynamic interactions, but also on effects arising from highly inelastic collisions between particles. Collisions may result in aggregation of single particles to doublets and, to a lesser extent, triplets. Vand[54] presents a treatment which attempts to take into account the collision effect for a suspension of spheres, neglecting, in the first approximation, the effect of Brownian motion. After colliding, it is assumed that the particles roll around each other until they reach positions where they can be disengaged by shear of the fluid suspending them. It is thus possible to estimate the number of doublets, triplets, etc., present at any time in a given suspension. It is further hypothesized that one may compute the intrinsic viscosity of an aggregate by assuming that the intrinsic viscosity ratio is in proportion to the volume of the spheres plus immobilized solvent to the original volume of unagglomerated spheres, corrected by a small factor to allow for nonsphericity of the particle aggregates. The intrinsic viscosity of a suspension is a dimensionless constant defined as follows:

$$[\mu] = \lim_{\phi \to 0} \frac{(\mu/\mu_0) - 1}{\phi} \tag{9–3.12}$$

Thus $[\mu] = 2.5$ for singlets, according to the Einstein relationship. For a doublet, the volume of particles plus immobilized solvent is assumed equal to the volume of the spherocylinder formed from the two spheres. Vand thus obtains a value 3.175 for $[\mu]_2$. $[\mu]_2$ is a characteristic constant expressing the effect of doublets on intrinsic viscosity, in the same sense that $[\mu]_1$ applies to singlets. The effective intrinsic viscosity of the suspension is then taken to be the mean of the intrinsic viscosity contributions due to all species in suspension:

$$[\bar{\mu}] = \frac{[\mu]\phi_1 + [\mu]_2\phi_2 + \cdots}{\phi} \tag{9–3.13}$$

Vand employs this relationship in conjunction with a treatment of hydrodynamic interaction using the method of reflections, resembling that of Guth and Simha[15]. His final result for relative viscosity is as follows:

$$\frac{\mu}{\mu_0} = 1 + 2.5\phi + 7.349\phi^2 + O(\phi^3) \tag{9–3.14}$$

Vand's treatment must be regarded as largely qualitative. The assumptions previously outlined do not correspond to a specific mathematical boundary-value problem, nor do they seem justifiable by any other rigorous standards. Nevertheless, the treatment does illustrate the potential problems inherent in a hydrodynamic treatment taking collisions into account.

Kynch[29], in a treatment devoted largely to higher-order concentration effects, developed an equation for dilute systems similar to that given by Guth and Simha, and Vand:

$$\frac{\mu}{\mu_0} = 1 + 2.5\phi + 7.5\phi^2 + \cdots \tag{9–3.15}$$

Simha's[48] cell-model treatment of concentrated suspensions (see Section 9–4) may also be specialized to examine the coefficient of the ϕ^2 term in these relationships. He obtains the coefficient $15.6/f^3$, where f is a somewhat arbitrarily determined constant. A coefficient of 7–8 for the quadratic term in ϕ would correspond to $f \approx 1.31–1.25$, which Simha believes to be a reasonable value for this interaction parameter.

Since Eqs. (9–3.13)–(9–3.15) each involve a coefficient of 2.5 for the linear term, and since the validity of this constant is not wholly free from doubt, it is obviously difficult to be certain of the quadratic coefficient on the basis of either theoretical or experimental considerations. This is especially true at concentrations where the effect of the ϕ^2 term becomes appreciable, since other higher-order terms may then also contribute substantially to the relative viscosity.

It may be concluded, in general, that fundamental treatments of first-order concentration effects based on hydrodynamic considerations are few and that, even granting the validity of the Einstein equation for the zeroth-order approximation, they do not give definitive solutions to the effects in question. Even when collision effects are neglected, the mathematics becomes very involved if one wishes to take proper account of both particle-particle and particle-wall interactions to a first order.

9-4 Concentrated Systems

To apply hydrodynamic principles to more concentrated suspensions of spherical particles, it is again necessary to assume idealized systems, in which effects arising from collisions and aggregation are ignored. In such systems, the method of reflections, though applicable in principle, has thus far proved too complicated to carry out in practice. It should be noted, however, that when correctly applied, the method of reflections permits the over-all conservation of volume condition to be satisfied in any finite domain. This requires that the field originated by each particle be carried to the walls of the containing vessel. The Smoluchowski treatment, discussed in the preceding chapter [see also the discussion preceding Eq. (9–3.1)], applies only in an infinite medium, and thus will not correctly yield the over-all flow pattern. This has been discussed by Simha[48] as a "shielding effect" due to finite particle size. The use of the Smoluchowski treatment for calculation of particle interaction is visualized in this discussion as equivalent to a point center of dilation, although in fact a finite sphere radius is employed. True, a particle of radius a will produce the same disturbance to fluid motion as a point if the fluid domain is infinite in extent, but it is possible in using the method of reflections to employ a finite domain. It is thus thought that this method has possibilities for applications which have not yet been fully exploited.

A cell model provides a simpler, albeit less rigorous, approach to the viscosity of concentrated systems. By their nature, such models are limited to situations where the ratio of particle area to wall area is very great. Otherwise wall effects may contravene. The definition of relative viscosity for such models involves the ratio of dissipation of energy per unit volume of suspension compared with that of the suspending fluid alone. This definition is not as satisfactory as the operational definition given by Eq. (9–1.1). It is forced upon us by our inability to reconcile cell models with the existence of boundaries constraining the flow.

Simha[48] applied such a model to a calculation of the viscosity of concentrated suspensions. The cell in this case consists of a rigid concentric spherical enclosure containing a representative sphere at its center. Perturbations of the flow caused by the presence of other particles outside the cell cannot influence the dilational flow within it. Denoting the radius of the spherical shell or cage by b, one assumes that the effect of all other particles in the suspension being sheared is such as to cause the disturbance velocity to the dilational flow to vanish at the surface of the shell. Such a simplified model stresses the interaction between the central particle and its immediate neighbors. Inside the cage, $a < r < b$, the solvent flow satisfies the creeping motion equations. The hydrodynamics of this simplified model may be obtained in closed form. We omit mathematical details, as these may be inferred from Happel's solution for a similar cell model, discussed in the sequel. One ultimately finds for the relative viscosity

$$\frac{\mu}{\mu_0} = 1 + \frac{5}{2}\lambda(\gamma)\phi \tag{9–4.1}$$

where $\gamma = a/b$. $\lambda(\gamma)$ is a steeply increasing function of γ, measuring the interaction between the central particle and its surroundings. It is given by the expression,

$$\lambda(\gamma) = \frac{4(1-\gamma^7)}{4(1+\gamma^{10}) - 25\gamma^3(1+\gamma^4) + 42\gamma^5} \tag{9–4.2}$$

The concentration dependence of suspension viscosity arises because the radius of the outer shell varies with concentration, becoming smaller the closer particles come together. Further assumptions are required to establish the size of this outer shell. In dilute solutions this poses a problem since, in fact, it is conceptually difficult to visualize a single solid shell as replacing the actual surfaces of solid particles which the central particle "sees." Simha introduces arguments permitting him to assign plausible values to γ. If R_{12} is taken as the average distance between particles, we can assume that the radius $r = b$ of the cell will vary with concentration in such a way that b is proportional to R_{12} or, at higher concentrations, to $R_{12} - a$. Corresponding to the preceding proportionalities one may assume

$$\gamma^3 = \frac{\phi}{f^3} \tag{9–4.3}$$

or

$$\gamma^3 = \frac{\phi}{f^3[1 + (\phi^{1/3}/f)]^3} \tag{9–4.4}$$

In Eq. (9–4.4), the maximum concentration corresponding to close packing and $\gamma = 1$ is given by $\phi_{\max} = f^3/8$. Thus, for an hexagonal or simple cubic arrangement corresponding to $\phi_{\max}$, one has $f = 1.81$ or 1.61, respectively, if the shell b is identified with the wall of the cage formed by the nearest neighbors, $b = R_{12} - a$. If these relationships are inserted into Eq. (9–4.2) and the latter expanded for low concentrations, Simha finds, respectively:

$$\frac{\mu}{\mu_0} = 1 + \frac{5}{2}\phi\left(1 + \frac{25}{4f^3}\phi - \frac{21}{2f^5}\phi^{5/3} + \frac{625}{16f^6}\phi^2 + \cdots\right) \tag{9–4.5}$$

$$\frac{\mu}{\mu_0} = 1 + \frac{5}{2}\phi\left(1 + \frac{25}{4f^3}\phi + \frac{75}{4f^4}\phi^{4/3} + \frac{27}{f^5}\phi^{5/3} + \frac{785}{16f^6}\phi^2 + \cdots\right) \tag{9–4.6}$$

For high concentrations, corresponding to Eq. (9–4.4), the limiting approximation is

$$\lim_{\phi \to \phi_{max}} \frac{\mu}{\mu_0} = \frac{54}{5f^3} \frac{\phi^2}{[1 - (\phi/\phi_{max})]^3} \tag{9–4.7}$$

Because of the arbitrary nature of the location of the cell wall, it is possible to obtain varying results with these formulas.

The restriction of no disturbance velocity at a cell wall seems too limiting an assumption. In an alternative treatment of the concentric sphere cell model, Happel[16] introduced what appears to be a more plausible hypothesis as to the behavior of the disturbance velocity at $r = b$. This was supplemented with a more rational proposal for relating the latter radius to the solids concentration. In particular, in place of Simha's solid spherical cage, Happel assumed that the disturbance due to the presence of each sphere is confined to a cell of fluid bounded by a frictionless envelope on whose surface the normal velocity vanishes. The outer fluid shell is assumed to enclose an amount of fluid such that the fluid-solid volume ratio in the cell is identical to that in suspension. This "free surface" model is the same model employed in the preceding chapter, in connection with the calculation of sedimentation velocity and pressure drop for flow through suspensions of spherical particles. Details of the treatment now follow.

In the absence of suspended particles, the flow field is assumed to consist of a simple shearing motion with a constant velocity gradient $q = \partial u/\partial y$ in the xy plane. Since only spherical particles are considered, the rotational part of this field is not disturbed by the presence of a sphere at the origin of the coordinate system. Only the dilational component, $\mathbf{v}^{(0)}$, need be considered. Thus,

$$u^{(0)} = \frac{q}{2}y, \qquad v^{(0)} = \frac{q}{2}x, \qquad w^{(0)} = 0 \tag{9–4.8}$$

where $u^{(0)}$, $v^{(0)}$, and $w^{(0)}$ are the components of $\mathbf{v}^{(0)}$ in the x, y, and z directions, respectively. In view of the spherical boundaries involved it is convenient to employ spherical coordinates (r, θ, ϕ), where $x = r\cos\theta$, $y = r\cos\phi\sin\theta$, and $z = r\sin\phi\sin\theta$. Expressed in these coordinates, Eq. (9–4.8) becomes

$$\begin{aligned} v_r^{(0)} &= qr\cos\phi\sin\theta\cos\theta \\ v_\theta^{(0)} &= \tfrac{1}{2}qr\cos\phi(\cos^2\theta - \sin^2\theta) \\ v_\phi^{(0)} &= \tfrac{1}{2}qr\sin\phi\cos\theta \end{aligned} \tag{9–4.9}$$

where $v_r^{(0)}$, $v_\theta^{(0)}$, and $v_\phi^{(0)}$ are the appropriate components of $\mathbf{v}^{(0)}$.

The velocity field $\mathbf{v}$, which is required to estimate the apparent viscosity, consists of two parts: the undisturbed field $\mathbf{v}^{(0)}$, and the disturbance $\mathbf{v}^{(1)}$ due to the presence of the particles. Because of the linearity of the creeping motion equations, the resultant field may be written as

$$\mathbf{v} = \mathbf{v}^{(0)} + \mathbf{v}^{(1)}, \qquad p = p^{(0)} + p^{(1)}$$

At the solid sphere of radius a, the field $\mathbf{v}^{(1)}$ must cancel the initial field $\mathbf{v}^{(0)}$

in order to satisfy the no-slip condition, $\mathbf{v} = \mathbf{0}$. At the boundary of the fluid envelope of radius b, we require that the normal component of the perturbation velocity $\mathbf{v}^{(1)}$ vanish in order to confine the disturbance to the interior of the cell. The condition of no friction on the envelope boundary due to the disturbance corresponds to the vanishing of the tangential stress components[35] $\Pi_{r\theta}^{(1)}$ and $\Pi_{r\phi}^{(1)}$. The complete boundary value problem is expressed as follows:

$$\mu_0 \nabla^2 \mathbf{v}^{(1)} = \nabla p^{(1)} \tag{9–4.10}$$

$$\nabla \cdot \mathbf{v}^{(1)} = 0 \tag{9–4.11}$$

$$\left.\begin{aligned} v_r^{(1)} &= -qr\cos\phi\sin\theta\cos\theta \\ v_\theta^{(1)} &= -\tfrac{1}{2}qr\cos\phi(\cos^2\theta - \sin^2\theta) \\ v_\phi^{(1)} &= -\tfrac{1}{2}qr\sin\phi\cos\theta \end{aligned}\right\} \text{ at } r = a \tag{9–4.12}$$

and

$$\left.\begin{aligned} v_r^{(1)} &= 0 \\ \Pi_{r\theta}^{(1)} &= \mu_0\left(\frac{\partial v_r^{(1)}}{r\partial\theta} + \frac{\partial v_\theta^{(1)}}{\partial r} - \frac{v_\theta^{(1)}}{r}\right) = 0 \\ \Pi_{r\phi}^{(1)} &= \mu_0\left(\frac{\partial v_r^{(1)}}{r\sin\theta\,\partial\phi} + \frac{\partial v_\phi^{(1)}}{\partial r} - \frac{v_\phi^{(1)}}{r}\right) = 0 \end{aligned}\right\} \text{ at } r = b \tag{9–4.13}$$

Lamb's general solution of the creeping motion equations in spherical coordinates, Eqs. (9–2.4) and (9–2.5), may be employed to determine $\mathbf{v}^{(1)}$. A closed solution is obtained by setting $\chi_n = 0$ and retaining harmonics of orders -3 and $+2$. The solution takes the form

$$\begin{aligned} v_r^{(1)} &= \left(6Ar^3 + 2Cr + \frac{6E}{r^2} - \frac{3G}{r^4}\right)\cos\phi\sin\theta\cos\theta \\ v_\theta^{(1)} &= \left(5Ar^3 + Cr + \frac{G}{r^4}\right)\cos\phi(\cos^2\theta - \sin^2\theta) \\ v_\phi^{(1)} &= \left(-5Ar^3 - Cr - \frac{G}{r^4}\right)\sin\phi\cos\theta \end{aligned} \tag{9–4.14}$$

$$p^{(1)} = \left(42Ar^2 + \frac{12E}{r^3}\right)\mu_0\cos\phi\sin\theta\cos\theta \tag{9–4.15}$$

The constants are determined from the boundary conditions. They are given below in terms of the parameters $\gamma = a/b$ and $\omega = (10 + 4\gamma^7)/[10(1 - \gamma^{10}) - 25\gamma^3(1 - \gamma^4)]$:

$$\begin{aligned} A &= -\frac{5q}{2a^2}\left(\frac{\gamma^7}{10 + 4\gamma^7}\right)\omega \\ C &= \frac{5q}{4}\left(\frac{4 + 10\gamma^7}{10 + 4\gamma^7}\right)\omega - \frac{q}{2} \\ E &= -\frac{5qa^3}{12}\omega \\ G &= -\frac{5a^5}{10 + 4\gamma^7}\omega \end{aligned} \tag{9–4.16}$$

If $b \to \infty$, then $\gamma \to 0$ and $\omega \to 1$. In this limiting case, the preceding solution reduces to a form equivalent to that originally obtained by Einstein —cf. Eq. (9–3.2)—for the perturbed field due to the presence of a sphere in a dilatant field which extends to infinity. Einstein, however, employed the following boundary condition at $b = \infty$:

$$v_r^{(1)} = v_\theta^{(1)} = v_\phi^{(1)} = 0 \tag{9–4.17}$$

rather than the "free-surface" condition expressed by Eq. (9–4.13). Simha's treatment, which uses the same boundary condition (9–4.17) at $r = b$, also reduces to the Einstein solution when $b \to \infty$. As noted later, however, the energy dissipation for the two cell models is not the same—even in the limit where $b \to \infty$.

The energy dissipated per unit time within a cell may be computed via the surface integral in Eq. (9–1.5). The integration in this case need be extended over the outer surface only, since $\mathbf{v} = \mathbf{0}$ at the particle surface. The field $\mathbf{v} = \mathbf{v}^{(0)} + \mathbf{v}^{(1)}$ required in the integral represents the entire motion. Evaluation of the integral yields

$$E = \frac{4}{3}\pi b^3 \mu_0 q^2 + \frac{4}{3}\pi b^3 \mu_0 q \left(\frac{24}{5} b^2 A - \frac{36E}{5b^3} + 2C + \frac{6G}{b^5}\right) \tag{9–4.18}$$

This may be expressed in terms of the parameter $\gamma = a/b$ and the volume of a cell, $V = (4/3)\pi b^3$, as follows:

$$E = Vq^2\mu_0(1 + 5.5\phi\psi) \tag{9–4.19}$$

where

$$\phi = \gamma^3 = \left(\frac{a}{b}\right)^3 \tag{9–4.20}$$

and

$$\psi = \frac{4\gamma^7 + 10 - (84/11)\gamma^2}{10(1 - \gamma^{10}) - 25\gamma^3(1 - \gamma^4)} \tag{9–4.21}$$

It is assumed that since $v_r^{(1)} = 0$ at $r = b$, the perturbed field $\mathbf{v}$ experiences the same average rate of shear as the undisturbed field $\mathbf{v}^{(0)}$. The energy dissipation for the unperturbed flow in a unit cell is $E^{(0)} = Vq^2\mu_0$ since the interaction factor ψ reduces to unity as $\gamma \to 0$. From Eq. (9–1.28) we then obtain

$$\frac{\mu}{\mu_0} = \frac{E}{E^{(0)}} = \frac{Vq^2\mu_0(1 + 5.5\phi\psi)}{Vq^2\mu_0} = 1 + 5.5\phi\psi \tag{9–4.22}$$

Here, we have assumed that the domain of integration is the unit cell rather than the apparatus volume.

Table 9–4.1 gives values for several of the variables in Eq. (9–4.22).

Note that, at low concentrations, Eq. (9–4.22) reduces to Eq. (9–2.17), whose significance we have already discussed. The table is not extended beyond $\phi = 0.50$ because the loose packed condition for a bed of uniform spheres occurs at $\phi \approx 0.52$. It seems clear that at concentrations approaching this, friction due to interparticle contacts may become very important, in

TABLE 9–4.1
EFFECT OF CONCENTRATION ON VISCOSITY—FREE SURFACE MODEL

Solids Concentration, Volume Fraction ϕ	Interaction Factor ψ	Relative Viscosity μ/μ_0
0	1.000	1.000
0.05	1.022	1.281
0.10	1.100	1.605
0.15	1.208	1.997
0.20	1.351	2.486
0.25	1.538	3.115
0.30	1.783	3.942
0.35	2.110	5.062
0.40	2.555	6.621
0.45	3.176	8.861
0.50	4.071	12.20

which case the resistance to shear will be much larger than predicted on the basis of purely hydrodynamic considerations.

Kynch[28] developed a mathematical treatment of the viscosity of dilute suspensions of particles, and later applied it to more concentrated systems.[29] In consequence of the relation

$$\nabla^2 p = 0 \tag{9–4.23}$$

there is a simple particular integral of the equations of motion of the form

$$\mathbf{v} = \frac{\mathbf{r}p}{2\mu_0} \tag{9–4.24}$$

where $\mathbf{r}$ is a position vector. Thus we may write

$$\mathbf{v} = \frac{\mathbf{r}p}{2\mu_0} + \mathbf{u} \tag{9–4.25}$$

where the field $\mathbf{u}$ satisfies the equations

$$\nabla^2 \mathbf{u} = 0 \tag{9–4.26}$$

$$\begin{aligned} -2\mu_0 \nabla \cdot \mathbf{u} &= 3p + \mathbf{r} \cdot \nabla p \\ &= 3p + r\frac{\partial p}{\partial r} \end{aligned} \tag{9–4.27}$$

If $\mathbf{v}_b$ is the velocity on any of the boundaries, the boundary condition imposed on $\mathbf{u}$ is

$$\mathbf{u} = \mathbf{v}_b - \frac{\mathbf{r}p}{2\mu_0} \quad \text{on the boundary} \tag{9–4.28}$$

The solution of Eq. (9–4.26) employed by Kynch involves a Green's function $G_P(\mathbf{r})$, satisfying the equation $\nabla^2 G = 0$, vanishing on the boundaries, and reducing to the form $G = (4\pi R)^{-1}$ near the point P, where R is the distance from P. The solution of Eq. (9–4.26) is then

$$\mathbf{u}_P = \int \left(\mathbf{v}_b - \frac{\mathbf{r}p}{2\mu_0} \right) \frac{\partial G_P}{\partial n} \, dS \qquad (9\text{–}4.29)$$

The integration extends over all the elements dS of the boundaries; $\partial G_P/\partial n$ is the derivative of G normal to these boundaries into the fluid. Similarly, because of Eq. (9–4.23), the pressure is given by

$$p_P = \int p \frac{\partial G_P}{\partial n} \, dS \qquad (9\text{–}4.30)$$

For any prescribed $\mathbf{v}_b$, Kynch's method of solution requires that we express $\mathbf{u}$ in terms of p, using Eq. (9–4.29), and then adjust p so that Eq. (9–4.27) is satisfied. This adjustment is accomplished by a method of successive approximations. It should, however, be noted that no matter where the origin is chosen, the term in $\mathbf{r}p$ in Eq. (9–4.25) will become indefinitely large as we move away from the origin. Kynch overcomes this difficulty by noting, in the usual apparatus, that we do not deal with an infinite medium, but rather with one possessing finite boundaries. The foregoing equations are then taken to refer to the *additional* pressure p and *additional* velocity $\mathbf{v}$ due to the presence of the spheres, the velocity field due to the apparatus boundary being added in separately. This, in effect, amounts to a *first reflection* approximation.

In view of the large number of particle boundaries present, the Green's function is advantageous because the total velocity is expressed as a sum of contributions from each boundary, due to the motion of each particle. It is only possible to establish the Green's function readily, however, if each particle is considered to be a point disturbance. This is equivalent to assuming that the particle-to-wall area ratio is very small, a postulate we have seen to be characteristic of the Einstein-type treatment.

Kynch's treatment gives rise to a set of simultaneous equations containing numerical factors as functions of the positions of the particles. Solution of these equations yields the fluid velocity and pressure fields. For example, with the velocity field given by Eq. (9–4.8), the extra pressure near each particle has the form:

$$p = \frac{5\mu_0 q}{a^2} \left(A_1 + A_2 \frac{a^5}{r^5} \right) xy \qquad (9\text{–}4.31)$$

This field satisfies $\nabla^2 p = 0$, and constitutes an elementary extension of the pressure existing near a single sphere. The center of the particle is chosen as the origin. The pressure at the particle surface is

$$p = \frac{5\mu_0 q}{a^2} Axy, \qquad A = A_1 + A_2 \qquad (9\text{–}4.32)$$

Utilizing this expression for the pressure, the velocity field can be deduced by substitution into Eq. (9–4.29). In order to calculate the constant A, it is necessary to make a further important assumption; namely, that the value of A for any particle is independent of the position of the other particles. In fact, this assumption is made everywhere up to the walls of the apparatus containing the flow. This is the same basic assumption made in the Einstein-type derivations; namely, that the shape of the containing vessel and the location of the particle within it have no effect on the rate of energy dissipation. In dilute solutions, the result obtained is identical to Einstein's viscosity formula, Eq. (9–2.15), as might be expected from the similarity of the basic assumptions employed.

Kynch's detailed mathematical treatment is rather complicated, involving methods successfully employed to compute the potential of a point electric charge in the presence of a large number of earthed conducting spheres. It is assumed that the zone around each particle is divided into a cavity immediately adjacent to it, occupied by no particles, and an isotropic continuum outside of the cavity. The value of the cavity radius is taken as $b = 2a$, the collision diameter of the particles, until the mean distance b_0 between particles becomes less than $2a$, in which case it is assumed that $b = b_0$. Viscosity is computed using this "statistical" model by first calculating the velocity, which is accurately determined near a particle surface. Details of the final calculations are not given in the original paper but some numerical values are reported. These are reproduced in Table 9–4.2.

Kynch also discusses Simha's[48] cage model and states that his own method can be employed to give results very similar to Simha's when using the same basic assumptions. The cell-type models of Simha[48] and Happel[16] are intended to provide a reasonable approximation to the velocity distribution inside the unit cell. This, in turn, is used to calculate the rate of viscous dissipation of energy and hence to determine the effective viscosity. Kynch developed his statistical treatment to calculate the fluid velocity as accurately as possible near the particle boundaries. On the other hand, he is of the opinion that it is more appropriate to calculate the effective viscosity from the rate of shear at the walls. It does not seem possible to reconcile the conceptual differences between the two theories. Moreover, it is equally unclear that the insertion of a large sphere in the suspension will correspond to the presence of a wall.

TABLE 9–4.2
EFFECT OF CONCENTRATION ON VISCOSITY—KYNCH MODEL

Solids Concentration, Volume Fraction ϕ	Relative Viscosity μ/μ_0
0	1
0.05	1.16
0.10	1.38
0.15	1.70
0.25	2.72

The theories of Kynch and Simha, since they reduce to the Einstein relationship in dilute systems, predict viscosities lower than the Happel relationship. This effect continues up to moderate concentrations; for example, at $\phi = 0.20$, Happel predicts $\mu/\mu_0 = 2.486$ whereas the value predicted by Kynch is about 15 per cent lower. The value predicted by Simha is over 30 per cent lower, when Eq. (9–4.4) is used to evaluate γ and ϕ_{max} is taken as 0.74, corresponding to hexagonal packing, as Simha suggests. Kynch does not give any values for viscosity above 25 per cent concentration of solids. Simha's relationship, however, shows a very rapid rise in suspension viscosity as the concentration increases, so that the predicted relative viscosity becomes equal to that predicted by Happel at $\phi = 0.38$, corresponding to $\mu/\mu_0 = 5.7$. Comparison of these relationships with experimental data is deferred to a later section in this chapter, which also includes discussion of several empirical relationships (see Section 9–6).

9-5 Nonspherical and Nonrigid Particles

For shapes other than spheres, theory, owing to the complexities which arise, has not progressed beyond treatments for dilute systems. Thus, when a dilute suspension of particles is subjected to shear, the particles undergo rotation and translation. If deformable, they will also undergo changes in shape. Recall, too, that the rate of energy dissipation incurred by a non-spherical particle depends on the orientation of the particle relative to the principal axes of shear. If the particle rotates, this rate will vary with time. Translational movements of a freely suspended particle in a shear field may cause even uniform-sized spheres to collide with one another. In the case of extremes in shape these collision effects may be enhanced. In certain instances aggregates may be formed or particles may coalesce. Brownian motion effects may serve to further complicate the problem.

Jeffery[21] investigated the motion of a rigid ellipsoidal particle in a shear flow on the basis of the creeping motion equations. For a simple shear flow in the xy plane, $u^{(0)} = qy$, $v^{(0)} = 0$, $w^{(0)} = 0$; it is now necessary to consider both rotational and dilational contributions to the flow field, in contrast to the case for a sphere. The stresses at the ellipsoid surface are found to be equivalent to the sum of two couples. One tends to make an elongated particle (prolate ellipsoid of revolution) adopt the same rotation as the undisturbed flow, so that the particle axis tends to lie parallel to the z axis. The other tends to make the particle set its axis within the xy plane. The combined effect in the absence of external couples is that, for intermediate positions, the ends of the particle describe a spherical ellipse, given in spherical coordinates by the following expression:

$$\tan^2 \theta = \frac{a^2 b^2}{k^2(a^2 \cos^2 \phi + b^2 \sin^2 \phi)} \tag{9–5.1}$$

Here, for an ellipsoid of revolution with the surface of the particle described by

$$\frac{x^2}{a^2} + \frac{y^2}{b^2} + \frac{z^2}{c^2} = 1 \tag{9–5.2}$$

we take $b = c$. k is a constant of integration which depends on the particle orientation: when $k = 0$, the major axis (a axis) rotates in the xy plane; when $k = \infty$, the particle remains parallel to the z axis. For any intermediate value of k, the particle also spins axially on the x axis, at an angular velocity

$$\omega = \frac{q}{2} \cos \theta$$

For $k = \infty$, this reduces to $\omega = q/2$; for $k = 0$, the spin ω is zero. For rigid spheres, $a/b = 1$, the angular velocity becomes $\omega = q/2$. Corresponding to $a = 0$, it is found that a circular disk may take any position in which its faces are entirely composed of streamlines of the undisturbed fluid motion. It thus moves through the fluid edgewise with one axis along the x direction and the other making an arbitrary angle with the z direction. Letting $b = c$ approach zero gives the case of a thin rod, which it is found can set itself in any direction in a plane parallel to the xz plane. The investigation, however, reveals no tendency on the part of an ellipsoid to set its axis in any particular direction with regard to the undisturbed motion of the fluid.

Effects like this, in which no preferred orientations or positions exist, have previously been noted to be consequences of the creeping motion equations. Thus, for the case of a translating ellipsoid (see Section 5–11), it was found that no couple is exerted on the ellipsoid, regardless of its orientation relative to its direction of motion through the fluid. Similarly, a sphere settling axially in an eccentric position within a cylinder (see Section 7–3) experiences no force tending to move it in a radial direction.

To calculate the viscosity of a suspension of ellipsoidal particles, Jeffery assumes the ellipsoid to be surrounded by a large sphere whose center coincides with that of the ellipsoid. The disturbance created by the presence of the ellipsoid is assumed to vanish at the surface of this sphere, and the method of reflections is employed to compute the reflection of this disturbance from the surrounding spherical envelope. The additional energy dissipation due to the presence of the particle, $E^* = E - E^{(0)}$, is then computed by equating it to the work done by the additional stresses acting over the surface of the large sphere. Jeffery notes that the additional energy dissipation calculated from the *original* disturbance created by the ellipsoid is only one-fifth that obtained by including the field represented by the reflection from the surrounding spherical envelope, even though the radius of the latter is finally assumed to extend to infinity. A similar variation in energy dissipation was noted and discussed following Eq. (9–4.17) in connection with the development of the free surface model. In that case a difference in boundary conditions rather than shape was the cause. Jeffery obtains a complicated expression for energy dissipation which, as expected, depends upon the constant of integration, k. The motion which gives the minimum average dissipation of energy corresponds to $k = \infty$ for a prolate spheroid. In this case it spins about its axis, which lies in the z direction. The case of minimum energy dissipation for an oblate spheroid corresponds to $k = 0$.

Jeffery also calculated the maximum and minimum energy dissipation rates. Since, from Eq. (9–1.29), $\mu/\mu_0 = 1 + (E^*/E^{(0)})$, and since $E^{(0)} = Vq^2\mu_0$, our knowledge of the additional dissipation rates permits us to calculate the suspension viscosity,

$$\frac{\mu}{\mu_0} = 1 + \nu\phi \tag{9–5.3}$$

where ν is a factor which depends on the geometry of the particular ellipsoid involved.

TABLE 9–5.1
VISCOSITY CONSTANT FOR PROLATE SPHEROIDS

Ellipticity $\epsilon = (a - b)/a$	Minimum ν	Maximum ν
0	2.5	2.5
0.1	2.431	2.540
0.2	2.361	2.586
0.3	2.295	2.645
0.4	2.232	2.719
0.5	2.174	2.819
0.6	2.120	2.958
0.7	2.073	3.170
0.8	2.035	3.548
0.9	2.010	4.485
1.0	2.000	∞

TABLE 9–5.2
VISCOSITY CONSTANT FOR OBLATE SPHEROIDS

Ellipticity $\epsilon = (b - a)/b$	Minimum ν	Maximum ν
0	2.5	2.5
0.1	2.464	2.582
0.2	2.426	2.683
0.3	2.388	2.818
0.4	2.348	3.003
0.5	2.306	3.267
0.6	2.262	3.670
0.7	2.216	4.354
0.8	2.168	5.744
0.9	2.116	9.960
1.0	2.061	∞

The maximum and minimum values of ν for spheroids of different shapes are given in Tables 9–5.1 and 9–5.2. They are reported by Jeffery in terms of the "ellipticity" ϵ of the meridian section of the particle, defined to be the difference of the greatest and least diameters divided by the greatest. Thus $\epsilon = (a - b)/a$ for a prolate spheroid and $\epsilon = (b - a)/b$ for an oblate spheroid.

Jeffery hypothesized that ellipsoidal particles would ultimately tend to adopt that orbital motion which corresponds to the least dissipation of energy. Taylor[52], in experiments on a single aluminum spheroid immersed in a water-glass solution between the two vertical cylinders of a Couette apparatus, found the type of periodic rotation about a vertical axis predicted by Jeffery. This gradually passed over into a definite orientation, with the axis vertical in the case of a prolate spheroid and with an equatorial diameter vertical in the case of an oblate spheroid. About one hundred periods were required for the motion to reach these final states corresponding to minimum energy dissipation.

Eisenschitz[10], on the other hand, assumed that the behavior of a suspension of ellipsoids corresponds to the case that originally the axes are distributed at random, and that throughout the subsequent motion each particle retains its original orbit. Mason and Manley[33], in experimental investigations, found a distribution of orbits which lies between those computed on the basis of Jeffery's and Eisenschitz' assumptions.

In all these calculations Brownian motion has been neglected. If one is concerned with very small particles, such as macromolecules, Brownian motion will be the main factor influencing particle orientation. Brownian motion increases viscosity by bringing the particles into a more disorientated position with respect to fluid motion. Such orientations relative to the principal axes of shear do not correspond to minimization of the energy dissipation. Frisch and Simha[13] review a number of treatments involving the case of small elongated prolate spheroids subject to Brownian motion. In all these expressions the specific viscosity is, for a given ratio of particle dimensions, proportional to the total volume, as in Einstein's equation. For ellipsoids of revolution of major axis a_1 and minor axis a_2 (axial ratio $p = a_1/a_2$) Kuhn and Kuhn[25] developed the following approximate expressions:

$$[\mu] = 2.5 + \frac{32}{15}\left(\frac{1}{p} - 1\right) - 0.628\,\frac{(1/p) - 1}{(1/p) - 0.075} \qquad (0 < p < 1) \qquad (9\text{–}5.4a)$$

$$[\mu] = 2.5 + 0.4075(p - 1)^{1.508} \qquad (1 < p < 15) \qquad (9\text{–}5.4b)$$

$$[\mu] = 1.6 + \frac{p^2}{5}\left[\frac{1}{3(\ln 2p - 1.5)} + \frac{1}{\ln 2p - 0.5}\right] \qquad (p > 15) \qquad (9\text{–}5.4c)$$

where $[\mu]$ is the intrinsic viscosity, defined in Eq. (9–3.12). The effect of Brownian movement on viscosity is very large. Thus, for prolate spheroids, on Jeffery's assumption $[\mu] = 2.0$ (for spheres, $[\mu] = 2.5$). If the Eisenschitz hypothesis is adopted with an axis ratio of 50 to 1, the intrinsic viscosity is 3.97 for no Brownian motion and 178 for complete Brownian motion according to Eq. (9–5.4).

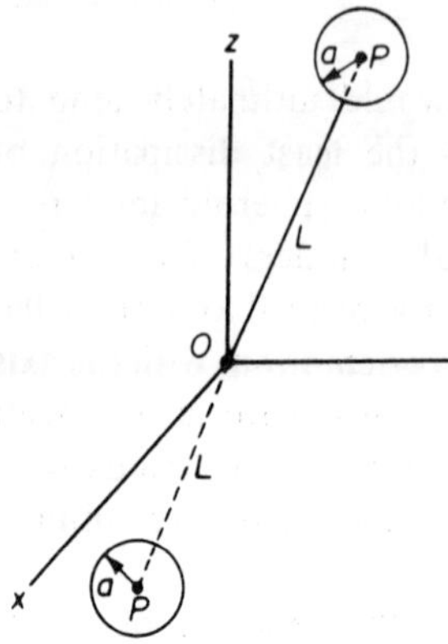

Figure 9-5.1. Dumbbell model for viscosity.

Another basic approach to the effects of particle shape on viscosity has been developed in some detail by Simha[49] using a dumbbell model (two rigidly connected spherical particles), originally considered by Kuhn[26] in the latter's treatment of intrinsic viscosity. Simha shows that, for the calculation of energy dissipation, it can be assumed that the translational contribution will be much larger than that due to dilation or rotation—see Eq. (9–1.21). On the basis of this assumption, Simha concludes that for a single dumbbell, Fig. 9–5.1, the energy dissipation may be written:

$$\left(\frac{dW}{dt}\right)_1 = \text{force on sphere } P \times \text{velocity of } P + \text{force on sphere } Q \times \text{velocity of } Q \quad (9\text{–}5.5)$$

Thus, $(dW/dt)_N$, the energy dissipated by the motion of N independent, infinitely separated suspended dumbbells must be proportional to $\mu_0 q^2 a L^2 N$. Since the volume fraction of suspended particles is $\phi = 2(\frac{4}{3}\pi a^3 N)/V$, where V is the volume in which the N particles are suspended, the dissipation per unit volume is then proportional to $\mu_0 q^2 (L/a)^2 \phi$. On the assumption of a completely random orientation due to Brownian movement, the following relationship is ultimately obtained:

$$\frac{\mu}{\mu_0} = 1 + \frac{3}{2}\left(\frac{L}{a}\right)^2 \phi \quad (9\text{–}5.6)$$

This result, obtained on the basis of Stokes' law of resistance, must be modified at higher concentrations by superposition of translational contributions owing to the presence of other particles, as was done in the sedimentation problems discussed in the previous chapter (Section 8–3).

The exact analysis for the first-order interaction effect is then taken into account in the following equation:

$$\frac{\mu}{\mu_0} = 1 + \frac{3}{2}\left(\frac{L}{a}\right)^2 \phi + \frac{87}{50}\left(\frac{L}{a}\right)^4 \phi^2 \quad (9\text{–}5.7)$$

Theoretical results for the viscosity of spherical particles in the form of a power series in ϕ, for example, Eqs. (9–3.11), (9–3.14), and (9–3.15), can be generalized to a form which may then be compared with Eq. (9–5.7). Thus, we write

$$\begin{aligned} \mu_{sp}/\phi &= [\mu] + a_2\phi + a_3\phi^2 + \cdots \\ &= [\mu] + k_1[\mu]^2\phi + k_2[\mu]^3\phi^2 + \cdots \end{aligned} \tag{9–5.8}$$

where $\mu_{sp} = (\mu/\mu_0) - 1$ is the specific viscosity and $[\mu]$ is the intrinsic viscosity, defined by Eq. (9–3.12). The intrinsic viscosity is equal to 2.5 for spheres according to the Einstein relationship. Thus for Eq. (9–3.11), $k_1 = 2.26$; for Eq. (9–3.14), $k_1 = 1.18$; and for Eq. (9–3.15), $k_1 = 1.20$. Some investigators[13] have suggested that Eq. (9–5.8) may be applicable to suspensions of nonspherical particles and polymer solutions.

Simha's investigation of dumbbell-shaped particles provides interesting confirmation of this equation. Thus, if Eq. (9–5.7) is used to evaluate the constants in Eq. (9–5.8), we find for dumbbell-shaped particles that $k_1 = 58/75 = 0.77$. Compact particles, such as spheres, thus give higher values for k_1 than are obtained for more extended shapes. Frisch and Simha[13] give more detailed results for other similar investigations.

Equation (9–5.7) is subject to the usual limitation of treatments involving single particles, which do not take the wall effect into consideration; namely, the question as to whether the "Stokes" energy dissipation is uniquely determined without specifying the shape of the outer boundary and position of the particle considered within the boundary. There is also the question of the applicability of the Smoluchowski linear approximation to allow for interaction of the two spheres comprising the dumbbell when the former are close together. Also, of course, inertial effects are neglected when the creeping motion equations are employed.

Debye and Bueche[8] have treated intrinsic viscosity of polymer molecules in solution by means of a generalization of Einstein's theory for spheres. For the coiled polymer molecule a sphere is substituted which hinders the liquid flow through its interior only to a degree, depending on the average density in space of the polymer molecule in solution. The amount of shielding of the liquid flow which is introduced in this way determines the exponent in the customary empirical exponential relation between intrinsic viscosity and molecular weight, M, namely,

$$[\mu] = AM^n \tag{9–5.9}$$

An alternative theory of intrinsic viscosity has been developed by Kirkwood and Riseman[24] on the basis of a random coil model consisting of a string of beads. Account is taken of the hydrodynamic interaction of the monomer units of the molecule and of inhibited flow through the chain. This theory leads to results which are qualitatively similar to those of Debye. Data are not presently available to decide which theory comes closer to agreement with the facts.

Taylor[53] extended Einstein's analysis of the viscosity of a suspension of rigid spherical particles to the case of small immiscible fluid spheres. The viscous stresses exerted by the sheared medium surrounding the particles bring about circulation currents within them, provided that no formation of a rigid film around the particles occurs to reduce the transmission of tangential stresses to the interior of the particles. If it is also assumed that the interfacial tension is sufficiently great to preserve a spherical shape, the following relationship is obtained for the viscosity μ of very dilute emulsions:

$$\frac{\mu}{\mu_0} = 1 + 2.5\left[\frac{\mu' + (2/5)\mu_0}{\mu' + \mu_0}\right]\phi \tag{9–5.10}$$

where μ' is the viscosity of the suspended liquid particles and μ_0 the viscosity of the continuous phase. If μ'/μ_0 is large, Eq. (9–5.10) reduces to the Einstein relationship, Eq. (9–2.15). On the other hand, if this ratio is very small, as in the case of gas bubbles, the factor multiplying ϕ becomes unity.

Similarly, the effect of boundary slip on viscosity of suspensions of rigid particles may be developed following the general procedure of Einstein[9] but substituting the slip boundary condition at the surface of the sphere. In this case the result obtained is

$$\frac{\mu}{\mu_0} = 1 + 2.5\left(\frac{\beta r + 2\mu_0}{\beta r + 5\mu_0}\right)\phi \tag{9–5.11}$$

where β is the coefficient of sliding friction at the boundary of each sphere. For complete slip, $\beta = 0$ and, as in the case of very low viscosity suspended particles, the Einstein factor is unity.

In the case of fluid particles the circulation currents inside the drops have been calculated theoretically by Bartok and Mason[2].

9-6 Comparison with Data

We shall consider the agreement with experimental data (or lack thereof) of the various theories discussed, in the same order as they were presented, beginning with the dilute system treatment. Philippoff's book[40] reviews experimental data obtained before 1942 in attempts to verify the Einstein relationship. The investigations cited show values of the constant varying from 2.5 to 4.75. Results of more recent investigators show similar discrepancies, indicating the difficulty in obtaining accurate experimental data in the very dilute region. Thus, Cheng and Schachman[6], experimenting with dilute suspensions of polystyrene latex particles, concluded that the Einstein formula with a constant of 2.5 was valid up to volumetric concentrations of 2 per cent. On the other hand, Ward and Whitmore[55] cite experimental data to show that, with uniform-sized spheres, the Einstein constant assumes a value of 4.0, but with a particle size ratio of 3:1 the proportionality constant will be reduced to 1.9.

Segré and Silberberg[47] made viscosity determinations with suspensions of spherical particles of substantially identical size in a capillary viscometer, obtaining values of the Einstein constant below 1.5. The results are believed due to particle segregation effects, discussed in Section 8–3. Thus, it appears that even for dilute suspensions there is no absolute agreement on the validity of the constant 2.5.

A very complete survey and discussion of the many equations which have been proposed for describing the relationship between relative viscosity and concentration over a wide range has been published by Rutgers[42], bringing up to date the earlier surveys by Philippoff[40] and Frisch and Simha[13]. Rutgers' selection of an average curve to represent the relationship between viscosity and concentration is more or less arbitrary, but he concludes that one of the most useful formulas is that given by Mooney[37], which is discussed later—see Eq. (9–6.4). At high solids concentrations, Rutgers' correlation shows a somewhat more rapid rise in viscosity than would be predicted by the free-surface relationship, which takes only hydrodynamic interaction effects into consideration.

Happel[16] has reviewed available data for suspensions of spheres of uniform size at higher concentrations, and finds that they may be reasonably represented by the free surface model, Eq. (9–4.22) and Table 9–4.1. Figure 9–6.1 shows a comparison of the Happel equation, as well as the relationships proposed by Simha and Kynch, with the data selected. The Simha

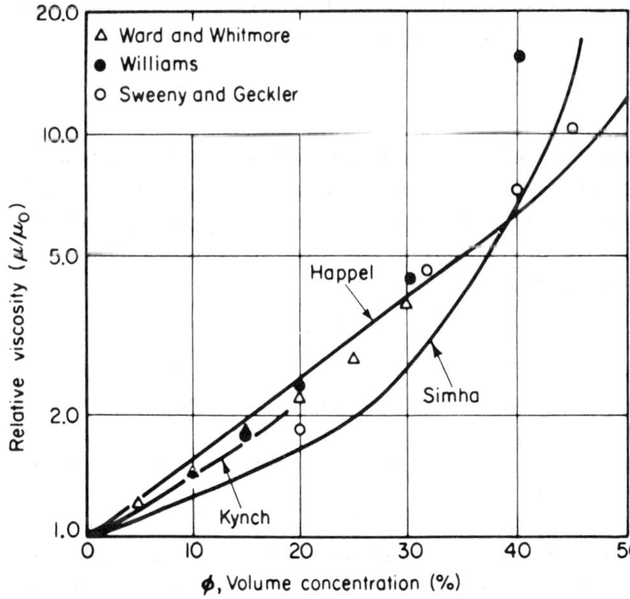

Figure 9-6.1. Comparison of theoretical viscosity relationships with data for uniform spheres.

formula plotted is taken from Eqs. (9–4.2) and (9–4.4). The factor γ was evaluated by assuming $\phi_{max} = 0.74$, corresponding to hexagonal packing. The graphical presentation of Kynch's results is taken from Table 9–4.2. Simha's relationship appears to predict values of relative viscosity lower than experimental data in the intermediate concentration range, but agrees with the rapid rise of viscosity exhibited by experimental data for concentrations of solids over 50 per cent by volume.

A number of semiempirical relationships have been developed to predict the concentration dependence of the viscosity of suspensions of spherical particles in newtonian fluids. Many of these are expressible in the form of a power series [cf. Eq. (9–5.8)]:

$$\mu_r = 1 + \alpha_0\phi + \alpha_1\phi^2 + \alpha_2\phi^3 + \cdots \tag{9–6.1}$$

where μ_r the relative viscosity, μ/μ_0. In these developments, α_0 is invariably taken as the Einstein constant of 2.5, more generally referred to as *intrinsic viscosity* in Eq. (9–5.8). Obviously the value of α_1, as determined from experimental data, will depend strongly upon the value chosen for α_0. A review by Ford[12] summarizes some of these relationships and observes that better agreement with data is obtained by expressing the power series in terms of the relative fluidity $1/\mu_r$ rather than in terms of the relative viscosity. On this basis, we have

$$\frac{1}{\mu_r} = 1 - 2.5\phi \tag{9–6.2}$$

in place of the usually quoted Einstein relationship, which, if the quadratic term in ϕ in Eq. (9–6.2) is negligible, yields

$$\mu_r = \frac{1}{1 - 2.5\phi} = 1 + 2.5\phi + 6.25\phi^2 + \cdots \tag{9–6.3}$$

This coefficient of 6.25 compares reasonably well with comparable values reported in the literature. It also appears to be in conformity with data on small glass spheres as well as milk fat, latex, and asphalt emulsions.

Another type of relationship employed to correlate data at higher concentrations is that due to Mooney[37],

$$\ln \mu_r = \frac{2.5\phi}{1 - c\phi} \tag{9–6.4}$$

where c is a factor to allow for hydrodynamic interaction, to be determined experimentally. The constant 2.5 is again based on the Einstein equation for dilute solutions. This can be expanded in a form similar to Eq. (9–6.1) to give

$$\mu_r = 1 + 2.5\phi + (3.125 + 2.5c)\phi^2 + \cdots \tag{9–6.5}$$

Mooney sets the value of c for monodisperse systems roughly between the limits 1.35 and 1.91, the reciprocals of the volume fractions of uniform spheres in close packing and cubical packing, respectively. Saunders[45] compares data for monodisperse polystyrene latexes with Mooney's equation and finds good agreement, with the interaction coefficient c varying from 1.357 (with particle diameters of 990 Å) to 1.118 (with particle diameters of 8710 Å). Sweeney and Geckler[50] correlated their data, plotted in Fig. 9–6.1, using Mooney's equation. They found that c varied from 1.000 to 1.470, the larger-sized particles giving the lower c values. In the case of Saunders' data, the apparent coefficient of the linear term in Eq. (9–6.5) was 2.642 with the 990 Å particles. Saunders argued that the constant 2.5 could be recovered by assuming an adsorbed layer of constant thickness of emulsifier surrounding each particle. Another explanation for these effects would be to assume that, with smaller particles, interaction effects persist even to the linear term in the empirical power series representation of μ_r.

In this connection a somewhat different empirical equation of the type $\ln \mu_r = f(\phi)$ appears to have merit, since it enables a different type of effect of the empirically established constant to be used. The simplest of these relationships is that proposed originally by Arrhenius[1], before Einstein's work:

$$\ln \mu_r = k\phi \tag{9–6.6}$$

In this case, k is not necessarily taken to be equal to the Einstein constant 2.5, and no constant involving the crowding or packing effect appears as in Mooney's equation. Such an additional constant would, of course, take into account that interparticle friction becomes important as concentration becomes high. Empirical introduction of such a constant will undoubtedly give better agreement with data.

An inspection of the data plotted in Fig. 9–6.1, where the ordinate is a logarithmic scale, indicates that, for certain data at least, a fairly good straight-line relationship of the form of Eq. (9–6.6) exists. If Eq. (9–4.22) is used to evaluate the proportionality constant over the range $\phi = 0.05$ to 0.30, the following relationship results:

$$\ln \mu_r = 4.58\phi \tag{9–6.7}$$

As the constant appears to depend on the relative particle size as related to viscometer dimensions, it is possible that other values of k ranging between the theoretical values of 2.5 and 5.5 could be obtained, depending on the system studied. Equations of the Arrhenius form are, of course, expandable as power series in ϕ. The higher-order coefficients of ϕ are then no longer independent, but depend rather on the value chosen for the linear constant. Thus, from Eq. (9–6.6), we obtain

$$\mu_r = 1 + k\phi + \frac{k^2}{2!}\phi^2 + \frac{k^3}{3!}\phi^3 + \cdots \tag{9–6.8}$$

Kunitz[27] proposed an empirical relationship for spheres which is in close agreement with Eq. (9–6.8), using $k = 4.15$. Johnson[22] applied the Kunitz relationship with success to describe the rheological behavior of beds in which particles were fluidized by liquids.

Mori and Ototake[38], in a review of previous theories and data, proposed a somewhat different functional relationship. They derive a relationship which extends an earlier investigation by Robinson[41] in which he considers that the specific viscosity at higher concentrations is proportional not only to the volume fraction of solids, but also to the reciprocal of the fraction of free liquid volume. In the special case where the suspended particles are spherical and of equal size, their formula reduces to the following, provided that the maximum attainable solids fraction is 0.52, which is the volume fraction of solids for cubical packing of spheres of equal size:

$$\mu_r = 1 + \frac{3}{(1/\phi) - (1/0.52)} \tag{9–6.9}$$

This formula agrees reasonably well with the observed values of several experimenters, which the authors compare graphically with their formula over a wide range of concentrations of spherical particles. Of course, at the lower limit of concentration, Eq. (9–6.9) is equivalent to using the power series type expansion with $k = 3.0$. Near the upper concentration limit of $\phi = 0.52$ it asymptotically approaches an infinite relative viscosity, which may be physically interpreted in terms of interparticle frictional effects. Values predicted for the range below $\phi = 0.4$ lie between those shown on Fig. 9–6.1 for the Happel and Simha relationships. At concentrations above $\phi = 0.4$ the predicted relative viscosity rises steeply.

Several investigators have derived semiempirical theories of the viscosity of suspensions based on similarities with sedimentation. Thus, the studies of Hawksley[19], based on Vand's[54] treatment of viscosity, lead to the relationship

$$\mu_r = \frac{(1 - \phi)^2}{U/U_0} \tag{9–6.10}$$

whereas Kynch[30], based on different reasoning, derives

$$\mu_r = \frac{1 - \phi}{U/U_0} \tag{9–6.11}$$

Here, U/U_0 is the ratio of the sedimentation velocity U of the suspension at the concentration ϕ to the settling velocity U_0 of an isolated particle. These relationships may be conveniently compared to the more usual plots of μ_r versus ϕ by using appropriate sedimentation data in the form of U/U_0 versus ϕ. The method selected here utilizes the consistent sedimentation[17] and viscosity[16] treatments represented by the free surface model, for these have been shown to be consistent with a considerable body of experimental data. Table 9–6.1 gives a comparison over a range of concentrations.

TABLE 9–6.1
RELATIONSHIP BETWEEN RELATIVE SEDIMENTATION VELOCITY AND RELATIVE VISCOSITY

Volume Fraction of Solids ϕ	Relative Settling Velocity* U/U_0	Relative Viscosity, $\mu_r = \mu/\mu_0$		
		$\frac{(1-\phi)^2}{U/U_0}$	$\frac{(1-\phi)}{U/U_0}$	Theory, Eq. (9–4.22)
0.00	1.000	1.00	1.00	1.00
0.05	0.453	1.99	2.10	1.281
0.10	0.3215	2.52	2.80	1.605
0.20	0.1773	3.60	4.50	2.486
0.30	0.0969	5.04	7.24	3.94
0.40	0.0529	6.80	11.4	6.62
0.50	0.02638	9.48	18.9	12.20

*See Table 8–4.1.

Hawksley's equation does indeed give evidence of a reasonable correspondence in the higher concentration range. In the low concentration range it must be stated that both sedimentation and viscosity data show some fluctuation, so that comparison with the theory expressed by the free surface model[16,17] may not be valid in many instances. Thus the self-consistent data of Cheng and Schachman[6] on sedimentation and viscosity of polystyrene latex suspensions confirm the Einstein relationship, whereas for sedimentation they obtain an empirical relationship with the limiting form

$$\frac{U}{U_0} = \frac{1}{1 + 4.06\phi} \tag{9–6.12}$$

From Eq. (9–6.12) we find for $\phi = 0.01$ that $U/U_0 = 0.962$. Thus the relative viscosity based on Hawksley's relationship, Eq. (9–6.10), would be $\mu_r = (0.99)^2/0.962 = 1.02$. From the Kynch relationship, Eq. (9–6.11), one would obtain $\mu_r = 0.99/0.962 = 1.03$. The Einstein equation for this case gives, of course, $\mu_r = 1.025$. Thus, it appears that in the dilute region these relationships may be of value, though there is not good agreement among data on sedimentation velocity at low concentrations.

In view of the apparent importance of particle size in the determination of intrinsic viscosity, it seems desirable to examine data which might be pertinent. The viscosity of aqueous solutions of sucrose has been accurately determined over a wide range of concentrations. Sucrose molecules represent a lower limit of size at which a continuum theory might be expected to apply. Einstein's original work, in fact, used data on sugar solutions as a method for estimating the size of the sugar molecule. He noted that, experimentally, the partial specific volume of sugar in solution is the same as for solid sugar, and assumed that the molecules would form a suspension approximating small spherical particles. He found an intrinsic viscosity equal to 4.0 instead of 2.5. Einstein explained this discrepancy by assuming that the sugar molecules present in solution limit the mobility of the water immediately adjacent, so that a quantity of water, whose volume is approximately one-half the volume of the sugar molecule, is bound on to the sugar molecule (4.0/2.5 = 1.6). An equally tenable explanation would appear to be that the correct value of the Einstein constant for such small particles may not be as low as 2.5.

Aside from the explanation, it is of interest to check as precisely as possible, from presently available data, the true value of the constant for very dilute sugar solutions. Jones and Tilley[23], in an investigation which extended down to concentrations of 0.004 volume per cent of sucrose, obtained a viscosity correlation which is equivalent to

$$\mu_r = 1 + 4.18\phi \tag{9–6.13}$$

Zakin[59], in a recent study, obtained the data on the viscosity of dilute sugar solutions tabulated in Table 9–6.2.

TABLE 9–6.2
VISCOSITY OF DILUTE SUCROSE SOLUTIONS

Volume per cent Sugar, $100\,\phi$	μ_r	$(\mu_r - 1)/\phi$	$(\ln \mu_r)/\phi$
0.655	1.030	4.5	4.46
1.262	1.063	5.0	4.86

Thus it would appear that for particles as small as sugar molecules (diameter $\approx$ 8–9 Å) a first-order constant greater than 4.0 is definitely established. The values of $(\ln \mu_r)/\phi$ lend experimental support to Eq. (9–6.7).

A further point of interest is the behavior of some concentrated sugar solutions as predicted by Eq. (9–6.7). Thus, consider the viscosity of a 20 weight per cent sucrose solution. This is given as $\mu_r = 1.941$ at 20°C by Swindells, *et al.*[51]. (Note, however, that their viscosity values extrapolated to low concentrations are not in exact agreement with the other sucrose data noted above.) Taking the specific volume of sugar as approximately 0.613 cc/gram we obtain $\phi = 0.133$, which corresponds to a value of $\mu_r = 1.84$ according to Eq. (9–6.7), within 5 per cent of the experimental value. At higher concentrations the theory predicts still lower values of viscosity as compared with the experimental, the discrepancy being about 22 per cent at a sugar concentration of 30 per cent by weight. With an irregularly shaped molecule like that of sucrose, one could expect interference between molecules at substantially less than the concentration corresponding to a packing of smooth spheres. It is felt that the excellent agreement up to 20 per cent by weight lends support to the idea that the viscosity of suspensions of very small particles can be explained hydrodynamically, rather than by the assumption of adsorbed layers of solvent on the particles.

It is of some practical interest that certain types of suspensions exhibit newtonian behavior, and that their viscosities can be reasonably well predicted on the basis of existing theoretical treatment, for example, Eq. (9–4.22), even though the particles are neither spherical nor uniform in size. Comparative values for two of these materials—human red blood corpuscles and rubber latexes—are shown in Table 9–6.3.

TABLE 9–6.3.
RELATIVE VISCOSITY OF NATURAL SUSPENSIONS

Particle Concentration, Volume per cent, 100 ϕ	Relative Viscosity, μ_r		
	Theory, Eq. (9–4.22)	Blood	Latex
20	2.49	3.3	2.5
30	3.94	4.4	4.0
40	6.62	5.8	5.5
50	12.2	7.5	7.5
60	25.8	9.0	27.0

The data on blood are taken from Coulter and Pappenheimer[7] on red corpuscles supended in brine. The authors state that if capillary viscometers of over 3 mm bore are employed, blood of less than 55 per cent hemocrit may be regarded as a newtonian fluid. The hemocrit is a standard centrifuging procedure which, according to Leeson and Reeve[32], results in 5 per cent of the packed cell volume separated from normal blood as due to trapped plasma. The concentrations shown in Table 9–6.3 are corrected so that they refer to the actual concentration of red blood cells. Much work has been done in the field of physiology pertaining to blood flow, starting with the classic work on this subject by Jean Louis Marie Poiseuille (1799–1869). A good summary is presented in J. F. Fulton, ed., *A Textbook of Physiology*, especially chapter 31, "Hemodynamics"[14].

The data on latex are taken from H. L. Fisher's *Chemistry of Natural and Synthetic Rubbers*[11]. Natural rubber contains about 35 per cent of rubber hydrocarbons together with a low percentage of nonrubber constituents. The hydrocarbon is insoluble in water and is dispersed in particles not greater than 3 microns in diameter, the average size being 0.5 micron; many are much smaller. The particles are either spherical or pear-shaped, resulting from the attachment of one to another.

9-7 Non-newtonian Behavior

Newtonian behavior of a suspension is taken to mean that if the entire suspension is treated as a continuum the stress will be proportional to rate of deformation, as in the case of a newtonian fluid. Even if the suspending fluid is newtonian in behavior it does not always follow that a suspension will also obey this relationship. If the particles are spherical and uniform in size, hydrodynamic treatments and data indicate that the suspension will be newtonian provided that particles are uniformly distributed and the suspension is not near the upper limit ($\phi = 0.52$) of loose-packed bed concentration.

As indicated in Section 9–5, changes in particle shape cause non-newtonian effects in the apparent viscosity of a suspension of small particles or macromolecules, even assuming that they are suspended in a liquid which exhibits newtonian flow properties. One of the simplest models that can be chosen to account for differences in behavior due to shape is that of an ellipsoid. The orientation of the particles relative to each other will be dependent on the competing effects introduced by the orienting effect of the shear field and the disorienting effect of the superimposed random Brownian motion. The Kuhn and Kuhn[25] theories, summarized in Eq. (9–5.4), indicate that the rheological behavior of a dilute suspension of ellipsoids can be represented in terms of an apparent viscosity for the case where Brownian movement is predominant and orientation effects negligible. Brodnyan[4] has extended this theory to more concentrated systems by the same arguments utilized by Mooney[37] in developing his viscosity relationship from the original Einstein formula. Thus, for the relative viscosity of concentrated solutions of ellipsoidal particles, he obtains

$$\ln \mu_r = \frac{2.5\phi + 0.399(p - 1)^{1.48}\phi}{1 - k\phi} \tag{9–7.1}$$

The constants 0.399 and 1.48 are based upon experimental data on suspensions of ellipsoids available to Mooney. These values are within 2 per cent of the values derived theoretically by Kuhn and Kuhn and given in Eq. (9–5.4b), namely 0.4075 and 1.508, respectively. The data of Yang[58] were considered best for testing Eq. (9–7.1). In comparing the latter with Yang's data, Brodnyan finds that the experimental viscosity does not increase quite as rapidly with increasing concentration as the equation would predict. This would indicate the existence of shear-dependent orientation effects. Yang's study indicates that in dilute suspensions of ellipsoids the viscosity will vary with the rate of shear in the manner predicted by Scheraga's[46] numerical results, based on the Jeffery-Peterlin-Saito[44] treatment. It seems likely that Eq. (9–7.1) could be used to extend the dilute ellipsoid treatment to more concentrated assemblages, but data are lacking at present to confirm this.

At high rates of shear, the data of a number of investigators[57] indicate divergence from simple newtonian behavior for suspensions containing distributions of particle sizes. The apparent viscosity can be correlated approximately by the Mooney-type equation which predicts that the viscosity of a suspension having a given percentage of solids may be decreased by using selected blends of particle sizes in place of a single-particle-sized fraction.

Various investigators of the rheology of polymer solutions have employed empirical relationships with more arbitrary constants than the Mooney equation. In general, the values of the constants employed have depended on the particular polymer-solvent system studied.[56] In the cases studied[56], the rate of shear was sufficiently low that newtonian viscosity was approximately applicable. Concentrated solutions of high polymers at higher rates of shear, and under shear rates varying with time, exhibit more complicated behavior.[39] In the case of flexible macromolecules, mechanical interactions, such as temporary association and entanglement of two or more molecules, or a change in the diameter of coiled macromolecules in different solvents with concentration and temperature, must also be considered.

In questions pertaining to suspension viscosity, there is the basic question of the limit of validity of the creeping motion equations themselves. For a pure liquid undergoing shear in a capillary or Couette type of viscometer, inertial forces do not come into play. With particles immersed in a shear field, however, inertial effects do arise owing to the three-dimensional nature of the resulting flow. Saito[44] has proposed that suspension viscosity be evaluated using an Oseen type of approximation in order to assess the importance of these effects, but no such solution is available.

Another factor entering into the determination of flow properties of suspensions is the influence of the type of viscometer employed. When the viscosity of a suspension of spherical particles uniformly distributed in a newtonian liquid is measured in a capillary tube viscometer, a wall effect, not found with coaxial rotating cylindrical viscometers of the Couette type, is observed. Maude and Whitmore[34] suggest that this may be caused by a redistribution of a portion of the suspended particles on entering the tube of the viscometer, leading to a reduction in the concentration and, hence, in the apparent viscosity of the suspension near the walls of the tube. Data on flow of blood[14] confirm these observations.

BIBLIOGRAPHY

1. Arrhenius, S., Z. Physik. Chem. **1** (1887), 285.
2. Bartok, W., and S. G. Mason, J. Colloid. Sci. **13** (1958), 293.
3. Brenner, H., Phys. Fluids **1** (1958), 338.
4. Brodnyan, J. G., Trans. Soc. Rheology **3** (1959), 61.
5. Burgers, J. M., chap. 3 in *Second Report on Viscosity and Plasticity*. Amsterdam: North Holland, 1938.
6. Cheng, P. Y., and H. K. Schachman, J. Polymer Sci. **16** (1955), 19.
7. Coulter, N. A., and J. R. Pappenheimer, Amer. J. Physiol. **159** (1949), 401.
8. Debye, P., and A. M. Bueche, J. Chem. Phys. **16** (1948), 573.

9. Einstein, A., Ann. Physik **19** (1906), 289; **34** (1911), 591; for English translations see Einstein, A., *The Theory of Brownian Movement.* New York: Dover, 1956.

10. Eisenschitz, R., Z. Physik. Chem. **A158** (1932), 85.

11. Fisher, H. L., *Chemistry of Natural and Synthetic Rubbers.* New York: Reinhold, 1957.

12. Ford, T. F., J. Phys. Chem. **64** (1960), 1168.

13. Frisch, H. L., and R. Simha, in *Rheology—Theory and Applications*, Vol. I. F. R. Eirich, ed. New York: Academic Press, 1956. chap. 14.

14. Fulton, J. F. (ed.), *A Textbook of Physiology*, 17th ed. New York: Saunders, 1955; chap. 31, H. Lamport, "Hemodynamics," and other chapters.

15. Guth, E., and R. Simha, Kolloid Z. **74** (1936), 266.

16. Happel, J., J. Appl. Phys. **28** (1957), 1288.

17. ———, A. I. Ch. E. Jour. **4** (1958), 197.

18. ———, and R. C. Lee, unpublished investigation (1962).

19. Hawksley, P. G. W., Paper 7 "Some Aspects of Fluid Flow," Inst. Phys., London: Arnold, 1951.

20. Hermans, J. J. (ed.), *Flow Properties of Disperse Systems.* New York: Interscience Publishers, 1953.

21. Jeffery, G. B., Proc. Roy. Soc. (London) **A102** (1922), 161.

22. Johnson, E., "A Theory of Fluidization," Inst. Gas Eng. *Publ.* 378, London: 1951.

23. Jones, G., and A. K. Tilley, J. Amer. Chem. Soc. **55** (1933), 624.

24. Kirkwood, J. G., and J. Riseman, J. Chem. Phys. **16** (1948), 565.

25. Kuhn, W., and H. Kuhn, Helv. Chim. Acta **28** (1945), 97.

26. ———, Z. Physik. Chem. **A161** (1932), 1.

27. Kunitz, M., J. Gener. Physiol. **9** (1926), 715.

28. Kynch, G. J., Brit. J. Appl. Phys. **3** (1954), S5.

29. ———, Proc. Roy. Soc. (London) **A237** (1956), 90.

30. ———, Nature **184** (1959), 1311.

31. Lamb, H., *Hydrodynamics.* New York: Cambridge Univ. Press, 1932; reprint, New York: Dover Publications, 1945.

32. Leeson and Reeve, J. Physiol. **115** (1951), 129.

33. Mason, S. G., and R. St. J. Manley, Proc. Roy. Soc. (London) **A238** (1956), 117.

34. Maude, A. D., and R. L. Whitmore, Brit. J. Appl. Phys. **7** (1956), 98.

35. Milne-Thomson, L. M., *Theoretical Hydrodynamics*, 3rd ed. New York: Macmillan, 1955.

36. Mooney, M., J. Appl. Phys. **25** (1954), 406.

37. ———, J. Colloid. Sci. **6** (1951), 162.

38. Mori, Y., and N. Ototake, J. Chem. Eng. (Tokyo) **20** (1956), 488.

39. Onogi, S., I. Hamana, and H. Hirai, J. Appl. Phys. **29** (1958), 1503.

40. Philippoff, W., *Viskosität der Kolloide*. Dresden and Leipzig: Theodor Steinkopff, 1942.

41. Robinson, J., J. Phys. Colloid. Chem. **53** (1949), 1042.

42. Rutgers, R., Rheologica Acta **2** (1962), 202, 305.

43. Saito, N., J. Phys. Soc. Japan **7** (1952), 447.

44. ———, and M. Sugita, J. Phys. Soc. Japan **7** (1952), 554.

45. Saunders, F. L., J. Colloid Sci. **16** (1961), 13.

46. Scheraga, H. A., J. Chem. Phys. **23** (1955), 1526.

47. Segré, G., and A. Silberberg, J. Colloid. Sci. **18** (1963), 312.

48. Simha, R., J. Appl. Phys. **23** (1952), 1020.

49. ———, J. Res. Nat. Bur. Stand. **42** (1949), 409.

50. Sweeny, K. H., and R. D. Geckler, J. Appl. Phys. **25** (1954), 1135.

51. Swindells, J. F., C. F. Snyder, R. C. Hardy, and P. E. Golden, "Viscosities of Sucrose Solutions at Various Temperatures: Tables of Recalculated Values," Suppl. Nat. Bur. Stand. *Circular* 440, 1958.

52. Taylor, G. I., Proc. Roy. Soc. (London) **A103** (1923), 58.

53. ———, Proc. Roy. Soc. (London) **A138** (1932), 41.

54. Vand, V., J. Phys. Colloid. Chem. **52** (1948), 277.

55. Ward, S. G., and R. L. Whitmore, Brit. J. Appl. Phys. **1** (1950), 286.

56. Weissberg, S. G., R. Simha, and S. Rothman, J. Res. Nat. Bur. Stand. **47** (1951), 298.

57. Williams, P. S., J. Appl. Chem. **3** (1953), 120.

58. Yang, J. T., J. Amer. Chem. Soc. **80** (1958), 1783.

59. Zakin, J. L., private communication (1958).

Orthogonal Curvilinear Coordinate Systems A

A-1 Curvilinear Coordinates

The location of a point in three-dimensional space (with respect to some origin) is usually specified by giving its three cartesian coordinates (x, y, z) or, what is equivalent, by specifying the position vector $\mathbf{R}$ of the point. It is often more convenient to describe the position of the point by another set of coordinates more appropriate to the problem at hand, common examples being spherical and cylindrical coordinates. These are but special cases of *curvilinear coordinate systems*, whose general properties we propose to examine in detail.

Suppose that q_1, q_2, and q_3 are independent functions of position such that

$$q_1 = q_1(x, y, z), \qquad q_2 = q_2(x, y, z), \qquad q_3 = q_3(x, y, z) \tag{A-1.1}$$

or, in terms of the position vector,

$$q_k = q_k(\mathbf{R}) \quad (k = 1, 2, 3)$$

In regions where the Jacobian determinant,

$$\frac{\partial(q_1, q_2, q_3)}{\partial(x, y, z)} = \begin{vmatrix} \frac{\partial q_1}{\partial x} & \frac{\partial q_1}{\partial y} & \frac{\partial q_1}{\partial z} \\ \frac{\partial q_2}{\partial x} & \frac{\partial q_2}{\partial y} & \frac{\partial q_2}{\partial z} \\ \frac{\partial q_3}{\partial x} & \frac{\partial q_3}{\partial y} & \frac{\partial q_3}{\partial z} \end{vmatrix} \tag{A-1.2}$$

is different from zero this system of equations can be solved simultaneously for x, y, and z, giving

$$x = x(q_1, q_2, q_3), \qquad y = y(q_1, q_2, q_3), \qquad z = z(q_1, q_2, q_3) \quad \text{(A-1.3)}$$

or

$$\mathbf{R} = \mathbf{R}(q_1, q_2, q_3)$$

A vanishing of the Jacobian implies that q_1, q_2, and q_3 are not independent functions but, rather, are connected by a functional relationship of the form $f(q_1, q_2, q_3) = 0$.

In accordance with (A-1.3), the specification of numerical values for q_1, q_2, and q_3 leads to a corresponding set of numerical values for x, y, z; that is, it locates a point (x, y, z) in space. Thus, we come to regard the set of three numbers (q_1, q_2, q_3) as the *curvilinear coordinates* of a point in space. It is natural in dealing with physical problems to restrict attention to systems of curvilinear coordinates in which each point in space may be represented at least once by letting q_1, q_2, and q_3 vary over all possible values.

Curvilinear coordinates have a simple geometric interpretation. If for the moment we ascribe some constant value to q_k, we have

$$q_k(x, y, z) = \text{constant} \quad (k = 1, 2, 3)$$

which describes a *surface* in space. By assigning a series of different values to q_k, we generate a *family of surfaces* on which q_k is constant. If the functions have been properly chosen there is at least one surface belonging to each of the three families which passes through any arbitrary point P in space. Thus, a point in space is characterized by the intersection of the three surfaces, $q_1 =$ constant, $q_2 =$ constant, $q_3 =$ constant (see Fig. A-1.1), termed *coordinate surfaces*. The coordinate surface is named for that coordinate which is constant, the other two coordinates being variable along

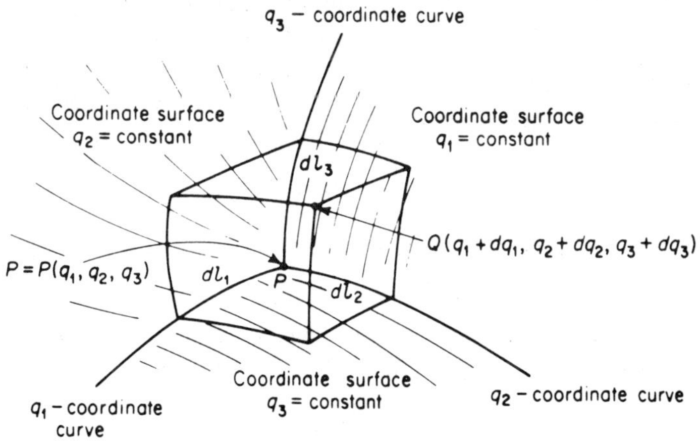

Figure A-1.1. Curvilinear coordinates.

that surface. The intersection of any two coordinate surfaces results in a skew curve termed a *coordinate curve*. For example, the intersection of the q_2- and q_3-coordinate surfaces results in the coordinate curve labeled q_1. Since this curve lies simultaneously on the surfaces q_2 = constant and q_3 = constant, only q_1 varies as we move along the curve; hence, the designation q_1-coordinate curve.

In the special case of cartesian coordinates the coordinate surfaces consist of three mutually perpendicular planes; the coordinate curves consist of three mutually perpendicular lines.

In cartesian coordinates the differentials dx, dy, and dz correspond to distances measured along each of the three cartesian coordinate curves. The analogous differentials in curvilinear coordinates, dq_1, dq_2, and dq_3, do not necessarily have a similar interpretation. As in Fig. A-1.1, let dl_1 be the distance measured along the q_1-coordinate curve from the point $P(q_1, q_2, q_3)$ to the neighboring point $(q_1 + dq_1, q_2, q_3)$. Similar definitions apply to dl_2 and dl_3. We define the three quantities*

$$h_k = \left|\frac{dq_k}{dl_k}\right| > 0 \qquad (k = 1, 2, 3) \tag{A–1.4}$$

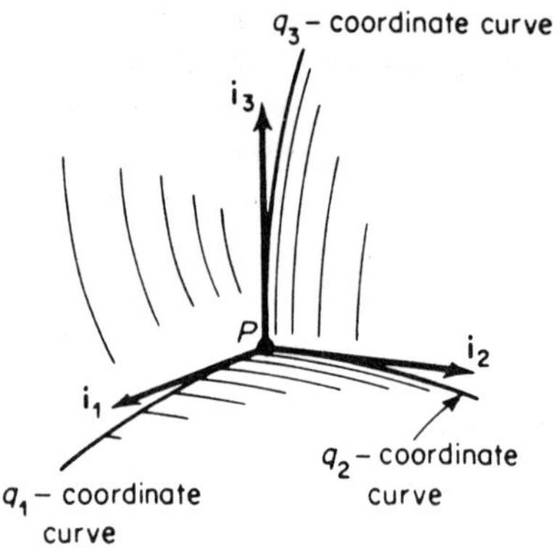

Figure A-1.2. Unit tangent vectors.

These quantities (or simple variations thereof) are termed *metrical coefficients*. They are intrinsic properties of any particular system of curvilinear coordinates and, in general, are functions of position,

$$h_k = h_k(q_1, q_2, q_3)$$

Let $\mathbf{i}_1$, $\mathbf{i}_2$ and $\mathbf{i}_3$ be unit vectors drawn tangent to the q_1-, q_2- and q_3-coordinate curves, respectively, in the directions of algebraically *increasing* q_k's (Fig. A-1.2). It is evident that these unit tangent vectors are given by

$$\mathbf{i}_k = \frac{\partial \mathbf{R}}{\partial l_k} \qquad (k = 1, 2, 3) \tag{A–1.5}$$

Whereas the *magnitudes* of these unit vectors are necessarily constant,

$$\mathbf{i}_k \cdot \mathbf{i}_k = 1 \quad \text{or} \quad |\mathbf{i}_k| = 1 \tag{A–1.6}$$

it does not follow that their *directions* remain constant from point to point, so the unit vectors are, in general, functions of position,

$$\mathbf{i}_k = \mathbf{i}_k(q_1, q_2, q_3)$$

*Some authors prefer to define these symbols to be the reciprocals of the values given here.

The three non-coplanar unit vectors $\mathbf{i}_1$, $\mathbf{i}_2$, $\mathbf{i}_3$ are said to constitute a *base* set of unit vectors for the particular system of curvilinear coordinates. Any arbitrary vector, $\mathbf{u}$, may be uniquely expressed in terms of them by a relation of the form $\mathbf{u} = \mathbf{i}_1 u_1 + \mathbf{i}_2 u_2 + \mathbf{i}_3 u_3$. A simple calculation shows that

$$u_1 = \frac{\mathbf{u}\cdot\mathbf{i}_2 \times \mathbf{i}_3}{\mathbf{i}_1\cdot\mathbf{i}_2 \times \mathbf{i}_3}, \qquad u_2 = \frac{\mathbf{i}_1\cdot\mathbf{u} \times \mathbf{i}_3}{\mathbf{i}_1\cdot\mathbf{i}_2 \times \mathbf{i}_3}, \qquad u_3 = \frac{\mathbf{i}_1\cdot\mathbf{i}_2 \times \mathbf{u}}{\mathbf{i}_1\cdot\mathbf{i}_2 \times \mathbf{i}_3}$$

Upon combining Eqs. (A-1.4) and (A-1.5), we obtain

$$\mathbf{i}_k = h_k \frac{\partial \mathbf{R}}{\partial q_k} \tag{A–1.7}$$

which provides an alternative formula for determining the metrical coefficients,

$$h_k = \frac{1}{|\partial \mathbf{R}/\partial q_k|} \tag{A–1.8}$$

The utility of this particular expression lies in the calculation of metrical coefficients for systems of curvilinear coordinates defined *explicitly* by Eq. (A-1.3). Thus, if we put

$$\mathbf{R} = \mathbf{i}x + \mathbf{j}y + \mathbf{k}z$$

it is evident that

$$\frac{\partial \mathbf{R}}{\partial q_k} = \mathbf{i}\frac{\partial x}{\partial q_k} + \mathbf{j}\frac{\partial y}{\partial q_k} + \mathbf{k}\frac{\partial z}{\partial q_k}$$

and hence,

$$\frac{1}{h_k^2} = \left(\frac{\partial x}{\partial q_k}\right)^2 + \left(\frac{\partial y}{\partial q_k}\right)^2 + \left(\frac{\partial z}{\partial q_k}\right)^2 \quad (k = 1, 2, 3) \tag{A–1.9}$$

For example, in cartesian coordinates where $q_1 = x, q_2 = y,\ q_3 = z$, we find that $h_1 = h_2 = h_3 = 1$.

A-2 Orthogonal Curvilinear Coordinates

If (q_1, q_2, q_3) are the curvilinear coordinates of a point P whose position vector is $\mathbf{R}$ and $(q_1 + dq_1, q_2 + dq_2, q_3 + dq_3)$ those of a neighboring point Q whose position vector is $\mathbf{R} + d\mathbf{R}$, then

$$\begin{aligned}\overrightarrow{PQ} = d\mathbf{R} &= \frac{\partial \mathbf{R}}{\partial q_1} dq_1 + \frac{\partial \mathbf{R}}{\partial q_2} dq_2 + \frac{\partial \mathbf{R}}{\partial q_3} dq_3 \\ &= \mathbf{i}_1 \frac{dq_1}{h_1} + \mathbf{i}_2 \frac{dq_2}{h_2} + \mathbf{i}_3 \frac{dq_3}{h_3}\end{aligned} \tag{A–2.1}$$

Thus, the distance dl between these adjacent points is given by

$$\begin{aligned}dl^2 = |d\mathbf{R}|^2 = \frac{dq_1^2}{h_1^2} + \frac{dq_2^2}{h_2^2} + \frac{dq_3^2}{h_3^2} + 2(\mathbf{i}_1\cdot\mathbf{i}_2)\frac{dq_1\,dq_2}{h_1 h_2} \\ + 2(\mathbf{i}_2\cdot\mathbf{i}_3)\frac{dq_2\,dq_3}{h_2 h_3} + 2(\mathbf{i}_3\cdot\mathbf{i}_1)\frac{dq_3\,dq_1}{h_3 h_1}\end{aligned} \tag{A–2.2}$$

When the system of curvilinear coordinates is such that the three coordinate surfaces are mutually perpendicular at each point, it is termed an *orthogonal* curvilinear coordinate system. In this event the unit tangent vectors to the coordinate curves are also mutually perpendicular at each point and thus

$$\mathbf{i}_j \cdot \mathbf{i}_k = 0 \quad (j, k = 1, 2, 3) \quad (j \neq k) \tag{A-2.3}$$

whereupon

$$dl^2 = \frac{dq_1^2}{h_1^2} + \frac{dq_2^2}{h_2^2} + \frac{dq_3^2}{h_3^2} \tag{A-2.4}$$

Therefore, an essential attribute of orthogonal systems is that mixed terms of the form $dq_j dq_k$ $(j \neq k)$ do not appear in the expression for the distance dl. This condition is not only necessary for orthogonality but is sufficient as well; for q_1, q_2, and q_3 in Eq. (A-2.2) are *independent* variables.

In consequence of Eq. (A-1.7) the necessary and sufficient conditions for orthogonality may also be expressed by the relations

$$\frac{\partial \mathbf{R}}{\partial q_j} \cdot \frac{\partial \mathbf{R}}{\partial q_k} = 0 \quad (j, k = 1, 2, 3) \quad (j \neq k) \tag{A-2.5}$$

or, putting $\mathbf{R} = \mathbf{i}x + \mathbf{j}y + \mathbf{k}z$,

$$\frac{\partial x}{\partial q_j}\frac{\partial x}{\partial q_k} + \frac{\partial y}{\partial q_j}\frac{\partial y}{\partial q_k} + \frac{\partial z}{\partial q_j}\frac{\partial z}{\partial q_k} = 0 \quad (j, k = 1, 2, 3) \quad (j \neq k) \tag{A-2.6}$$

which provides a useful test of orthogonality for systems of curvilinear coordinates defined explicitly by the relations

$$x = x(q_1, q_2, q_3), \qquad y = y(q_1, q_2, q_3), \qquad z = z(q_1, q_2, q_3)$$

If, instead, the system of coordinates is defined explicitly by the equations

$$q_1 = q_1(x, y, z), \qquad q_2 = q_2(x, y, z), \qquad q_3 = q_3(x, y, z)$$

the computation of the partial derivatives required in Eq. (A-2.6) can be a tedious chore, and it is best to proceed somewhat differently. In consequence of the general properties of the gradient operator, each of the vectors ∇q_k $(k = 1, 2, 3)$ is necessarily perpendicular to the corresponding coordinate surface $q_k =$ constant. Thus, the necessary and sufficient conditions for orthogonality are equally well expressed by the relations

$$\nabla q_j \cdot \nabla q_k = 0 \quad (j, k = 1, 2, 3) \quad (j \neq k) \tag{A-2.7}$$

or, expressing ∇ in cartesian coordinates,

$$\frac{\partial q_j}{\partial x}\frac{\partial q_k}{\partial x} + \frac{\partial q_j}{\partial y}\frac{\partial q_k}{\partial y} + \frac{\partial q_j}{\partial z}\frac{\partial q_k}{\partial z} = 0 \qquad (j, k = 1, 2, 3) \quad (j \neq k) \tag{A-2.8}$$

These may be contrasted with Eqs. (A-2.6).

If the system of coordinates does prove orthogonal we can avail ourselves of still another method for computing the metrical coefficients. For, in this event, the q_2- and q_3-coordinate surfaces are perpendicular to the q_1-coordinate surfaces. But since the q_1-coordinate curves lie simultaneously on each of the former surfaces, these curves must be perpendicular to the surfaces q_1 = constant. In general, then, the q_k-coordinate curves lie *normal* to the surfaces on which q_k is constant. The general properties of the ∇ operator are such that the vector ∇q_k is normal to the surfaces on which q_k is constant and points in the direction of increasing q_k. Consequently, the unit tangent vector, $\mathbf{i}_k$, to the q_k-coordinate curve passing through a particular point in space is identical to the unit normal vector, $\mathbf{n}_k$, to the q_k-coordinate surface passing through the point in question. Since, from the general properties of the ∇ operator,

$$\mathbf{n}_k = \frac{\nabla q_k}{|dq_k/dl_k|}$$

we have that

$$\mathbf{i}_k = \frac{\nabla q_k}{h_k} \tag{A-2.9}$$

This makes

$$h_k = |\nabla q_k| \tag{A-2.10}$$

or, again writing ∇ in cartesian coordinates,

$$h_k^2 = \left(\frac{\partial q_k}{\partial x}\right)^2 + \left(\frac{\partial q_k}{\partial y}\right)^2 + \left(\frac{\partial q_k}{\partial z}\right)^2 \quad (k = 1, 2, 3) \tag{A-2.11}$$

In comparing this with Eq. (A-1.9) it should be borne in mind that Eqs. (A-2.11) hold only for orthogonal systems.

We shall say nothing further about nonorthogonal coordinate systems, for these find no application in conventional hydrodynamic problems.

For vector operations involving cross products, it is convenient to order the orthogonal curvilinear coordinates (q_1, q_2, q_3) in such a way that the base unit vectors $\mathbf{i}_1$, $\mathbf{i}_2$, $\mathbf{i}_3$ form a *right-handed* system of unit vectors; that is,

$$\mathbf{i}_1 \times \mathbf{i}_2 = \mathbf{i}_3, \qquad \mathbf{i}_2 \times \mathbf{i}_3 = \mathbf{i}_1, \qquad \mathbf{i}_3 \times \mathbf{i}_1 = \mathbf{i}_2 \tag{A-2.12}$$

With the aid of the properties of the scalar triple product, it is simple to show that these three relations are all satisfied by ordering q_1, q_2, and q_3 so as to satisfy the relation

$$\mathbf{i}_1 \cdot \mathbf{i}_2 \times \mathbf{i}_3 = +1 \tag{A-2.13}$$

Inasmuch as the metrical coefficients are essentially positive, it follows from Eq. (A-1.7) that this is equivalent to choosing the sequence of coordinates in such a way that

$$\frac{\partial \mathbf{R}}{\partial q_1} \cdot \frac{\partial \mathbf{R}}{\partial q_2} \times \frac{\partial \mathbf{R}}{\partial q_3} > 0 \tag{A-2.14}$$

or, setting $\mathbf{R} = \mathbf{i}x + \mathbf{j}y + \mathbf{k}z$,

$$\begin{vmatrix} \dfrac{\partial x}{\partial q_1} & \dfrac{\partial x}{\partial q_2} & \dfrac{\partial x}{\partial q_3} \\ \dfrac{\partial y}{\partial q_1} & \dfrac{\partial y}{\partial q_2} & \dfrac{\partial y}{\partial q_3} \\ \dfrac{\partial z}{\partial q_1} & \dfrac{\partial z}{\partial q_2} & \dfrac{\partial z}{\partial q_3} \end{vmatrix} = \frac{\partial(x, y, z)}{\partial(q_1, q_2, q_3)} > 0 \tag{A-2.15}$$

There is only one way of ordering the coordinates to make this Jacobian determinant positive.

Alternatively, a right-handed system is obtained when

$$\nabla q_1 \cdot \nabla q_2 \times \nabla q_3 > 0 \tag{A-2.16}$$

or, expressing ∇ in terms of cartesian coordinates,

$$\begin{vmatrix} \dfrac{\partial q_1}{\partial x} & \dfrac{\partial q_1}{\partial y} & \dfrac{\partial q_1}{\partial z} \\ \dfrac{\partial q_2}{\partial x} & \dfrac{\partial q_2}{\partial y} & \dfrac{\partial q_2}{\partial z} \\ \dfrac{\partial q_3}{\partial x} & \dfrac{\partial q_3}{\partial y} & \dfrac{\partial q_3}{\partial z} \end{vmatrix} = \frac{\partial(q_1, q_2, q_3)}{\partial(x, y, z)} > 0 \tag{A-2.17}$$

With the curvilinear coordinates arranged in proper order we have

$$\frac{\partial(x, y, z)}{\partial(q_1, q_2, q_3)} = \frac{1}{[\partial(q_1, q_2, q_3)/\partial(x, y, z)]} = \frac{1}{h_1 h_2 h_3} > 0 \tag{A-2.18}$$

A-3 Geometrical Properties

When the curvilinear coordinates are orthogonal, the infinitesimal volume depicted in Fig. A-1.1 is a rectangular parallepiped the dimensions of which are dl_1, dl_2, and dl_3 or, employing Eq. (A-1.4), dq_1/h_1, dq_2/h_2, and dq_3/h_3, respectively. Thus, if dS_k is an element of surface area lying on the coordinate surface $q_k =$ constant, we have

$$\begin{aligned} dS_1 &= dl_2 dl_3 = \frac{dq_2\, dq_3}{h_2 h_3} \\ dS_2 &= dl_3 dl_1 = \frac{dq_3\, dq_1}{h_3 h_1} \\ dS_3 &= dl_1 dl_2 = \frac{dq_1\, dq_2}{h_1 h_2} \end{aligned} \tag{A-3.1}$$

Furthermore, an element of volume is given by

$$dV = dl_1\, dl_2\, dl_3 = \frac{dq_1\, dq_2\, dq_3}{h_1 h_2 h_3} \tag{A-3.2}$$

A-4 Differentiation of Unit Vectors

As previously remarked, the base unit vectors associated with a particular system of orthogonal curvilinear coordinates are vector functions of position. In this connection it is a matter of some importance to determine the form taken by the nine possible derivatives $\partial \mathbf{i}_k/\partial q_l$ $(k, l = 1, 2, 3)$. These vectors do not, in general, have the same direction as the unit vectors themselves. This contention is easily demonstrated by observing that $\mathbf{i}_k \cdot \mathbf{i}_k = 1$, from which it follows that

$$\frac{\partial}{\partial q_l}(\mathbf{i}_k \cdot \mathbf{i}_k) = 0 = 2\mathbf{i}_k \cdot \frac{\partial \mathbf{i}_k}{\partial q_l}$$

Thus, the vector $\partial \mathbf{i}_k/\partial q_l$ is either zero or else it is *perpendicular* to $\mathbf{i}_k$.

To evaluate these derivatives we proceed as follows. For an orthogonal system, Eq. (A-1.7) yields

$$\frac{\partial \mathbf{R}}{\partial q_k} \cdot \frac{\partial \mathbf{R}}{\partial q_l} = 0 \quad (k \neq l)$$

Hence, for any j,

$$\frac{\partial}{\partial q_j}\left(\frac{\partial \mathbf{R}}{\partial q_k} \cdot \frac{\partial \mathbf{R}}{\partial q_l}\right) = 0 \quad (k \neq l)$$

or, performing the differentiations,

$$\frac{\partial^2 \mathbf{R}}{\partial q_j\, \partial q_k} \cdot \frac{\partial \mathbf{R}}{\partial q_l} + \frac{\partial^2 \mathbf{R}}{\partial q_j\, \partial q_l} \cdot \frac{\partial \mathbf{R}}{\partial q_k} = 0 \quad (k \neq l) \tag{A-4.1a}$$

Inasmuch as j, k, and l are *dummy* indices, we obtain upon interchanging j and k in Eq. (A-4.1a),

$$\frac{\partial^2 \mathbf{R}}{\partial q_k\, \partial q_j} \cdot \frac{\partial \mathbf{R}}{\partial q_l} + \frac{\partial^2 \mathbf{R}}{\partial q_k\, \partial q_l} \cdot \frac{\partial \mathbf{R}}{\partial q_j} = 0 \quad (j \neq l) \tag{A-4.1b}$$

Similarly, interchanging j and l in Eq. (A-4.1a),

$$\frac{\partial^2 \mathbf{R}}{\partial q_l\, \partial q_k} \cdot \frac{\partial \mathbf{R}}{\partial q_j} + \frac{\partial^2 \mathbf{R}}{\partial q_l\, \partial q_j} \cdot \frac{\partial \mathbf{R}}{\partial q_k} = 0 \quad (k \neq j) \tag{A-4.1c}$$

If $j \neq k \neq l$, all three of these relations are valid. Since $\partial^2 \mathbf{R}/\partial q_m\, \partial q_n = \partial^2 \mathbf{R}/\partial q_n\, \partial q_m$, we find upon subtracting Eq. (A-4.1c) from Eq. (A-4.1b) that

$$\frac{\partial^2 \mathbf{R}}{\partial q_k\, \partial q_j} \cdot \frac{\partial \mathbf{R}}{\partial q_l} - \frac{\partial^2 \mathbf{R}}{\partial q_l\, \partial q_j} \cdot \frac{\partial \mathbf{R}}{\partial q_k} = 0 \quad (j \neq k \neq l)$$

Adding this to Eq. (A-4.1a) then yields

$$\frac{\partial \mathbf{R}}{\partial q_l} \cdot \frac{\partial^2 \mathbf{R}}{\partial q_j\, \partial q_k} = 0 \quad (j \neq k \neq l) \tag{A-4.2}$$

But, from Eq. (A-1.7),

$$\frac{\partial \mathbf{R}}{\partial q_l} = \frac{\mathbf{i}_l}{h_l} \quad \text{and} \quad \frac{\partial \mathbf{R}}{\partial q_k} = \frac{\mathbf{i}_k}{h_k}$$

Equation (A-4.2) may therefore be written as

$$\mathbf{i}_l \cdot \frac{\partial}{\partial q_j}\left(\frac{\mathbf{i}_k}{h_k}\right) = 0 \quad (j \neq k \neq l)$$

that is,

$$\mathbf{i}_l \cdot \left[\mathbf{i}_k \frac{\partial}{\partial q_j}\left(\frac{1}{h_k}\right) + \frac{1}{h_k}\frac{\partial \mathbf{i}_k}{\partial q_j}\right] = 0 \quad (j \neq k \neq l)$$

Since k and l are different, $\mathbf{i}_l \cdot \mathbf{i}_k = 0$; hence, the preceding equation becomes

$$\mathbf{i}_l \cdot \frac{\partial \mathbf{i}_k}{\partial q_j} = 0 \quad (j \neq k \neq l)$$

This means that the vector $\partial \mathbf{i}_k/\partial q_j$ does not have an l component. But, since $\mathbf{i}_k \cdot \mathbf{i}_k = 1$, we find by differentiation with respect to q_j that

$$\mathbf{i}_k \cdot \frac{\partial \mathbf{i}_k}{\partial q_j} = 0 \tag{A–4.3}$$

from which it is clear that $\partial \mathbf{i}_k/\partial q_j$ does not have a k component either. We therefore obtain the important result that, for $j \neq k$, $\partial \mathbf{i}_k/\partial q_j$ has, at most, a component in the j direction; that is,

$$\frac{\partial \mathbf{i}_k}{\partial q_j} \parallel \mathbf{i}_j \quad (j \neq k) \tag{A–4.4}$$

To obtain this component, observe that since the order of differentiation is immaterial,

$$\frac{\partial}{\partial q_j}\left(\frac{\partial \mathbf{R}}{\partial q_k}\right) = \frac{\partial}{\partial q_k}\left(\frac{\partial \mathbf{R}}{\partial q_j}\right) \quad \text{or} \quad \frac{\partial}{\partial q_j}\left(\frac{\mathbf{i}_k}{h_k}\right) = \frac{\partial}{\partial q_k}\left(\frac{\mathbf{i}_j}{h_j}\right)$$

Differentiation gives

$$\frac{1}{h_k}\frac{\partial \mathbf{i}_k}{\partial q_j} + \mathbf{i}_k \frac{\partial}{\partial q_j}\left(\frac{1}{h_k}\right) = \frac{1}{h_j}\frac{\partial \mathbf{i}_j}{\partial q_k} + \mathbf{i}_j \frac{\partial}{\partial q_k}\left(\frac{1}{h_j}\right) \tag{A–4.5}$$

Suppose in this equation that $j \neq k$. Then, since $\partial \mathbf{i}_k/\partial q_j$ has only a j component—see Eq. (A-4.4)—it immediately follows upon equating j components in the foregoing that

$$\frac{\partial \mathbf{i}_k}{\partial q_j} = \mathbf{i}_j h_k \frac{\partial}{\partial q_k}\left(\frac{1}{h_j}\right) \quad (j \neq k) \tag{A–4.6}$$

The preceding relation gives an explicit formula for differentiating unit vectors when $j \neq k$. The corresponding relations for the case $j = k$ may be obtained as follows: Suppose $j \neq k \neq l$ and that $[jkl]$ are arranged in right-handed cyclic order, that is, [123], [231], or [312]. Then

$$\mathbf{i}_j = \mathbf{i}_k \times \mathbf{i}_l$$

Differentiate with respect to q_j and obtain

$$\frac{\partial \mathbf{i}_j}{\partial q_j} = \frac{\partial \mathbf{i}_k}{\partial q_j} \times \mathbf{i}_l + \mathbf{i}_k \times \frac{\partial \mathbf{i}_l}{\partial q_j}$$

Utilizing Eq. (A-4.6) this becomes

$$\frac{\partial \mathbf{i}_j}{\partial q_j} = (\mathbf{i}_j \times \mathbf{i}_l) h_k \frac{\partial}{\partial q_k}\left(\frac{1}{h_j}\right) + (\mathbf{i}_k \times \mathbf{i}_j) h_l \frac{\partial}{\partial q_l}\left(\frac{1}{h_j}\right)$$

But $\mathbf{i}_j \times \mathbf{i}_l = -\mathbf{i}_k$ and $\mathbf{i}_k \times \mathbf{i}_j = -\mathbf{i}_l$. Thus, we finally obtain

$$\frac{\partial \mathbf{i}_j}{\partial q_j} = -\mathbf{i}_k h_k \frac{\partial}{\partial q_k}\left(\frac{1}{h_j}\right) - \mathbf{i}_l h_l \frac{\partial}{\partial q_l}\left(\frac{1}{h_j}\right) \tag{A-4.7}$$

Written out explicitly, Eqs. (A-4.6) and (A-4.7) yield the desired nine derivatives,

$$\begin{aligned}
\frac{\partial \mathbf{i}_1}{\partial q_1} &= -\mathbf{i}_2 h_2 \frac{\partial}{\partial q_2}\left(\frac{1}{h_1}\right) - \mathbf{i}_3 h_3 \frac{\partial}{\partial q_3}\left(\frac{1}{h_1}\right) \\
\frac{\partial \mathbf{i}_1}{\partial q_2} &= \mathbf{i}_2 h_1 \frac{\partial}{\partial q_1}\left(\frac{1}{h_2}\right) \\
\frac{\partial \mathbf{i}_1}{\partial q_3} &= \mathbf{i}_3 h_1 \frac{\partial}{\partial q_1}\left(\frac{1}{h_3}\right) \\
\frac{\partial \mathbf{i}_2}{\partial q_1} &= \mathbf{i}_1 h_2 \frac{\partial}{\partial q_2}\left(\frac{1}{h_1}\right) \\
\frac{\partial \mathbf{i}_2}{\partial q_2} &= -\mathbf{i}_3 h_3 \frac{\partial}{\partial q_3}\left(\frac{1}{h_2}\right) - \mathbf{i}_1 h_1 \frac{\partial}{\partial q_1}\left(\frac{1}{h_2}\right) \qquad \text{(A-4.8)} \\
\frac{\partial \mathbf{i}_2}{\partial q_3} &= \mathbf{i}_3 h_2 \frac{\partial}{\partial q_2}\left(\frac{1}{h_3}\right) \\
\frac{\partial \mathbf{i}_3}{\partial q_1} &= \mathbf{i}_1 h_3 \frac{\partial}{\partial q_3}\left(\frac{1}{h_1}\right) \\
\frac{\partial \mathbf{i}_3}{\partial q_2} &= \mathbf{i}_2 h_3 \frac{\partial}{\partial q_3}\left(\frac{1}{h_2}\right) \\
\frac{\partial \mathbf{i}_3}{\partial q_3} &= -\mathbf{i}_1 h_1 \frac{\partial}{\partial q_1}\left(\frac{1}{h_3}\right) - \mathbf{i}_2 h_2 \frac{\partial}{\partial q_2}\left(\frac{1}{h_3}\right)
\end{aligned}$$

A-5 Vector Differential Invariants

To establish the form taken by the vector operator ∇ in orthogonal curvilinear coordinates we proceed as follows. The quantity $\nabla\psi$ is *defined* by the relation

$$d\psi = d\mathbf{R}\cdot\nabla\psi$$

for an arbitrary displacement $d\mathbf{R}$. The function ψ may be a scalar, vector, or polyadic. Expanding the left-hand side and employing the expression for $d\mathbf{R}$ given in Eq. (A-2.1) we obtain

$$\frac{\partial\psi}{\partial q_1}dq_1 + \frac{\partial\psi}{\partial q_2}dq_2 + \frac{\partial\psi}{\partial q_3}dq_3 = \frac{\mathbf{i}_1\cdot\nabla\psi}{h_1}dq_1 + \frac{\mathbf{i}_2\cdot\nabla\psi}{h_2}dq_2 + \frac{\mathbf{i}_3\cdot\nabla\psi}{h_3}dq_3$$

Since q_1, q_2, q_3 are independent variables, it follows that

$$\frac{\partial\psi}{\partial q_1} = \frac{\mathbf{i}_1\cdot\nabla\psi}{h_1}, \qquad \frac{\partial\psi}{\partial q_2} = \frac{\mathbf{i}_2\cdot\nabla\psi}{h_2}, \qquad \frac{\partial\psi}{\partial q_3} = \frac{\mathbf{i}_3\cdot\nabla\psi}{h_3}$$

and thus, in consequence of the orthogonality of the unit vectors,

$$\nabla\psi \equiv \text{grad}\,\psi = \mathbf{i}_1 h_1 \frac{\partial\psi}{\partial q_1} + \mathbf{i}_2 h_2 \frac{\partial\psi}{\partial q_2} + \mathbf{i}_3 h_3 \frac{\partial\psi}{\partial q_3} \tag{A-5.1}$$

For the vector differential operator ∇ alone, we then have

$$\nabla = \mathbf{i}_1 h_1 \frac{\partial}{\partial q_1} + \mathbf{i}_2 h_2 \frac{\partial}{\partial q_2} + \mathbf{i}_3 h_3 \frac{\partial}{\partial q_3} \tag{A-5.2}$$

In conjunction with the formulas of the previous section for the curvilinear derivatives of unit vectors, this expression permits a straightforward—if somewhat lengthy—calculation of the various ∇ operations.

For example, if $\mathbf{u}$ is a vector function whose components are given by

$$\mathbf{u} = \mathbf{i}_1 u_1 + \mathbf{i}_2 u_2 + \mathbf{i}_3 u_3$$

it is relatively simple to demonstrate that

$$\nabla\cdot\mathbf{u} \equiv \text{div}\,\mathbf{u} = h_1 h_2 h_3\left[\frac{\partial}{\partial q_1}\left(\frac{u_1}{h_2 h_3}\right) + \frac{\partial}{\partial q_2}\left(\frac{u_2}{h_3 h_1}\right) + \frac{\partial}{\partial q_3}\left(\frac{u_3}{h_1 h_2}\right)\right] \tag{A-5.3}$$

and

$$\begin{aligned}\nabla\times\mathbf{u} \equiv \text{curl}\,\mathbf{u} = {}& \mathbf{i}_1 h_2 h_3\left[\frac{\partial}{\partial q_2}\left(\frac{u_3}{h_3}\right) - \frac{\partial}{\partial q_3}\left(\frac{u_2}{h_2}\right)\right] \\ &+ \mathbf{i}_2 h_3 h_1\left[\frac{\partial}{\partial q_3}\left(\frac{u_1}{h_1}\right) - \frac{\partial}{\partial q_1}\left(\frac{u_3}{h_3}\right)\right] \\ &+ \mathbf{i}_3 h_1 h_2\left[\frac{\partial}{\partial q_1}\left(\frac{u_2}{h_2}\right) - \frac{\partial}{\partial q_2}\left(\frac{u_1}{h_1}\right)\right]\end{aligned} \tag{A-5.4}$$

or, what is equivalent,

$$\nabla\times\mathbf{u} = h_1 h_2 h_3\begin{vmatrix}\dfrac{\mathbf{i}_1}{h_1} & \dfrac{\mathbf{i}_2}{h_2} & \dfrac{\mathbf{i}_3}{h_3} \\ \dfrac{\partial}{\partial q_1} & \dfrac{\partial}{\partial q_2} & \dfrac{\partial}{\partial q_3} \\ \dfrac{u_1}{h_1} & \dfrac{u_2}{h_2} & \dfrac{u_3}{h_3}\end{vmatrix} \tag{A-5.5}$$

The Laplace operator may be transformed to orthogonal curvilinear coordinates by substituting $\mathbf{u} = \nabla\psi$ in Eq. (A-5.3) and observing from Eq. (A-5.1) that this makes $u_k = h_k\,\partial\psi/\partial q_k$ $(k = 1, 2, 3)$. Hence,

$$\nabla^2\psi = \nabla\cdot\nabla\psi = h_1h_2h_3\left[\frac{\partial}{\partial q_1}\left(\frac{h_1}{h_2h_3}\frac{\partial\psi}{\partial q_1}\right) + \frac{\partial}{\partial q_2}\left(\frac{h_2}{h_3h_1}\frac{\partial\psi}{\partial q_2}\right) + \frac{\partial}{\partial q_3}\left(\frac{h_3}{h_1h_2}\frac{\partial\psi}{\partial q_3}\right)\right] \tag{A-5.6}$$

which gives the Laplacian of a scalar function. For the Laplace operator alone we have

$$\nabla^2 = h_1h_2h_3\left[\frac{\partial}{\partial q_1}\left(\frac{h_1}{h_2h_3}\frac{\partial}{\partial q_1}\right) + \frac{\partial}{\partial q_2}\left(\frac{h_2}{h_3h_1}\frac{\partial}{\partial q_2}\right) + \frac{\partial}{\partial q_3}\left(\frac{h_3}{h_1h_2}\frac{\partial}{\partial q_3}\right)\right] \tag{A-5.7}$$

The computation of the vector $\nabla^2\mathbf{u}$ in curvilinear coordinates can be carried out with the aid of Eq. (A-5.7) and the formulas of Section A-4 for the differentiation of unit vectors. Although the calculation is straightforward in principle, it is rather lengthy in practice. A less tedious method of calculation utilizes the vector identity $\nabla^2\mathbf{u} = \nabla(\nabla\cdot\mathbf{u}) - \nabla\times(\nabla\times\mathbf{u})$. If, in Eq. (A-5.1), we put $\psi = \nabla\cdot\mathbf{u}$ and employ the expression for the divergence of $\mathbf{u}$ given in Eq. (A-5.3), a simple calculation yields

$$\nabla(\nabla\cdot\mathbf{u}) = \mathbf{i}_1 h_1\frac{\partial}{\partial q_1}\left\{h_1h_2h_3\left[\frac{\partial}{\partial q_1}\left(\frac{u_1}{h_2h_3}\right) + \frac{\partial}{\partial q_2}\left(\frac{u_2}{h_3h_1}\right) + \frac{\partial}{\partial q_3}\left(\frac{u_3}{h_1h_2}\right)\right]\right\} + \cdots \tag{A-5.8}$$

Likewise, if we obtain $\nabla\times\mathbf{A}$ from Eq. (A-5.4) and put $\mathbf{A} = \nabla\times\mathbf{u}$, there is no difficulty in obtaining the relation

$$\begin{aligned}\nabla\times(\nabla\times\mathbf{u}) = \mathbf{i}_1h_2h_3\Big\langle&\frac{\partial}{\partial q_2}\left\{\frac{h_1h_2}{h_3}\left[\frac{\partial}{\partial q_1}\left(\frac{u_2}{h_2}\right) - \frac{\partial}{\partial q_2}\left(\frac{u_1}{h_1}\right)\right]\right\}\\ &- \frac{\partial}{\partial q_3}\left\{\frac{h_3h_1}{h_2}\left[\frac{\partial}{\partial q_3}\left(\frac{u_1}{h_1}\right) - \frac{\partial}{\partial q_1}\left(\frac{u_3}{h_3}\right)\right]\right\}\Big\rangle + \cdots\end{aligned} \tag{A-5.9}$$

Combining these two results in accordance with the vector identity previously cited, and simplifying by means of Eq. (A-5.6), the expression for the Laplacian of a vector function may be put in the form

$$\begin{aligned}\nabla^2\mathbf{u} = \mathbf{i}_1\Bigg[&\nabla^2u_1 - \frac{u_1}{h_1}\nabla^2h_1\\ &+ \frac{u_1}{h_1}h_1\frac{\partial}{\partial q_1}(\nabla^2q_1) + \frac{u_2}{h_2}h_1\frac{\partial}{\partial q_1}(\nabla^2q_2) + \frac{u_3}{h_3}h_1\frac{\partial}{\partial q_1}(\nabla^2q_3)\\ &- 2h_1^2\frac{\partial h_1}{\partial q_1}\frac{\partial}{\partial q_1}\left(\frac{u_1}{h_1}\right) - 2h_2^2\frac{\partial h_1}{\partial q_2}\frac{\partial}{\partial q_1}\left(\frac{u_2}{h_2}\right) - 2h_3^2\frac{\partial h_1}{\partial q_3}\frac{\partial}{\partial q_1}\left(\frac{u_3}{h_3}\right)\\ &+ h_1\frac{\partial h_1^2}{\partial q_1}\frac{\partial}{\partial q_1}\left(\frac{u_1}{h_1}\right) + h_1\frac{\partial h_2^2}{\partial q_1}\frac{\partial}{\partial q_2}\left(\frac{u_2}{h_2}\right) + h_1\frac{\partial h_3^2}{\partial q_1}\frac{\partial}{\partial q_3}\left(\frac{u_3}{h_3}\right)\Bigg] + \cdots\end{aligned} \tag{A-5.10}$$

The sixth and ninth terms in brackets cancel one another. They are retained here only to indicate clearly the form taken by the $\mathbf{i}_2$ and $\mathbf{i}_3$ components of $\nabla^2\mathbf{u}$, these being readily obtained by permuting the appropriate subscripts.

A-6 Relations between Cartesian and Orthogonal Curvilinear Coordinates

In most problems involving fluid motion, cartesian coordinates are best suited to the formulation of boundary conditions. On the other hand, the partial differential equations describing the motion are usually more conveniently solved in some other system of orthogonal curvilinear coordinates, characteristic of the geometrical configuration of the fluid domain. This suggests the potential value of certain general relations which enable us to convert handily from one system of coordinates to the other.

We shall assume that the equations describing the coordinate system transformations are given *explicitly* by

$$x = x(q_1, q_2, q_3), \qquad y = y(q_1, q_2, q_3), \qquad z = z(q_1, q_2, q_3) \quad \text{(A–6.1)}$$

The metrical coefficients are then most readily calculated via the relations

$$\frac{1}{h_k^2} = \left(\frac{\partial x}{\partial q_k}\right)^2 + \left(\frac{\partial y}{\partial q_k}\right)^2 + \left(\frac{\partial z}{\partial q_k}\right)^2 \quad (k = 1, 2, 3) \qquad \text{(A–6.2)}$$

(i) *Transformation of partial derivatives:* To express the partial differential operators $\partial/\partial x$, $\partial/\partial y$, and $\partial/\partial z$ in curvilinear coordinates we note that

$$\frac{\partial}{\partial x} = \mathbf{i}\cdot\nabla = (\nabla x)\cdot\nabla, \cdots$$

Upon writing the nabla operator in orthogonal curvilinear coordinates, there is no difficulty in obtaining the relations

$$\frac{\partial}{\partial x} = \sum_{k=1}^{3} h_k^2\left(\frac{\partial x}{\partial q_k}\right)\frac{\partial}{\partial q_k}, \cdots \qquad \text{(A–6.3)}$$

which give the desired expressions.

By allowing these derivatives to operate on q_l ($l = 1, 2, 3$) we are led to interesting relations of the form

$$\frac{\partial q_k}{\partial x} = h_k^2 \frac{\partial x}{\partial q_k}, \cdots \quad (k = 1, 2, 3) \qquad \text{(A–6.4)}$$

These might also have been obtained by equating Eqs. (A-1.7) and (A-2.9). They are of particular value in evaluating the derivatives $\partial x/\partial q_k, \ldots$, required in Eq. (A-6.3) and the sequel, when the transformation equations are given *explicitly* by

$$q_1 = q_1(x, y, z), \qquad q_2 = q_2(x, y, z), \qquad q_3 = q_3(x, y, z)$$

rather than Eq. (A-6.1), as assumed.

The transformation of partial derivatives inverse to Eq. (A-6.3) is simply obtained by application of the "chain rule" for partial differentiation,

$$\frac{\partial}{\partial q_k} = \frac{\partial x}{\partial q_k}\frac{\partial}{\partial x} + \frac{\partial y}{\partial q_k}\frac{\partial}{\partial y} + \frac{\partial z}{\partial q_k}\frac{\partial}{\partial z} = \nabla q_k \cdot \nabla \quad (k = 1, 2, 3) \qquad \text{(A–6.5)}$$

(ii) *Transformation of unit vectors:* Writing $\mathbf{R} = \mathbf{i}x + \mathbf{j}y + \mathbf{k}z$ in Eq. (A-1.7), we obtain

$$\mathbf{i}_k = h_k\left(\mathbf{i}\frac{\partial x}{\partial q_k} + \mathbf{j}\frac{\partial y}{\partial q_k} + \mathbf{k}\frac{\partial z}{\partial q_k}\right) \quad (k = 1, 2, 3) \qquad \text{(A–6.6)}$$

which gives the transformation of unit vectors from curvilinear to cartesian coordinates.

The inverse transformation of unit vectors follows immediately from the relations

$$\mathbf{i} = \nabla x, \cdots$$

by expressing ∇ in curvilinear coordinates. Thus,

$$\mathbf{i} = \sum_{k=1}^{3} \mathbf{i}_k h_k \frac{\partial x}{\partial q_k}, \cdots \qquad \text{(A–6.7)}$$

with analogous formulas for $\mathbf{j}$ and $\mathbf{k}$.

Inasmuch as for $k = 1, 2, 3$

$$\cos(x, q_k) = \frac{\partial x/\partial q_k}{[(\partial x/\partial q_k)^2 + (\partial y/\partial q_k)^2 + (\partial z/\partial q_k)^2]^{1/2}} = h_k \frac{\partial x}{\partial q_k}, \cdots$$

these formulas have an obvious geometric interpretation.

(iii) *Transformation of vector components:* If we put

$$\mathbf{u} = \mathbf{i}u_x + \mathbf{j}u_y + \mathbf{k}u_z = \mathbf{i}_1 u_1 + \mathbf{i}_2 u_2 + \mathbf{i}_3 u_3$$

then scalar multiplication of $\mathbf{u}$ with Eqs. (A-6.6) and (A-6.7), respectively, gives

$$u_k = h_k\left(u_x\frac{\partial x}{\partial q_k} + u_y\frac{\partial y}{\partial q_k} + u_z\frac{\partial z}{\partial q_k}\right) \quad (k = 1, 2, 3) \qquad \text{(A–6.8)}$$

and

$$u_x - \sum_{k=1}^{3} u_k h_k \frac{\partial x}{\partial q_k}, \cdots \qquad \text{(A–6.9)}$$

These permit rapid conversion of vector components between cartesian and curvilinear coordinates.

(iv) *Position vector:* Writing $\mathbf{R}$ in cartesian coordinates and employing Eq. (A-6.7) results in

$$\mathbf{R} = \sum_{k=1}^{3} \mathbf{i}_k h_k\left(x\frac{\partial x}{\partial q_k} + y\frac{\partial y}{\partial q_k} + z\frac{\partial z}{\partial q_k}\right)$$
$$= \frac{1}{2}\sum_{k=1}^{3} \mathbf{i}_k h_k \frac{\partial}{\partial q_k}(x^2 + y^2 + z^2) = \frac{1}{2}\nabla r^2 \qquad \text{(A–6.10)}$$

where we have set $r^2 = x^2 + y^2 + z^2$.

A-7 Dyadics in Orthogonal Curvilinear Coordinates

The most general dyadic may be written in the form*

$$\boldsymbol{\phi} = \sum_{j=1}^{3} \sum_{k=1}^{3} \mathbf{i}_j \mathbf{i}_k \phi_{jk} \tag{A–7.1}$$

where $(\mathbf{i}_1, \mathbf{i}_2, \mathbf{i}_3)$ are unit vectors in a curvilinear coordinate system. It should be clearly understood that the nine scalar numbers ϕ_{jk} $(j, k = 1, 2, 3)$ depend upon the particular system of curvilinear coordinates under discussion. We shall restrict our attention only to *orthogonal* systems.

The transpose or conjugate of the dyadic $\boldsymbol{\phi}$ is

$$\boldsymbol{\phi}^{\dagger} = \sum_{j=1}^{3} \sum_{k=1}^{3} \mathbf{i}_k \mathbf{i}_j \phi_{jk} = \sum_{j=1}^{3} \sum_{k=1}^{3} \mathbf{i}_j \mathbf{i}_k \phi_{kj} \tag{A–7.2}$$

That the two summations are identical depends on the fact that the indices j and k are dummy indices.

(i) *Idemfactor:* Since the idemfactor is given by the expression

$$\mathbf{I} = \nabla \mathbf{R} \tag{A–7.3}$$

we find from Eq. (A-5.2) that

$$\mathbf{I} = \mathbf{i}_1 h_1 \frac{\partial \mathbf{R}}{\partial q_1} + \mathbf{i}_2 h_2 \frac{\partial \mathbf{R}}{\partial q_2} + \mathbf{i}_3 h_3 \frac{\partial \mathbf{R}}{\partial q_3}$$

But, from Eq. (A-1.7), $\mathbf{i}_k = h_k \partial \mathbf{R} / \partial q_k$, whence

$$\mathbf{I} = \mathbf{i}_1 \mathbf{i}_1 + \mathbf{i}_2 \mathbf{i}_2 + \mathbf{i}_3 \mathbf{i}_3 \tag{A–7.4}$$

which may also be expressed as

$$\mathbf{I} = \sum_j \sum_k \mathbf{i}_j \mathbf{i}_k \delta_{jk} \tag{A–7.5}$$

where

$$\delta_{jk} = \begin{cases} 1 & \text{if } j = k \\ 0 & \text{if } j \neq k \end{cases} \tag{A–7.6}$$

is the Kronecker delta.

(ii) *Gradient of a vector:* An important dyadic in physical applications arises from forming the gradient of a vector function. By writing [see Eq. (A-5.2)]

$$\nabla = \sum_{j=1}^{3} \mathbf{i}_j h_j \frac{\partial}{\partial q_j}, \qquad \mathbf{u} = \sum_{k=1}^{3} \mathbf{i}_k u_k$$

we find, bearing in mind that $\mathbf{i}_k$ is in general a variable vector,

*In actual manipulations it is convenient to introduce the Einstein summation convention, whereby the summation symbols are suppressed, it being understood that repeated indices are to be summed. With this convention, Eq. (A-7.1) may be written

$$\boldsymbol{\phi} = \mathbf{i}_j \mathbf{i}_k \phi_{jk}$$

We shall, however, not use this convention here.

$$\nabla \mathbf{u} = \sum_{j=1}^{3}\sum_{k=1}^{3} h_j \mathbf{i}_j\left(\mathbf{i}_k \frac{\partial u_k}{\partial q_j} + u_k \frac{\partial \mathbf{i}_k}{\partial q_j}\right)$$

Using the formulas in Eq. (A-4.8) for the derivatives of the unit vectors, $\partial \mathbf{i}_k/\partial q_j$, we obtain

$$\begin{aligned}
\nabla \mathbf{u} = {} & \mathbf{i}_1\mathbf{i}_1 h_1\left[\frac{\partial u_1}{\partial q_1} + h_2 u_2 \frac{\partial}{\partial q_2}\left(\frac{1}{h_1}\right) + h_3 u_3 \frac{\partial}{\partial q_3}\left(\frac{1}{h_1}\right)\right] \\
& + \mathbf{i}_1\mathbf{i}_2 h_1\left[\frac{\partial u_2}{\partial q_1} - h_2 u_1 \frac{\partial}{\partial q_2}\left(\frac{1}{h_1}\right)\right] \\
& + \mathbf{i}_1\mathbf{i}_3 h_1\left[\frac{\partial u_3}{\partial q_1} - h_3 u_1 \frac{\partial}{\partial q_3}\left(\frac{1}{h_1}\right)\right] \\
& + \mathbf{i}_2\mathbf{i}_1 h_2\left[\frac{\partial u_1}{\partial q_2} - h_1 u_2 \frac{\partial}{\partial q_1}\left(\frac{1}{h_2}\right)\right] \\
& + \mathbf{i}_2\mathbf{i}_2 h_2\left[\frac{\partial u_2}{\partial q_2} + h_3 u_3 \frac{\partial}{\partial q_3}\left(\frac{1}{h_2}\right) + h_1 u_1 \frac{\partial}{\partial q_1}\left(\frac{1}{h_2}\right)\right] \\
& + \mathbf{i}_2\mathbf{i}_3 h_2\left[\frac{\partial u_3}{\partial q_2} - h_3 u_2 \frac{\partial}{\partial q_3}\left(\frac{1}{h_2}\right)\right] \\
& + \mathbf{i}_3\mathbf{i}_1 h_3\left[\frac{\partial u_1}{\partial q_3} - h_1 u_3 \frac{\partial}{\partial q_1}\left(\frac{1}{h_3}\right)\right] \\
& + \mathbf{i}_3\mathbf{i}_2 h_3\left[\frac{\partial u_2}{\partial q_3} - h_2 u_3 \frac{\partial}{\partial q_2}\left(\frac{1}{h_3}\right)\right] \\
& + \mathbf{i}_3\mathbf{i}_3 h_3\left[\frac{\partial u_3}{\partial q_3} + h_1 u_1 \frac{\partial}{\partial q_1}\left(\frac{1}{h_3}\right) + h_2 u_2 \frac{\partial}{\partial q_2}\left(\frac{1}{h_3}\right)\right]
\end{aligned} \tag{A-7.7}$$

(iii) *Divergence of a dyadic:* Using Eqs. (A-5.2) and (A-7.1), we have

$$\nabla = \sum_{j=1}^{3} \mathbf{i}_j h_j \frac{\partial}{\partial q_j}, \quad \text{and} \quad \boldsymbol{\phi} = \sum_{k=1}^{3}\sum_{l=1}^{3} \mathbf{i}_k \mathbf{i}_l \phi_{kl}$$

whence

$$\nabla \cdot \boldsymbol{\phi} = \sum_{j=1}^{3}\sum_{k=1}^{3}\sum_{l=1}^{3} h_j \mathbf{i}_j \cdot \left[\left(\frac{\partial \mathbf{i}_k}{\partial q_j}\right)\mathbf{i}_l \phi_{kl} + \mathbf{i}_k\left(\frac{\partial \mathbf{i}_l}{\partial q_j}\right)\phi_{kl} + \mathbf{i}_k \mathbf{i}_l \frac{\partial \phi_{kl}}{\partial q_j}\right]$$

Using Eq. (A-4.8) to differentiate the unit vectors, and bearing in mind that $\mathbf{i}_j \cdot \mathbf{i}_k = \delta_{jk}$, we eventually obtain

$$\begin{aligned}
\nabla \cdot \boldsymbol{\phi} = \mathbf{i}_1 \Bigg[& h_1 h_2 h_3 \left\{\frac{\partial}{\partial q_1}\left(\frac{\phi_{11}}{h_2 h_3}\right) + \frac{\partial}{\partial q_2}\left(\frac{\phi_{21}}{h_3 h_1}\right) + \frac{\partial}{\partial q_3}\left(\frac{\phi_{31}}{h_1 h_2}\right)\right\} \\
& + h_1 h_1 \phi_{11} \frac{\partial}{\partial q_1}\left(\frac{1}{h_1}\right) + h_1 h_2 \phi_{12} \frac{\partial}{\partial q_2}\left(\frac{1}{h_1}\right) + h_1 h_3 \phi_{13} \frac{\partial}{\partial q_3}\left(\frac{1}{h_1}\right) \\
& - h_1 h_1 \phi_{11} \frac{\partial}{\partial q_1}\left(\frac{1}{h_1}\right) - h_1 h_2 \phi_{22} \frac{\partial}{\partial q_1}\left(\frac{1}{h_2}\right) - h_1 h_3 \phi_{33} \frac{\partial}{\partial q_1}\left(\frac{1}{h_3}\right)\Bigg] + \cdots
\end{aligned} \tag{A-7.8}$$

The fourth and seventh terms cancel one another, and are included only to show the structure of the general formula. The two remaining components of the preceding vector can be written down by appropriate permutation of the subscripts.

A-8 Cylindrical Coordinate Systems (q_1, q_2, z)

In this section we deal with the special class of curvilinear coordinate systems defined by the equations

$$x = x(q_1, q_2), \qquad y = y(q_1, q_2), \qquad z = q_3 \tag{A–8.1}$$

Since q_1 and q_2 depend only on x and y they can be regarded as curvilinear coordinates in a *plane* perpendicular to the z axis. The curves in this plane on which q_1 and q_2 are constant are then the generators of the coordinate surfaces q_1 = constant and q_2 = constant, obtained by moving the plane perpendicular to itself. These coordinate surfaces have the shape of cylinders.

We shall further restrict ourselves to situations in which the system of curvilinear coordinates defined in Eq. (A-8.1) is both orthogonal and right-handed. In the present instance,

$$\frac{\partial z}{\partial q_1} = 0, \qquad \frac{\partial z}{\partial q_2} = 0, \qquad \frac{\partial z}{\partial q_3} = 1$$

Thus, in accordance with Eq. (A-2.6), the stipulation that (q_1, q_2, q_3) form an *orthogonal* system requires only that

$$\frac{\partial x}{\partial q_1}\frac{\partial x}{\partial q_2} + \frac{\partial y}{\partial q_1}\frac{\partial y}{\partial q_2} = 0 \tag{A–8.2}$$

Furthermore, expansion of the Jacobian determinant in Eq. (A-2.15) shows that the system will be *right-handed* if we order the curvilinear coordinates defined in Eq. (A-8.1) so as to satisfy the inequality

$$\frac{\partial x}{\partial q_1}\frac{\partial y}{\partial q_2} - \frac{\partial x}{\partial q_2}\frac{\partial y}{\partial q_1} = \frac{\partial(x, y)}{\partial(q_1, q_2)} > 0 \tag{A–8.3}$$

The metrical coefficients are now given by

$$\frac{1}{h_1^2} = \left(\frac{\partial x}{\partial q_1}\right)^2 + \left(\frac{\partial y}{\partial q_1}\right)^2, \qquad \frac{1}{h_2^2} = \left(\frac{\partial x}{\partial q_2}\right)^2 + \left(\frac{\partial y}{\partial q_2}\right)^2, \qquad h_3 = 1 \tag{A–8.4}$$

The majority of important cylindrical coordinate systems fulfilling these criteria can be generated via the theory of analytic functions. This is discussed at further length in Section A–10 and examples given in Sections A-11–A-13. *Circular* cylindrical coordinates, discussed in Section A-9, constitute an important exception to this generalization.

A-9 Circular Cylindrical Coordinates (ρ, ϕ, z)

[Figs. A-9.1 (a), (b), (c)]

Circular cylindrical coordinates,

$$q_1 = \rho, \qquad q_2 = \phi, \qquad q_3 = z \tag{A–9.1}$$

are defined by the relations

$$x = \rho \cos\phi, \qquad y = \rho \sin\phi, \qquad z = z \tag{A–9.2}$$

These may be solved explicitly for the curvilinear coordinates,

$$\rho = (x^2 + y^2)^{1/2}, \qquad \phi = \tan^{-1}\frac{y}{x}, \qquad z = z \qquad \text{(A-9.3)}$$

This system of coordinates is depicted from various points of view in Figs. A-9.1(a), (b), (c).

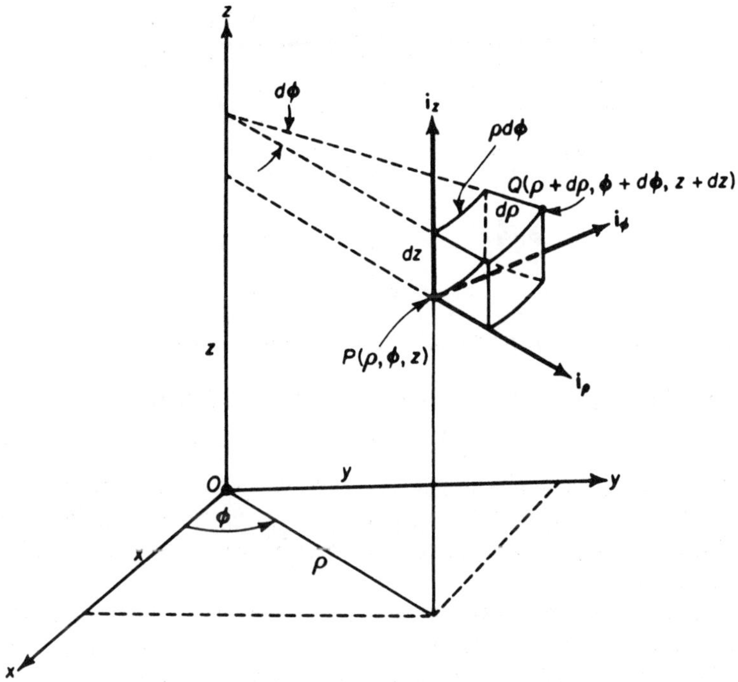

Figure A-9.1(a). Circular cylindrical coordinates.

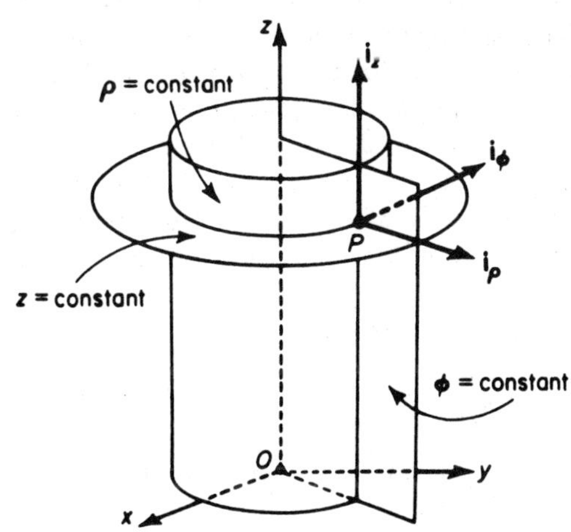

Figure A-9.1(b). Circular cylindrical coordinate system surfaces.

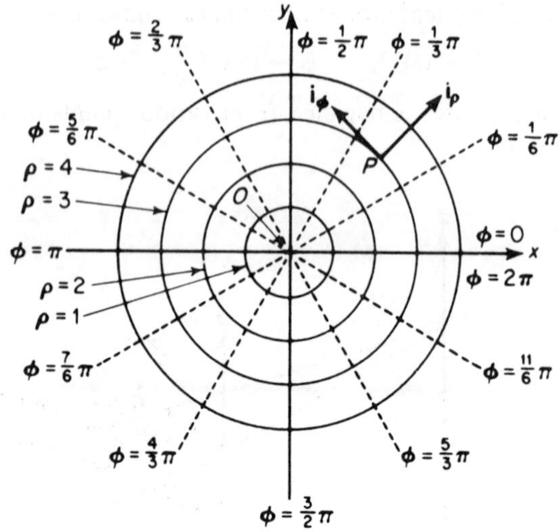

Figure A-9.1(c). Circular cylindrical coordinate system curves.

By restricting the ranges of these coordinates as follows:

$$0 \leq \rho < \infty, \qquad 0 \leq \phi < 2\pi, \qquad -\infty < z < \infty \tag{A-9.4}$$

each point in space is given once and only once with the exception of those points along the z axis, for which ϕ is undetermined. It is understood that when applying Eq. (A-9.3), ϕ (radians) is to be measured in the quadrant in which the point lies.

The family of coordinate surfaces $\rho =$ constant are concentric cylinders whose longitudinal axes coincide with the z axis. The coordinate surfaces $\phi =$ constant are vertical half-planes containing, and terminating along, the z axis. If $\phi = \phi_o =$ constant ($<\pi$) is one of these semi-infinite planes, the extension of this plane across the z axis corresponds to the value $\phi = \phi_o + \pi =$ constant. The coordinate surfaces $z =$ constant are horizontal planes.

The ρ coordinate curves, formed by the intersection of the planes $\phi =$ constant and $z =$ constant, are horizontal rays issuing from the z axis. The ϕ coordinate curves, produced by the intersection of the cylinders $\rho =$ constant with the planes $z =$ constant, are concentric, horizontal circles having the z axis at their center. The z coordinate curves, resulting from the intersection of the cylinders $\rho =$ constant with the planes $\phi =$ constant, are vertical lines.

Upon calculating the derivatives,

$$\frac{\partial x}{\partial q_1} = \cos \phi, \qquad \frac{\partial x}{\partial q_2} = -\rho \sin \phi$$

$$\frac{\partial y}{\partial q_1} = \sin \phi, \qquad \frac{\partial y}{\partial q_2} = \rho \cos \phi$$

from Eq. (A-9.2), and employing the formulas of Section A-8, it follows that circular cylindrical coordinates constitute a right-handed system of orthogonal curvilinear coordinates whose metrical coefficients are given by

$$h_1 = 1, \qquad h_2 = \frac{1}{\rho}, \qquad h_3 = 1 \tag{A–9.5}$$

We tabulate here for reference some of the more important properties of these coordinates:

$$x + iy = \rho e^{i\phi} = e^{(\ln\rho)+i\phi} \tag{A–9.6}$$

$$\begin{gathered} dl_1 = d\rho, \qquad dl_2 = \rho\, d\phi, \qquad dl_3 = dz \\ dl^2 = (d\rho)^2 + \rho^2 (d\phi)^2 + (dz)^2 \end{gathered} \tag{A–9.7}$$

$$dS_1 = \rho\, d\phi\, dz, \qquad dS_2 = d\rho\, dz, \qquad dS_3 = \rho\, d\rho\, d\phi \tag{A–9.8}$$

$$dV = \rho\, d\rho\, d\phi\, dz \tag{A–9.9}$$

$$\begin{aligned} &\mathbf{i}_\rho \cdot \mathbf{i}_\rho = 1, \qquad \mathbf{i}_\phi \cdot \mathbf{i}_\phi = 1, \qquad \mathbf{i}_z \cdot \mathbf{i}_z^* = 1 \\ &\mathbf{i}_\rho \cdot \mathbf{i}_\phi = 0, \qquad \mathbf{i}_\phi \cdot \mathbf{i}_z = 0, \qquad \mathbf{i}_z \cdot \mathbf{i}_\rho = 0 \\ &\mathbf{i}_\rho \times \mathbf{i}_\phi = \mathbf{i}_z, \qquad \mathbf{i}_\phi \times \mathbf{i}_z = \mathbf{i}_\rho, \qquad \mathbf{i}_z \times \mathbf{i}_\rho = \mathbf{i}_\phi \end{aligned} \tag{A–9.10}$$

$$d\mathbf{R} = \mathbf{i}_\rho\, d\rho + \mathbf{i}_\phi\, \rho\, d\phi + \mathbf{i}_z\, dz \tag{A–9.11}$$

$$\mathbf{R} = \mathbf{i}_\rho \rho + \mathbf{i}_z z \tag{A–9.12}$$

All partial derivatives of unit vectors are zero except for

$$\frac{\partial \mathbf{i}_\rho}{\partial \phi} = \mathbf{i}_\phi, \qquad \frac{\partial \mathbf{i}_\phi}{\partial \phi} = -\mathbf{i}_\rho \tag{A–9.13}$$

$$\nabla\psi = \mathbf{i}_\rho \frac{\partial \psi}{\partial \rho} + \mathbf{i}_\phi \frac{1}{\rho}\frac{\partial \psi}{\partial \phi} + \mathbf{i}_z \frac{\partial \psi}{\partial z} \tag{A–9.14}$$

$$\nabla^2\psi = \frac{1}{\rho}\frac{\partial}{\partial \rho}\left(\rho \frac{\partial \psi}{\partial \rho}\right) + \frac{1}{\rho^2}\frac{\partial^2 \psi}{\partial \phi^2} + \frac{\partial^2 \psi}{\partial z^2} \tag{A–9.15}$$

$$\mathbf{u} = \mathbf{i}_\rho u_\rho + \mathbf{i}_\phi u_\phi + \mathbf{i}_z u_z \tag{A–9.16}$$

$$|\mathbf{u}|^2 = u_\rho^2 + u_\phi^2 + u_z^2 \tag{A–9.17}$$

$$\nabla \cdot \mathbf{u} = \frac{1}{\rho}\frac{\partial}{\partial \rho}(\rho u_\rho) + \frac{1}{\rho}\frac{\partial u_\phi}{\partial \phi} + \frac{\partial u_z}{\partial z} \tag{A–9.18}$$

$$\nabla \times \mathbf{u} = \mathbf{i}_\rho\left(\frac{1}{\rho}\frac{\partial u_z}{\partial \phi} - \frac{\partial u_\phi}{\partial z}\right) + \mathbf{i}_\phi\left(\frac{\partial u_\rho}{\partial z} - \frac{\partial u_z}{\partial \rho}\right) + \mathbf{i}_z\left[\frac{1}{\rho}\frac{\partial}{\partial \rho}(\rho u_\phi) - \frac{1}{\rho}\frac{\partial u_\rho}{\partial \phi}\right] \tag{A–9.19}$$

$$\nabla^2 \mathbf{u} = \mathbf{i}_\rho\left(\nabla^2 u_\rho - \frac{2}{\rho^2}\frac{\partial u_\phi}{\partial \phi} - \frac{u_\rho}{\rho^2}\right) + \mathbf{i}_\phi\left(\nabla^2 u_\phi + \frac{2}{\rho^2}\frac{\partial u_\rho}{\partial \phi} - \frac{u_\phi}{\rho^2}\right) + \mathbf{i}_z \nabla^2 u_z \tag{A–9.20}$$

*To be consistent we are here using the symbol $\mathbf{i}_z$ in place of $\mathbf{k}$.

$$\frac{\partial}{\partial x} = \cos\phi\frac{\partial}{\partial\rho} - \frac{\sin\phi}{\rho}\frac{\partial}{\partial\phi}, \qquad \frac{\partial}{\partial y} = \sin\phi\frac{\partial}{\partial\rho} + \frac{\cos\phi}{\rho}\frac{\partial}{\partial\phi} \tag{A-9.21}$$

$$\frac{\partial}{\partial\rho} = \cos\phi\frac{\partial}{\partial x} + \sin\phi\frac{\partial}{\partial y}, \qquad \frac{\partial}{\partial\phi} = -\rho\sin\phi\frac{\partial}{\partial x} + \rho\cos\phi\frac{\partial}{\partial y} \tag{A-9.22}$$

$$\mathbf{i} = \mathbf{i}_\rho\cos\phi - \mathbf{i}_\phi\sin\phi, \qquad \mathbf{j} = \mathbf{i}_\rho\sin\phi + \mathbf{i}_\phi\cos\phi \tag{A-9.23}$$

$$\mathbf{i}_\rho = \mathbf{i}\cos\phi + \mathbf{j}\sin\phi, \qquad \mathbf{i}_\phi = -\mathbf{i}\sin\phi + \mathbf{j}\cos\phi \tag{A-9.24}$$

$$u_x = u_\rho\cos\phi - u_\phi\sin\phi, \qquad u_y = u_\rho\sin\phi + u_\phi\cos\phi \tag{A-9.25}$$

$$u_\rho = u_x\cos\phi + u_y\sin\phi, \qquad u_\phi = -u_x\sin\phi + u_y\cos\phi \tag{A-9.26}$$

$$\begin{aligned}\nabla\mathbf{u} = {} & \mathbf{i}_\rho\mathbf{i}_\rho\frac{\partial u_\rho}{\partial\rho} + \mathbf{i}_\rho\mathbf{i}_\phi\frac{\partial u_\phi}{\partial\rho} + \mathbf{i}_\rho\mathbf{i}_z\frac{\partial u_z}{\partial\rho} \\ & + \mathbf{i}_\phi\mathbf{i}_\rho\frac{1}{\rho}\left(\frac{\partial u_\rho}{\partial\phi} - u_\phi\right) + \mathbf{i}_\phi\mathbf{i}_\phi\frac{1}{\rho}\left(\frac{\partial u_\phi}{\partial\phi} + u_\rho\right) + \mathbf{i}_\phi\mathbf{i}_z\frac{1}{\rho}\frac{\partial u_z}{\partial\phi} \\ & + \mathbf{i}_z\mathbf{i}_\rho\frac{\partial u_\rho}{\partial z} + \mathbf{i}_z\mathbf{i}_\phi\frac{\partial u_\phi}{\partial z} + \mathbf{i}_z\mathbf{i}_z\frac{\partial u_z}{\partial z}\end{aligned} \tag{A-9.27}$$

and

$$\begin{aligned}\nabla\cdot\boldsymbol{\tau} = {} & \mathbf{i}_\rho\left[\frac{1}{\rho}\frac{\partial}{\partial\rho}(\rho\tau_{\rho\rho}) + \frac{1}{\rho}\frac{\partial\tau_{\phi\rho}}{\partial\phi} + \frac{\partial\tau_{z\rho}}{\partial z} - \frac{\tau_{\phi\phi}}{\rho}\right] \\ & + \mathbf{i}_\phi\left[\frac{1}{\rho}\frac{\partial}{\partial\rho}(\rho\tau_{\rho\phi}) + \frac{1}{\rho}\frac{\partial\tau_{\phi\phi}}{\partial\phi} + \frac{\partial\tau_{z\phi}}{\partial z} + \frac{\tau_{\phi\rho}}{\rho}\right] \\ & + \mathbf{i}_z\left[\frac{1}{\rho}\frac{\partial}{\partial\rho}(\rho\tau_{\rho z}) + \frac{1}{\rho}\frac{\partial\tau_{\phi z}}{\partial\phi} + \frac{\partial\tau_{zz}}{\partial z}\right]\end{aligned} \tag{A-9.28}$$

where $\boldsymbol{\tau}$ is the dyadic

$$\begin{aligned}\boldsymbol{\tau} = {} & \mathbf{i}_\rho\mathbf{i}_\rho\tau_{\rho\rho} + \mathbf{i}_\rho\mathbf{i}_\phi\tau_{\rho\phi} + \mathbf{i}_\rho\mathbf{i}_z\tau_{\rho z} \\ & + \mathbf{i}_\phi\mathbf{i}_\rho\tau_{\phi\rho} + \mathbf{i}_\phi\mathbf{i}_\phi\tau_{\phi\phi} + \mathbf{i}_\phi\mathbf{i}_z\tau_{\phi z} \\ & + \mathbf{i}_z\mathbf{i}_\rho\tau_{z\rho} + \mathbf{i}_z\mathbf{i}_\phi\tau_{z\phi} + \mathbf{i}_z\mathbf{i}_z\tau_{zz}\end{aligned}$$

A-10 Conjugate Cylindrical Coordinate Systems

Transformations of the type

$$x + iy = f(q_1 + iq_2), \quad z = q_3 \tag{A-10.1}$$

generate systems of cylindrical curvilinear coordinate systems, for upon equating real and imaginary parts we obtain

$$x = x(q_1, q_2), \qquad y = y(q_1, q_2), \qquad z = q_3$$

It follows from the Cauchy-Riemann equations applied to Eq. (A-10.1) that

$$\frac{\partial x}{\partial q_1} = \frac{\partial y}{\partial q_2} \quad \text{and} \quad \frac{\partial x}{\partial q_2} = -\frac{\partial y}{\partial q_1} \tag{A-10.2}$$

Coupled with Eqs. (A-8.2), (A-8.3), and (A-8.4), these are sufficient to show that the system of curvilinear coordinates generated by Eq. (A-10.1),

(q_1, q_2, q_3)—in that order—form a right-handed system of orthogonal, curvilinear coordinates whose metrical coefficients are given by

$$\frac{1}{h_1^2} = \frac{1}{h_2^2} = \left(\frac{\partial x}{\partial q_1}\right)^2 + \left(\frac{\partial y}{\partial q_1}\right)^2 = \cdots = \left|\frac{d(x+iy)}{d(q_1+iq_2)}\right|^2 \tag{A–10.3}$$

and

$$h_3 = 1 \tag{A–10.4}$$

Coordinate systems of the type under discussion are termed *conjugate* systems, for obvious reasons. Important examples of these are elliptic cylinder coordinates, bipolar cylinder coordinates, and parabolic cylinder coordinates, discussed in Sections A-11–A-13.

A-11 Elliptic Cylinder Coordinates (ξ, η, z) (Fig. A–11.1)

The transformation

$$x + iy = c \cosh(\xi + i\eta) \tag{A–11.1}$$

$c > 0$, yields, upon expanding the right-hand side and equating real and imaginary parts,

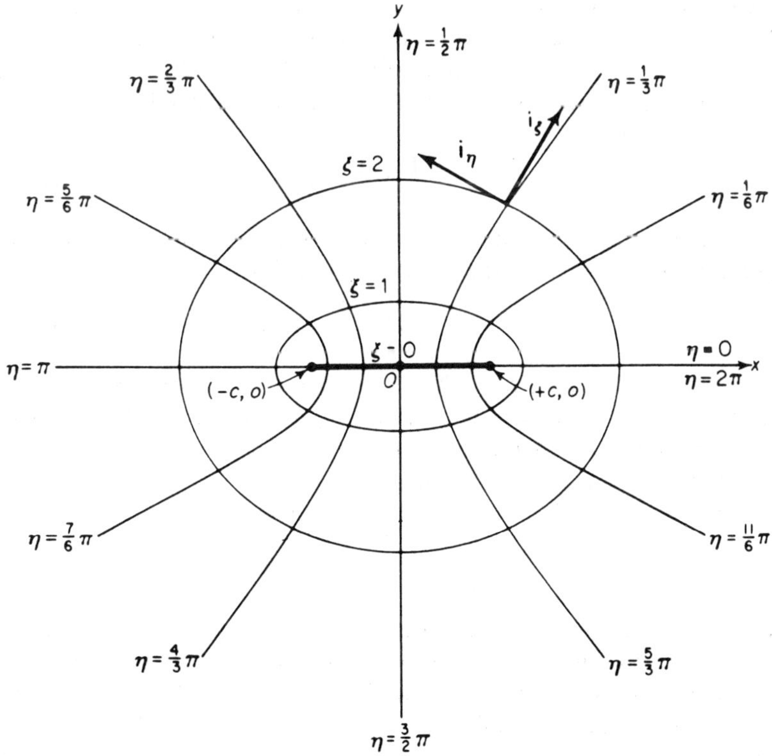

Figure A-11.1. Elliptic cylinder coordinates.

$$x = c \cosh \xi \cos \eta, \qquad y = c \sinh \xi \sin \eta \tag{A–11.2}$$

Referring to Fig. A-11.1 it is clear that if we restrict the *elliptic coordinates* (ξ, η) to the ranges

$$0 \leq \xi < \infty, \qquad 0 \leq \eta < 2\pi \tag{A–11.3}$$

each point (x, y) in a plane $z =$ constant is represented at least once and, with the exception of the points $(-c < x < c, y = 0)$ (for which η is doubled-valued), only once.

Eliminating η from Eq. (A-11.2) gives

$$\frac{x^2}{c^2 \cosh^2 \xi} + \frac{y^2}{c^2 \sinh^2 \xi} = 1 \tag{A–11.4}$$

Thus, the family of curves in the xy plane characterized by the parameters $\xi =$ constant are ellipses having their centers at the origin. In addition, since $\cosh \xi > \sinh \xi \geq 0$, the major and minor semiaxes, a_o and b_o, respectively, of a typical ellipse, $\xi = \xi_o =$ constant, are

$$a_o = c \cosh \xi_o, \qquad b_o = c \sinh \xi_o \tag{A–11.5}$$

These lie along the x and y axes, respectively. From Eq. (A-11.5)

$$a_o^2 - b_o^2 = c^2 \tag{A–11.6}$$

from which it follows that the family of ellipses are confocal; that is, every ellipse of the family has the same foci. The two foci are on the x axis at the points $(x = \pm c, y = 0)$, corresponding to the values $(\xi = 0, \eta = 0)$ and $(\xi = 0, \eta = \pi)$, respectively. Upon eliminating c from Eq. (A-11.5) we eventually obtain

$$\xi_o = \frac{1}{2} \ln \frac{a_o + b_o}{a_o - b_o} \tag{A–11.7}$$

which expresses the parameter ξ_o in terms of the lengths of the semiaxes. The eccentricity of the ellipse $\xi = \xi_o =$ constant is

$$e_o = \left[1 - \left(\frac{b_o}{a_o}\right)^2\right]^{1/2} = \operatorname{sech} \xi_o \tag{A–11.8}$$

Equation (A-11.2) shows that when $\xi_o = 0$ the ellipse is degenerate and corresponds to that segment of the x axis lying between the two foci; that is, the line is composed of the points $(-c \leq x \leq c, y = 0)$. As $\xi_o \to \infty$ the ellipse approaches a circle of infinite radius.

Upon eliminating ξ from Eq. (A-11.2) we find

$$\frac{x^2}{c^2 \cos^2 \eta} - \frac{y^2}{c^2 \sin^2 \eta} = 1 \tag{A–11.9}$$

The family of curves in the xy plane corresponding to different constant values of the parameter η are, therefore, hyperbolas whose principal axes coincide with the x axis. Closer inspection of Eq. (A-11.2) reveals that each curve $\eta =$ constant is actually only one-quarter of a hyperbola; if $\eta = \eta_o$ $=$ constant $< \pi/2$ is that branch of the hyperbola which lies in the first quadrant, the values of η corresponding to those branches of the same hyperbola which lie in the second, third, and fourth quadrants are $\pi - \eta_o$, $\pi + \eta_o$, and $2\pi - \eta_o$, respectively. The major and minor semiaxes, A_o and B_o, of a typical hyperbola, $\eta = \eta_o =$ constant, are

$$A_o = c\,|\cos\eta_o|, \qquad B_o = c\,|\sin\eta_o| \tag{A–11.10}$$

so that

$$A_o^2 + B_o^2 = c^2 \tag{A–11.11}$$

from which it follows that the family of hyperbolas are confocal, having the same foci as the confocal ellipses. In terms of the semiaxes, the parameter η_o is given by

$$\eta_o = \tan^{-1}\frac{B_o}{A_o} \tag{A–11.12}$$

When $\eta_o = 0$, we find from Eq. (A-11.2) that the corresponding hyperbola is degenerate, reducing to the straight line which extends from the focus $x = c$ to $x = +\infty$ along the x axis. Likewise, for $\eta = \pi$, the hyperbola again becomes a straight line, extending along the x axis from the focus $x = -c$ to $x = -\infty$. For $\eta = \pi/2$ and $3\pi/2$, the hyperbola becomes the upper and lower halves of the y axis, respectively.

It is possible to ascribe a geometric significance to the elliptic coordinates ξ and η. This can be established without difficulty from the geometric interpretation given to *prolate spheroidal coordinates* in Section A-17.

If we put

$$q_1 = \xi, \qquad q_2 = \eta, \qquad q_3 = z \tag{A–11.13}$$

the coordinate surfaces $\xi =$ constant are confocal elliptic cylinders, whereas the surfaces $\eta =$ constant are confocal hyperbolic cylinders. *Elliptic cylinder coordinates* (ξ, η, z) constitute a right-handed system of orthogonal curvilinear coordinates whose metrical coefficients are

$$h_1 = h_2 = \frac{1}{c(\sinh^2\xi + \sin^2\eta)^{1/2}} \quad \text{or} \quad \frac{2^{1/2}}{c(\cosh 2\xi - \cos 2\eta)^{1/2}}$$

$$h_3 = 1 \tag{A–11.14}$$

Typical unit vectors are shown in Fig. A-11.1, the unit vector $\mathbf{i}_z$ being directed out of the page at the reader.

A-12 Bipolar Cylinder Coordinates (ξ, η, z)

[Figs. A-12.1 (a), (b)]

Bipolar coordinates (ξ, η) in a plane are defined by the transformation

$$x + iy = ic\cot\tfrac{1}{2}(\xi + i\eta) \tag{A–12.1}$$

$c > 0$, from which we obtain

$$x = c\frac{\sinh\eta}{\cosh\eta - \cos\xi}, \qquad y = c\frac{\sin\xi}{\cosh\eta - \cos\xi} \tag{A–12.2}$$

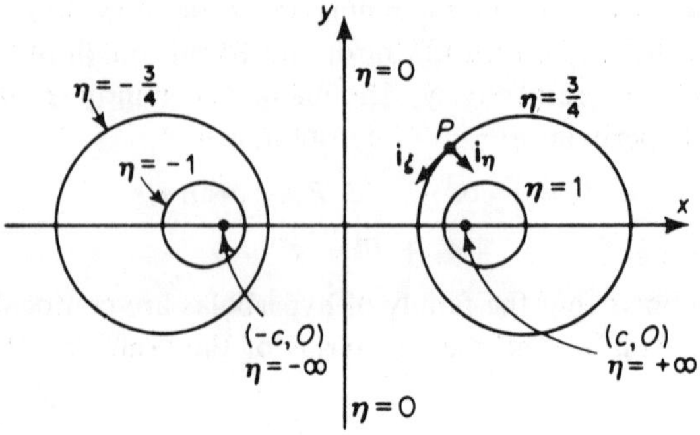

Figure A-12.1(a). Bipolar coordinates; η = constant.

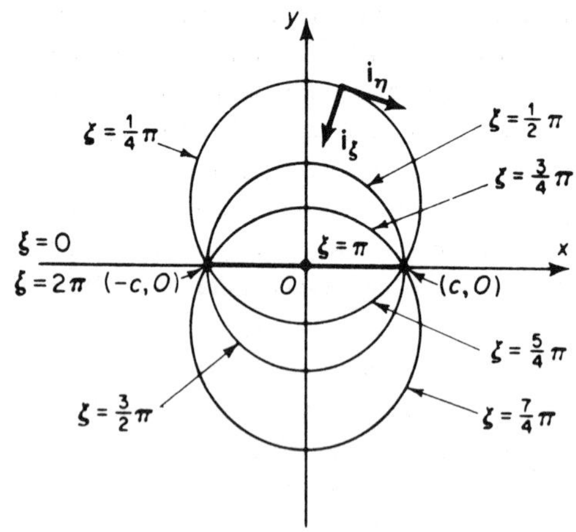

Figure A-12.1(b). Bipolar coordinates; ξ = constant.

The foregoing denominators are essentially positive. As is evident from Figs. A-12.1(a), (b), each point (x, y) in the xy plane is represented at least once by limiting ξ and η to the ranges

$$0 \leq \xi < 2\pi, \qquad -\infty < \eta < \infty \tag{A–12.3}$$

With the exception of the two points $(x = \pm c, y = 0)$, for which ξ is infinitely many-valued, there is now a one-to-one correspondence between the cartesian coordinates (x, y) and the bipolar coordinates (ξ, η).

When ξ is eliminated from Eq. (A-12.2) we obtain

$$(x - c \coth \eta)^2 + y^2 = c^2 \operatorname{csch}^2 \eta \tag{A–12.4}$$

In the xy plane, the curves given by the parameter η = constant are, therefore, a family of nonintersecting circles whose centers all lie along the x

axis (coaxial circles). The center of a typical circle $\eta = \eta_0 =$ constant is located at the point ($x = c \coth \eta_0$, $y = 0$) and its radius is $c|\operatorname{csch} \eta_0|$. For $\eta_0 > 0$, the circle lies entirely to the right of the origin; for values of $\eta_0 < 0$ the circle is to the left of the origin. The value $\eta_0 = 0$ generates a circle of infinite radius whose center is at either ($x = \pm\infty$, $y = 0$) so that $\eta_0 = 0$ corresponds to the entire y axis. When $\eta_0 = \pm\infty$ the radius is zero and the centers are located at the points ($x = \pm c$, $y = 0$), respectively. As η_0 varies from $+\infty$ to 0 the radius of the circle corresponding to this value of η_0 increases from 0 to ∞ and the center moves from ($x = c$, $y = 0$) to ($x = \infty$, $y = 0$) along the x axis. On the other hand, as η_0 varies from $-\infty$ to 0 the center moves from ($x = -c$, $y = 0$) to ($x = -\infty$, $y = 0$) along the x axis.

Elimination of η from Eq. (A-12.2) results in

$$x^2 + (y - c \cot \xi)^2 = c^2 \csc^2 \xi \tag{A–12.5}$$

Thus, the family of curves in the xy plane which arise by assigning different constant values to the parameter ξ appear to be intersecting circles whose centers all lie along the y axis. Every circle of the family passes through the *limiting points* of the system, ($x = \pm c$, $y = 0$). The center of a typical circle $\xi = \xi_0 =$ constant is at ($x = 0$, $y = c \cot \xi_0$) and its radius is $c \csc \xi_0$. More careful examination of Eq. (A-12.2) shows, however, that the curves characterized by $\xi =$ constant are not complete circles but, rather, are *circular arcs* terminating on the x axis at the limiting points of the system. If the circular arc $\xi = \xi_0 = \text{constant} < \pi$ is that part of the circle lying above the x axis, its extension below the x axis is given by the value $\xi = \xi_0 + \pi$. When $\xi_0 = \pi$, the arc is degenerate and corresponds to that portion of the x axis lying between the limiting points of the system. The value $\xi_0 = 0$ is a circular arc of infinite radius with center at $+\infty$ on the y axis, and gives the entire x axis with the exception of those points between ($x = \pm c$, $y = 0$), for which $\xi_0 = \pi$. The values $\xi_0 = \pi/2$ and $3\pi/2$ are semicircles of radii c, having their centers at the origin.

The geometric significance of the bipolar coordinates ξ and η can be established without difficulty from the analogous discussion of three-dimensional bipolar coordinates in Section A-19.

If we select

$$q_1 = \xi, \qquad q_2 = \eta, \qquad q_3 = z \tag{A–12.6}$$

the conjugate system of curvilinear coordinates thereby obtained constitutes a right-handed, orthogonal system whose metrical coefficients are

$$h_1 = h_2 = \frac{1}{c}(\cosh \eta - \cos \xi), \qquad h_3 = 1 \tag{A–12.7}$$

The unit vectors $\mathbf{i}_\xi$ and $\mathbf{i}_\eta$ are depicted in Figs. A-12.1(a), (b). The unit vector $\mathbf{i}_z$ is directed out of the page at the reader.

A-13 Parabolic Cylinder Coordinates (ξ, η, z) (Fig. A–13.1)

Parabolic coordinates (ξ, η) in a plane are defined by the transformation

$$x + iy = c(\xi + i\eta)^2 \tag{A–13.1}$$

$c > 0$, whereupon

$$x = c(\xi^2 - \eta^2), \qquad y = 2c\xi\eta \tag{A–13.2}$$

Each point in space is represented at least once by allowing the parabolic coordinates to vary over the ranges

$$-\infty < \xi < \infty, \qquad 0 \leq \eta < \infty \tag{A–13.3}$$

Other interpretations are possible.

From Eq. (A-13.2) we obtain

$$y^2 = 4c\xi^2(c\xi^2 - x) \tag{A–13.4}$$

so that in the xy plane the family of curves $\xi =$ constant are confocal parabolas having their foci at the origin. These parabolas open to the left along the x axis.

In a similar manner, Eq. (A-13.2) yields, upon elimination of ξ,

$$y^2 = 4c\eta^2(c\eta^2 + x) \tag{A–13.5}$$

The curves $\eta =$ constant are therefore confocal parabolas having their foci at the origin. These parabolas open to the right along the x axis.

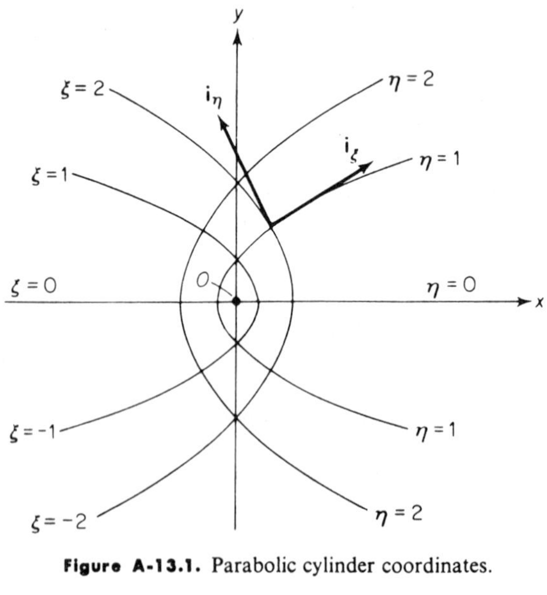

Figure A-13.1. Parabolic cylinder coordinates.

With the choice of curvilinear coordinates

$$q_1 = \xi, \qquad q_2 = \eta, \qquad q_3 = z \tag{A–13.6}$$

parabolic cylinder coordinates (ξ, η, z) form a right-handed, orthogonal system whose metrical coefficients are

$$h_1 = h_2 = \frac{1}{2c(\xi^2 + \eta^2)^{1/2}}, \qquad h_3 = 1 \tag{A–13.7}$$

The coordinate surfaces ξ = constant and η = constant are each confocal parabolic cylinders. In addition to the unit vectors shown in Fig. A-13.1, the unit vector $\mathbf{i}_z$ is directed out of the page at the reader.

Other properties of parabolic cylinder coordinates may be deduced from the properties of *paraboloidal coordinates*, discussed in Section A-21. For example, the analog of Eq. (A-21.7) is, in the present instance,

$$\xi = \left(\frac{\rho}{c}\right)^{1/2} \cos\frac{\phi}{2}, \qquad \eta = \left(\frac{\rho}{c}\right)^{1/2} \sin\frac{\phi}{2} \tag{A–13.8}$$

A-14 Coordinate Systems of Revolution (q_1, q_2, ϕ) (Fig. A–14.1)

The properties of orthogonal curvilinear coordinate systems associated with *bodies of revolution* may be systematically developed by specializing the general formulas of Sections A-1–A-7.

Consider the class of curvilinear coordinate systems defined explicitly by relations of the form

$$z = z(q_1, q_2), \qquad \rho = \rho(q_1, q_2), \qquad \phi = q_3 \tag{A–14.1}$$

where (ρ, ϕ, z) are circular cylindrical coordinates. The cartesian coordinates x and y may be obtained from the relations

$$x = \rho \cos\phi, \qquad y = \rho \sin\phi \tag{A–14.2}$$

In this way we are led to consider the relations

$$x = \rho(q_1, q_2) \cos q_3, \qquad y = \rho(q_1, q_2) \sin q_3, \qquad z = z(q_1, q_2) \tag{A–14.3}$$

which give (x, y, z) in terms of (q_1, q_2, q_3).

Differentiating Eq. (A-14.3) we find that

$$\begin{gathered}
\frac{\partial x}{\partial q_1} = \cos q_3 \frac{\partial \rho}{\partial q_1}, \qquad \frac{\partial y}{\partial q_1} = \sin q_3 \frac{\partial \rho}{\partial q_1} \\
\frac{\partial x}{\partial q_2} = \cos q_3 \frac{\partial \rho}{\partial q_2}, \qquad \frac{\partial y}{\partial q_2} = \sin q_3 \frac{\partial \rho}{\partial q_2} \\
\frac{\partial x}{\partial q_3} = -\rho \sin q_3, \qquad \frac{\partial y}{\partial q_3} = \rho \cos q_3, \qquad \frac{\partial z}{\partial q_3} = 0
\end{gathered} \tag{A–14.4}$$

Thus, in place of Eq. (A-2.6), the necessary and sufficient conditions for the orthogonality of q_1, q_2, q_3 now require only that

$$\frac{\partial \rho}{\partial q_1}\frac{\partial \rho}{\partial q_2} + \frac{\partial z}{\partial q_1}\frac{\partial z}{\partial q_2} = 0 \tag{A–14.5}$$

In addition, if we substitute the derivatives appearing in Eq. (A-14.4) into Eq. (A-2.15) and expand the resulting determinant, we obtain

$$\frac{\partial(x, y, z)}{\partial(q_1, q_2, q_3)} = \rho\frac{\partial(z, \rho)}{\partial(q_1, q_2)}$$

Since $\rho > 0$, it follows that the system of curvilinear coordinates defined by Eq. (A-14.1) forms a right-handed system whenever

$$\frac{\partial(z, \rho)}{\partial(q_1, q_2)} = \frac{\partial z}{\partial q_1}\frac{\partial \rho}{\partial q_2} - \frac{\partial z}{\partial q_2}\frac{\partial \rho}{\partial q_1} > 0 \tag{A–14.6}$$

Finally, substituting Eq. (A-14.4) into Eq. (A-1.9) yields the following expressions for the metrical coefficients:

$$\frac{1}{h_1^2} = \left(\frac{\partial \rho}{\partial q_1}\right)^2 + \left(\frac{\partial z}{\partial q_1}\right)^2, \quad \frac{1}{h_2^2} = \left(\frac{\partial \rho}{\partial q_2}\right)^2 + \left(\frac{\partial z}{\partial q_2}\right)^2, \quad h_3 = \frac{1}{\rho(q_1, q_2)} \tag{A–14.7}$$

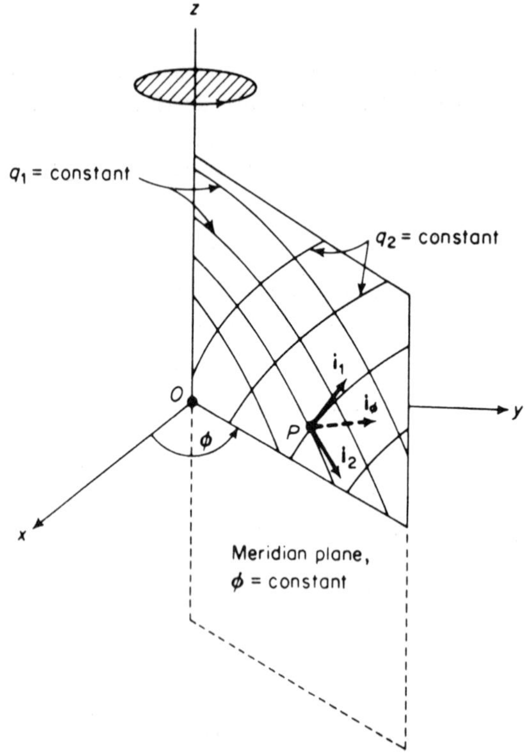

Figure A-14.1. Curvilinear coordinate systems of revolution.

We shall term any plane for which ϕ is constant a *meridian plane*. On the assumption that Eqs. (A-14.5) and (A-14.6) are both satisfied, the curves $q_1 =$ constant and $q_2 =$ constant intersect each other orthogonally in such a meridian plane, as in Fig. A-14.1. These curves are termed the *generators* of the coordinate surfaces $q_1 =$ constant and $q_2 =$ constant, these being *surfaces of revolution* obtained by rotating the corresponding curves in a meridian plane about the z axis.

It is often convenient in applications involving bodies of revolution to be able to convert functions readily from orthogonal curvilinear coordinate systems of revolution to *circular cylindrical coordinates* and vice versa. The relations to be developed are analogous to those discussed in Section A-6, except that the present results are limited to the special class of orthogonal curvilinear coordinate systems defined by Eq. (A-14.1).

(i) *Transformation of partial derivatives:* To express the partial differential operators $\partial/\partial\rho$ and $\partial/\partial z$ in orthogonal curvilinear coordinates of revolution, observe that

$$\frac{\partial}{\partial\rho} = (\nabla\rho)\cdot\nabla, \qquad \frac{\partial}{\partial z} = (\nabla z)\cdot\nabla$$

Thus, writing ∇ in curvilinear coordinates, we obtain the desired transformations,

$$\frac{\partial}{\partial\rho} = \sum_{k=1}^{2} h_k^2 \frac{\partial\rho}{\partial q_k}\frac{\partial}{\partial q_k}, \qquad \frac{\partial}{\partial z} = \sum_{k=1}^{2} h_k^2 \frac{\partial z}{\partial q_k}\frac{\partial}{\partial q_k} \tag{A-14.8}$$

By permitting these partial differential operators to operate on q_1 and q_2 we find

$$\frac{\partial q_k}{\partial\rho} = h_k^2\frac{\partial\rho}{\partial q_k}, \qquad \frac{\partial q_k}{\partial z} = h_k^2\frac{\partial z}{\partial q_k} \quad (k = 1, 2) \tag{A-14.9}$$

which are the analogs of Eqs. (A-6.4).

The transformation of partial derivatives inverse to Eq. (A-14.9) are easily obtained by application of the "chain-rule,"

$$\frac{\partial}{\partial q_k} = \frac{\partial\rho}{\partial q_k}\frac{\partial}{\partial\rho} + \frac{\partial z}{\partial q_k}\frac{\partial}{\partial z} \quad (k = 1, 2) \tag{A-14.10}$$

(ii) *Transformation of unit vectors:* Since $\mathbf{i}_k = (\nabla q_k)/h_k$ we obtain, upon writing ∇ in cylindrical coordinates and employing Eq. (A-14.9),

$$\mathbf{i}_k = h_k\left(\mathbf{i}_\rho\frac{\partial\rho}{\partial q_k} + \mathbf{i}_z\frac{\partial z}{\partial q_k}\right) \quad (k = 1, 2) \tag{A-14.11}$$

The inverse transformation is obtained by utilizing the relations

$$\mathbf{i}_\rho = \nabla\rho, \qquad \mathbf{i}_z = \nabla z$$

and expressing ∇ in curvilinear coordinates; whence

$$\mathbf{i}_\rho = \sum_{k=1}^{2} \mathbf{i}_k h_k \frac{\partial\rho}{\partial q_k}, \qquad \mathbf{i}_z = \sum_{k=1}^{2} \mathbf{i}_k h_k \frac{\partial z}{\partial q_k} \tag{A-14.12}$$

(iii) *Transformation of vector components:* Set

$$\mathbf{u} = \mathbf{i}_\rho u_\rho + \mathbf{i}_\phi u_\phi + \mathbf{i}_z u_z = \mathbf{i}_1 u_1 + \mathbf{i}_2 u_2 + \mathbf{i}_3 u_3$$

and multiply scalarly with Eqs. (A-14.11) and (A-14.12), respectively. One obtains

$$u_k = h_k\left(u_\rho \frac{\partial \rho}{\partial q_k} + u_z \frac{\partial z}{\partial q_k}\right) \quad (k = 1, 2) \tag{A–14.13}$$

and

$$u_\rho = \sum_{k=1}^{2} u_k h_k \frac{\partial \rho}{\partial q_k}, \qquad u_z = \sum_{k=1}^{2} u_k h_k \frac{\partial z}{\partial q_k} \tag{A–14.14}$$

which permit an easy transformation of vector components between the two systems of coordinates.

Most important orthogonal coordinate systems of revolution encountered in applications can be generated by simple application of the theory of analytic functions. This technique is discussed further in Section A-16 and examples given in Sections A-17–A-21. *Spherical coordinates*, discussed in the next section, constitute an important exception to this generalization.

A-15 Spherical Coordinates (r, θ, ϕ)

[Figs. A–15.1 (a), (b)]

Spherical coordinates

$$q_1 = r, \qquad q_2 = \theta, \qquad q_3 = \phi \tag{A–15.1}$$

are defined by the relations

$$\rho = r \sin\theta, \qquad z = r\cos\theta \tag{A–15.2}$$

This makes

$$x = r\sin\theta\cos\phi, \qquad y = r\sin\theta\sin\phi, \qquad z = r\cos\theta \tag{A–15.3}$$

Equations (A-15.2) can be solved explicitly for the spherical coordinates

$$r = (\rho^2 + z^2)^{1/2}, \qquad \theta = \tan^{-1}\frac{\rho}{z} \tag{A–15.4}$$

This system of coordinates is depicted from various points of view in Figs. A-15.1(a), (b).

By restricting the ranges of these coordinates as follows:

$$0 \leq r < \infty, \qquad 0 \leq \theta \leq \pi, \qquad 0 \leq \phi < 2\pi \tag{A–15.5}$$

each point in space is represented once and only once, with the exception of the points along the z axis, for which ϕ is undetermined.

The coordinate surfaces $r =$ constant, $\theta =$ constant, and $\phi =$ constant are, respectively: concentric spheres with center at the origin; right-circular cones with apex at the origin, having the z axis as their axis of revolution; vertical half-planes.

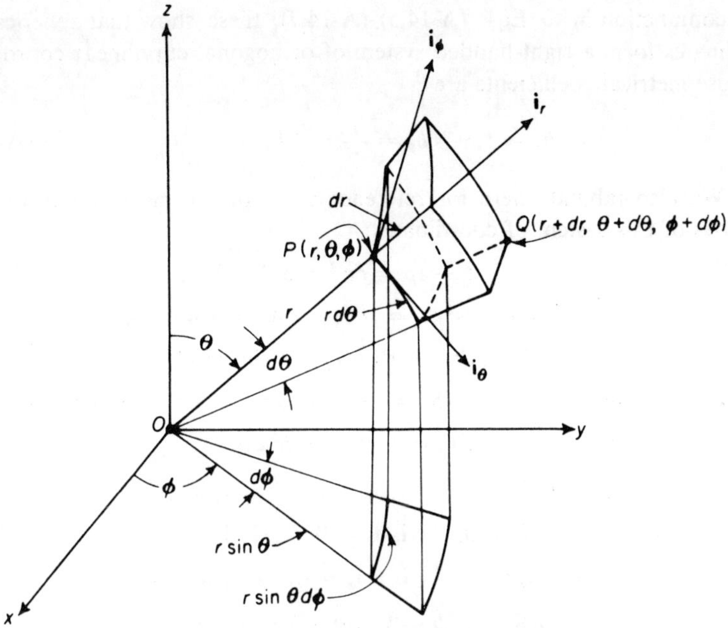

Figure A-15.1(a). Spherical coordinates.

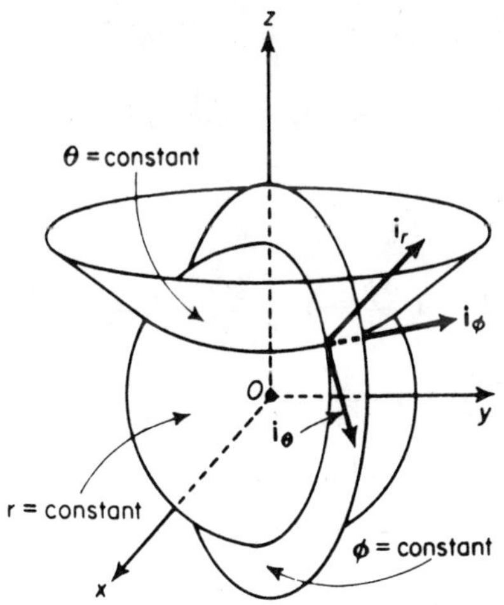

Figure A-15.1(b). Spherical coordinate system surfaces.

From Eqs. (A-15.1) and (A-15.2) we have

$$\frac{\partial \rho}{\partial q_1} = \sin\theta, \qquad \frac{\partial \rho}{\partial q_2} = r\cos\theta$$

$$\frac{\partial z}{\partial q_1} = \cos\theta, \qquad \frac{\partial z}{\partial q_2} = -r\sin\theta$$

In conjunction with Eqs. (A-14.5)–(A-14.7), these show that spherical coordinates form a right-handed system of orthogonal curvilinear coordinates whose metrical coefficients are

$$h_1 = 1, \qquad h_2 = \frac{1}{r}, \qquad h_3 = \frac{1}{r \sin\theta} \tag{A-15.6}$$

We also tabulate here for reference some of the more important properties of this system of coordinates:

$$z + i\rho = re^{i\theta} = e^{(\ln r)+i\theta} \tag{A-15.7}$$

$$\begin{gathered} dl_1 = dr, \qquad dl_2 = rd\theta, \qquad dl_3 = r\sin\theta\, d\phi \\ dl^2 = (dr)^2 + r^2(d\theta)^2 + r^2\sin^2\theta\,(d\phi)^2 \end{gathered} \tag{A-15.8}$$

$$dS_1 = r^2\sin\theta\, d\theta\, d\phi, \qquad dS_2 = r\sin\theta\, dr\, d\phi, \qquad dS_3 = r\, dr\, d\theta \tag{A-15.9}$$

$$dV = r^2 \sin\theta\, dr\, d\theta\, d\phi \tag{A-15.10}$$

$$\begin{gathered} \mathbf{i}_r\cdot\mathbf{i}_r = 1, \qquad \mathbf{i}_\theta\cdot\mathbf{i}_\theta = 1, \qquad \mathbf{i}_\phi\cdot\mathbf{i}_\phi = 1 \\ \mathbf{i}_r\cdot\mathbf{i}_\theta = 0, \qquad \mathbf{i}_\theta\cdot\mathbf{i}_\phi = 0, \qquad \mathbf{i}_\phi\cdot\mathbf{i}_r = 0 \\ \mathbf{i}_r\times\mathbf{i}_\theta = \mathbf{i}_\phi, \qquad \mathbf{i}_\theta\times\mathbf{i}_\phi = \mathbf{i}_r, \qquad \mathbf{i}_\phi\times\mathbf{i}_r = \mathbf{i}_\theta \end{gathered} \tag{A-15.11}$$

$$d\mathbf{R} = \mathbf{i}_r\, dr + \mathbf{i}_\theta r\, d\theta + \mathbf{i}_\phi r\sin\theta\, d\phi \tag{A-15.12}$$

$$\mathbf{R} = \mathbf{i}_r r \tag{A-15.13}$$

$$\begin{gathered} \frac{\partial \mathbf{i}_r}{\partial r} = \mathbf{0}, \qquad \frac{\partial \mathbf{i}_r}{\partial \theta} = \mathbf{i}_\theta, \qquad \frac{\partial \mathbf{i}_r}{\partial \phi} = \mathbf{i}_\phi \sin\theta \\ \frac{\partial \mathbf{i}_\theta}{\partial r} = \mathbf{0}, \qquad \frac{\partial \mathbf{i}_\theta}{\partial \theta} = -\mathbf{i}_r, \qquad \frac{\partial \mathbf{i}_\theta}{\partial \phi} = \mathbf{i}_\phi \cos\theta \\ \frac{\partial \mathbf{i}_\phi}{\partial r} = \mathbf{0}, \qquad \frac{\partial \mathbf{i}_\phi}{\partial \theta} = \mathbf{0}, \qquad \frac{\partial \mathbf{i}_\phi}{\partial \phi} = -\mathbf{i}_r \sin\theta - \mathbf{i}_\theta\cos\theta \end{gathered} \tag{A-15.14}$$

$$\nabla\psi = \mathbf{i}_r\frac{\partial\psi}{\partial r} + \mathbf{i}_\theta\frac{1}{r}\frac{\partial\psi}{\partial\theta} + \mathbf{i}_\phi\frac{1}{r\sin\theta}\frac{\partial\psi}{\partial\phi} \tag{A-15.15}$$

$$\nabla^2\psi = \frac{1}{r^2}\left[\frac{\partial}{\partial r}\left(r^2\frac{\partial\psi}{\partial r}\right) + \frac{1}{\sin\theta}\frac{\partial}{\partial\theta}\left(\sin\theta\frac{\partial\psi}{\partial\theta}\right) + \frac{1}{\sin^2\theta}\frac{\partial^2\psi}{\partial\phi^2}\right] \tag{A-15.16}$$

$$\mathbf{u} = \mathbf{i}_r u_r + \mathbf{i}_\theta u_\theta + \mathbf{i}_\phi u_\phi \tag{A-15.17}$$

$$|\mathbf{u}|^2 = u_r^2 + u_\theta^2 + u_\phi^2 \tag{A-15.18}$$

$$\nabla\cdot\mathbf{u} = \frac{1}{r^2}\frac{\partial}{\partial r}(r^2u_r) + \frac{1}{r\sin\theta}\frac{\partial}{\partial\theta}(\sin\theta\, u_\theta) + \frac{1}{r\sin\theta}\frac{\partial u_\phi}{\partial\phi} \tag{A-15.19}$$

$$\begin{aligned} \nabla\times\mathbf{u} = {} & \mathbf{i}_r\frac{1}{r\sin\theta}\left[\frac{\partial}{\partial\theta}(\sin\theta\, u_\phi) - \frac{\partial u_\theta}{\partial\phi}\right] + \mathbf{i}_\theta\frac{1}{r}\left[\frac{1}{\sin\theta}\frac{\partial u_r}{\partial\phi} - \frac{\partial}{\partial r}(r\, u_\phi)\right] \\ & + \mathbf{i}_\phi\frac{1}{r}\left[\frac{\partial}{\partial r}(ru_\theta) - \frac{\partial u_r}{\partial\theta}\right] \end{aligned} \tag{A-15.20}$$

$$\nabla^2\mathbf{u} = \mathbf{i}_r\left[\nabla^2 u_r - \frac{2}{r^2}\left(u_r + \frac{\partial u_\theta}{\partial\theta} + u_\theta\cot\theta - \frac{1}{\sin\theta}\frac{\partial u_\phi}{\partial\phi}\right)\right]$$
$$+ \mathbf{i}_\theta\left[\nabla^2 u_\theta + \frac{1}{r^2}\left(2\frac{\partial u_r}{\partial\theta} - \frac{u_\theta}{\sin^2\theta} - \frac{2\cos\theta}{\sin^2\theta}\frac{\partial u_\phi}{\partial\phi}\right)\right] \quad \text{(A–15.21)}$$
$$+ \mathbf{i}_\phi\left[\nabla^2 u_\phi + \frac{1}{r^2}\left(\frac{2}{\sin\theta}\frac{\partial u_r}{\partial\phi} + \frac{2\cos\theta}{\sin^2\theta}\frac{\partial u_\theta}{\partial\phi} - \frac{u_\phi}{\sin^2\theta}\right)\right]$$

$$\frac{\partial}{\partial\rho} = \sin\theta\frac{\partial}{\partial r} + \frac{\cos\theta}{r}\frac{\partial}{\partial\theta}, \qquad \frac{\partial}{\partial z} = \cos\theta\frac{\partial}{\partial r} - \frac{\sin\theta}{r}\frac{\partial}{\partial\theta} \quad \text{(A–15.22)}$$

$$\frac{\partial}{\partial r} = \sin\theta\frac{\partial}{\partial\rho} + \cos\theta\frac{\partial}{\partial z}, \qquad \frac{\partial}{\partial\theta} = r\cos\theta\frac{\partial}{\partial\rho} - r\sin\theta\frac{\partial}{\partial z} \quad \text{(A–15.23)}$$

$$\mathbf{i}_r = \mathbf{i}_\rho\sin\theta + \mathbf{i}_z\cos\theta, \qquad \mathbf{i}_\theta = \mathbf{i}_\rho\cos\theta - \mathbf{i}_z\sin\theta \quad \text{(A–15.24)}$$

$$\mathbf{i}_\rho = \mathbf{i}_r\sin\theta + \mathbf{i}_\theta\cos\theta, \qquad \mathbf{i}_z = \mathbf{i}_r\cos\theta - \mathbf{i}_\theta\sin\theta \quad \text{(A–15.25)}$$

$$u_r = u_\rho\sin\theta + u_z\cos\theta, \qquad u_\theta = u_\rho\cos\theta - u_z\sin\theta \quad \text{(A–15.26)}$$

$$u_\rho = u_r\sin\theta + u_\theta\cos\theta, \qquad u_z = u_r\cos\theta - u_\theta\sin\theta \quad \text{(A–15.27)}$$

The following relations connect unit vectors in spherical and cartesian coordinates:

$$\mathbf{i}_r = \mathbf{i}\sin\theta\cos\phi + \mathbf{j}\sin\theta\sin\phi + \mathbf{k}\cos\theta$$
$$\mathbf{i}_\theta = \mathbf{i}\cos\theta\cos\phi + \mathbf{j}\cos\theta\sin\phi - \mathbf{k}\sin\theta \quad \text{(A–15.28)}$$
$$\mathbf{i}_\phi = -\mathbf{i}\sin\phi + \mathbf{j}\cos\phi$$

The relations inverse to these are

$$\mathbf{i} = \mathbf{i}_r\sin\theta\cos\phi + \mathbf{i}_\theta\cos\theta\cos\phi - \mathbf{i}_\phi\sin\phi$$
$$\mathbf{j} = \mathbf{i}_r\sin\theta\sin\phi + \mathbf{i}_\theta\cos\theta\sin\phi + \mathbf{i}_\phi\cos\phi \quad \text{(A–15.29)}$$
$$\mathbf{k} = \mathbf{i}_r\cos\theta - \mathbf{i}_\theta\sin\theta$$

If $\mathbf{u}$ is the vector given by Eq. (A-15.17), then

$$\nabla\mathbf{u} = \mathbf{i}_r\mathbf{i}_r\frac{\partial u_r}{\partial r} + \mathbf{i}_r\mathbf{i}_\theta\frac{\partial u_\theta}{\partial r} + \mathbf{i}_r\mathbf{i}_\phi\frac{\partial u_\phi}{\partial r}$$
$$+ \mathbf{i}_\theta\mathbf{i}_r\frac{1}{r}\left(\frac{\partial u_r}{\partial\theta} - u_\theta\right) + \mathbf{i}_\theta\mathbf{i}_\theta\frac{1}{r}\left(\frac{\partial u_\theta}{\partial\theta} + u_r\right) + \mathbf{i}_\theta\mathbf{i}_\phi\frac{1}{r}\frac{\partial u_\phi}{\partial\theta}$$
$$+ \mathbf{i}_\phi\mathbf{i}_r\frac{1}{r}\left(\frac{1}{\sin\theta}\frac{\partial u_r}{\partial\phi} - u_\phi\right) + \mathbf{i}_\phi\mathbf{i}_\theta\frac{1}{r}\left(\frac{1}{\sin\theta}\frac{\partial u_\theta}{\partial\phi} - \cot\theta\, u_\phi\right) \quad \text{(A–15.30)}$$
$$+ \mathbf{i}_\phi\mathbf{i}_\phi\frac{1}{r}\left(\frac{1}{\sin\theta}\frac{\partial u_\phi}{\partial\phi} + u_r + \cot\theta\, u_\theta\right)$$

If $\boldsymbol{\tau}$ is the dyadic

$$\boldsymbol{\tau} = \mathbf{i}_r\mathbf{i}_r\tau_{rr} + \mathbf{i}_r\mathbf{i}_\theta\tau_{r\theta} + \mathbf{i}_r\mathbf{i}_\phi\tau_{r\phi}$$
$$+ \mathbf{i}_\theta\mathbf{i}_r\tau_{\theta r} + \mathbf{i}_\theta\mathbf{i}_\theta\tau_{\theta\theta} + \mathbf{i}_\theta\mathbf{i}_\phi\tau_{\theta\phi}$$
$$+ \mathbf{i}_\phi\mathbf{i}_r\tau_{\phi r} + \mathbf{i}_\phi\mathbf{i}_\theta\tau_{\phi\theta} + \mathbf{i}_\phi\mathbf{i}_\phi\tau_{\phi\phi}$$

then its divergence is

$$\nabla \cdot \boldsymbol{\tau} = \mathbf{i}_r \left[\frac{1}{r^2} \frac{\partial}{\partial r}(r^2 \tau_{rr}) + \frac{1}{r \sin \theta} \frac{\partial}{\partial \theta}(\sin \theta \tau_{\theta r}) + \frac{1}{r \sin \theta} \frac{\partial \tau_{\phi r}}{\partial \phi} - \frac{\tau_{\theta\theta}}{r} - \frac{\tau_{\phi\phi}}{r} \right]$$
$$+ \mathbf{i}_\theta \left[\frac{1}{r^2} \frac{\partial}{\partial r}(r^2 \tau_{r\theta}) + \frac{1}{r \sin \theta} \frac{\partial}{\partial \theta}(\sin \theta \tau_{\theta\theta}) + \frac{1}{r \sin \theta} \frac{\partial \tau_{\phi\theta}}{\partial \phi} + \frac{\tau_{\theta r}}{r} - \frac{\cot \theta}{r} \tau_{\phi\phi} \right]$$
$$+ \mathbf{i}_\phi \left[\frac{1}{r^2} \frac{\partial}{\partial r}(r^2 \tau_{r\phi}) + \frac{1}{r \sin \theta} \frac{\partial}{\partial \theta}(\sin \theta \tau_{\theta\phi}) + \frac{1}{r \sin \theta} \frac{\partial \tau_{\phi\phi}}{\partial \phi} + \frac{\tau_{\phi r}}{r} + \frac{\cot \theta}{r} \tau_{\phi\theta} \right] \tag{A-15.31}$$

A-16 Conjugate Coordinate Systems of Revolution

Transformations of the type

$$z + i\rho = f(q_1 + iq_2), \qquad \phi = q_3 \tag{A-16.1}$$

generate curvilinear coordinate systems of rotation; for, upon equating real and imaginary parts, we obtain

$$z = z(q_1, q_2), \qquad \rho = \rho(q_1, q_2), \qquad \phi = q_3$$

which is of precisely the form discussed in Section A-14. Furthermore, the systems of coordinates defined by Eq. (A-16.1) are, of necessity, of the right-handed, orthogonal type. This follows by observing that Eqs. (A-14.5) and (A-14.6) are automatically satisfied by virtue of the applicability of the Cauchy-Riemann equations,

$$\frac{\partial z}{\partial q_1} = \frac{\partial \rho}{\partial q_2}, \qquad \frac{\partial z}{\partial q_2} = -\frac{\partial \rho}{\partial q_1} \tag{A-16.2}$$

to Eq. (A-16.1).

The metrical coefficients of the present system (q_1, q_2, q_3), obtained from Eq. (A-14.7), adopt the particularly simple forms

$$\frac{1}{h_1^2} = \frac{1}{h_2^2} = \left| \frac{d(z + i\rho)}{d(q_1 + iq_2)} \right|^2, \qquad h_3 = \frac{1}{\rho(q_1, q_2)} \tag{A-16.3}$$

where

$$\left| \frac{d(z + i\rho)}{d(q_1 + iq_2)} \right|^2 = \left(\frac{\partial \rho}{\partial q_1} \right)^2 + \left(\frac{\partial z}{\partial q_1} \right)^2 = \left(\frac{\partial \rho}{\partial q_2} \right)^2 + \left(\frac{\partial z}{\partial q_2} \right)^2 = \cdots, \text{ etc.}$$

Important examples of *conjugate* systems, characterized by Eq. (A-16.1), are prolate spheroidal coordinates, oblate spheroidal coordinates, bipolar coordinates, toroidal coordinates, and paraboloidal coordinates. These are discussed at length in the following sections.

A-17 Prolate Spheroidal Coordinates (ξ, η, ϕ)
[Figs. A-17.1(a), (b), (c)]

The transformation

$$z + i\rho = c \cosh (\xi + i\eta) \tag{A-17.1}$$

$c > 0$, gives rise to the relations

$$z = c \cosh \xi \cos \eta, \qquad \rho = c \sinh \xi \sin \eta \tag{A-17.2}$$

Every point in space is represented at least once and, with the exceptions to be cited, only once by restricting the ranges of the prolate spheroidal coordinates (ξ, η, ϕ) as follows:

$$0 \le \xi < \infty, \qquad 0 \le \eta \le \pi, \qquad 0 \le \phi < 2\pi \tag{A-17.3}$$

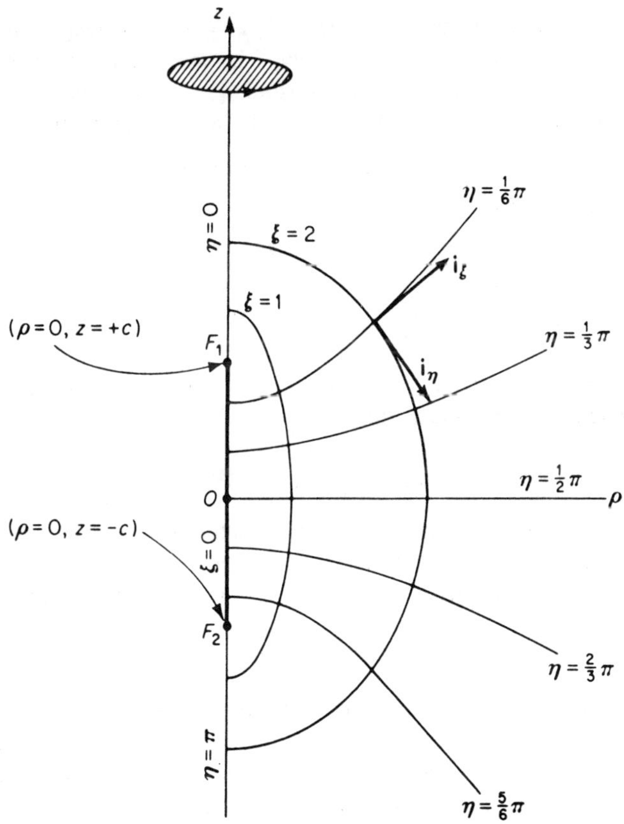

Figure A-17.1(a). Prolate spheroidal coordinates in a meridian plane.

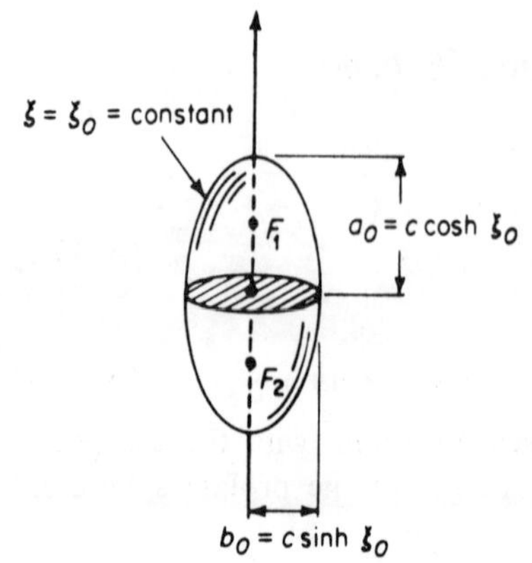

Figure A-17.1(b). Prolate spheroid.

Eliminating η from Eq. (A-17.2) yields

$$\frac{z^2}{c^2 \cosh^2 \xi} + \frac{\rho^2}{c^2 \sinh^2 \xi} = 1 \quad \text{(A–17.4)}$$

Since $\cosh \xi \geq \sinh \xi$, the coordinate surfaces ξ = constant are a confocal family of prolate spheroids having their common center at the origin. Spheroids of this type are also referred to as ovary, egg-shaped, or elongated ellipsoids, and are generated by the rotation of an ellipse about its major axis—in this case the z axis—as indicated in Figs. A-17.1(a), (b). The foci, F_1 and F_2, of the confocal system are located on the z axis at the points* $\{\rho = 0, z = \pm c\}$ corresponding to the values $\{\xi = 0, \eta = 0 \text{ and } \pi\}$, respectively. The major and minor semiaxes, a_o and b_o, respectively, of a typical ellipsoid, $\xi = \xi_o$ = constant, lie along the z axis and in the plane $z = 0$, respectively, and are given by

$$a_o = c \cosh \xi_o, \qquad b_o = c \sinh \xi_o \quad \text{(A–17.5)}$$

We note from Eq. (A-17.5) that

$$c^2 = a_o^2 - b_o^2 \quad \text{(A–17.6)}$$

and

$$\xi_o = \frac{1}{2} \ln \frac{a_o + b_o}{a_o - b_o} \quad \text{(A–17.7)}$$

which give the parameters c and ξ_o in terms of the lengths of the semiaxes. The eccentricity e_o of a typical ellipsoid is

$$e_o = \left[1 - \left(\frac{b_o}{a_o}\right)^2\right]^{1/2} = \operatorname{sech} \xi_o \quad \text{(A–17.8)}$$

The value $\xi_o = 0$ is a degenerate ellipsoid which reduces to the line segment $-c \leq z \leq c$ along the z axis, connecting the foci.

When ξ is eliminated between z and ρ in Eq. (A-17.2), one obtains

$$\frac{z^2}{c^2 \cos^2 \eta} - \frac{\rho^2}{c^2 \sin^2 \eta} = 1 \quad \text{(A–17.9)}$$

The coordinate surfaces characterized by η = constant are, therefore, a confocal family of two-sheeted hyperboloids of revolution having the z axis as their axis of rotation—Fig. A-17.1(c). The foci of this family are the same as those of the corresponding spheroids. It is evident from Eq. (A-17.2) that

*Where the value of ϕ is not uniquely determined we shall write the point (ρ, ϕ, z) as $\{\rho, z\}$.

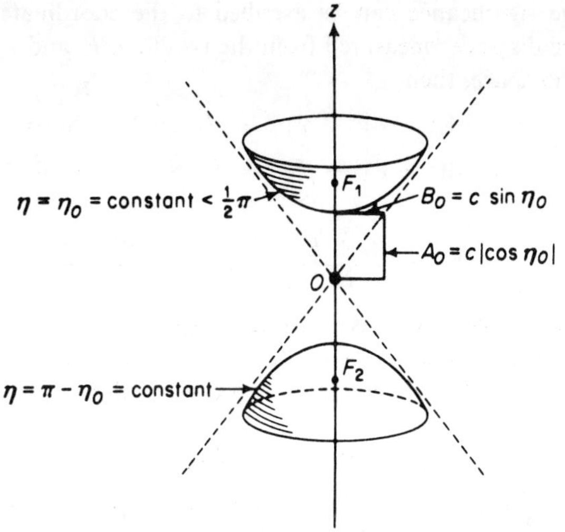

Figure A-17.1(c). Two-sheeted hyperboloid of revolution.

z is positive for the values $0 \leq \eta < \pi/2$ and negative for the values $\pi/2 < \eta \leq \pi$. In general, then, if $\eta = \eta_o$ = constant $< \pi/2$ gives that sheet of the hyperboloid lying above the plane $z = 0$, the value $\eta = \pi - \eta_o$ = constant gives the corresponding sheet lying below this plane. The major and minor semiaxes, $|A_o|$ and B_o, respectively, of a typical hyperboloid, $\eta = \eta_o$ = constant $< \pi/2$, are

$$A_o = c \cos \eta_o \qquad B_o = c \sin \eta_o \tag{A–17.10}$$

whence, in terms of these semiaxes,

$$c^2 = A_o^2 + B_o^2 \tag{A–17.11}$$

and

$$\eta_o = \tan^{-1} \frac{B_o}{A_o} \tag{A–17.12}$$

The values $\eta_o = 0$ and $\eta_o = \pi$ are the two halves of a degenerate hyperboloid and reduce to those segments of the z axis consisting of the points $\{\rho = 0, c \leq z < +\infty\}$ and $\{\rho = 0, -\infty < z \leq -c\}$, respectively. $\eta_o = \pi/2$ is again a degenerate hyperboloid, the two sheets coinciding to give the entire plane $z = 0$.

The distance from the origin to any point is

$$r = (\rho^2 + z^2)^{1/2} = c(\sinh^2 \xi + \cos^2 \eta)^{1/2} = \frac{c}{2^{1/2}} (\cosh 2\xi + \cos 2\eta)^{1/2} \tag{A–17.13}$$

the coordinates of the origin being $\{\xi = 0, \eta = \pi/2\}$. It follows from the preceding that large distances from the origin are equivalent to large values of ξ, and that as $\xi \to \infty$, $r \to \frac{1}{2}ce^{\xi}$.

Geometric significance can be ascribed to the coordinates ξ and η. If R_1 and R_2 are distances measured from the two foci, F_1 and F_2, respectively, to a point P in space, then

$$R_1 = [(z - c)^2 + \rho^2]^{1/2} = c(\cosh \xi - \cos \eta) \tag{A–17.14}$$

and

$$R_2 = [(z + c)^2 + \rho^2]^{1/2} = c(\cosh \xi + \cos \eta) \tag{A–17.15}$$

so that

$$\cosh \xi = \frac{R_1 + R_2}{2c}, \qquad \cos \eta = \frac{R_2 - R_1}{2c} \tag{A–17.16}$$

Thus, ξ and the angle η are easily determined from the triangle whose sides are R_1, R_2, and $2c$, the latter being the distance between foci. Finally, with the help of Eq. (A–17.5), the equation of a typical ellipsoid may be expressed in the form

$$R_1 + R_2 = 2a_o \tag{A–17.17}$$

whereas, from Eq. (A–17.10), the equation of a typical hyperboloid may be written as

$$|R_1 - R_2| = 2|A_o| \tag{A–17.18}$$

If we put

$$q_1 = \xi, \qquad q_2 = \eta, \qquad q_3 = \phi \tag{A–17.19}$$

then prolate spheroidal coordinates form a right-handed system of orthogonal curvilinear coordinates whose metrical coefficients are

$$h_1 = h_2 = \frac{1}{c(\sinh^2 \xi + \sin^2 \eta)^{1/2}} = \frac{2^{1/2}}{c(\cosh 2\xi - \cos 2\eta)^{1/2}} \tag{A–17.20}$$

and

$$h_3 = \frac{1}{c \sinh \xi \sin \eta} \tag{A–17.21}$$

Typical unit vectors are depicted in Fig. A–17.1(a). The unit vector $\mathbf{i}_\phi$ is directed into the page.

This system of coordinates constitutes a special case of *ellipsoidal* coordinates in which, of the three axes of the general ellipsoid, the two smallest are equal.

A-18 Oblate Spheroidal Coordinates (ξ, η, ϕ)

[Figs. A–18.1(a), (b), (c)]

The transformation

$$z + i\rho = c \sinh (\xi + i\eta) \tag{A–18.1}$$

$c > 0$, leads to the relations

$$z = c \sinh \xi \cos \eta, \qquad \rho = c \cosh \xi \sin \eta \tag{A–18.2}$$

Each point in space is obtained once and, with minor exceptions, only once by limiting the ranges of the oblate spheroidal coordinates (ξ, η, ϕ) in the following manner:

$$0 \le \xi < \infty, \qquad 0 \le \eta \le \pi, \qquad 0 \le \phi < 2\pi \tag{A-18.3}$$

Elimination of η from Eq. (A-18.2) results in

$$\frac{z^2}{c^2 \sinh^2 \xi} + \frac{\rho^2}{c^2 \cosh^2 \xi} = 1 \tag{A-18.4}$$

from which it is readily established that the coordinate surfaces $\xi =$ constant are a confocal family of oblate spheroids having their geometric center at the origin. Spheroids of this type are also termed planetary, disk-shaped, or flattened ellipsoids and are generated by rotation of an ellipse about its minor axis (in this instance the z axis) as indicated in Figs. A-18.1(a), (b).

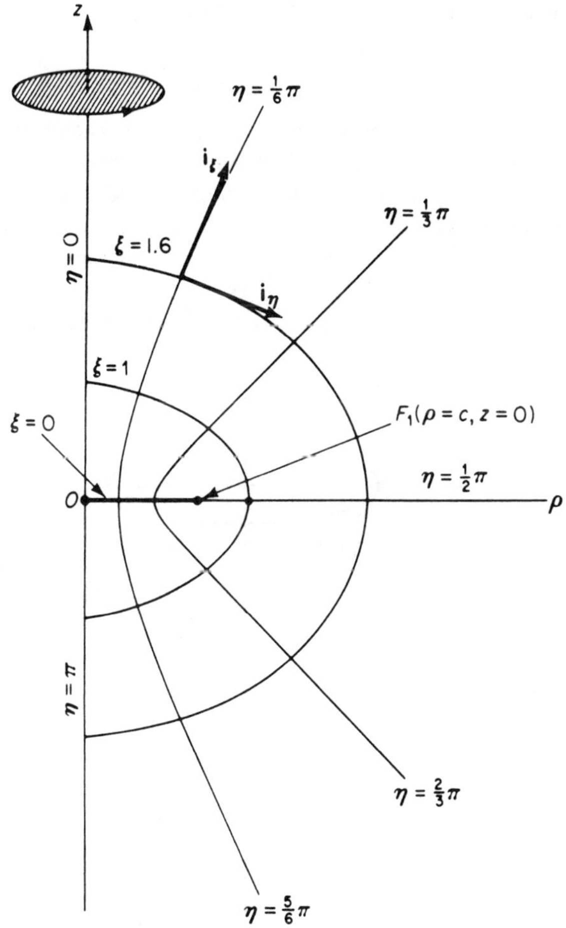

Figure A-18.1(a). Oblate spheroidal coordinates in a meridian plane.

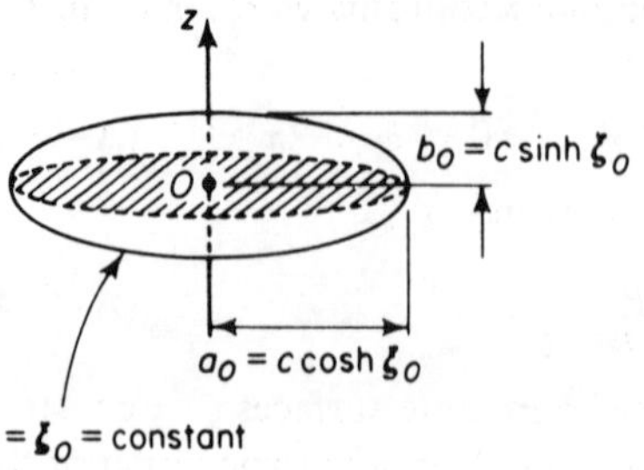

Figure A-18.1(b). Oblate spheroid.

The *focal circle* of the confocal family lies in the plane $z = 0$ and corresponds to the circle $\rho = c$. The major and minor semiaxes, a_0 and b_0, respectively, of a typical oblate spheroid, $\xi = \xi_0 =$ constant lie in the plane $z = 0$ and along the z axis, respectively. They are given by Eq. (A–17.5). Equations (A-17.6)–(A-17.8) are also applicable in the present instance. The ellipsoid given by $\xi_0 = 0$ is degenerate and corresponds to that portion of the plane $z = 0$ inside the focal circle, that is, $0 \leq \rho \leq c$.

When ξ is eliminated from Eq. (A–18.2) we find

$$-\frac{z^2}{c^2 \cos^2 \eta} + \frac{\rho^2}{c^2 \sin^2 \eta} = 1 \tag{A–18.5}$$

The coordinate surfaces given by $\eta =$ constant are, therefore, a family of confocal hyperboloids of revolution of one sheet having as their axis of rotation the z axis—Fig. A-18.1(c). These hyperboloids have the same focal circle as the family of oblate spheroids. Equations (A-17.10)–(A-17.12), involving the semiaxes of the hyperboloid, are also applicable here. As is evident from Eq. (A-18.2), values of η between 0 and $\pi/2$ correspond to the region $z > 0$, whereas values of η between $\pi/2$ and π belong to the region

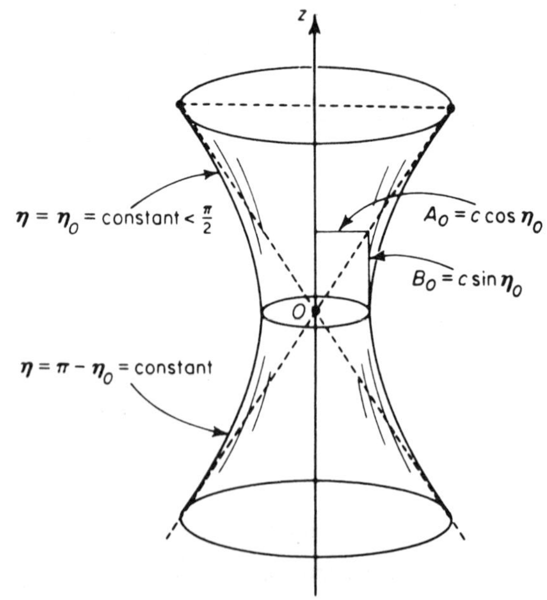

Figure A-18.1(c). One-sheeted hyperboloid of revolution.

$z < 0$. Thus, if $\eta = \eta_0 = \text{constant} < \pi/2$ gives the points on that portion of the hyperboloid situated above the plane $z = 0$, then $\eta = \pi - \eta_0$ gives the points on the same hyperboloid lying below the plane $z = 0$. For $\eta_0 = 0$ and π, the hyperboloid degenerates into the positive and negative z axes, respectively. The value $\eta_0 = \pi/2$ is also a degenerate hyperboloid, corresponding to that part of the plane $z = 0$ external to the focal circle, that is, $c \leq \rho < \infty$.

The distance from the origin to any point is

$$r = (\rho^2 + z^2)^{1/2} = c(\sinh^2\xi + \sin^2\eta)^{1/2} = \frac{c}{2^{1/2}}(\cosh 2\xi - \cos 2\eta)^{1/2} \quad \text{(A-18.6)}$$

the coordinates of the origin being either $\{\xi = 0, \eta = 0\}$ or $\{\xi = 0, \eta = \pi\}$. Large distances from the origin correspond to large values of ξ; as $\xi \to \infty$, $r \to \frac{1}{2}ce^{\xi}$.

A geometric interpretation of the coordinates ξ and η is possible. Let F_1 and F_2, respectively, be the points formed by the intersection of the two half-planes $\phi = \phi_0 = \text{constant} < \pi$ and $\phi = \pi - \phi_0$ with the focal circle. Then F_1 and F_2 lie at either end of a diameter of the focal circle and are separated by a distance of $2c$. If R_1 and R_2 are distances in the plane formed from ϕ_0 and $\pi - \phi_0$, measured from F_1 and F_2, respectively, then

$$R_1 = [z^2 + (\rho - c)^2]^{1/2} = c(\cosh\xi - \sin\eta) \quad \text{(A-18.7)}$$

and

$$R_2 = [z^2 + (\rho + c)^2]^{1/2} = c(\cosh\xi + \sin\eta) \quad \text{(A-18.8)}$$

These give

$$\cosh\xi = \frac{R_1 + R_2}{2c}, \qquad \sin\eta = \frac{R_2 - R_1}{2c} \quad \text{(A-18.9)}$$

from which ξ and the angle η are easily obtained from the triangle whose sides are R_1, R_2, and $2c$. Employing Eqs. (A-17.5) and (A-17.10), we are thus led to

$$R_1 + R_2 = 2a_0 \quad \text{and} \quad R_2 - R_1 = 2B_0 \quad \text{(A-18.10)}$$

as the equations for a typical ellipsoid and hyperboloid, respectively, of the present system.

Upon putting

$$q_1 = \xi, \qquad q_2 = \eta, \qquad q_3 = \phi \quad \text{(A-18.11)}$$

it follows that oblate spheroidal coordinates constitute a right-handed system of orthogonal, curvilinear coordinates having the metrical coefficients

$$h_1 = h_2 = \frac{1}{c(\cosh^2\xi - \sin^2\eta)^{1/2}} = \frac{2^{1/2}}{c(\cosh 2\xi + \cos 2\eta)^{1/2}} \quad \text{(A-18.12)}$$

and

$$h_3 = \frac{1}{c\cosh\xi\sin\eta} \quad \text{(A-18.13)}$$

Typical unit vectors are depicted in Fig. A-18.1(a). The unit vector $\mathbf{i}_\phi$ is directed into the page.

Oblate spheroidal coordinates are, again, a special case of ellipsoidal coordinates in which, of the three axes of the general ellipsoid, the two largest are equal.

A-19 Bipolar Coordinates (ξ, η, ϕ)

[Figs. A–19.1(a), (b), (c)]

Upon setting

$$z + i\rho = ic \cot \tfrac{1}{2}(\xi + i\eta) \tag{A–19.1}$$

$c > 0$, we obtain

$$z = c\frac{\sinh \eta}{\cosh \eta - \cos \xi}, \qquad \rho = c\frac{\sin \xi}{\cosh \eta - \cos \xi} \tag{A–19.2}$$

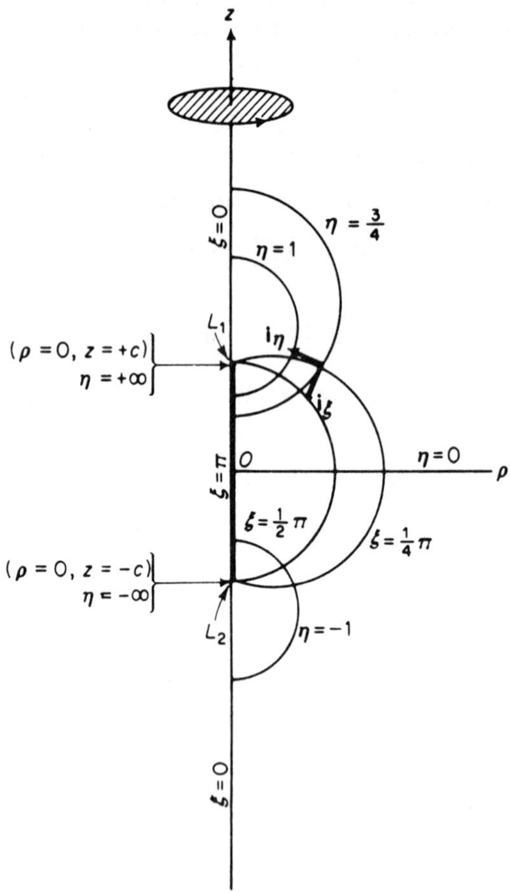

Figure A-19.1(a). Bipolar coordinates in a meridian plane.

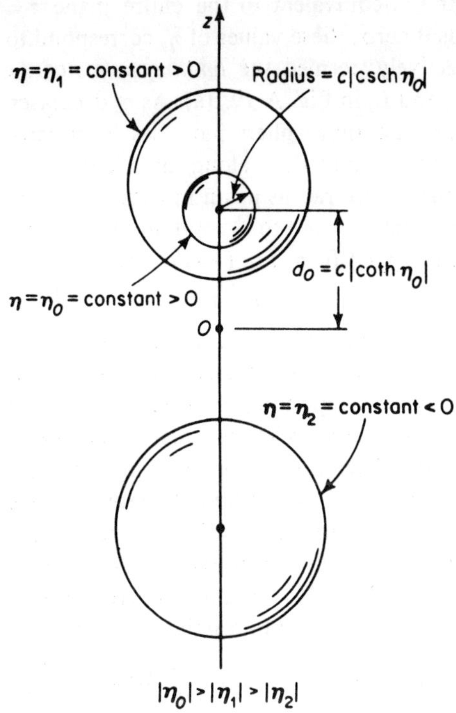

Figure A-19.1(b). Coaxial spheres.

The foregoing denominators are essentially positive. Each point in space is represented once and, with minor exceptions, only once by restricting the range of the coordinates to the following intervals:

$$0 \leq \xi \leq \pi, \qquad -\infty < \eta < \infty, \qquad 0 \leq \phi < 2\pi \tag{A–19.3}$$

Upon eliminating ξ in Eq. (A-19.2), we obtain

$$(z - c \coth \eta)^2 + \rho^2 = c^2 \operatorname{csch}^2 \eta \tag{A–19.4}$$

The coordinate surfaces $\eta =$ constant are, therefore, a family of non-intersecting, coaxial spheres whose centers lie along the z axis. A typical sphere, $\eta = \eta_o =$ constant, has its center at the point $\{\rho = 0, z = c \coth \eta_o\}$ and has a radius of $c\,|\operatorname{csch} \eta_o|$. It follows from Eq. (A-19.2) that, if $\eta_o > 0$, the sphere lies entirely above the plane $z = 0$. Conversely, for $\eta_o < 0$, the sphere is situated below this plane. The value $\eta_o = 0$ is a sphere of

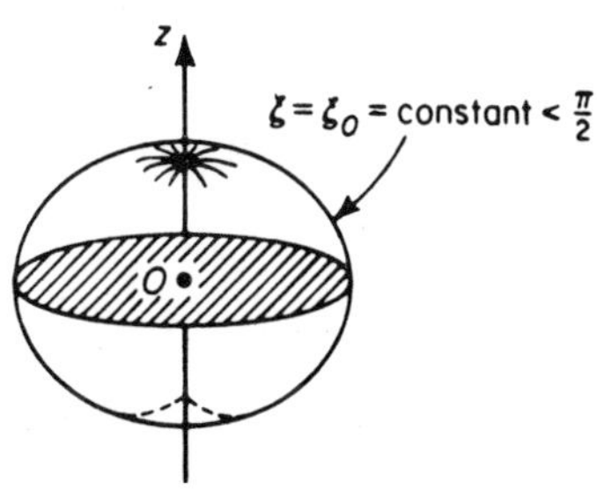

Figure A-19.1(c). Surfaces $\xi =$ constant.

infinite radius and is equivalent to the entire plane $z = 0$. For $\eta_0 = \pm\infty$ the sphere radius is zero; these values of η_0 correspond to the points $\{\rho = 0, z = \pm c\}$, respectively, termed the *limiting points* of the system. These are designated by L_1 and L_2 in Fig. A-19.1(a). As η_0 decreases from $+\infty$ to 0, the radius of the corresponding sphere increases from zero to infinity and the center moves from $z = c$ to $+\infty$ along the z axis. Likewise, as η_0 increases from $-\infty$ to 0 the sphere radius again increases from zero to infinity, while the center moves from $z = -c$ to $-\infty$ along the z axis.

If η is now eliminated from Eq. (A–19.2) we obtain

$$z^2 + (\rho - c \cot \xi)^2 = c^2 \csc^2 \xi \tag{A–19.5}$$

In a meridian plane, the *curves* $\xi =$ constant are arcs of circles, terminating at the limiting points of the system and having their centers in the plane $z = 0$. Circular arcs corresponding to values of ξ_0 between 0 and $\pi/2$ are greater in length than semicircles, whereas values between $\pi/2$ and π are less than semicircles. Therefore, upon rotating these arcs about the z axis, the coordinate *surfaces* $\xi =$ constant thereby obtained are spindle-like surfaces of revolution. The value $\xi = 0$ corresponds to the two segments of the z axis which lie above L_1 and below L_2. For $\xi_0 = \pi$ we obtain the line segment between L_1 and L_2. $\xi_0 = \pi/2$ is a sphere of radius c.

The distance from the origin to a point in space is

$$r = (\rho^2 + z^2)^{1/2} = c\left(\frac{\cosh \eta + \cos \xi}{\cosh \eta - \cos \xi}\right)^{1/2} \tag{A–19.6}$$

the origin being given by $\{\xi = \pi, \eta = 0\}$. We note that $\cosh \eta - \cos \xi \geq 0$, the value zero being attained only when both $\xi = 0$ and $\eta = 0$. These, then, are the values corresponding to $r = \infty$.

To secure a geometric interpretation of bipolar coordinates, denote by R_1 and R_2 the distances measured to a point P from the limiting points L_1 and L_2, respectively. Thus,

$$R_1^2 = (z - c)^2 + \rho^2 = \frac{2c^2 e^{-\eta}}{\cosh \eta - \cos \xi} \tag{A–19.7}$$

$$R_2^2 = (z + c)^2 + \rho^2 = \frac{2c^2 e^{\eta}}{\cosh \eta - \cos \xi} \tag{A–19.8}$$

from which we obtain

$$\eta = \ln \frac{R_2}{R_1} \tag{A–19.9}$$

and

$$\cos \xi = \frac{R_1^2 + R_2^2 - (2c)^2}{2R_1 R_2} \tag{A–19.10}$$

But $2c$ is the distance between the limiting points of the system. Thus, in the triangle whose sides are R_1, R_2, and $2c$, ξ is the subtended angle $\widehat{L_1PL_2}$. That this is so is equally evident from Fig. A–19.2.

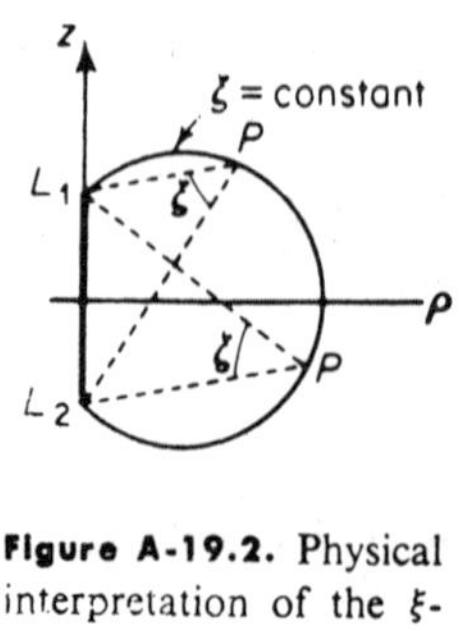

Figure A-19.2. Physical interpretation of the ξ-coordinate.

Upon arranging that

$$q_1 = \xi, \qquad q_2 = \eta, \qquad q_3 = \phi \tag{A–19.11}$$

it follows that bipolar coordinates are a right-handed system of orthogonal, curvilinear coordinates whose metrical coefficients are

$$h_1 = h_2 = \frac{\cosh \eta - \cos \xi}{c}, \qquad h_3 = \frac{\cosh \eta - \cos \xi}{c \sin \xi} \tag{A–19.12}$$

Typical unit vectors are displayed in Fig. A–19.1(a). The unit vector $\mathbf{i}_\phi$ is directed into the page.

A-20 Toroidal Coordinates (ξ, η, ϕ)

[Figs. A–20.1(a), (b), (c)]

Toroidal coordinates are generated by the transformation

$$z + i\rho = ic \coth \tfrac{1}{2}(\xi + i\eta) \tag{A–20.1}$$

$c > 0$, from which we obtain the relations

$$z = c\frac{\sin \eta}{\cosh \xi - \cos \eta}, \qquad \rho = c\frac{\sinh \xi}{\cosh \xi - \cos \eta} \tag{A–20.2}$$

By permitting the coordinates to range over the values

$$0 \leq \xi < \infty, \qquad 0 \leq \eta < 2\pi, \qquad 0 \leq \phi < 2\pi \tag{A–20.3}$$

each point in space is represented at least once and, with minor exceptions, only once.

Elimination of η from Eq. (A–20.2) yields

$$z^2 + (\rho - c \coth \xi)^2 = c^2 \operatorname{csch}^2 \xi \tag{A–20.4}$$

In a meridian plane the *curves* $\xi =$ constant are, therefore, nonintersecting, coaxial circles having their centers in the plane $z = 0$. A typical circle, $\xi = \xi_0 =$ constant, has its center at a distance $c \coth \xi_0$ from the origin and has a radius of $c \operatorname{csch} \xi_0$. Upon rotation about the z axis these circles generate an eccentric family of toruses (anchor-rings). A typical toroidal coordinate surface, $\xi = \xi_0$, is depicted in Fig. A–20.1(b). The value $\xi = 0$ corresponds to the entire z axis, whereas the value $\xi = \infty$ gives the points lying on the circle $\rho = c$ in the plane $z = 0$.

When ξ is eliminated from Eq. (A–20.2) there results

$$(z - c \cot \eta)^2 + \rho^2 = c^2 \csc^2 \eta \tag{A–20.5}$$

In a meridian plane the *curves* $\eta =$ constant are, therefore, circular arcs beginning on the z axis and terminating at the plane $z = 0$. Revolving these arcs about the z axis, the coordinate surfaces $\eta =$ constant thereby obtained are lenses or spherical caps having their centers along the z axis. A typical cap is shown in Fig. A–20.1(c). This family of spherical caps intersect in a common circle, $\rho = c$, lying in the plane $z = 0$. For $0 < \eta_0 < \pi/2$ the cap

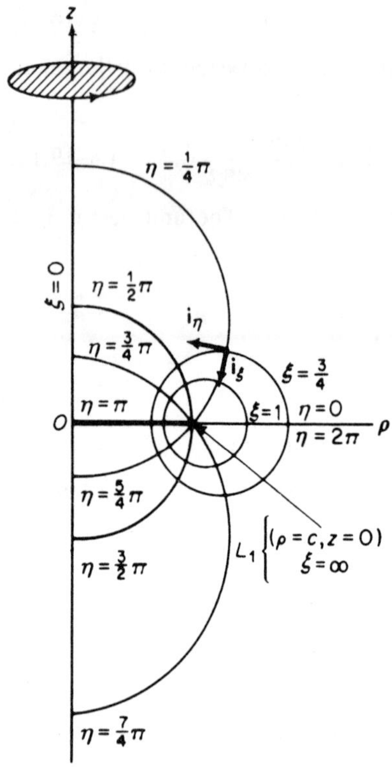

Figure A-20.1(a). Toroidal coordinates in a meridian plane.

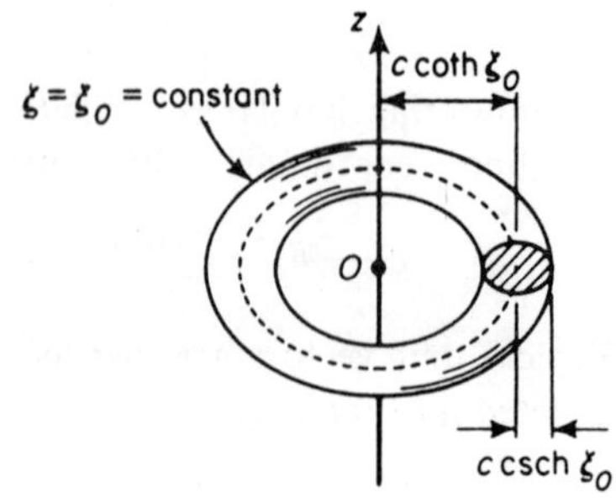

Figure A-20.1(b). Anchor rings.

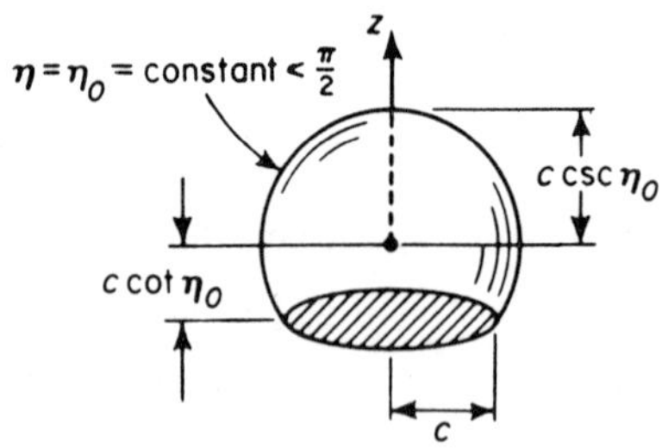

Figure A-20.1(c). Spherical caps (or lenses).

is greater than a hemisphere and lies above the plane $z = 0$. Likewise, for values of $\pi/2 < \eta_o < \pi$ the cap is less than a hemisphere and has the form of a curved diaphragm. $\eta_o = \pi/2$ is exactly a hemisphere. In general, if $\eta = \eta_o =$ constant $< \pi$ gives the surface of the cap lying above the plane $z = 0$, the extension of the spherical cap below this plane is given by $\eta = \pi + \eta_o$. The value $\eta = 0$ yields those points in the plane $z = 0$ which lie outside the circle $\rho = c$, whereas $\eta = \pi$ gives those points in the plane $z = 0$ which lie inside this circle.

The distance from the origin to a point in space is given by

$$r^2 = (\rho^2 + z^2)^{1/2} = c\left(\frac{\cosh \xi + \cos \eta}{\cosh \xi - \cos \eta}\right)^{1/2} \tag{A–20.6}$$

the origin having the coordinates $\{\xi = 0, \eta = \pi\}$. $r = \infty$ is characterized by the "point"

$$\{\xi = 0, \eta = 0\} \tag{A–20.7}$$

To interpret the present system of coordinates geometrically, let L_1 and L_2 be the points obtained by the intersection of the two parallel meridian planes $\phi = \phi_o = \text{constant} < \pi$ and $\phi = \pi - \phi_o$, respectively, with the circle $\rho = c$ lying in the plane $z = 0$. L_1 and L_2 thus lie at opposite ends of the circle, being separated by a distance $2c$. If a point P lies in one or the other of these two meridian planes, and if R_1 and R_2, respectively, are distances measured from L_1 and L_2 to P, then

$$R_1^2 = z^2 + (\rho - c)^2 = \frac{2c^2 e^{-\xi}}{\cosh \xi - \cos \eta} \tag{A–20.8}$$

$$R_2^2 = z^2 + (\rho + c)^2 = \frac{2c^2 e^{\xi}}{\cosh \xi - \cos \eta} \tag{A–20.9}$$

These combine to give

$$\xi = \ln \frac{R_2}{R_1} \tag{A–20.10}$$

and

$$\cos \eta = \frac{R_1^2 + R_2^2 - (2c)^2}{2R_1 R_2} \tag{A–20.11}$$

In consequence of these, η is the subtended angle $\widehat{L_1 P L_2}$ in the triangle whose sides are R_1, R_2, and $2c$.

Choosing

$$q_1 = \xi, \qquad q_2 = \eta, \qquad q_3 = \phi \tag{A–20.12}$$

the system of toroidal coordinates forms a right-handed system of orthogonal, curvilinear coordinates with metrical coefficients

$$h_1 = h_2 = \frac{\cosh \xi - \cos \eta}{c}, \qquad h_3 = \frac{\cosh \xi - \cos \eta}{c \sinh \xi} \tag{A–20.13}$$

Typical unit vectors are shown in Fig. A–20.1(a). The unit vector $\mathbf{i}_\phi$ is directed into the page.

A-21 Paraboloidal Coordinates (ξ, η, ϕ)
[Figs. A 21.1(a), (b), (c)]

Paraboloidal coordinates arise from the transformation

$$z + i\rho = c(\xi + i\eta)^2 \tag{A–21.1}$$

$c > 0$, whereupon

$$z = c(\xi^2 - \eta^2), \qquad \rho = 2c\xi\eta \tag{A–21.2}$$

Each point in space is represented at least once by letting the paraboloidal coordinates (ξ, η, ϕ) range over the values

$$0 \le \xi < \infty, \qquad 0 \le \eta < \infty, \qquad 0 \le \phi < 2\pi \tag{A–21.3}$$

From Eq. (A–21.2) we obtain

$$\rho^2 = 4c\xi^2(c\xi^2 - z) \tag{A–21.4}$$

so that the coordinate surfaces $\xi =$ constant are confocal paraboloids of revolution having the z axis as their axis of rotation and their foci at the origin. These paraboloids open in the direction of z negative.

In a similar manner Eq. (A–21.2) yields

$$\rho^2 = 4c\eta^2(c\eta^2 + z) \qquad \text{(A–21.5)}$$

whence the coordinate surfaces $\eta =$ constant are also confocal paraboloids of revolution having the z axis as their axis of rotation and their foci at the origin. This family of paraboloids, however, opens along the positive z axis.

The distance from the origin to any point in space is

$$r = (\rho^2 + z^2)^{1/2} = c(\xi^2 + \eta^2) \qquad \text{(A–21.6)}$$

It is useful in some applications to employ the relations

$$\xi = \left(\frac{r}{c}\right)^{1/2} \cos\frac{\theta}{2}, \qquad \eta = \left(\frac{r}{c}\right)^{1/2} \sin\frac{\theta}{2} \qquad \text{(A–21.7)}$$

(r, θ, ϕ) being spherical coordinates.

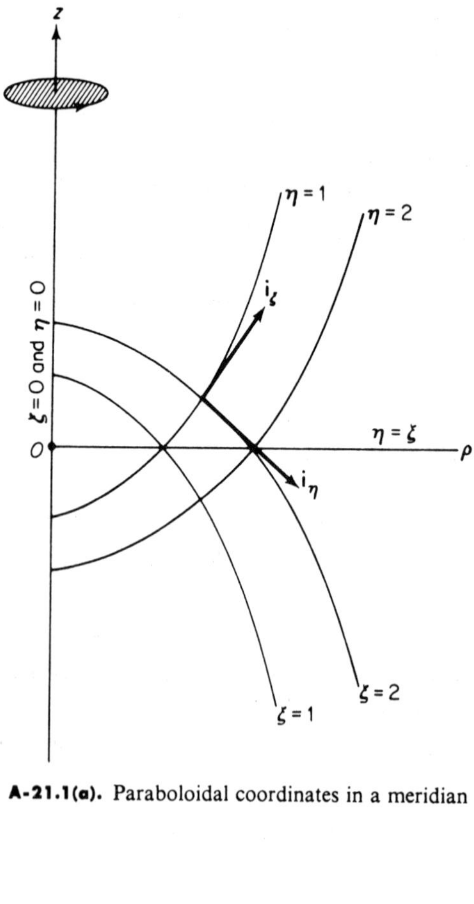

Figure A-21.1(a). Paraboloidal coordinates in a meridian plane.

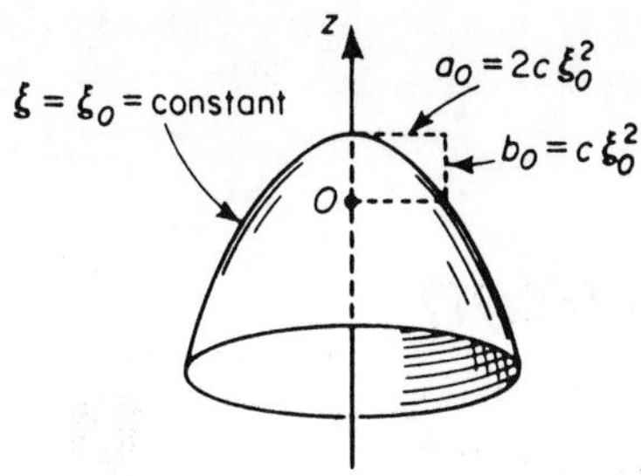

Figure A-21.1(b). Paraboloid of revolution ξ = constant.

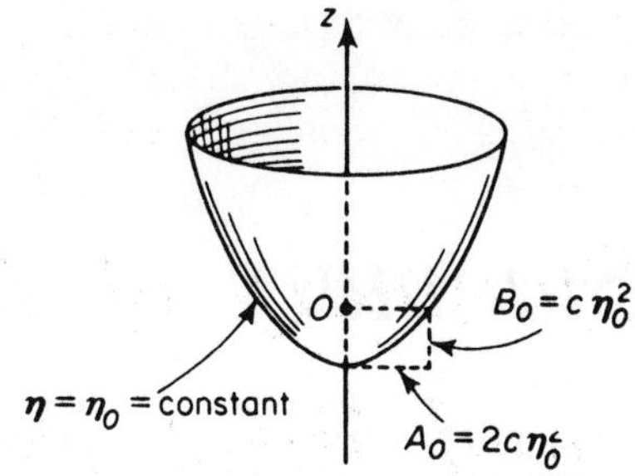

Figure A-21.1(c). Paraboloid of revolution η = constant.

With the choice of curvilinear coordinates,

$$q_1 = \xi, \qquad q_2 = \eta, \qquad q_3 = \phi \tag{A-21.8}$$

the system of orthogonal, curvilinear coordinates is right-handed and has metrical coefficients whose values are

$$h_1 = h_2 = \frac{1}{2c(\xi^2 + \eta)^{1/2}}, \qquad h_3 = \frac{1}{2c\xi\eta} \tag{A-21.9}$$

A set of unit vectors is shown in Fig. A–21.1(a). The unit vector $\mathbf{i}_\phi$ is directed into the page.

Paraboloidal coordinates may also be obtained as a limiting case of prolate spheroidal coordinates. This technique is useful in obtaining solutions to various problems involving paraboloids of revolution when the solution to the corresponding problem is known for a prolate spheroid. In Section A–17 replace z by $z + 2ck^2$, c by $2ck^2$, and ξ and η by ξ/k and η/k, respectively. Thus, prolate spheroidal coordinates are now given by the transformation

$$(z + 2ck^2) + i\rho = 2ck^2 \cosh\left(\frac{\xi + i\eta}{k}\right)$$

This gives

$$z + i\rho = 4ck^2 \sinh^2\left(\frac{\xi + i\eta}{2k}\right)$$

As we let $k \to \infty$ and expand the hyperbolic sine term for small values of its argument, we obtain in the limit

$$z + i\rho = c(\xi + i\eta)^2$$

which is precisely the definition of paraboloidal coordinates.

Summary of Notation and Brief Review of Polyadic Algebra B

The vector, dyadic, polyadic, and tensor notation used in this book follows customary American usage, being derived from the work of Gibbs[4]. Extensions of Gibbs' notation to polyadics of ranks greater than 2 (that is, dyadics) is discussed in Drew's *Handbook of Vector and Polyadic Analysis*[3]. For our immediate purposes, however, the formidable general symbolism developed by Drew is unnecessary. Block[1] has produced a very brief and readable textbook on the relationship between polyadics and tensors. Milne's[5] book affords an excellent example of the physical insights afforded by polyadic symbolism in physical problems. The reader should be cautioned, however, that Milne utilizes the "nesting convention" of Chapman and Cowling[2] with regard to multiplication of polyadics, rather than the original notation of Gibbs[4].

In this text, physical quantities encountered are distinguished as being scalars, vectors, and polyadics. They are distinguished, wherever feasible,* by differences in type as follows:

*Thus, boldface Greek symbols may represent *either* vectors or polyadics.

s = scalar (lightface italic)
$\mathbf{v}$ = vector (boldface roman)
$\mathbf{i}, \mathbf{e}$ = unit vectors (boldface roman)
T = polyadic (boldface sans serif)

When, without specific designation, the same letter appears in the same context both as a vector (for example, $\mathbf{v}$) and as a scalar (for example, v), the scalar is the magnitude of the vector. When necessary for clarity $|\mathbf{v}|$ is used to denote the magnitude of the vector $\mathbf{v}$.

Vectors and polyadics are often conveniently expressed in terms of their components in some particular system of curvilinear coordinates (q_1, q_2, q_3), for example, cartesian coordinates (x, y, z), spherical coordinates (r, θ, ϕ), cylindrical coordinates (ρ, ϕ, z). In this text we work only with *orthogonal* curvilinear coordinates, of which the preceding systems are examples. If $(\mathbf{i}_1, \mathbf{i}_2, \mathbf{i}_3)$ are a right-handed* triad of *unit* base vectors in such a system, for example $(\mathbf{i}, \mathbf{j}, \mathbf{k})$ in cartesian coordinates or $(\mathbf{i}_r, \mathbf{i}_\theta, \mathbf{i}_\phi)$ in spherical coordinates, then any vector $\mathbf{v}$ may be expressed in the form

$$\mathbf{v} = \mathbf{i}_1 v_1 + \mathbf{i}_2 v_2 + \mathbf{i}_3 v_3 \tag{B–1}$$

where (v_1, v_2, v_3) are the components of the vector $\mathbf{v}$ in the particular coordinate system. This relation may also be written concisely as

$$\mathbf{v} = \sum_{j=1}^{3} \mathbf{i}_j v_j \tag{B–2}$$

We shall review briefly some of the more important properties of polyadics. The most general dyadic can be expressed in the nonion form

$$\mathsf{D} = \sum_{j=1}^{3} \sum_{k=1}^{3} \mathbf{i}_j \mathbf{i}_k D_{jk} \tag{B–3}$$

Written out explicitly this is

$$\begin{aligned} \mathsf{D} = {} & \mathbf{i}_1\mathbf{i}_1 D_{11} + \mathbf{i}_1\mathbf{i}_2 D_{12} + \mathbf{i}_1\mathbf{i}_3 D_{13} \\ & + \mathbf{i}_2\mathbf{i}_1 D_{21} + \mathbf{i}_2\mathbf{i}_2 D_{22} + \mathbf{i}_2\mathbf{i}_3 D_{23} \\ & + \mathbf{i}_3\mathbf{i}_1 D_{31} + \mathbf{i}_3\mathbf{i}_2 D_{32} + \mathbf{i}_3\mathbf{i}_3 D_{33} \end{aligned} \tag{B–4}$$

The set of nine scalar numbers D_{jk} $(j, k = 1, 2, 3)$ are the components of the dyadic. Though their numerical values depend on the particular system of coordinates (q_1, q_2, q_3) employed, the dyadic D itself has a significance which transcends any particular choice of coordinates. If the unit vectors in Eq. (B–4) are suppressed, one may regard the dyadic D as the 3×3 matrix

*The system is right-handed in the cyclic order

$$\begin{array}{ccc} & 3 & \\ \swarrow & & \nwarrow \\ 1 & \rightarrow & 2 \end{array}$$

if, for the scalar triple product,

$$\mathbf{i}_1 \cdot \mathbf{i}_2 \times \mathbf{i}_3 = +1$$

$$(\mathbf{D}) \equiv \begin{pmatrix} D_{11} & D_{12} & D_{13} \\ D_{21} & D_{22} & D_{23} \\ D_{31} & D_{32} & D_{33} \end{pmatrix} \tag{B-5}$$

The determinant of the dyadic $\mathbf{D}$ is the scalar

$$\det \mathbf{D} = \begin{vmatrix} D_{11} & D_{12} & D_{13} \\ D_{21} & D_{22} & D_{23} \\ D_{31} & D_{32} & D_{33} \end{vmatrix} \tag{B-6}$$

whose value can be shown to be an invariant which is independent of the particular system of coordinates employed.

The transpose (or conjugate) of the dyadic $\mathbf{D}$ is denoted by the symbol $\mathbf{D}^{\dagger}$ (some authors use $\widetilde{\mathbf{D}}$, $\mathbf{D}_c$, $\mathbf{D}^t$) and may be defined as the dyadic obtained by interchanging the order of the unit vectors in Eq. (B–3). Thus,

$$\mathbf{D}^{\dagger} = \sum_{j=1}^{3} \sum_{k=1}^{3} \mathbf{i}_k \mathbf{i}_j D_{jk} \tag{B-7}$$

or, since the indices j and k are *dummy* indices, we have, upon interchanging j and k, that

$$\mathbf{D}^{\dagger} = \sum_{j=1}^{3} \sum_{k=1}^{3} \mathbf{i}_j \mathbf{i}_k D_{kj} \tag{B-8}$$

This operation can be shown to have an invariant meaning. It is also equivalent to the usual transposition operation with matrices, where the rows and columns in Eq. (B–5) are interchanged.

A dyadic is said to be *symmetric* if it is equal to its own transpose, that is,

$$\mathbf{D} = \mathbf{D}^{\dagger} \tag{B-9}$$

From Eqs. (B–3) and (B–8), this is equivalent to the three scalar equations

$$D_{jk} = D_{kj} \quad (j, k = 1, 2, 3) \tag{B-10}$$

which, when written out explicitly, requires that

$$D_{12} = D_{21}, \qquad D_{23} = D_{32}, \qquad D_{31} = D_{13} \tag{B-11}$$

A symmetric dyadic thus possesses only six independent components. Any symmetric dyadic can be written in the diagonal form

$$\mathbf{D} = \mathbf{e}_1\mathbf{e}_1 D_1 + \mathbf{e}_2\mathbf{e}_2 D_2 + \mathbf{e}_3\mathbf{e}_3 D_3 \tag{B-12}$$

where $(\mathbf{e}_1, \mathbf{e}_2, \mathbf{e}_3)$ are a particular system of mutually perpendicular unit vectors called the *eigenvectors* (characteristic vectors) of the symmetric dyadic $\mathbf{D}$. The three scalars D_1, D_2, D_3 are called its *eigenvalues* (principal values, characteristic values, characteristic roots). For a given symmetric dyadic $\mathbf{D}$, the problem of establishing its eigenvalues and eigenvectors is equivalent to that involved in diagonalizing the matrix in Eq. (B–5).

A dyadic $\mathbf{D}$ is antisymmetric if it is equal to the negative of its transpose, that is, if

$$\mathbf{D} = -\mathbf{D}^{\dagger} \tag{B-13}$$

From Eqs. (B–3) and (B–8), this requires that

$$D_{11} = D_{22} = D_{33} = 0$$

and $$D_{12} = -D_{21}, \quad D_{23} = -D_{32}, \quad D_{31} = -D_{13} \tag{B–14}$$

An antisymmetric dyadic thus possesses only three independent components.

Any dyadic can be uniquely expressed as the sum of a symmetric and antisymmetric dyadic as follows:

$$\mathbf{D} = \tfrac{1}{2}(\mathbf{D} + \mathbf{D}^{\dagger}) + \tfrac{1}{2}(\mathbf{D} - \mathbf{D}^{\dagger}) \tag{B–15}$$

the first term on the right being symmetric and the second antisymmetric.

A particularly important dyadic is the *idemfactor* or unit dyadic. This may be written in the form*

$$\mathbf{I} = \sum_{j=1}^{3}\sum_{k=1}^{3} \mathbf{i}_j\mathbf{i}_k\delta_{jk} \tag{B–16}$$

where $$\delta_{jk} = \begin{cases} 1 & \text{if } j = k \\ 0 & \text{if } j \neq k \end{cases} \tag{B–17}$$

is the Kronecker delta. Hence, an equivalent form of the idemfactor is

$$\mathbf{I} = \mathbf{i}_1\mathbf{i}_1 + \mathbf{i}_2\mathbf{i}_2 + \mathbf{i}_3\mathbf{i}_3 \tag{B–18}$$

The most general triadic, say $\mathbf{T}$, can be expressed in the form†

$$\mathbf{T} = \sum_{j=1}^{3}\sum_{k=1}^{3}\sum_{l=1}^{3} \mathbf{i}_j\mathbf{i}_k\mathbf{i}_l T_{jkl} \tag{B–19}$$

and thus has 27 independent components, T_{jkl}. In applications one must consider two possible transposition operations:

pre-transposition—

$$\begin{aligned} {}^{\dagger}\mathbf{T} &= \sum_{j=1}^{3}\sum_{k=1}^{3}\sum_{l=1}^{3} \mathbf{i}_k\mathbf{i}_j\mathbf{i}_l T_{jkl} \\ &= \sum_{j=1}^{3}\sum_{k=1}^{3}\sum_{l=1}^{3} \mathbf{i}_j\mathbf{i}_k\mathbf{i}_l T_{kjl} \end{aligned} \tag{B–20}$$

post-transposition—

$$\begin{aligned} \mathbf{T}^{\dagger} &= \sum_{j=1}^{3}\sum_{k=1}^{3}\sum_{l=1}^{3} \mathbf{i}_j\mathbf{i}_l\mathbf{i}_k T_{jkl} \\ &= \sum_{j=1}^{3}\sum_{k=1}^{3}\sum_{l=1}^{3} \mathbf{i}_j\mathbf{i}_k\mathbf{i}_l T_{jlk} \end{aligned} \tag{B–21}$$

An especially useful triadic is the unit isotropic triadic (alternating triadic, alternator),

$$\boldsymbol{\epsilon} = \sum_{j=1}^{3}\sum_{k=1}^{3}\sum_{l=1}^{3} \mathbf{i}_j\mathbf{i}_k\mathbf{i}_l\epsilon_{jkl} \tag{B–22}$$

*Some authors prefer $\mathbf{U}$ for the unit dyadic.

†Since polyadics of order greater than 2 appear only infrequently in the text, we do not use any special type style to distinguish the different orders. If necessary, one can attach the affix n to indicate the order of the polyadic. Thus ${}^{n}\mathbf{A}$ is an n-adic; for example, ${}^{2}\mathbf{A}$ is a dyadic, ${}^{3}\mathbf{A}$ a triadic, etc.

where ϵ_{jkl} is the permutation symbol, having the following properties: it is zero if any two of the three indices are equal; it has the value $+1$ if (j, k, l) is an even cyclic permutation of the integers (1, 2, 3); it has the value -1 if (j, k, l) is an odd cyclic permutation of the integers (1, 2, 3). Thus, written out explicitly we have

$$\begin{aligned}\boldsymbol{\epsilon} = \mathbf{i}_1\mathbf{i}_2\mathbf{i}_3 - \mathbf{i}_1\mathbf{i}_3\mathbf{i}_2 \\ + \mathbf{i}_2\mathbf{i}_3\mathbf{i}_1 - \mathbf{i}_2\mathbf{i}_1\mathbf{i}_3 \\ + \mathbf{i}_3\mathbf{i}_1\mathbf{i}_2 - \mathbf{i}_3\mathbf{i}_2\mathbf{i}_1\end{aligned} \tag{B–23}$$

The most general polyadic of rank n in three-dimensional space is the n-adic

$$^n\mathbf{P} = \sum_{k_1=1}^{3} \sum_{k_2=1}^{3} \cdots \sum_{k_n=1}^{3} \mathbf{i}_{k_1}\mathbf{i}_{k_2} \cdots \mathbf{i}_{k_n} P_{k_1k_2\cdots k_n} \tag{B–24}$$

which has 3^n components.

There exist several different types of "multiplication" pertaining to vectors and polyadics. Since all such entities may be expressed in the form of Eq. (B–24), the multiplication rules may be conveniently expressed in terms of operations on the unit vectors in Eq. (B–24), at least for the orthogonal systems of interest to us in this text. Attention is confined to those multiplicative operations which appear explicitly in this book.

For the dot multiplication of vectors, we have (for $j, k = 1, 2, 3$)

$$\mathbf{i}_j \cdot \mathbf{i}_k = \delta_{jk} \tag{B–25}$$

whereas for cross multiplication of vectors

$$\mathbf{i}_j \times \mathbf{i}_k = \sum_{l=1}^{3} \epsilon_{jkl}\mathbf{i}_l \tag{B–26}$$

Dot and cross multiplications may be applied to polyadics of any order by invoking the convention that the operation denoted by the dot or cross is to be performed on the vectors appearing immediately on either side of the operational symbol. For example, the two possible dot products of a dyadic with a vector are

$$\begin{aligned}\mathbf{D}\cdot\mathbf{v} &= \Big(\sum_j \sum_k \mathbf{i}_j\mathbf{i}_k D_{jk}\Big)\cdot \sum_l \mathbf{i}_l v_l \\ &= \sum_j \sum_k \sum_l \mathbf{i}_j(\mathbf{i}_k\cdot\mathbf{i}_l) D_{jk} v_l \\ &= \sum_j \sum_k \sum_l \mathbf{i}_j \delta_{kl} D_{jk} v_l \\ &= \sum_j \sum_k \mathbf{i}_j D_{jk} v_k \quad \text{(a vector)}\end{aligned} \tag{B–27}$$

and

$$\begin{aligned}\mathbf{v}\cdot\mathbf{D} &= \Big(\sum_j \mathbf{i}_j v_j\Big)\cdot\Big(\sum_k \sum_l \mathbf{i}_k\mathbf{i}_l D_{kl}\Big) \\ &= \sum_j \sum_k \sum_l (\mathbf{i}_j\cdot\mathbf{i}_k)\mathbf{i}_l D_{kl} v_j \\ &= \sum_j \sum_k \sum_l \mathbf{i}_l \delta_{jk} D_{kl} v_j \\ &= \sum_k \sum_l \mathbf{i}_l D_{kl} v_k \\ &= \sum \sum \mathbf{i}_j D_{kj} v_k \quad \text{(a vector)}\end{aligned} \tag{B–28}$$

the latter equation being obtained by replacing the dummy index l with j. We observe that

$$\mathbf{v}\cdot\mathbf{D} \neq \mathbf{D}\cdot\mathbf{v} \tag{B–29}$$

unless $\mathbf{D}$ is symmetric, in which case $D_{jk} = D_{kj}$. Observe, however, that it is always true that

$$\mathbf{v}\cdot\mathbf{D} = \mathbf{D}^{+}\cdot\mathbf{v} \quad \text{and} \quad \mathbf{v}\cdot\mathbf{D}^{+} = \mathbf{D}\cdot\mathbf{v} \tag{B–30}$$

One can dot multiply two polyadics of any order; for example, for a triadic $\mathbf{T}$ and dyadic $\mathbf{D}$,

$$\begin{aligned}\mathbf{T}\cdot\mathbf{D} &= (\sum_j \sum_k \sum_l \mathbf{i}_j\mathbf{i}_k\mathbf{i}_l T_{jkl})\cdot(\sum_m \sum_n \mathbf{i}_m\mathbf{i}_n D_{mn})\\ &= \sum_j \sum_k \sum_l \sum_m \sum_n \mathbf{i}_j\mathbf{i}_k(\mathbf{i}_l\cdot\mathbf{i}_m)\mathbf{i}_n T_{jkl} D_{mn}\\ &= \sum_j \sum_k \sum_l \sum_m \sum_n \mathbf{i}_j\mathbf{i}_k\mathbf{i}_n \delta_{lm} T_{jkl} D_{mn}\\ &= \sum_j \sum_k \sum_l \sum_n \mathbf{i}_j\mathbf{i}_k\mathbf{i}_n T_{jkl} D_{ln} \quad \text{(a triadic)}\end{aligned} \tag{B–31}$$

As examples of cross products we have, with the aid of Eq. (B–26),

$$\begin{aligned}\mathbf{D}\times\mathbf{v} &= (\sum_j \sum_k \mathbf{i}_j\mathbf{i}_k D_{jk}) \times (\sum_l \mathbf{i}_l v_l)\\ &= \sum_j \sum_k \sum_l \mathbf{i}_j(\mathbf{i}_k \times \mathbf{i}_l) D_{jk} v_l\\ &= \sum_j \sum_k \sum_l \sum_m \mathbf{i}_j\mathbf{i}_m \epsilon_{klm} D_{jk} v_l \quad \text{(a dyadic)}\end{aligned} \tag{B–32}$$

and

$$\begin{aligned}\mathbf{v}\times\mathbf{D} &= (\sum_j \mathbf{i}_j v_j) \times (\sum_k \sum_l \mathbf{i}_k\mathbf{i}_l D_{kl})\\ &= \sum_j \sum_k \sum_l (\mathbf{i}_j \times \mathbf{i}_k)\mathbf{i}_l D_{kl} v_j\\ &= \sum_j \sum_k \sum_l \sum_m \mathbf{i}_m\mathbf{i}_l \epsilon_{jkm} D_{kl} v_j\\ &= \sum_j \sum_k \sum_l \sum_m \mathbf{i}_j\mathbf{i}_m \epsilon_{lkj} D_{km} v_l \quad \text{(a dyadic)}\end{aligned} \tag{B–33}$$

which we have obtained from the preceding equation by the substitutions $j \rightarrow l$, $l \rightarrow m$, $m \rightarrow j$. Because of the relations

$$\epsilon_{jkl} = \epsilon_{ljk} = \epsilon_{klj} = -\epsilon_{jlk} = -\epsilon_{lkj} = -\epsilon_{kjl} \tag{B–34}$$

it is possible to express these relations in a variety of equivalent forms. We note that

$$\mathbf{D}\times\mathbf{v} \neq \mathbf{v}\times\mathbf{D} \tag{B–35}$$

Cross products may also be formed from higher-order polyadics; for example, $\mathbf{D} \times \mathbf{T} =$ a tetradic.

Another form of multiplication is direct multiplication, in which no operational symbol is employed. For example,

$$\begin{aligned}\mathbf{D}\mathbf{v} &= (\sum_j \sum_k \mathbf{i}_j\mathbf{i}_k D_{jk})(\sum_l \mathbf{i}_l v_l)\\ &= \sum_j \sum_k \sum_l \mathbf{i}_j\mathbf{i}_k\mathbf{i}_l D_{jk} v_l \quad \text{(a triadic)}\end{aligned} \tag{B–36}$$

It is sometimes convenient to use multiple operational symbols. The only such operator employed extensively in this book is the double-dot multiplication of Gibbs. In Gibbs' notation, if **a**, **b**, **c**, **d** are any vectors, then*

$$\mathbf{ab}:\mathbf{cd} = (\mathbf{a}\cdot\mathbf{c})(\mathbf{b}\cdot\mathbf{d}) \tag{B–37}$$

In particular, if these be unit vectors

$$\mathbf{i}_j\mathbf{i}_k:\mathbf{i}_l\mathbf{i}_m = (\mathbf{i}_j\cdot\mathbf{i}_l)(\mathbf{i}_k\cdot\mathbf{i}_m) = \delta_{jl}\,\delta_{km} \tag{B–38}$$

For example,

$$\begin{aligned}\mathbf{D}^{(1)}:\mathbf{D}^{(2)} &= (\sum_j\sum_k \mathbf{i}_j\mathbf{i}_k D^{(1)}_{jk}):(\sum_l\sum_m \mathbf{i}_l\mathbf{i}_m D^{(2)}_{lm})\\ &= \sum_j\sum_k\sum_l\sum_m (\mathbf{i}_j\mathbf{i}_k:\mathbf{i}_l\mathbf{i}_m)\,D^{(1)}_{jk}D^{(2)}_{lm}\\ &= \sum_j\sum_k\sum_l\sum_m (\mathbf{i}_j\cdot\mathbf{i}_l)(\mathbf{i}_k\cdot\mathbf{i}_m)\,D^{(1)}_{jk}D^{(2)}_{lm}\\ &= \sum_j\sum_k\sum_l\sum_m \delta_{jl}\,\delta_{km}\,D^{(1)}_{jk}D^{(2)}_{lm}\\ &= \sum_j\sum_k D^{(1)}_{jk}D^{(2)}_{jk} \quad \text{(a scalar)}\end{aligned} \tag{B–39}$$

Similarly,

$$\begin{aligned}\mathbf{T}:\mathbf{D} &= (\sum_j\sum_k\sum_l \mathbf{i}_j\mathbf{i}_k\mathbf{i}_l T_{jkl}):(\sum_m\sum_n \mathbf{i}_m\mathbf{i}_n D_{mn})\\ &= \sum_j\sum_k\sum_l \mathbf{i}_j T_{jkl}D_{kl} \quad \text{(a vector)}\end{aligned} \tag{B–40}$$

Another useful multiple operation is the double cross-product, which in Gibbs' notation is defined by the relation

$$\mathbf{ab} \overset{\times}{\times} \mathbf{cd} = (\mathbf{a}\times\mathbf{c})(\mathbf{b}\times\mathbf{d}) \quad \text{(a dyadic)} \tag{B–41}$$

In a sequence involving more than two operations, parentheses may be required to define the operations unambiguously. For example,

$$(\mathbf{v}^{(1)}\times\mathbf{v}^{(2)})\cdot\mathbf{D} \neq \mathbf{v}^{(1)}\times(\mathbf{v}^{(2)}\cdot\mathbf{D})$$

whereas

$$\mathbf{v}^{(1)}\times(\mathbf{D}\cdot\mathbf{v}^{(2)}) = (\mathbf{v}^{(1)}\times\mathbf{D})\cdot\mathbf{v}^{(2)} = \mathbf{v}^{(1)}\times\mathbf{D}\cdot\mathbf{v}^{(2)}$$

so that no parentheses are necessary to specify the order of the operation.

The idemfactor has the property that if $\boldsymbol{\Psi}$ is any vector or polyadic, then

$$\mathbf{I}\cdot\boldsymbol{\Psi} = \boldsymbol{\Psi}\cdot\mathbf{I} = \boldsymbol{\Psi} \tag{B–42}$$

Other useful properties possessed by the idemfactor are

$$\mathbf{I}:\mathbf{v}^{(1)}\mathbf{v}^{(2)} = \mathbf{v}^{(1)}\cdot\mathbf{v}^{(2)} \tag{B–43}$$

where $\mathbf{v}^{(1)}$ and $\mathbf{v}^{(2)}$ are any vectors. As particular examples we have

$$\mathbf{I}:\nabla\mathbf{v} = \nabla\cdot\mathbf{v} \tag{B–44}$$

and

$$\mathbf{I}:\nabla\,\nabla = \nabla\cdot\nabla = \nabla^2 \tag{B–45}$$

where ∇^2 is the Laplace operator.

*Some authors[2,5] define

$$\mathbf{ab}:\mathbf{cd} = (\mathbf{a}\cdot\mathbf{d})(\mathbf{b}\cdot\mathbf{c})$$

The unit alternating triadic has the useful property that

$$\boldsymbol{\epsilon} : \mathbf{v}^{(1)}\mathbf{v}^{(2)} = \mathbf{v}^{(1)} \times \mathbf{v}^{(2)} \tag{B–46}$$

In particular

$$\boldsymbol{\epsilon} : \nabla \mathbf{v} = \nabla \times \mathbf{v} \tag{B–47}$$

We note that

$$\boldsymbol{\epsilon} = -\mathbf{I} \times \mathbf{I} \tag{B–48}$$

If $\mathbf{D}^{(1)}$ and $\mathbf{D}^{(2)}$ are dyadics and $\mathbf{v}^{(1)}$ and $\mathbf{v}^{(2)}$ are *arbitrary* vectors, then if

$$\mathbf{v}^{(1)} \cdot \mathbf{D}^{(1)} \cdot \mathbf{v}^{(2)} = \mathbf{v}^{(1)} \cdot \mathbf{D}^{(2)} \cdot \mathbf{v}^{(2)}$$

the principle of equality of dyadics permits us to conclude that

$$\mathbf{D}^{(1)} = \mathbf{D}^{(2)}$$

The determinant of a dyadic may be expressed in the invariant form

$$\det \mathbf{D} = \tfrac{1}{6}(\mathbf{D} \overset{\times}{\times} \mathbf{D}) : \mathbf{D} \tag{B–49}$$

If the determinant of a dyadic $\mathbf{D}$ is different from zero, then the dyadic possesses a unique inverse $\mathbf{D}^{-1}$ defined either by the relation

$$\mathbf{D} \cdot \mathbf{D}^{-1} = \mathbf{I}$$

or

$$\mathbf{D}^{-1} \cdot \mathbf{D} = \mathbf{I} \tag{B–50}$$

The inverse possesses the properties that

$$(\mathbf{D}_{(1)} \cdot \mathbf{D}_{(2)})^{-1} = \mathbf{D}_{(2)}^{-1} \cdot \mathbf{D}_{(1)}^{-1} \tag{B–51}$$

and

$$(\mathbf{D}^{-1})^{\dagger} = (\mathbf{D}^{\dagger})^{-1} \tag{B–52}$$

The inverse or reciprocal dyadic may be computed from the relation

$$\mathbf{D}^{-1} = \frac{\tfrac{1}{2}(\mathbf{D} \overset{\times}{\times} \mathbf{D})^{\dagger}}{\det \mathbf{D}} \tag{B–53}$$

The relationship

$$\mathbf{D} \cdot \mathbf{v} = 0$$

for $\mathbf{v}$ an *arbitrary* vector requires that $\mathbf{D} = \mathbf{O}$. On the other hand, if $\mathbf{v}$ is a *given* vector the preceding relation requires that $\mathbf{D} = \mathbf{O}$ if, and only if, $\det \mathbf{D} \neq 0$. Conversely, if $\mathbf{v}$ is a *given* vector and $\det \mathbf{D} = 0$ then $\mathbf{D}$ need not be zero. For example, if we let $\mathbf{i}_3$ be a unit vector parallel to the given vector $\mathbf{v}$, that is, $\mathbf{v} = \mathbf{i}_3 v$, then the relation $\mathbf{D} \cdot \mathbf{v} = \mathbf{0}$ is clearly satisfied by any dyadic of the general form

$$\begin{aligned} \mathbf{D} = {} & \mathbf{i}_1\mathbf{i}_1 D_{11} + \mathbf{i}_1\mathbf{i}_2 D_{12} \\ & + \mathbf{i}_2\mathbf{i}_1 D_{21} + \mathbf{i}_2\mathbf{i}_2 D_{22} \\ & + \mathbf{i}_3\mathbf{i}_1 D_{31} + \mathbf{i}_3\mathbf{i}_2 D_{32} \end{aligned}$$

In such cases for which $\det \mathbf{D} = 0$, $\mathbf{D}$ is said to be an *incomplete* dyadic. Conversely, $\mathbf{D}$ is a *complete* dyadic if $\det \mathbf{D} \neq 0$.

A vector $\mathbf{v}$ for which the vector $\mathbf{D}\cdot\mathbf{v}$ is parallel to $\mathbf{v}$ is an eigenvector of $\mathbf{D}$. If, for such vectors $\mathbf{v}$ we write

$$\mathbf{D}\cdot\mathbf{v} = \lambda\mathbf{v}$$

then λ is called the eigenvalue associated with $\mathbf{v}$. The foregoing may be written in the form

$$(\mathbf{D} - \mathbf{I}\lambda)\cdot\mathbf{v} = \mathbf{0}$$

so that the eigenvalues are the roots of the characteristic (secular) equation

$$\det(\mathbf{D} - \mathbf{I}\lambda) = 0 \tag{B–54}$$

In the particular case where $\mathbf{D}$ is a *symmetric* dyadic, there are in general three real roots λ_i ($i = 1, 2, 3$), not necessarily distinct, of the cubic equation

$$\begin{vmatrix} D_{11} - \lambda & D_{12} & D_{13} \\ D_{12} & D_{22} - \lambda & D_{23} \\ D_{13} & D_{23} & D_{33} - \lambda \end{vmatrix} = 0 \tag{B–55}$$

If the roots are distinct, then the corresponding three eigenvectors $\mathbf{v}_i$, defined by

$$\mathbf{D}\cdot\mathbf{v}_i = \lambda_i\mathbf{v}_i \quad (i = 1, 2, 3) \tag{B–56}$$

form a mutually perpendicular triad of vectors. The preceding equation remains unaltered if $\mathbf{v}_i$ is multiplied by a constant, say c_i:

$$\mathbf{D}\cdot(c_i\mathbf{v}_i) = \lambda_i(c_i\mathbf{v}_i)$$

It is convenient to choose c_i such that the vector $c_i\mathbf{v}_i$ is normalized to unity, that is, $|c_i\mathbf{v}_i| = 1$. Thus the *normalized* eigenvectors, $\mathbf{e}_i = \mathbf{v}_i/v_i$, satisfy $|\mathbf{e}_i| = 1$, and the equation

$$\mathbf{D}\cdot\mathbf{e}_i = \lambda_i\mathbf{e}_i \quad (i = 1, 2, 3) \tag{B–57}$$

Since they are mutually perpendicular, then

$$\mathbf{e}_j\cdot\mathbf{e}_k = \delta_{jk} \tag{B–58}$$

By utilizing these eigenvalues and normalized eigenvectors the symmetric dyadic $\mathbf{D}$ may be written in the form of Eq. (B–12), where $\lambda_i = D_i$.

A symmetric dyadic is said to be *positive definite* if for all non-zero vectors $\mathbf{u}$, the scalar

$$\mathbf{u}\cdot\mathbf{D}\cdot\mathbf{u} > 0 \tag{B–59}$$

A necessary and sufficient condition that $\mathbf{D}$ be a positive definite symmetric dyadic is that its three eigenvalues $\lambda_1, \lambda_2, \lambda_3$ each be positive scalars; for if we write the symmetric dyadic in the form of Eq. (B–12) and note that $\lambda_i = D_i$, we obtain

$$\mathbf{u}\cdot\mathbf{D}\cdot\mathbf{u} = \lambda_1 u_1^2 + \lambda_2 u_2^2 + \lambda_3 u_3^2 \tag{B–60}$$

where u_i is the component of $\mathbf{u}$ in the direction of $\mathbf{e}_i$.

If we restrict ourselves to situations in which the unit vectors ($\mathbf{i}_1, \mathbf{i}_2, \mathbf{i}_3$) in Eqs. (B–3) and (B–24) are the constant *cartesian* unit vectors ($\mathbf{i}, \mathbf{j}, \mathbf{k}$), the calculus of polyadics can be made equivalent to ordinary scalar calculus. For example, by writing

$$\nabla = \mathbf{i}_1 \frac{\partial}{\partial x_1} + \mathbf{i}_2 \frac{\partial}{\partial x_2} + \mathbf{i}_3 \frac{\partial}{\partial x_3} = \sum_{j=1}^{3} \mathbf{i}_j \frac{\partial}{\partial x_j} \tag{B–61}$$

where (x_1, x_2, x_3) are cartesian coordinates, the divergence of a dyadic may be written in cartesian form as follows:

$$\begin{aligned}\nabla \cdot \mathbf{D} &= \left(\sum_{j=1}^{3} \mathbf{i}_j \frac{\partial}{\partial x_j}\right) \cdot \left(\sum_{k=1}^{3} \sum_{l=1}^{3} \mathbf{i}_k \mathbf{i}_l D_{kl}\right) \\ &= \sum_j \sum_k \sum_l (\mathbf{i}_j \cdot \mathbf{i}_k) \mathbf{i}_l \frac{\partial D_{kl}}{\partial x_j} \\ &= \sum_j \sum_k \sum_l \mathbf{i}_l \, \delta_{jk} \frac{\partial D_{kl}}{\partial x_j} \\ &= \sum_k \sum_l \mathbf{i}_l \frac{\partial D_{kl}}{\partial x_k}\end{aligned} \tag{B–62}$$

Written out explicitly this is

$$\begin{aligned}\nabla \cdot \mathbf{D} = \mathbf{i}_1 &\left(\frac{\partial D_{11}}{\partial x_1} + \frac{\partial D_{21}}{\partial x_2} + \frac{\partial D_{31}}{\partial x_3}\right) \\ + \mathbf{i}_2 &\left(\frac{\partial D_{12}}{\partial x_1} + \frac{\partial D_{22}}{\partial x_2} + \frac{\partial D_{32}}{\partial x_3}\right) \\ + \mathbf{i}_3 &\left(\frac{\partial D_{13}}{\partial x_1} + \frac{\partial D_{23}}{\partial x_2} + \frac{\partial D_{33}}{\partial x_3}\right)\end{aligned} \tag{B–63}$$

In noncartesian coordinate systems, the unit vectors are not constant, but are themselves functions of position (see Sections A–4 and A–7). Hence, in such cases,

$$\frac{\partial}{\partial x_j}(\mathbf{i}_k Y) \neq \mathbf{i}_k \frac{\partial Y}{\partial x_j}$$

which shows why Eq. (B–62) is not valid except when (x_1, x_2, x_3) and $(\mathbf{i}_1, \mathbf{i}_2, \mathbf{i}_3)$ refer to cartesian coordinates.

If we confine ourselves solely to cartesian coordinates one may, in a sense, ignore the unit vectors and summation signs in equations such as (B–2), (B–3), (B–16), (B–24), and (B–62), and write

$$\mathbf{v} \equiv v_j \tag{B–2$'$}$$

$$\mathbf{D} \equiv D_{jk} \tag{B–3$'$}$$

$$\mathbf{I} \equiv \delta_{jk} \tag{B–16$'$}$$

$$^{n}\mathbf{P} \equiv P_{k_1 k_2 \cdots k_n} \tag{B–24$'$}$$

$$\nabla \cdot \mathbf{D} \equiv \frac{\partial D_{kl}}{\partial x_k} \tag{B–62$'$}$$

so that an obvious correspondence can be made to exist between polyadics and cartesian tensors. Polyadics, however, are clearly much more general entities than are cartesian tensors, for it is obviously not essential that the unit vectors be cartesian unit vectors. The distinction is, of course, of significance only in the calculus of polyadics and tensors—not in their algebras.

With regard to the integral calculus of polyadics, the only formula we shall mention explicitly is the analog of Gauss' divergence theorem,

$$\int_S d\mathbf{S}\cdot{}^n\mathbf{P} = \int_V \nabla\cdot({}^n\mathbf{P})\,dV \tag{B–64}$$

where S is a closed surface completely bounding the volume V, $d\mathbf{S}$ is a directed element of surface area pointing out of the volume V, ${}^n\mathbf{P}$ is a polyadic of any rank, and dV is an element of volume. Since, in general

$$d\mathbf{S}\cdot{}^n\mathbf{P} \neq {}^n\mathbf{P}\cdot d\mathbf{S}$$

it is important to maintain the proper ordering of the directional quantities in the integral theorem.

The significance of polyadic integrals is, perhaps, most readily grasped by expressing them in terms of cartesian unit vectors. For example, if $\mathbf{T}$ is the triadic

$$\mathbf{T} = \sum_k \sum_l \sum_m \mathbf{i}_k\mathbf{i}_l\mathbf{i}_m T_{klm}$$

and we write

$$d\mathbf{S} = \sum_j \mathbf{i}_j\, dS_j$$

and

$$\nabla = \sum_n \mathbf{i}_n \frac{\partial}{\partial x_n}$$

then Eq. (B–64) may be written

$$\sum_k \sum_l \sum_m \mathbf{i}_l\mathbf{i}_m \int_S dS_k T_{klm} = \sum_k \sum_l \sum_m \mathbf{i}_l\mathbf{i}_m \int_V \frac{T_{klm}}{\partial x_k}\,dV$$

which is a dyadic equation, equivalent to nine scalar relations. In differentiating the unit vectors and in bringing them through the integration sign, we have explicitly utilized the fact that they are constants, independent of position. Thus, the preceding relation written out in component form is valid only for cartesian systems. The relation from whence it emanated,

$$\int_S d\mathbf{S}\cdot\mathbf{T} = \int_V \nabla\cdot\mathbf{T}\,dV$$

is, of course, an invariant relation and holds true in any system of coordinates.

BIBLIOGRAPHY

1. Block, H. D., *Introduction to Tensor Analysis.* Columbus, Ohio: Merrill, 1962.
2. Chapman, S., and T. G. Cowling, *The Mathematical Theory of Non-Uniform Gases*, 2nd ed. Cambridge: Cambridge Univ. Press, 1961.
3. Drew, T. B., *Handbook of Vector and Polyadic Analysis.* New York: Reinhold, 1961.
4. Gibbs, J. W., and E. B. Wilson, *Vector Analysis* (reprint). New York: Dover, 1960.
5. Milne, E. A., *Vectorial Mechanics.* London: Methuen, 1957.

Indices

Name Index

[Names listed here refer only to the text itself. Additional names appear in the Bibliography at the end of each chapter.]

O

P

R

S

T

Subject Index

A

B

F